Handbook of Research on Implementing Digital Reality and Interactive Technologies to Achieve Society 5.0

Francesca Maria Ugliotti
Politecnico di Torino, Italy

Anna Osello
Politecnico di Torino, Italy

A volume in the Advances in Human and Social Aspects of Technology (AHSAT) Book Series

Published in the United States of America by
IGI Global
Information Science Reference (an imprint of IGI Global)
701 E. Chocolate Avenue
Hershey PA, USA 17033
Tel: 717-533-8845
Fax: 717-533-8661
E-mail: cust@igi-global.com
Web site: http://www.igi-global.com

Library of Congress Cataloging-in-Publication Data

Names: Ugliotti, Francesca Maria, 1987- editor. | Osella, Anna, editor.
Title: Handbook of research on implementing digital reality and interactive technologies to achieve Society 5.0 / Francesca Maria Ugliotti and Anna Osello, editors.
Description: Hershey PA : Information Science Reference, [2023] | Includes bibliographical references and index. | Summary: "The book opens a new interpretation to the digital initiatives experienced in the different sectors of construction, healthcare and smart city to promote human-centred services as physical and virtual space become more interconnected to create highly inclusive communication networks and social frameworks"-- Provided by publisher.
Identifiers: LCCN 2022023495 (print) | LCCN 2022023496 (ebook) | ISBN 9781668448540 (hardcover) | ISBN 9781668448557 (ebook)
Subjects: LCSH: Virtual reality--Case studies. | Human services--Simulation methods. | Society 5.0. | Smart cities.
Classification: LCC QA76.9.C65 H3465 2023 (print) | LCC QA76.9.C65 (ebook) | DDC 006.8--dc23/eng/20220810
LC record available at https://lccn.loc.gov/2022023495
LC ebook record available at https://lccn.loc.gov/2022023496

This book is published in the IGI Global book series Advances in Human and Social Aspects of Technology (AHSAT) (ISSN: 2328-1316; eISSN: 2328-1324)

British Cataloguing in Publication Data
A Cataloguing in Publication record for this book is available from the British Library.

For electronic access to this publication, please contact: eresources@igi-global.com.

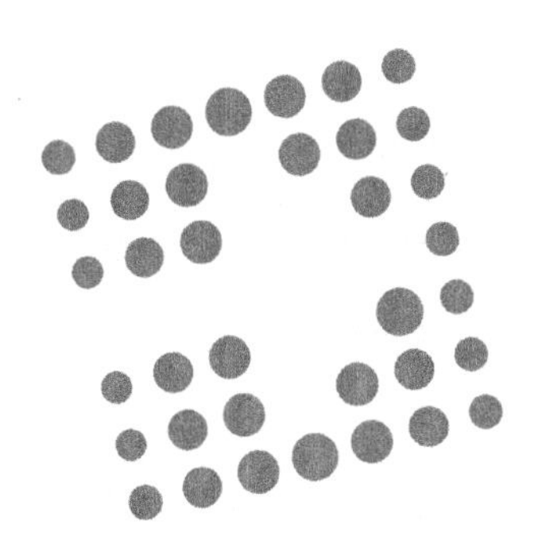

Advances in Human and Social Aspects of Technology (AHSAT) Book Series

Mehdi Khosrow-Pour, D.B.A.
Information Resources Management Association, USA

ISSN:2328-1316
EISSN:2328-1324

Mission

In recent years, the societal impact of technology has been noted as we become increasingly more connected and are presented with more digital tools and devices. With the popularity of digital devices such as cell phones and tablets, it is crucial to consider the implications of our digital dependence and the presence of technology in our everyday lives.

The **Advances in Human and Social Aspects of Technology (AHSAT) Book Series** seeks to explore the ways in which society and human beings have been affected by technology and how the technological revolution has changed the way we conduct our lives as well as our behavior. The AHSAT book series aims to publish the most cutting-edge research on human behavior and interaction with technology and the ways in which the digital age is changing society.

Coverage

- ICTs and human empowerment
- Philosophy of technology
- Technology and Social Change
- Information ethics
- Human Rights and Digitization
- Activism and ICTs
- Human-Computer Interaction
- Cyber Behavior
- Human Development and Technology
- ICTs and social change

IGI Global is currently accepting manuscripts for publication within this series. To submit a proposal for a volume in this series, please contact our Acquisition Editors at Acquisitions@igi-global.com or visit: http://www.igi-global.com/publish/.

The Advances in Human and Social Aspects of Technology (AHSAT) Book Series (ISSN 2328-1316) is published by IGI Global, 701 E. Chocolate Avenue, Hershey, PA 17033-1240, USA, www.igi-global.com. This series is composed of titles available for purchase individually; each title is edited to be contextually exclusive from any other title within the series. For pricing and ordering information please visit http://www.igi-global.com/book-series/advances-human-social-aspects-technology/37145. Postmaster: Send all address changes to above address.

Titles in this Series

For a list of additional titles in this series, please visit: www.igi-global.com/book-series

Technological Influences on Creativity and User Experience
Joshua Fairchild (Creighton University, USA)
Information Science Reference • © 2022 • 305pp • H/C (ISBN: 9781799843542) • US $195.00

Machine Learning for Societal Improvement, Modernization, and Progress
Vishnu S. Pendyala (San Jose State University, USA)
Engineering Science Reference • © 2022 • 290pp • H/C (ISBN: 9781668440452) • US $270.00

The Digital Folklore of Cyberculture and Digital Humanities
Stamatis Papadakis (University of Crete, Greece) and Alexandros Kapaniaris (Hellenic Open University, Greece)
Information Science Reference • © 2022 • 361pp • H/C (ISBN: 9781668444610) • US $215.00

Multidisciplinary Perspectives Towards Building a Digitally Competent Society
Sanjeev Bansal (Amity University, Noida, India) Vandana Ahuja (Jaipuria Institute of Management, Ghaziabad, India) Vijit Chaturvedi (Amity University, Noida, India) and Vinamra Jain (Amity University, Noida, India)
Engineering Science Reference • © 2022 • 335pp • H/C (ISBN: 9781668452745) • US $270.00

Analyzing Multidisciplinary Uses and Impact of Innovative Technologies
Emiliano Marchisio (Giustino Fortunato University, Italy)
Information Science Reference • © 2022 • 275pp • H/C (ISBN: 9781668460153) • US $240.00

Handbook of Research on Applying Emerging Technologies Across Multiple Disciplines
Emiliano Marchisio (Giustino Fortunato University, Italy)
Information Science Reference • © 2022 • 548pp • H/C (ISBN: 9781799884767) • US $270.00

Handbook of Research on Digital Violence and Discrimination Studies
Fahri Özsungur (Mersin University, Turkey)
Information Science Reference • © 2022 • 837pp • H/C (ISBN: 9781799891871) • US $270.00

Opportunities and Challenges for Computational Social Science Methods
Enes Abanoz (Ondokuz Mayıs University, Turkey)
Information Science Reference • © 2022 • 277pp • H/C (ISBN: 9781799885535) • US $215.00

Handbook of Research on Promoting Economic and Social Development Through Serious Games

701 East Chocolate Avenue, Hershey, PA 17033, USA
Tel: 717-533-8845 x100 • Fax: 717-533-8661
E-Mail: cust@igi-global.com • www.igi-global.com

Editorial Advisory Board

List of Contributors

Table of Contents

Detailed Table of Contents

Touch is fundamental to create our perception of reality and to allow fulfilling social experiences, such as those at the basis of metaverses. In order to be accurately reproduced, a number of scientific and technological aspects should be considered. In this chapter, the authors highlight the relevance of the tactile modality in eliciting 'presence' in virtual reality interactions. They also discuss the neuroscientific foundation of our bodily interactions and the fact that they are based on a number of receptors and neural circuits that contribute to the complexity of our perceptions. The available technological devices for the reproduction of touch in virtual environments and their limitations are also described. They suggest that virtual interactions should include more of this sensory modality and that attempts should be made to go beyond the actual approach to 'mimicking reality'. In particular, future simulations should consider the perspective of creative 'hyper-sensations' including 'hyper-touch' on the basis of our psychological and neuroscientific knowledge.

The chapter presents research on the relationship between the human body and the space implemented by data and digital interfaces. In this relationship, technology plays a mediating role. The research introduces the concept of a digital threshold to an interactive space that has the capacity to preserve the cognitive well-being of users and invite interaction. To do this, some characteristics are identified that can be used in the design with the aim of relating the body to the devices in the space. Pressure stimuli, rhythm, and body symmetry are the components of a natural language capable of activating a natural motorial reaction mechanism. The details of the experimentation carried out and the processing of the data collected through data visualisation are provided to support the argument.

Section 2
Smart Environments and Systems

The research is based on the hypothesis that integrating site-specific and global data into the design process requires a methodological design approach, which connects local to global systems and extends the application of available predefined algorithmic scripts and singular solutions. These tools allow the designer to apprehend and simulate possible future scenarios with unparalleled precision and speed. Computational design thinking will help us master increasingly complex design challenges as well as build a profound theoretical knowledge base to meaningfully integrate current and future technologies. After re-evaluating the principles of the computational pioneers, computationally driven methods for pressing urban challenges through data-informed design speculations are discussed. Cutting-edge design speculations aim to open up new immersive design simulation and participatory processes in environmental design and urban development and give sustainable answers to societal and environmental challenges, ultimately shaping our future world.

This research experiments the theme of cultural heritage (CH) in architectural/engineering fields, located in urban space. Primary sources and new tactics for digital reconstruction allow interactive contextualization-access to often inaccessible data creating pedagogical apps for spreading. Digital efforts are central, in recent years based on new technological opportunities that emerged from big data, Semantic Web technologies, and exponential growth of data accessible through digital libraries – EUROPEANA. Also, the use of data-based BIM allowed the gaining of high-level semantic concepts. Then, interdisciplinary collaborations between ICT and humanities disciplines are crucial for the advance of workflows that allow research on CH to exploit machine learning approaches. This chapter traces the visualizing cities progress, involving Duke and Padua University. This initiative embraces the analysis of urban systems to reveal with diverse methods how documentation/understanding of cultural sites complexities is part of a multimedia process that includes digital visualization of CH.

This chapter aims to describe the multidimensional virtual reality tools applied to healthcare: in particular the comparison between virtual reality traditional tolls and the 360° videos. The VR traditional devices could differ in terms of specific graphics (2D/3D), display devices (head mounted display), and tracking/sensing tools. Although they are ecological tools, they have several problems such as cybersickness, high-cost software, and psychometric issues. Instead, the 360° videos can be described as an extension of virtual reality technology: they are immersive videos or spherical videos that give the opportunity to immerse the subject in authentic natural environments, being viewed via an ordinary web browser in that a user can pan around by clicking and dragging. The comparison between those two technologies stems from the question if 360° videos could solve and overcome the problems related to virtual reality and be an effective and more ecological alternative.

The development of serious games has enabled new challenges for the healthcare sector in psychological, cognitive, and motor rehabilitation. Thanks to virtual reality, stimulating and interactive experiences can be reproduced in a safe and controlled environment. This chapter illustrates the experimentation conducted in the hospital setting for the non-pharmacological treatment of cognitive disorders associated with dementia. The therapy aims to relax patients of the agitation cluster through a gaming approach through the immersion in multisensory and natural settings in which sound and visual stimuli are provided. The study is supported by a technological architecture including the virtual wall system for stereoscopic wall projection and rigid body tracking.

Technological progress must aim at creating Society 5.0 by developing tools to support people. This contribution aims to show how modern technologies and their integration into society can support people with fragility. In particular, the authors present the prototype of a technology that the Turin Polytechnic has developed to provide an IoT device control tool for people with motor neuron degeneration. This, through the use of eye-trackers and building information models (BIM), allows the navigation of models in virtual reality and interaction with different devices and services. Furthermore, the use of micro-

Preface

TOWARD A PEOPLE-CENTERED FUTURE

The great challenges of the 21st century constantly push and stretch the frontiers of human knowledge. We are always looking for high-performance, digital, and connected systems. Whether we are talking about problems such as climate change, energy, regions and cities, mobility, healthcare, manufacturing and services, technology is almost always at the center. All spheres of human life are thus influenced by the achievements in the information and knowledge segments, which are the raw materials and main products of our society. However, the quality of life will trust not only information. How will a person be able to exploit it? How will each user be able to satisfy its own needs in terms of information products or services within the established informative space? These simple reflections can completely change the point of view and reinforce the objectives shifting the focus from developing devices to disseminating well-being. Hence, the idea of Society 5.0 (Cabinet Office, 2016) has been conceived to harness the enormous potential expressed by the innovations of the fourth industrial revolution toward solving society's most significant problems and social life. The critical question is, therefore, what to use technologies for. A necessary condition is that the users accept innovation: everyone is called upon to participate in pursuing the values of sustainability, inclusiveness, and welfare. Thus the super-intelligent society (Cabinet Office, 2016) contributes to the vision of the ideal city by rediscovering the anthropocentrism of needs. Within the Society 5.0 paradigm, people and machines establish a positive relationship to balance economic advancement with the resolution of social issues for every possible end-user - from the child to the elderly, from students to professionals, and from citizen to the fragile person – considering the different contexts of everyday life – i.e. homes, schools, workplaces, natural environment, and care facilities. This perspective establishes a strong interconnection between physical (real) space and cyberspace (virtual space), making the user an active player in a knowledge-intensive society. A future of interconnectedness between individuals through a network of 3D virtual worlds centered on social connection is looked for in the Metaverse. Research on Digital Reality – modeling real-world issues in various way, and Interactive Technologies – facilitating the interaction between people and establishing an information flow between the user and the technology, have been extensive in recent years, covering a wide range of topics and leading to new ways to approach and deal with complex situations. In this widely debated scenario, the book looks at experiences and approaches that place people at the center enabling multi-dimensional strategies and additional levels of interdisciplinary collaboration to create a highly inclusive communication network and social framework.

CONSIDERING SUPER-SMART SOCIETY'S NEEDS

The blueprint of Society 5.0 strategically supports the United Nations 2030 Agenda development target and Sustainable Development Goals aimed at creating a sustainable future on a global scale and change in mindset by a shared common direction. Therefore, the topics covered in the book are of international interest as they reflect on some of the most critical challenges of our contemporary times linked to social aspects. In particular, diverse human needs concerning the individual and community dimensions are tackled in the following areas.

Education

Digital Transformation requires digital and transversal skills by companies, public administrations, citizens, and students to work more collaboratively and benefit from new services. Automation, in tandem with the Covid-19 recession, creates a double-disruption scenario for workers. Moreover, according to the World Economic Forum, 65% of children enrolled in primary school today will work in jobs that do not yet exist (World Economic Forum, 2020). Education and training will therefore play a decisive role in the following years to bridge the gap and ease the transition of workers into more sustainable job opportunities. The bounty of technological innovation can be leveraged to unleash human potential. New tools and approaches allow for better interaction and learning in an increasingly inclusive and collaborative manner, considering the users' psycho-emotional well-being. The goal is not just to collect knowledge but to develop problem-solving skills.

Environment and Cities

Half of the world's population lives in urban areas, and the percentage is expected to rise to almost 70 percent by 2050 (United Nations, 2018). Thus, cities have become the crucial reference for sustainable development, facing priorities related to climate change and increasingly limited resources in a context often characterized by growing urban and infrastructural decay. The systemic strategy is centered on increasingly advanced monitoring systems and the strengthening of interdisciplinary interactions: between the environment and networks, between network and network, and between the city and the end users so that the city can be more adapted to the needs of the citizen, and the citizen more active in creating a resilient environment (Mitchell, 2001). Urban information should be integrated through Information Technology to result in the collective optimization of services and develop new businesses. The integration of spatial and temporal information represents the initial approach to the architecture of social and technical integration (Hitachi-UTokyo Laboratory, 2020).

Cultural Enrichment and Accessibility

It is recognized that Cultural Heritage, in its tangible or intangible forms, acts as a resource for creating social cohesion, educating society, and enhancing the territory. Therefore, the relevance of accessibility emerges as a fundamental requirement for transmitting society's values and overcoming social imbalances and inequality. However, this possibility is affected by environmental, social, geographical, political, religious, physical, or cognitive impairments and security-related factors. A further condition of deprivation has been experienced in the last few years within the Covid-19 pandemic by the need for social distancing. The democratization of heritage can arise by ensuring the accessibility of the digital replica, the Digital Heritage, for disseminating knowledge and achieving users' involvement and awareness. This way, physically distant sites can be brought closer in virtual space, enabling virtuous behavioral mechanisms.

Healthcare Services

Pressured by a growing and aging population and constraints on public spending, the healthcare sector requires a comprehensive re-thinking to seek the machine's efficiency and to improve the quality of healthcare treatments and services delivery for the patient's well-being. The idea is to achieve a Smart Hospital with a digital brain and systems that can connect with users' needs and meet them in every respect by combining advanced medical concepts and state-of-the-art devices and technologies. Solutions to bring the world into hospitals or private homes are also desirable for patients' therapy and entertainment to overcome physical and spatial barriers.

LOOKING FOR SOLUTIONS

The handbook proposes an interdisciplinary approach based on matching frontier technologies to enable a network among data, information, and all the city stakeholders. The following represents the key drivers to frame innovative solutions addressing social needs. The synoptic overview below shows their use in the different chapters.

- **Digital Twin (DT).** Virtual representation of a physical asset or process integrating multi-source data operating symbiotically in real-time. Querying the digital replica contributes to a dynamic view and monitoring of a system by enabling decision-making processes by managers and owners.
- **Geographic Information System (GIS) and Building Information Modeling (BIM).** Database containing geographic and building data combined with tools for storing, managing, analyzing, and visualizing that information. They support engineers in describing built environment phenomena for cyberspace-based data-driven planning and can be used for interactive knowledge experiences for citizens.
- **Unmanned Aerial Vehicles (UAVs).** Drones are aerial vehicles operated by humans from a remote location or autonomously. They are widespread for three-dimensional mapping activities and reaching inaccessible places, having great potential for monitoring activities on territory, historical heritage and infrastructure, catastrophic events, surveillance, and rescue operations.
- **Digital Reality (DR).** Augmented Reality (AR) refers to the real world overlapping digital information. Virtual Reality (VR) is a computer-generated technology that allows users to be completely immersed in a virtual environment. With Mixed Reality (MR), real and virtual worlds overlap with interactive superimposition of digital objects. 360-degree videos are spherical recordings captured by sophisticated cameras with omnidirectional lenses that can collect images from all over the scene. They enrich the way of approaching information and extracting value from it for a pool of users.

- **Big Data (BD) and Internet of Things (IoT).** BD refers to large or complex data sets to be handled by traditional data-processing application software. IoT identifies the ability to connect every object to the Internet. Together they maximize data availability and utilization towards greater digitalization and process automation.
- **Artificial Intelligence (AI) and Machine Learning (ML).** AI broadly refers to any human-like behavior displayed by a machine or system and responds to the need for lightning-fast responses to automate and speed up decision-making processes. ML technique develops pattern recognition by intelligent algorithms to continuously learn and make predictions based on data available or collected. The potential lies in analyzing masses of data with varying degrees of autonomy to successfully react to unfolding scenarios based on previous outcomes.
- **Robotics.** Designing machines that can help and assist humans. While they reduce jobs, they also improve the quality of life for performing sensitive, routine, strenuous, or risky tasks.

BUILDING THE SOCIAL NETWORK OF THE FUTURE

The Handbook provides an overview of methods, processes, and tools adopted to achieve super-smart society needs by exploiting Digital Reality and Interactive Technologies. It includes use-cases that illustrate applications that place people's quality of life at the center of digitalization, accessing and managing various information and data domains. Due to its cross-cutting approach, the book aims to bring together different types of actors interacting with the functions considering diverse and complementary points of view. Indeed, Society 5.0 is not something to come but something to co-create (Keidanren, 2018). This work can represent a stimulating reference resource addressed to the following.

- Researchers from engineering, computing, medicine, psychology, and sociology with the aim of underscoring the potential for cross-fertilization and collaboration among these communities.
- Students who can, on the one hand, benefit from advanced didactic experimentation proposing collaborative tools establishing interactive virtual learning environments and, on the other hand, can develop a strategic attitude and working method to approaching complex problems.
- Professionals of the same fields as researchers (i.e., Engineers, Managers, Technicians, Health staff, Heritage curators, Cultural institutions, Educator) to exploit and make operative the scientific and academic outcomes and contribute to the digital transition process by formulating new products and services to be placed on the market (i.e., Multinational companies, ICT developers).
- Citizens who can be made aware of and engaged with various subjects by accessing information and cultural content more effectively and inclusively, playing an active role in the digital social network.
- Patients and fragile people who can understand how modern technology can play an essential role in supporting their daily activities and improving their quality of life through increasingly personalized and cutting-edge entertainment and therapy opportunities.

ORGANIZATION OF THE HANDBOOK

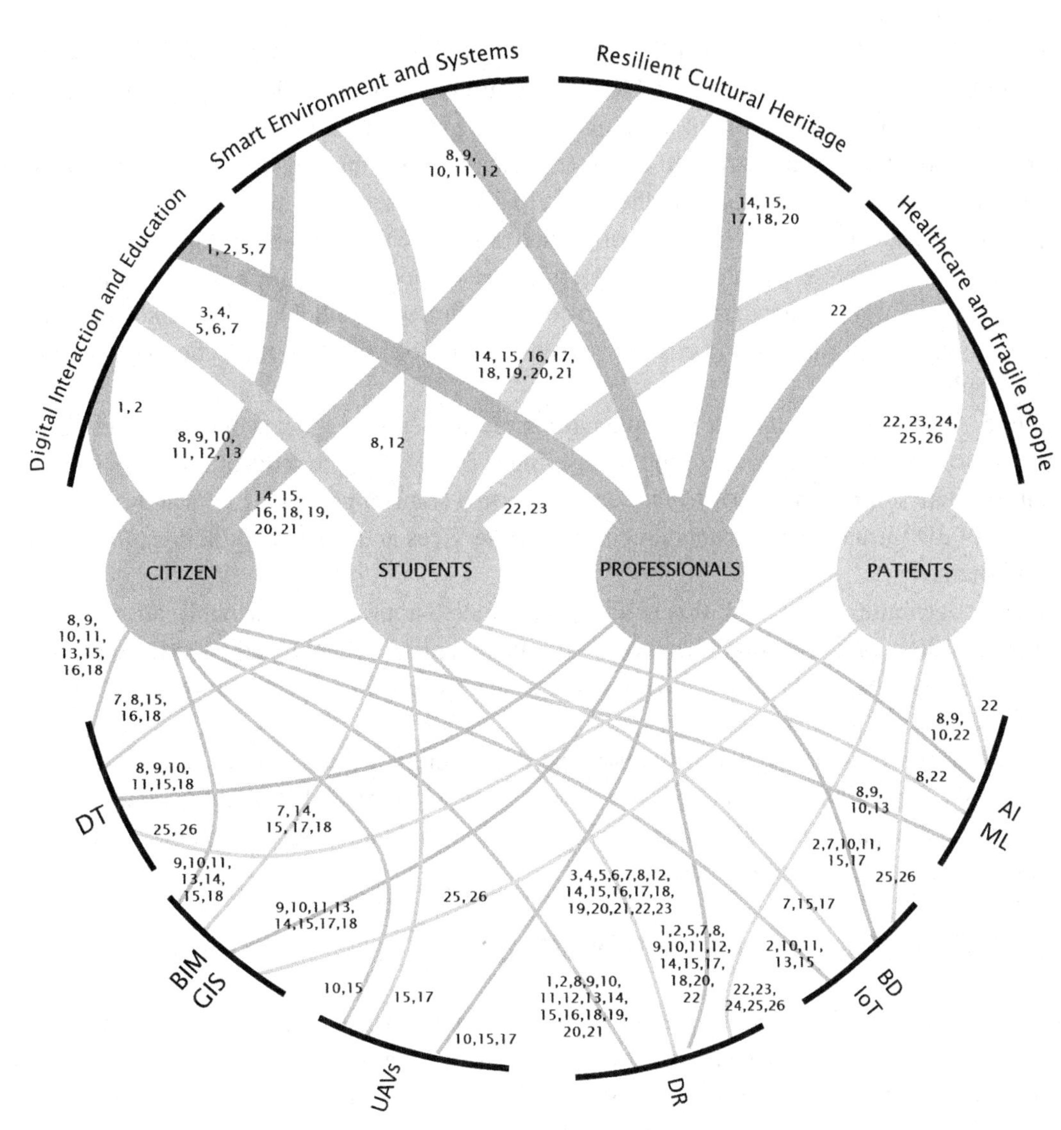

The book includes 26 chapters covering four application areas linked to the social needs outlined above. A brief description of each chapter follows.

Section 1: Digital Interaction and Education

Chapter 1 analyzes the importance of the sense of touch and its study from a neuroscientific point of view to reproducing credible and fulfilling experiences in Virtual Reality, starting with a reflection on what is "real" for us. The chapter discusses how this sensory modality contributes to the sense of presence and its fundamental role in reproducing virtual social interactions (such as those occurring in Metaverses). The current technical limitations and future potentialities involved in its application are also discussed.

Chapter 2 discusses the application opportunities of threshold design in Mixed Reality scenarios. These thresholds identify steps beyond which the human body has the ability to access the content. The experiment is described, and the data recorded through multiple visualizations are discussed to support the conceptual and bibliographical work.

Chapter 3 described the virtual world challenges to overcoming common learning disabilities in children and the possibilities of Augmented Reality within learning disabilities. Virtual Reality has not only proven to be a corrective action for learning disorders but has also succeeded in educating and curing them. Currently, the virtual world can discover the signs and symptoms of learning disorders early, making it easier for special educators and therapists to tackle these problems.

Chapter 4 presents an Android-based learning application related to the digital laboratory concept for phytochemical screening. The chapter considers the users' needs for remote practicum in vocational colleges during the Covid-19 pandemic. Through collaboration with Poltekkes Kemenkes Jakarta II and UIN Syarif Hidayatullah Jakarta, the chapter presents the design of learning technology for building interactive applications.

Chapter 5 presents the results of a multiple case study that explores the affordances and limitations of using 3D virtual environments for group counseling services in higher education. The chapter compares multiple data sources from face-to-face and 3D virtual counseling groups. The authors discuss the implications of 3D virtual environments for counseling services and provide guidelines for researchers and practitioners.

Chapter 6 contributes to the definition of a human-centered approach to Virtual Reality in university teaching. The authors propose a digital game environment applied to health and nutrition education and provide experimental evidence that its effectiveness is strictly related to an educational model based on the specific features of presence and immersion.

Chapter 7 briefly presents digital innovation and the enabling technologies as engines for applying new educational approaches. It introduces two examples of Master's and Bachelor's courses related to BIM and algorithmic parametric modeling that integrates several tools and technologies, such as Cloud Computing, Big Data, and Machine Learning.

Section 2: Smart Environment and Systems

Chapter 8 discusses computationally driven methods for wicked urban challenges through data-informed design speculations. The introduced cutting-edge design speculations conducted at Aalto University (Finland) aim to open up new ways of immersive design simulation and participatory processes in environmental design and urban development to shape the future and give sustainable answers to societal and environmental challenges.

Chapter 9 is focused on the communication of the history of architecture and the city to a broader public with the aid of new tools, underlining how these same tools intervene in the study of the city compared to more traditional, no longer effective today. The challenge is to address issues difficult to narrate with the traditional urban history tools (archive research, bibliographic, iconographic, cartographic): an example is how to deal with not only historical dynamics of transformation (flows of goods and people, movements, sounds) of the life of a city. Therefore, multimedia tools - configured with Machine Learning and Artificial Intelligence - are configured as a moment not only for innovation but also for the development of further research.

Chapter 10 analyzes the current process for evaluating the digital readiness of construction companies starting from the report on the current technological trends in the construction sector. The study highlights the need to integrate the existing methods proposing a dedicated assessment of the level of digitalization of construction companies to provide a common ground of analysis that can be used to identify and develop the best strategies for digital innovation.

Chapter 11 presents the outcomes of the POR-FESR "eBIM: existing Building Information Modeling for the management of the intervention on the built environment" project in terms of developing skills, models, and solutions related to the conservation and enhancement of the built heritage. BIM methodology, implemented on open standard dedicated IT platforms, is used to identify and characterize building material.

Chapter 12 presents the first results of a research program that aims to improve the designer's interpretation of fire evolution and its effects. In particular, the development of an application is proposed that allows the real-time monitoring of fire effects, occupants, signage systems, and fire scenarios. The method represents an aid to research tools related to the world of Fire Safety Engineering and Virtual Reality, as it can fully immerse the user in the environment and evaluate its performance from the point of view of fire prevention.

Chapter 13 illustrates integrated applications for monitoring and managing a sport facility as an information base to involve spectators by recording their behavior, movements, and preferences through sensors, Big Data, and Artificial Intelligence capable of prefiguring customized and fan-engaging solutions.

Section 3: Resilient Cultural Heritage

Chapter 14 explores new directions Digital Reality is currently taking in the field of Cultural Heritage. New needs to visit, enjoy, understand and preserve cultural assets through digital fruition are described, as well as an overview of some significant projects, methodological approaches, and applications in the field of Semantic Web technologies. The topic of the role of digital technologies in safeguarding endangered Cultural Heritage is faced as well, by shortly presenting the EU project 4CH and other related activities.

Chapter 15 presents an innovative approach using HBIM technology as an essential tool for bringing Society 5.0 closer to the concept of tangible Smart Heritage. The final goal is to clarify the impact of ICT during the pandemic and post-pandemic era. In addition, the chapter describes the creation of a virtual tour of the Baia Castle in the south of Italy to break down architectural barriers and allow access to all users as a benefit of dematerializing the architectural asset.

Chapter 16 addresses the topic of Accessibility of Cultural Heritage for educational purposes, enhancing the involving mechanisms that can be triggered from the communication of Digital Heritage to societal users. The design of multiple strategies for presenting Virtual Heritage Environments, and supporting prototypes, to citizens is considered in three different solutions adopted during the European Research Night in Pavia from 2019 to 2021, within the digital data from the documentation of Architectural sites in the H2020 project PROMETHEUS.

Chapter 17 analyzes how digital technologies could provide a new human centrality in interpreting and presenting Cultural Heritage, realizing an inclusive and engaging phygital environment. The contribution focuses on the relationship between heritage and communities, starting from recognizing the progressive settlement of the conception of Cultural Heritage through the years. It reflects the opportunities that technological innovations can open in this field, considering tools for digitization and the enjoyment of heritage and the ICT infrastructure needed to achieve this process given a cultural-based smart society.

Chapter 18 investigates the digitization process's role in the valorization and dissemination of Cultural Heritage. This chapter also aims to highlight the numerous positive impacts that this operation produces, both from a social and economic point of view, both on individual and collective levels. The main goal is to illustrate and critically analyze the research carried out on the Farnese Theatre in Parma, Italy. Particular attention is paid to the methodological choices for creating an extremely versatile three-dimensional model for its possible uses.

Chapter 19 describes the adoption of digital applications in the cultural and entertainment sector, where Gamification and Serious Games are exploited to transfer reality into new virtuality. Starting with the digitalization of the Theatre system, the accessibility of content and the sustainability of events is pursued by creating an Interactive Digital Service Model to expand the offer of digital services to an increasingly broad and heterogeneous audience.

Chapter 20 focuses on how real-time computer visualization technology proves to be highly efficient in disseminating cultural values inferred by ancient drawings, manuscripts, and objects belonging to museums and cultural institutions. Following the knowledge-intensive society paradigm, some experiences producing digital replicas are introduced, highlighting opportunities and criticalities to be considered in adopting custom processes and applications for the Cultural Heritage.

Chapter 21 shows some outcomes of a research project begun in 2021 in collaboration with the Archivio Progetti IUAV of Venice to disseminate the preserved drawings, documents, and projects. The chapter deepens the topic of the unbuilt Venice, circumscribing the investigation of the project by Luciano Semerani for the international competition held in Venice in 1978. The discussion on the Venetian structural system, the urban trace, and the architectural configuration is re-established in a dialogue between history and contemporaneity from digital models and virtual tours integrating information actions concerning the qualities of architectures and urban spaces.

Section 4: Healthcare and Fragile People

Chapter 22 summarizes the current application of Mixed Medical Reality within neurosurgery, encompassing the treatment of several neurological pathologies, including brain tumors, traumatic brain injuries, and spine disorders. The chapter also illustrates the pilot study investigating the benefits of using Mixed Reality to reduce exposure time to Covid-19 ICU red zone doctors conducted at the Covid-19 Intensive Care Unit of the University Clinical-Hospital Centre "Dr. Dragiša Mišović-Dedinje" in Belgrade, Serbia.

Chapter 23 examines the benefits and drawbacks of the Virtual Reality range of technologies, highlighting where VR potential ends and 360-degree video potential begins. 360-degree movies, often known as immersive videos or spherical videos, are a contemporary rising trend. They may be a more inventive, cost-effective, and realistic tool than VR devices: they allow users to immerse in truly natural surroundings using a standard web browser or mobile device to pan around the area by clicking and dragging.

Chapter 24 illustrates the application of Virtual Reality and Serious Games to improve the quality of life of Dementia patients through the implementation of non-pharmacological treatment for cognitive disorders. The semi-immersion in Snoezelen and natural settings are experienced to offer a relaxation/rehabilitation therapy besides an entertainment experience with multisensory stimuli.

Chapter 25 discusses how modern technologies and the concept of virtuality, in Digital Twin and Metaverse forms, can help frail people increase their autonomy and quality of life. An analysis is provided on the leading technologies and research projects concerning the use of the digital social world to support frail persons, and the SIRIO solution, developed by Politecnico di Torino, to support people with Amyotrophic Lateral Sclerosis is presented.

Chapter 26 investigates the benefits related to the adoption of innovative methods and tools in the Digital Health field for the development of a Digital Twin for Amyotrophic Lateral Sclerosis with a human-centered approach. Through the development of virtual environments, the goal is to evaluate the involvement of ALS patients who succeed in increasing their autonomy in daily actions. The ongoing research aims to evaluate the benefits of BIM-IoT-VR interaction through innovative graphical interfaces with an interdisciplinary approach, assessing user response by increasing user engagement.

SOCIETY 5.0 WILL CHANGE THE WORLD

The search for solutions to achieve Society 5.0 will be a challenging topic in the coming decades; therefore, this volume will be a trailblazer in this line of investigation, establishing a precise identity and affirming its high interest. It opens a new interpretation key to the digital initiatives experienced in the different sectors promoting human-centered services. The case studies presented and how technologies are used are highlighted as models capable of generating a positive impact. Technological development will always go further, but in the future, humans will require imagination to change the world and creativity to materialize their ideas (Keidanren, 2018). This book has a significant impact on the issue of Society 5.0 and on how technology should be used in the service of people, helping to stimulate stakeholders to seek new solutions, even starting from the examples described in the different chapters.

Francesca Maria Ugliotti
Politecnico di Torino, Italy

Anna Osello
Politecnico di Torino, Italy

REFERENCES

Cabinet Office. (2016). *The 5th Science and Technology Basic Plan.* https://www8.cao.go.jp/cstp/kihonkeikaku/5basicplan_en.pdf

Deloitte. (2022). *Digital Reality.* https://www2.deloitte.com/us/en/pages/consulting/topics/digital-reality.html

Hitachi-UTokyo Laboratory. (2020). Society 5.0. A People-centric Super-smart Society. *Springer Singapore, 177.* Advance online publication. doi:10.1007/978-981-15-2989-4

Keidanren (Japan Business Federation). (2018). *Society 5.0 - Co-creating the future.* http://www.keidanren.or.jp/en/policy/2018/095.html

Mitchell, W. J. (2001). *Smart Cities.* https://smartcities.media.mit.edu/

United Nations. (2018). *World Urbanization Prospects. The 2018 Revision.* https://population.un.org/wup/publications/Files/WUP2018-Report.pdf

World Economic Forum (WEF). (2020). *The Future of Jobs Report 2020.* https://www3.weforum.org/docs/WEF_Future_of_Jobs_2020.pdf

Section 1

Digital Interaction and Education

Chapter 1
Haptic Interaction in Virtual Reality:
Are We Ready for the Metaverse? Neuroscientific and Behavioral Considerations

Alberto Gallace
University of Milano Bicocca, Italy

ABSTRACT

Touch is fundamental to create our perception of reality and to allow fulfilling social experiences, such as those at the basis of metaverses. In order to be accurately reproduced, a number of scientific and technological aspects should be considered. In this chapter, the authors highlight the relevance of the tactile modality in eliciting 'presence' in virtual reality interactions. They also discuss the neuroscientific foundation of our bodily interactions and the fact that they are based on a number of receptors and neural circuits that contribute to the complexity of our perceptions. The available technological devices for the reproduction of touch in virtual environments and their limitations are also described. They suggest that virtual interactions should include more of this sensory modality and that attempts should be made to go beyond the actual approach to 'mimicking reality'. In particular, future simulations should consider the perspective of creative 'hyper-sensations' including 'hyper-touch' on the basis of our psychological and neuroscientific knowledge.

INTRODUCTION

The technological advances of the last few decades have allowed virtual and augmented reality to enter people's lives for the first time. Most recently, the term 'metaverse' (taken from the 1992 sci-fi novel "Snow Crash" by Neal Stephenson), referred to virtual worlds shared over the internet, would seem more popular than ever before and millions of dollars are invested into its development (Kye, Han, Kim, Park, & Jo, 2021; Gallace, 2022). The success of companies such as Oculus (now Meta) or HTC certainly

DOI: 10.4018/978-1-6684-4854-0.ch001

testifies this growing interest in the world of virtuality. As a matter of fact, these technological devices are certainly not 'new' and virtual reality has experienced several 'waves' of popularity from the 1960s to our days. It seems almost prehistoric to think about Morton Helig's 'Sensorama', dating back to 1957, one of the first commercial attempt to simulate an hedonic/pleasant complex experience (i.e. the first flight simulators for military purposes were implemented between the 1910s and 1930s), such as driving a motorcycle. Interestingly, these early examples of virtual reality already embedded some form of tactile stimulation in the experience. That is, since the very beginning of virtual reality it was very clear that the world had to be simulated taking into account all sensory modalities that our brain can process, in order to be perceived as a fulfilling and enjoyable experience. This trend never stopped since then, even if the available technologies and scientific knowledge somehow affected its speed.

The popular way in which most people today think about virtual reality is through headsets (devices that can be worn on a person face, as a sort of heavy pair of glasses) and their hand controllers. From this perspective, the Sutherland's 'Sword of Damocles', one of the first head mounted display anchored to the ceiling due to its weight and dated back to 1968, might look now to most people as a peculiar experimental attempt to create wearable displays, from a bygone era. However, once again such device and many others not mentioned here, although still embryonic, have certainly laid the foundations for 'portable' virtual reality as we know it today. Even the interfaces developed by VPL research (the company founded by Jaron Lanier, considered one of the fathers of Virtual Reality) at the dawn of the 80s - including headsets, whole body suits and haptic gloves - were strikingly similar, at least in their basic operating principles, to those available nowadays (Lanier, 2017)! However, if the technological and theoretical systems at the foundation of VR were already well underway in the 80s, why after more than 50 years from the first studies in this field, we still don't use these technologies as an integral part of our daily activities?

Certainly part of the answer to these questions is linked to a technical/computational aspect: Moore's famous law, according to which the number of transistors in microprocessors doubles every 12 months (value then changed to 18 months) and their price is halved. In other words, the computing capabilities of electronic computers (microprocessors), such as those necessary for the interactive reproduction of complex three-dimensional graphic environments, have increased considerably since the 1980s, making computations once impossible or too slow, now possible in faster times. In fact, we should consider that the reproduction through a visual medium (e.g., a display) of an interactive three-dimensional content (e.g., an avatar or object) depends on the number of polygons (simplified geometrical structures used for 3D graphic rendering) of which it is composed. More polygons, better graphics quality, but also greater computational demands. The reader belonging to generation X (born between 1965 and 1980) could think here about the first version of 'Lara Croft', the fictional main character of the 1996 popular game 'Tomb Raider'. The features of the first versions of Lara, as well as of the remaining characters of the game, were based on a limited number of polygons, the only ones which could be managed by the Home-PC processors of the time. By contrast, the youngest readers belonging to the millennial generation, are certainly accustomed to the highest graphic quality of games such as League of Legends or Fortnight, where the characters are created by means of millions of polygons. They will certainly be horrified by the visual simplification at the basis of the early Lara's details! However, so far we have only spoken about the visual aspect of early videogames, but we might wonder about what kind of advancements (if any) were made since the 80s on the reproduction of other sensations beside the visual one. After all, the early videogame players were by no mean supposed to physically touch any element within the Tomb Raider game or any other game (just as no haptic interaction was still provided in the 2002 social game 'Second Life', a sort of ante-litteram metaverse by Linden Lab)! In a few words, is there anything more than vision to care about in digital interactions? More specifically, did we make enough progress in the delivery of haptic sensations in virtual reality? and are these sensations really relevant for virtual interactions and economical investments in this field to take place? Unfortunately, when we exit the realm of vision, the story may become a bit more complicated. In particular, with touch we enter a word made of physical forces, physiology, engineering, emotions and social relationships (and the interaction between all of them).

In order to answer the questions highlighted above we must understand what are the main problems related to the reproduction of touch in VR. A first point that we should make clear here, is that the main issue is certainly not only one of computational power. A set of further technical limitations, theoretical knowledge and appropriate business models has certainly contributed to the lack of diffusion of tactile technologies in the past, but it may not be so in the near future. In this chapter, we will focus on an element that we believe central to the future development of virtual reality, namely, the limitations in our ability to reproduce the direct contact between our body and the world in simulated environments. We will thus analyze the importance to define, from a neuroscientific point of view what is reality and how touch can contribute to its perception. We will also define the importance of touch in our everyday interactions and how bodily sensations could be reproduced with the available technologies. We will finally highlight that from individual interactions with objects to more social forms of relationships (human to human, human to AI, and metaverses), the study of tactile information processing should be a fundamental aspect of the 'virtual revolution' that we are witnessing in our era.

TOUCH AND REALITY

Before we can even think of being able to reproduce our real experiences in a virtual world we must reflect on the very concept of reality. What is real? How do we define 'real' and, above all, how do we discriminate something real from something that is not? The whole essence of the concept of virtual reality lies in the answer to these questions, because reality in digital worlds needs to be 'rebuilt' from scratch. Unfortunately, however, the answer to this question is not as simple as one might think and hundreds of scientists, philosophers and researchers have come up against this important aspect. Let's start with a question as old as the history of Western thought: does the sound of a tree falling in an uninhabited forest really exist? This early question by George Berkely in 1710 summarizes very well the quest to understand how 'real' a phenomenon can be, in the absence of a human observer who is there to perceive it. Today, with the development of psychology and brain sciences we can try to provide a scientific answer. As strange as it may seem to the reader, the answer is no, there is no sound without a listener, but this does not necessarily mean that the observer is fundamental in order to create the reality of a phenomenon. The noise of the tree does not exist as 'noise', simply because what we define as noise implies a transduction of one form of energy (that produced by the fall of the tree that propagates as a shock wave in the air) into another (electrochemical signals) by our central nervous system. In other words, a falling tree produces a vibration in a specific range of frequencies, some of which are translated by our inner ear systems into neural impulses. These impulses activate certain areas of our brain responsible for analyzing and interpreting the stimuli and producing the sound experiences that we are used to hear (i.e., a noise in this case). Without a receiving organ, a transmission system and a central operating structure, therefore, we cannot have 'sounds and noises' but only an energy vibration in the air.

The same happens for visual sensations. In nature, electromagnetic energy can travel through photons whose frequencies of vibration are translated by the visual apparatus into electrochemical signals. Their neural processing in specific areas of the brain gives rise to all colors of the visible spectrum.

All of this implies that our perception of reality is very different from the reality of the physical world (interestingly, it is worth noting here that even theoretical physics, especially after the development of quantomechanics, has started to challenge the claim that reality exist regardless of observation and devices created to ask questions about nature; see Wheeler, 1989, for a discussion on this point). Coming back to the main topic of this chapter, we can ask ourself if and how our sense of touch can contribute to define what we perceive as 'reality'. Interestingly, it is worth noting that we recur to the sense of touch, every time we are uncertain about the physical existence of a given stimulus (see Gallace & Spence, 2014). That is, if we are uncertain if something is unreal, we can always extend our hand and try to touch it. This point – touch as a proof of reality (see also the Saint Thomas in the Bible, reproduced in the famous painting by Caravaggio) – should then be considered very important in virtual simulations. Whenever we interact with a well reproduced virtual reality environment and we try to grab objects, we are likely to experience a sense of disappointment in noticing that our hand trespasses the object without giving us that important perception of a world that exists beyond the boundaries of our mind (we observe this every day in our laboratories, whenever people try VR for the first time!). In this case, a breach in the sense of 'presence' (the feeling of being there; see Slater & Wilbur, 1995; Bystrom, Barfield, & Hendrix, 1999; see also Gallace, Ngo, Sulaitis, & Spence, 2012) occurs and the realism of the situation is compromised. This is a very important problem that should be seriously considered if one's aim is to make people behave in a virtual reality environment just as they will in its real counterpart. Think about a simulation whose main purpose is to learn a manual job (one requiring visuo-motor coordination). The gestures that need to be performed in order to safely work often depends on the structure, shape and affordances given by the tools and instruments that we need to operate. What kind of learning can we hope to obtain when the virtual objects can be trespassed without any apparent consequence on our performance? Imagine an engine that needs to be mounted in a certain way. A visual only simulation does not allow to experience the difficulties in reaching (without bumping against physical structures), the inner and most difficult-to-see parts of the structure. It is not simple to anticipate how this lack of tactile information can compromise the final performance/learning by the users (see Nazir, Gallace, Manca, & Overgard, 2016, for the virtual training of chemical plant field operators).

It is also worth noticing that, at least from a scientific point of view, in the last few decades our concept of learning has changed a lot. More specifically, a number of theories, such as the one known as 'embodied cognition', have started to suggest that we learn with our body and its physicality (e.g, Varela, Thompson, & Rosch, 1991), and not only by means of language-based contents. This concept is not far from the initials intuitions that resulted into the success of Montessori approach to learning and teaching. That is, touch and the involvement of our body makes things real and help to understand better the properties of the world where we live. This cannot be neglected in any serious approach to virtual reality development.

THE IMPORTANCE OF TOUCH IN OUR LIFE

The importance of touch in our everyday interactions is certainly underestimated. Very often we are tempted to believe that vision is our most important sensory channel, and that most resources (economical and computational) should be dedicated to recreating high definition visual content in virtual interactions. This approach is simply wrong and probably due to the lack of technological and scientific knowledge on how to reproduce tactile sensations in virtual reality. In fact, just as Tiffany Field is used to say, 'nothing can arouse you like touch' (Field, 2001; see also Etzi & Gallace, 2016). As a matter of fact, we use touch and proprioception (the perception of body movement and position; e.g., Canzoneri, Ferrè, & Haggard, 2014) for a lot of everyday activities, even if we are often unaware of the processing of such information by our central nervous system. For example, we use touch and proprioception to adjust our posture whenever we walk, sit or lie down. In addition to this, the information coming from our body is used to organize movements and interact with the environment. People who suffer from a rare condition, known as 'deafferentation', that prevents them from using tactile and proprioceptive information, experience huge difficulties in performing even the most simple actions such as standing or eating. This is due to the fact that these systems continuously inform our motor system regarding the current outcome of motor commands and the actual body position in space (see Gallace & Spence, 2014).

We also use touch for food perception. In fact, our sense of taste depends on a huge amount of information that is not only provided by the chemical receptors on our tongue, but also by the tactile (mechanoceptors) sensors present in our mouth. It is not difficult to think here about how different can be our gustative experiences when the exact same or very similar ingredients are used to create food having different textures or shapes (gelled, liquidized, dehydrated, moussed, etc.; see Spence, Hobkinson, Gallace, A. et al., 2013, for a review). Most importantly, touch is used to enrich our social experiences. This aspect is more relevant now than ever before while talking about virtual reality, because our society is starting to get serious about the development and use of shared digital environments (metaverses). We should then ask ourself how fulfilling social experiences (comprehending romantic an sexual interactions) might be within the metaverses without their tactile components (see Gallace & Girondini, 2022).

Touch strongly shapes our interactions with other people and it can be considered a biological part of our very nature (see Gallace & Spence, 2010, 2016). We can only mention here that the tactile system is the very first to develop in an embryo and at birth we can interact with objects and people using touch, even at an age where vision is still particularly poor. The grabbing reflex that we can observe in infants when objects are placed in the palm of their hands, has probably developed by nature in order to establish a first and safe interaction with the world, a fundamental aspect to guarantee the survival of a specie that is extremely depended on other individuals for several years after birth. Interestingly, the presence of social touch profoundly affects our behavior. For example, it has been shown that social contact can enhance compliance to requests, a phenomenon known as the 'Midas touch effect' (see Crusco & Wetzel, 1984). That is, in a now famous experiment, real costumers in a restaurant were randomly touched on the shoulder by the waitress while they received the bill. Surprisingly, those who received the touch provided the highest tip, as compared to those who were not touched. This occurred independently from the fact that the costumers were able or not to remember any touch that occurred during this situation. The effect was replicated in many other situations leading to the exact same result: social touch, at least under certain circumstances, can lead to an increase of compliance and positive evaluations.

Another interesting effect of touch would seem to be related to empathy and/or the susceptibility to authority. In particular, in another series of famous studies, it was investigated how random people can, apparently blindly, submit to authority, also under conditions where they could harm other individuals (Milgram, 1974). In these experiments participants were requested to provide a painful electric shock to other participants (actually actors who did not receive any arm) whenever requested by the experimenter, due to their errors in performing a simple task. The large majority of the participants continued to administer these shocks (up to a hypothetical level of 400 volts!) even when they could hear the complaints and even screams by the other individuals. The experimenters tried to understand what elements could mitigate this complete submission to the established authority and to modulate different aspects of the experiment (e.g., where it was performed, whether or not the experimenters wear the white coat and so one). The most effective of all changes in reducing submission, was actually the possibility of touching the other participant and share with him/her for a few seconds the same physical space. This finding can be perhaps taken to suggest an important role of touch in activating some forms of empathic behavior (but maybe also in creating bonds between people, that could be useful in defying authority!).

Certainly, touch plays an important role in shaping our emotions (see Gallace, 2012; Gallace & Spence, 2010). Emotional experiences in humans should be considered the product of a complex interaction occurring in the nervous systems between physiological, sensorial, motor and cognitive information (see Schachter & Singer, 1962). As far as physiological arousal is considered - an aspect that determines the degree of activation of our autonomic system responsible for basic reactions to the environment (i.e., flight or fight response) – research has shown that touch is very effective in modulating our bodily signals (e.g., sweeting of the skin, heartbeat, respiration rate etc; e.g., Etzi, Carta, & Gallace, 2018). Unfortunately, however our capability of generating long distance emotional content through haptic devices is still far from been fully effective (see the next section; see also Gallace & Girondini, 2022).

Social touch would also seem at the very basis of our wellbeing and health. In particular, touch has been shown to be involved in the release of hormones like oxytocin (responsible of creating bonds between individuals), but also endorphins (substances responsible for pleasant sensations). Social tactile interactions have also been proven to reduce other neurochemicals, such as cortisol and adrenocorticotropin, related to our stress response; see Morhenn, Beavin, & Zak, 2012). It is thus very surprising that, despite this scientifically-proven importance, the sense of touch is now more than ever before taken away from our everyday interactions, progressively leading to 'touch free societies'. This trend has already been shown to have detrimental effects on our wellbeing, a phenomenon known as 'skin hunger' (e.g., Field, 2001; see also Gallace, 2012). The question becomes then if we do really want to create touch free metaverses through virtual realities or rather addressing more seriously the technical and scientific difficulties in introducing touch in virtual environments.

HOW TO REPRODUCE TOUCH IN VR

Neuroscientific Aspects

Taken for granted the importance of touch in virtual interactions, it is now relevant to understand what kind of sensations need to be reproduced and how. From a neuroscientific point of view, understanding how the tactile system works is certainly a great challenge, and this difficulty has partially justified in the past the lower number of scientific papers dedicated to this sensory modality as compared to those addressed to vision and audition (see Gallace & Spence, 2014). One of the first problem in understanding touch is due to the fact that this system continuously interacts with other neural systems (e.g., such as those involved in proprioception, motion, balance, body representation, etc.) in order to provide its main functions. There are certainly a number of specific tactile receptors that would seem to respond to highly specific forms of stimulation (see Gallace & Spence, 2014). For example, the Pacinian corpuscle receptors, present in the deepest part of the skin respond to limited deformations of the skin surface and contribute to the perception of pressure changes and vibrations. Other receptors, such as the Meissner corpuscles, are present in the most superficial part of the skin and are responsible of the perception of textures (such as in Braille reading) and fine movements on the skin (see Iggo & Andres, 1982). Each receptors has a specific pattern of response (a sort of 'language' that they use to communicate with the brain); some send impulses in very rapid sequences only when a stimulus is applied on the skin and then they become silent even if the stimulus is still there (rapidly adapting receptors), while others continue to respond until the stimulus is removed from the body (slowly adapting receptors). However, even these highly specialized receptors cannot work in isolation or without the input provided by other sensory modalities. For example, the perception of an object shape is also determined by proprioceptors (receptors that are connected to our muscles, tendons and joints and determine the position of body parts and the degree of muscle flexion). Similarly, there are perceptions that are totally due to the integration of signals from two different classes of receptors. Think for example about the sensation of 'wetness'. In order to obtain such feeling (very hard to reproduce in virtual reality), we need the integration between the information provided by thermal receptors (that inform about the presence of stimuli that differ from body temperature) and movement receptors (sensors often associated to body hairs, that detect variations on a stimulus movement over the skin). Without such combination we can't get this sensation. That is, information provided by proprioceptors, mechanoreceptors (sensors tuned to different forces applied to the skin), and thermal receptors (not to mention 'nociceptors' for pain sensations) continuously interact in our brain with the output commands sent to our muscle and with information provided by our vestibular system (responsible for the perception of the gravity force), in order to recreate the complexity of our tactile world. Amazingly, most of this processing is done by our brain unconsciously!

From a neuroscientific point of view the complexity the tactile system is even greater if one considers our social interactions. In particular, it has been shown that humans, just like other animal species who live in societies, have a neural fiber system totally dedicated to some sort of 'social touch'. In particular, the non-glabrous part of our skin (the one covered by body hairs) is innervated by a class of neural fibers, known as 'CT', that respond to stimuli moving on the skin surface at a specific velocity and administered with a specific level of pressure; something that can be described as a caress! These fibers, would seem to be part of a neural network responsible for establishing comfort, pleasure and wellbeing through tactile social interactions (e.g., Löken, Wessberg, Morrison, McGlone, Olausson, 2009; Manzotti, Cerritelli, Esteves et al. 2019; see also Gallace & Spence, 2010, 2016). It is finally important to notice here, that the way in which our brain analyses information coming from our body and interpret it, is not unidirectional (i.e., given a certain stimulus the perception elicited is always the same). We now know that the multisensory context where the stimulation occurs contributes to our sensations and interpretations of reality. For example, we know that the same stroke received on the skin elicits different perceptions and neural activity in the somatosensory cortex (one of the first structures responsible to process information coming from the body surface), as a function of the visual available information about the material touched or the person who delivered the touch (e.g., Gazzola, Spezio, Etzel, Castelli, Adolphs, & Keysers, 2012). Once again, it cannot be conceivable to reproduce credible and fulfilling multiuser interactions in VR (e.g., metaverses) without considering these important neuroscientific discoveries (see Gallace & Spence, 2008, 2014; Moseley, Gallace, & Spence, 2012, for reviews on the neuroscientific aspects of touch).

Technical and Behavioral Aspects

Having seen that there are many different signals used by our brain in order to perceive touch, we can now approach the technical issues about how to reproduce them in virtual reality, and the effects that any potential hardware limitation can have on our behavior as users/individuals. Several devices are now available in order to reproduce some forms of tactile sensations and each of them has some serious issues which still need to be overcome (see Sreelakshmi & Subash, 2017). Just as it occurs for other sensory modalities, in virtual reality interfaces and devices need to reproduce those physical energies that our senses can translate into the electrochemical signals used to lie the foundations of our perceptual world. First and most importantly, as previously mentioned, the relationship between the physical world and our perceptual world is not a simple one, and certainly not even a geometrically linear one. That is, there are some forms of energies that results into no activation of our neural system, others that can activate our neural system but are processed under the level of consciousness, and some others that produce some form of conscious perception (end eventually behaviors). That is, in order to develop efficient instruments to recreate virtual tactile interactions these relationships need to be known, and the science than analyses and discover them is known as 'psychophysics'. Instruments need then to comply with the law of psychophysics whenever available (consider that these laws are expressed in a mathematical terms and this certainly can help engineers and designers).

What kind of devices are now available to reproduce those forms of energy that our tactile neural system can interpret as 'real' or 'veridical'? A first hardware classification can be based on the fact that the devices can or cannot generate forces able to stimulate not only tactile receptors on the skin, but also our proprioceptors (that inform us about the tension of our muscles and/or the position of our body in space). The first class of devices comprehend vibrotactile, electromagnetic and ultrasound stimulators, while the latter are generally known as 'force feedback devices'. Surface skin receptors can be activated by vibrators (just as those mounted inside our mobile phones) or by ultrasound force fields generated in mid-air (e.g., Carter, Seah, Long, Drinkwater, Subramanian, 2013; Sand, Rakkolainen, Isokoski, Kangas, Raisamo, & Palovuori, 2015; see Figure 1). Vibrations are certainly useful to simulate simple forms of contact with objects. However, the kind of sensations provided by them should be considered more of a 'proxy' of the real interactions than an actual simulation of touch. That is, the vibration provided by these devices is something that act as some form of tactile feedback regarding the presence/absence of an object in VR, rather than as an equivalent of the full tactile qualities of that object in the simulated environment. A similar line of reasoning can be applied to mid-air interfaces, that use ultrasounds for the stimulation of the skin (generally the fingertips, because they are the most sensitive part of the body). These devices generally couple a hand tracking system (with gesture/posture recognition) with an arrays of ultrasound loudspeakers. The coordination between the position of the hand within a given area and the activation of the loudspeakers allows to generate sensations limited to certain points of stimulation on the hands. The advantage of this system is that nothing needs to be attached to the skin in order to generate sensations (from this feature derives the name of 'touchless' interfaces often given to these device). So far, the intensity of the sensations generated by these device is rather limited, and so is the ability to generate stimulations that can reach the boundary of our tactile acuity (the minimum distance between two stimulated points in order to be perceived as separate sensations; a parameter that varies for different regions of the skin).

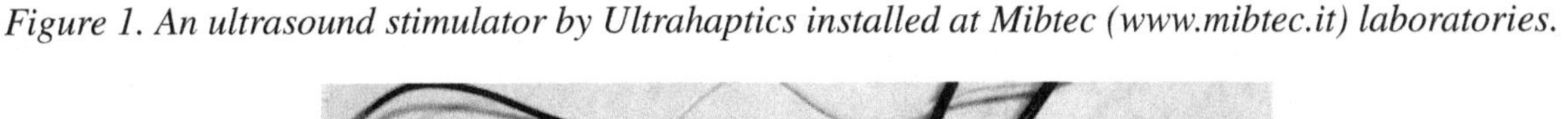

Figure 1. An ultrasound stimulator by Ultrahaptics installed at Mibtec (www.mibtec.it) laboratories.

None of the devices mentioned above can allow users to touch virtual objects without trespassing them. For this reason, they cannot be used to make participants perceive in a very accurate way complex shapes, weight or the solidity of virtual objects. As said, in order to achieve these aspects, we need force feedback devices (see for example Massie & Salisbury, 1994; see Figure 2). These instruments need to apply forces that counteract the tension exert by our body muscles. In order to achieve those forces there are at least two possibilities (others require more complex interactions with the body physiology). One relates to some sort of exoskeletons (more or less extended to different body parts). In this case, some effectors mounted on a sort of robotic structure counteract the force exert by our muscles giving us the feeling of holding objects having real weight and shape. The problem with this hardware is that they are often cumbersome (especially considering that physical forces need to be 'grounded' in some way) and that they may reduce the degrees of freedom necessary to explore objects by means of complex movements (not to mention the issues related to the friction produced by the mechanical movements of the device). Note also that exoskeletons cannot stimulate very accurately all the receptors present in the skin, reducing for example the capability of reproducing fine textures in virtual reality. A specific kind of force feedback device are the magnetic levitation interfaces. In this case a tridimensional object (a sphere) is inserted into a magnetic field and grabbed by the user. Changing the intensity of the magnetic field along different spatial axes allows to generate forces that can counteract those exert by the user's muscles. So far this class of device has not found a large field of application and experimentation is still under way (see also Gallace & Girondini, 2022).

Figure 2. A force feedback stimulator by 3D systems installed at Mibtec (www.mibtec.it) laboratories.

It is worth noting here that all of the devices mentioned so far cannot reproduce the complexity of our tactile world, but are tailored on the basis of specific user cases or scenarios. That is, no device so far can be used interchangeably to reproduce virtual texture, contact with objects, or social interactions within different environments (and certainly not for the whole body surface). The trend is now to try and reproduce some of the most important tactile aspects of a given and highly specific situation and within a very limited space of action. Other strategies adopted for encompassing the limitations of the available devices, are related to the attempt to provide some forms of multisensory compensation for the very impoverish quality of the virtual touch reproduced (e.g., using sound to improve the perception of tactile textures; e.g., Etzi, Ferrise, Bordegoni, Zampini, & Gallace, 2018). All of these technical limitations certainly affect our experience of simulated touch, and also our behavior in virtual environments. For example, the results in attempting to reproduce the Midas Touch effect described above by means of mediated devices in virtual interactions were not incredibly exciting, even if certainly promising (e.g., Haans & IJsselsteijn, 2009). Only the interaction between engineers, psychologists and neuroscientists will allow to develop better systems for the reproduction and the multisensory integration of tactile sensations within virtual reality environments in the future.

CONCLUSION

This overview has analyzed the important role played by tactile sensations in virtual interactions with the purpose of improving our knowledge on this fundamental aspect. We have seen that the sense of touch contributes to determine our perception of reality, but also the quality of our social interactions. Touch is a powerful sensory modality and its lack might determine the failure or success of a certain technological approach to virtual interactions, such as those involved in metaverses, in the future. In this chapter we mainly concentrated on the reproduction of touch under conditions where virtual reality is considered a sort of 'digital twin' of reality (a sort of training camp where to exercise for the real world). A few considerations on this aspect should be drawn at this point. The first relates to the fact that VR is not and should not be considered a complete substitution of reality. From that point of view, touch is fundamental in VR only when it is functional to the given situation to be reproduced (e.g., for learning how to operate a device/machine). However, not necessarily we need to accurately reproduce all the stimuli involved in the real word for a simulation to be effective. A simplified version of our tactile reality might be sufficient in some cases (and not in others). The second point relates to the fact that whenever VR is not considered a 'training camp' of reality (for safety or economic reasons), we should be a bit more creative about its use. That is, VR can offer the opportunity to go well beyond the limitations of the real world and create experiences that might extend our current mental and physical capabilities (e.g., by playing with the laws of physics and of our imagination). The sense of touch is not exempt from this view and we should try to achieve new uses for tactile sensations in VR (or even generate 'hyper-tactile' experiences). This approach might result in an interesting evolutionary cultural challenge for our specie, one that is not far from our grasp.

REFERENCES

Berkeley, G. (1710). A Treatise Concerning the Principles of Human Knowledge. Academic Press.

Bystrom, K. E., Barfield, W., & Hendrix, C. (1999). A conceptual model of the sense of presence in virtual environments. *Presence (Cambridge, Mass.)*, *8*(2), 241–244. doi:10.1162/105474699566107

Canzoneri, E., Ferrè, E. R., & Haggard, P. (2014). Combining proprioception and touch to compute spatial information. *Experimental Brain Research*, *232*(4), 1259–1266. doi:10.100700221-014-3842-z PMID:24468725

Carter, T., Seah, S. A., Long, B., Drinkwater, B., & Subramanian, S. (2013). UltraHaptics: Multi-point mid-air haptic feedback for touch surfaces. *UIST 2013 - Proceedings of the 26th Annual ACM Symposium on User Interface Software and Technology*, 505–514.

Crusco, A. H., & Wetzel, C. G. (1984). The Midas Touch: The Effects of Interpersonal Touch on Restaurant Tipping. *Personality and Social Psychology Bulletin*, *10*(4), 512–517. doi:10.1177/0146167284104003

Etzi, R., Ferrise, F., Bordegoni, M., Zampini, M., & Gallace, A. (2018). The Effect of Visual and Auditory Information on the Perception of Pleasantness and Roughness of Virtual Surfaces. *Multisensory Research, 31*(6), 501–522. doi:10.1163/22134808-00002603 PMID:31264615

Etzi, R., & Gallace, A. (2016). The arousing power of everyday materials: An analysis of the physiological and behavioral responses to visually and tactually presented textures. *Experimental Brain Research, 2016*(234), 1659–1666. doi:10.100700221-016-4574-z PMID:26842855

Field, T. (2001). *Touch.* MIT Press. doi:10.7551/mitpress/6845.001.0001

Gallace, A. (2012). Living with touch: Understanding tactile interactions. *The Psychologist, 25*, 3–5.

Gallace, A. (2022). Cervelli reali in mondi virtuali: psicologia e neuroscienze del metaverso [Real brains in virtual worlds: psychology and neuroscience of metaverses]. In Metaverso. Mondadori Editore.

Gallace, A., & Girondini, M. (2022). Social touch in virtual reality. *Current Opinion in Behavioral Sciences, 43*, 249–254. doi:10.1016/j.cobeha.2021.11.006

Gallace, A., Ngo, M. K., Sulaitis, J., & Spence, C. (2012). Multisensory presence in virtual reality: Possibilities & limitations. In G. Ghinea, F. Andres, & S. Gulliver (Eds.), *Multiple sensorial media advances and applications: New developments in MulSeMedia* (pp. 1–40). IGI Global. doi:10.4018/978-1-60960-821-7.ch001

Gallace, A., & Spence, C. (2008). The cognitive and neural correlates of "tactile consciousness": A multisensory perspective. *Consciousness and Cognition, 17*(1), 370–407. doi:10.1016/j.concog.2007.01.005 PMID:17398116

Gallace, A., & Spence, C. (2010). The science of interpersonal touch: An overview. *Neuroscience and Biobehavioral Reviews, 34*(2), 246–259. doi:10.1016/j.neubiorev.2008.10.004 PMID:18992276

Gallace, A., & Spence, C. (2014). *Touch with the Future: The Sense of Touch from Cognitive Neuroscience to Virtual Reality.* Oxford University Press. doi:10.1093/acprof:oso/9780199644469.001.0001

Gallace, A., & Spence, C. (2016). Social touch. In *Affective Touch and the Neurophysiology of CT Afferents* (pp. 227–238). Springer New York. doi:10.1007/978-1-4939-6418-5_14

Gallace, A., Tan, H. Z., & Spence, C. (2007). The body surface as a communication system: The state of the art after 50 years. *Presence (Cambridge, Mass.), 16*(6), 655–676. doi:10.1162/pres.16.6.655

Gazzola, V., Spezio, M. L., Etzel, J. A., Castelli, F., Adolphs, R., & Keysers, C. (2012). Primary somatosensory cortex discriminates affective significance in social touch. *Proceedings of the National Academy of Sciences of the United States of America, 2012*(109), 9688. doi:10.1073/pnas.1113211109 PMID:22665808

Haans, A., & IJsselsteijn, W. A. (2009). The Virtual Midas Touch: Helping Behavior After a Mediated Social Touch. *IEEE Transactions on Haptics, 2*(3), 136–140. doi:10.1109/TOH.2009.20 PMID:27788077

Iggo, A., & Andres, K. H. (1982). Morphology of cutaneous receptors. *Annual Review of Neuroscience, 5*(1), 1–31. doi:10.1146/annurev.ne.05.030182.000245 PMID:6280572

Kye, B., Han, N., Kim, E., Park, Y., & Jo, S. (2021). Educational applications of metaverse: Possibilities and limitations. *Journal of Educational Evaluation for Health Professions*, *18*, 32. doi:10.3352/jeehp.2021.18.32 PMID:34897242

Lanier, J. (2017). *Dawn of the New Everything: A Journey Through Virtual Reality*. Henry Holt and Co., Inc.

Löken, L. S., Wessberg, J., Morrison, I., McGlone, F., & Olausson, H. (2009). Coding of pleasant touch by unmyelinated afferents in humans. *Nature Neuroscience*, *12*(5), 547–548. doi:10.1038/nn.2312 PMID:19363489

Manzotti, A., Cerritelli, F., Esteves, J. E., Lista, G., Lombardi, E., La Rocca, S., Gallace, A., McGlone, F. P., & Walker, S. C. (2019, October). Dynamic touch reduces physiological arousal in preterm infants: A role for c-tactile afferents? *Developmental Cognitive Neuroscience*, *39*, 100703. doi:10.1016/j.dcn.2019.100703 PMID:31487608

Massie, T. H., & Salisbury, K. J. (1994). PHANToM haptic interface: a device for probing virtual objects. American Society of Mechanical Engineers, Dynamic Systems and Control Division.

Milgram, S. (1974). *Obedience to Authority; An Experimental View*. Harpercollins.

Morhenn, V., Beavin, L. E., & Zak, P. J. (2012). Massage increases oxytocin and reduces adrenocorticotropin hormone in humans. *Alternative Therapies in Health and Medicine*, *18*(6), 11–18. PMID:23251939

Moseley, G. L., Gallace, A., & Spence, C. (2012). Bodily illusion in health and disease: Physiological and clinical perspectives and the concept of a cortical body matrix. *Neuroscience and Biobehavioral Reviews*, *36*(1), 34–46. doi:10.1016/j.neubiorev.2011.03.013 PMID:21477616

Nazir, S., Gallace, A., Manca, D., & Overgard, K. I. (2016). Immersive virtual environment or conventional training? Assessing the effectiveness of different training methods on the performance of industrial operators in an accident scenario. In P. M. Ferreira Martins Arezes & P. V. Rodrigues de Carvalho (Eds.), *Ergonomics and Human Factors in Safety Management*. CRC Press.

Sand, A., Rakkolainen, I., Isokoski, P., Kangas, J., Raisamo, R., & Palovuori, K. (2015). Head-mounted display with mid-air tactile feedback. In *Proceedings of the ACM Symposium on Virtual Reality Software and Technology, VRST*. Association for Computing Machinery. 10.1145/2821592.2821593

Schachter, S., & Singer, J. (1962). Cognitive, social, and physiological determinants of emotional state. *Psychological Review*, *69*(5), 379–399. doi:10.1037/h0046234 PMID:14497895

Slater, M., & Wilbur, S. (1995). Through the looking glass world of presence: A framework for immersive virtual environments. In Five (Vol. 95, pp. 1-20). Academic Press.

Spence, C., Hobkinson, C., Gallace, A., & Fiszman, B. P. (2013). A touch of gastronomy. *Flavour (London)*, *2*(1), 14. doi:10.1186/2044-7248-2-14

Sreelakshmi, M., & Subash, T. D. (2017). Haptic Technology: A comprehensive review on its applications and future prospects. *Materials Today: Proceedings*, *2017*(2), 4182–4187. doi:10.1016/j.matpr.2017.02.120

Varela, F. J., Thompson, E., & Rosch, E. (1991). *The embodied mind: Cognitive science and human experience*. MIT Press. doi:10.7551/mitpress/6730.001.0001

Wheeler, J. A. (1989) Information, Physics, Quantum the Search for Links. *The 3rd International Symposium Foundations of Quantum Mechanics*, 310-336.

KEY TERMS AND DEFINITIONS

Autonomic Nervous System: Is a component of the peripheral nervous system that regulates involuntary physiologic processes, such as heart rate, blood pressure, skin sweating, respiration, and arousal.

Avatar: The embodiment of a person within a digital/virtual representation.

Cortex: The outer layer of the brain, composed of folded grey matter.

Digital Twin: A virtual representation that serves as the digital counterpart of a physical object or process.

Haptics: The perception of objects by touch and proprioception, involves sensory processing and body movement.

Psychophysics: The branch of psychology that investigates the relations between physical stimuli and mental phenomena.

Quantomechanic: A description of the physical properties of nature at the scale of atoms and subatomic particles.

Realism: The idea that nature exists independently of man's mind.

Somatosensory Cortex: THE area of the brain, responsible for the processing of information provided by tactile receptors. It contains a complete—although distorted—representation of the entire body surface, known as 'homunculus', with a point-to-point correspondence to actual body parts.

Chapter 2
Design Elements for the Implementation of Threshold Crossing In and Out of Mixed Reality

Giorgio Dall'Osso
Alma Mater Studiorum – University of Bologna, Italy

Michele Zannoni
https://orcid.org/0000-0003-2703-772X
Alma Mater Studiorum – University of Bologna, Italy

Ami Licaj
Alma Mater Studiorum – University of Bologna, Italy

ABSTRACT

The chapter presents research on the relationship between the human body and the space implemented by data and digital interfaces. In this relationship, technology plays a mediating role. The research introduces the concept of a digital threshold to an interactive space that has the capacity to preserve the cognitive well-being of users and invite interaction. To do this, some characteristics are identified that can be used in the design with the aim of relating the body to the devices in the space. Pressure stimuli, rhythm, and body symmetry are the components of a natural language capable of activating a natural motorial reaction mechanism. The details of the experimentation carried out and the processing of the data collected through data visualisation are provided to support the argument.

DOI: 10.4018/978-1-6684-4854-0.ch002

INTRODUCTION

Mixed Reality is a multidisciplinary field of study, in which macro and micro scenarios using *Virtual Reality* and *Augmented Reality* as seamless solutions are described. *Mixed Reality* is applied inside architectural or natural spaces where human users carry out their daily activities.

This research aims to explore the current context of *Mixed Reality* to understand how the integration of *wearable* devices has triggered the interaction on the body, linked to the attentional processes through the visual, tactile, and pressure feedback.

In this framework, the primary relationship investigated by scientific research is between the human body and space. In the contemporary world, space is not only describable through the sensoriality of the body; digital information referenced both to places (Zannoni, 2018) and to the actions performed by the user are anchored in space. When this information becomes part of perception, it contributes to defining the situational characteristics in which the human body lives; moreover, by adding to the pre-existing sensory reality, it consciously or unconsciously modifies the behaviors of humans. Scientific and applicative research devotes continuous work to the aim of increasing the quality of the mediation performed by technology in the relationship between the body of the individual user and the digital information referenced to space.

The paper presented explores some characteristics of body-space mediation devices. It identifies the threshold of entry and exit from an immersive experience as a design objective that can be developed through haptic feedback elements. The areas in which a potential spillover of the study is imagined are marked by spaces in which it is possible to enter and exit mixed reality experiences such as could be workplaces or cultural fruition environments.

BACKGROUND

Embodiment theories emphasize the close interconnection between body and mind; what happens on one level always affects the other and vice versa.

The motor and sensory capacities and past experiences that every human has had in the natural environment define how humans think. These theories show the way forward for researchers and designers: working from the human body means knowing the most subtle mechanisms that regulate it and designing with great respect for the balances at play.

Every extension of the body and every piece of information that manifests itself on or around it always generates complex effects; these act on both perception and behavior. For this reason, a primary research objective is to develop mediating technologies that are able to manage the wellbeing of the human body.

The advances in ergonomics indicate well-being as the sum of all the levels that define the human factor (anatomy, physiology, cognitive resources, pleasantness, etc.) and the environmental context of reference (Sicklinger, 2020). Measuring well-being in the contemporary world is possible through the analysis of qualitative data and quantitative data (De Luca, 2016). Technological developments in sensors and artificial intelligence allow the collection of data that detail the motorial behaviors in space and track the emotional states of users. Reference technologies are those derived from human face analysis (*face tracking*), attention analysis (*eye tracking*, *body orientation tracking*), detection of physiological parameters related to stress levels (skin conductance EDA, heart rate variability HRV, etc.).

Technology as a technical element supporting the dialogue between body and digital information is a field that is particularly addressed with the terminology prosthesis. An extensive contribution in this sense can be found in the work of Maldonado (1997) in which the author identifies four types of prosthesis: motorial, sensory-perceptual, intellectual and syncretic. Motorial prostheses include objects designed to increase the subjects' capacity for movement or strength. Sensory-perceptual prostheses include the types of projects that reinforce or replace human sensory capacities. Intellectual prostheses are characterized by the ability to memorize information. Finally, syncretic prostheses are those that combine the capacities of the previous categories.

With the aim of focusing the field of prosthetic research on digital objects, additional types of prostheses were subsequently introduced by Zannoni (2018), in particular passive and interactive ones. In passive prostheses the body receives information from the outside, in interactive prostheses the user is also able to interact by controlling the functioning of the prosthesis.

In application areas, the topic of prosthetic technology finds many case studies in the scenarios triggered by wearable devices. In recent years, this product sector has continued to expand considerably, as highlighted by data showing a growth of 20% in the calendar year 2021 alone (International Data Corporation, 2022).

Users have identified several characteristics that are considered most important in this product sector and that differ from those of other portable devices such as laptops or smartphones. The importance of criteria such as convenience, lightness, durability, good looks and effortless use outweighs the ability to share data (Rantakari et al., 2016). These characteristics, combined with human morphology and physiology, determine the achievement of good integration between user and activities performed. Modes and types of intervention of wearables are therefore related to both specific characteristics of the human body and those of the activities; literature research shows numerous application guidelines of wearables both in positioning on the body and in the modes of use (Zeagler, 2017). The wearables market is therefore very sensitive to functional body/object integration and it is evident how devices are chosen based on the degree of quality with which they manage to integrate into the end user's body patterns (Buiatti, 2014).

One of the ways in which scientific research tries to find answers to the usability of human/machine languages are Natural User Interfaces (NUI) based on natural languages. The idea behind NUIs is that in a scenario of ubiquitous computing, dialogue with technology is facilitated when the modes of interaction habitually used by humans in tangible reality are recalled.

Jetter et al. (2014) take up the concept linked to Dourish's (2004) embodied interaction and state that the key principle of natural interaction is as follows:

Instead of drawing on artifacts in the everyday world, it draws on the way the everyday world works or, perhaps more accurately, the ways we experience the everyday world.

Designers and researchers can achieve effective interactions within scenarios populated by heterogeneous digital interfaces (Jetter et al., 2014) using common knowledge of the laws of physics governing the behaviors of things (speed, gravity, friction, etc.), awareness of human capabilities related to the body (Shi, 2018) and space, and experience of existing social principles.

The *Blended Interaction* framework helps to define patterns of use of natural languages in interfaces (Jetter et al., 2014). It explains how *image schemas* and *affordances* guide user behavior when interacting with a new interface. *Image schemas* are the behavioral schemas linked to sensorimotor experiences that humans transfer from physical to virtual space and which, in addition to guiding us on basic interactions, are used for the construction of more complex concepts. *Affordances* are invitations to interaction with the body that humans, based on their own experience, identify in their relationship with objects (Norman, 1988).

The *Blended Interaction* framework further defines four domains that design should pay attention to when intervening in post-WIMP *ubiquitous computing* scenarios. These domains are: the development of individual interaction through multitouch and multimodal interfaces, the design of social interaction capable of reading and using existing social models and protocols, the design of workflows, and finally the design of the physical environment in which displays and technologies will coexist.

These domains introduce the second component of the technology-mediated relationship: space. The *Blended Interaction* framework emphasizes space both as a domain of interaction design and as a container for natural social interactions. Technologies and sociality coexist in space.

The study of the distribution of relationships in space finds its greatest contributions in the large field of proxemics first defined in the mid-1960s and then developed in relation to architecture and ergonomics (Hall, 1966; Panero & Zelnik, 1979). The use of intimate, personal, social and public proxemic areas has been used extensively in the study of *ambient displays* (Jakobsen et al. 2013). The aim of using proxemics is to design digital systems capable of autonomously transforming in response to social situations or personal configurations that happen between people and in spaces.

Using the rules of proxemics, areas have been defined for specific types of interaction with ambient displays. These can help in defining what should be a good interaction flow within mixed reality scenarios. In detail, four areas have been identified: *ambient display*, *implicit interaction*, *subtle interaction* and *personal interaction* (Vogel & Balakrishnan, 2004).

The first area of interaction highlights a neutral state of technology capable of relating to the architectural space with a calm aesthetics characterized by slowly changing elements. The second area is where digital systems detect the behavioral characteristics of the body and interpret the user's intention to interact or not. The third area is the one intended for a conscious but fast interaction, based on macro information that can be managed with an interaction based on body language. The last area is that of personal interaction in which the previous interactive typologies are valid and further more detailed ones are added, intended to manage personal information.

In mixed reality contexts, spaces are characterized by numerous possibilities of interaction. Multiple interactive systems, technically and morphologically different, coexist in the same environment.

In ambient display areas, users walk through without focusing their attention on the interactive objects. Users are aware of the presence of objects because the characteristics of objects belong to the perceptual field of space. The next area, the area of implicit interaction, is reactive: as users cross the threshold, digital systems detect and read the behavior of the body and implement an automatic transformation in response.

The transition from one area to the next must be managed through changes that make the transition smooth and coherent. The user must therefore be invited and not forced to enter into a relationship with the objects; moreover, he/she must be able to effectively access a higher level of detail or easily exit the interaction flow.

These characteristics are consistent with the *Person Center Design* approach (Bagnara & Pozzi, 2014, p. 54) which insists on the need to equip interaction flows with moments of pause, reflection and serendipity so that people can remember the steps of their overall experience.

The transition between the area of ambient display and that of Implicit interaction is the threshold where the user's body enters into a dialogical relationship with the object of interaction. The body is oriented and muscles are stretched, attention can vary, emotions surface; all this without dragging the user into an obligatory and constrained interaction.

The application context of this research highlights how the triple connection between digital technologies, human body, and common space opens to new design scenarios in which human actions are naturally integrated in between real spaces and mixed reality. The fluid transition through the material and digital dimensions is a critical threshold that needs attention throughout design processes. Building a framework to support the development of these new tools, for the perception and representation of reality, is a significant phase for a new and innovative design process.

In summary, this study investigates the principles and methods through which a person can perceive access to an area dedicated to an augmented reality or virtual reality interaction within a *Person Center Design* interaction flow.

After defining the broad scientific framework of reference, an investigation was carried out on the existing languages between body and technologies, verifying application aspects related to them, particularly in the field of wearables. With this aim, the available enabling technologies and the ergonomic requirements to make these interventions effective were analyzed.

Subsequently, the research attempted to construct a reflection on the transformation of digital information systems in relation to changed living spaces and the proliferation of geo-referenced data. In this scenario, it was indicated some ways in which the studied thresholds can effectively respond to human needs and behaviors related to individuals, groups of people and communities.

MAIN FOCUS OF THE CHAPTER

Nowadays, the transition between the real and virtual dimension takes place in specific spaces, in which the spatial rim of experimentation is explicit and fixed. However, considering the spread of georeferenced data in urban areas relevant and more pervasive (Zannoni & Formia, 2018), it is necessary to balance the interaction experience on these new devices. This kind of interaction could give to the user a mediation system that underlines the thresholds between the real and the digital space. One possible developer solution is to improve the mode of switching between the two different phases of immersion and emersion; it could help people have a good experience in mixed reality use.

Research on the transition between real and virtual will make the designing process more natural and open, towards a reflection on the features of these embodied media, to enable the body in this perception of the difference between real vision and augmented things.

In this context, the role of wearables devices emerges, improving a bidirectional form of communication between humans and machines enhanced by their interface transduction with the body. This role as a mediator among devices permits also to mediates between data and actions in the virtual and material dimensions, expanding the interactive components in the augmented vision experience.

The wide range of actions that humans can perform when interacting in the spaces and with the objects, depends on the ability of any single person to understand the opportunities for use of each artifact. On this matter, there are many researchers in different academic fields who have expressed their scientific positions. These studies have arrived at a scientific conceptualization identified by the term affordance. This word indicates the capability of the artifacts to offer some indication of their possible use to the people that interact with them. The theoretical studies on the concept of affordance started in the early twentieth century with the Gestalt studies of the early twentieth century and evolve into many interpretations depending on the disciplinary areas addressed. A great contribution was offered by the cognitive sciences and in the studies of human-computer interaction of the 1980s. A more extensive formulation of Affordance was introduced in the scientific literature by the American psychologist James Gibson in his work *The Theory of Affordances* (1977) and later in the book *The ecological approach to visual perception* (1979). He formulated a theory based on the evidence that human beings with their senses are able to perceive an environmental context and the possibilities it offers. According to Gibson, the environment is a key factor for humans to understand the possible use of objects.

According to another important American psychologist Donald Norman, author of the book *The psychology of everyday things* (1988) who has worked in the fields of human-computer interaction and artifact design for the past 40 years, there is greater complexity beyond the studies of Gibson formulated ten years earlier. Norman argues that there is a real affordance and a perceived affordance. The first one is the set of actions we can really do with an artifact, the second one is what we can perceive starting from the cultural and semiotic factors that permit us to recognize a series of "signifiers" that a particular object offers us.

In contrast with Gibson, Norman interprets the concept of affordance shifted to an experiential level, changing his theory over the years and focusing on the idea that a correct affordance of an object is mainly a result of a good design process. Attention to the morphological and visual "signifiers" of any part of an object in relation to the person's previous experience allows him to perceive a possible use, that can hardly be univocal but plausibly recognizable. According to Norman, affordance is not an intrinsic property of an object but a result of a relationship:

We are used to thinking that properties are associated with objects. But affordance is not a property. An affordance is a relationship. Whether an affordance exists depends upon the properties of both the object and the agent. (Norman, 2013, p.11)

Starting from a classic example in the affordance literature, we can observe that it is not the shape of a door handle that affords its correct use, but it is the set of tactile and experiential knowledge involved in the relational experience that suggests its use. In the second instance also seeing other people use it allows us to understand how to use it. It is not severely correct to consider that the geometries or the visual elements are universally recognizable by all the people in the same mode. For example, a handle of a door does not communicate its affordance itself but is the personal information that a person has collected in his life experience that allows him to understand how to use it. At the same time, shapes and signs do not acquire meaning until them will start a semiotic association process that allows people to understand them. The context in which the forms and the artifacts are seen is the primary element that activates the cognition process in perception giving hierarchy, dimensions, and semantic relations to the space and other objects. Norman said that he chose the term signifiers to help designers to understand the concept that affordance is generated by a communicative process. In the last edition of his book *The psychology of everyday things*, updated in 2013, he wrote:

[...] designers needed a word to describe what they were doing, so they chose affordance. What alternative did they have? I decided to provide a better answer: signifiers. Affordances determine what actions are possible. Signifiers communicate where the action should take place. We need both." (Norman, 2013, p. 14)

In relation to these matters, a question arises about how in the complex use scenarios of the mixed reality, can the exponential grown visual elements added to the augmented vision influence the perception of affordance. When the experience becomes total and immersive, the body becomes an important sensorial interface element through making experience and interacting with physical or virtual spaces and their objects. The possibility of extensive communication between people, space, and objects evolves into a very complex synesthetic experience that invites reflections and analytical evaluations on how motor and sensory factors could be changed.

In the original of Gibson's theoretical vision, he argues how affordance is linked exclusively to the environment and is not property inside of the objects. It does not dependent on its morphology, but it is only related to the context that allows us to understand it.

In summary, for Gibson, affordance is linked to a context that influences the possible actions of things inside it. The senses help animals and humans in this action allowing them to percept the "structural invariants" (Gibson, 1974, 1979) available in the environment. These characteristics of visual elements always remain the same despite the people changing their point of vision and distance to an object.

I have described the invariants that enable a child to perceive the same solid shape at different points of observation and that likewise enable two or more children to perceive the same shape at different points of observation. These are the invariants that enable two children to perceive the common affordance of the solid shape despite the different perspectives, the affordance of a toy, for example. Only when each child perceives the values of things for others as well as for herself does she begin to be socialized. ~(Gibson, 1986, 141)

In summary, Gibson said that the affordances are what they can really provide to humans without interpretations and therefore without any high-level cognitive processes. They are not the result of analysis and reflection that involve memory, but they are a result of basic processes that do not change in relation to people's mental abilities:

The affordances of the environment are what it offers the animal, what it provides or furnishes, either for good or ill. The verb to afford is found in the dictionary, but the noun affordance is not. I have made it up. I mean by it something that refers to both the environment and the animal in a way that no existing term does. ~(Gibson, 1986, p. 127)

In the last fifty years, this Gibson's absolute position reported above has been a matter of complex debate in the scientific community and in some cases, misinterpreted. Specifically, in the field of design, it has often leaned towards Norman's position which has shifted the debate on the perception of the observer who interprets signifiers that afford a possible use of an object based on people's previous experience.

Starting from these two approaches, in accord with the fact that the body is a complex system characterized by physicality and sensorially both directly connected to the brain and its mnemonic and cognitive abilities, we could argue that the perception of an affordance goes beyond the range of visual perception and involve the wider sphere and senses of the body up to proprioception. In this context, the technology as a mediator between body and space can also assume a role as a tool that builds relationships and signifiers in which the human with his body can read through some haptic stimulations possible threshold variations between one place to an another, between one object to an another. The sensory information collected by the body that interacts with technology should be read as a factor that helps people to find a relationship between themself and virtual or real space. The question that emerges from this issue is if the cognitive processes activated by our sensory system in virtual o extended reality are similar to real experience. This dilemma has been debated in the last ten years and the study of the affordance of objects in virtual reality remains a field of experiments and evaluation. For example, the recent studies carried out by Tony Regia-Corte and other colleagues from INRIA in Rennes France arrived at interesting results. In several experiments conducted within a virtual environment with a selected group of people, it was demonstrated that when asked to different subjects to rate their balance on different types of inclined surfaces, were obtained values comparable to similar studies previously conducted in real environments (Regia-Corte et al., 2013). The work on the Perception of Affordances and Experience of Presence in Virtual Reality by Paweł Grabarczyk and Marek Pokropski (2016) also confirms these results. Research shows that there is a minimum threshold for activation of affordance processes detectable by the fact that the user begins to perceive the use of objects in a virtual space. Similar to the previous research are the experiments conducted by researchers at RMIT University in Melbourne in Australia led by Jules Moloney. The research group analyzed forms of representation of big data in virtual environments. They verified that the affordances offered by an immersive 3D system can expand the range of the information that a person could perceive giving him a possibility to access to very high data complexity in different ways from the traditional visual tools (Moloney et al., 2018).

All these studies show how visual perception and embodied cognition concur in a virtual context and how all sensory stimuli can also contribute to developing affordance processes in relation to physical and virtual artifacts. We presume in a short-term scenario that the tactile stimulus generated by wearable devices on the body skin could contribute to activating a cognitive flow in mixed reality experience that could help to perceive changes in spatial thresholds between different content such as exhibitions or working environments or correct walking direction and body posture.

If we consider wearable devices as a medium, we could suppose, as a first assessment, that the requirements of this new class of objects are plausibly similar to the same type of product currently on the market for several years. The second assessment is that, in order to be accepted and integrated into human actions, they first of all must be comfortable for the body and have "good-looking" (Rantakari et al., 2016). They must also be able to communicate directly with the body without reducing cognitive resources from the primary functions dedicated to visual attention and avoiding harmful interference with other activities the user is carrying out (Nicoletti & Borghi, 2007).

A large area of research emerging in the scientific literature underlines that these multiple classes of feedback triggered by these technological devices may communicate with the body according to different forms of attention. Shifting the user towards a focused level of attention is an action that can be implemented through a type of impactful notification (e.g. alarm) that interrupts the actions in progress. Opposite, in a situation where attention is divided on several activities, notifications will have the purpose of making the user aware of specific information or its variation over time and should not affect pre-existing activities. Finally, notifications can be sent to the user without the user being consciously aware of their arrival; in this case we refer to an inattentive state and notifications are designed to induce an unconscious change in perceptual or motorial behavior. Lastly, it is also argued that they may also be a vehicle for transition from one form of attention to another. (Matthews et al., 2004).

It is therefore necessary to develop a design awareness of the notifications that technology transmits to the human body. These must take into account the prevailing activities the user is performing and choose the sensory channel, shape and power with which to manifest in the users' experience.

The experimentations carried on in the last years demonstrate promising applications of haptic pressure stimulation on the surfaces of the body to notify structured information making human's attention locked to other activities (Löffler et al., 2019). Of particular interest are whether these actions may tell us how the whole epidermic surface could be used as a medium for tactile information. Furthermore, it is possible that these interactive processes on the skin could be capable of manifesting themselves with greater or lesser emphasis, according to how stimuli are executed (Karuei et al., 2011).

The research proposes the development of a design tool capable of improving the user's experience when crossing the threshold of implicit interaction with digital objects in the surrounding space. To this end, experimentation has been carried out on some types of stimuli that can be ascribed to natural languages. These stimuli aim to provoke an instinctive behavioral reaction in the body of the users.

In order to elaborate an effective NUI with inter-subjective effects, three elements deeply linked to the way the human body relates perceptually to the environment around it were chosen and added together. These three elements belong to the human cognitive experience of everyone and for this reason they have a greater possibility of being read as natural language in the human-computer dialogue relationship. They are: the tactile stimulus of surface pressure; the rhythm, that is the structuring of stimuli in time; finally, the symmetry of the human body with respect to the sagittal plane used by it to perceive the world with opposite and symmetrical parts (eyes, hands, feet, nostrils, etc.).

The use of touch as a channel for communicating thresholds, and more specifically the pressure stimulus, was identified in the literature review. The studies that consider human experiences as whole as multisensory activities underlines how the sensory channel that brings the lowest intensity variation in the subject becomes a guide to the qualitative perception of information detected by the other senses (Buiatti, 2014, p. 158). The somatosensory system is frequently the least stressed sensory channel and therefore central in guiding overall stimulus perception.

Designing stimuli acting on the tactile channel is possible through vibration, temperature, pressure, pain. Pain is a type of stimulus that is difficult to apply and certainly not accepted by users. Temperature is an interesting stimulus, but its variation is slow and difficult to detect except with strong contrasts (Matthies et al., 2013). Actuators used for vibratory stimulation allow complex applications with little impact on volumes and energy consumption (Song et al., 2015). Currently, actuators that act through pressure on the body have higher technological and application complexities. Research, however, shows how the communicative potential of them is in fact equivalent to those related to vibration (Song et al., 2015; Pohl et al., 2017). Studies using vibration patterns to create information also suggest guidelines for the application of pressure stimuli. In the context of postural correction, for example, the body is described as being able to read up to four different levels of vibratory intensity, but above all it has a very good ability to detect a change in stimulus information and not a specific intensity. Pressure stimuli, unlike vibration stimuli, are read by the human's body with greater familiarity (Kettner et al., 2017) which makes them more readily acceptable when worn frequently on the body.

Pressure stimuli, moreover, have an obvious link to the human emotional component. Several tests show that these stimuli cause similar feelings in the subjective (Hertenstein et al., 2006). Some studies, in particular, use periodic stimuli characterized by; slow or fast rhythms; symmetric or asymmetric waveform types (Baumann et al., 2010); variation of parameters such as frequency and intensity (Suzuki et al., 2015). Pressure in the literature is effective for subtle communication (Baumann et al., 2010) that can be used to send information while respecting privacy in public places. Furthermore, it can create synergies with other stimuli (vision or touch) to enhance or ensure the arrival of information (Lederman & Klatzky, 2009). These stimuli can support orientation and navigation actions in space (Frey, 2007;

Velazquez et al., 2009) and finally send spatial information to the user as experienced in automotive (Ploch et al., 2016). In the latter area, it has been witnessed how a pressure response on the body can suggest virtuous behaviors in the management of the vehicle (e.g. Bosch "Smart Gas Pedal").

Pressure is therefore a stimulus that manifests itself privately on the body of individuals without generating cognitive imbalances and respecting the tasks that users are already performing. For this reason, the use of pressure is a useful tool to communicate the transition from a classical real space to a space in which a mixed reality content is placed.

The analysis of the rhythm component is infrequent in the scientific literature associated with digital technologies. However, it is present and often mentioned among the factors that can be used in the development of haptic languages based on vibration, pressure or hybrid systems. On a scientific level, studies supporting the identification of rhythm as a structural element for the generation of natural languages are found within multiple fields: figurative art (Klee, 1959), music, neuroscience (Mallgrave, 2015, p. 113), cognitive, technological, sociology (Apolito, 2014), architecture and urbanism (Lefebvre, 2019), etc.

The use of rhythm as a structural element of a natural language is due to three processes that the body naturally implements when it enters a relationship with rhythm: synchrony, rhythmic imitation and rhythmic entrainment.

Synchrony occurs when perception consciously identifies two stimuli with a common rhythm. Synchrony is an element that often evokes moments of sudden amazement in humans, as Strogatz (2003) well describes in his book dedicated to studies of it in different scientific fields. A second element with which the body manages to naturally enter a relationship with rhythm is rhythmic imitation, that is, the ability to become aware of a rhythm and manage to reproduce it through different sensory channels or systems. The last phenomenon highlighted differs from the previous ones in that it acts in the unconscious of human behavior, it is called rhythmic entrainment. People have a natural tendency to reproduce in their behavior the rhythms perceived externally (Trost et al., 2017), as if the human body were in constant search of synchrony with external elements. This process is used, for example, to increase and decrease the speed of walking by modifying the speed of the music being listened to (Moens et al., 2010).

These three processes support the claim that rhythm is a fundamental variable in natural languages as it can be read consciously or unconsciously.

The human body is divided on the sagittal plane into two equal and opposite parts called antimere. This characteristic is intrinsically linked to the way in which man develops his perception of the world in that he walks with two feet moving alternately, grasps objects with two hands, observes through two eyes, listens to environments with two ears, and so on.

The sum of the stimuli collected by the antimeres is not arithmetic; thanks to the synergy of the opposite elements, humans grasp the depth of spaces and become aware of surfaces that are not touched but lie between the ends of the object held in their hands.

In the scientific bibliography, some indications of the potential of applying haptic technology to different areas of the body emerge. For example, it is highlighted how the perceptual threshold of touch is different in different areas of the body (Lederman & Klatzky, 2009) and there are points where it is more effective to transmit punctual stimuli. This parameter has to compromise with the opportunities related to comfort and usability of wearable systems.

Another important piece of information is that single-site interfaces are perceived more safety than multi-site interfaces, which on the other hand cause users to feel surprised (Karuei et al., 2011).

The conscious use of symmetrical body parts is of great interest for the development of natural languages that can adequately relate to the morphological and perceptual characteristics of the human body.

In order to investigate the effectiveness of a human/machine language composed of these three elements, a series of experiments have been started based on the effects of stimuli on the static and moving body (Dall'Osso, 2021). The collection of quantitative data related to human behavior in a real space is a complex challenge. Every space, in fact, through the permanent or temporary characteristics that identify it, generates specific effects on the behavior of those who live in it.

The tests investigating the impact of rhythmic and pressure stimuli on the postural oscillation of the human body, described in the following paragraphs, focused primarily on achieving suitable control of the space around the subjects.

EXPERIMENTATION METHODOLOGY

In order to experimentally verify the effects of rhythmic pressor stimuli on behavior, the area of the plantar arch was chosen as the point of application of the prototypes.

This choice is due to several factors: the ease of application of the technical apparatuses in the volumes of the footwear, the perceptive qualities of touch in the soles and finally, the natural function that the feet play in the mediation of information from the environment with the needs of the body.

In the tests carried out, the technological system for sending the stimuli was created by means of two servomotors located in the footwear (Figure 1). These move a plastic tongue which, moving from bottom to top, creates pressure in the plantar arch. The maximum intensity of the pressure can be adjusted by the subjects by varying the length of the cable connecting the servomotor to the tongue. The system is battery-powered and remotely controlled via Wi-Fi.

Figure 1. Prototype

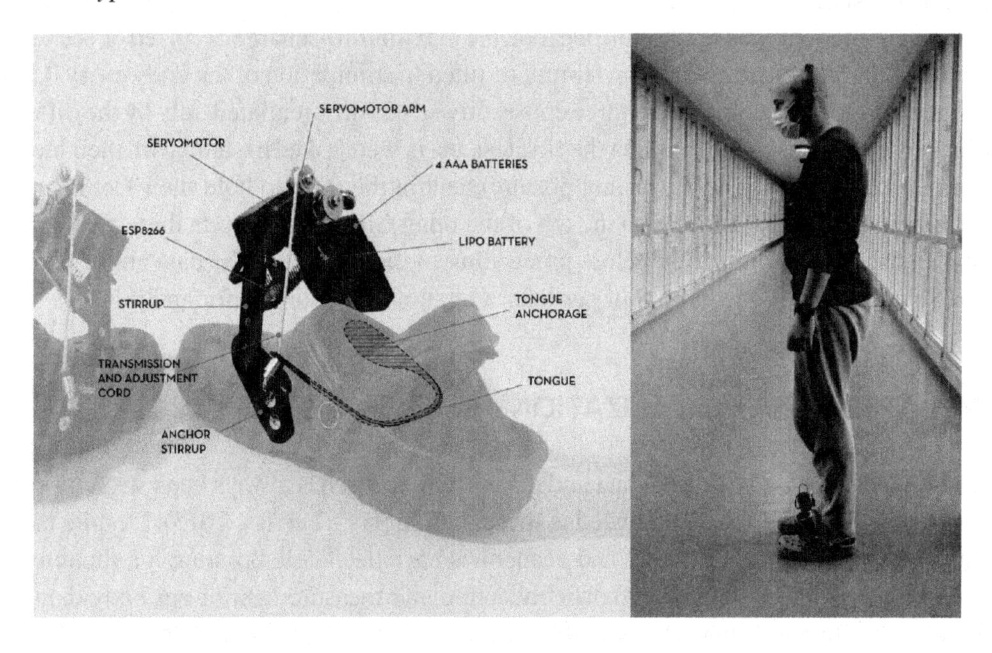

During the tests, the movement detection system mHealth Technologies called SWAY (Palmerini et al., 2011) was used, based on a pod placed at the level of the lumbar vertebrae, inside which there is an accelerometer. During the tests, data on skin conductance and heart rate variability were also collected through the E4 wristband wearable device by Empatica.

Seventeen testers were involved, eight women and nine men, aged between 19 and 66. Of these, five were left-handed and the remainder right-handed. All testers, at the time of test, were healthy subjects in terms of motor health status. The test site was a straight corridor with homogeneous and repetitive surface characteristics. Each tester wore headphones for the duration of the tests, with white noise to eliminate the noise produced by the environment and the prototypes.

Before the start of each test, each tester was given stimuli in order to adapt and prepare the body for what would happen during the exercise.

Each tester was asked to remain stationary in a balanced, upright position for a total of 60 seconds. During this time, the subjects had to look continuously at a fixed point at eye level on a wall at a distance of approximately 1.5 meters.

Stimuli transmitted were divided into 4 parts of equal duration (15 s), constant speed and intervals:

- **Synchronous** (simultaneous) pressure stimuli on both feet.
- **Monopod** pressure stimulus on the right.
- **Monopod** pressure stimulus on the left.
- **Alternating** pressure stimuli (not simultaneous) on both feet.

The sinusoid of stimulus intensity over time was symmetrical. The maximum intensity of stimulus pressure was regular throughout the test and occurred with a temporal distance between one stimulus and the next of one second.

During transmission of the alternating stimulus, the body received stimuli more quickly because while the maximum pressure peak was reached on one foot, the maximum discharge occurred on the second. The phase-shifting of the waveforms of the two stimuli resulted in stimulation of the body every 0.5 seconds.

The designed protocol was applied in three consecutive trials differentiated only by the different positions in which users were asked to stand: In the first test, users were asked to stand with their feet together, in the second to hold their feet open (resting posture), and in the third to hold their feet in tandem (the heel of the dominant foot in contact with the toe of the other). In-between tests there was a short break during which participants could relax before proceeding to the next one. The data collected thus relate to a total of 51 tests each divided into four parts for a total of 204 data sets for analysis.

BODY-RELATED DATA VISUALIZATION

The relationship between body-related data and data visualization has always been strongly explored in the field of Quantified Self, or 'self-knowledge through numbers' (Ferriss, 2013). Despite the fact that major companies, professionals, athletes and academics have been collaborating for about ten years to make devices more and more sensitive to data collection and measurement of our body data, there are still, as some tests with users show, many problems related to the representation of this data in order to make it usable and readable. The problems that have emerged are varied: from the lack of user literacy, to the lack of storytelling, to the small size of screens (if we are talking about wearables), to information overload. (Rooksby et al., 2014; Parnow, 2015; Rapp & Cena, 2016; Liçaj, 2018; Zannoni, 2018; Dall'Osso, 2021; Pronzati, 2021).

In addition to these issues, the present experimentation revealed a further problematic issue concerning how to simultaneously represent a micro-data (such as the movement of the center of gravity) on a macro scale, such as that of a whole-body posture.

The data visualization approach in this scientific work aims to investigate forms of representation of the analytical information that emerged in the experimentation, constructing a language that takes into account the limitations just expressed, with which to contextualize the measurements of body movement data according to easily interpretable models. Thus previously invisible phenomena become evident signs and movements, intelligible objects for multidisciplinary dialogue.

The first aim in representing the experimental data sets was to respect the body-space-time relationship by working with very small units (micro units).

The first choice was therefore to understand which visualizations might be the best to explore the data referring to the space and time variables through the available data: the points on the X-axis, i.e. the displacements of the center of gravity to the left or right in space, and the points on the Y-axis, i.e. the displacements of the center of gravity forwards or backwards in space.

There are mainly two types of representations suitable for this purpose. The contour-plot (or isarithms), to be created for each rhythm of each subject, with a distributive function, capable of highlighting precise position relationships between the data of the barycenter. It relates the data of the X with the data of the Y (the punctual position of the barycenter), generating proximity areas in which we can immediately understand the zones, in relation to the axes, in which the body has lingered the most (Figure 2). (Bertin, 1968) The second type of graph identified was the multiple scatterplots, where each scatterplot coincides with one of the four rhythms of each test (feet together, open, tandem). This typology of visualization, distributive and correlative, combines the X and Y variables of the dataset as well as a third variable that determines the color of each point on the Cartesian plane and therefore adds a temporal value of the barycentric movement from start to finish (Figure 3).

Figure 2. Contour-plot Tester n.2

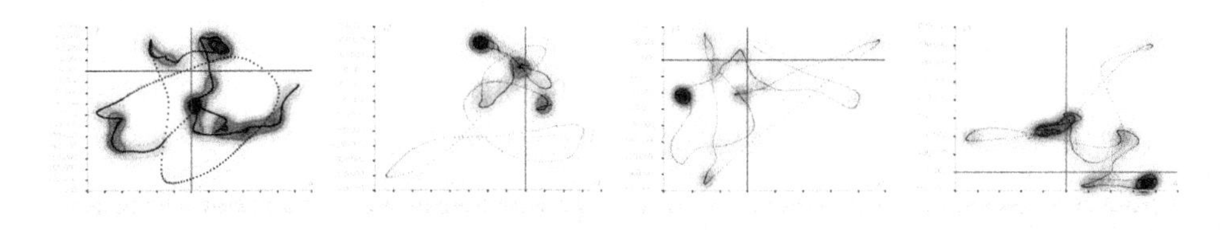

Figure 3. Scatterplot Tester n.2

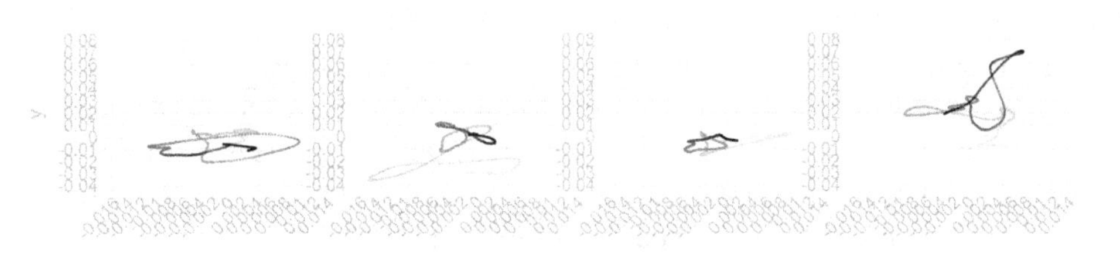

Operationally, contour-plots of each of the four rhythms were made for each of the 17 subjects in the three feet positions, for a total of 204 graphs. The same procedure was applied to the structuring of the multiple scatterplots where each group of scatterplots containing the four rhythms was created for each subject and for each position of the feet. Methodologically, the choice was made not to scale the graphs according to the same units of measurement for the X-axis and Y-axis, since at this stage of analysis it was the direction rather than the size of the movement that was of interest. Surely the next steps will also involve other types of analysis of the effective range of movement, as a future development of the experimentation in progress.

The first reading of the results was therefore the comparison of the various visualizations. Observing the numerous graphs, the first question was about the incidence of the different rhythms in the displacement of the barycenter, and therefore of the body. We wanted to understand how, and if, the different rhythms trigger a movement and towards which direction in the axes. It was therefore decided to overlay, as a first attempt, all the synchronous rhythms with feet together (Figure 5) and it was immediately evident the effectiveness of the visualization in showing how generally for the subjects, in the first fifteen seconds of the test, there was a determined tendency of movement in a certain portion of the area of the Cartesian plane. It was therefore decided to repeat the procedure with the following three rhythms, and with the other foot positions, thus obtaining twelve visualizations (Figure 6) that were certainly more usable and comparable with respect to the complexity of the 204 initial graphs (Figure 4).

The analysis of multiple scatterplot results showed how it was necessary to simplify them in order to deduce conclusions about the characteristics of the displacement of the barycenter in time. What is important in these graphs, in order to understand the real incidence of the stimulus on the body movement, is the starting point, the path taken by the centre of gravity and where it ends. Therefore, the decision was made to mark these three phases with areas of three different gradations (Figure 7). The three areas thus marked allow determining three important types of information immediately:

1. the main direction of the movement according to the X-axis or the Y-axis, and thus whether vertical or horizontal;
2. the width of the area defines the size of the displacement of the center of gravity, if it is narrow the center of gravity has had less displacement, if it is wide it has had a greater displacement;
3. the type of movement given by the order of the three areas: if the movement was linear, the distribution of the areas goes in an orderly manner from the lightest to the darkest, but if, on the other hand, the movement was irregular, the distribution of the areas does not go in ascending gradient order but is mixed, indicating a very turbulent movement of the barycenter. Through the division into areas it was immediately possible to determine schematically the type of movement over time, and to make everything even more immediately usable, symbols were added to each graph to indicate in a simplified way the type of movement - if linear, simple arrow, if complex, circular arrow - and the direction - right, left, high, low (Figure 8, 9).

Once the various visualizations through the two types of graphs have been completed, and subsequent post-productions, it can be seen that the contour-plot immediately reveals the relationships between body and space, while the scatterplot allows us to better understand the relationships between body, space and time. In this phase comes the moment of greatest complexity regarding the processing of data visualization, because what was missing from this great work of data processing and representation was the most important element, which is much talked about in this text: the body.

Figure 4. The complexity of the 204 contour-plot graphs.

The visualization with the element of the human body has the role of making people perceive how this data was effectively related to the real, physical element. So, the challenge in this case was to bring the microscale into the representation of the macroscale. The body here becomes a key element for a multi-level narrative: emotional and technical (Figure 10) (Bihanic, 2015). The visualization was structured to have two main levels of readability. The first level has the duty, in the rules of communication, to self-explain the most important messages in a direct way, in a few seconds. The second one, now that the observer has been prepared for the contents, leads to the side messages, which have the task of enriching and detailing. Moreover, the more complex the themes are, the more useful it is to combine a qualitative, evocative and direct representation with a quantitative, numerical and accurate one. Qualitative representation, due to its simplicity, has the power to be narratively faster than a complex quantitative visualization. The choice of keeping both visualizations is an action of attention and respect towards the observer (Licaj, 2018).

Figure 5. Overlay of contour-plot graphs made for the synchronous stimulus in the feet together test.

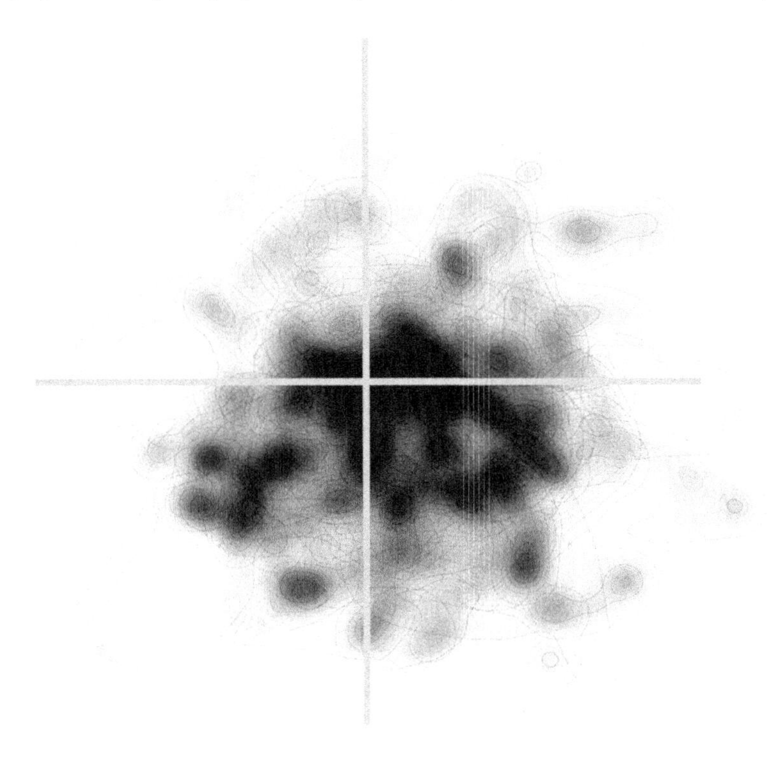

Figure 6. Overlay of the contour-plot graphs made for each type of stimulus and exercise.

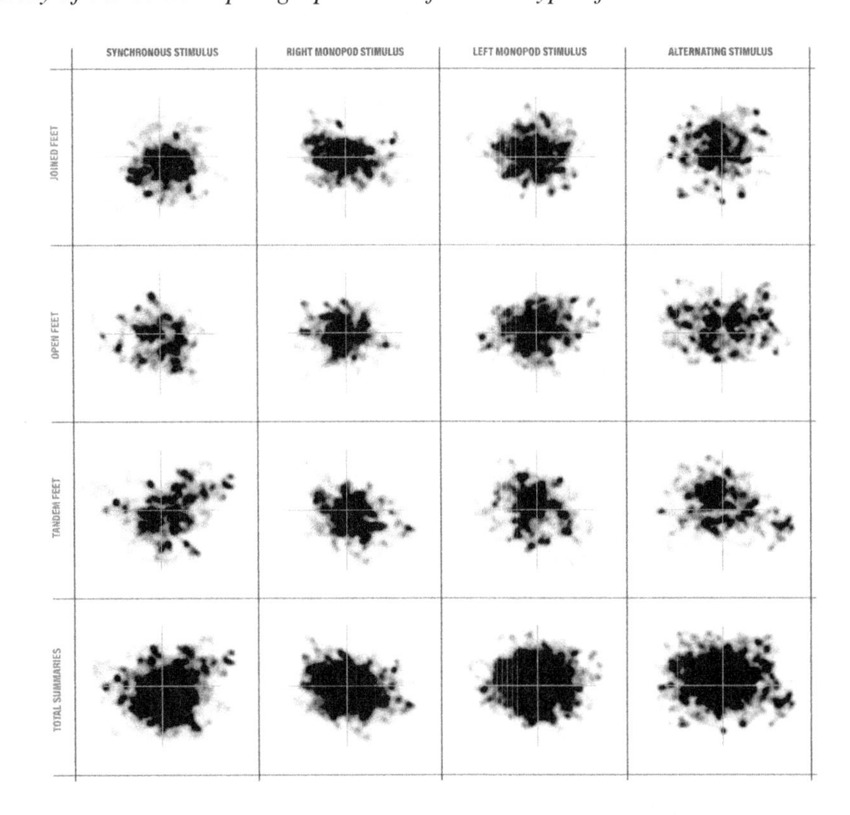

Figure 7. Division of the quadrants into the three areas: beginning, middle and end of the test.

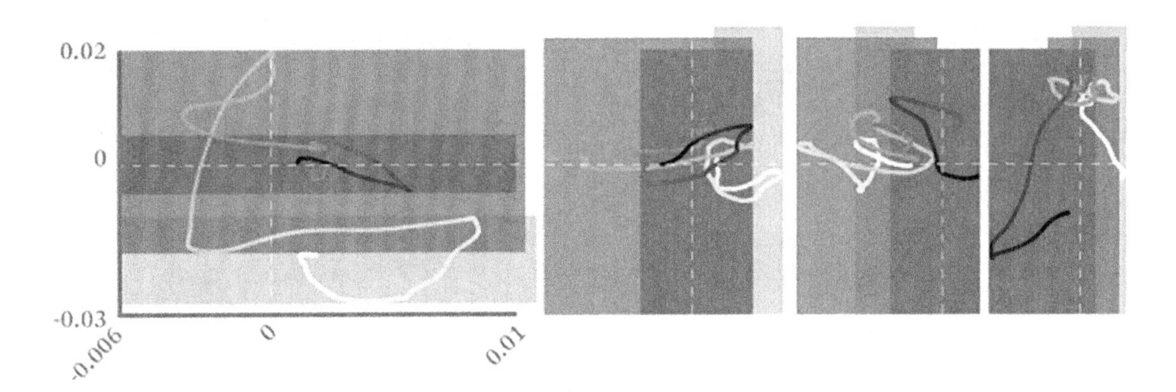

In this case, the emotional/suggestive representation must make the observer understand the more general impacts of the test on the body of the testers. It is therefore forgiven for being approximate to provide a more qualitative but direct representation. Its aim is to show how the body really tended to move in a certain direction during the various tests. On the contrary, the second level representations have the duty to represent the data in a quantitative way and to make people understand with scientific rigor, putting aside the emotional side, the elements that influenced the movement of the body, i.e. the displacement of the barycenter in relation to the rhythms of the stimuli.

The aim of this visualization is to represent the relationships between the four types of stimuli of the feet and their incidence on the human body of a typical subject with feet together. After having therefore positioned the body in the drawing with the correct posture, it was decided to place a zoom of the pelvic bones into which could be positioned the contour-plot visualization. The double micro/macro scale overlaid in this way allows to immediately understand the first reading level. Visually, the red band not only creates a zoom zone but also divides the representation into two parts: at the top is the qualitative part and at the bottom the quantitative one.

In the quantitative section we find the representation of the rhythm of the stimuli on the feet - synchronous, right monopod, left monopod, alternating - and its correspondence with a new type of graph that has not yet been addressed. This is a column chart that responds to the need to correlate the rhythm of the stimuli with the actual displacement of the barycenter on the X and Y axis. In order to obtain a consequential temporal correspondence of the X and Y values, numerical values corresponding to the sequence of values have been placed on the ordinate, while the X and Y values of the test have been distributed on the abscissa. With this type of graph we can therefore see the trend of X and Y over time, throughout the test, divided by the four rhythms. We can observe how the various directions of the displacements coincide with the contour-plot and how the effects of the four stimuli on the movements of the barycenter are different. In conclusion, going up again, we can also understand with the approximate representation of the displacement of the skeleton, the effective incidence of the stimuli on the final posture of the body, where in fact for each of the four phases we observe a different position of the body.

Figure 8. Scatter plot of the feet together test with addition of the movement symbol for each phase.

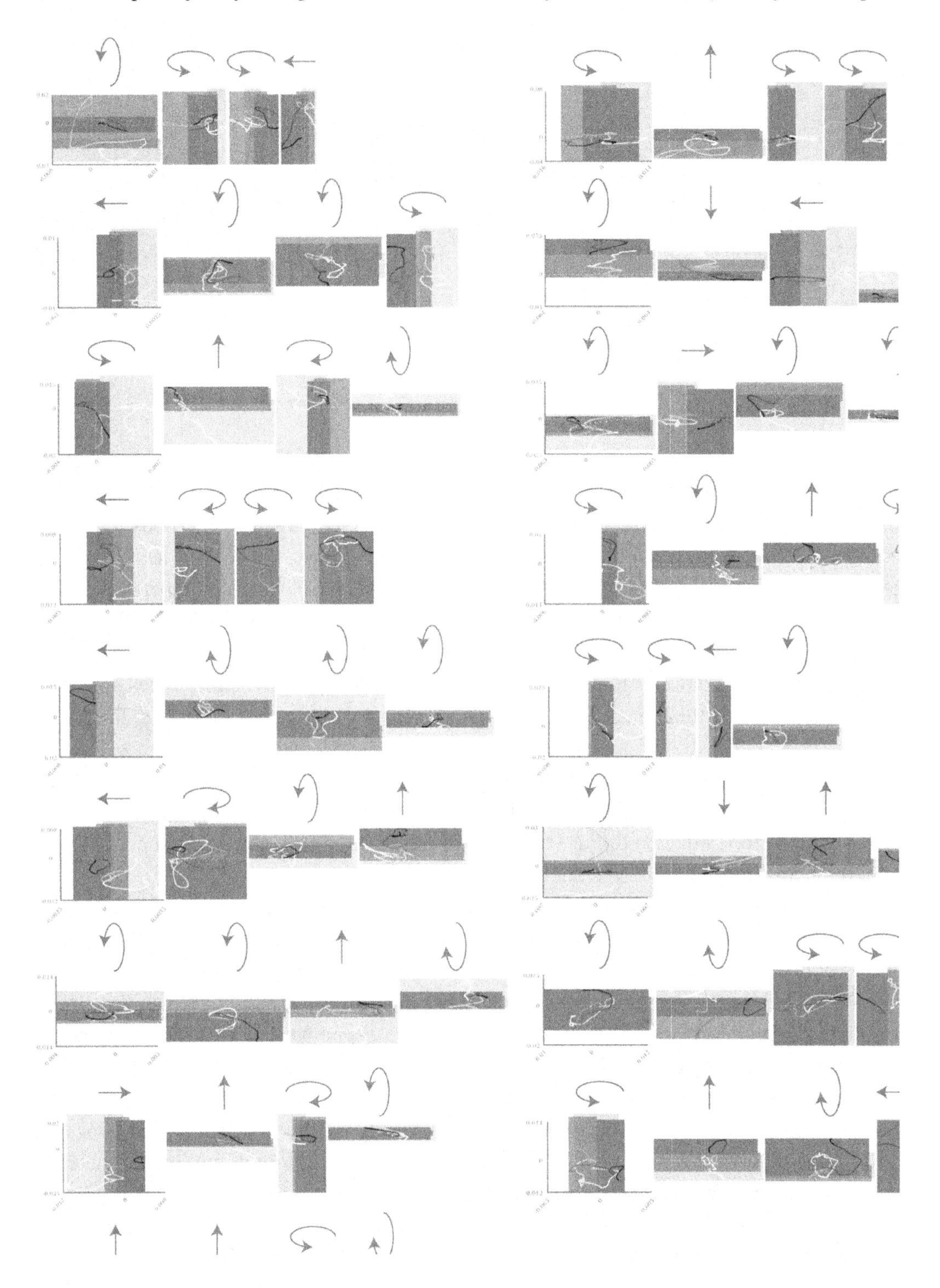

Figure 9. Simplification of movement typology during feet together test.

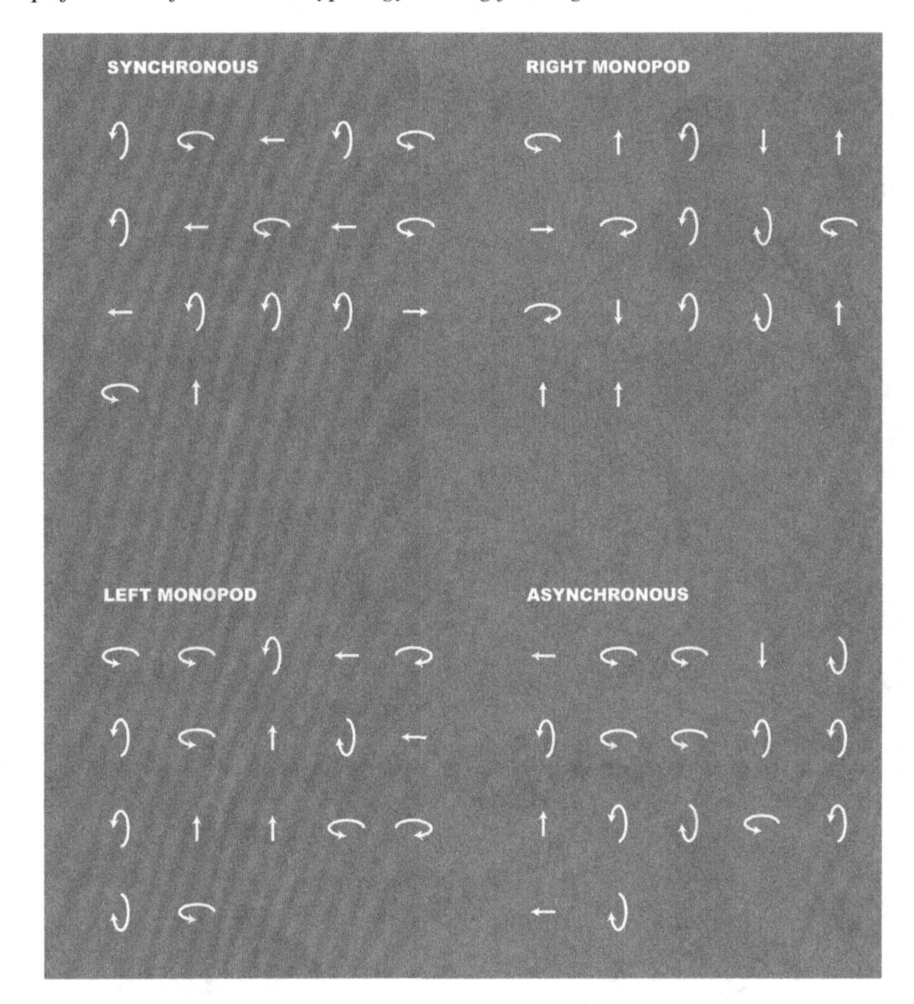

DISCUSSION

The data analysis on the movement of the barycenter describes the relationship between stimuli and motorial behavior. As in everyday experience, to an external tactile stimulus, the body reacts with an immediate motor behavior.

For this reason, the 15-second time span over which the same stimulus is transmitted is large compared to the observation of body reaction. During this time the body will implement a series of steps: the immediate reaction to the stimulus, the adaptation to the stimulus, the achievement of a balance between the behavior and the stimulus. The exact duration of these phases is not defined in this research and will be the subject of future studies.

The summations of the contour-plots shown in Figure 4 show, in a grid, the general motor behavior that the testers had during the tests. On the rows we find the results divided according to the type of posture while on the columns we find the results divided according to the type of stimulus transmitted.

Figure 10. Relationship between rhythmic pressor stimulus and motorial behaviour. Subject n.1, Feet together Test.

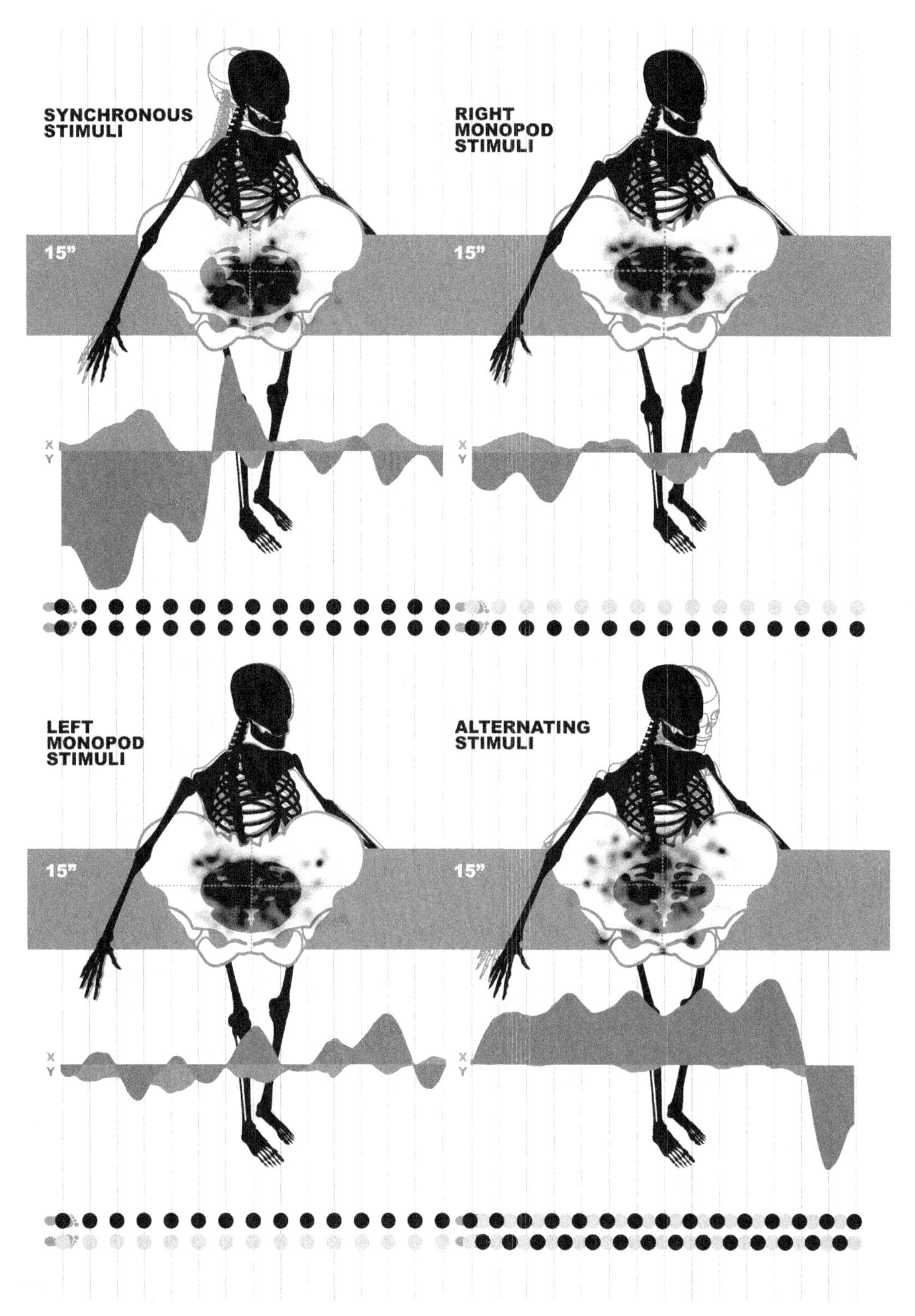

Observation of the graphs allows some notes:

- The two rhythmic monopod types generated similar quadrant incidence graphs. The points at which the center of gravity most frequently transited are mainly located near the origin of the graph. The distribution of X and Y points is equally comparable in all four quadrants, and thus there is no preponderance of one quadrant with more points than others. This is particularly true if we look at the graphs for the opened feet test and the foot together test. The test with tandem feet shows an unbalance towards the lower part of the graph (right-monopod stimulus) and towards the upper part of the graph (left-monopod stimulus).
- The two rhythmic monopods generated graphs with a higher concentration around the origin. In exercises with alternating and synchronous stimuli there is greater heterogeneity of distribution between tests. This is particularly true for the right-monopod stimulus.
- Alternating and synchronous rhythmic provided opposite results.
    - In the tests carried out with synchronous stimulus the center of gravity moved mainly in the lower part of the graph. Also in this case the tandem graphs are in countertendency showing a lateral shift to the right on the x-axis.
    - In the tests carried out with alternating stimuli the center of gravity passed more in the upper parts of the graphs.
- The total summations of the graphs for each rhythm show three trends: synchronous rhythms generate a motorial behavior that shifts the center of gravity backwards; monopod rhythms do not show a prevalent behavior oriented towards a specific area of the graph; alternating rhythms generate a motorial behavior that shifts the center of gravity forwards.

The division of the scatterplot graphs into three areas (Figure 6) highlights the need to further detail the moments that occurred during the exercises. The simplifications by means of directions and vectors (Figures 7, 8) highlight the tendency of the center of gravity to general circular motor behavior during the duration of the test. The center of gravity during each step of the test tends to describe a trajectory whose end point is close to the initial point.

From the considerations some reflections emerge both on the postural positions required and on the effects of the stimuli on the behavior.

The observations allow to identify a design mode linked to the use of pressor rhythmic stimuli that are able to exploit the symmetry of the body on the sagittal axis. The data analysis shows how a type of rhythmic stimulus transmitted symmetrically on the two antimeres invites the body to stretch and unload the weight backwards with respect to the ideal center. In an opposite way, an alternating stimulus causes a forward tension of the body.

- A few final notes at the end of the discussion of data processing highlight points that should be reconsidered in the protocols of subsequent experiments:
- The test performed with tandem feet generated data in contrast to the previous ones, this could have been determined by the precarious position in which the subjects found themselves. The complexity of maintaining this position may therefore have prevailed over the type of stimulus transmitted.
- The stimuli were always transmitted in the following sequence: synchronous, right monopod, left monopod and alternating. There was no pause between the different types of stimuli so each

stimulus always had the same position during the test. Even if the time was short enough not to influence the tiredness of the subjects' bodies, an error could have occurred due to specific behaviors that the body performs differently at the end and beginning of the test.
- An adaptation phase to the stimulus was performed for each test during the pre-test phases, but a postural survey of the condition without plantar stimulus was not carried out. Doing this would have allowed to generate benchmark graphs to be used in the comparison table.
- Graph overlays were made without unifying the scales. The choice depended on the assumption that the introduction of a common scale would reduce the detail of barycenter motion in the graphs. However, we know that an investigation that takes this into account will help uncover new trends of interest in the evidence related to the range of motion factors and not just direction.

FUTURE RESEARCH DIRECTIONS

Experimentally, the results show a relationship between rhythmic pressor stimuli and involuntary body behavior. This relationship needs to be strengthened with further experimental test protocols. In these, it is hypothesized that the positioning of stimuli in other areas of the body can be varied while maintaining the qualities highlighted: rhythm, pressure and body symmetry. The use of virtual reality experiments could help the realization of tests that are not constrained by the architectural context.

CONCLUSION

The analysis of the data collected during the test highlights a close relationship between rhythmic pressor stimuli and human body behavior. The activation of a posture leaning forward or retracted backwards, following the transmission of specific stimuli in the feet, suggests the existence of a motorial reaction mechanism that the human body implements automatically in response. This implicit behavior is not sharp but is evident in the data collected in the tests.

The activated relationship between body and stimulus invites the body to assume a specific position and to tense its muscles according to a preconfigured layout.

In this sense we think that the pressor/rhythmic stimulus can be placed within the semantic typology of affordances and used in the future as a basis for research on thresholds that identify an area of afference to a place of interaction of *mixed reality*.

Starting from the results, the research identified possible multidisciplinary approaches to design users' behaviors with these new wearable devices that assumes the role of mediators between real spaces and mixed reality. Possible areas of application for this research could be mixed reality projects and tools existing or in advanced testing. Therefore, there are two design areas to be investigated: workspaces where digital information helps to create and understand tangible objects, and the field of cultural heritage such as museums or any other kind of exhibition.

In the workspaces, characterized by mixed interaction elements and undefined architectural spaces, it will be possible to highlight the access to an implicit interaction through a haptic threshold perceived by the subjects. These stimuli could also be useful to manage the transition between one workspace and another within virtual reality.

Within the exhibit scenarios, it will be interesting to place these studies in relation to the multimedia and multisensory moments proposed to the users. Complex interactive experiences are increasingly used as exhibition systems and often involve fast and consecutive fruition in which the quality of immersion in the proposed experience decreases each time. Somatosensory stimulation could be an answer to activate and involve the body, as a whole, during specific moments in the paths.

REFERENCES

Apolito, P. (2014). *Ritmi di festa. Corpo, danza, socialità*. Il Mulino.

Bagnara, S., & Pozzi, S. (2014). Interaction design e riflessione. In Le ragioni del design. FrancoAngeli.

Baumann, M. A., MacLean, K. E., Hazelton, T. W., & McKay, A. (2010). Emulating human attention-getting practices with wearable haptics. *2010 IEEE Haptics Symposium*, 149–156. 10.1109/HAPTIC.2010.5444662

Bertin, J. (2010). *Semiology of Graphics: Diagrams, Networks, Maps* (1st ed.). Esri Press.

Bihanic, D. (2015). New Challenges for Data Design. Springer.

Buiatti, E. (2014). *Forma Mentis. Neuroergonomia sensoriale applicata alla progettazione*. FrancoAngeli.

Dall'Osso, G. (2021). Haptic Rhythmics for Mediation Design Between Body and Space. *DIID*, *74*(74). Advance online publication. doi:10.30682/diid7421d

De Luca, V. (2016, luglio). Oltre l'interfaccia: Emozioni e design dell'interazione per il benessere. *MD Journal,* 106–119.

Dourish, P. (2004). *Where the action is: the foundations of embodied interaction*. MIT Press.

Fass, D. (2015). Affordances and Safe Design of Assistance Wearable Virtual Environment of Gesture. *Procedia Manufacturing*, *3*, 866–873. doi:10.1016/j.promfg.2015.07.343

Ferriss, T. (2013, April 3). *The First-Ever Quantified Self Notes (Plus: LSD as Cognitive Enhancer?).* The Blog of Author Tim Ferriss. https://tim.blog/2013/04/03/the-first-ever-quantified-self-notes-plus-lsd-as-cognitive-enhancer

Frey, M. (2007). CabBoots: Shoes with integrated guidance system. *Proceedings of the 1st International Conference on Tangible and Embedded Interaction - TEI '07*, 245. 10.1145/1226969.1227019

Gibson, J. J. (1974). *The perception of the visual world.* Greenwood Press.

Gibson, J. J. (1977). The Theory of Affordances. In *R. E. Shaw & J. Bransford (A c. Di), Perceiving, Acting, and Knowing*. Lawrence Erlbaum Associates.

Gibson, J. J. (1979). The ecological approach to visual perception [Un approccio ecologico alla percezione visiva]. Houghton Mifflin.

Gibson, J. J. (1986). *The ecological approach to visual perception*. L. Erlbaum.

Grabarczyk, P., & Pokropski, M. (2016). Perception of Affordances and Experience of Presence in Virtual Reality. *Avant (Torun)*, *7*(2), 25–44. doi:10.26913/70202016.0112.0002

Hall, E. T. (1966). *The Hidden Dimension*. Anchor Books.

Hertenstein, M. J., Keltner, D., App, B., Bulleit, B. A., & Jaskolka, A. R. (2006). Touch communicates distinct emotions. *Emotion (Washington, D.C.)*, *6*(3), 528–533. doi:10.1037/1528-3542.6.3.528 PMID:16938094

International Data Corporation. (2022). *Wearables Deliver Double-Digit Growth for Both Q4 and the Full Year 2021, According to IDC*. IDC. https://www.idc.com/getdoc.jsp?containerId=prUS48935722

Jakobsen, M. R., Sahlemariam Haile, Y., Knudsen, S., & Hornbæk, K. (2013). Information Visualization and Proxemics: Design Opportunities and Empirical Findings. *IEEE Transactions on Visualization and Computer Graphics*, *19*(12), 2386–2395. doi:10.1109/TVCG.2013.166 PMID:24051805

Jetter, H. C., Reiterer, H., & Geyer, F. (2014). Blended Interaction: Understanding natural human–computer interaction in post-WIMP interactive spaces. *Personal and Ubiquitous Computing*, *18*(5), 1139–1158. doi:10.100700779-013-0725-4

Karuei, I., MacLean, K. E., Foley-Fisher, Z., MacKenzie, R., Koch, S., & El-Zohairy, M. (2011). Detecting vibrations across the body in mobile contexts. *Proceedings of the SIGCHI Conference on Human Factors in Computing Systems*, 3267–3276. 10.1145/1978942.1979426

Kettner, R., Bader, P., Kosch, T., Schneegass, S., & Schmidt, A. (2017). Towards pressure-based feedback for non-stressful tactile notifications. *Proceedings of the 19th International Conference on Human-Computer Interaction with Mobile Devices and Services - MobileHCI '17*, 1–8. 10.1145/3098279.3122132

Klee, P. (1959). *Teoria della forma e della figurazione* (Vol. 1). Feltrinelli.

Lederman, S. J., & Klatzky, R. L. (2009). Haptic perception: A tutorial. *Attention, Perception & Psychophysics*, *71*(7), 1439–1459. doi:10.3758/APP.71.7.1439 PMID:19801605

Lefebvre, H. (2019). *Elementi di Ritmanalisi. Introduzione alla conoscenza dei ritmi*. LetteraVentidue.

Licaj, A. (2018). *Information Visualization. Disciplina liquida intersoggettiva* [Doctoral dissertation]. University of Genoa. doi:10.15167/licaj-ami_phd2018-05-09

Löffler, D., Tscharn, R., Schaper, P., Hollenbach, M., & Mocke, V. (2019). Tight Times: Semantics and Distractibility of Pneumatic Compression Feedback for Wearable Devices. *Proceedings of Mensch Und Computer 2019 on - MuC'19*, 411–419. doi:10.1145/3340764.3340796

Maldonado, T. (1997). *Critica alla ragione informatica*. Feltrinelli.

Mallgrave, H. F. (2015). *L'empatia degli spazi*. Raffaello Cortina Editore.

Matthews, T., Dey, A. K., Mankoff, J., Carter, S., & Rattenbury, T. (2004). A toolkit for managing user attention in peripheral displays. *Proceedings of the 17th Annual ACM Symposium on User Interface Software and Technology - UIST '04*, 247. 10.1145/1029632.1029676

Matthies, D. J. C., Müller, F., Anthes, C., & Kranzlmüller, D. (2013). ShoeSoleSense: Proof of concept for a wearable foot interface for virtual and real environments. *Proceedings of the 19th ACM Symposium on Virtual Reality Software and Technology - VRST '13*, 93. 10.1145/2503713.2503740

Moens, B., van Noorden, L., & Leman, M. (2010). D-Jogger: Syncing Music with Walking. *Proceedings of SMC Conference 2010,* 451–456. http://hdl.handle.net/1854/LU-1070528

Moloney, J., Spehar, B., Globa, A., & Wang, R. (2018). The affordance of virtual reality to enable the sensory representation of multi-dimensional data for immersive analytics: From experience to insight. *Journal of Big Data*, *5*(1), 53. doi:10.118640537-018-0158-z

Nicoletti, R., & Borghi, A. M. (2007). *Il controllo motorio.* Il Mulino.

Norman, D. (1988). The psychology of everyday things [Psicopatologia degli oggetti quotidiani]. Basic Books.

Norman, D. (2013). The psychology of everyday things [Il design degli oggetti quotidiani]. Basic Books.

Palmerini, L., Rocchi, L., Mellone, S., Valzania, F., & Chiari, L. (2011). Feature Selection for Accelerometer-Based Posture Analysis in Parkinson's Disease. *IEEE Transactions on Information Technology in Biomedicine*, *15*(3), 481–490. doi:10.1109/TITB.2011.2107916 PMID:21349795

Panero, J., & Zelnik, M. (1979). *Human Dimension and Interior Space. A Source Book of Design Reference Standards*. Whitney Library of Design.

Parnow, J. (2015). *Micro Visualization.* Retrieved from https://microvis.info/

Ploch, C. J., Bae, J. H., Ju, W., & Cutkosky, M. (2016). Haptic skin stretch on a steering wheel for displaying preview information in autonomous cars. *2016 IEEE/RSJ International Conference on Intelligent Robots and Systems (IROS)*, 60–65. 10.1109/IROS.2016.7759035

Pohl, H., Brandes, P., Ngo Quang, H., & Rohs, M. (2017). Squeezeback: Pneumatic Compression for Notifications. *Proceedings of the 2017 CHI Conference on Human Factors in Computing Systems - CHI '17,* 5318–5330. 10.1145/3025453.3025526

Pronzati, A. (2021). *Micro-viz: modelli per la progettazione grafica di visualizzazioni dati in spazi ridotti.* https://www.politesi.polimi.it/handle/10589/175020?mode=complete

Rantakari, J., Inget, V., Colley, A., & Häkkilä, J. (2016). Charting Design Preferences on Wellness Wearables. *Proceedings of the 7th Augmented Human International Conference 2016 on - AH '16,* 1–4. 10.1145/2875194.2875231

Rapp, A., & Cena, F. (2016). Personal informatics for everyday life: How users without prior self-tracking experience engage with personal data. *International Journal of Human-Computer Studies*, *94*, 1–17. doi:10.1016/j.ijhcs.2016.05.006

Regia-Corte, T., Marchal, M., Cirio, G., & Lécuyer, A. (2013). Perceiving affordances in virtual reality: Influence of person and environmental properties in perception of standing on virtual grounds. *Virtual Reality (Waltham Cross)*, *17*(1), 17–28. doi:10.100710055-012-0216-3

Rooksby, J., Rost, M., Morrison, A., & Chalmers, M. (2014). Personal tracking as lived informatics. *Proceedings of the SIGCHI Conference on Human Factors in Computing Systems*. 10.1145/2556288.2557039

Shi, Y. (2018). Interpreting User Input Intention in Natural Human Computer Interaction. *Proceedings of the 26th Conference on User Modeling, Adaptation and Personalization*, 277–278. 10.1145/3209219.3209267

Sicklinger, A. (2020). *Design e Corpo umano. Cenni storici di Ergonomia, Antropometria e Movimento Posturale*. Maggioli Editore.

Song, S., Noh, G., Yoo, J., Oakley, I., Cho, J., & Bianchi, A. (2015). Hot & tight: Exploring thermo and squeeze cues recognition on wrist wearables. *Proceedings of the 2015 ACM International Symposium on Wearable Computers - ISWC '15*, 39–42. 10.1145/2802083.2802092

Strogatz, S. (2003). *Sincronia. I ritmi della natura, i nostri ritmi*. Rizzoli.

Suzuki, Y., Suzuki, R., Watanabe, J., Yoshida, A., & Shigeru, S. (2015). Haptic vibrations for hands and bodies. *SIGGRAPH Asia 2015 Haptic Media and Contents Design on - SA '15*, 1–3. doi:10.1145/2818384.2818389

Trost, W., Labb, C., & Grandjean, D. (2017). Rhythmic entrainment as a musical affect induction mechanism. *Neuropsychologia*, *96*, 96–110. doi:10.1016/j.neuropsychologia.2017.01.004 PMID:28069444

Velazquez, R., Bazan, O., & Magana, M. (2009). A shoe-integrated tactile display for directional navigation. *2009 IEEE/RSJ International Conference on Intelligent Robots and Systems*, 1235–1240. 10.1109/IROS.2009.5354802

Vogel, D., & Balakrishnan, R. (2004). Interactive public ambient displays: Transitioning from implicit to explicit, public to personal, interaction with multiple users. *Proceedings of the 17th annual ACM symposium on User interface software and technology*, 137–146. 10.1145/1029632.1029656

Zannoni, M. (2018). *Progetto e interazione. Il design degli ecosistemi interattivi*. Quodlibet. https://www.quodlibet.it/libro/9788822901668

Zannoni, M., & Formia. (2018). *"Geo-media" e Data Digital Humanities*. Academic Press.

Zeagler, C. (2017). Where to wear it: Functional, technical, and social considerations in on-body location for wearable technology 20 years of designing for wearability. *Proceedings of the 2017 ACM International Symposium on Wearable Computers*, 150–157. 10.1145/3123021.3123042

KEY TERMS AND DEFINITIONS

Antimere: Homotypic and opposing parts that determine the symmetry of the body with respect to the sagittal axis.

Haptic: Field of research that investigates the potential of using the tactile and proprioceptive system of the human body in relation to digital technologies.

Mixed Reality: The design field in which human actions can be carried out in continuity between virtual reality and augmented reality systems.

Natural Language: Languages used in interaction design drawing on the natural rules of the physical world on which each person has based their experience.

Proxemics: Study of the influence of space on social relations between humans.

Rhythm: The layout on which sensory stimuli are positioned temporally and/or spatially.

Threshold: Element of discontinuity in the space that allows to perceive the access from one area to another.

Chapter 3
Virtual World in Learning Disability

Anjana Prusty
https://orcid.org/0000-0002-6205-4712
SR University, India

Priyaranjan Maral
https://orcid.org/0000-0002-5266-8681
Central University of Rajasthan, India

ABSTRACT

In terms of technological advantages, virtual reality or augmented reality remains less popular within the field of learning disabilities. Research shows that children with learning disabilities face various challenges in their day-to-day lives dealing with these disorders, demanding massive solutions. This chapter will address the pros and cons of virtual reality in learning disabilities across different age groups by combining theories of virtual worlds and learning disorders. Exciting research in virtual reality focuses on finding out how psychotherapies have benefits in learning and education. Upon review, it becomes evident that research in the virtual world along with learning disabilities has not yet been examined from a cohesive perspective, illustrating a lack of alliance that determines a more global understanding of the technological advantages of disabilities. Thus, this chapter aims to provide educators with an overview of explanations of the virtual world and to ensure appropriate development of VR/AR applications and special assistance for learning disabilities.

INTRODUCTION

"A hero is an ordinary individual who finds the strength to persevere and endure in spite of overwhelming obstacles." - Christopher Reeve

DOI: 10.4018/978-1-6684-4854-0.ch003

Figure 1. Experiencing virtual world through virtual reality headset (Ashworth, 2020)

The term "*virtual world*" needs a simple, meaningful understanding of what it means and what it means to have a virtual world. Mostly, it needs to underrate the relevance of the virtual world in the disabled sector, which can be used to help the individual live. There are a lot of schools of thought on virtual worlds, but unfortunately, they have not yet brought the concept to clarity to clarify the benefits for the disabled. If technology can do everything, which is unbelievable, then it is also essential to see in the other possible directions. Through this, disabled people will benefit, along with special educators and scientists. So far, the conceptual meaning or definition of a virtual world has not been well-defined. The ability to predict a technical definition has its own benefits and gives users a wide range of experience. However, because a virtual world is defined by a mix of different technologies, it makes it hard to tell which technologies have similar features. For example, a smartphone has multiple advantages with different technologies. In this section, we examined technology and its advancement in the field of learning disabilities (LD). According to one of the eminent researchers on learning disabilities, "*Learning disabilities are not a prescription for failure. With the right kinds of instruction, guidance and support, there are no limits to what individuals with LD can achieve" said by* Sheldon H.Horowiz, 2014 (p3) (Cortiella & Horowiz, 2014).

According to the National Center for Education Statistics report (2020), from 2018 to 2019, no less than 33% of students have specific learning disabilities. Learning disabled students are likely to drop out of school three times more often than other dropout students. In line with Butterworth & Kovas (2013) said that students with high IQ may fail to understand the standard mathematical curriculum. Similarly, another report reported that with the help of smartphones, children can learn and interact with their environment easily using augmented and virtual reality (Panwala et al., 2017). And also, it has been proven that the more student interaction is involved in learning technology, the more they can enhance their learning ability (Blaster et al., 2016). Image processing technique has been proved a method for the interactive learning through different applications in higher education (Yaman & Karakose, 2016).

VIRTUAL WORLD IN EDUCATION

Advantages of virtual world is an umbrella term, from them the virtual reality, sometimes known as VR, is becoming increasingly popular in the educational sector, with more and more schools embracing the technology. VR gives pupils the opportunity to go to different parts of the world without ever having to physically leave the classroom. Envisage a classroom in which the students are free to investigate any topic while remaining seated (Figure 1) and virtual reality enables this kind of learning in education. VR is something that most people have heard of, but several people do not understand what VR is or how it may be used to enhance learning and education. The term "virtual reality" refers to interactive media like photos or movies that let the viewer move around in a setting in all 360 degrees (Immersion VR, 2022).

The advanced benefits of technologies such as the internet, e-mail, and video teleconferencing are becoming familiar methods for diagnosis, therapy, education, and training. In particular, the use of virtual reality and computerized therapies is high in homework assignments, relapse prevention, and problem-solving techniques. It has been seen that the use of virtual reality is rapidly growing in the field of education, in which students are encouraged to explore and experiment in order to form new ideas and concepts. VR is a multisensory interactive medium that replaces the normal sensory inputs of the subject with artificial sensory signals generated by a computer. These particular aspects of VR make users initiate similar behaviors in learning as they do in the real world. It is observed that the method we use to process information from the environment forms the concept of multisensory in nature, in which different sense organs work together to reduce ambiguity. From the literature, it is found that multisensory inputs allow for disambiguation in real life. It also provides a comfortable environment to form the concepts. The aspects of multisensory visualization, complex and abstract concepts, are used to teach through virtual reality easily (Christou, 2010). Moreover, the nature of VR supports the constructive approach to learning (Winn, 1993). In psychology, constructivism is a theory of knowledge acquisition that states that humans construct knowledge by learning from their experiences. The model theory of Jean Piaget states that the learner attempts to assimilate new experiences within their environment itself. If the learner cannot successfully assimilate new details, they change their world view to accommodate the new experience (Baron et al., 2009). Learning is a form of active and constructive processing that accumulates or accepts facts. Similarly, virtual reality provides an environment for this particular active process of learning. The use of virtual reality in the classroom can enhance students' learning by giving them opportunities to participate in unique, life-changing experiences that would not be accessible in any other setting. In addition to this, everything can occur within the framework of the classroom. Every student has access to virtual reality, and it is simple for teachers to keep an eye on their progress. Students can be engaged and inspired in a way that is truly one of a kind and potent through the use of virtual experiences.

VIRTUAL WORLD IN HIGHER EDUCATION

When we think about higher education, then the probability of usability of virtual world is not too much common among people. After pandemic people have realised that the need of virtual world for providing education on time to everyone without any interruptions. People are now able to learn at a distance, on their own time, and at their own speed thanks to online learning, which has become a supplement to traditional teaching methods over the course of the past few years and has fundamentally altered the educational landscape. It has evolved into a strong instrument that enables teachers to broaden the scope of their instruction while also giving students the opportunity to obtain an education without physically attending classes. The use of remote learning allows for all of this to be accomplished in a way that is both convenient and frugal. Presently the VR technology has highly appeared in different fields and sectors (Asnar, et al., 2018) such as in surgical education uses it in terms of intervention (Kyaw et al., 2019; Yognathan, et al., 2018), heritage education (Ibanes-Etxeberria, et al., 2020), sports training (Panchuk, et al., 2018) also it is useful to treat the psychological disorder of stage fright and anxiety disorder (Stupar-Rutenfrans, et al., 2017; Cardos, et al., 2017)) and language learning (Parmaxi, 2020). Educators are discovering that, contrary to popular belief, it is possible to cover any topic in this environment, even though the potential benefits of virtual world education appear to be greatest for certain fields of study. The study of history appears to offer the most potential benefits; for instance, one can re-enact events such as the American Civil War and offer students the opportunity to participate in virtual field trips that take them to the battlefields, allow them to march in rapid-fire formation through the South, and even allow them to meet George Washington. All of the technology that promote education, including recording lectures, can be used in a virtual classroom, allowing students to return to the lecture later, replay the sections they need to hear again, and not rely on lengthy notes. Since classes can be accessed from anywhere with an Internet connection, students save significantly on travel and living expenses. The finest of the physical world, the best of the internet and online apps, and the best of the virtual world technology will come together in the form of virtual learning, allowing students to acquire an education using the most cutting-edge methods possible.

DEVELOPMENT OF VIRTUAL WORLD IN DIFFERENT AGE GROUPS

It is very essential to address the advantages of a virtual environment in different age groups, especially in school to be useful efficiently. Teaching and learning methods have been continuously changing in the last few decades. Superscape, a software tool used with people with learning disabilities and on the autism spectrum, is one of the software tools used by the VIRART group. The result suggested that individuals with ADS (Vasudevan et al., 2022), low IQ, and weak executive abilities require more facilitated virtual environment software to access the task accurately (Parsons et al., 2005). Similarly, virtual software developed by the University of Valencia called Virtual Supermarket found very authentic tools to generalize and acquire knowledge (Herrera et al., 2005). Recent reports from NMC/CoSN Horison K-12 (Freeman et al., 2017) and EDUCAUSE 2020 horizon (Brown et al., 2020) said that VR has the most impact in educational settings. It is also proved by the recent findings of Villena-Taranilla et al. (2022) with K-6 students. VR promotes greater learning in students. There are a lot of studies that have proved virtual simulation has a great impact on learning in K-12 (Merchant et al., 2014) students as well as students in higher education with VR theologies (Maral et al., 2020). A very useful and recent study by Vasuden et al. (2022) on ASD children with the implementation of virtual reality games tested effectiveness, and they found that virtual reality is capable of diagnosing the ASD. Higher education in terms of Internet Virtual Reality (IVR) proposes academic motivational improvements and achievements (Hamilton et al., 2021; Radianti et al., 2020). Similarly, Pellas et al., 2021 found that non-immersive VR has a positive effect on students' motivation and learning. Another review of 18 research papers by Di Natale et al., 2020 suggested that IVR has positive outcomes for academic performance and motivation towards learning in students.

Subsequently, technology's involvement likes VR in eliminating barriers that cause cognitive deficiencies to become intellectual disabilities. The realizationthat virtual reality can play a part in the cognitive rehabilitation of individuals with disabilities has opened the door to the possibility of developing systems that simultaneously assess and address the educational, rehabilitative, and therapeutic requirements of the population that stands to gain a great deal from advancements in technology (Standen & Brown, 2006).

Exciting researchers have developed varieties of applications for general population to acknowledge their need; unfortunately, very rare applications have developed for special population. Below the table 1 is describing different types of applications for different age groups of abled and disabled populations.

Table 1. Virtual applications for both abled/disabled populations

Sl. No.	Application	Author	Year	User	Age Group
1	Heromask VR Headset + Mathematics Games	Edevento	2018	Abled	5-12
2	Discovery Kids VR Goggles	Jackie Cucco	2017	Abled	7+
3	Oculus Go Standalone Virtual Reality Headset - 32gb	Huggo Barra	2018	Abled/Disabled	13+
4	The Number Race	Wilson et al.	2016	Disabled (Dyscalculia)	7-9
5	Grapho-Game Maths	Gowswami & Fiona	2008	Disabled (Dyscalculia/Dyslexia)	1-8
6	RobotLAB Autism VR	Elad Inbar	2007	Disabled (Autism)	K-2
7	Kobi360	Oyedele & Ozor	2019	Disabled (Dyslexia)	5+
8	AR-DAWE	Khan et al.	2017	Disabled (Dysgraphia)	7+
9	Imaginator	Rossi et al.	2018	Disabled (Auditory Processing Disorder)	5+
10	EEG-Based Serious Game	Alchalabi et al.	2018	Disabled (ADHD)	5+

LEARNING DISABILITY (LD)

Learning disability, also known as a learning disorder, begins at a young age. It is an umbrella term for having a wide range of learning problems. A learning disability does not mean a problem with intelligence, creativity, or motivation. Those children who have learning disabilities deal with the problem of understanding, not counted as lazy or dumb. People with learning disabilities may be exceptionally intelligent and creative, such as Sir Winston Leonard Spencer Churchill (Novel Prize winner), Wernher von Braun (Father of Rocketry), and Thomas Alva Edition (indigenous American Inventor) (Prusty et al., 2021). Simply put, people with learning disabilities have different ways of observing, hearing, and understanding things. This aspect makes it difficult for them to learn new information and improve their silks for proper use (Kemp et al., 2020).

The severity of a learning issue can range from a) Mild (i.e., some learning challenges in one or two academic areas, although the student may be able to compensate for these issues.), b) Moderate (i.e., significant challenges with learning that require some specialist instruction in addition to some accommodations or supportive services.), and c) Severe (i.e., significant challenges with learning, affecting multiple subject areas and necessitating continued participation in intensive, specialised instruction).

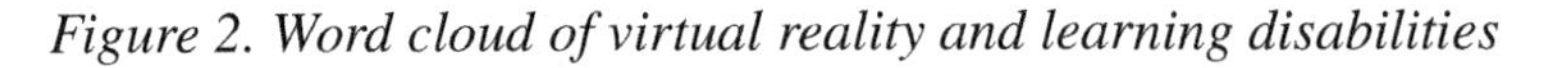

Figure 2. Word cloud of virtual reality and learning disabilities

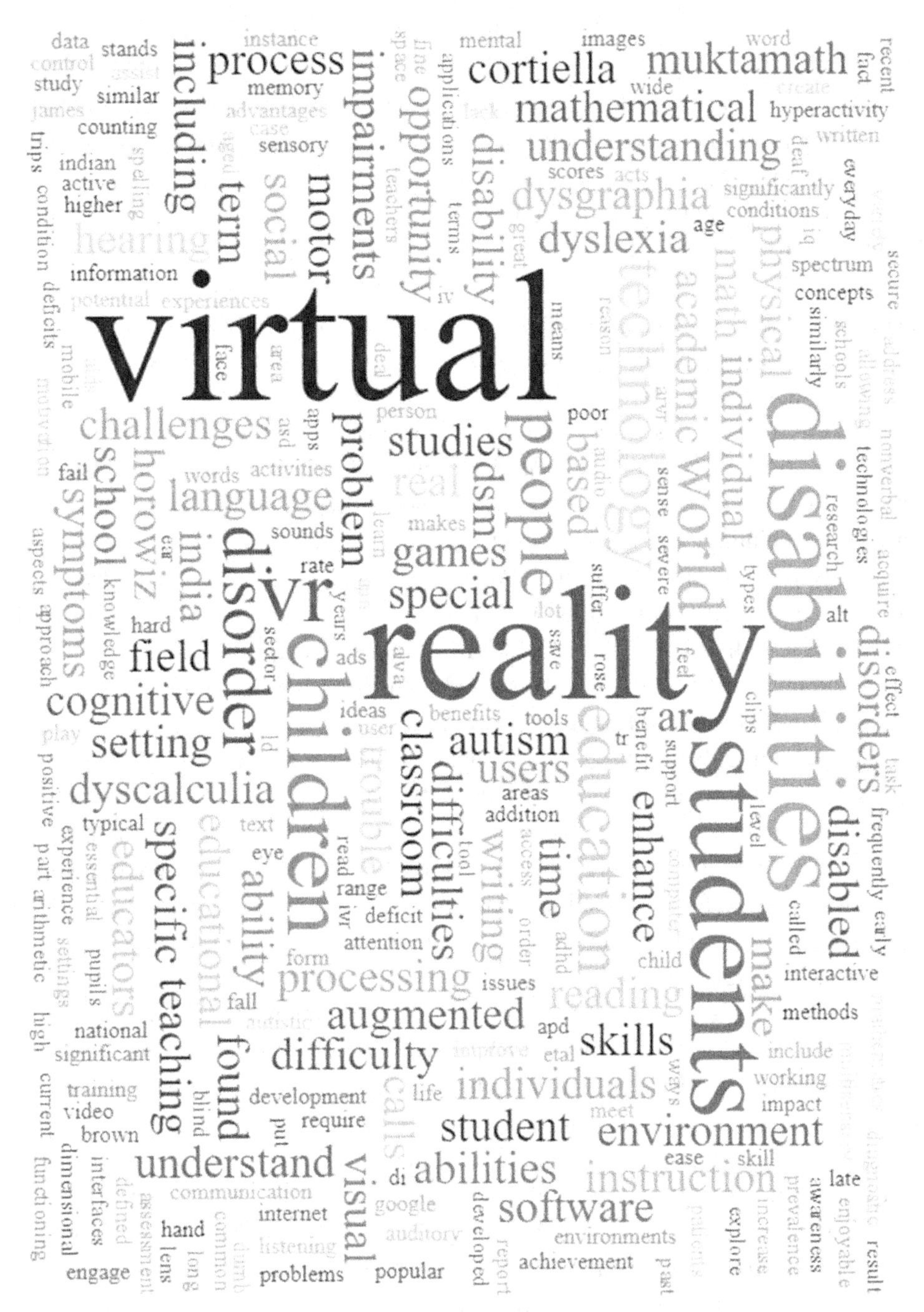

Many kinds of learning difficulties fall into one of these three categories: dyslexia, dysgraphia, or dyscalculia as mentioned in above (Figure 2). Learning disabilities encompass a wide range of conditions. The most common types of learning disabilities in individuals are problem with reading calls; dyslexia problem with writing class; dysgraphia, problem with math and reasoning calls; dyscalculia, problem with listening and speaking calls: autism. Other types of learning disabilities are such as problem in motor skill calls; dyspraxia, problem with language calls; aphasia/dysphasia, problem with listening calls; auditory processing disorder, problem with visualization calls visual processing disorder (Kemp et al., 2020).

Reading, writing, and overall understanding can all be affected by the language processing condition known as dyslexia. People who have dyslexia may have trouble decoding words or with phonemic awareness, which is the process of distinguishing particular sounds contained inside words. The condition known as dyslexia frequently goes undiagnosed for a significant amount of time, which frequently leads to difficulties with reading, grammar, reading comprehension, and several other language skills.

1. Dyslexia: It is a term that predates LD and refers especially to difficulties with correct or fluent word identification, poor spelling, and weaknesses in coding abilities. LD is an acronym that stands for learning disability (International Dyslexia Association, 2015). It is still utilised in therapeutic settings as well as research settings, and it is incorporated into the overarching diagnostic of LD that is provided by the DSM-5. (Cortiella & Horowiz, 2014; Muktamath et al., 2021).
2. Dysgraphia: People who suffer from dysgraphia have a difficult time putting their thoughts down on paper or in a painting. The inability to write legibly is a hallmark of dysgraphia, but it is by no means the only symptom of the condition. Patients have difficulty putting their ideas down on paper, whether that's with their spelling, their grammar, their vocabulary, their ability to think critically, or their memory. Individuals who suffer from dysgraphia may have issues with letter spacing, poor motor planning and spatial awareness, and difficulty thinking and writing at the same time. (Cortiella & Horowiz, 2014; Muktamath et al., 2021).
3. Dyscalculia: Learning difficulties that are connected to mathematical computations are referred to collectively as dyscalculia. People who have dyscalculia have difficulty understanding mathematical ideas, numbers, and logical reasoning. People who have what is sometimes called "math dyslexia" may have trouble reading clocks to tell time, counting money, seeing patterns, recalling arithmetic information, and completing mental math problems. Other symptoms of math dyslexia include difficulties recognising patterns. (Cortiella & Horowiz, 2014; Muktamath et al., 2021).
4. Auditory processing disorder (APD): Patients suffering from auditory processing disorder have trouble processing noises in their environment. People who have APD might get the order of sounds mixed up or be unable to distinguish between various noises, such a teacher's speech and environmental noise. In APD, the information that is received and processed by the brain from the ear is erroneously interpreted. (Cortiella & Horowiz, 2014; Muktamath et al., 2021).
5. Language processing disorder: Language processing disorder is a subtype of auditory processing disorder that occurs when an individual has specific difficulties in processing spoken language. This disorder can have an effect on both the individual's ability to understand and to convey language. Language processing disorder is characterised by "difficulty attaching meaning to sound groups that create words, phrases, and tales," as stated by the Learning Disabilities Association of America (LDA, 2018).
6. Nonverbal learning disabilities (NVLD): They are characterised by difficulty in deciphering nonverbal behaviours or social cues, despite the fact that the term "nonverbal learning disabilities" (NVLD) may give the impression that they are associated with a person's incapacity to speak. Patients who suffer from NVLD often have difficulty comprehending the nonverbal parts of communication, such as body language, facial expressions, and tone of voice. (Cortiella & Horowiz, 2014; Muktamath et al., 2021).

7. Visual perceptual/visual motor deficiency: People who have a visual perceptual or visual motor deficit have poor hand-eye coordination; frequently lose their position when reading, and struggle with pencils, crayons, glue, scissors, and other activities that require fine motor skills. In addition to this, they may have issues distinguishing between letters that seem alike, have difficulty navigating their environment, or exhibit odd eye activity when reading or completing their responsibilities. (Cortiella & Horowiz, 2014; Muktamath et al., 2021).
8. Attention deficit hyperactivity disorder (ADHD): Individual faces problem with inattention, hyperactivity and distractibility (Cortiella & Horowiz, 2014; Muktamath et al., 2021).
9. Executive Functioning Deficits: Individual faces problem with planning, organising, remembering and strategies, manage time and space efficiently (Cortiella & Horowiz, 2014; Muktamath et al., 2021).
10. Autism: Inability to communicate behaviour as per situation (National Institute of Mental Heath, 2018)
11. Dyspraxia: Immaturity of the organisation of movement (Thomas et al., 2015).
12. Aphasia: inability to perform and understand language to speech (Ardila, 2014).

The Learning Impairments Association of America and a large number of other professionals in the field of mental health consider the seven diseases described above to be examples of distinct learning disabilities. They identify autism spectrum disorder (ASD) and ADHD as linked conditions that influence learning, but they do not recognise particular learning disorders. It is essential for all those interested in professions in psychology to have an understanding of learning difficulties. People who have ASD and ADHD may have coexisting problems, including learning impairments. When professionals in the field of psychology have an understanding of the neurodiversity and learning characteristics of their patients, they are able to become more inclusive, empathic, and effective contributors to the populations that they serve.

Sign and Symptoms of Learning Disabilities

- Difficulties with academic achievement and progress
- Discrepancies exist between a person's potential for learning and what that person actually learns
- An uneven pattern of development (language development, physical development, academic development, and/or perceptual development)
- Mental retardation or emotional disturbance
- To read, write, speak, spell, compute math, and reason
- Attention, memory, coordination, social skills, and emotional maturity
- Individual differing capabilities, with difficulties in certain academic areas but not in others (National Association of Special Education Teachers, 2022)
- Have trouble learning the alphabet, rhyming words, or matching letters to their sounds
- Make many mistakes when reading aloud, and repeat and pause often
- Not understand what he or she reads
- Have real trouble with spelling
- Have very messy handwriting or hold a pencil awkwardly
- Struggle to express ideas in writing
- Learn language late and have a limited vocabulary

- Have trouble remembering the sounds that letters make, or in hearing slight differences between words
- Have trouble understanding jokes, comic strips, and sarcasm
- Have trouble following directions
- Mispronounce words or use a wrong word that sounds similar
- Have trouble organizing what he or she wants to say or not be able to think of the word needed for writing or conversation
- Not follow the social rules of conversation, such as taking turns, and may stand too close to the listener
- Confuse math symbols and misread numbers
- Not be able to retell a story in order (what happened first, second, third)
- Not know where to begin a task or how to go on from there (National Association of Special Education Teachers, 2022)

Learning Disabilities in DSM-5

The Diagnostic and Statistical Manual of Mental Disorders (DSM) is "a classification system of mental disorders with associated criteria fashioned to facilitate more reliable diagnoses of psychiatric disorders." Diagnoses can be made using the DSM descriptions of symptoms and other criteria. It establishes consistent and reliable diagnoses that may be utilised in research on mental diseases and gives physicians a common vocabulary to speak about their patients. It was established by the American Psychiatric Association (APA) in 1952 in order to facilitate the diagnosis of mental diseases in the United States. The Diagnostic and Statistical Manual of Mental Disorders (DSM) has undergone five revisions since its initial release (including the DSM-III-R, DSM-IV-TR). The most recent revision of DSM-5 was unveiled in 2013. As part of a case formulation assessment, qualified clinicians use DSM-5 to help them identify their patients' mental disorders and devise a treatment plan that is tailored to their specific needs. The DSM-5 is founded on explicit disorder criteria, which, when taken as a whole, represent a "nomenclature" of mental diseases. In addition, the DSM-5 includes a comprehensive explanatory text that, for the first time, is completely referenced in the electronic edition of this DSM (Maral, 2020).

DSM-5 has made significant changes to the classification of LDs by consolidating all the subtypes into a single overall diagnosis. It does not restrict the diagnostic to reading, mathematics, or written expression, but rather describes issues in academic skills more generally, with the possibility of specifying the more traditional domains if necessary (American Psychological Association, 2013). A clinical evaluation of an individual's history, as well as reports from teachers and academic records, as well as responses to interventions, are all taken into consideration when determining a diagnosis. The complications should continue over a period of time, the scoring must be significantly lower than the range on the applicable measures, and the issues cannot be better addressed by other disorders. There must be a considerable impediment existing in the individual's ability to achieve, work, or carry out activities necessary for everyday existence (Tannock, 2013). The following are the symptoms that must be present in order to diagnose a specific learning disorder using DSM-5 criteria.

- First, long-lasting difficulty in reading, writing, arithmetic, or the ability to reason mathematically during the years spent formally enrolled in an educational programme. Imprecise reading, reading that is slow and laborious, poor written expression that lacks precision, trouble in recalling mathematical concepts, and erroneous mathematical reasoning are some of the markers that may be present.

- Second, the applicant's new academic skills need to score significantly lower than the typical range of scores on tests of reading, writing, or mathematics that are culturally and linguistically adequate. As a consequence of this, a person who has dyslexia needs to put in significantly more effort when they read, and they cannot read in the same way that typical readers do.
- Third, challenges in learning manifest themselves when the child is still in school.
- Fourth, the individual's challenges cannot be explained simply by developmental, neurological, sensory (vision or hearing), or motor disorders, and they must significantly interfere with academic achievement, occupational performance, or activities of daily living in order for the individual to be eligible for special education services (APA, 2013).
- It is important to note that the fourth version of the DSM (also known as DSM-IV-TR) did not include a general category of LD; rather, it featured various diagnoses that were specific to deficits in reading, arithmetic, and written language (APA, 2000). The DSM-IV-TR method acknowledges three diagnostic categories that are specifically defined: reading disorders, mathematical disorders, and disorders of written expression. There is also a residual category available, which is for learning difficulties that cannot be otherwise identified. In federal rules, these phrases are often used in place of the term "learning disability." The DSM-IV approach, which has been around for more than 20 years, was described and evaluated using methods that were based on discrepancy scores. This means that a learning difficulty was said to exist in a specific area, such as reading, when the scores in that particular area were significantly below what would be expected judging by the individual's overall cognitive ability. Currently, the DSM-V approach is based on methods that rely on mean scores. The definition of these conditions in the International Classification of Diseases, Ninth Revision, is quite comparable to one another, but it contains an explicit provision that the child's educational setting must be suitable to the child's capacity to acquire the skill. It is possible to have a sensory deficiency, but a long-term learning problem won't be found until the delays in getting things done are much bigger than would be expected.

Prevalence of Learning Disabilities in Worldwide

The current prevalence rate of specific learning disabilities in the global population is 15.17%, whereas the dyscalculia rate is 10.5% (Kuriyan & James, 2018). Similarly, and globally, around 7% to 8% of children aged between 3 and 17 years old have difficulties with learning or suffer from learning disability (National Survey of Children's Health, 2007a, b). There are various types of learning disabilities, including dyscalculia, dysgraphia, dyslexia, and autism. Each of these manifested different signs and symptoms in children, such as counting, spelling, writing, and listening. Learning disabilities in mathematics include deficits in multiple mathematical abilities, depending on the student's age and education. Children face problems with counting, understanding number patterns, and operations that include addition, subtraction, division, and multiplication, along with the measurement of area, shape, and size (American Psychological Association, 2000; Santos, 2009). In accordance with the international context, there are some mobile apps developed for learning disabilities. The majority of them (approximately 250) have been developed for dyslexia, 14 for dysgraphia, and only 49 for dyscalculia children (Ariffin, et al., 2019). This indicates that there is still a lack of awareness of dyscalculia in the field of technology.

There are several studies on technology-based learning tools to support learning disabled students (Weng & Taber-Dought, 2015; Zain, et al., 2013). Many VR games and applications have been developed for the purpose of educational goals to support the disability sector. Several studies have shown positive results for children with learning disabilities in reading with the use of virtual environments (Nicolson, et al., 2000; Macaruso, et al., 2006). It has been proven that children can teach themselves concepts of mathematics along with developing their specific skills by playing games (Mendes & Grando, 2008). Computer games can encourage children's engagement in the learning process through games because the games hold challenges that can motivate children to search for a solution to the presented mathematical problem (Papastergiou, 2009). Children can also build logical-mathematical thinking processes when it involves fun (Mendes & Garando, 2008). As a result, the more inspiring the challenges, the more children will be interested in the game (Wilson & Dahaens, 2007). Technology-based games develop self-rectification by allowing the child to see the result of their actions immediately, which helps them to understand their mistakes. Therefore, technology-based games can be a way to support people with learning disabilities (Bottino et al., 2007). According to Shin & Bryant, (2015) and Ok et al., (2016) reports, learning disabilities can be dealt with technology based learning tools such as mobile applications because they have specific strategic instruction, colourful backgrounds (Nagavali & Juliet, 2015) enjoyable sounds, different language options (Poobrasert & Gestubtim, 2013), and also a multisensory aspect of learning (Skiada, et al., 2014). In the same way, Ariffin, et al. (2017) said that children with dyscalculia should utilize mobile app learning to make the learning process easier. Furthermore, Skiada, et al. (2014) suggested that mobile applications have supported children with learning disabilities to improve number series and also enhance their mind in mapping capabilities. Instructive assessment with fun in mobile apps can enhance the dyscalculia mind in a better way (Ok et al., 2016).

Prevalence of Learning Disabilities in India

In India, the current prevalence rate of learning disability is 10% among children (Kuriyan & James, 2018). There are a lot of theoretical studies conducted on learning disabilities in India, but this lack of studies has still not detected the advantages of new technology in an Indian context. Despite this, Indian schools are not given preference or advancement. Although the Indian market (disability sector) is still at an early stage in the adoption of AR/VR technologies, many start-ups in the country are presently working on finding solutions for the Indian special education sector using VR technology. Academic experts predict that the demand for virtual reality and augmented reality solutions to make education more engaging will increase with time. In that case, the role of AR/VR in the education system will redefine India's teaching-learning space (Kulkarni, 2019). Both AR and VR increase different skills, motivation, and memory. Basically, AR/VR teaching lessons are becoming more enjoyable in the field of special education. The most effect of AR/VR use improves communication skills on hearing problem students, for acoustic student VR increases interaction, and for learning-disabled individuals, AR can improve writing, reading, and math skills through gamification. Previously, in India, few educators have created an interactive textbook with 3D images, audio clips, and videos to explain the text. But nowadays, educators are moving to take advantage of AR/VR tools to explain and save time for special children (Pixelplex, 2020). A long decade of literature suggests that new technology improves the learning effectiveness of special education. Yadav (2004) found that in terms of achievement, the students had a significant gain through an IT-enabled instructional package. Similarly, Vaishnav & Parage (2013) found computer aided instruction is very effective in terms of student achievement. Anilakumari (2012) used a technology she made and found that a multimedia remedial tracking package can help reduce the symptoms of dysgraphia.

VIRTUAL REALITY IN SPECIAL EDUCATION

We always talk about learning disabilities in terms of their symptoms, types, and approaches because their causes are biological, developmental, or environmental. However, instead of describing this level, we have taken 'learning disabilities' to be very positive rather than pejorative in the sense that they can be overcome by using a particular strategy in early life. Nowadays, gaming theory or strategies are very popular with lots of technology use. Some AR games are non-educational and many of them are educational games (non 3D). After COVID-19, there are a lot of software programs with limited access that can be used to keep learning. It is an even greater lack of facilities within special education. The possibilities for use are practically endless as virtual reality and augmented reality continue to advance. For instance, a group of researchers at the University of Michigan are working on a project called iGYM, which is an augmented reality system that will teach youngsters who are confined to wheelchairs how to play community-level sports. For users with physical disabilities, this means lighter headsets and more user-friendly controllers; for children who are blind, appropriate colour selection, audio descriptions, and text and image magnification; and for users who are deaf or hard of hearing, clear transcripts and closed captioning. Developers are also starting to prioritize accessibility during design, which has resulted in lighter headsets and more user-friendly controllers, among other gear. In the end, teachers will be able to use virtual reality and augmented reality together to create classrooms that can be changed to meet the needs of each student (Gurwin, 2019).

Students, who have trouble studying, may get benefit greatly from virtual reality. Students are able to acquire knowledge regarding the world around them in an atmosphere that is both secure and supervised. Immersive classrooms are typically the most popular form of virtual reality for teaching children who are having difficulty studying, similar to how they are in primary school. Students frequently report experiencing difficulties when attempting to use virtual reality headgear. Teachers are able to conduct the lesson in a lot more typical manner when they have access to an immersive classroom. The students can still explore around and investigate their virtual environment, but they can do it in a manner that is familiar to them and does not make them feel as though they are being confined. Students diagnosed with autism may benefit from participating in VR activities. A study conducted by Strickland et al. (2007) found that children diagnosed with autism were able to transfer abilities that they had acquired in a virtual setting to the real world. This would include knowledge of how to stay safe in the event of a fire and on the street or road.

The use of assistive technology, such as computer software, communication devices, tablets, and virtual reality, is a new and innovative trend among educators. The tremendous aids of virtual reality help students with disabilities, including sensory impairment, autism, learning disabilities, attention deficit disorder, behavioural disorders, and severe cognitive disabilities, particularly in special needs classrooms, where the technology has the potential to make a real difference. Children with varying literacy, physical, language, and cognitive levels can be accommodated in virtual environments. Disabled children can explore and create new ideas and manipulate objects through new design tools. Basically, virtual reality allows children to control their senses over the environment, and they can actively participate, focus on their abilities (Sims, 1994). Even VR/VEs improve individuals' strengths rather than limiting their capabilities (Greenleaf & Tovar, 1994). Benefits have also been noted for the use of virtual reality in rehabilitation, such as for applications for cognitive assessment (Rose, et al., 1996; Casey 1995), physical assessment (Greenleaf & Tovar, 1994), and training in motor (Inman et al., 1997; Wilson, et al., 1997) or cognitive functioning (Rose, et al., 1996). Wilson et al. (1997), on the other hand, said that the user's sense of freedom and well-being in virtual environments makes them shy away from social situations in real life. They do this because they are almost addicted to this artificial reality.

There is a lot of research that has been using AR/VR to understand learning disabilities. According to one of the important studies by Joseph (2020), AR/VR can enhance children's learning ability by using an interactive method and making it enjoyable. Sanchez & Saenz (2006) found that the use of 3D sound enhances the problem-solving skills of children with visual impairments. Standen & Brown (2006) explain that virtual reality learning environments remove the barrier for individuals with moderate to severe disabilities. Through AR/VR, autistic children can build everyday routines and social skills such as turn taking, imitation, and play. Research studies by Herrera, et al. (2006) and Cobb (2007) pointed out those virtual learning environments can be used as a tool to support communication skills for the autistic, thus enabling independence. Grynszpan, et al. (2008) pointed out that multimedia interfaces for users with high functioning autism and concluded that richer multimedia learning interfaces. Self et al. (2007) said that autistic (6–12-year-old) students can learn fire and tornado safety skills through a virtual learning environment.

Usability of Virtual Reality in Learning Disabilities Children

Virtual reality puts a person in a three-dimensional environment that stimulates their senses of hearing, touch, smell, and taste. Students have the option of interacting with virtual reality software on a conventional desktop computer or by wearing a head-mounted display (HMD) and data glove. AR adds three-dimensional effects to real-world content, allowing users to keep their position as objective observers while observing the augmented world through applications like Google Lens. Virtual reality and augmented reality face primarily logistical and technological obstacles in their implementation. Virtual reality calls for users to have sufficient space, qualified personnel, and the ability to avoid distractions. Additionally, users will require instruction in digital skills. There are data privacy and security concerns associated with using apps. Instruments might also be hefty. But even the most basic AR and VR technology can be used in schools, especially to help kids with special educational needs. Students with special needs can take advantage and resonate themselves from the advantages of VR and AR. In the past few years, virtual and augmented reality has assisted educators in educating, motivating, and enhancing the teaching process and abilities for students of all ages. This has been accomplished by making learning more approachable, remembered, applicable, and compelling (Jeffs, 2009).

Research has demonstrated that augmented and virtual reality can boost students' motivation, improve their ability to communicate with one another, strengthen cognitive skills, increase short-term memory, and make classroom experiences more entertaining. The largest impact will be seen in enhanced communication abilities, particularly among children who struggle with hearing impairments. Virtual reality appears to be helpful for autistic students in terms of facilitating social engagement. There are numerous instances of virtual reality and augmented reality being used to aid students with disabilities. Veronica Lewis, a teacher at a school for the visually impaired, makes use of Google Chromecast to enlarge images for her students, and she also employs VR screen readers such as Voice Over and Talk Back, which provide a description of the environment based on the information contained in the alt text of the images and videos. Sue Parton, a researcher at Morehead State University, has demonstrated how beneficial Google Glass as well as video and 2D barcode camera phone scanning may be for students who are deaf. The faculty at The Deaf and Dumb School in Gujarat, India, where some of the pupils are unable to recognize their own names, teaches them using virtual reality photos that have been processed by a program called Foton. Multiple studies have proven that virtual and augmented reality can help those on the autistic spectrum improve their social skills and detect facial expressions (Joseph, 2020). After COVID-19 pandemic, virtual reality got more recognition and important in the application of education field (as mentioned in Figure 3) such as encourage creativity in with or without special need children, make it more convenient for children to get education through distance learning programs, save time and money of children through virtual field trips, allow to present information without any distractor disturbance, increase the engagement of all children with lots of motivation and trained them with these types of high technology skills (Organization for Economic Co-operation Development, OECD-2016).

Figure 3. Benefits of virtual reality in education

AR can help people with learning disabilities enhance their vocabulary through the use of game mechanics. Educators in India have developed an interactive textbook that explains text through the use of three-dimensional pictures, audio clips, and video clips. Other studies have shown that virtual reality can be an effective tool for treating a variety of disabilities, including social anxiety, language deficiencies, attention deficit hyperactivity disorder, physical or motor disability, cognitive deficits, dyslexia, and Down syndrome, amongst others. Megan Rierdon, an educator who works with students who have special needs at the 53rd St. School in Milwaukee, uses Google Earth VR on field trips. Rierdon told the Milwaukee Journal Sentinel that the children took part in a virtual reality experience in which they "sat down in a chair, put on a virtual reality helmet, and witnessed a tour strolling around a whole greenhouse." They were putting their hands on the ground to feel the dirt and waving to the individuals they saw." Students are better prepared to address their limitations with potential employers after participating in AI-based Training with Molly Porter. Students at the Perkins School for the Blind are helped to feel more at ease by seeing calming 360-degree movies of either inanimate or living items or locales (Yoganathan et al., 2018)

Virtual Reality Assists Special Need Students

Children who have impairments can benefit immensely from using technology in the classroom. Because of assistive technologies, teachers of pupils with special needs can now engage their students on a deeper and more complete level than ever before. For instance, virtual reality is currently making its way into schools for special education, where it gives pupils the opportunity to manipulate their environments without actually moving any physical things. The use of virtual reality as a teaching tool in special education is becoming increasingly common. It is essential to have a solid understanding of the advantages that virtual reality may provide for individuals who are currently working in the field of special education or who are going toward a master's degree in the subject. In addition to providing students with a secure setting in which they can practice important skills, it also gives students the opportunity to learn from one another socially, gives them a sense of control over the surrounding environment, and gives them the ability to modify the program to better suit their individual user experiences. Students who require special education may have physical restrictions that limit the amount of control they are able to exercise over their everyday circumstances. On the other hand, virtual reality enables students with impairments to exert influence over their settings in a variety of different ways, using a variety of different movements. Students engage in physical activities that assist them in working toward their own recovery objectives. Those students who, in other aspects of their lives, do not feel as though they are in charge of their lives are provided with a sense of skill mastery through this activity. In addition, people with impairments have the opportunity to practice skills necessary for "real world" situations in a protected setting thanks to virtual reality. They have the opportunity to train themselves in a variety of real-world scenarios using virtual reality, including everyday activities such as getting dressed, obeying traffic signals while crossing the street, and shopping in a grocery store. Furthermore, they might even be able to go on virtual field trips that take them to places in their own towns that they wouldn't normally be able to see (University of Lumar, 2016).

Interfaces for virtual reality can be customized to meet the requirements of each individual child. Students who have sensory impairments, such as not being able to manage a great quantity of stimulus, for instance, are able to use the virtual reality interface in a customized fashion that displays fewer audio or visual clues. This makes the virtual reality experience more accessible to these students. There are certain virtual reality programs that allow users to network with one another. This affords students with disabilities the chance to engage in conversation with their classmates, both those who also have disabilities and those who do not, providing them with an invaluable opportunity for socialization that is not usually available to them in the classroom setting. Students are frequently given the opportunity to select an avatar, which is a cartoon figure that acts as a representation of the student within the realm of virtual reality. Students who have disabilities can benefit from concentrating on how they see themselves in contrast to those limitations when using this avatar (University of Lumar, 2016).

CONCLUSION

Virtual reality may be more important because of a wide range of factors, from teaching to learning and treatment all these included with virtual now. However, a subset of research and books not given a clear conscience about possible uses to find out learning disability, clinician will have opportunities to create new psychotherapies. This paper will discuss characteristics, pros and cons, classroom uses, and different types of virtual reality prevailing among different types of learning disabilities and applications benefits in different age groups. While the context provides new ideas and benefits to plan to diagnose learning disability, this chapter will be presenting a plan with a primary goal to ensure that different learning disorders characteristics by age will be described systematically. The description of this plan will be helpful to set tools to improve approaches by age groups of individuals, as well as will be made available to a wider community of researchers. The chapter concludes with a summary, a balanced assessment of the contribution of virtual reality and a roadmap for future directions of disabilities. It is very important that educators recognize the importance of virtual reality in learning disability early so remedial actions can be taken to help individuals to succeed in educating and healing. And also virtual reality will help to point out the signs and symptoms of learning disability in early age. In this paper, people will have more insight into the uses of virtual reality in learning disability and the necessary procedures that can also be taken to help in disability sectors. Professionals, therapists, psychologists, school teachers, and parents will be benefited to assist their children appropriately.

ACKNOWLEDGMENT

This research received no specific grant from any funding agency.

REFERENCES

Alchalabi, A. E., Shirmohammadi, S., Eddin, A. N., & Elsharnouby, M. (2018). FOCUS: Detecting ADHD Patients by an EEG-Based Serious Game. *IEEE Transactions on Instrumentation and Measurement, 67*(7), 1512–1520. doi:10.1109/TIM.2018.2838158

American Psychiatric Association. (2000). Diagnostic and statistical manual of mental disorders, DSM-5 (4th ed.). doi:10.1176/appi.books.9780890423349

American Psychiatric Association. (2013). Diagnostic and statistical manual of mental disorders, DSM-5 (5th ed.). doi:10.1176/appi.books.9780890425596

Anilakumari, M. C. (2012). *Developing a multimedia remedial tracking package for Dysgraphia among primary school students with Specific learning disabilities*. School of Pedagogical Sciences, Mahatma Gandhi University. http://hdl.handle.net/10603/25931

Ardila, A. (2014). *Aphasia Handbook*. Department of Communication Sciences and Disorders, Florida International University.

Ariffin, M. M., Halim, F. A. A., Arshad, N. I., Mehat, M., & Hashim, A. S. (2019). Calculic kids© mobile app: the impact on educational effectiveness of dyscalculia children. *International Journal of Innovative Technology and Exploring Engineering, 8*(8S), 701-705. https://H11200688S1919©BEIESP

Ariffin, M. M., Halim, F. A. A., & Sugathan, S. K. (2017). Mobile application for Dyscalculia children in Malaysia. *Proceedings of the 6th International Conference on Computing and Informatics-ICOCI*, 467-472. https://repo.uum.edu.my/id/eprint/22891

Ashworth, J. (2020). *Virtual Reality for Family Education, Exercise and Entertainment.* https://www.parentmap.com/article/virtual-reality-family-education-exercise-and-entertainment

Asnar-Díaz, I., Rodríguez-García, A. M., & Romero-Rodríguez, J. M. (2018). La tecnología móvil de Realidad Virtual en educación: unarevisión del estado de la literatura científica en España [Virtual Reality mobile technology in education: a review of the state of the scientific literature in Spain]. *Revista de Educación Mediática y TIC, 7*(1), 256–274. doi:10.21071/edmetic.v7i1.10139

Baron, R. A., Branscombe, N. R., & Byrne, D. (2009). *Social psychology* (12th ed.). Pearson.

Blaster, B., Ladner, R., & Burgstahler, S. (2016). *Lesson Learned: Engaging Students With Disabilities on a National Scale*. University of Washington.

Bottino, R. M., Ferlino, L., Ott, M., & Tavella, M. (2007). Developing strategic and reasoning abilities with computer games at primary school level. *Computers & Education, 49*(4), 1272–1286. doi:10.1016/j.compedu.2006.02.003

Brown, M., McCormack, M., Reeves, J., Brooks, C., & Grajek, S. (2020). *2020 EDUCAUSE Horizon report, teaching and learning edition*. EDUCAUSE. https://www.educause.edu/horizon-report-2020

Butterworth, B., & Kovas, Y. (2013). Understanding neurocognitive developmental disorders can improve education for all. *Science, 19*(340), 300–305. doi:10.1126cience.1231022 PMID:23599478

Cardos¸, R. A. I., David, O. A., & David, D. O. (2017). Virtual reality exposure therapy in flight anxiety: A quantitative meta-analysis. *Computers in Human Behavior, 72*, 371–380. doi:10.1016/j.chb.2017.03.007

Casey, J. A. (1995). Developmental issues for school counselors using technology. *Elementary School Guidance & Counseling, 30*(1), 26-34. https://www.jstor.org/stable/42871189

Christou, S. (2010). Virtual Reality in Education. In *Affective, Interactive and Cognitive Methods for E-Learning Design: Creating an Optimal Education Experience*. IGI Global. doi:10.4018/978-1-60566-940-3.ch012

Cobb, S. V. G. (2007). Virtual environments supporting learning and communication in special needs education. *Topics in Language Disorders*, *27*(3), 211–225. doi:10.1097/01.TLD.0000285356.95426.3b

Cortiella, C., & Horowitz, S. H. (2014). *The State of Learning Disabilities: Facts, Trends and Emerging Issues* (3rd ed.). National Center for Learning Disabilities. https://www.ncld.org/wp-content/uploads/2014/11/2014-State-of-LD

Di Natale, A. F., Repetto, C., Riva, G., & Villani, D. (2020). Immersive virtual reality in K-12 and higher education: A 10-year systematic review of empirical research. *British Journal of Educational Technology*, *51*(6), 2006–2033. doi:10.1111/bjet.13030

Freeman, A., Adams Becker, S., Cummins, M., Davis, A., & Hall Giesinger, C. (2017). *NMC/CoSN Horizon Report: 2017 K-12 Edition*. New Media Consortium; Consortium for School Networking. https://files.eric.ed.gov/fulltext/ED588803.pdf

Greenleaf, W. J., & Tovar, M. A. (1994). Augmenting reality in rehabilitation medicine. *Artificial Intelligence in Medicine, 6*, 289-299.

Grynszpan, O., Martin, J.-C., & Nadel, J. (2008). Multimedia interfaces for users with high functioning autism: An empirical investigation. *International Journal of Human-Computer Studies*, *66*(8), 628–639. doi:10.1016/j.ijhcs.2008.04.001

Gurwin, G. (2019). *AR Technology is Letting Children With Disabilities Play Sports*. VR Fitness Insider. https://www.vrfitnessinsider.com/ar-technology-is-letting-children-with-disabilities-play-sports/

Hamilton, D., McKechnie, J., Edgerton, E., & Wilson, C. (2021). Immersive virtual reality as a pedagogical tool in education: A systematic literature review of quantitative learning outcomes and experimental design. *Journal of Computers in Education*, *8*(1), 1–32. doi:10.100740692-020-00169-2

Herrera, G., Alcantua, F., Jordan, R., Blanquer, A., Labajo, G., & De Pablo, C. (2005). Development of Symbolic play through the use of Virtual Reality tools in children with Autistic Spectrum Disorders: Two case studies. *Autism*. Advance online publication. doi:10.1177/1362361307086657 PMID:18308764

Herrera, G., Jordan, R., & Vera, L. (2006). Abstract concept and imagination teaching through virtual reality in people with autism spectrum disorders. *Technology and Disability*, *18*(4), 173–180. doi:10.3233/TAD-2006-18403

Ibanes-Etxeberria, A., G'omez-Carrasco, C. J., Fontal, O., & García-Ceballos, S. (2020). Virtual environments and augmented reality applied to heritage education. An evaluative study. *Applied Sciences (Basel, Switzerland)*, *10*(7), 2352. doi:10.3390/app10072352

ImmersionVR. (2022). *VR for Education - The Future of Education*. https://immersionvr.co.uk/about-360vr/vr-for-education/

Inman, D. P., Loge, K., & Leavens, J. (1997). VR education and rehabilitation. *Communications of the ACM*, *40*(8), 53–58. doi:10.1145/257874.257886

Jeffs, T. L. (2009). Virtual reality and special needs. *Themes in Science and Technology Education, 2*(1-2), 253-268. https://eric.ed.gov/?id=EJ1131319

Joseph, C. (2020). Augmented reality and virtual reality to aid students with learning disability: A review. *International Journal of Scientific & Technology Research*, *9*(2), 6475–6478. http://www.ijstr.org/paper-references.php?ref=IJSTR-1219-26850

Kemp, G., Smith, M., & Segal, J. (2020). *Learning Disabilities and Disorders*. HelpGuide, Trusted non-profit guide to mental health & wellness. https://www.helpguide.org/articles/autism-learning-disabilities/learning-disabilities-and-disorders.htm

Khan, M. F., Hussain, M. A., Ahsan, K., Saeed, M., Naddem, A., & Ali, S. A. (2017). Augmented reality based spelling assistance to dysgraphia students. *Journal of Basic and Applied Sciences*, *13*, 500–507. doi:10.6000/1927-5129.2017.13.82

Kulkarni, S. (2019). *Virtual And Augmented Realities Landscape In Indian Higher Education*. Adaption of social media in academia. https://www.asmaindia.in/blog/virtual-and-augmented-reality-landscape-in-indian-higher-education

Kuriyan, N. M., & James, J. (2018). Prevalence of learning disability in India: A need for mental health awareness programme. *Conference: First National Conference on Mental Health Education*. doi: 10.4103/0253-

Kyaw, B. M., Saxena, N., Posadzki, P., Vseteckova, J., Nikolaou, C. K., George, P. P., Divakar, U., Masiello, I., Kononowicz, A. A., Zary, N., & Tudor Car, L. (2019). Virtual reality for health professions education: Systematic review and meta-analysis by the digital health education collaboration. *Journal of Medical Internet Research*, *21*(1), 12959. doi:10.2196/12959 PMID:30668519

Learning Disabilities Association of America. (2018). *Types of Learning Disabilities*. LDA: Core Principles for the Identification and Support of Individuals with Learning Disabilities. https://ldaamerica.org/types-of-learning-disabilities/

Macaruso, P., Hook, P. E., & McCabe, R. (2006). The efficacy of computer-based supplementary phonics programs for advancing reading skills in at-risk elementary students. *Journal of Research in Reading*, *29*(2), 162–172. doi:10.1111/j.1467-9817.2006.00282.x

Maral, P., & Pande, N. (2020). Progressive development of posttraumatic stress disorder and its holistic evolution of natural treatments. In R. Nicholson (Ed.), *Natural Healing as Conflict Resolution* (1st ed., pp. 73–99). IGI Global.

Mendes, R. N., & Grando, R. C. (2008). The computer game SimCity 4 and its pedagogical potential in math classes. *Revista Zetetiké*, *2*(16), 118–176.

Merchant, Z., Goetz, E. T., Cifuentes, L., Keeney-Kennicutt, W., & Davis, T. J. (2014). Effectiveness of virtual reality-based instruction on students' learning outcomes in K-12 and higher education: A meta-analysis. *Computers & Education*, *70*, 29–40. doi:10.1016/j.compedu.2013.07.033

Muktamath, V. U., Priya, R. H., & Chand, S. (2021). Types of specific learning disability. *IntechOpen*, 1-20. doi:10.5772/intechopen.100809

National Association of Special Education Teachers. (2022). *Introduction to Learning Disabilities*. https://www.naset.org/fileadmin/user_upload/LD_Report/LD_Report_1_Intro_to_LD.doc.pdf

National Center for Education Statistic. (2020). *Students with Disabilities*. The Condition of Education. https://nces.ed.gov/programs/coe/indicator_cgg.asp

National Institute of Mental Health. (2018). *Autism Spectrum Disorder*. U.S. Department of Health and Human Services. https://www.nimh.nih.gov/sites/default/files/documents/health/publications/autism-spectrum-disorder/19-mh-8084-autismspectrumdisorder.pdf

National Survey of Children's Health. (2007). *Data query from the child and adolescent health measurement initiative*. http://childhealthdata.org/browse/survey/results?q=1219

Nicolson, R. I., Fawcett, A. J., & Nicolson, M. K. (2000). Evaluation of a computer-based reading intervention in infant and junior schools. *Journal of Research in Reading*, *23*(2), 194–209. doi:10.1111/1467-9817.00114

Ok, M. W., Kim, M. K., Kang, E. Y., & Bryant, B. R. (2016). How to find good apps: An evaluation rubric for instructional apps for teaching students with learning disabilities. *Intervention in School and Clinic*, *51*(4), 244–252. doi:10.1177/1053451215589179

Organisation for Economic Co-operation and Development. (2016). *Innovating Education and Educating for Innovation: The Power of Digital Technologies and Skills*. OECD Publishing. doi:10.1787/9789264265097-en

Panchuk, D., Klusemann, M. J., & Hadlow, S. M. (2018). Exploring the effectiveness of immersive video for training decision-making capability in elite, youth basketball players. *Frontiers in Psychology*, *9*(27), 2315. doi:10.3389/fpsyg.2018.02315 PMID:30538652

Panwala, S., Shaikh, A. S., Ghare, A., Kazi, S., Khan, M., & Rangwala, M. (2017). *Augmented Realilty for Educational Enhancement*. University of Mumbai. http://ir.aiktclibrary.org:8080/xmlui/handle/123456789/2052

Papastergiou, M. (2009). Digital game-based learning in high school Computer Science education: Impact on educational effectiveness and student motivation. *Computers & Education, 1*(52), 1-12. doi:10.1016/j.compedu.2008.06.004

Parmaxi, A. (2020). Virtual reality in language learning: A systematic review and implications for research and practice. *Interactive Learning Environments*, 1–13, 1049–4820. doi:10.1080/10494820.2020.1765392

Parsons, S., Mitchell, P., & Leonard, A. (2005). Do adolescents with autistic spectrum disorders adhere to social conventions in virtual environments? *Autism: an International Journal of Research and Practise.*, *9*(1), 95–117. doi:10.1177/1362361305049032 PMID:15618265

Pellas, N., Mystakidis, S., & Kazanidis, I. (2021). Immersive virtual reality in K-12 and higher education: A systematic review of the last decade scientific literature. *Virtual Reality (Waltham Cross)*, *25*(3), 835–861. doi:10.100710055-020-00489-9

Pixelplex. (2020). *VR/AR in Education and Training*. https://pixelplex.io/blog/ar-and-vr-in-education-and-training/

Poobrasert, O., & Gestubtim, W. (2013). Development of assistive technology for students with dyscalculia. *2nd International Conference in E-Learning and E-Technologies Education,* 60-63. 10.1109/ICeLeTE.2013.6644348

Prusty, A., Yeh, C. J., Sengupta, R., & Smith, A. (2021). Dyscalculia: Difficulties in Making Arithmetical Calculation. In Handbook of Research on Critical Issues in Special Education for School Rehabilitation Practices. IGI Global. doi:10.4018/978-1-7998-7630-4.ch023

Radianti, J., Majchrzak, T. A., Fromm, J., & Wohlgenannt, I. (2020). A systematic review of immersive virtual reality applications for higher education: Design elements, lessons learned, and research agenda. *Computers & Education, 147*, 103778. doi:10.1016/j.compedu.2019.103778

Rose, F. D., Attree, E. A., & Johnson, D. A. (1996). Virtual reality: An assistive technology in neurological rehabilitation. *Current Opinion in Neurology, 9*(6), 461-467. doi:https://pubmed.ncbi.nlm.nih.gov/9007406/

Rossi, H. S., Santos, S. M., Prates, R., & Ferreira, R. A. C. (2018). Imaginator: A virtual reality based game for the treatment of sensory processing disorders. *IEEE 6th International Conference on Serious Games and Applications for Health (SeGAH).* 10.1109/SeGAH.2018.8401355

Sanchez, J., & Saenz, M. (2006). 3D sound interactive environments for blind children problem solving skills. *Behaviour & Information Technology, 25*(4), 367–378. doi:10.1080/01449290600636660

Santos, V. M. (2009). The relationship and difficulties of students with math: A topic of study. *Revista Zetetiké, 32*(17), 1744.

Self, T., Scudder, R. R., Weheba, G., & Crumrine, D. (2007). A virtual approach to teaching safety skills to children with autism spectrum disorder. *Topics in Language Disorders, 27*(3), 242–253. doi:10.1097/01.TLD.0000285358.33545.79

Shin, M., & Bryant, D. P. (2015). A synthesis of mathematical and cognitive performances of students with mathematics learning disabilities a synthesis of mathematical and cognitive performances of students with mathematics learning disabilities. *Journal of Learning Disabilities, 48*(1), 96–112. doi:10.1177/0022219413508324 PMID:24153404

Sims, D. (1994). Multimedia camp empowers disabled kids. *IEEE Computer Graphics and Applications, 14*(1), 13–14. doi:10.1109/38.250912

Skiada, R., Soroniati, E., Gardeli, A., & Zissis, D. (2014). EasyLexia: A mobile application for children with learning difficulties. *Procedia Computer Science, 27*, 218–228. doi:10.1016/j.procs.2014.02.025

Standen, P. J., & Brown, D. J. (2006). Virtual reality and its role in removing the barriers that turn cognitive impairments into intellectual disability. *Virtual Reality (Waltham Cross), 10*(3), 241–252. doi:10.100710055-006-0042-6

Strickland, D. C., McAllister, D. F., Coles, C., & Osborne, S. (2007). An evolution of virtual reality training designs for children with autism and fetal alcohol spectrum disorders. *PubMed, 27*(3), 226–241. doi:10.1097/01.TLD.0000285357.95426.72 PMID:20072702

Stupar-Rutenfrans, S., Ketelaars, L. E. H., & van Gisbergen, M. S. (2017). Beat the fear of public speaking: Mobile 360o video virtual reality exposure training in homeenvironment reduces public speaking anxiety. *Cyberpsychology, Behavior, and Social Networking*, *20*(10), 624–633. doi:10.1089/cyber.2017.0174 PMID:29039704

Tannock, R. (2013). Specific Learning Disabilities in DSM-5: are the changes for better or worse? *International Journal of Research in Learning Disabilities, 1*(2), 2-30. https://eric.ed.gov/?id=EJ1155677

Thomas, P. T., Goswami, S. P., & Samasthitha, S. (2015). Developmental coordination disorder (dyspraxia). *Asian Journal of Cognitive Neurology*, *3*(1), 41–43. https://www.academia.edu/29076970/developmental_coordination_disorder_dyspraxia

University of Lumar. (2016). *Can Virtual Reality Assist Special Needs Students?* https://degree.lamar.edu/articles/education/can-virtual-reality-assist-special-needs-students

Vaishnav, R., & Parage, P. (2013). Innovative instructional strategies interactive multimedia instruction and computer aided instruction for teaching biology. *Voice of Research, 2*(2), 1-4. http://www.voiceofresearch.org/doc/sep-2013/sep-2013_1.pdf

Vasudevan, S. K., Saravanan, P. G., John, M. K., & Sasidharan, A. (2022). Virtual reality-based real-time solution for children with learning disabilities and slow learners - an innovative attempt. *International Journal of Medical Engineering and Informatics*, *14*(2), 165–175. doi:10.1504/IJMEI.2022.121131

Villena-Taranilla, R., Tirado-Olivares, S., C'ozar-Guti'errez, R., & Gonz'alez-Calero, J. A. (2022). Effects of virtual reality on learning outcomes in K-6 education: A meta-analysis. *Educational Research Review*, *35*(11), 100434. doi:10.1016/j.edurev.2022.100434

Weng, P. L., & Taber-Doughty, T. (2015). Developing an app evaluation rubric for practitioners in special education. *Journal of Special Education Technology*, *30*(1), 43–58. doi:10.1177/016264341503000104

Wilson, A. J., & Dehaene, S. (2007). *Number sense and developmental dyscalculia, in human behavior, learning and the developing brain: A typical development*. Guilford Press.

Wilson, P. N., Foreman, N., & Stanton, D. (1997). Virtual reality, disability and rehabilitation. *Disability and Rehabilitation*, *19*(6), 213–220. doi:10.3109/09638289709166530 PMID:9195138

Winn, W. (1993). *A Conceptual Basis for Educational Applications of Virtual Reality*. Human Interface Technology Laboratory, Washington Technology Center, University of Washington. http://www.hitl.washington.edu/projects/learning_center/winn/winn-paper.html

Yadav, K. (2004). *Development of an IT enabled Instructional Package for Teaching English medium students of Vadodara city* [M.eD Dissertation]. CASE, The MS University of Boroda, Borodara.

Yaman, O., & Karaköse, M. (2016). Development of image processing based methods using augmented reality in higher education. *2016 15th International Conference on Information Technology Based Higher Education and Training (ITHET)*, 1-5. 10.1109/ITHET.2016.7760723

Yoganathan, S., Finch, D. A., Parkin, E., & Pollard, J. (2018). 360o virtual reality video for the acquisition of knot tying skills: A randomised controlled trial. *International Journal of Surgery*, *54*, 24–27. doi:10.1016/j.ijsu.2018.04.002 PMID:29649669

Zain, N. Z. M., Mahmud, M., & Hassan, A. (2013). Utilization of mobile apps among student with learning disability from Islamic perspective. *5th International Conference Information Community Technology Muslim World*, 1-4. 10.1109/ICT4M.2013.6518889

KEY TERMS AND DEFINITIONS

Attention Deficit Hyperactive Disorder (ADHD): A neurological disorder affects to sustain attention, hyperactivity, and impulsive behaviour.

Augmented Reality (AR): Experiences of real-world objects which integrated in real time use of information in the form of audio, graphic, and text.

Autism Spectrum Disorder (ASD): A developmental disorder affects how to interact, communicate, and behave with others.

Gamification: A simple technique to insert game plays elements in a non-gaming setting.

Immersive Experience: An experience which simulates the realistic feeling of being present in the visual space. An example such as exploring the surface of platens in the classroom with mixed virtual reality.

Learning Disabilities (LD): A kind of disorder that roots from brain causes difficulties in reading, writing, and processing information.

Multimedia Interfaces: A digital interface that has capacity of transmitting uncompressed audios and videos data in devices.

Three-Dimensional Effects: It is an effect that produces in depth perception of an image. For example, cubes, pyramids, cones, and cylinders.

Virtual Reality: An environment is created through computer to feel the objects are surrounded and appear to be real.

Visual Motor Deficiency: Inability to observe and use visual information in the form of shapes, figures, and size.

Chapter 4
A Design of Interactive Learning Applications for Phytochemical Screening in Vocational College: Augmented Reality and Gamification

Ruth Elenora Kristanty
Poltekkes Kemenkes Jakarta II, Indonesia

Maulita Prima Sari
Poltekkes Kemenkes Jakarta II, Indonesia

Purnama Fajri
Poltekkes Kemenkes Jakarta II, Indonesia

Nashrul Hakiem
UIN Syarif Hidayatullah Jakarta, Indonesia

Hana Relita
Poltekkes Kemenkes Jakarta II, Indonesia

Yollan Gusnanda Setiawan
UIN Syarif Hidayatullah Jakarta, Indonesia

ABSTRACT

The challenges of practicum learning for the vocational institute are increasingly prominent. Innovation is needed to utilize technology and learning media to support distance learning and adaptive learning. Phytochemistry Practicum, a course given in the third semester of the Pharmaceutical and Food Analysis Department of Poltekkes Kemenkes Jakarta II, provides knowledge and skills to analyze chemical compounds in plants. This study aimed to develop interactive learning media for remote practicum of phytochemical screening materials at the Pharmaceutical and Food Analysis Department of Poltekkes Kemenkes Jakarta II. The methods used in this study were descriptive exploratory for laboratory experiment, multi-media development life cycle (MDLC) for AR development, and game development life cycle (GDLC) for building the gamification system. The augmented reality application and education game have been published in Playstore under the name AR Fitokimia and Virtual Lab Fitokimia. Both of these products were able to be accessed easily through mobile devices.

DOI: 10.4018/978-1-6684-4854-0.ch004

INTRODUCTION

The phytochemical subject is a compulsory course for second-year diploma students (third semester) in the Pharmaceutical and Food Analysis Department of Poltekkes Kemenkes Jakarta II. One of the learning materials is the phytochemical screening of Simplicia, such as alkaloid screening, glycoside screening, flavonoid screening, saponin screening, steroids-triterpenoids, and tannin screening.

Mobile learning application in the phytochemical area is limited. Research on AR technology in the learning of phytochemicals had not been conducted. It is essential to develop intelligent learning objects to assist students in practical laboratories from the experimental results of phytochemical screening by designing a mobile AR application (Srivastava, 2016).

This research aimed to enrich teaching materials for Phytochemical Practicum courses in Pharmaceutical and Food Analysis Department of Poltekkes Kemenkes Jakarta II by applying Augmented Reality (AR) technology and optimizing smartphone usage. In particular, this study aimed to obtain interactive learning media in the form of marker-based mobile applications and to obtain the results of testing trials for learning Phytochemical Practicum in Poltekkes Kemenkes Jakarta II. The AR technology-based learning media were expected to introduce experimental results interactively and efficiently so that students could observe in detail and make remote practicum easier and fun.

The authors tried to study the format of microlearning media that teachers can use as a form of innovation in the learning process. Based on the results of the preliminary data survey held in 2020 and 2022 as explained in this chapter, the authors needed to develop learning applications with various features to meet the needs of remote practicum learning. These studies are in line with the research topic required in the research roadmap of the team for the period 2020-2024, which is the development of learning technology (Figure 1). Completing the research roadmap for the innovation of phytochemical practicum learning is the background of the research team to develop two Android-based applications that are proposed to students in the same course.

The utilization of gamification in learning phytochemical practicums can hopefully facilitate students in understanding the phytochemical screening procedures for alkaloid compounds, glycosides, flavonoids, saponins, and tannins in mobile technology and virtual laboratory environments and can improve student learning outcomes. Marker information for the augmented reality application is added to the game's material features to enrich the understanding of phytochemical screening results. The research program aimed to design virtual laboratory applications based on gamification for the phytochemical screening materials.

Figure 1. State of the art chart and research roadmap

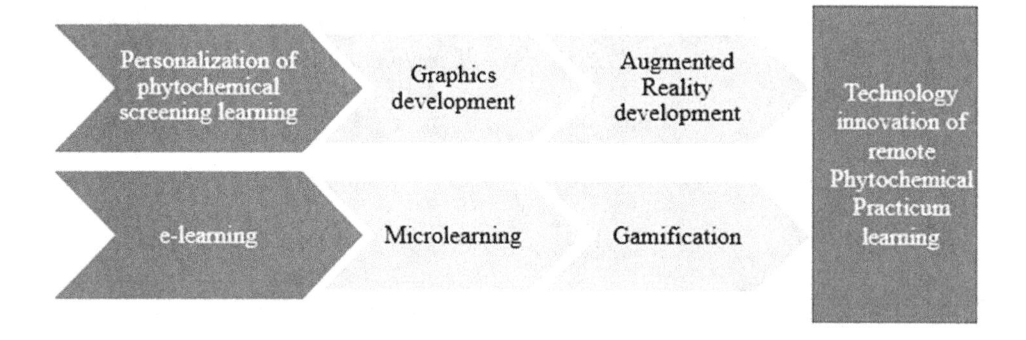

BACKGROUND

With the use of increasingly massive digital technology coupled with the Covid-19 pandemic that has lasted over two years, the challenges of the health and education sectors are increasing. The Covid-19 pandemic encourages educators to be ready and able to teach by integrating digital technology in today's science era. Online learning has become an alternative used to solve conventional learning problems. Today's students prefer to self-study by viewing interactive videos, games, projects, and more (Ng, 2012; Thomas, 2011).

The Indonesian Internet Service Providers Association survey for the second quarter of 2020 recorded that Indonesia's Internet users increased by 23.5 million or 8.9% compared to 2018. Based on the empirical evidence, the use of gadgets and the internet during the pandemic increased as distance learning. Gadget devices used by society today have diverse functions; some use them for business purposes, office work, entertainment media, learning media, or just for communication. The most common gadget device encountered in people's lives is a smartphone. Almost everyone has a smartphone, from office workers, teenagers, and children to the elderly. In addition to being easy to carry anywhere, smartphones today can support various activities and ease accessing the internet.

The results of the Ministry of Communication and Informatics survey in 2017 showed that 66.31% of Indonesian people already had a smartphone. The productive age of internet users (20-29 years) was higher than the others, 83.97% came from a diploma or undergraduate educational background. By 2025, at least 89.2% of Indonesia's population will have taken advantage of smartphones. Smartphone penetration in the country will grow by 25.9% in six years since 2019 (KOMINFO, 2017; Pusparisa, 2020).

Modern students need to digitally process information instead of conventional materials' delivery and packaging methods. In the learning process, innovations are required, especially in the utilization of technology and media development, so that students can understand the learning environment in the new prospect and most effective way (Setyawan et al., 2019).

An innovative learning environment is a form of personalized adaptive learning (Peng et al., 2019). It is built with intelligent device technologies and smart technologies. The features of its adaptive function and natural interaction would bring students more adaptation, flexibility, and engagement (Huang et al., 2013; Peng et al., 2019; Zhu et al., 2016).

Online learning must be optimized with various modifications and innovations to achieve the institution's vision as a technology reference. Problems related to the implementation of remote practicum can be solved by developing digital simulation technology and a virtual laboratory so students can feel the laboratory atmosphere of the laboratory in distance learning (Frima, 2020).

The latest technological development is the increased use of gadgets, especially smartphones, in daily life, including education (Setyawan et al., 2019). Changes in learning and teaching methods have led to the widespread use of tools and technologies. There has been a lot of research related to gamification and Augmented Reality technology. The design of interactive learning media using Augmented Reality (AR) technology has developed the field of chemistry and other subjects, including chemical reactions between two elements (Maier et al., 2009), chemical structures in three dimensions (Nuñez et al., 2008), children of early childhood learning groups (Saurina, 2016), and animal recognition for Android-based devices (Indriani et al., 2016).

In 2020, more than 60% of students from the Pharmaceutical and Food Analysis Department of Poltekkes Kemenkes Jakarta II were interested in the idea of AR for the visualization of phytochemical screening observations. The respondents were interested in AR ideas in developing the design of the teaching material for phytochemical screening.

The phytochemical analysis is a part of pharmacognosy science that studies the method or method of calculating chemical content contained in plants or animals as a whole or its components, including isolation or separation. The area of attention of phytochemicals is a variety of organic compounds formed and stored by plants, namely their chemical structure, biosynthesis, changes and metabolism, their scientific spread, and biological function (Frima, 2020). Phytochemical screening is a way to identify bioactive that can quickly separate between natural materials that consist of certain phytochemicals. In general, phytochemical screening of Simplicia powders and extracts included examining the group of alkaloidal compounds, glycosides, flavonoids, terpenoids/steroids, tannins, and saponins (Tukiran et al., 2016). Phytochemical screening can be done with tube tests and chromatography tests. The method is primarily a reaction of color testing with a color reagent because it is simple, fast, designed with minimal equipment, selective for the group of compounds identified, and can be semiquantitative to provide additional information to the group compounds studied (Owolabi et al., 2017; Sasidharan et al., 2011). In performing phytochemicals screening, a method to analyze the suitable compound is needed because each plant has different chemical structural properties (Sahira Banu & Cathrine, 2015). Solvent selection and extraction methods were essential in phytochemical screening (Maier et al., 2009).

Educational or learning activities have been widely used on smartphones but have not been maximal in using chat, social media, etc. To obtain data on the system needs for building a game education in phytochemistry subject, the authors conducted a preliminary data survey through the distribution of online questionnaires for the vocational students of the Pharmaceutical and Food Analysis Department of Poltekkes Kemenkes Jakarta II at the beginning of 2022. Figure 2 showed that the students needed applications that have features in gamification to assist them in understanding phytochemical screening materials during practicum learning interactively and adaptively.

Microlearning can be used as a strategy to present practicum material or content in e-learning in small, focused segments (Tinacci et al., 2022). Content that is easily absorbed and remembered by students needs to be optimized to achieve e-learning goals amid the many distractions that can shift focus while learning. One distraction that can be utilized to innovate learning content is gaming. Based on a study, gamification learning approaches are more effective than social networks(De-Marcos et al., 2016).

One of the processes that need to be considered in building an online learning system is processing teaching materials with sorting techniques into small parts. This technique then becomes a small piece of teaching material called Microlearning. Microlearning is learning/training presented in small sizes and gives learners control over what they learn (Noriska et al., 2021).

Serving with a microlearning strategy results in a short and practical type of content. Microlearning helps learners learn the material in a short period. The available range is already smaller and more focused so that the amount of information needed is appropriate to help learners achieve learning goals.

At the university, information and communication technologies should not be an obstacle. The development of teaching materials based on information and communication technologies is one of them through the development of micro-learning teaching materials. Microlearning is referred to as a small-scale learning method, in which learning is designed in small segments across different media formats so that available information is converted into "concise content" which enables quick understanding of content and can be learned anywhere, anytime through technology, information and communication devices (H. Nugraha et al., 2021).

Figure 2. Survey results in data

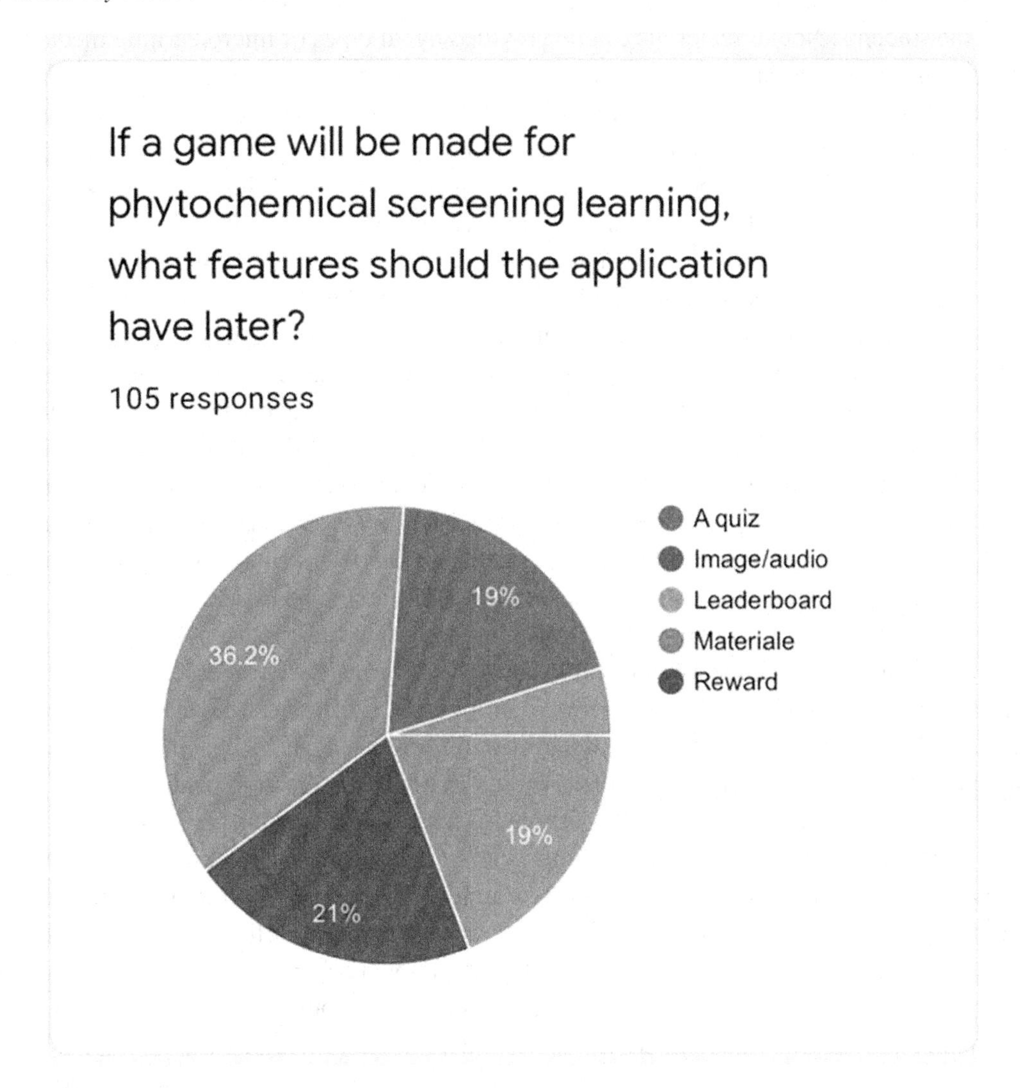

Some studies showed how effective Microlearning is in the learning process. Microlearning can make learning content easier to understand and memorable for a long time, while Microlearning can improve the effectiveness and efficiency of the learning process (Mohammed et al., 2018). In addition, other studies also state that Microlearning and the development of modern technology can provide sophisticated practical solutions to training and education problems (Alsehri, 2021). Microlearning may be one of the efforts universities can make as a form of innovation to break the saturation of the online learning process amid the Covid-19 pandemic. The development of micro-learning teaching materials is expected to bring a breath of fresh air to teachers and students to create more effective and efficient learning for teachers and students during the online learning process.

Pandey (2018) explains that there are types of media formats that can be used in Microlearning, namely infographics, interactive infographics, PDF, interactive PDF, e-books and flipbooks, animated videos, whiteboard animation, kinetic animation, explainer videos, interactive videos, webcasts or podcasts, expert videos or webinars or recorded webinars, and mobile apps. Each microlearning media format has characteristics and needs to be matched between the media format and the content delivered by the teachers to the students. Not all materials can be provided by all types of microlearning media formats, so the ability of teachers to analyze the kind of material adequately provided by each microlearning media format is required (H. Nugraha et al., 2021).

During the new normal, entertainment sources are centered on the digital world, one of which is online gaming. The growing number of people working from home and learners engaging in online learning during the new normal is causing Internet access and computer games (PCs, macOS, and other high-resolution desktop platforms with high graphics) to increase by more than 38% in 2020. Gaming hours have shifted to weekdays, and this trend persists until now (Lestiyo, 2021).

Three types of immersive technology can be utilized in education: game-based, problem-based, and inquiry-based active learning tools. The success of immersive learning technology depends on the ability of the learning program designer to build a learning environment that contains as many elements as possible (Blessinger & Wankel, 2012; Graesser, 2017). Nugraha et al (2017) designed educational games as a learning medium for digital game engineering practice courses that have the advantage that games have been through material tests, media tests, and student assessments. Windawati and Koeswanti (2021) developed android-based educational games for elementary school students, having the advantage that the game has been through the validity test of material experts and media experts, while the drawback is that it can only be used on smartphones and requires a large enough data quota to install applications. The development of Educational Game Media as an Alternative Information System Ice Breaking Learning in pandemic times researched by Santoso and Hastutiningtyas (2021) has the advantage of being accessed through PC / laptop and smartphone. It has gone through the R &D stage, but the form of games in simple PowerPoint is only equipped with animations, action, and selection pane.

Educational games have been specifically designed to teach people about a particular subject, expand concepts, strengthen development, understand an event, or help them learn a skill (Ahmad, 2020). It is a program that creates capabilities in the game environment. Games are given to motivate students and get them to go through game procedures meticulously to develop their abilities (Batubara et al., 2017).

There are four main features of a game related to the educational aspect, namely the existence of goals, rules, feedback systems, and contributions. Each game contains all four of these features. In addition to these four main features. These features are diverse media (multimedia), realistic nature (realism), image display, challenges to move, adaptivity, feedback, interactivity, modeling, collaboration, competition, reflection, fantasy, and narrative. These features are mainly found in digital games. Additional features push and confirm all four basic features (Blessinger & Wankel, 2012; McGonigal, 2011).

Gamification is a learning approach that uses elements in the game to motivate students in the learning process and maximize the feeling of enjoyment and participation in the learning process; also, this media can be used to capture things that interest students and inspire students to continue learning. Gamification uses game mechanics to provide practical solutions by building the engagement of certain groups (Jusuf, 2016). Gamification aims to change non-game contexts (examples: teaching-learning, marketing, etc.) into much more interesting by integrating game thinking, game design, and game mechanics. Gamification in e-learning is enough to influence users in studying material (Handani et al., 2016).

The use of digital technology is increasingly and massively used as a result of the development of industrial revolution 4.0 and the concept of society 5.0 (H. Nugraha et al., 2021). To achieve the vision of Poltekkes Kemenkes Jakarta II, the proper facilities and infrastructures are needed to support the improvement of practicum learning with the use of appropriate technology. Since the government implemented the policy of distance learning programs as a priority agenda during the Covid-19 pandemic, online learning (e-learning) in a learning system, commonly known as LMS (Learning Management System), has become a choice of learning method without having to face-to-face on campus. However, the results obtained from web-based platforms such as Google Classroom and Poltekkes Jakarta II E-Learning have not been adequate to facilitate online practicum learning. Therefore, there needs to be sustainable development of practicum learning media following the needs of students in the new normal era so that competence can be achieved.

Learning digital games based on media can be one of the main attractions that allow learners to remember by playing, unlike the conventional way the subject sits down and reads a book while studying (Ninaus et al., 2017). Gamification of digital generation enables a learning process that follows the personal characteristics of learners (Clark, 2006; Prensky, 2001).

A game can have the same role as books, movies, and museums. Games can be utilized as entertainment and as a reliable means to increase learning effectiveness. When playing educational games, learners can learn according to their character and understand the scope of the material in more detail. Through a virtual lab designed by a team of gamification-based researchers, it is hoped that many people can learn about phytochemical screening.

Numerous studies have shown that video games positively affect mental health and well-being. Psychology Today states that video games like first-person shooters help improve hand-eye coordination and quick reflexes among players. Also, choice-based games give players a better sense of self and improved emotional stability. Not to mention that schools and universities have used games such as Portal and Minecraft to teach physics to even chemistry subjects. Games could be a valuable tool for the education system.

In this phytochemical screening microlearning, the application is designed to have gamification features. The leaderboard feature is for the students to have the realm of competition in the learning environment. The reward feature is for the results achieved and book material to answer quiz questions so that students can understand the fabric following the guidelines of phytochemical practicum learning in the institution. The authors developed practicum learning objects in infographics of various animations, graphic illustrations, and colors that add realism to phytochemical screening material and some Simplicia powder by combining elements in the game simulation genre.

VIRTUAL LABORATORY FOR PHYTOCHEMICAL SCREENING

Issues and Problems

A laboratory is a place for students to conduct experiments from theories that have been given in the classroom. The function of the investigation itself is the learning support to improve learners' understanding of a material that has been studied. The laboratory has been at the heart of science education. The laboratory has several essential functions, namely as a source of learning related to the cognitive, affective, and psychomotor realms by conducting experiments, and as a means of research, namely the place of various studies so that the personal form of learners who behave scientifically (Wibawanto, 2020).

Digital technology facilitates the learning process by presenting new formats to support the learning process. The digital age is driving innovations where technology has opened up the possibility of new science learning processes in the form of virtual laboratories.

A virtual laboratory is a learning experience that simulates an authentic laboratory. Laboratories are simulated and visualized in digital format, and then students can use them to explore concepts and theories. Virtual laboratories can be defined as multisensory software with the interactivity to simulate certain practicums by replicating conventional laboratories. Virtual laboratories allow students to learn through a case study approach, interact with laboratory equipment, conduct experiments, and analyze experiments while evaluating processes performed. The possibility to explore, experiment, and learn becomes more dynamic with virtual laboratories.

Virtual laboratories require hardware that supports specific inputs from their users. The types of hardware used to operate virtual laboratories are increasingly diverse as technology evolves, ranging from computers, gadgets, and virtual reality devices. This immersive experience puts the users in a virtual environment to create the feeling they are in it and a part of it (Wibawanto, 2020).

The problem that will be examined is how the form of virtual laboratory application would develop the learning quality of Phytochemical Practicum content. The application was designed with AR technology and a gamification-based strategy for learning phytochemical screening in a virtual laboratory.

Augmented Reality

This study covered two stages: phytochemical screening experiment, which was descriptively exploratory, and AR development. AR system development used the Multimedia Development Life Cycle (MDLC) method, a multimedia development method with six characteristics, namely concept, design, material collecting, assembly, testing, and distribution (Lusa et al., 2020).

The research used four types of Simplicia powder samples from the Bogor area, Indonesia: brotowali stems, ginger rhizomes, katuk leaves, and guava leaves, as listed in the Phytochemistry Practice Manual of the Pharmaceutical and Food Analysis Department. The samples were determined at the LIPI Bogor Plant and Botanical Garden Conservation Research Center. The phytochemical screening was performed on samples using several chemical reagents for alkaloids, flavonoids, triterpenoids/steroids, glycosides, saponins, and tannins/polyphenols.

The concept stage was used to determine the system specifications and analyze the needs. The data obtained was in the form of application specifications, what 3D objects would be created, audio for the material explanation, and the text of the material to be used in the application. The design stage applied concepts in the flowchart, the appearance of the layout, and the material needs of the system. The collecting material stage collects the materials used in the manufacture of applications, namely photos of the objects of phytochemical screening test results as 3D objects with Blender 3D software, audio used for explanations in the application, and text used to provide specific procedures. Other needs, such as AR markers and main menu displays, were created using Adobe Photoshop. The author button design was edited using Adobe Illustrator. At this stage, the author used Unity 3D software and used the Vuforia SDK to implement Augmented Reality technology into applications. The application was then named AR Fitokimia and used in the Android platform.

This study developed application testing based on ISO 25010 on functional suitability and usability. The authors tested operational suitability in a questionnaire stating application function scenarios tested by an expert in software development from UIN Jakarta and a lecturer from Poltekkes Kemenkes Jakarta II to see the suitability of existing functions. Usability testing of AR Fitokimia application involved diploma students from the Department of Pharmaceutical and Food Analysis of Poltekkes Kemenkes

Jakarta II (40 respondents) through online questionnaires. Both the application and the test used the native language of Indonesia.

Gamification-Based Microlearning

The method used consists of data collection methods and software system development methods. The data collection method was library studies and field studies by conducting an online questionnaire deployment. System development methods design android-based phytochemical screening interactive learning applications using the GDLC (Game Development Life Cycle) method. The stages of GDLC were initiation, pre-production, production, testing, and application distribution.

The development of these virtual laboratories falls into the category of product-based development research and tests the effectiveness of those products. Related to the development of virtual laboratories as a medium of learning, the generally used development model is the Dick and Carrey procedural model, which emphasizes applying development design principles tailored to sequential steps.

Microlearning applications are designed and built using Unity 3D software and Google play games service as a leaderboard database. The authors used the phytochemical practicum manual of the Department of Pharmaceutical and Food Analysis Department of Poltekkes Kemenkes Jakarta II and Herbal Pharmacopeia as a reference for the content of materials and quizzes.

Gamification plans in phytochemical practicums are:

a) Understand the target audience and context, namely students participating in a phytochemical practicum in the Pharmaceutical and Food Analysis Department of Poltekkes Kemenkes Jakarta II.
b) Determining learning goals is achieving phytochemical screening competencies and completing assignments, quizzes, exams, and other projects.
c) Organizing the Embassy that students must understand the concept of phytochemical screening learning and implement it into practice.
d) Identify resources, i.e., *applying game* mechanics, determining level achievement, and implementing game rules.
e) Apply gamification elements, namely points, badges, time limits, leaderboards, and social/community elements.

Components available in the gamification of this program include:

a) Level described as completing the mission of each chapter of practicum learning.
b) Challenges in the form of quizzes, task completion, participation/liveliness in discussions, video conferences, and problem exercises.
c) The score obtained by students each time they complete a challenge.
d) A leaderboard showing the comparison of mission completion between students.
e) Progress bar showing the mission completion process.

Model outline/feature map in the app:

a. Login using an account.

b. View and select practicum materials the authors will do in the virtual lab.
c. Introduction of Phytochemical Laboratory *(tour lab).*
d. Completion of missions/challenges in the form of simulations.

SOLUTIONS AND RECOMMENDATIONS

The phytochemical screening experiment as the first stage of this research started from the determination of the sample. The results of the samples' identity were in Table 1. Brotowali stems were tested to obtain positive observations on the screening of alkaloids, glycosides, and saponins: katuk leaves in the screening of flavonoid and guava leaves to determine tannin levels ginger rhizomes in the determination of steroids-triterpenoids. The positive results were selected as 3D objects, the materials for AR development (Table 2).

Table 1. Phytochemical screening results of Simplicia

No.	Chemical Group	Experiment Results			
		Tinospora crispa **(L.) Hook. F.& Thomson Stems**	*Psidium guajava L.* **Leaves**	*Zingiber officinale Roscoe* **Rhizomes**	*Sauropus androgynus (L.) Merr.* **Leaves**
1.	Alkaloid	+	-	-	-
2.	Glycoside	+	-	-	-
3.	Flavonoid	-	-	-	+
4.	Saponin	+	-	-	-
5.	Tannin	-	+	-	-
6.	Steroid-Triterpenoid	-	+	+	+

Table 2. Objectives results for 3D models

Chemical Groups		Observation	Results
Alkaloid	Mayer	White precipitates	+
	Dragendorf	Brown precipitates	+
	Phosphomolibdic acid Phosphotungstate acid Silicotungstate acid Sulfuric acid P Nitric acid P Frohde LP Erdman LP	Orange precipitates White precipitates White precipitates Brownish green Yellowish Brownish green Brownish	+ + + + + + +
Glycoside	Molisch Lieberman-Bourchard	Purple ring formed Green and blue	+ +
Flavonoid		Orange	+
Saponin		Foam more than 1 cm	+
Tannin		Golden yellow at the endpoint of a titration	+

Figure 3 and Figure 4 showed the system flowchart and use case diagram of the AR Fitokimia app. The programming started by creating the main menu (Figure 5) by importing the design developed into the Unity 3D project. The interface design for the main menu view was created using Adobe Photoshop. Each button was given to work when pressing it. Programming continued until all designs could run on smartphone devices in their original language, which is Indonesian.

Figure 3. Flowchart of the AR Fitokimia application

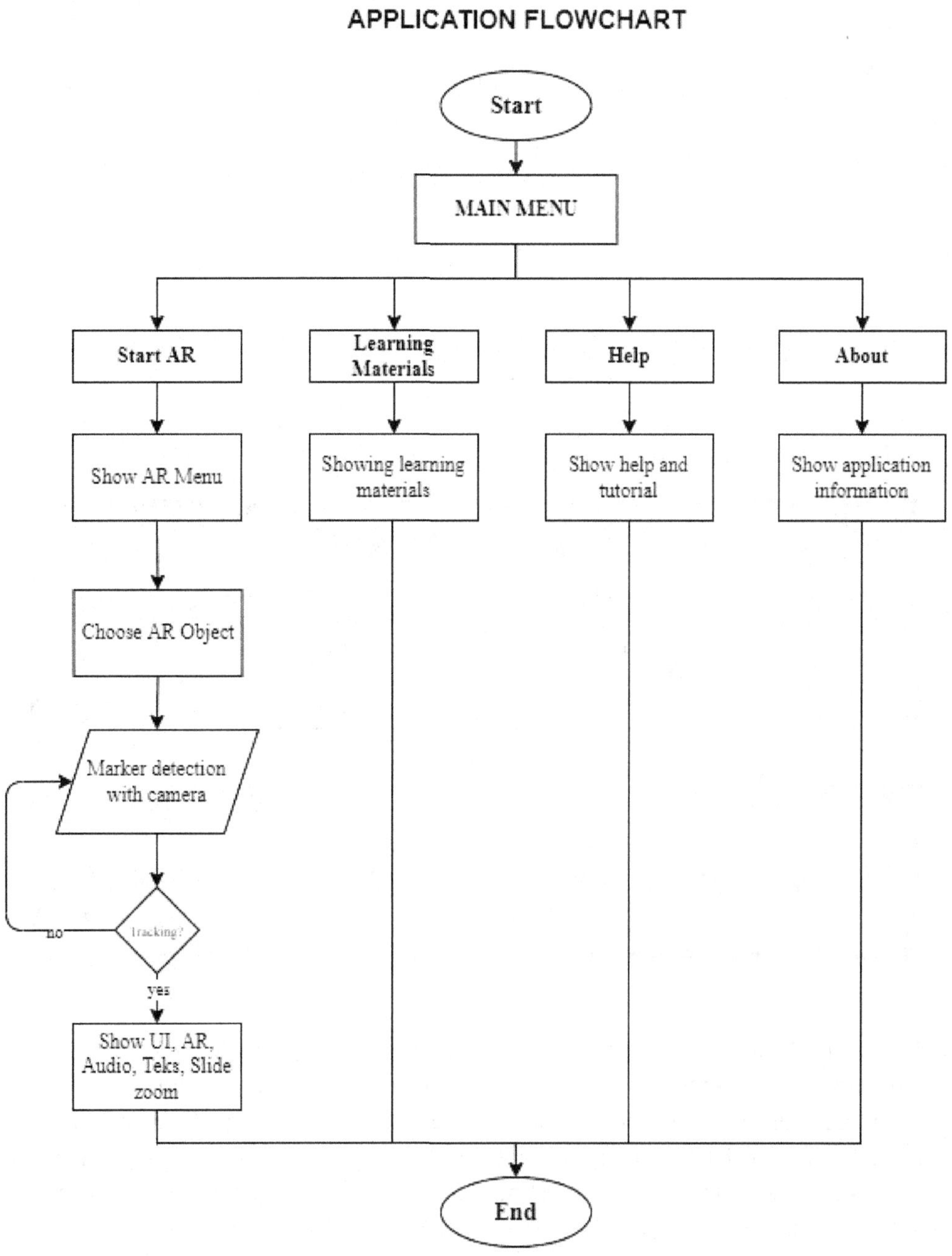

Figure 4. Use Case Diagram of the AR Fitokimia application

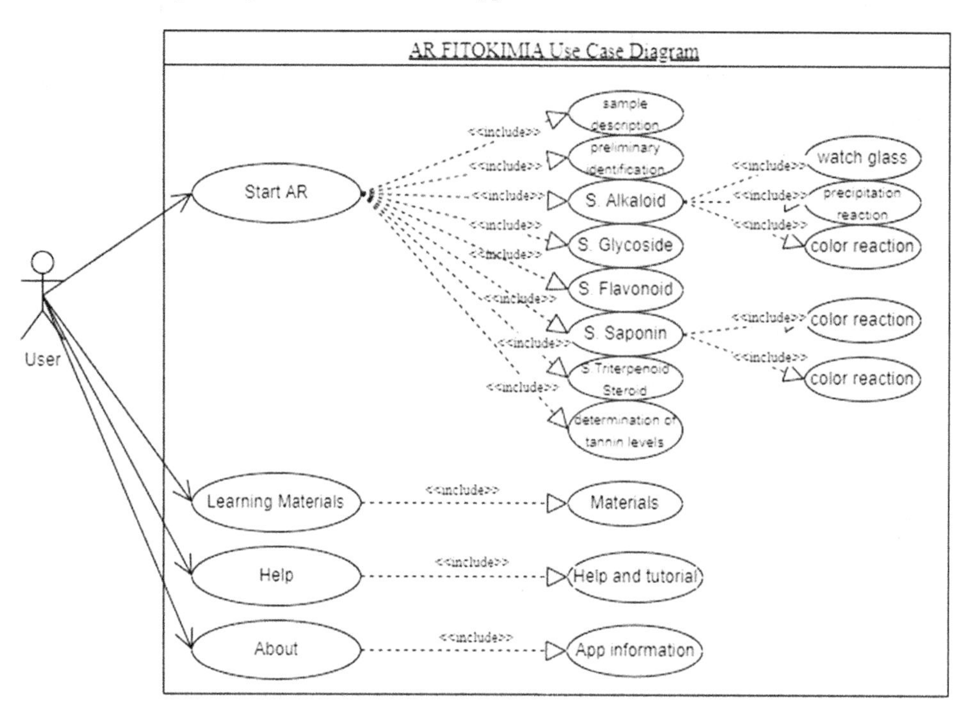

Figure 5. Main menu view in the AR Fitokimia application

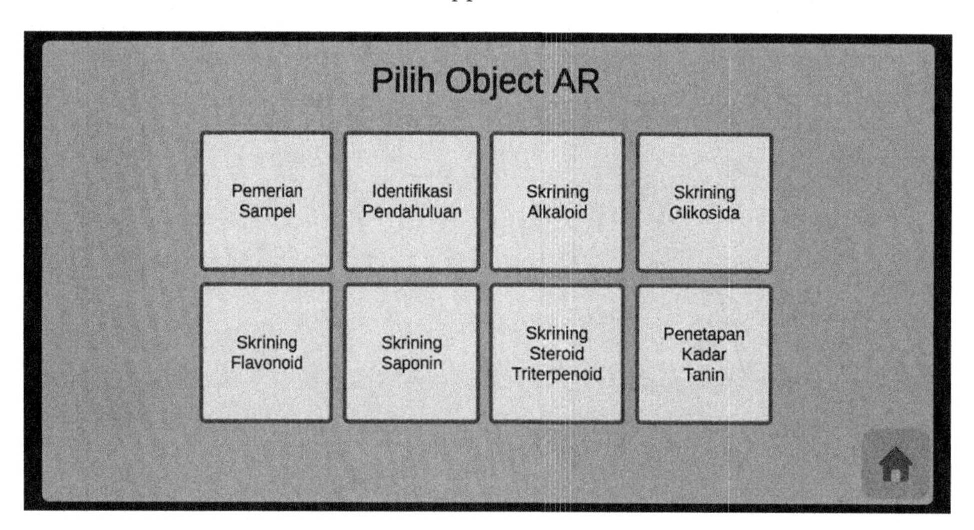

After clicking the “Start” button on the Main Menu, the system would go to the menu “Select Object AR,” which then displayed 8 (eight) options buttons, as shown in Figure 6. Alkaloid screening and saponin screening had more than one 3D object, making a submenu on each screening (Figure 7). In addition, the AR application was designed interactively in displaying 3D objects in the alkaloid, saponin, and PK Tanin screening menus that showed the stages of testing steps that the users and the animation could select.

Figure 6. Select Object in the AR Fitokimia menu view

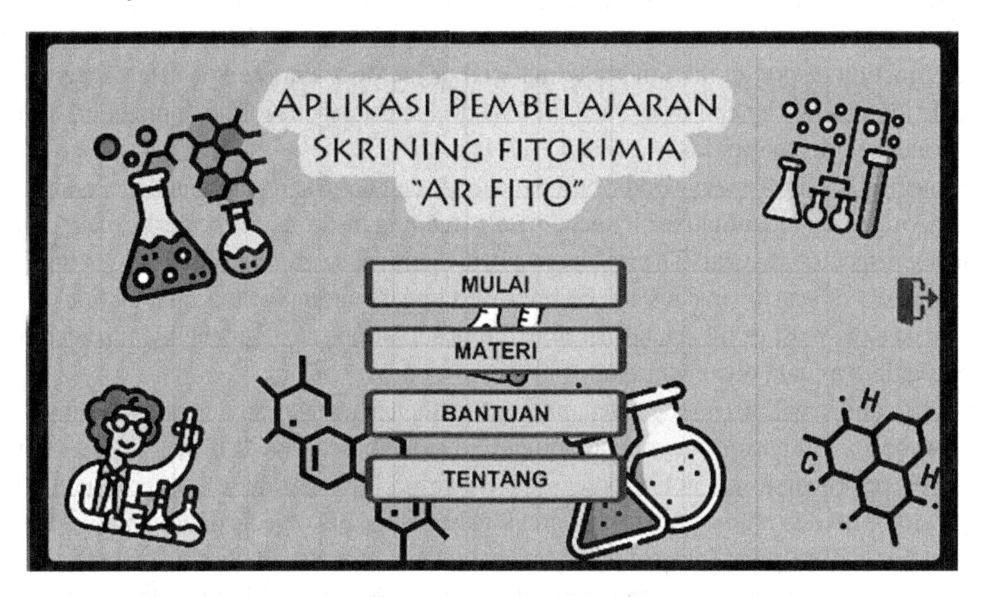

Figure 7. Alkaloid and saponin screening sub-menu

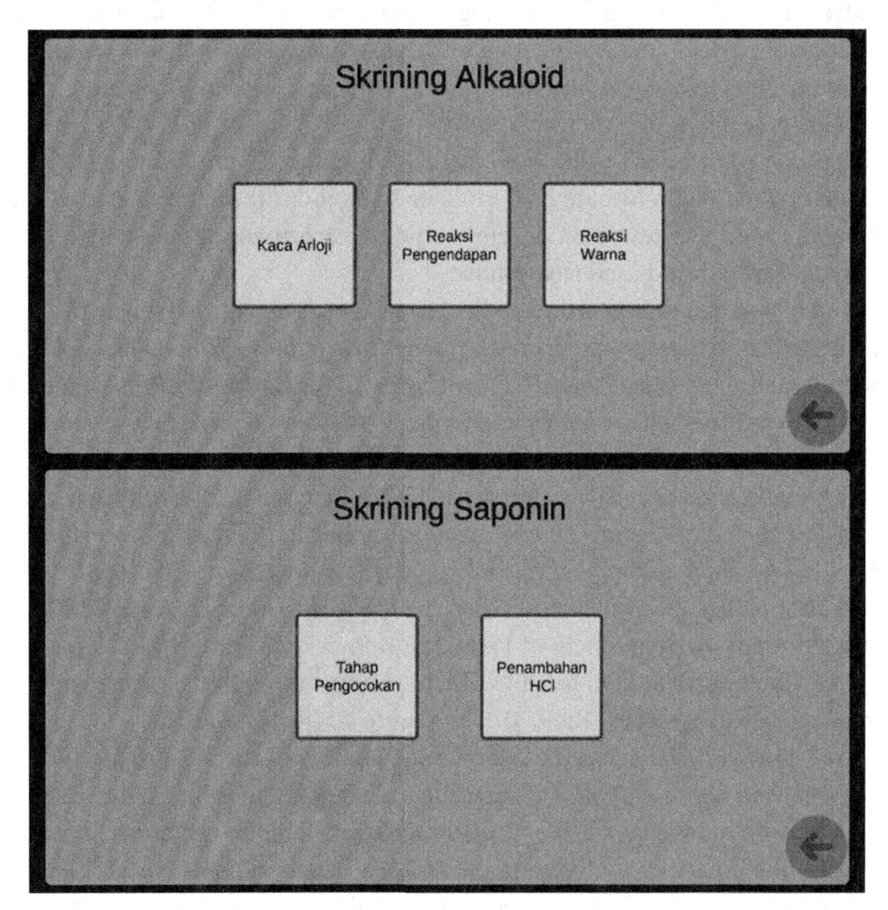

The 'Material' menu included preliminary identification procedures, alkaloid screening, glycosides, flavonoids, saponins, steroid-triterpenoids, and tannin rate determination. The next and previous buttons of the swipe function displayed the following material or previous material. A 'Help' menu explained how to use the application to end-users, and the About menu introduced the application development team from Poltekkes Kemenkes Jakarta II and UIN Jakarta.

The Augmented Reality system worked based on image detection and automated the learning through data. Qualitative and semiquantitative phytochemical screening methods can be performed through color reactions using a specific reagent. The results of phytochemical screening observations that have only existed in photo documents or images can be improved understanding of the imagination with the use of AR. AR makes the most of the data. Therefore, all observations of phytochemical screening experiments were required as data documentation.

The device's camera will detect the given marker, and then after recognizing and marking the marker pattern, the webcam will perform a comparison with the database owned. If the database is not available, the marker information will not be processed. However, if the database is appropriate, the marker information will be used to render and display previously created 3D objects or animations. The creation of this application will help improve students' understanding and participation in distance learning.

One of the ideas to optimize the usefulness of smartphones, especially in distance practical learning, is to create an interactive learning medium using AR technology that can provide more perspective for users. The advantages offered by this technology include being able to bring up objects in 3D, text, sound, and animation of phytochemical screening results in a natural environment. Android can support various system operations to customize digital devices (without the need for special instruments). No specialized hardware is required for Android users.

In the final stage, the application testing was completed through the programming process by debugging the application on the Android-based smartphone to find out the success rate of the application when used. The application was tested on different types of smartphones to test their reliability when used in various kinds of Android-based smartphones.

In AR trials, smartphone cameras were directed to the marker to be identified (Figure 8). The application then displayed the object of laboratory experiments on the smartphone screen in interactive 3D visualizations with audio, text, and animation (Figure 9). The trial tested the functions of the buttons, markers, and the application itself. It needed a phone with a minimum of Android specification version 7.1.1 and 2 GB RAM. The authors conducted an Android-based AR Fitokimia performance evaluation of ISO 25010, a standard for testing software quality. The aspects tested were functional suitability and usability.

The results of the black-box testing on each scenario indicated that all the functions and buttons contained in the AR Fitokimia application run well as expected, without any errors. After passing the black-box testing, the application trial continued on the usability aspect by Android users in the Pharmaceutical and Food Analysis Department of Poltekkes Kemenkes Jakarta II. Each participant was allowed to try the AR Fitokimia and then provided an assessment through the questionnaire. The results were calculated using the Likert scale (Table 3). There were 40 (forty) students who filled out an online questionnaire to see the opinion and accuracy of the functionality of the AR Fitokimia. The instrument used consisted of 7 (seven) questions with 4 (four) criteria, including usability value, ease of use, ease of understanding, and satisfaction value (Table 4). After the questionnaire results were obtained, the next step was to calculate the points by multiplying the number of respondents by the Likert score. Then, calculate the percentage, sum all points received and divided by maximum points, and multiply by 100 percent. Maximum points are 40 students x 5 points = 200 points.

Figure 8. The marker of the AR Fitokimia application

Figure 9. The visualization of AR Fitokimia on an Android-based device

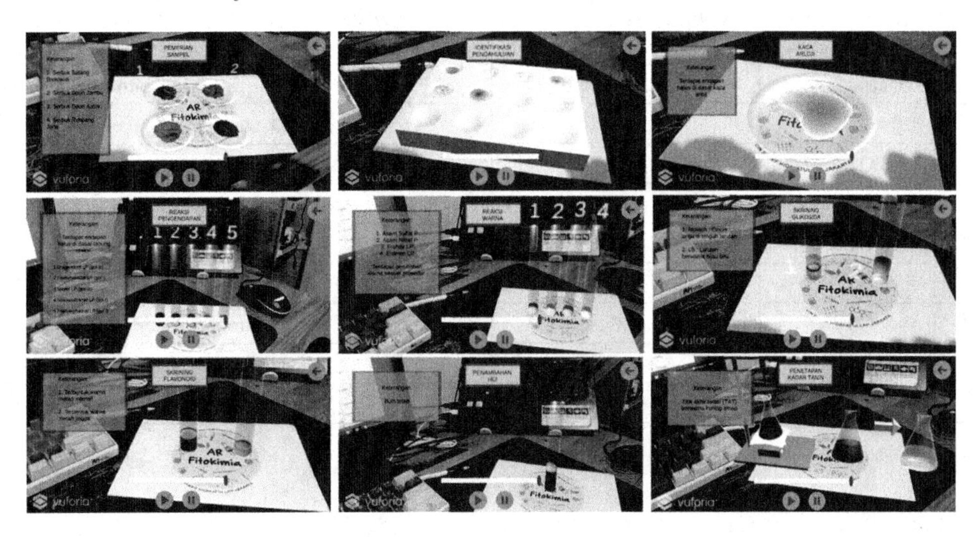

Table 3. Interpretation score based on the Likert score

Likert Score	Interpretation of the score with interval = 20	Options
1	0% - 19.99%	Strongly Disagree
2	20% - 39.99%	Disagree
3	40% - 59.99%	Sufficient
4	60% - 79.99%	Agree
5	80% - 100%	Strongly Agree

Table 4. Questionnaire for the 'AR Fitokimia' app

No	Questions	SD	D	S	A	SA
1	Do you think the AR Fitokimia application is easy to use?	31	9	0	0	0
2	Are all menus in the AR Fitokimia application accessible?	28	12	0	0	0
3	Is the AR Fitokimia application able to display phytochemical screening materials properly?	28	12	0	0	0
4	Can the Augmented Reality feature in the AR Fitokimia application work properly?	26	11	3	0	0
5	Is the appearance and design of the AR Fitokimia application attractive?	27	12	1	0	0
6	Can the AR Fitokimia application be used as a learning aid?	33	7	0	0	0
7	Do you think the AR Fitokimia application is feasible to apply in learning?	28	12	0	0	0

The questionnaire assessment results (Table 5) showed that 95.5% of the respondents strongly agreed that the AR Fitokimia application was easy to use, 94% strongly agreed that the menu could be accessed, and 94% strongly agreed that the application could display phytochemical screening materials well. 91.5% strongly agree that Augmented Reality on the application can work well, and 93% strongly agree that the appearance and design of the AR Fitokimia application were suitable for learning.

Table 5. Percentage of AR Fitokimia's respondents after multiplied by the Likert score

No	Questions	Level accepted	Percentage	Indicator of Category
1	Q1	191	95.5%	Strongly Agree
2	Q2	188	94%	Strongly Agree
3	Q3	188	94%	Strongly Agree
4	Q4	183	91.5%	Strongly Agree
5	Q5	186	93%	Strongly Agree
6	Q6	193	96.5%	Strongly Agree
7	Q7	188	94%	Strongly Agree
Average		188	94%	Strongly Agree

The results of the questionnaire assessment showed that the average percentage of each question reached 94% with the strongly agree category. These results showed that the AR Fitokimia application is suitable for learning. Marker-based AR Fitokimia application can be an alternative solution for adaptive and interactive distance practicum learning in virtual form to support college students' activities in the Pharmaceutical and Food Analysis Department during the pandemic.

In Microlearning development, some stages must be passed so that phytochemical screening material can be utilized to the maximum. The adapted settings are related to content analysis, media production, and dissemination (H. Nugraha et al., 2021). All these stages are done so that the content presented in Microlearning follows the purpose of phytochemical screening learning.

The initial stage of this gamification design is to define the game genre in the form of educational game modes presented to students in vocational higher education environments, especially in the Pharmaceutical and Food Analysis Department Poltekkes Kemenkes Jakarta II. The application was created in a virtual lab system that features guidance for users to complete missions and quiz questions, tool and material pick-up features, progress storage, leveling, and reward systems. The visualization of the room was not static, so users could learn and play by exploring the contents of the laboratory while answering questions or challenges given in the game.

Storyboard design was done to determine the appearance of the application. Figure 10 showed the system flowchart of the Virtual Lab Fitokimia app. The Storyboard included the login menu, the home menu, theory, leaderboard, mission, score, level, progress, and the end of the game. When the user opened the game, the user could see some options to choose from, Material, Mission, Help Menu, and Leaderboard (Figure 11). The user was directed to do some personal preparations in-Home menu before entering the laboratory. After completing those steps, the laboratory door would be opened. Then the user was allowed to enter and explore the lab virtually (Figure 12). The user could choose the theory menu to learn about the phytochemical screening material before starting the mission and see the available options, as shown in Figure 13. Besides material, there was also a piece of information about the AR Fitokimia location on the Google Playstore and AR marker image (Figure 14) so it would be clear that the AR Fitokimia and Virtual Lab Fitokimia app were independent. The user could scan the picture to help visualize the results of phytochemical screening observations after installing the AR Fitokimia application as instructed. On the mission menu, the player would see the available missions to complete in each laboratory area (Figure 15). The user must choose the correct answer to some quiz questions and get the score based on the time running. The total points would be displayed when the player accomplished all of the questions in a mission. The score would be sent to the leaderboard, and the user could select the following task (Figure 16). The game ended when the entire mission had been completed (Figure 17). Similar to AR Fitokimia app, the Virtual Lab Fitokimia app also used Indonesian language.

The application production started with collecting the assets needed in the game, coding, developing, and integrating the leaderboard system into the game. Some gaming assets came from the authors themselves and various sources on the internet that are open access and license-free. The application used GooglePlayGamesPlugin-010.14 (source: github.com) for connection to the leaderboard database, OpenSans Family (source: Google fonts) as text material, and the design for the display of virtual laboratory images and other objects was based on laboratory room data of the Pharmaceutical and Food Analysis Department of Poltekkes Kemenkes Jakarta II. Once the app was completed, it was further integrated with the Google Play Games leaderboard. The installation of the SDK plugin aimed to make it easier to integrate the Google Play Games API system. The leaderboard data and scores were automatically updated by Google. The system only needed to send the updated score to Google's leaderboard data by using the internet. The app used background music from an internet source and is royalty-free.

Upon starting the first mission, each user or player is given a limited time to get a high score after analyzing and answering all of the questions displayed. The system would not provide the right option when the user responded to the question incorrectly. If the user receives a value of < 60, the user must repeat the quiz from the beginning to strengthen understanding. Players would get the point based on the scoring system. Points would not increase if the answer were wrong, and the system would proceed to the next question. Each question in the quiz is taken based on the phytochemical screening material. The material available in the game is based on the phytochemical practicum guidelines in the Pharmaceutical and Food Analysis Department of Poltekkes Kemenkes Jakarta II, Herbal Pharmacopoeia, and AR application.

Figure 10. Flowchart of the Virtual Lab Fitokimia application

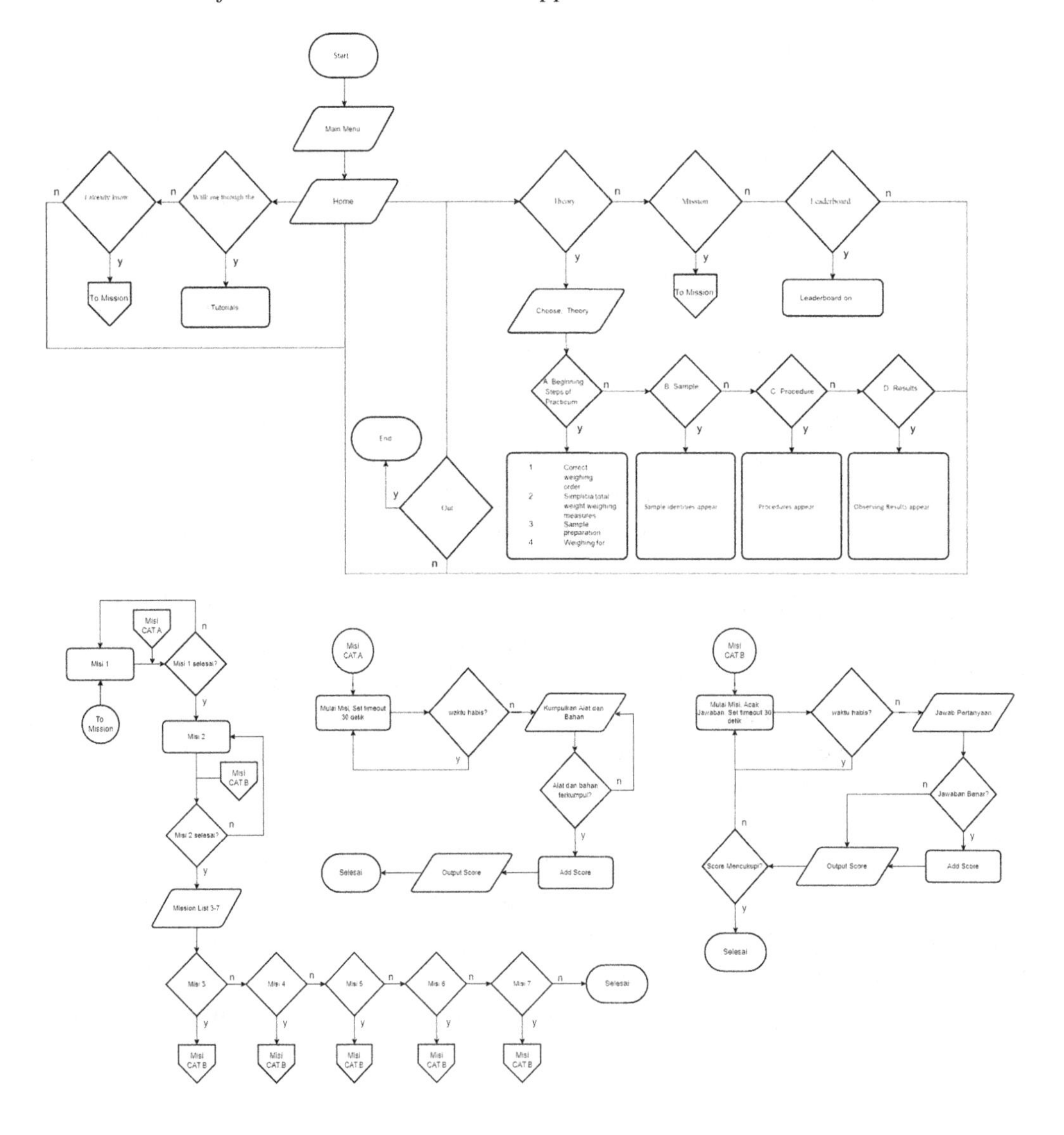

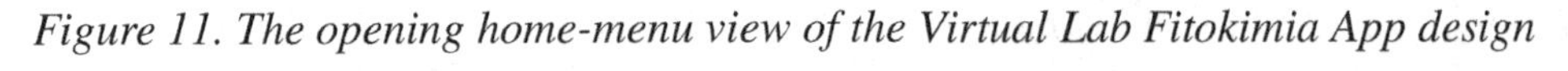

Figure 11. The opening home-menu view of the Virtual Lab Fitokimia App design

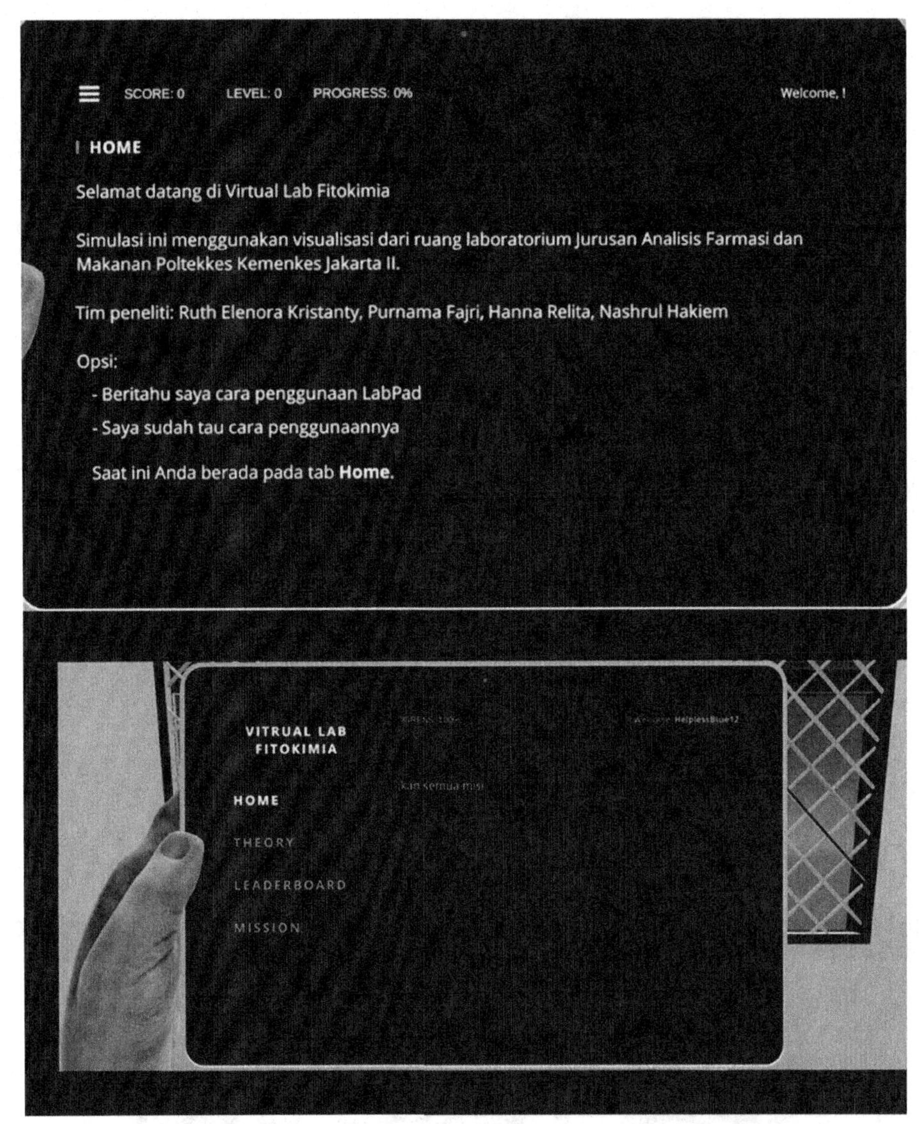

For the second mission, there is no timeout set in the system. Users were allowed to answer without time restrictions. The time length user used to reply would affect the score obtained. The user only gets 1 point after answering for more than 60 seconds. The scoring system was built as follows (running time backward from 60s-0s):

- >48s score +8
- >36s score +6
- >24s score +4
- >12s score +2
- else score +1

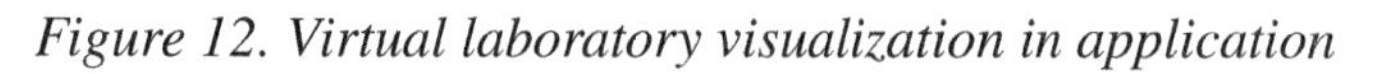
Figure 12. Virtual laboratory visualization in application

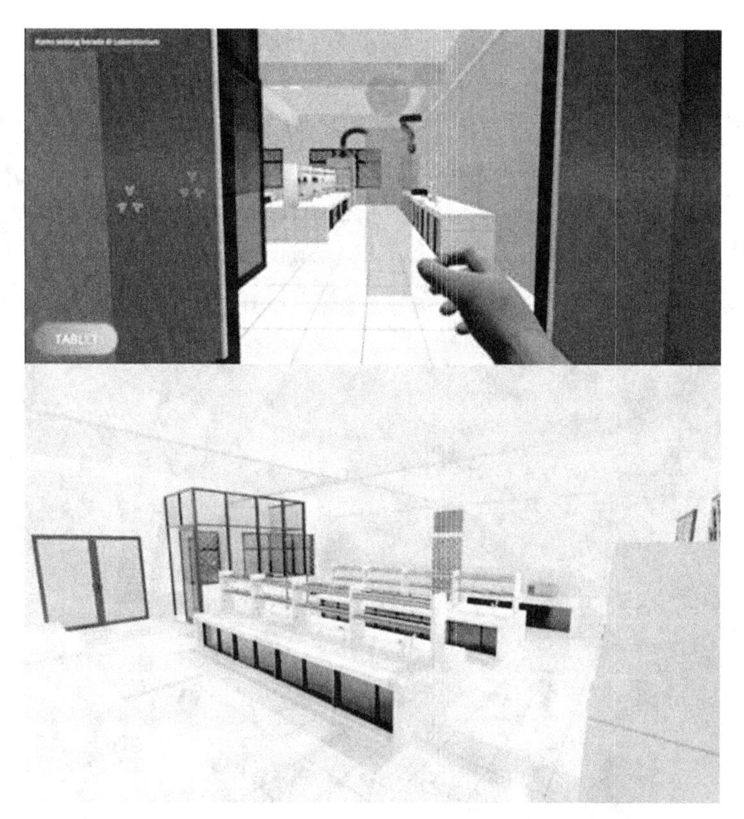

Figure 13. The available options in the theory menu

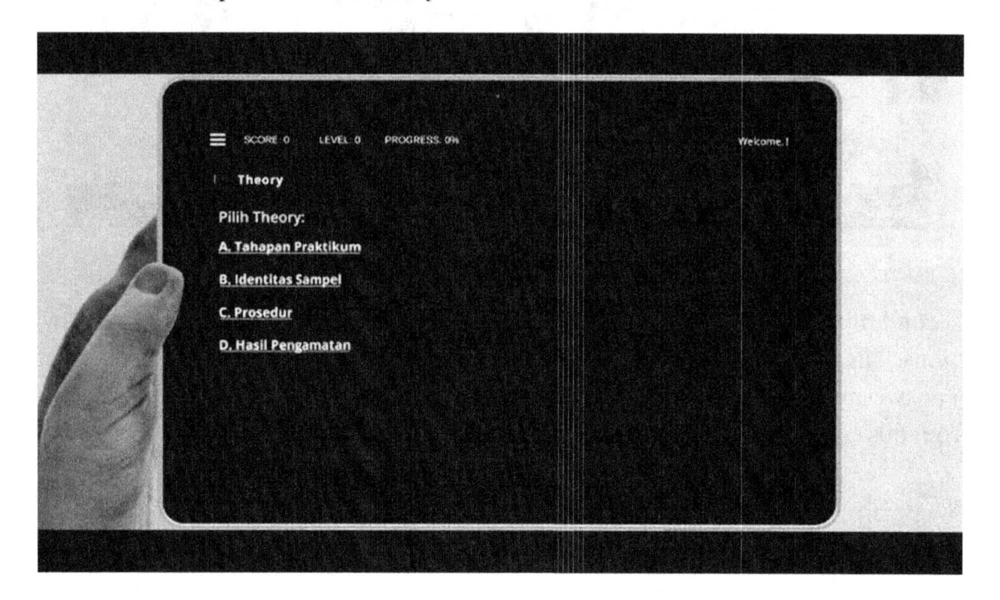

Figure 14. The AR Fitokimia information in the theory menu

Figure 15. The available missions to complete in each laboratory area

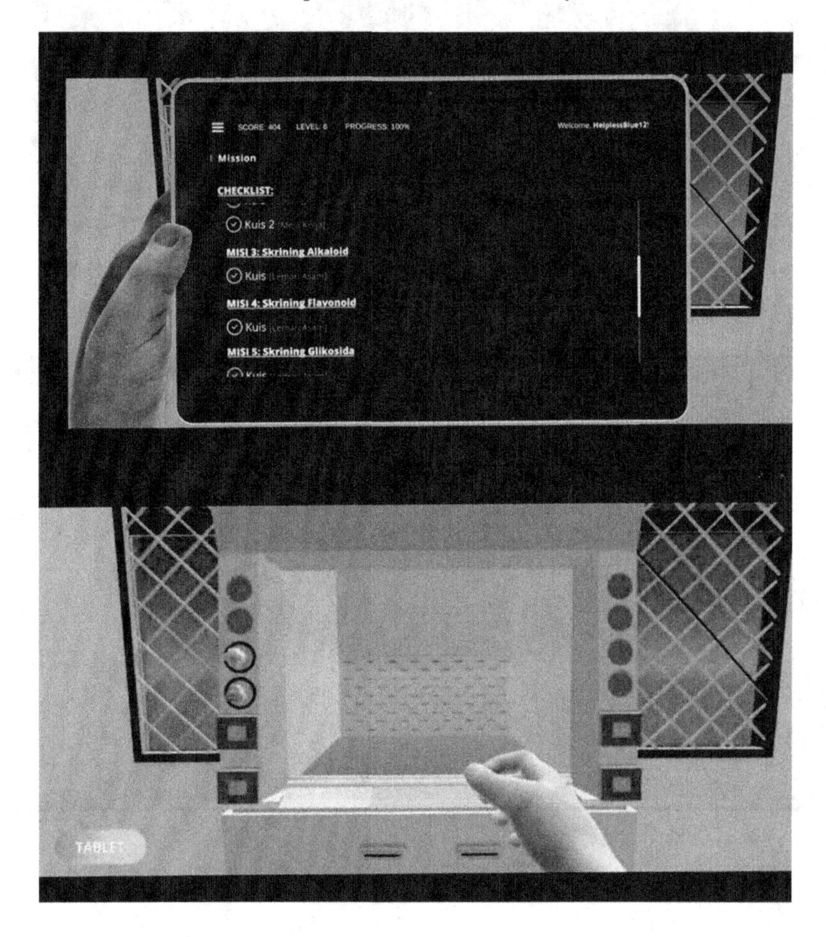

Figure 16. Instruction to the next mission as the score sent to the leaderboard

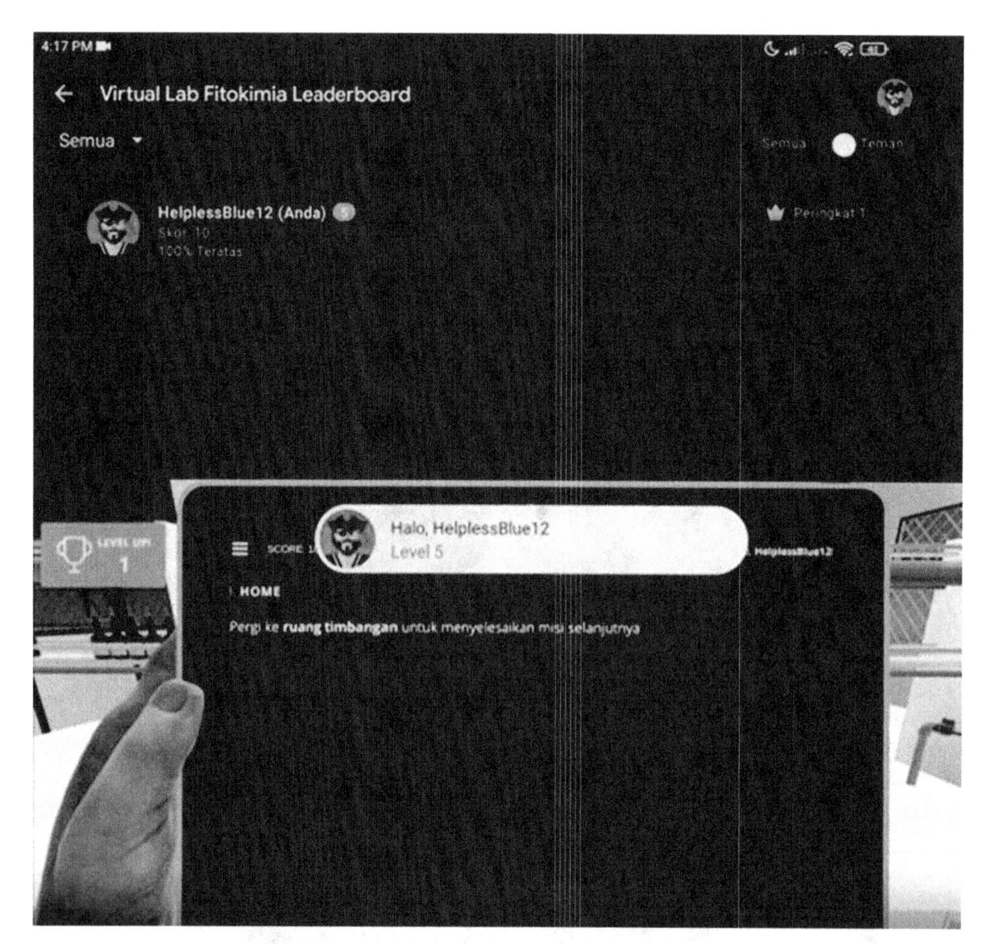

Figure 17. Missions accomplished

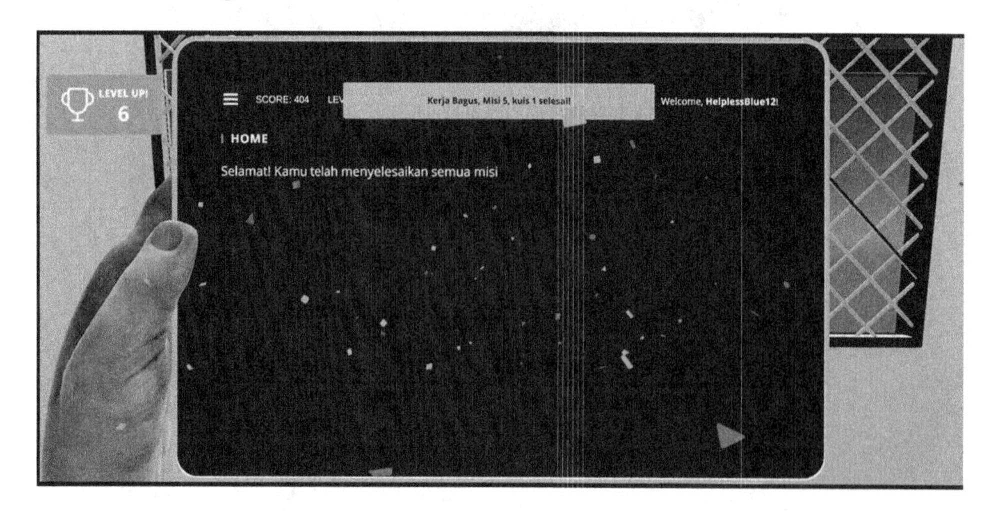

The system is the same as the second mission for the first mission. If the user chooses the wrong answer, then the system will add no point for the user. The scoring system was built as follows (time running backward from 60s-0s):

- >240s score +10
- >180s score +8
- >120s score +6
- >60s score +4
- >20s score +2
- else score +1

The user must answer correctly in the first and second missions to proceed to the next task. For essay questions on third until seventh missions, question sentences are displayed in the form of right or wrong conclusion options. If the users have completed the task, then the progress will be saved automatically so that the user can exit the game application at any time. The user cannot repeat the game if he has completed all the missions. Users can repeat the game and create a leaderboard from scratch again with a new account.

The gamification strategy rewards the player for completing a given task or making the player compete with other players. The bonuses provided were points, badges, levels, progress bars, virtual objects, or achievements that other players could see. The ranking or leaderboard was used so that players would feel like they were competing with other players.

The concept of gamification is concerned with interaction and participation. When applied in the learning process, students as players must follow the rules for goals (to achieve specific scores). In gamification, the elements contained in the application are implemented to increase motivation and involvement in learning (Alsawaier, 2018). Some features in a game, such as standings board, test report, and time restriction, create interest in gamification in the learning process (Pitoyo et al., 2020). This makes sense of pleasure and curiosity to continue trying. Seeing students' interest in Poltekkes Kemenkes Jakarta II towards the mobile application of phytochemical screening, the teachers can apply gamification methods to this learning content.

The features in this application are phytochemical screening practicum material, 7 (seven) missions, and 7 (seven) quizzes—the course of consecutive missions from the first mission to the second mission. Completing the third to seventh mission (alkaloid screening, glycosides, flavonoids, saponins, and tannin rate determination) is done according to the user's choice.

In the mission 2, the user is faced with four types of quizzes that must be completed in sequence, namely:

1. visualization of weight space (two question instructions appear: arrange the correct sequence of steps in the form of a choice of sentence arrangement2)
2. visualization of the desk (1 question instruction appears: arranges the correct steps in the form of a choice of sentence arrangements)
3. The user returns to the scale room (visualization of the scale room) and then appears two quiz questions (questions and answer options as I include in the soft file)
4. The user returns to the desk (table visualization and reagent) and completes the quiz question (1 question).

5. The user makes experimental conclusions (reward: level up and can continue the next mission)

The user could open the materials any time they want to learn before answering the quiz. Missions were created to obtain a score. Scores were calculated based on the total amount of correct answers on the examination referred to by the scoring system. This application had no life feature because there was no correlation with the virtual lab. The length of time answering questions was a challenging factor based on the scoring system.

The scoring system for the first and second missions was installed to get rewarded level and achieve the total score required to open the next mission (i.e., mission 3 to mission 7). The 3rd to 7th missions are given their respective scores attached to the leaderboard so that the teaching team can see which users managed to complete answering all questions correctly and which users still did not understand the material. The scores obtained in each of the 3rd mission to the 7th mission is a maximum score of 100 so that it can simultaneously be a task score to be designed to be the final score of the practicum (averaged together with test scores).

The advantages of online applications lied in the storage of progress, the existence of a leaderboard, dynamic content, and the ability to see user activity. However, an admin must be in charge of updating content and requires officers to perform server maintenance. Due to cost and time limitations, this phytochemical screening game application is made in a simple online form. Any updated content can be input into the application. The leaderboard can still be added using the Google play games feature.

This game education was placed as prerequisites for practicum examinations to obtain task scores while reviewing practicum material (briefing materials) and to strengthen or affirm students' understanding. It could be done at once and gradually until all missions are carried out. The score achieved in the screenshot was then sent to the practicum coordinator (offline) or recorded directly on the leaderboard on the server (online).

The application was equipped with background sound because the music element or background sound became one of the fundamental aspects in assessing a game in addition to storyline, gameplay, and graphics. Music played an essential role in bringing the game to life, building the atmosphere, and providing more experience while playing because the audio can affirm better than the visuals.

The Storyboard contained a visualization of ideas for the application to be built. A storyboard can also be a scenario that will be used to outline a project. Storyboards are more widely used to create interactive media frameworks in the design/design stage. The system is spelled out by creating work points that help identify what materials must be completed or arranged for the application to run. The information listed in the Storyboard is sketches or screen/page images, vivid color and size, original text, font size, narration, video, audio, animation, and interaction.

Testing of the 'Virtual Lab Fitokimia' app started from the function of the button to the gamification function included in the game by one expert in the field of software development from UIN Jakarta and one lecturer of Phytochemistry courses from the Pharmaceutical and Food Analysis Department of Poltekkes Kemenkes Jakarta II. After passing this first testing stage, the next test involved end users (30 respondents) to see the level of satisfaction in using this game application as an additional supporting media in learning phytochemical practicum for vocational students. The results can be seen in Table 6 and Table 7.

Table 6. Questionnaire for the 'Virtual Lab Fitokimia'app

No	Questions	SD	D	S	A	SA
1	'Virtual Lab Fitokimia' app is easy to install on Android smartphones	29	1	0	0	0
2	All menus on the app are well accessible	25	5	0	0	0
3	All icons and functions in the application are appropriate and understandable	23	7	0	0	0
4	The features provided in the application are appropriate and complete	26	2	2	0	0
5	The look and design of the app are already attractive	24	5	1	0	0
6	Augmented Reality features in the form of markers embedded in the application can run well	25	5	0	0	0
7	The quiz feature in the 'Virtual Lab Fitokimia' application is following the phytochemical screening material in practicum learning	27	3	0	0	0
8	The scoring and leaderboard features in the 'Virtual Lab Fitokimia' application are appropriate	27	3	0	0	0
9	The game element in the 'Virtual Lab Fitokimia' application can increase understanding and interest in phytochemical screening learning	26	4	0	0	0
10	The 'Virtual Lab Fitokimia' application can be used as a medium to help learn phytochemical screening	29	1	0	0	0
11	The 'Virtual Lab Fitokimia' application is feasible to be applied in phytochemical practicum learning, especially as evaluation material	28	2	0	0	0

Table 7. Percentage of the Virtual Lab Fitokimia's respondents after multiplied by the Likert score

No	Questions	Level accepted	Percentage	Indicator of Category
1	Q1	149	99.3%	Strongly Agree
2	Q2	145	96,7%	Strongly Agree
3	Q3	143	94%	Strongly Agree
4	Q4	144	95.3%	Strongly Agree
5	Q5	143	94%	Strongly Agree
6	Q6	145	96,7%	Strongly Agree
7	Q7	147	98%	Strongly Agree
8	Q8	147	98%	Strongly Agree
9	Q9	146	97,3%	Strongly Agree
10	Q10	149	99,3%	Strongly Agree
11	Q11	148	98,6%	Strongly Agree
Average		146	97,02%	Strongly Agree

Lecture activities in the college environment, especially the vocational world, during the Covid-19 pandemic experienced many changes. Learning innovation is needed to overcome the saturation of online learning. Microlearning can be one of the most effective efforts for "refreshing" learning content. Its short presentation makes it easy to understand when used to deliver learning content. The use of Microlearning can increase the effectiveness and efficiency of the learning process. In addition, the knowledge gained from Microlearning can also last longer in memory. The authors applied gamification in the form of challenges to the virtual application of the laboratory so that users honed their ability to think quickly and responsively in answering and honed their understanding of questions about phytochemical screening.

Practicum learning in vocational institutions will run well if talented resources and infrastructures are supported to become competent and technologically responsive analysts and adapt to the new normal. Things to consider when selecting learning media are learning objectives, practical, easy to obtain, learners, use, not rigid, cost, and quality (Mustaqim & Kurniawan, 2017; Pradibta et al., 2016). In this case, the creation of learning media using Augmented Reality and gamification meets all these considerations. AR and gamification-based learning media can cover the lack of practicum modules used simultaneously, reach all students, last longer, be easy to use and understand, and be an alternative media of better quality.

The practicum will become more engaging by bringing virtual objects through the application. Practicum preparation will also be more efficient because teachers only prepare intelligent devices. The institution can reduce the cost of practicum implementation because innovative technology can replace consumable practicum materials. AR can address the digitalization needs for practicum learning to promote adaptive learning in the laboratory.

The learning innovation produced by this research followed the purpose of the study and contributed to the achievement of the institution's vision as a reference for health technology and graduated analysts that would be technology adaptive. These efforts had been made to overcome the learning boredom created by the pandemic. The mobile platforms as the research outcome can be the appropriate technology for public users.

Both of these applications had been tested by internal parties and beta testers and released free on the Google Playstore under the name AR Fitokimia and Virtual Lab Fitokimia. AR Fitokimia's marker was used as one of the material menus in the Virtual Lab Fitokimia application to display the results of phytochemical screening observation for users who want to play the game. The results of the assessments showed that the average percentage of each question was higher than 90% with the strongly agree category. These results showed that the AR Fitokimia and Virtual Lab Fitokimia applications are suitable for vocational students in learning phytochemical screening. Students of the Pharmaceutical and Food Analysis Department of Poltekkes Kemenkes Jakarta II are currently using the applications to learn phytochemical screening. The reviews showed that both applications received a 5-star rating from users (Figure 18).

FUTURE RESEARCH DIRECTIONS

The phytochemistry course in vocational college needs microlearning by using interactive technology to improve remote practicum learning. The concept of microlearning leads to knowledge acquisition as well as skills growth, based on user thinking, application processing methods, and electronic devices. The development of mobile applications can be an advantage in the field of education to support the teaching and learning process of students.

There are some suggestions to note for the prospects' development, as follows:

1. The 'Virtual Lab Fitokimia' application update is needed to add features and functions that do not yet exist. The app's current version does not have an 'exit' feature to exit the game at certain moments and a back button to return to the main page before completing the mission.

2. Adding more animations to make the 'AR Fitokimia application more engaging and interactive is necessary.
3. The scope of the applications' test is still limited to the community in Poltekkes Kemenkes Jakarta II.

Figure 18. Rating review of the learning applications in Google Playstore

CONCLUSION

Based on the results of AR technology and gamification development, these studies concluded as follows:

1. The design of an interactive learning application for phytochemical screening material in Poltekkes Kemenkes Jakarta II resulted in two Android-based applications, the 'AR Fitokimia' and 'Virtual Lab Fitokimia'.
2. Black-box testing and user acceptance test (UAT) showed that all functions in both applications could work properly, and Poltekkes Kemenkes Jakarta II could apply the applications for remote practicum.
3. The augmented reality and educational game applications that had been created to approach microlearning were useful to enhance students' understanding of phytochemical screening in the Pharmaceutical and Food Analysis Department of Poltekkes Kemenkes Jakarta II.

ACKNOWLEDGMENT

This study was carried out in the Pharmaceutical and Food Analysis Department of Poltekkes Kemenkes Jakarta II and collaborated with the Faculty of Science and Technology of UIN Syarif Hidayatullah Jakarta.

REFERENCES

Ahmad, M. (2020). Categorizing Game Design Elements into Educational Game Design Fundamentals. In I. Deliyannis (Ed.), *Game Design and Intelligent Interaction*. IntechOpen. doi:10.5772/intechopen.89971

Alsawaier, R. S. (2018). The effect of gamification on motivation and engagement. *The International Journal of Information and Learning Technology*, *35*(1), 56–79. doi:10.1108/IJILT-02-2017-0009

Alsehri, A. (2021). The Effectiveness of a Micro-Learning Strategy in Developing the Skills of Using Augmented Reality Applications among Science Teachers in Jeddah. *International Journal of Educational Research Review*, *6*(2), 176–183. doi:10.24331/ijere.869642

Batubara, M. H., Mesran, Sihite, A. H., & Saputra, I. (2017). Aplikasi Pembelajaran Teknik Mesin Otomotif Kendaraan Ringan Dengan Metode Computer Assisted Instruction. *Informasi Dan Teknologi Ilmiah*, *12*(2), 266–270.

Blessinger, P., & Wankel, C. (2012). Innovative Approaches in Higher Education: An Introduction to Using Immersive Interfaces. In C. Wankel & P. Blessinger (Eds.), *Increasing Student Engagement and Retention Using Immersive Interfaces: Virtual Worlds, Gaming, and Simulation* (Vol. 6, Part C, pp. 3–14). Emerald Group Publishing Limited. doi:10.1108/S2044-9968(2012)000006C003

Clark, D. (2006). *Games and E-Learning*. Caspian Learning Ltd.

De-Marcos, L., Garcia-Lopez, E., & Garcia-Cabot, A. (2016). On the effectiveness of game-like and social approaches in learning: Comparing educational gaming, gamification & social networking. *Computers & Education*, *95*, 99–113. doi:10.1016/j.compedu.2015.12.008

Frima, F. K. (2020). Penerapan Praktikum Jarak Jauh Pada Topik Pertumbuhan Mikroba Dalam Masa Darurat Covid-19 Di Institut Teknologi Sumatera. *Jurnal Pendidikan Sains (Jps)*, *8*(2), 102. doi:10.26714/jps.8.2.2020.102-109

Graesser, A. C. (2017). Reflections on Serious Games BT - Instructional Techniques to Facilitate Learning and Motivation of Serious Games. Springer International Publishing. doi:10.1007/978-3-319-39298-1_11

Handani, S. W., Suyanto, M., & Sofyan, A. F. (2016). Penerapan Konsep Gamifikasi pada E-Learning Untuk Pembelajaran Animasi 3 Dimensi. *Jurnal Telematika*, *9*(1), 42–53. doi:10.2214/ajr.181.6.1811716b

Huang, R., Yang, J., & Zheng, L. (2013). The Components and Functions of Smart Learning Environments for Easy, Engaged and Effective Learning The Demands on Rebuilding Learning Environments in Information Society. *International Journal for Educational Media and Technology*, *7*(1), 4–14.

Indriani, R., Sugiarto, B., & Purwanto, A. (2016). Pembuatan Augmented Reality Tentang Pengenalan Hewan Untuk Anak Usia Dini Berbasis Android Menggunakan Metode Image Tracking Vuforia. *Seminar Nasional Teknologi Informasi Dan Multimedia*, 73–78.

Jusuf, H. (2016). Penggunaan Gamifikasi dalam Proses Pembelajaran [The Use of Gamification in the Learning Process]. *Jurnal TICOM*, *5*(1), 1–6.

KOMINFO. (2017). *Survei Penggunaan TIK Serta Implikasinya terhadap Aspek Sosial Budaya Masyarakat 2017*. BPPSDM KOMINFO.

Lestiyo, I. (2021). *Gaming Report*. Unity Technologies.

Lusa, S., Rahmanto, Y., & Priyopradono, B. (2020). The Development Of Web 3d Application For Virtual Museum Of Lampung Culture. *Psychology and Education Journal*, *57*(9), 188–193.

Maier, P., Klinker, G., & Tonis, M. (2009). *Augmented reality for teaching spatial relations*. Conference Ofthe International Journal Of Arts & Sciences.

McGonigal, J. (2011). *Reality is Broken : Why games Make Us Better and How They Can Change the World*. The Pinguin Press.

Mohammed, G. S., Wakil, K., & Nawroly, S. S. (2018). The Effectiveness of Microlearning to Improve Students' Learning Ability. *International Journal of Educational Research Review*, *3*(3), 32–38. doi:10.24331/ijere.415824

Mustaqim, I., & Kurniawan, N. (2017). Pengembangan Media Pembelajaran Berbasis Augmented Reality [Augmented Reality-Based Learning Media Development]. *Edukasi Elektro*, *1*(1), 26–48.

Ng, W. (2012). Can we teach digital natives digital literacy? *Computers & Education*, *59*(3), 1065–1078. doi:10.1016/j.compedu.2012.04.016

Ninaus, M., Moeller, K., McMullen, J., & Kiili, K. (2017). Acceptance of Game-Based Learning and Intrinsic Motivation as Predictors for Learning Success and Flow Experience. *International Journal of Serious Games*, *4*(3), 15–30. doi:10.17083/ijsg.v4i3.176

Noriska, N. J., Widyaningrum, R., & Nursetyo, K. I. (2021). Pengembangan Microlearning pada Mata Kuliah Difusi Inovasi Pendidikan di Prodi Teknologi Pendidikan [Microlearning Development in Diffusion of Educational Innovation Courses in Educational Technology Study Program]. *Jurnal Pembelajaran Inovatif*, *4*(1), 100–107. doi:10.21009/JPI.041.13

Nugraha, A. C., Khairudin, M., & Hertanto, D. B. (2017). Rancang Bangun Game Edukasi Sebagai Media Pembelajaran Mata Kuliah Praktik Teknik Digital [Design of Educational Games as Learning Media for Digital Engineering Practice Courses]. *Jurnal Edukasi Elektro*, *1*(1), 92–98. doi:10.21831/jee.v1i1.15121

Nugraha, H., Rusmana, A., Khadijah, U., & Gemiharto, I. (2021). Microlearning Sebagai Upaya dalam Menghadapi Dampak Pandemi pada Proses Pembelajaran [Microlearning as an Effort to Deal with The Pandemic Impact on The Learning Process]. *Kajian Dan Riset Dalam Teknologi Pembelajaran*, *8*(3), 225–236. doi:10.17977/um031v8i32021p225

Nuñez, M., Quirós, R., Nuñez, I., Carda, J. B., & Camahort, E. (2008). Collaborative Augmented Reality for Inorganic Chemistry Education. *WSEAS International Conference. Proceedings. Mathematics and Computers in Science and Engineering*, *5*(January), 271–277.

Owolabi, M. S., Omowonuola, A. A., Lawal, O. A., Dosoky, N. S., Collins, J. T., Ogungbe, I. V., & Setzer, W. N. (2017). Phytochemical and bioactivity screening of six Nigerian medicinal plants. *Journal of Pharmacognosy and Phytochemistry*, *6*(6), 1430–1437.

Pandey, A. (2018). *15 Types Of Microlearning For Formal And Informal Learning In The Workplace*. https://elearningindustry.com/types-of-microlearning-formal-informal-learning-workplace-15

Peng, H., Ma, S., & Spector, J. M. (2019). Personalized Adaptive Learning: An Emerging Pedagogical Approach Enabled by a Smart Learning Environment. In Lecture Notes in Educational Technology (pp. 171–176). Springer International Publishing. doi:10.1007/978-981-13-6908-7_24

Pitoyo, M. D., Sumardi, S., & Asib, A. (2020). Gamification-based assessment: The washback effect of quizizz on students' learning in higher education. *International Journal of Language Education*, *4*(1), 1–10. doi:10.26858/ijole.v4i2.8188

Pradibta, H., Harijanto, B., & Wibowo, D. W. (2016). Penerapan Augmented Reality Sebagai Alternatif Media Pembelajaran. *Smartics, 2*(2), 43–48. https://ejournal.unikama.ac.id/index.php/jst/article/view/1693

Prensky, M. (2001). *Digital Game-based Learning*. McGraw-Hill.

Pusparisa, Y. (2020). *Pengguna Smartphone diperkirakan Mencapai 89% Populasi pada 2025 | Databoks*. https://databoks.katadata.co.id/datapublish/2020/09/15/pengguna-smartphone-diperkirakan-mencapai-89-populasi-pada-2025

Sahira Banu, K., & Cathrine, L. (2015). General Techniques Involved in Phytochemical Analysis. *International Journal of Advanced Research in Chemical Science*, *2*(4), 25–32. www.arcjournals.org

Santoso, T. N. B., & Hastutiningtyas, K. N. (2021). Pengembangan Media Game Edukasi Sebagai Sistem Informasi Alternatif Ice Breaking Pembelajaran Di Masa Pandemi [Development of Educational Game Media as an Alternative Information System for Ice Breaking Learning During a Pandemic]. *Ecodunamika : Jurnal Pendidikan Ekonomi*, *4*(1), 1–6.

Sasidharan, S., Chen, Y., Saravanan, D., Sundram, K. M., & Yoga Latha, L. (2011). Extraction, Isolation and Characterization of Bioactive Compounds from Plants' Extracts. *African Journal of Traditional, Complementary, and Alternative Medicines*, *8*(1), 1. doi:10.4314/ajtcam.v8i1.60483 PMID:22238476

Saurina, N. (2016). Pengembangan Media Pembelajaran Untuk Anak Usia Dini Menggunakan Augmented Reality [Learning Media Development for Early Childhood Using Augmented Reality]. *Jurnal IPTEK*, *20*(1), 95. doi:10.31284/j.iptek.2016.v20i1.27

Setyawan, B., Rufii, N., & Fatirul, A. N. (2019). Augmented Reality Dalam Pembelajaran IPA Bagi Siswa SD [Augmented Reality in Natural Science Learning for Elementary School Students]. *Kwangsan: Jurnal Teknologi Pendidikan*, *7*(1), 78–90. doi:10.31800/jtp.kw.v7n1.p78--90

Srivastava, A. (2016). Enriching student learning experience using augmented reality and smart learning objects. *ICMI 2016 - Proceedings of the 18th ACM International Conference on Multimodal Interaction, October 2016*, 572–576. 10.1145/2993148.2997623

Thomas, M. (2011). Technology, Education, and the Discourse of the Digital Native: Between Evangelists and Dissenters. In M. Thomas (Ed.), *Deconstructing Digital Natives: Young People, Technology, and The New Literacies* (pp. 1–14). Routledge. doi:10.4324/9780203818848

Tinacci, L., Guardone, L., Giusti, A., Pardini, S., Benedetti, C., Di Iacovo, F., & Armani, A. (2022). Distance Education for Supporting "Day One Competences" in Meat Inspection: An E-Learning Platform for the Compulsory Practical Training of Veterinarians. *Education Sciences*, *12*(1), 24. Advance online publication. doi:10.3390/educsci12010024

Tukiran, Pramudya, A., Wardana, Nurlaila, E., Santi, A. M., & Hidayati, N. (2016). Analisis Awal Fitokimia pada Ekstrak Metanol Kulit Batang Tumbuhan Syzygium (Myrtaceae) [Preliminary Phytochemical Analysis on Methanol Extract of Syzygium (Myrtaceae)]. *Prosiding Seminar Nasional Kimia Dan Workshop*, (September), 2–8.

Wibawanto, W. (2020). Laboratorium Virtual Konsep Dan Pengembangan Simulasi Fisika LPPM UNNES.

Windawati, R., & Koeswanti, H. D. (2021). Pengembangan Game Edukasi Berbasis Android untuk Meningkatkan Hasil Belajar Siswa di Sekolah Dasar [Android-Based Educational Game Development to Improve Student Learning Outcomes in Elementary Schools]. *Journal Basicedu*, *5*(2), 1028–1038.

Zhu, Z. T., Yu, M. H., & Riezebos, P. (2016). A research framework of smart education. *Smart Learning Environments*, *3*(1), 4. Advance online publication. doi:10.118640561-016-0026-2

KEY TERMS AND DEFINITIONS

AR Fitokimia: An android-based interactive learning application that focuses on phytochemical practicum in the Pharmaceutical and Food Analysis Department of Poltekkes Kemenkes Jakarta II. This application uses a marker-based augmented reality technology to help understand phytochemical screening material in distance learning.

Digital Learning: An educational method that uses digital devices with a system in which students learn using online resources to adapt technology in the class.

Microlearning: A learning strategy to present teaching material in a small-scale learning process.

Phytochemical Screening: The practice of examining plant secondary metabolites to find out the presence of various biologically active compounds.

Remote Practicum: The part of a course in college or university consisting of practical work in a particular field that is capable of being operated from a distance with electronic devices and interactive technology.

Simplicia: Natural ingredients for medicine that are generally in the form of dried materials. The plant parts used comprised rhizomes, leaves, stems, bark, thallus, fruit, flowers, herbs, and seeds.

Virtual Lab Fitokimia: An android-based interactive learning application based on simulation of the phytochemical laboratory in the Pharmaceutical and Food Analysis Department of Poltekkes Kemenkes Jakarta II to help the students practice phytochemical screening and explore the learning in the virtual world.

Chapter 5
Exploring Affordances and Limitations of 3D Virtual Worlds in Psychoeducational Group Counseling

Abdulmenaf Gul
https://orcid.org/0000-0002-3683-8441
Hakkari University, Turkey

Saniye Tugba Tokel
Middle East Technical University, Turkey

ABSTRACT

The purpose of this study was to explore the potential affordances and challenges of 3D virtual environments in psychoeducational group counseling. The research design was based on multiple case study methodology. Face-to-face and 3D virtual psychoeducational counseling groups were formed that focused on procrastination, and multiple forms of data were collected from both groups' participants. The study's results revealed that perceived affordances of the 3D environment for group counseling were similar in both groups, with self-disclosure, anonymity, convenience, interactive environment, and accessible content as the emerged affordances. However, the study also revealed mixed results in terms of perceived challenges. While interaction issues, multitasking, lack of social interaction, and trust concerns emerged as common to both groups, factors such as technical issues and negative attitudes towards virtual intervention were revealed as divergent themes. Intervention outcome results revealed similar patterns in terms of procrastination behavior change in both groups.

INTRODUCTION

Education's role is no longer perceived solely as academic development, with social and psychological development having also become a key area of responsibility. The college learning period, in particular,

DOI: 10.4018/978-1-6684-4854-0.ch005

constitutes a unique life stage during which many individuals experience a significant level of change. As such, they are required to cope with issues of identity, socialization, as well as adaptation to university life and an academic career. These challenges have led to increasing demands for psychological support among university students (Erkan et al., 2011). A growing body of literature in recent years has identified the most significant problems faced by students, and their patterns of seeking help in order to deal with these issues (Aluede et al., 2006; Çebi, 2009; Erkan et al., 2012; Güneri et al., 2003; Koydemir et al., 2010; Rickwood et al., 2005; Schwitzer, 2005). Schwitzer (2005) surveyed undergraduate students and revealed their major concerns to be academic issues, followed by emotional problems, and career-related concerns. Similarly, a number of studies have reported that academic, career, emotional, and social issues are the highest-ranking problems amongst college students (Atik & Yalçın, 2010; Erkan et al., 2012). Therefore, higher education institutions bear a responsibility to develop appropriate interventions for their students, and to actively provide support so as to help students deal with these difficulties (Richards, 2009). Many universities have made significant efforts to develop counseling and guidance services in response to the mental health and academic development needs of their students.

Despite the growing importance of college student counseling services (Güneri, 2006), many universities struggle with the logistical challenges of meeting the increasing demand (Kincade & Kalodner, 2004; Riva & Haub, 2004). The evolution of information and communication technologies has greatly impacted on educational reforms, and has raised the question of whether or not computer mediated communication (CMC) tools can be appropriately utilized in student counseling. Travers and Benton (2014) asserted that today's rapidly advancing technology may offer promising options for the development of innovative and effective interventions in universities. Furthermore, students' usage of CMC tools for communication and collaboration is increasing exponentially, which could therefore be leveraged in order to help students master the complex psychosocial skills of today's information age.

BACKGROUND

Online Counseling

Mallen and Vogel (2005) defined online counseling as;

Any delivery of mental and behavioral health services, including but not limited to therapy, consultation, and psychoeducation, by a licensed practitioner to a client in a non-FtF setting through distance communication technologies such as the telephone, asynchronous e-mail, synchronous chat, and videoconferencing. (p. 764)

The evolution of communication technologies has had a significant impact on CMC in both online counseling practice and academic research. This trend has affected both counselors' and their clients' attitudes towards online counseling (Richards & Viganó, 2013), with technology-based communication becoming an indispensable part of daily life. Research comparing online and face-to-face counseling has revealed that the virtual intervention modality is just as effective as the archetypal face-to-face form of counseling (Pordelan et al., 2018).

A considerable volume of theoretical and empirical literature has been published on the affordances and challenges of online counseling (Baker & Ray, 2011; Rochlen et al., 2004). Some of the highlighted advantages are a higher degree of self-disclosure (Suler, 2004), better accessibility (Mallen, Vogel, & Rochlen, 2005), convenience (Richards, 2009), cost-effectiveness (Colon & Stren, 2011), and the opportunity for richer and more accessible multimedia content (Mallen & Vogel, 2005). However, certain challenges have also been identified, including ethical issues (Richards & Viganó, 2013), privacy and security concerns (Young, 2005), lack of nonverbal cues, technical and logistical problems (Mallen, Vogel, & Rochlen, 2005), and deficiencies in counselors' computer-based skills pertinent to offering counseling through the online medium (Fenichel et al., 2002). Although some of the challenges identified stem from the communication modality itself and the technology employed, some relate to the nature of the distance between counselor and client.

Despite the criticisms and arguments regarding the challenges of online counseling, the global COVID-19 pandemic brought this form of counseling to the fore as a prominent alternative to face-to-face counseling. A recent study offered online group career counseling during the pandemic, and revealed that this type of intervention was found to be effective in terms of supporting young unemployed adults' future career goals (Santilli et al., 2021). The same study also reported participants being satisfied with the online group counseling intervention. In a similar study, online counseling during the COVID-19 pandemic was offered to individuals with an at-risk mental state, and was reported to be effective as the mental health of the participants showed significant signs of improvement (Tsai et al., 2021). In another study, researchers investigated the experiences of professionals having offered counseling during the pandemic, and although the counselors reportedly found their clients to be disengaged and less comfortable, they perceived the online counseling to be just as effective as the face-to-face method (Barker & Barker, 2021). Furthermore, researchers noted that the online intervention's effectiveness was significantly related to the participants' characteristics as well as the type of intervention. These recent studies have clearly demonstrated that online counseling can be an effective method, especially when offered during special circumstances such as a global pandemic, although it may not be appropriate for all clients and all types of intervention.

Various technologies and communication modalities have been utilized in online counseling. Most of the current research has focused upon traditional communication technologies such email, web pages, forums, and video conferencing. However, immersive 3D virtual world (3D VW) is one of the technologies that has attracted increased attention in recent years. As this technology has promising affordances in terms of communication, interaction, and user representation, it may be considered a viable method of delivering a new form of counseling service (Zack, 2011). It is therefore crucial that both researchers and practitioners investigate and understand the unique affordances of and issues related to this medium of counseling. The current study therefore aims to identify the affordances and challenges of 3D VWs in terms of group counseling, and to identify the similarities and differences between 3D VWs and the face-to-face setting.

3D Virtual World and Its Implications for Counseling

Bartle (2009) defined 3D VW as "an automated, shared, persistent environment with and through which people can interact in real time by means of a virtual self" (p. 24). However, it should be acknowledged that this covers a broad spectrum of virtual environments developed for various target audiences (Tokel & Topu, 2016). For instance, some are developed specifically for teenagers, whilst others are aimed at business and industry. Similarly, some virtual world platforms are offered as open-source applications, whilst others are proprietary software or offered as a paid online service. Although 3D VW applications come in different forms and offer various different features, the literature defines certain recurrent features. According to Dickey (2005), illusion of 3D space, avatar representation, and an interactive chat environment are the three typical features of 3D VWs. Furthermore, Warburton (2009) defined the features of persistence, simultaneous shared user space, virtual embodiment as an avatar, interaction between users and objects, immediacy of actions, and the illusion of being there through actions similar to the real world as the most common features on offer.

There has been a growing interest amongst both researchers and practitioners from a range of disciplines to leverage the interactive and immersive virtual space of 3D VWs. This type of virtual environment offers important implications for counseling interventions. First, the use of avatars ensures the visual anonymity of clients, which may help to make them feel more comfortable and secure (Evans, 2009). Second, 3D VWs are highly immersive environments, which offers another potential area of benefit in terms of the counseling process as the user's sense of physical and social presence is significantly promoted. As such, an increased feeling of presence may help to decrease the user's sense of there being "distance" between client and counselor, which may eventually eliminate any notable difference between real life (face-to-face) counselling and the virtual medium (Holmes & Foster, 2012). Level of presence also has an effect on intimacy (Kang & Gratch, 2010), commitment (Gorini et al., 2008), and the therapeutic relationship between a counselor and their clients (Riva, 2005; Suler, 2011). Third, rich and interactive 3D environments enable the utilization of a variety of multimedia formats that can help make the environment more dynamic and engaging. In the context of counseling, the use of varied media can make the whole process not only more attractive, but may also enhance the effectiveness of the intervention (Abbott et al., 2008; Barak et al., 2008). Interactive environments enable counselors to conduct either individual or group activities as a means to increasing the effectiveness of learning content, facilitating collaboration amongst group members, and in fostering increased feelings of connectedness (Barak et al., 2009). Finally, effective communication is crucial in counseling, and as such the most obvious and significant difference between virtual (online) counseling and face-to-face counseling is the lack of nonverbal cues (Barak et al., 2008; Fenichel, 2011; Mallen & Vogel, 2005; Suler, 2011). However, compared to other technology-based communication mediums, 3D VWs offer rich verbal and nonverbal communication capabilities via the use of avatars.

Despite the asserted implications of using 3D VWs for online counseling services, there has been limited research that has investigated the potential affordances and limitations in an applied setting. Therefore, the aim of the current study is to explore the affordances and limitations of the 3D virtual environment for counseling services from both the virtual and face-to-face participants' perspective. Based on the purpose of the study, the following research questions were investigated;

1. What are the perceived *affordances* of 3D VWs for group counseling?
2. What are the perceived *limitations* of 3D VWs for group counseling?
3. What are the similarities and differences between 3D and face-to-face groups in terms of *group outcomes*?

METHOD

Research Model

In this study, qualitative research methodology guided the research process as this method is considered appropriate when the aim is to understand people's perceptions about their experiences (Merriam, 2009; Stake, 2010), where no form of manipulation is applied to the research setting (Yıldırım & Şimşek, 2013), and the investigated problem lacks distinct variables that require exploration (Creswell, 2007). More specifically, the multiple case-study methodology, one of the most common types of qualitative research (Yıldırım & Şimşek, 2013) was applied, since it enables the researcher to investigate a phenomenon in depth using multiple data collection strategies and rich description of the findings (Merriam, 2009). The current study included two psychoeducational counseling groups, with the first case being a 3D virtual group whilst the second was face-to-face (FtF).

Research Context

The study was conducted in Ankara, Turkey, at the Middle East Technical University (METU), one of the country's largest public universities. Two counselors working at the university's Learning and Student Development Office (LSD) led the two counseling groups, with each having more than 2 years' experience counseling individuals and groups.

Psychoeducational groups can focus on various topics such as grief, interpersonal communication, forgiveness, or academic support, etc. In the current study, the psychoeducational group was formed to provide support for those who procrastinate. This decision was based on a needs analysis that was conducted and on meetings held with experts from the field. In addition, due to the high demand for support related to procrastination among university students, the topic was considered to be sufficiently appropriate for adequate participant numbers to be recruited to the study.

The group counseling program lasted for a period of 8 weeks for both groups, with both groups operating in parallel. The group sessions for the FtF group were conducted within the LSD's physical office space. For the virtual group, a 3D virtual counseling environment was specifically developed using OpenSim, a free and open-source system. The design and development of the environment involved researchers and counselors. Following the best practices for designing 3D virtual environment, researchers designed the environment with collaboration of counselors based on the group intervention curriculum.

Developed virtual environment was composed of these four main components;

- OpenSim Server: This component was the main server that clients connected. It was also the management tool for all server operations.
- Database Server: This component was responsible for storing all data used by the system such as regions, objects, users, groups etc. MySQL Server was used as database server.
- 3D Viewer: Viewer is the software that users used to connect to the main server and interact with 3D environment. User were required to install this software on their personal computers for connecting to the virtual world. 3D Viewer had many elements that users could use to communicate with other avatars and interact with 3D environment. The most frequently used components were movement control, communication tools, avatar appearance, maps and inventory.

- Web Panel: This web-based interface was used for managing users. In addition, users created and managed their accounts using this component.

Among the various available virtual platforms, OpenSim was selected mainly due to its flexibility and ethical concerns. First, in the provision of counseling services, client privacy and confidentiality is paramount (Gorini et al., 2008). In order achieve this, the system was managed by a researcher with control over the system's functionality. With the OpenSim platform only being open to the participants of the study, a private and confidential environment was maintained. Furthermore, as a self-hosted and self-managed platform, the researchers were able to add additional functionality to the system as required. A group session area for virtual group is shown in Figure 1.

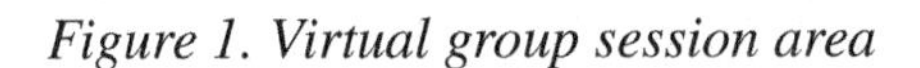

Figure 1. Virtual group session area

Participants

The study's participant selection process started with the announcement of the forming of the procrastination psychoeducational groups across the university campus, and through its website and social media accounts. A total of 48 undergraduate students applied for the program via an online application. The final participant selection strategy for the study was based on maximum variation purposeful sampling (Patton, 2002), which aims to obtain information about the phenomenon under investigation with as much diversity as possible. In total, 30 students applied for the virtual form of the program. Among these 30, nine participants were selected for the study from different departments and different academic years. For the FtF group, a total of 18 students applied. However, following the initial informative session, 12 students opted to continue attending the program, and then three left the group after the third week mainly due to scheduling conflicts. Therefore, a total of nine participants regularly attended the FtF group sessions and completed the program.

Examination of the participant demographics (see Table 1) show that both groups were balanced in terms of gender and study year. Similarly, participants from both groups were active Internet users and their average usage time were shown to be similar across both groups. Three students from each group had some previous counseling experience, whilst one participant from each group had prior experience of 3D virtual worlds. The only notable demographic difference between the two groups was with regards to the participants' active computer/video gaming experience; with six participants from the 3D group having played computer-based games, whilst only three from the FtF group had similar experience.

Table 1. Participant demographics for both groups

Demographic	Frequency (*f*)	
	3D Group	FtF Group
Gender		
Female	5	4
Male	4	5
Level		
Prep. Class	1	2
Grade 1	2	1
Grade 2	2	3
Grade 3	2	2
Grade 4	2	1
Average Weekly Internet Usage		
1-5 hours	1	2
6-10 hours	1	1
11-15 hours	2	2
16-20 hours	2	-
20+ hours	3	4
Played computer/video games	6	3
Previous counseling experience	3	3
Previous 3D VW experience	1	1

Data Collection Tools

In order to answer the study's research questions and realize an in-depth understanding of the phenomenon under investigation, semi-structured interviews were held and a questionnaire was applied as the two primary data sources.

Semi-structured interview protocols were developed for each of the two groups. Prior to employing the interview protocols, two methods were applied to enhance the instruments' credibility. First, the interview protocols were reviewed by two experts for their content validity, organization, clarity of language used, and for ethical considerations. Second, the interview forms were tested using think-aloud protocol with two undergraduate students who were not participants of the study, but were from a similar demographic background. The final version interview protocol for the 3D group consisted of 14 questions, whilst there were 11 questions for the FtF group.

The questionnaire was developed in order to collect the information regarding the participants' demographics, and also their procrastination scores using the Tuckman Procrastination Scale (TPS) (Uzun Özer et al., 2013). The original scale was developed in the English language and consisted of 16 items. A Turkish translation and adaptation of the original instrument was conducted by Uzun Özer et al. (2013), with the adapted scale consisting of 14 items. According to Uzun Özer et al. (2013), the validity and reliability measures of the scale were established through exploratory and confirmatory factor analyses which supported the scale's validity. Both the Cronbach internal consistency coefficient ($\alpha = .90$) and Pearson coefficient of stability ($r = .80$) indicated the scale to have a high level of reliability.

Data Analysis

Prior to conducting the qualitative data analyses, the audio-recorded interviews were transcribed verbatim. Qualitative data analysis was then applied according to the inductive approach (Patton, 2002). As outlined by Yıldırım and Şimşek (2013), the data analysis process followed four main steps; (1) coding data, (2) constructing themes, (3) organizing codes and themes, and (4) presenting and interpreting findings. The data analysis was conducted using MAXQDA.

In order to ensure trustworthiness of the qualitative data analysis results, several methods were followed. First, the research process was discussed with two researchers familiar with the topic of research, and they were consulted on a regular basis throughout the study. Second, for interrater reliability of the coded data, two independent experts who were familiar with the research topic coded the data, and their results were then compared to those of the researchers using MAXQDA's "intercoder agreement" feature. The intercoder level of agreement was revealed as 86% for the 3D group and 92% for the FtF group. The researchers then discussed the differences that had been revealed until reaching full consensus.

Data from the study's questionnaire were analyzed using nonparametric testing due to the small sample size. Although nonparametric tests are considered to be statistically less powerful, they have less or sometimes no assumptions regarding the data distribution, and are considered appropriate for analyzing data from a small-sized sample (Field, 2009). The Friedman test, a nonparametric equivalent of the repeated-measures analysis of variance (ANOVA), was applied in order to identify whether or not a statistically significant difference was found to exist among the participants' procrastination scores based on the pretest, posttest, and follow-up test scores for both groups.

Ethics

The first ethical consideration was to obtain approval from the university's Institutional Review Board (IRB), which was granted by the METU Ethics Committee (IRB approval: 28620816/100). Second, at the beginning of the program all of the study's participants were openly and clearly informed about the program so as to ensure that they were all fully informed prior to confirming their voluntary participation. All of the participants signed an informed consent form prior to joining the study. Third, during the data collection and data analyses processes, ethical issues were considered based on guidelines provided by Creswell (2009). During the data analysis and presentation of the study's findings, the participants' anonymity was fully maintained. Finally, in addition to general research design-related ethical standards, ethical concerns related to online counseling (Mallen, Vogel, & Rochlen, 2005; Richards & Viganó, 2013) were also carefully considered and adhered to.

FINDINGS

Affordances of 3D Virtual VWs

The affordances of 3D VWs were investigated based on the 3D group members' experience of virtual group counselling and also from the FtF group members' perception of 3D VWs. The findings revealed that both groups identified similar affordances; *comfort of self-disclosure and sharing*, *anonymity*, *interactive and rich environment*, *convenience*, and *accessible content* were the themes that emerged. The number of students and code frequencies for each perceived affordance are presented in Table 2.

Table 2. Frequencies of perceived affordances of 3D VWs by groups

3D Group ($N = 9$)		**FtF Group ($N = 9$)**	
	f		*f*
Comfort of self-disclosure / sharing	9	Comfort of self-disclosure / sharing	7
Convenience	9	Anonymity	7
Anonymity	8	Convenience	6
Interactive and rich environment	8	Interactive and rich environment	5
Accessible and reachable content	5	Accessible and reachable content	3

Comfort of Self-Disclosure and Sharing

The findings from the 3D case revealed that comfort of self-disclosure and level of sharing experiences was considered the most significant affordance of the 3D VWs, with all members of the 3D group having mentioned this factor. The participants expressed that the virtual group environment increased their comfort for self-disclosure. This affordance was considered very important, especially by those participants who felt shy or experienced a stigma for seeking psychological help. For instance, one member of the virtual group expressed that;

I am a very shy person. I care too much about what other people think and what they will think about me. I needlessly care too much, but the virtual environment helped me feel more comfortable. (3DMember1)

Similar to the 3D group, the majority of participants ($n = 7$) in the FtF group perceived comfort of self-disclosure and level of sharing experience as an affordance of the virtual group environment. The FtF group participants perceived that a virtual environment would help to increase their self-disclosure level. One participant asserted this as;

I think that group members would definitely be more comfortable there [virtual environment]. Whilst there would be no difference for me, there would be in general, because people really express things more comfortably in virtual environments that they could not otherwise express in real life [face-to-face]. (FtFMember6)

Convenience

Convenience is a well-known advantage associated with online counseling. As expected, this factor was identified as a major affordance of 3D VWs. All members of the 3D group (n = 9) and more than half of the FtF group (n = 6) shared the same opinion about the convenience of counseling within a virtual environment. Further analysis of the results showed that convenience was mentioned in three aspects; time, being able to connect from anywhere, and the comfort of where they connected from. Two of the 3D group's participants shared their experience about the comfort of connecting from their own home. They stated that it felt good to connect virtually from their home as it was considered more comfortable compared to attending face-to-face. The participants' comments about this convenience aspect were as follows;

I think there are many people that would really prefer it [virtual group] because of the time aspect. Those who do not have much spare time would likely choose to attend like that [virtual group] just because of this. (3DMember6)

Virtual is more favorable because of the time factor. For example, you can organize it at any time you want; so, in my opinion this is important. (FtFMember1)

Anonymity

Anonymity, in general, was the second most mentioned affordance within the 3D group, with the majority of participants (n = 8) considering this as a significant factor. They considered anonymity to be positive as it helped decrease their level of anxiety, helped them adapt more quickly, and did not experience any prejudice within the group environment. One interesting result revealed in the FtF group was that although the participants had no experience of anonymous counseling, most (n = 7) perceived that this factor would be considered beneficial for group counseling. Considering that virtual environments provide a unique user experience in terms of avatar representation and the ability to use nicknames, this was considered to afford complete anonymity whereby participants' visual and identity anonymity were ensured. One of the virtual group participants expressed the effect of anonymity on potential prejudices as follows;

It is a big advantage of the virtual environment, as nobody holds any prejudice simply due to my name. Because I was there from the beginning, there is no information known about me in anyone's mind. (3DMember8)

It was expressed that virtual counseling has the potential to ease the adaptation process, and to make participants more comfortable from the outset. One member of the FtF group pointed out the perceived importance of physical anonymity in the same manner as the virtual group, saying that;

It could be easier to adapt [within a virtual environment] since people aren't looking directly at you while you are speaking. This is somewhat discomforting [within a face-to-face environment]; I mean, you tend to keep moving your hands as you don't know where to place them. (FtFMember8)

Interactive and Rich Environment

The 3D interactive space is one of the virtual world's unique characteristics in comparison to traditional mediums. Users can interact with objects within the virtual environment or even manipulate them. In the current study, eight participant group members who experienced the 3D group counseling process considered this to be an affordance of the virtual environment. Further analysis showed that the interactive and visually rich 3D environment was considered to contribute to the psychoeducational counseling process in terms of creating interactive materials and activities which had a therapeutic effect on the participants. On this, two of the 3D group's participants stated:

There are many possibilities in the virtual environment in terms of materials. It is possible to open many things in a computer, with tables, surveys, and interactive boards. (3DMember2)

In the [virtual] environment, especially when on an island, a person feels good and relaxed... I don't know, even though we are not actually there, seeing it makes one feel happy. In fact, sometimes I logged in and walked around by myself for no reason. I can relate this to my liking of the [virtual] environment. (3DMember6)

Interestingly, the frequency of participants who identified this factor in the FtF group was also high; with five participants having perceived the interactive and rich environment characteristic as an affordance of the virtual environment.

It [virtual environment] could be more interactive...There are unlimited resources there, whereas it was a bit limited here [face-to-face environment]. More effective things could be conducted there [virtual environment]. (FtFMember2)

Accessible and Reachable Content

The accessible and reachable nature of the virtual environment was a theme that emerged in both cases in the current study. While in the 3D group more than half of the participants ($n = 5$) noted this as a benefit of the 3D virtual environment, only three participants from the FtF group perceived this as an affordance. Regarding this affordance, one member of each group highlighted the benefits of reachable content and activities in the same manner.

The fact that activities and presentations were there [in the virtual environment] always helped me a lot. When I missed something, it helped me as I could recheck them after the session. (3DMember5)

It can be more permanent in 3D. Here [face-to-face environment], it is happening instantaneously; we are here for 1.5 hours, it is like the process flows faster. Maybe they could reach the content more easily in 3D. (FtFMember9)

Challenges of 3D VWs

The perceived challenges of 3D VWs in the group counseling intervention context were also investigated in the study, based on the 3D group members' virtual counseling experiences and the FtF group members' perceptions. The results of the study's findings showed that whilst there were commonalities identified by the participants of both groups, some of the challenges were only mentioned by one of the groups (see Table 3). Interaction issues, lack of socialization, multitasking and trust concerns were the common emerged themes. On the other hand, technical issues and "gaming" 3D environment were themes noted only by the participants of the 3D group, whilst negative attitude towards virtual counseling and less dedication were challenges mentioned only by the FtF group's participants. Another interesting finding was that the frequency of these challenges varied between the two groups. For instance, while *lack of socialization* in the virtual environment was perceived as a major challenge by some participants in the FtF group, the same challenge was considered less significant by those in the 3D group.

Table 3. Frequencies of perceived challenges of 3D VWs by groups

3D Group (N = 9)		**FtF Group (N = 9)**	
	f		f
Technical issues	9	Lack of socialization	7
Interaction issues	8	Interaction issues	7
Lack of socialization	5	*Negative attitudes for virtual counseling*	4
Multitasking	5	Multitasking	4
Trust concerns	4	*Less dedication*	3
"Gaming" 3D Environment	3	Trust concerns	2

Technical Issues

The findings clearly showed that technical issues were the main perceived challenge of the 3D environment in terms of psychoeducational group counseling. Voice problems (n = 8), Internet connectivity problems (n = 7), and computer performance issues related to rendering the 3D environment (n = 4) were the technical issues mentioned by the participants. Although the participants did not experience these issues during every session, those participants who frequently experienced technical problems noted these issues as having affected their experience, with frustration, barrier to adopt, and feeling of isolation having been mentioned. One participant (3DMember3) stated having been less motivated to attend the group session during the week when technical problems had been experienced. Similarly, another participant highlighted the effect that such problems had on adaptation, stating;

Technical problems delayed my adaptation to the environment. (3DMember4)

Interaction Issues

The second most commented challenge related to 3D VWs was concerning interaction. Almost all of the participants (n = 8) stated experiencing interaction challenges during their counseling. These issues were mostly related to technical limitations and a lack of nonverbal cues.

The study's results showed that six of the participants considered lack of nonverbal cues as a challenge of 3D VWs, while even more participants perceived this phenomenon as a challenge in the FtF case. Based on the comments of the participants, for both cases it may be concluded that communication deficiency was the main challenge, and that it was due to a lack of nonverbal cues.

For example, while talking to you now, we have eye contact and I can see your mimics and gestures. This helps me to relax, and I consider that we are communicating better. (3DMember6)

I think that communication would be limited [within a virtual environment]. I would feel limited ability to express myself, and limited to understanding people properly because I cannot see their gestures and mimics, and I consider this very important for me. (FtFMember4)

Multitasking

Computer environments in general offer various forms of distraction for users. Unlike individual counseling sessions, the focus within group sessions is not always on the counselee. This is valid for both FtF and virtual group programs, and therefore, virtual participants may opt to multitask during their counseling sessions.

Five of the 3D group's participants reported multitasking and distractions in the computer environment as challenges of conducting psychoeducational counseling in 3D VWs. Based on their comments, playing games, talking with friends, checking social media accounts, and organizing books were the multitasking activities. Similar to the 3D group, multitasking and distractions in online environments were also perceived as challenges by the FtF group's participants. Four participants identified this challenge in the same manner; noting distraction factors as social media and gaming. On this, the following participant underlined the risk of losing focus when attending virtual counseling;

Sometimes I lost focus, it is easier to lose focus due to smartphone notifications etc. In the face-to-face environment, we cannot go out to check our smartphone. However, in the virtual environment, we could more easily lose focus with such minor distractions. (3DMember5)

Lack of Socialization

Participants in the 3D group used avatars and nicknames during their counselling sessions. It is clear that distance communication over a technological medium is quite different to that of face-to-face communication. Although anonymous socialization is a construct within some virtual worlds, this was not evident in the current study. Moreover, socialization was not an intended goal of the program, and this factor was verbalized only by members who had certain social expectations. As expected, some members of the FtF group expressed that they had certain socialization expectations from attending a group counseling program.

The study's results showed that five of the 3D group participants noted lack of socialization as a challenge of 3D VWs. Interestingly, this factor was the most perceived challenge by the FtF group's participants. Due to the social relationships constructed in the FtF environment, seven members of this group expressed concerns related to this factor over other challenges. The following comments portrayed that such concerns were similarly held by the participants of both groups.

In the physical environment there would be the same content, and you always know who you are talking to. That second factor [lack of socialization] did not happen for me. In the face-to-face environment, we would gain 10 more friends after the program had ended; however, in the [3D] interactive environment, this did not happen because nobody actually knows each other. (3DMember5)

It would not be easy to establish friend relationships there [virtual environment]. It would be binding for me in terms of social perspective. I was coming here for socialization as well, but there is no such thing there [virtual environment]. (FtFMember2)

"Gaming" 3D Environment

One interesting finding was that some members of the 3D group ($n = 3$) considered the 3D environment to be highly interactive and that this prompted them to experience "gaming" within the environment. While two of the participants had prior experience using a virtual environment for gaming and rarely as a medium of procrastination, the following participant mentioned having become too immersed in the environment that their focus on the subject was lost.

No be honest, for me this experience was transformed by a fascination for the program's virtual environment. At some point, I moved away from the procrastination topic...I saw the procrastination topic as being of lesser importance as the program's [3D VW] attractiveness and gamification aspect became more valued. (3DMember9)

Trust Concerns

The findings revealed that trust concern was a perceived challenge in terms of the virtual environment for the participants of both groups. In the 3D group this was stated by almost half of the participants ($n = 4$), while it was mentioned less ($n = 2$) in the FtF group. Visual and identity anonymity led some of the participants to feel a level of uncertainty, and experience doubts about some of the other participants' actions. For instance, one participant (3DMember5) stated having doubts about other members' shared information during the first week. One key finding was that trust-related concerns existed at the beginning of the process.

I did not know who I was talking to, and this was a disadvantage at the beginning. After a while this changed and I started to trust others. (3DMember5)

Participants of the FtF group shared similar concerns, and mentioned that trust concerns could lead to miscommunication and misbehavior within the virtual group environment. For example, FtFMember9 said that "There can be communication and trust problems. People are sometimes behaving disrespectfully in the virtual environment".

Negative Attitudes for Virtual Counseling

Four participants from the FtF group highlighted the importance of attitudes towards virtual communication. They expressed that some people dislike virtual communication and would not feel comfortable within such an environment in terms of receiving counseling. However, they also highlighted the importance of personal preferences, and that some people may actually prefer virtual communication. On this, one participant commented that;

Even though I am new generation, I am against virtual things, I don't like them. I don't even like text messaging. (FtFMember1)

Interestingly, one participant from the 3D group explained their attitude from both before and during the program.

If it [FtF group] fitted my schedule, I most probably would not choose the virtual environment as an option. However, what I came across invalidated the prejudice I held. The result, even for a person like me who approached the program with some prejudice, made me realize that it was okay, that it can be good, and that it can be done. (3DMember7)

Less Dedication

This factor emerged only from the FtF group' participants. Unlike those from the 3D group, they perceived that virtual counseling participants may not feel an adequate level of responsibility to fully participate. Three members of the FtF group perceived this factor as a challenge of the 3D VWs. On this, FtFMember4 underlined the relationship between procrastination and responsibility; saying that the virtual environment could be seen as less serious, which would then decrease dedication to the program.

I think, my enthusiasm for the program would decrease within a virtual environment. I could procrastinate about my attending, and I could underestimate its benefits. Ours was a face-to-face environment, and we felt a sense of responsibility towards each other. (FtFMember4)

Group Outcomes

In order to assess the effectiveness of the counseling intervention on the participants' behavior, both groups were evaluated based on changes in their participants' procrastination behaviors according to three measures (pretest, posttest, and a follow-up test). The first Friedman test was conducted with the nine members of the 3D group, and the results revealed no significant differences in the repeated measures of their procrastination scores over time, $\chi^2(2, N = 9) = 5.200, p = .087$. Similarly, the results for the FtF group revealed that their procrastination scores also did not significantly change across the three time periods, $\chi^2(2, N = 9) = 5.543, p = .059$. The score changes for the three measure points is illustrated in

Figure 2. Although there were no statistically significant decreases in the participants' procrastination scores in either group, there was a notable similarity in the descending trend for both groups.

Figure 2. Procrastination score changes in pretest, posttest, and follow-up

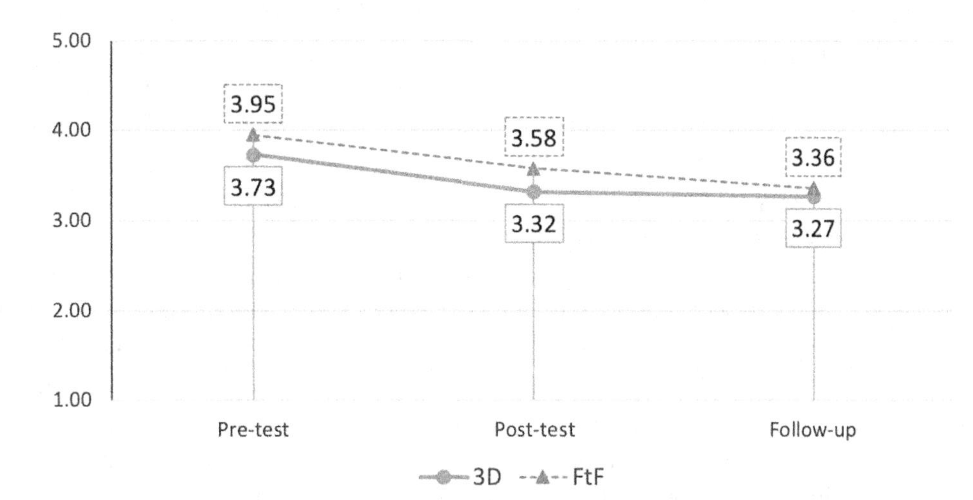

DISCUSSION AND CONCLUSION

In this study the potential affordances and challenges of 3D WVs were investigated in terms of online group counseling based on a multiple case study approach. Two counselling groups were organized as interventions, with one held face-to-face and a second as a 3D virtual group. Analysis of data collected from multiple sources was then used to address the study's research questions.

The affordances of 3D virtual environments identified for group counseling were found to parallel those of previous studies in the literature related to online counseling where these affordances were explored using other mediums of communication. The current study's results showed that self-disclosure, convenience, anonymity, interactive environment and accessible content were the most prominent affordances identified. Notably, these aspects were similarly considered by the participants of both the 3D group and those from the FtF group.

The first major affordance identified was comfort of self-disclosure and sharing within a 3D virtual environment. Previous online counseling research has explored that people feel more comfortable in revealing their thoughts, emotions, feelings, and concerns in online communication. Suler (2004, 2005) coined the term "disinhibition effect" for this phenomenon in virtual communication. In the current study, it was notable that all of the participants from the 3D group identified this particular factor. It was found that shy or timid group members especially benefitted from this characteristic of the virtual environment. Some of the participants shared their experiences in terms of comfort of self-disclosure during the counseling process that they would not otherwise have shared within the face-to-face setting. It is worth mentioning here that those participants who stated that they would also feel the same comfort level in a face-to-face environment mentioned this affordance as well. It is somewhat surprising, however, that the number of FtF group participants who identified this factor was quite high. This factor has crucial and clear implications for the counseling process, as comfort of self-disclosure can see the realization of an honest and more expressive level of communication being established between counselors and clients.

Anonymity was the second major affordance of 3D VWs that the participants identified, and by almost all of the 3D group members. Surprisingly, anonymity was also perceived as an affordance by the majority of the FtF group as well, even though they did not personally experience the anonymous counseling process. The level of anonymity, however, depends largely upon the technological medium employed, as well as the specific setup of the system. For instance, participants in video conferencing are naturally not afforded the aspect of physical anonymity. Therefore, in order to achieve full anonymity, the participants in the current study used avatars and nicknames so as to ensure they had anonymity both in physical terms and identity. One interesting finding was that visual anonymity came to prominence more than identity anonymity. The participants found anonymity to be beneficial as it helped manage their anxiety levels, made them feel somewhat safer, and helped to eliminate the potential for others to exhibit prejudices against them. This finding corroborates the arguments put forward by Richards and Viganó (2013), who stated that visual anonymity allows clients to feel safer and to express themselves more freely due to the lack of nonverbal cues indicating disapproval or judgment. This finding also corroborates Miller and Gergen's (1998) arguments; with some of the participants having highlighted the benefits of visual anonymity for eliminating potential prejudice due to a lack of social markers, especially in terms of physical attributes. Finally, physical anonymity was deemed beneficial in order to isolate negative feelings and situations from within the group environment. This finding was also considered in line with previous studies that investigated attitudes towards help-seeking and barriers to counseling services in the Turkish context, whereby a stigma was revealed against seeking psychological help or support (Güneri, 2006; Koydemir et al., 2010). In their study, Koydemir et al. (2010) found that university students in Turkey associated seeking psychological help with "sickness," and perceived it to be a sign of weakness and deemed socially unacceptable. These findings support the assertion that anonymity can decrease group members' prejudices and prejudgments towards each other in the group counseling environment.

The findings of the current study also revealed that the rich and interactive environment of 3D VWs was considered a major affordance in terms of group counseling practices. Almost all of the participants who experienced the 3D virtual environment concurred on this opinion. It is somewhat surprising, however, that more than half of the FtF participants also noted this same affordance, even though they themselves lacked any direct experience in this virtual environment. However, in their detailed responses, it can be seen that the FtF participants only expressed this affordance in terms of content presentation such as the use of interactive slides, images, and videos. On the other side, the participants of the 3D group expressed that in addition to interactivity of content presentation, the virtual environment also presents considerable potential in terms of the environment's therapeutic effects. As has been discussed in the literature by other researchers (e.g., Abbott et al., 2008; Barak et al., 2008), these findings suggest that interactive 3D environments can be more effective than other CMC tools in terms of organizing effective group activities, as well as enhancing the learning experience in psychoeducational counseling groups. Furthermore, 3D virtual environments can be designed in such a way so as to serve as a therapeutic tool in the diagnosis and treatment of various disorders such as phobias, social anxiety, and stress disorders (Gorini et al., 2008).

There is no question that convenience can be considered as one of the most significant advantages of recent technological advancements. In fact, as asserted by Rochlen et al. (2004), convenience is one of the most frequently cited affordances associated with online counseling, with its facilitation of easy access for both counselors and their clients. The results of the current study showed that all of the 3D group's participants noted convenience as an affordance of 3D VWs. Similarly, and quite expectedly, it was also the most perceived affordance factor identified by the participants of the FtF group. The study's participants considered 3D VWs to be convenient due to time-related flexibility, the ability to connect from anyplace where the technical requirements are met, and in terms of the comfort from where they connected. These findings support the previous research in this area which found that convenience was a major factor associated with online counseling (Chester & Glass, 2006; Griffiths et al., 2006; Rochlen et al., 2004). Finally, given that logistical and physical space limitations were identified as a major issue for counseling services in both the school and university setting (Güneri, 2006; Riva & Haub, 2004), the use of 3D environments has the potential to enable counselors to create more accessible environments when the available physical facilities are deemed inadequate.

The second research question of the study investigated the limitations of 3D VWs in the context of online group counseling. The findings revealed that challenges linked to interaction, lack of socialization, multitasking, and trust concerns were the factors mentioned by both the virtual group's participants and also those in the FtF group. Unlike the affordances, both groups' participants diverged somewhat when it came to identifying these limitations. Whilst technical issues and online gaming emerged as limitations mentioned only by those from the virtual group, negative attitudes towards virtual counseling and less dedication were identified only by the FtF group.

Technological issues were considered in the current study as a major challenge to the realization of online counseling services in 3D VWs by the participants of the virtual group. Compared to other more common forms of CMC tools, 3D VWs are expected to be less robust as they require a higher level of computer hardware specification and performance, plus a reliable Internet connection with adequate bandwidth. Problems with Internet connectivity, voice problems, and computer performance issues were the most common technical challenges identified. Surprisingly, however, none of the FtF group participants perceived technical issues to be a challenge. One possible explanation for this finding may be that they responded in terms of considering their current daily use of CMC tools, which are predominantly more robust in terms of being prone to technical issues. These kinds of problems are considered critical for online counseling, since any sudden disconnection may cause clients to feel abandoned, and counselors to then panic due to the interrupted process (Mallen, Vogel, Rochlen, & Day, 2005). Therefore, it is crucial that counselors be prepared to deal with unexpected situations and to address the concerns of group members where a disconnection takes place, as well as to address the concerns of those remaining members in the group, and on the possible effects to the group counselling process (Page, 2004; Suler, 2011).

The second emerged challenge related to interaction issues, which was mentioned by almost all of the participants across both groups. The main cause of interaction issues was identified as the lack of nonverbal cues, and is the most commonly discussed challenge related to online counseling in the literature (Barak et al., 2008; Fenichel, 2011; Mallen & Vogel, 2005; Suler, 2011). Whilst this may not manifest as a challenge for all CMC tools (e.g., video conferencing), it can also be seen as a challenge for other mediums too. The current study's findings have shown that virtual communication faces certain challenges compared to face-to-face group counselling. However, it can also be argued that when compared to other communication mediums, interaction using avatars can be more effective in terms of nonverbal cues. Members of both groups and their counsellors interpreted the avatar's gaze and proximity as a form of nonverbal signal, which other CMC tools lack.

Although socialization is not the primary goal of psychoeducational counselling groups, the majority of the participants in the FtF group and about half of the participants in the virtual group considered this to be a limitation of 3D virtual counseling. Although socialization whilst maintaining identity anonymity has been achieved in some virtual worlds, this was not the case in the current study's virtual group. Whilst some of the participants, particularly those who had previous counseling experience and did not have any concerns over their anonymity, expressed a desire to meet with other group members, others did not seek this due to concerns over their anonymity and privacy. These findings indicate that expectations related to socialization may depend on each participant's previous counseling experience and their level of comfort related to maintaining a certain level of anonymity.

Despite not being considered to be a major factor, trust concerns were identified as a virtual environment challenge by the current study's participants. The anonymity of virtual communication led the participants to be skeptical at the beginning of the process. However, scholars' arguments about the relationship between anonymity and trust do not show a consensus. For instance, whilst Barak et al. (2008) asserted that the anonymity of online communication can foster trust and intimacy, Suler (2011) stated the exact opposite; having argued that the lack of physical presence might reduce trust, intimacy, and thereby the effectiveness of the therapeutic relationship. In an empirical study by Erdost (2004), it was revealed that trust and intimacy scores of those using computer mediated communication were significantly lower than in face-to-face communication. The findings of the current study were also found to be in line with Suler's (2011) argument in terms of decreased levels of trust and intimacy; however, the study's participants stated that such concerns were really only present at the beginning of the group process and that they diminished after group cohesion was established.

Negative attitudes towards online counseling and dedication concerns were two perceived challenges that emerged only in the FtF group. Previous research in this area has shown that college students are more favorable toward FtF counseling services than online forms (Bathje et al., 2014; Rochlen et al., 2004). However, other research has shown that college students have positive attitudes towards online counseling (Bathje et al., 2014; Travers & Benton, 2014). A similar pattern was also seen related to counselors (Tanrikulu, 2009); although it was noted that they preferred counseling face-to-face, with online counseling considered valuable only as an adjunct method. More research on this topic, particularly in terms of the cultural context of these studies, needs to be undertaken in order to better understand the findings revealed. When it comes to dedication and commitment, Kincade and Kalodner (2004) argued that college students might have reduced levels of commitment for group counseling. However, in contrast to these arguments and the concerns of the current study's FtF participants, dedication was not mentioned as a challenge by those from the 3D group. In fact, the 3D group participants' effort and statements revealed that they acted with increased responsibility and dedication in the 3D group.

The other two challenges identified were multitasking and the "gaming" of the environment. Nearly half of the participants from each group noted there being an association with multitasking during online sessions. In a previous study, Can and Zeren (2019) found Internet addiction to be moderately correlated with academic procrastination. As the computer environment offers various forms of distraction for users, and in group sessions the focus is not always maintained on one particular person being counseled, some participants may tend to drift towards multitasking during virtual counseling sessions. One minor challenge which emerged in the 3D virtual group was "gaming" or the gamification of the environment. Some participants became overly immersed in the environment, and had started to utilize the 3D environment's features more for gaming purposes. Some scholars have noted in the limited published research in this area a need to determine client characteristics appropriate to the type of online counseling on offer

(Barak et al., 2009; Barnett, 2005; Mallen & Vogel, 2005; Richards & Viganó, 2013). These findings indicate that the multitasking tendencies and self-control of the clients, and especially any history of addictive online gaming behaviors should be considered whilst selecting group participants for counselling in 3D VW environments.

The final finding revealed in the current study related to a comparison of intervention outcomes between the two participant groups. The results showed a declining trend in the procrastination scores for the majority of both 3D and FtF participants, from pretest to posttest and follow-up test. However, this decline was not found to be statistically significant for either group. One possible explanation for this finding may relate to behavioral changes requiring a longer period of intervention. It may therefore be suggested that despite some major challenges identified in terms of virtual communication, counseling within 3D VWs can yield promising outcomes that are similar to those seen in traditional face-to-face counselling.

FUTURE RESEARCH DIRECTIONS

Current study touched one major aspect of online psychoeducational group counseling; however, counseling is a complex process and needs further investigation. Based on the findings of current study several research topics are recommended for future studies. Firstly, this study investigated the process of psychoeducational group, thus, findings should not be generalized to all counseling groups. 3D VWs can be utilized for different counseling groups as they might reveal different results. For instance, further research should investigate whether avatar-based counseling is appropriate for therapy groups. Second, online group counseling process in 3D VWs can be compared with other CMC tools as each medium has different affordances and challenges. Such studies can provide invaluable insights for practitioners and administrator for making decisions about technology to use. Third, similar studies can be conducted with participants from different demographics. For instance, similar studies can be conducted with people who need counseling services but do not have that chance due to limited resources, geographical constraints or physical disabilities. Finally, future studies might incorporate emerging virtual reality technologies along with 3D VWs and investigate their affordances and challenges for psychoeducational groups.

ACKNOWLEDGMENT

This study was based on the first author's dissertation under supervision of the second author. This research received no specific grant from any funding agency in the public, commercial, or not-for-profit sectors.

REFERENCES

Abbott, J. M., Klein, B., & Ciechomski, L. (2008). Best practices in online therapy. *Journal of Technology in Human Services*, *26*(2–4), 360–375. doi:10.1080/15228830802097257

Aluede, O., Imhonde, H., & Eguavoen, A. (2006). Academic, career and personal needs of Nigerian university students. *Journal of Instructional Psychology*, *33*(1), 50–57.

Atik, G., & Yalçın, I. (2010). Counseling needs of educational sciences students at the Ankara University. *Procedia: Social and Behavioral Sciences*, *2*(2), 1520–1526. doi:10.1016/j.sbspro.2010.03.228

Baker, K. D., & Ray, M. (2011). Online counseling: The good, the bad, and the possibilities. *Counselling Psychology Quarterly*, *24*(4), 341–346. doi:10.1080/09515070.2011.632875

Barak, A., Hen, L., Boniel-Nissim, M., & Shapira, N. (2008). A comprehensive review and a meta-analysis of the effectiveness of Internet-based psychotherapeutic interventions. *Journal of Technology in Human Services*, *26*(2–4), 109–160. doi:10.1080/15228830802094429

Barak, A., Klein, B., & Proudfoot, J. G. (2009). Defining Internet-supported therapeutic interventions. *Annals of Behavioral Medicine*, *38*(1), 4–17. doi:10.100712160-009-9130-7 PMID:19787305

Barker, G. G., & Barker, E. E. (2021). Online therapy: Lessons learned from the COVID-19 health crisis. *British Journal of Guidance & Counselling*, *50*(1), 66–81. doi:10.1080/03069885.2021.1889462

Barnett, J. E. (2005). Online counseling: New entity, new challenges. *The Counseling Psychologist*, *33*(6), 872–880. doi:10.1177/0011000005279961

Bartle, R. A. (2009). From MUDs to MMORPGs: The history of virtual worlds. In J. Hunsinger, L. Klastrup, & M. Allen (Eds.), *International Handbook of Internet Research* (pp. 23–39). Springer Netherlands. doi:10.1007/978-1-4020-9789-8_2

Bathje, G. J., Kim, E., Rau, E., Bassiouny, M. A., & Kim, T. (2014). Attitudes toward face-to-face and online counseling: Roles of self-concealment, openness to experience, loss of face, stigma, and disclosure expectations among Korean college students. *International Journal for the Advancement of Counseling*, *36*(4), 408–422. doi:10.100710447-014-9215-2

Can, S., & Zeren, Ş. G. (2019). The role of Internet addiction and basic psychological needs in explaining the academic procrastination behavior of adolescents. *Çukurova Üniversitesi Eğitim Fakültesi Dergisi*, *48*(2), 1012–1040. 10.14812/cufej.544325

Çebi, E. (2009). *University students' attitudes toward seeking psychological help: Effects of perceived social support, psychological distress, prior help-seeking experience and gender* [Doctoral Dissertation]. Middle East Technical University. https://open.metu.edu.tr/handle/11511/18743

Chester, A., & Glass, C. A. (2006). Online counselling: A descriptive analysis of therapy services on the Internet. *British Journal of Guidance & Counselling*, *34*(2), 145–160. doi:10.1080/03069880600583170

Colon, Y., & Stren, S. (2011). Counseling groups online: Theory and framework. In R. Kraus, G. Stricker, & C. Speyer (Eds.), *Online counseling: A handbook for mental health professionals* (2nd ed., pp. 183–202). Academic Press. doi:10.1016/B978-0-12-378596-1.00010-1

Creswell, J. W. (2007). *Qualitative inquiry and research design: Choosing among five approaches* (2nd ed.). Sage.

Creswell, J. W. (2009). *Research design: Qualitative, quantitative, and mixed methods approaches* (3rd ed.). Sage.

Dickey, M. D. (2005). Brave new (interactive) worlds: A review of the design affordances and constraints of two 3D virtual worlds as interactive learning environments. *Interactive Learning Environments, 13*(1–2), 121–137. doi:10.1080/10494820500173714

Erdost, T. (2004). *Trust and self-disclosure in the context of computer mediated communication* [Doctoral Dissertation]. Middle East Technical University. https://open.metu.edu.tr/handle/11511/14126

Erkan, S., Çankaya, Z. C., Terzi, S., & Özbay, Y. (2011). Üniversite psikolojik danışma ve rehberlik merkezlerinin incelenmesi. *Mehmet Akif Ersoy Üniversitesi Eğitim Fakültesi Dergisi, 11*(22), 174–198. https://dergipark.org.tr/en/pub/maeuefd/issue/19395/206011

Erkan, S., Yaşar, Ö., Cihangir-Çankaya, Z., & Terzi, Ş. (2012). University students' problem areas and psychological help-seeking willingness. *Education in Science, 37*(164), 94–107. http://egitimvebilim.ted.org.tr/index.php/EB/article/view/402

Evans, J. (2009). Online counselling and guidance skills: A practical resource for trainees and practitioners. *Sage (Atlanta, Ga.)*. Advance online publication. doi:10.4135/9781446216705

Fenichel, M. A. (2011). Online behavior, communication, and experience. In R. Kraus, G. Stricker, & C. Speyer (Eds.), *Online counseling: A handbook for mental health professionals* (2nd ed., pp. 3–18). Academic Press. doi:10.1016/B978-0-12-378596-1.00001-0

Fenichel, M. A., Suler, J., Barak, A., Zelvin, E., Jones, G., Munro, K., Meunier, V., & Walker-Schmucker, W. (2002). Myths and realities of online clinical work. *Cyberpsychology & Behavior, 5*(5), 481–497. doi:10.1089/109493102761022904 PMID:12448785

Field, A. (2009). *Discovering statistics using SPSS* (3rd ed.). Sage.

Gorini, A., Gaggioli, A., Vigna, C., & Riva, G. (2008). A second life for eHealth: Prospects for the use of 3-D virtual worlds in clinical psychology. *Journal of Medical Internet Research, 10*(3), e21. doi:10.2196/jmir.1029 PMID:18678557

Griffiths, F., Lindenmeyer, A., Powell, J., Lowe, P., & Thorogood, M. (2006). Why are health care interventions delivered over the Internet? A systematic review of the published literature. *Journal of Medical Internet Research, 8*(2), e10. doi:10.2196/jmir.8.2.e10 PMID:16867965

Güneri, O. Y. (2006). Counseling services in Turkish universities. *International Journal of Mental Health, 35*(1), 26–38. doi:10.2753/IMH0020-7411350102

Güneri, O. Y., Aydin, G., & Skovholt, T. (2003). Counseling needs of students and evaluation of counseling services at a large urban university in Turkey. *International Journal for the Advancement of Counseling, 25*(1), 53–63. doi:10.1023/A:1024928212103

Holmes, C., & Foster, V. (2012). A preliminary comparison study of online and face-to-face counseling: Client perceptions of three factors. *Journal of Technology in Human Services, 30*(1), 14–31. doi:10.1080/15228835.2012.662848

Kang, S. H., & Gratch, J. (2010). Virtual humans elicit socially anxious interactants' verbal self-disclosure. *Computer Animation and Virtual Worlds, 21*(May), 473–482. doi:10.1002/cav.345

Kincade, E. A., & Kalodner, C. R. (2004). The use of groups in college and university counseling centers. In J. L. DeLucia-Waack, D. A. Gerrity, C. R. Kalodner, & M. T. Riva (Eds.), *Handbook of group counseling and psychotherapy* (pp. 366–377). Sage., doi:10.4135/9781452229683.n26

Koydemir, S., Erel, Ö., Yumurtacı, D., & Şahin, G. N. (2010). Psychological help-seeking attitudes and barriers to help-seeking in young people in Turkey. *International Journal for the Advancement of Counseling*, *32*(4), 274–289. doi:10.100710447-010-9106-0

Mallen, M. J., & Vogel, D. L. (2005). Introduction to the major contribution: Counseling psychology and online counseling. *The Counseling Psychologist*, *33*(6), 761–775. doi:10.1177/0011000005278623

Mallen, M. J., Vogel, D. L., & Rochlen, A. B. (2005). The practical aspects of online counseling. *The Counseling Psychologist*, *33*(6), 776–818. doi:10.1177/0011000005278625

Mallen, M. J., Vogel, D. L., Rochlen, A. B., & Day, S. X. (2005). Online counseling: Reviewing the literature from a counseling psychology framework. *The Counseling Psychologist*, *33*(6), 819–871. doi:10.1177/0011000005278624

Merriam, S. B. (2009). *Qualitative research: A guide to design and implementation*. Jossey-Bass.

Miller, J. K., & Gergen, K. J. (1998). Life on the line: The therapeutic potentials of computer-mediated conversation. *Journal of Marital and Family Therapy*, *24*(2), 189–202. doi:10.1111/j.1752-0606.1998.tb01075.x PMID:9583058

Page, B. J. (2004). Online group counseling. In J. L. DeLucia-Waack, D. A. Gerrity, C. R. Kalodner, & M. T. Riva (Eds.), *Handbook of group counseling and psychotherapy* (pp. 609–620). Sage. doi:10.4135/9781452229683.n43

Patton, M. Q. (2002). *Qualitative research and evaluation methods* (3rd ed.). Sage.

Pordelan, N., Sadeghi, A., Abedi, M. R., & Kaedi, M. (2018). How online career counseling changes career development: A life design paradigm. *Education and Information Technologies*, *23*(6), 2655–2672. doi:10.100710639-018-9735-1

Richards, D. (2009). Features and benefits of online counselling: Trinity College online mental health community. *British Journal of Guidance & Counselling*, *37*(3), 231–242. doi:10.1080/03069880902956975

Richards, D., & Viganó, N. (2013). Online counseling: A narrative and critical review of the literature. *Journal of Clinical Psychology*, *69*(9), 994–1011. doi:10.1002/jclp.21974 PMID:23630010

Rickwood, D., Deane, F. P., Wilson, C. J., & Ciarrochi, J. (2005). Young people's help-seeking for mental health problems. *Advances in Mental Health*, *4*(3), 218–251. doi:10.5172/jamh.4.3.218

Riva, G. (2005). Virtual reality in psychotherapy [Review]. *Cyberpsychology & Behavior*, *8*(3), 220–230. doi:10.1089/cpb.2005.8.220 PMID:15971972

Riva, M. T., & Haub, A. L. (2004). Group counseling in the schools. In J. L. DeLucia-Waack, D. A. Gerrity, C. R. Kalodner, & M. T. Riva (Eds.), *Handbook of group counseling and psychotherapy* (pp. 309–321). Sage. doi:10.4135/9781452229683.n22

Rochlen, A. B., Zack, J. S., & Speyer, C. (2004). Online therapy: Review of relevant definitions, debates, and current empirical support. *Journal of Clinical Psychology*, *60*(3), 269–283. doi:10.1002/jclp.10263 PMID:14981791

Santilli, S., Ginevra, M. C., Di Maggio, I., Soresi, S., & Nota, L. (2021). In the same boat? An online group career counseling with a group of young adults in the time of COVID-19. *International Journal for Educational and Vocational Guidance*. Advance online publication. doi:10.100710775-021-09505-z PMID:34642592

Schwitzer, A. M. (2005). Self-development, social support, and student help-seeking. *Journal of College Student Psychotherapy*, *2*(2), 29–52. doi:10.1300/J035v20n02_04

Stake, R. E. (2010). *Qualitative research: Studying how things work*. Guilford Press.

Suler, J. R. (2004). The online disinhibition effect. *Cyberpsychology & Behavior*, *7*(3), 321–326. doi:10.1089/1094931041291295 PMID:15257832

Suler, J. R. (2005). The online disinhibitation effect. *Journal of Applied Psychoanalytic Studies*, *2*(2), 184–188. doi:10.1002/aps.42

Suler, J. R. (2011). The psychology of text relationships. In R. Kraus, G. Stricker, & C. Speyer (Eds.), *Online counseling: A handbook for mental health professionals* (2nd ed., pp. 21–53). Academic Press. doi:10.1016/B978-0-12-378596-1.00002-2

Tanrikulu, İ. (2009). Counselors-in-training students' attitudes towards online counseling. *Procedia: Social and Behavioral Sciences*, *1*(1), 785–788. doi:10.1016/j.sbspro.2009.01.140

Tokel, S. T., & Topu, F. B. (2016). Üç boyutlu sanal dünyalar [Three dimensional virtual worlds]. In K. Çağıltay & Y. Göktaş (Eds.), *Öğretim Teknolojilerinin Temelleri: Teoriler, Araştırmalar, Eğilimler* [Foundations of Instructional Technologies: Theories, Research, Trends] (2nd ed., pp. 825–844). Pegem Akademi.

Travers, M. F., & Benton, S. (2014). The acceptability of therapist-assisted, Internet-delivered treatment for college students. *Journal of College Student Psychotherapy*, *28*(1), 35–46. doi:10.1080/87568225.2014.854676

Tsai, C.-L., Tu, C.-H., Chen, J.-C., Lane, H.-Y., & Ma, W.-F. (2021). Efficiency of an online health-promotion program in individuals with at-risk mental state during the COVID-19 pandemic. *International Journal of Environmental Research and Public Health*, *18*(22), 11875. Advance online publication. doi:10.3390/ijerph182211875 PMID:34831631

Uzun Özer, B., Saçkes, M., & Tuckman, B. W. (2013). Psychometric properties of the Tuckman Procrastination Scale in a Turkish sample. *Psychological Reports*, *113*(3), 874–884. doi:10.2466/03.20.PR0.113x28z7 PMID:24693816

Warburton, S. (2009). Second Life in higher education: Assessing the potential for and the barriers to deploying virtual worlds in learning and teaching. *British Journal of Educational Technology*, *40*(3), 414–426. doi:10.1111/j.1467-8535.2009.00952.x

Yıldırım, A., & Şimşek, H. (2013). *Sosyal bilimlerde nitel araştırma yöntemleri* [Qualitative research methods in the social sciences] (9th ed.). Seçkin Yayıncılık.

Young, K. S. (2005). An empirical examination of client attitudes towards online counseling. *Cyberpsychology & Behavior*, *8*(2), 172–177. doi:10.1089/cpb.2005.8.172 PMID:15938657

Zack, J. S. (2011). The technology of online counseling. In R. Kraus, G. Stricker, & C. Speyer (Eds.), *Online counseling: A handbook for mental health professionals* (2nd ed., pp. 67–84). Academic Press. doi:10.1016/B978-0-12-378596-1.00004-6

KEY TERMS AND DEFINITIONS

3D Virtual World: A three-dimensional virtual environment in which people are represented with avatars and interact through synchronous and asynchronous communication tools. Also, it simulates the real world in terms of navigation and basic rules of physics.

Anonymity: It is the condition when acting person's real identity is not known. In virtual environment, people usually use a nickname to ensure the anonymity.

Avatar: It is a visual representation of the self that increase the sense of embodiment in the virtual environment. In a typically 3D VW, a user can customize her/his avatar's appearance and perform various actions such as walking, running, dancing, rising hand or waving.

Computer Mediated Communication (CMC): A form of communication that occurs through networked devices.

Immersion: Perception of being physically present in a virtual environment.

Online Counseling: A form of counseling in which counselor and client are physically separated and computer mediated communication tools are utilized to conduct counseling including but not limited to therapy, psychoeducation, consolation, and guidance.

Presence: A psychological state in which one feels being in the place or virtual environment rather than in the immediate physical environment.

Psychoeducational Groups: A counseling group which has a significant educational component. The goal of psychoeducational groups is to increase awareness of group members about some certain life problems and help group members develop specific skills to cope with these problems.

Chapter 6
Immersive Virtual Reality as a Tool for Education:
A Case Study

Sara Ermini
University of Siena, Italy

Giulia Collodel
https://orcid.org/0000-0003-1587-0159
University of Siena, Italy

Alessandro Innocenti
University of Siena, Italy

Maurizio Masini
https://orcid.org/0000-0003-3129-9810
GTM & Partners, Italy

Elena Moretti
University of Siena, Italy

Vincenzo Santalucia
University of Siena, Italy

ABSTRACT

After introducing the topic of education in immersive virtual reality (iVR), the authors describe the methodology and procedure used to test an educational game in virtual reality. The objective of this chapter is to contribute to the definition of a format for the evaluation of educational experiences in VR by describing the methodology adopted in the mentioned case study. A group of 30 students completed a lesson in virtual reality, and their experience was evaluated through qualitative (questionnaires, thinking aloud, interviews) and quantitative (task completion and time) tools. The results show some need for improvement of the simulation, but subjects were immersed in the experience and scored highly on the final assessment on understanding the educational content.

INTRODUCTION

The challenges posed by the Covid-19 health emergency have led the academic institutions to start an innovative process with the aim of using all the available digital technologies to support the quality of teaching, research, and student services (Baran, E., AlZoubi, D., 2020).

DOI: 10.4018/978-1-6684-4854-0.ch006

In this perspective, attention has been given to immersive Virtual Reality (iVR) technologies. The Digital Agenda for Europe, one of the seven pillars of the European 2020 Strategy, states that VR is an innovative tool that, thanks to its multisensory and immersive nature, can satisfy the principles of active learning. Immersive virtual experiences foster, in fact, the sense of presence and embodiment, both key factors that can promote learning (Wiewiorra, L., Godlovitch, I., 2021). The use of immersive devices, with different types and different grades of involvement, is gaining growing interest for university education, in which students can no longer be considered recipients who acquire knowledge passively (Makransky, G et al. 2019). VR offers three main opportunities: it can change the abstract into the tangible, supports "doing" rather than just observing, can substitute methods that are desirable but practically infeasible even, if possible, in reality (Slater M., Sanchez-Vives M.V., 2016). In particular, many pivotal processes important in the teaching of Biology, Health Sciences, Medicine, Pharmacy, Biotechnology and Languages (Hein, R., et. al. 2001-2020) degree courses are difficult to visualize and iVR simulations can support students for a deeper understanding and easy learning of concepts.

The digital game environment has become an important tool for education and training, and evidence-based theories can be increasingly found on the educational benefits of interactive digital games related to the improvement of general cognitive skills (Johnson-Glenberg, M.C., 2018; Mayer, R. E., et.al., 2002) and motivation towards the content of learning (Roussou M, 2004; Checa, D., Bustillo A., 2019; Yildirim, B., et.al., 2020). When games are compared to conventional media, there are no results that indicate that they are generally inferior to traditional education (Mayer, R. E., et.al., 2002); especially when we look at case studies in the field of health and nutrition education (Ferguson B., 2012), we can argue that games can be as effective or more effective than traditional education for certain areas and learning objectives (Fox, J., Bailenson, J.N., 2009 ; Mayer, R. E., et.al., 2002).

This paper intends to contribute to the definition of a human-centered approach (Hassenzahl, M., 2010; IDEO, 2014; ISO, 2010) for the use of iVR experiences in teaching (Johnson-Glenberg, M.C., 2018) by describing a pilot conducted at the University of Siena at the LabVR UNISI with the collaboration of the Department of Molecular and Developmental Medicine of the University of Siena to measure the validity and usability of the educational game Oxistress (Collodel, G., et. al., 2019). The topic is male fertility; in men, infertility is related to poor seminal quality, often due to the presence of inflammatory states and an increase in free radicals (ROS) that damage sperm membranes. The ability to modulate the inflammatory process and ROS formation with dietary, non-pharmacological treatments could be a desirable goal for improving male reproductive efficiency.

Our main objective is an assessment of iVR to support learning outcomes in educational context (Di Natale et al., 2020). The basis of every immersive experience is the context in which the user's senses are effectively reached by the digitally reproduced artificial stimuli [Xie et al., 2021]. To date we are mainly talking about visual and auditory stimuli, but in the near future it will also be possible to provide feedback of a kinesthetic (or haptic) type. The environments or contexts that can be used for a lesson can be environments modeled in 3D or reproductions based on three-dimensional drawings produced by a technical artist, or a professional capable of modeling. The advantage of this approach is its extreme versatility, which allows you to reproduce at simplified resolutions and therefore with shorter times. It is also possible to reproduce environments acquired with 3D scans. 3D scanning and photogrammetry techniques today represent another way to obtain fast and high quality results. For example, in an archaeological site complex and irregular natural objects can be scanned with dedicated devices (3D scanner) or reconstructed from photographs (photogrammetry).

The virtual environments must also contain the objects of the lesson, also these three-dimensionally modeled or acquired by scanning. During modeling, the different parts are modeled and placed in a hierarchical relationship between them, allowing subsequent automatic framing and isolation of parts where multimedia and didactic elements are added. In addition, the various parts that make up an object can be "exploded" and "reassembled" automatically or manually, allowing for high interactivity. Also thanks to the three-dimensional geometry it is easy to create a variety of animations that illustrate mechanical, biological or logical processes. For the student, observing the interaction of the various parts in motion, from various points of view, is a more effective formative possibility than static images in sequence. Each identified and classified part of the lesson object can be enriched with other supporting multimedia elements, allowing students to learn necessary or optional insights (Olmos-Raya, et al., 2018). The virtual space lends itself particularly to the appearance of other images, videos, text and the simultaneous reproduction of the vocal notes that the teacher has disseminated along the way [Tolentino, l. et al., 2009].

In this perspective, the purpose of the project is to implement the Flipped Classroom, that is the pedagogical model that exploits the availability of advanced learning objects to ensure that the student can initially train independently and then share with the teacher the educational content.

Materials and Methods

Participants

The study group included 30 students (aged 17 year) from Siena High School. Only one participant was excluded because of anxiety problems. Student's parents signed a consent to participate to the project. All the anti-Covid procedures indicated by the Prevention and Protection Service of the University of Siena have been respected.

Protocol

Two areas were set up inside the LabVR UNISI. The first was equipped with four workstations to test the simulation, while the second with four desktop PCs to submit the pre- and post-experience anonymous questionnaires.

Each workstation had a PC connected to an Oculus Quest 2 Head Mounted Display (HMD) via cable and the Side Quest application was opened to allow video streaming to the desktop during the game experience.

Before starting the test session, each student filled the pre-experience questionnaire to understand the degree of knowledge in the field of nutrition and male reproduction and their interest in new technologies and VR.

Then, one researcher was paired up with a student. The participant sat on a chair positioned within the already drawn safe area boundaries and wore an HMD in which the Oxistress game had already been launched. The subject was asked to use the Thinking Aloud method [8] and to press the "start" button to enter the game.

The full-version simulation took between thirty and forty minutes to be completed. During this experience, by focusing on the behaviors and emotions of participants, researchers checked the storyboard and monitors the actions in the video streaming executed by the participant to attend in case of critical events and wrote a report.

At the end of the experience, the Oxistress game was self-evaluated by the students through two post-experiences questionnaires which focuses one mainly on the sickness, the sense of presence, and usability of the experiences, the other on biological knowledge.

The participant was accompanied in the second area of the LabVR UNISI by other researchers to complete the entire questionnaire in a quiet zone using the Google Form online tool.

Figure 1. The experimental setting

Oxistress

The game was developed by LabVR UNSI team using Unity development engine. The iVR experience run on Oculus Quest headsets which have a wireless six degrees of freedom (6DOF) of movement in 3D virtual environment.

In the game, the player is immersed in the epididymis and observes closer how dietary n-3 may change the sperm plasma membrane and consequently modify sperm traits.

The user played an interactive game in which the player, to improve the sperm health of a crew of astronauts, is asked to make food choices and then discover the consequences on the main biological aspects. In the game, the user, in a miniaturised spaceship, navigates inside the reproductive system and takes a close look at the male germ cells. Having identified the ROS responsible for inflammatory states, he will have to deactivate them by eating a diet rich in antioxidants. Subsequently, a diet rich in omega

3 will reverse the damage of ROS on the spermatozoa's plasma membrane, resulting in an increase in sperm parameters.

Post-experience Questionnaire

The post-experience self-reporting questionnaire was structured into modules to evaluate different aspects of the experience concerning sickness, presence, usability, and the general emotions involved in the experience. We investigated if there were effects of discomfort during the simulation (SSQ), what level of presence was experienced (PQ) and whether the simulation was perceived as more or less usable (SUS) and the level of acquired learning and finally subjects' opinion on the educational experience lived (think aloud, open ended questions and interviews).

First, we verified if the subject has reacted negatively to the exposure to iVR. The result of the SSQ was calculated on the following formula (Kennedy, R., et. al., 2003). Participants indicate for each symptom to what degree they experience 16 sickness symptoms by using a 4-point scale (None = 0 Slight = 1 Moderate = 2 Severe = 3). These 16 factors are divided into three sub-factors (nausea, oculomotor, and disorientation) to which scores are attributed. Nausea is the sum of the answers for general discomfort, increases salivation, nausea, difficulty concentrating, stomach awareness, burping. Oculomotor is the sum of general discomfort, fatigue, headache, eyestrain, difficulty focusing, difficulty concentrating, blurred vision. Disorientation is the result of addition of difficulty focusing, nausea, fullness of head, blurred vision, dizzy (eyes open), dizzy (eyes closed), vertigo. Normal Score: Nausea = 9.54, Oculomotor = 7.5 8, Disorientation = 13.9, Normal Total Score = Nausea + Oculomotor + Disorientation, 3.74. If participants expressed slight, moderate, or severe on any of the symptoms above, they stated if felt that way before using the system and if so, explained in the questionnaire how they felt worse after using the system (Bimberg, P., Weissker, T., & Kulik, A., 2020)

Secondly, we adopted the Witmer and Singer questionnaire (Bimberg, P., Weissker, T., & Kulik, A., 2020) to quantitatively measure both the degree of presence experienced within a virtual environment by each subject and how specific elements of the experience might have influenced the intensity of the experience [28]. The scale is among the most widely used (Katy Tcha-Tokey, et. al., 2016). The scale has 7 choices with different descriptive labels for each question and each question corresponds to one of the following items: "Realism", "Sensory Fidelity", "Immersion/Adaptation", "Interface Quality", "Sounds", "Haptic". To calculate the overall score of a category, all the questions in the questionnaire referable to that factor are added together to produce the category score for each item. The total sum of the averages of all the subcategories listed is the final attendance score according to the 4-factor model proposed by Witmer et al. (2014).

The third module was about assessing the usability of the system which is measured through the System Usability Scale Questionnaire (Brooke, J., 1996). This scale has been validated also in Italian. For each of the ten sentences (Appendix 1.c) the subject must choose what best describes his reaction to the experience using the 5-point scale, 1= Absolutely disagree; 2= Disagree; 3= Neutral; 4= Agree; 5= Strongly Agree. To calculate the scores obtained from the answers we considered separately questions: Odds (1, 3, 5, 7, and 9) and Even (2, 4, 6, 8, and 10). Then we added up the total score for all odd-numbered questions, and from the total obtained subtract 5 to get (X), and the total score for all even-numbered questions, and then subtract the resulting total value from 25 to get (Y).

Lastly, we added up the total score for the new values (X + Y) and multiplied by 2.5 to get the final SUS score. This score can range from a minimum of "0" to a maximum of "100". In literature, if the resulting SUS score is greater than 68, then the system is considered above average and user friendly.

RESULTS

The pre-experience questionnaire on student's interests in VR technologies showed the following results: really interested (13.8%), quite interested (41.4%), neutral interested (31%), little interested, no interested at all (13.8). Moreover, 51.7% of the participants had never tried VR experience, while 48.3% had previous experiences. Students involved were 75.9% without glasses, 24.1% with glasses and they used similar technologies about 4-8 hour a day.

The answers related to nutrition and male reproduction on average as follows:

1. Do you pay attention to your food choices? Average: 7.32
2. Do you think your diet affect your general health? Average: 8.61
3. Have you ever heard of omega 3? Yes 27/28
4. Do you know the basis of the Mediterranean diet? Correct answer: consumption of fruit, vegetables, legumes, whole grains, dried fruit, extra virgin olive oil and fish, 64%
5. Which foods do you think have the best nutritional properties? Correct answer: fish, 71%
6. Have you heard of saturated and trans fats? Yes, 85%
7. Can eating habits affect your ability to reproduce? Yes, 82%
8. Have you ever studied the spermatozoon? Yes, 67%
9. What do you think is a reactive oxygen species (ROS)? Correct answer: molecules with an unpaired electron produced by the body due to phenomena such as oxidative stress or normal oxidation-reduction reactions in the cell, 78%
10. Have you ever heard of antioxidants? Yes, 93%
11. What do you think is the correct definition of antioxidants? Correct answer: antioxidants are substances found inside the cell that we can take in through the diet and that can inhibit oxidation processes that damage cellular material, 82%
12. In your opinion, which group of macromolecules do n-3 belong to? Correct answer: fats 46%.

In relation to biological knowledge, we obtained the following percentage of correct answer:

Based on your experience, which of the following would you say is the correct definition of a free radical? Correct answer: these are molecules with a missing electron produced by our organism due to phenomena such as oxidative stress or the normal oxidation-reduction reactions of the cell which damage the phospholipid membranes of the spermatozoa, 90%

What are the parameters that allow us to understand if the spermatozoon is healthy and efficient? Correct answer: vitality, morphology, motility, 79%

Based on your experience just now, what would you try to take in your diet to counteract free radicals? Correct answer: antioxidants, 96%

Based on your experience, which of these can be defined as antioxidants? Correct answer: carotenoids, polyphenols, flavonoids, tocopherols and vitamin C, 96%

Table 1. Participants observations, in relation to the Oxistress experience students expressed the following observations.

Visibility of system status	Sometimes ROS were not found during navigation, because students did not understand the meaning of radar, and expressed the need for visual feedback to evidence the target item. During the task of repairing the damaged spermatozoa it would be useful to display the brushing interaction and feedback on the results of it.
Flexibility and efficiency of use	For the more experienced gamers it would have been useful to be able to customize the game speed and access an advanced level.
Aesthetic and minimalist design	Students generally appreciate the graphics and the VE.
User control and freedom	Students preferred to have a back button to return to the previous module or a cancel button to undo any bad executed action during the game. With the Oculus button it was always possible to exit the experience in case of necessity. Generally, participants preferred more control over their journey and to know exactly what they have done, which tasks they still had to complete and move from backward to repeat tasks.
Recognition rather than recall	In some case the game stopped because the subjects did not remember that it was necessary to close all the panels of information and their respective in-depth studies before entering action mode. This obligation imposed by the designer to oblige the student to receive the correct full training before acting, and in cases it was forgotten causing a stop in the game. In other cases, students did not remember that the button controller should be pressed to capture a target element.
Consistency and standards	Sometimes it was necessary, in some cases, to remind the subjects which interactions were necessary to shoot or grasp the control of the spacecraft.
Help and documentation	As far as interactions are concerned, students would have preferred that the main tutorial on the use and functions of the controller buttons was done before starting the game. Some students would have appreciated more contextual 'help and documentation' during the game. Even though the virtual guide indicated the objectives to be achieved to move to the next block, in some cases the subjects did not understand the concrete actions they would have to take to achieve them.
Match between the system and the real world	Students preferred to receive a confirmation for error-prone action such as answers and choices to respond to the doctor questions.
Error prevention	Students preferred to receive a confirmation for error-prone action such as answers and choices to respond to the doctor questions.
Help users recognize, diagnose, and recover from errors	Some subjects could not tell if they had achieved the goal, as to repair the damaged spermatozoa, and which error has been done or what they need to do to complete the required action. In these cases, an error message such as "Sorry! You have no more munitions, you must recharge, capturing other 7 antioxidants".

Why are fatty acids important for sperm health? Correct answer: because they make up the majority of the spermatozoa's plasma membrane, 54%, but another answer received a high percentage "because they counteract the action of ROS"42%

Based on your experience, which of the following are the most important fatty acids for sperm health? Correct answer, polyunsaturated fatty acids (omega 3 and omega 6), 93%

Where are omega 3 fatty acids found most? Correct answer: oily fish, dried fruit, and green leafy vegetables, 100%

The percentage of total correct answers is 87%.

The final score for Simulator Sickness Questionnaire (SSQ) score can range from 0 to 236, obtained for Oxistress is 121.4. According to the subscale value in *Tab 2 (b)*, while oculomotor symptoms are quite significant, disorientation and nausea symptoms are minimal.

Table 2(a). Simulator Sickness Questionnaire (SSQ) subscales' mean score and standard deviation.

Subscales of SSQ	Oculomotor	Disorientation	Nausea
Mean± Standard deviation (SD)	11.93±3.95	10.93±3.84	9.62±2.77

Table 2 (b). Simulator Sickness Questionnaire (SSQ) answers for each subscale.

Answers	Oculomotor	Disorientation	Nausea
Mean± Standard deviation (SD)	1.70±0.56	1.56±0.55	1.37±0.40

Table 3. Simulator Sickness Questionnaire (SSQ) categorization of results

SSQ total score	categorized
0	None symphtoms
<5	Negligible symphtoms
5>10	Minimal symphtoms
10-15	Significant symphtoms
15-20	Concerning symphtoms
>20	Simulation with problems

The total score for Presence Questionnaire (PQ) is 138.7 (24.9 SD) on a possible max of 203. Among the subfactors, involvement and immersive were the ones that scored the highest values.

Table 4. Presence Questionnaire (PQ) answers in each four factors, in a 7-point scale value.

Subfactors	Involvement	Adaptation Immersion	Sensors Fidelity	Interface Quality
Mean± Standard deviation (SD)	4.9±0.9	5.0±1.7	4.2±1.0	4.9±1.2

In the System Usability Scale (SUS) questionnaire, XXX obtained an average final score of 65. To read the SUS score as an evaluation adjective: Excellent (>80.3), Good (>68<80.3), Regular (<68), Pour (51<67), Terrible (<51).

Generally, female subjects experienced a higher level of sickness and instead scored lower on presence and usability. Subjects who had already tried VR had fewer side effects, higher levels of presence, and perceived the system as more usable. Participants with glasses reported fewer phenomena of discomfort but also less usability of the system, while there were no significant differences in the sense of presence.

The open-ended responses were analysed by going to see the most frequent words and correlations with them of other terms. The most frequent terms in the final comments were experience (21); interesting (16); wonderful (5); funny (4); useful (4); improvable (4); game (3); problem (3). The keyword analysis highlights that although the simulation needed some revisions of graphic elements and technical improvements, the experience was appreciated and perceived as positive by the participants. This sense of positive interest and involvement may have influenced the achievement of learning outcomes and the results of the final test on the biological knowledge (with an average of 87% of correct answers).

The final self-evaluation of the experience assessed a regular value for sickness symptoms, sense of presence and perceived usability.

DISCUSSION

Our project is based on the belief that to exploit the potential of iVR in learning it is necessary to define an educational model based on the specific features of iVR, among which presence and immersivity are the most relevant ones. The problem of the generalizability of iVR to the real world is commonly addressed on the double dimension of immersiveness, which is an objective parameter, i.e., depending on the technical characteristics of VR, and presence, which is a subjective variable depending on the user's perception.

The goal of our project was to assess the validity and usability of the virtual game Oxistress in order to test if the VR applied to scientific data could offer a credible alternative to the traditional didactic methods.

The group examined was homogeneous as school knowledge and age and showed a fair knowledge of biological topics. The user modulated the amount and interaction of FA, antioxidants and ROS and is encouraged to understand their role in the reproductive system to make informed and easy-to-remember choices. This project was based and grounded on scientific data obtained in vivo (Castellini, C., et. al., 2019) in an animal model (the rabbit) fed a diet with n-3 FA and the consequent design idea for the dissemination of scientific results (Collodel, G., et. al. 2019). The students highlighted that the information content was generally clear and easy to understand. Only some of the information on the game procedure was said to need more detail. For example, they indicated the exact number of antioxidants to be captured (seven elements) and had an automatic counting device to help them understand how far they were from reaching the goal.

The follow-up of information after the virtual experience showed that the students had increased their knowledge, in particular the importance of a diet rich in antioxidants and Ω3 had been clarified. Perhaps the different biological roles of antioxidants and Ω3 could be better interpreted, the former counterbalancing ROS, the latter essential for restoring the functionality of cell membranes. In any case, our results confirmed the role of virtual reality as a teaching tool and perhaps suggested implementing the relevance of some of the in-depth panels in the game.

The post experience questionnaire had three modules and the criteria used for the selection of the questionnaires regard the accessibility of the all-validation process, the frequency of use, the existence or easiness of Italian translation (Tcha-Tokey, K., et. al., 2015).

In the questionnaires of sickness, presence and usability, the simulation scored on average validating the game as sufficiently user friendly but with the need to be tested again after some improvement as indicated in the report. All data from the questionnaires to the final interviews confirm that the participants showed enthusiasm and engagement, although they highlighted some issues from the point of view of the usability.

Literature has identified suitable Nielsen's Heuristics to evaluate the iVR user experience (https://www.nngroup.com/articles/usability-heuristics-virtual-reality/) and to map the issues and comments (Wang, W., Cheng, J., & Guo, J.L., 2019). Regarding the first module, sometimes participants may exhibit symptoms that have been defined as cybersickness (McCauley, M. E., Sharkey, T. J., 1992) or more wisely sickness referring to all the random symptoms that tend to occur mainly during the first iVR experience and to disappear over time. To capture this variability with which the phenomenon occurs the most used method is the Kennedy Simulator-Sickness-Questionnaire (Kennedy, R., et. al., 1993; Davis, S., Nesbitt, K., Nalivaiko, E., 2014; Rebenitsch, L., Owen, C., 2014; Kim, H. K., et. al., 2018). For the second module, according to the literature (Grassini, S., Laumann, K., 2020; Hein, A., Böhm, M., Krcmar, H., 2018), we choose the most used questionnaire is the self-reporting Presence Questionnaire of Witmer et al. (2005) that measures the degree to which individuals experience presence in virtual environments and how the specific elements of the experience could have influenced the intensity of the experience. Data related to sickness and sense of presence allowed us to understand an intermediate degree of satisfaction.

In the third module a resulting score above 68 allows to consider the system above average and (Bangor, A., Kortum, P., Miller, J., 2009), otherwise it had some usability issues. These results may be used to structure further interviews and understand how user friendly the system is. In the System Usability Scale questionnaire, the Oxistress experience obtained an average final score of 65, almost a regular evaluation. These results and the difficulties found by the students during the game will suggest the ways of improving the experience (Ericsson, K. A., 2006).

In conclusions, this paper contributes to the literature providing a concrete assessment process of the iVR educational game Oxistress. The main question addressed was how the principles of human-centered design may contribute to a better understanding of how to evaluate the validity and usability of immersive simulations to improve teaching activity and students' learning. Secondly, we tried to understand what factors make training through simulations in a virtual environment effective.

After the UX evaluation process we reported student's requests including having the possibility to consult a panel to monitor actions done and to be done, always knowing where they were in their adventure, having more visual cues to identify relevant elements (e.g., light trail to find ROS), having feedback to understand when an action has been completed (e.g. you have deactivated 6 ROS, bravo you can proceed).

As further research, we would enrich the evaluation method presented in this paper with an automatic system for recording data related to actions, in terms of timing and completion of the educational play. For this purpose, we are designing a dashboard to automatically collect objective data on experiences, such as success rate on each task, the time a task requires, the error rate and completion time according to the usability metrics indicated by Jackob Nielsen (Nielsen, N., Budiu, R., 2001). Regarding knowledge retention (Huang, K. T., et al., 2019), we intend to obtain further evidence by comparing our results with those obtained with the standard methods of teaching (book, slides, desktop multimedia content).

REFERENCES

Bangor, A., Kortum, P., & Miller, J. (2009). Determining what individual SUS scores mean: Adding an adjective rating scale. *Journal of Usability Studies*, *4*, 114–123.

Baran, E., & AlZoubi, D. (2020). Human-Centered Design as a Frame for Transition to Remote Teaching during the COVID-19 Pandemic. *Journal of Technology and Teacher Education*, *28*(2), 365–372.

Bimberg, P., Weissker, T., & Kulik, A. (2020). On the Usage of the Simulator Sickness Questionnaire for Virtual Reality Research. *2020 IEEE Conference on Virtual Reality and 3D User Interfaces Abstracts and Workshops (VRW)*, 464-467. 10.1109/VRW50115.2020.00098

Brooke, J. (1996). SUS - A quick and dirty usability scale. *Usability Eval. Ind.*, *189*, 4–7.

Castellini, C., Mattioli, S., Signorini, C., Cotozzolo, E., Noto, D., Moretti, E., Brecchia, G., Dal Bosco, A., Belmonte, G., Durand, T., De Felice, C., & Collodel, G. (2019). Effect of Dietary n-3 Source on Rabbit Male Reproduction. *Oxidative Medicine and Cellular Longevity*, *16*(70), 531-549.

Checa, D., & Bustillo, A. (2019). A review of immersive virtual reality serious games to enhance learning and training. *Multimedia Tools and Applications*, *79*, 1–27.

Collodel, G., Masini, M., Signorini, C., Moretti, E., Castellini, C., Noto, D., & Innocenti, A. (2019). Antioxidants, Dietary Fatty Acids, and Sperm: A Virtual Reality Applied Game for Scientific Dissemination. Oxidative Medicine and Cellular Longevity.

Davis, S., Nesbitt, K., & Nalivaiko, E. (2014). A Systematic Review of Cybersickness. *Proceedings of the 2014 Conference on Interactive Entertainment*, 1–9.

Di Natale, A. F., Repetto, C., Riva, G., & Villani, D. (2020). Immersive virtual reality in K-12 and higher education: A 10-year systematic review of empirical research. *British Journal of Educational Technology*, *51*(6), 2006–2033. doi:10.1111/bjet.13030

Ericsson, K. A. (2006). Protocol Analysis and Expert Thought: Concurrent Verbalizations of Thinking during Experts' Performance on Representative Tasks. In K. A. Ericsson, N. Charness, P. J. Feltovich, & R. R. Hoffman (Eds.), *The Cambridge handbook of expertise and expert performance* (pp. 223–241). Cambridge University Press. doi:10.1017/CBO9780511816796.013

Ericsson, K. A., & Simon, H. A. (1993). *Protocol Analysis: Verbal Reports as Data*. MIT Press. doi:10.7551/mitpress/5657.001.0001

Ferguson, B. (2012). The emergence of games for health. *Games for Health Journal*, *1*(1), 1–2. doi:10.1089/g4h.2012.1010 PMID:26196423

Fox, J., & Bailenson, J. N. (2009). Virtual self–modeling: The effects of vicarious reinforcement and identification on exercise behaviors. *Media Psychology*, *12*(1), 1–25. doi:10.1080/15213260802669474

Grassini, S., & Laumann, K. (2020). Are modern head-mounted displays sexist? A systematic review on gender differences in HMD-mediated virtual reality. *Frontiers in Psychology*, *7*, 11. doi:10.3389/fpsyg.2020.01604 PMID:32903791

Hassenzahl, M. (2010). Experience design: Technology for all the right reasons. In J. M. Carroll (Ed.), *Synthesis lectures on human-centered informatics* (pp. 1–95). Morgan & Claypool. doi:10.1007/978-3-031-02191-6

Hein, A., Böhm, M., & Krcmar, H. (2018). *Platform Configurations within Information Systems Research: A Literature Review on the Example of IoT Platforms*. Multikonferenz Wirtschaftsinformatik.

Hein, R. M., Wienrich, C., & Latoschik, M. E. (2021). A systematic review of foreign language learning with immersive technologies. *AIMS Electronics and Electrical Engineering*, *5*(2), 117–145. doi:10.3934/electreng.2021007

Huang, K. T., Ball, C., Francis, J., Ratan, R., Boumis, J., & Fordham, J. (2019). Augmented versus virtual reality in education: An exploratory study examining science knowledge retention when using augmented reality/virtual reality Mobile applications. *Cyberpsychology, Behavior, and Social Networking*, *22*(2), 105–110. doi:10.1089/cyber.2018.0150 PMID:30657334

IDEO. (2014). *Design Thinking for Educators toolkit*. Retrieved March 10, 2020 from http://www.ideo.com/work/toolkit-for-educators

ISO. (2010) Human-centered design for interactive systems. ISO 9241-210:2010 (E).

Jerald, J. (2015). *The VR book: Human-centered design for virtual reality*. Morgan & Claypool Publishers. doi:10.1145/2792790

Johnson-Glenberg, M. C. (2018). Immersive VR and Education: Embodied Design Principles That Include Gesture and Hand Controls. *Frontiers in Robotics and AI*, *5*, 81. doi:10.3389/frobt.2018.00081 PMID:33500960

Kennedy, R. S., Drexler, J. M., Compton, D. E., Stanney, K. M., Lanham, D. S., & Harm, D. L. (2003). Configural scoring of simulator sickness, cybersickness, and space adaptation syndrome: Similarities and differences. In L. J. Hettinger & M. W. Haas (Eds.), *Virtual and adaptive environments: Applications, implications, and human performance issues* (pp. 247–278). Lawrence Erlbaum Associates Publishers.

Kennedy, R. S., Lane, N. E., Berbaum, K. S., & Lilienthal, M. G. (1993). Simulator sickness questionnaire: An enhanced method for quantifying simulator sickness. *The International Journal of Aviation Psychology*, *3*(3), 203–220. doi:10.120715327108ijap0303_3

Kim, H. K., Park, J., Choi, Y., & Choe, M. (2018). Virtual reality sickness questionnaire (VRSQ): Motion sickness measurement index in a virtual reality environment. *Applied Ergonomics*, *69*, 66–73. doi:10.1016/j.apergo.2017.12.016 PMID:29477332

Makransky, G., Terkildsen, T. S., & Mayer, R. E. (2019). Adding immersive virtual reality to a science lab simulation causes more presence but less learning. *Learning and Instruction*, *60*, 225–236. doi:10.1016/j.learninstruc.2017.12.007

Mayer, R. E., Mautone, P., & Prothero, W. (2002). Pictorial aids for learning by doing in a multimedia geology simulation game. *Journal of Educational Psychology*, *94*(1), 171–185. doi:10.1037/0022-0663.94.1.171

McCauley, M. E., & Sharkey, T. J. (1992). Cybersickness: Perception of Self-Motion in Virtual Environments. *Presence (Cambridge, Mass.)*, *1*(3), 311–318. doi:10.1162/pres.1992.1.3.311

Nielsen, N., & Budiu, R. (2001). *Success Rate: The Simplest Usability Metric*. N/G Nielsen Norman Group.

Olmos-Raya, E., Ferreira-Cavalcanti, J., Contero, M., Castellanos, M. C., Giglioli, I. A. C., & Alcañiz, M. (2018). Mobile virtual reality as an educational platform: A pilot study on the impact of immersion and positive emotion induction in the learning process. *Eurasia Journal of Mathematics, Science and Technology Education, 14*(6), 2045–2057.

Rebenitsch, L., & Owen, C. (2014). Individual variation in susceptibility to cybersickness. In *Proceedings of the 27th annual ACM symposium on User interface software and technology* (pp. 309-317). ACM. 10.1145/2642918.2647394

Roussou, M. (2004). Learning by doing and learning through play. *Computers in Entertainment, 2*(1), 2–10. doi:10.1145/973801.973818

Sanchez-Vives, M. V., & Slater, M. (2005). From presence to consciousness through virtual reality. *Nature Reviews. Neuroscience, 6*(4), 332–339. doi:10.1038/nrn1651 PMID:15803164

Slater, M., & Sanchez-Vives, M. V. (2016). Enhancing Our Lives with Immersive Virtual Reality. *Frontiers in Robotics and AI, 3*, 3–74. doi:10.3389/frobt.2016.00074

Tcha-Tokey, K., & Loup-Escande, E., Christmann, & O., Richir, S. (2016). A questionnaire to measure the user experience in immersive virtual environments. In *Proceedings of the 2016 Virtual Reality International Conference (VRIC '16)*. Association for Computing Machinery. 10.1145/2927929.2927955

Tcha-Tokey, K., Loup-Escande, E., Christmann, O., Canac, G., Fabien, F., & Richir, S. (2015). Towards a user experience in immersive virtual environment model: A review. In *Proceedings of the 27th Conference on l'Interaction Homme-Machine (IHM 15)*. Association for Computing Machinery.

Tolentino, L., Birchfield, D., Megowan-Romanowicz, C., Johnson-Glenberg, M. C., Kelliher, A., & Martinez, C. (2009). Teaching and learning in the mixed-reality science classroom. *Journal of Science Education and Technology, 18*(6), 501–1. doi:10.100710956-009-9166-2

Wang, W., Cheng, J., & Guo, J. L. (2019). Usability of Virtual Reality Application Through the Lens of the User Community: A Case Study. *Extended Abstracts of the 2019 CHI Conference on Human Factors in Computing Systems.*

Wiewiorra, L., & Godlovitch, I. (2021). The Digital Services Act and the Digital Markets Act – a forward-looking and consumer-centred perspective, Publication for the Committee on the Internal Market and Consumer Protection, Policy Department for Economic, Scientific and Quality of Life Policies. European Parliament.

Witmer, B. G., Jerome, C. J., & Singer, M. J. (2005). The Factor Structure of the Presence Questionnaire. *Presence (Cambridge, Mass.), 14*(3), 298–312. doi:10.1162/105474605323384654

Xie, B., Liu, H., Alghofaili, R., Zhang, Y., Jiang, Y., Lobo, F. D., Li, C., Li, W., Huang, H., Akdere, M., Mousas, C., & Yu, L.-F. (2021). A Review on Virtual Reality Skill Training Applications. *Frontiers in Virtual Reality, 2*, 1–19. doi:10.3389/frvir.2021.645153

Yildirim, B., Sahin-Topalcengiz, E., Arikan, G., & Timur, S. (2020). Using virtual reality in the classroom: Reflections of STEM teachers on the use of teaching and learning tools. *Journal of Education in Science, Environment and Health, 6*(3), 231–245. doi:10.21891/jeseh.711779

KEY TERMS AND DEFINITIONS

Classroom: A room in which a class of pupils or students is taught.

Digital Technologies: The branch of scientific or engineering knowledge that deals with the creation and practical use of digital or computerized devices, methods, systems.

Education: The process of receiving or giving systematic instruction, especially at a school or university.

Gamification: The application of typical elements of game playing to other areas of activity.

Learning: The acquisition of knowledge or skills through study, experience, or being taught.

Presence: The state or fact of existing, occurring, or being present.

Sickness: The feeling or fact of being affected with nausea or vomiting.

Virtual Reality: The computer-generated simulation of a three-dimensional image or environment that can be interacted with in a seemingly real or physical way by a person using special electronic equipment, such as a helmet with a screen inside or gloves fitted with sensors.

APPENDIX

In the following section, all the UX questionnaires used for the post-experience test:

1. **BACKGROUND QUESTIONNAIRE**
   Age
   Gender
   Please fill your level of interest in Virtual Reality:
   Please explain the answer to the previous question:
   Which aspects of Virtual Reality are you most interested in?
   Have you had any previous experience with Virtual Reality?
   Are you wearing glasses today?
   How many hours do you use computers, smartphones, tablets during the day?
   Do you have any physical or cognitive impairments that may possibly stop you from participating in this study?
   If you answered yes, please let me know of physical or cognitive impairments that might prevent you from completing this study:
2. **SICKNESS QUESTIONNAIRE**
   Fatigue (weariness or exhaustion of the body)
   Headache
   Eye strain (weariness or soreness of the eye)
   Difficulty focusing
   Increased salivation
   Sweating
   Nausea Difficulty concentrating Fullness of head (sinus pressure)
   Blurred Vision
   Dizziness with closed eyes
   Dizziness with open eyes
   Vertigo (surroundings seems to swirl)
   Stomach awareness (just a short feeling of nausea)
   Burping

If you expressed slight, moderate or severe on any of the questions above, please state if you felt that way before using the system and if so, explain how you felt worse after using the system.

3. **SUS QUESTIONNAIRE**
   I think that I would like to use this CAVE frequently
   I found this CAVE unnecessarily complex
   I thought this CAVE was easy to use
   I think that I would need assistance to be able to use this CAVE experience
   I found the various functions in this CAVE were well integrated
   I thought there was too much inconsistency in the CAVE
   I would imagine that most people would learn to use this CAVE very quickly

I found the CAVE very cumbersome/awkward to use 10_I feel confident using the CAVE
I needed to learn a lot of things before I could get going with this CAVE

4. **PRESENCE QUESTIONNAIRE**

How much were you able to control the events?
How responsive was the environment to actions that you initiated (or performed?)
How natural did your interactions with the environment seem?
How much did the visual aspects of the environment involve you?
How much did the auditory aspects of the environment involve you?
How natural was the mechanism which controlled changing between the different environments?
How compelling was your sense of objects moving through space?
How much did your experiences in the virtual environment seem consistent with your real world experiences?
Were you able to anticipate what would happen next in response to the actions that you performed?
How completely were you able to actively survey or search the environment using your vision?
How compelling was your sense of moving around inside the virtual environment?
How closely were you able to examine objects?
How well could you examine objects from multiple viewpoints?
How involved were you in the virtual environment experience?
How much delay did you experience between your actions and expected outcomes?
How quickly did you adjust to the virtual environment experience?
How proficient in moving and interacting with the virtual environment did you feel at the end of experience?
How much did the visual display quality interfere or distract you from performing assigned tasks or required activities?
How much did the control devices interfere with the performance of assigned tasks or with other activities?
How well could you concentrate on the assigned tasks or required activities rather than on the mechanisms used to perform those tasks or activities?
How completely were your senses engaged in this experience?
How easy was it to identify objects through physical interaction, like touching an object, walking over a surface, or bumping into a wall or object?
Were there moments during the virtual environment experience when you felt completely focused on the task or environment?
How easily did you adjust to the control devices used to interact with the virtual environment?
Was the information provided through different senses in the virtual environment (e.g., vision, hearing, touch) consistent?

5. **OPEN ENDED QUESTION**

If you have any additional comments or suggestions, please leave a comment below. We look forward to reading from you.

Chapter 7
Digital Innovation and Interactive Technologies:
Educating the Society 5.0

Arianna Fonsati
https://orcid.org/0000-0001-8566-3425
Politecnico di Torino, Italy

ABSTRACT

Is digital innovation helping towards achieving a higher level of education or not? Since the impact of technologies is affecting more of our society, it is also true that its use in education is still limited, even in university education, where it could have the real added value of experimenting with new approaches to didactics. Within this context, the chapter briefly presents digital innovation and the enabling technologies currently in use that are also producing new opportunities for the architectural, engineering, construction, and operation (AECO) sector. Furthermore, the chapter provides two examples of master and bachelor courses related to BIM and algorithmic parametric modelling that integrates several tools and technologies, such as cloud-computing, big data, and machine learning to add value to harnessing technologies so that digital innovation could truly improve the efficiency of the AECO sector.

INTRODUCTION

The impact of digital innovation has been increasing over the last years because of the fast development of new technologies and tools. However, such a growth did not always lead to positive effects in the society. Indeed, digital innovation is so fast that it is very complex for society to stay abreast of the times. The same applies to the education sector. To better understand the situation, it is of paramount importance to analyse the "enabling technologies" that are currently producing radical shifts in markets, business practice and society. Furthermore, it is also important to examine three concepts associated to digital innovation: (i) digitization, as the conversion of something non-digital into a digital representation or artifact; (ii) digitalization, as the improvement of processes by leveraging digital technologies and digitized data, meaning that digitalization precedes digitization; (iii) digital transformation, which

DOI: 10.4018/978-1-6684-4854-0.ch007

is actual business transformation enabled or forced by digitalization technologies (Gupta, 2020). The present study analyses part of the digital transformation process for the implementation of methods that involve the use of some of the available "enabling technologies". Such enabling technologies and tools are useful for educational purposes too, in order to prepare students to be competitive within the professional market.

The other side of the coin is represented by attempts to deploy digital technologies that were not successful, resulting in many examples of implementation failures and cost overruns. Such failures tend to be analysed focusing on the technology introduced, rather than on the role played by cultural factors and by organizations' willingness to embrace new approaches and working practice, individual characteristics of team members, team feelings and organizational governance, which result to be as essential to success as deploying the right Information and Communication Technologies (ICTs) (Mahroum, Ferchachi, & Gomes, 2018). For this reason, it is not only a question of choosing the right technologies, but the implementation method of such technologies establishes the real success of one over another.

In the past, the need for innovation required the development of new methods, which in turn involved the invention of new tools. Nowadays, the opposite is often proved, because it happens more frequently that first there is the invention and later new tools are applied to innovative methods in several fields of application. Digital transformation describes the deep-seated changes in industrial and organizational activities, processes, and competencies required to seize the opportunities and respond to the challenges engendered by the new digital paradigm, including a vast array of enabling technologies, such as the Internet of Things, Additive Manufacturing, Big Data, Artificial Intelligence, Cloud Computing, Augmented and Virtual Reality, and Blockchain (Rindfleisch, O'Hern, & Sachdev, 2017). Therefore, the time needed to develop and launch innovative technologies is decreasing the lifecycle of items and services, because customers and users are always looking for flexibility and personalization of products (Li, 2018). Furthermore, digital innovation requires a cross fertilization of knowledge in different fields, forcing users to implement technologies they are not always familiar with in a very short time and to step out of their comfort zone, possibly eliminating previous practices (Saarikko, Westergren, & Blomquist, 2020). Such an integration is paramount when considering the AECO sector, where several different disciplines are involved within the development of construction projects.

Digital innovation is fostered by technologies and innovation management too; within this context people have a leading role because they also have to introduce cultural shifts and transformation to working and educational procedures to succeed. Furthermore, the importance of digital assets and the way connectivity, data, AI etc. as well as basic and advanced digital skills sustain our economies and societies have been highlighted during COVID-19 pandemic. Digital systems allowed the work and the learning to continue, tracking the spread of the virus etc. and will also "play a key role in the economic recovery as the European Council and the Commission have undertaken to frame the support to the recovery along the twin transition to a climate neutral and resilient digital transformation" (European Commission, 2020). In this framework, monitoring such a performance in digital innovation is of paramount importance. Indeed, the European Commission uses the Digital Economy and Society Index (DESI) to monitor Europe's overall digital performance and track the progress of EU countries in their digital competitiveness (European Commission, 2020). The index is evaluated through indicators across five main dimensions: (i) Connectivity; (ii) Human capital; (iii) Use of internet services; (iv) Integration of digital technology; (v) Digital public services. In the 2020 edition of DESI rankings, Italy ranked 25th out of the twenty-eight EU countries, showing significant gaps in terms of Human capital, recording very low levels of basic and advanced digital skills, which in turn are reflected in the low use of Internet services (European Commission, 2020). Furthermore, results in terms of Integration of digital technologies are very low also, proving that Italian enterprises are still lag behind in the use of technologies such as cloud-computing, big data, etc. Also, the Italian "Piano Nazionale di Ripresa e Resilienza" (PNRR), prepared to have the economic resources from the NextGenerationEU available after COVID-19 pandemic, embraces digitalization and innovation among its missions. Indeed, digitalization represents a transversal necessity, as it concerns a continuous technological update of the production processes, concerning the educational sector also, which is required to be innovated and renovated very frequently. It must involve all kind of infrastructures, from energy and transport oness, where monitoring systems with sensors and data platforms represent an innovative archetype of quality and safety management of the assets (Governo Italiano, 2021). Among the consequences of digital transformation there is the transformation of work and accordingly educational methods. Indeed, as ICT systems facilitate the implementation and monitoring of processes, they also foster the capacity to optimize and invent better methods, so knowledged workers can prove their real added value because of the "return on talent" trend.

Within this context, education represents a key-factor towards the resolution of evident criticalities linked to the lack of professionals with high digital skills, currently required in the AECO context. Indeed, the digitalization process led to the evolution of educational courses requiring the introduction of new teaching approaches, mainly based on ICT. Therefore, ICT is visible as a school training instrument that could deliver a method to re-examine and modernize the educational frameworks and procedures. Indeed, the ICT implemented laboratories has also received overwhelming response from the students, teachers, parents and administration on several occasions, showing the great potential of ICT in creating engagement (Oiha, Jebelli, & Sharifironizi, 2023) (Pienimäki, Kinnula, & Livari, 2021) (Haleema, Javaid, Qadri, & Suman, 2022). For this reason, it is evident that the adoption of ICT has potential to improve the quality of education and productivity of AECO sector also. Students who learnt the potentialities of integrated technologies applications would not come back to "inefficient" systems. Therefore, building educational frameworks that show the effectiveness of applying digital innovation is of paramount importance in order to bridge the gap between the requests of the construction industry and the higher education sector. Anyway, various challenges must be faced during this implementation, starting from the opposition of some academics. For this reason, the present study aims at presenting the educational process applied to two master courses of the faculty of Architecture at the Politecnico di Torino. Such courses are based on Building Information Modelling (BIM), parametric algorithmic modelling and building and construction management. Advanced didactics experimentation using, for instance, Virtual Reality started to analyze and discuss the projects that the students were developing as a delivery for their final exam. Virtual reality was used not only as a substitute for traditional teaching, but as an added value, which allowed students to overcome physical barriers creating an effective interaction even at a distance.

ENABLING TECHNOLOGIES

The Business Dictionary defines the enabling technologies as "the invention or innovative system that, alone or in combination with associated technologies, provides the means to drive radical change in the capabilities of a user or culture". The examples of such enabling technologies are uncountable throughout history, such as glasses and ceramics in ancient and prehistorical eras or the printing press in the classic era, which ushered in the period of modernity. However, the meaning has evolved over time;

for instance, according to the European Commission definition, enabling technologies are "knowledge-intensive technologies associated with intensive R&D, rapid innovation cycles, significant investment costs and highly-qualified labour". For this reason, enabling technologies are able to drive innovation in processes, producing radical shifts in markets, business practice and society. To define the context of the present research, it was considered fundamental to introduce the three enabling technologies that enabled the diffusion of collaboration design methods such as Building Information Modelling (BIM) and that play a role in the definition of Digital Twins, which in turn represents the future of digital applications within the AECO sector: cloud-computing, IoT and Big Data.

The first technology that enabled great improvements in society is cloud-computing, which is described as "the delivery of computing services – including servers, storage, databases, networking, software, analytics and intelligence – over the Internet" (Microsoft, 2021). The introduction of cloud computing gave more flexibility in the way firms and individuals used storage capacity, processing power and software applications, such as Google Drive or Gmail. This way companies were able to provide flexibility without the need to dedicate resources and additional investments to increase their computational power and storage capacity. Therefore, cloud-computing increased market concentration in ICT vendors, with a few but very large players leading the market. Such a trend represented a peculiar trait of the internet society, influencing also the AECO sector, which has always been "dominated" by big software houses producing tools for different purposes related to specific disciplines, such as Autodesk or Bentley.

The second enabling technology is the Internet of Things (IoT) described as "the concept of gathering information from physical objects using computer networks or accelerated wireless connections" (Dilberoglu, Gharehpapagh, Yaman, & Dolen, 2017). IoT allows objects to be seized and controlled remotely across an existing network infrastructure, creating more opportunities for a more direct integration between the physical world and the computer based one. Within the AECO sector such infrastructure is extremely important because it represents the basis for specific analysis in several fields of application, such as Structural Health Monitoring (SHM). Indeed, monitoring existing infrastructures by equipping bridges and buildings with sensors represents the basis for real digital twins. Therefore, IoT enabled products to communicate with computers in a way that information is captured and directed where it can produce more economic value. IoT also allowed changes and innovation in business models in the way companies produce economic value for customers and make money. This trend raised new managerial challenges; IoT inside products often implies the increase of its mechanical complexity, whereas companies that were not specialized in information processes were required to become expert in data analysis and software in order to be competitive on the market. The same happened in the Construction Industry, where engineers and architects are currently required to have advanced ICT-related skills, forcing companies to make large investments in the training of their employees and in the hiring of new dedicated professionals.

A world where objects producing big amount of data can be monitored and sensed remotely is a world made of Big Data. This enabling technology has provided new opportunities including the following: the chance to observe products in use, which in the AECO sector is translated in the possibility to monitor buildings and infrastructures for several purposes, from structural health up to energy performance; the use of machine learning (ML), which encompasses a series of algorithms developed between 70s and early 90s when their applications were limited by constraints in storing and processing data. ML was defined as the study of computer algorithms able to automatically detect patterns in data, and then use the uncovered patterns to predict future data; the whole system of data prevision improves automatically through experience (Mitchell, 1997) (Kodratoff, 1989). Thanks to the three enabling technologies above mentioned, ML was provided with more support; persistence, big data and a lot more computer power made it possible for deep learning to make big breakthroughs (Sejnowski, 2018). Such a connection shows there is an important complementarity among enabling technologies; the development of one technology allows the for educational purposes could enhance the effectiveness of educations itself, thanks to the improvement of students' work.

Enabling technologies play a leading role in digital innovation of course, but as Kane, Palmer, Philips, Kiron, & Buckely (2015) said the real driver of digital transformation is strategy, not technology. Indeed, even though technical limits have gradually been overcome, technology does not automatically bring added value unless companies carefully consider how to implement it within the existing processes to gain benefits. For this reason, it is important to make students aware of how to use technology efficiently. Within this context, any company "seeking to make hay of digital technology must be willing to adapt its strategies and capabilities to accommodate new ways of perceiving and creating value" (Saarikko, Westergren, & Blomquist, 2020). For this reason, new challenges have been faced by managers to remain competitive on the market because of digital innovation. Furthermore, finding the right strategy should not be taken for granted. Indeed, especially in the AECO sector, there is still a lack of frameworks able to guide companies towards the creation of value and competitive advantage by using digital technologies within their innovation activity, as well as practical solutions on how companies should use digital technologies to transform their work practices and management of knowledge resources. Appio, Frattini, Petruzzelli, & Neirotti (2021) proposed an approach to organize the fragmented debate on digital transformation and innovation management by identifying three levels of analysis (i.e., macro, meso and micro), as shown in Fig. 1. The macro-level refers to the ability of digital transformation to influence industries organization and companies interconnection, considering contextual conditions which are in turn related to social, economic, political environments; the meso-level focuses on the way companies structure their processes and capabilities to embrace digital transformation; finally, the micro-level refers to the changes in micro-foundations of companies, investigating the way digital transformation influences routines and work practices underlying innovation processes (Appio, Frattini, Petruzzelli, & Neirotti, 2021).

Figure 1. Framing Digital Transformation and Innovation Management at Multiple Levels
Source: Appio, Frattini, Petruzzelli, & Neirotti (2021)

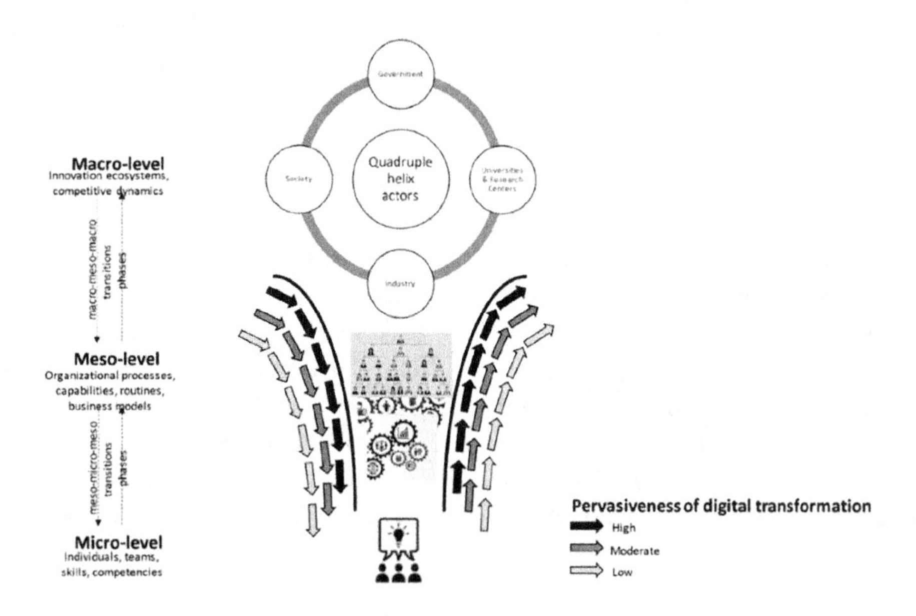

NEW OPPORTUNITIES IN THE AECO SECTOR

Within this context, the education sector should be prepared to offer a broader environment of possibilities in terms of applications useful to foster digital innovation since the very beginning of new professionals' careers. Even though innovation is a phenomenon affecting all sectors of society, the Construction industry is still lagging other fields, such as automotive or manufactoring industries. Such a gap influenced productivity also. The McKinsey Global Institute (MGI's) within the "Reinventing construction: A route to higher productivity" report, released in February 2017, highlighted the productivity problem of the AECO sector. Indeed, global labor-productivity growth in construction has averaged only 1% a year over the past two decades, against a growth of 2.8% in the world economy and 3.6% in manufacturing (McKinsey & Company, 2017). These results clearly prove that the AECO sector is underperforming, relying mainly on traditional methods for many projects rather than transforming itself to boost productivity as other sectors did in the past. For example, the US Construction industry has witnessed a decline in labor productivity since 1968, as shown in Fig. 3, while Agriculture, Manufacturing achieved quantum leaps. The compound annual growth rate between 1947 and 2010 was 4.5% for Agriculture, thanks to the land assembly, automation and advanced bioengineering to increase yields and 3.5% for Manufacturing, thanks to new concepts of flow, modularization and automation to increase production (McKinsey & Company, 2017). The report also highlights the major reasons for such a poor performance, including stringent regulations and dependence on public-sector demand, informality and high likelihood of corruption, fragmented nature of industry, inadequate design processes and underinvestment in skills development, R&D and innovation. This means that low investments in the educational sector strongly contribute to the low productivity of the sector. Furthermore, the AECO sector results one of the least digitized. Indeed, The Digital Economy and Society Index (DESI) Report 2018 Integration of Digital Technology reported that in the construction sector only 7.7% (European Commission, 2018) of enterprises have a high or very high Digital Intensity Index (DII), which measures the use of different digital technologies at company level (European Commission, 2020). Such negative trend cannot be contrasted only by digital innovation of course, but the integration of concept such as Digital Twin and processes optimization through the use of IoT, Big Data and other enabling technologies would certainly boost new opportunities for the AECO sector.

Digital Twin (DT) concept assumed more importance and meanings in recent times, shaping the future in the management of the Built Heritage. Amongst its definitions, a very interesting one is provided by Coupry et al. (2021): "A Digital Twin is a multi-scale representation of a whole consisting of a potential or existing system (physical product, user and activity) in the real environment, its virtual reflection in the digital space and the processes of automated exchange of data and information in real-time and using simulation algorithms and historical data or collected from smart sensors to predict the system's future state or its response to a given situation. A Digital Twin may also include the Digital Twins of its subsystems" (Coupry, Noblecourt, Richard, Baudry, & Bigaud, 2021). This definition takes inspiration from different observations and declination of DT (Grieves & Vickers, 2017) (Kritzinger, Karner, Traar, Henjes, & Sihn, 2018) (Aheleroff, Xu, Zhong, & Lu, 2021) (Glaessgen & Stargel, 2012) and focuses on DTs using predictive algorithms to envision the evolution of a system or its response to a specific situation. Also, Boje et al. (2020) discussed the composition of DT, introducing the DT paradigm which considers three main components interacting with each other (Fig. 3): (i) the Physical components; (ii) the Virtual models; (iii) the Data connecting the previous two elements. Within this context "the connection loop between the "Virtual-Physical" duality of the system is provided by the "Data" in its various forms" (Boje, Guerriero, Kubicki, & Rezgui, 2020).

Figure 2. Global labor-productivity growth in different sectors
Source: McKinsey & Company (2017)

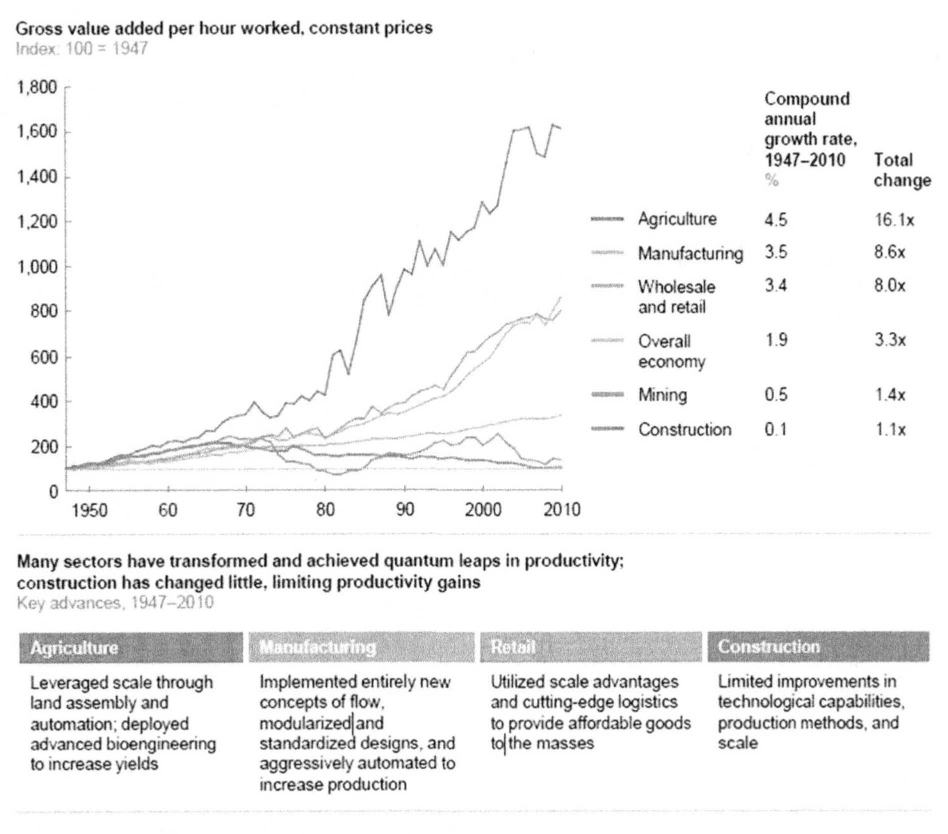

Within this context, data is the element connecting the two environments, "The Physical" and "The Digital" ones. This vision totally reflects the current trend of considering these two environments as pieces of a larger whole. A real connection between them optimizes the design and construction phases, and the asset and the management of facilities too. Furthermore, the immediate and great availability of real-time information gives the chance to use such a data in order to predict and anticipate how "The Physical" will behave on the basis of simulations performed on "The Virtual". One is linked to the other and needs its counterpart to "survive". However, despite of the potentialities of such integrated matching between physical and real, there are still some open questions on how to put it into practice, also because there is no agreement on how the different systems and technologies (BIM, InfraBIM, GIS etc.) should interact with each other.

Figure 3. The digital Twin Paradigm
Source: Adapted from Boje, Guerriero, Kubicki, & Rezgui (2020)

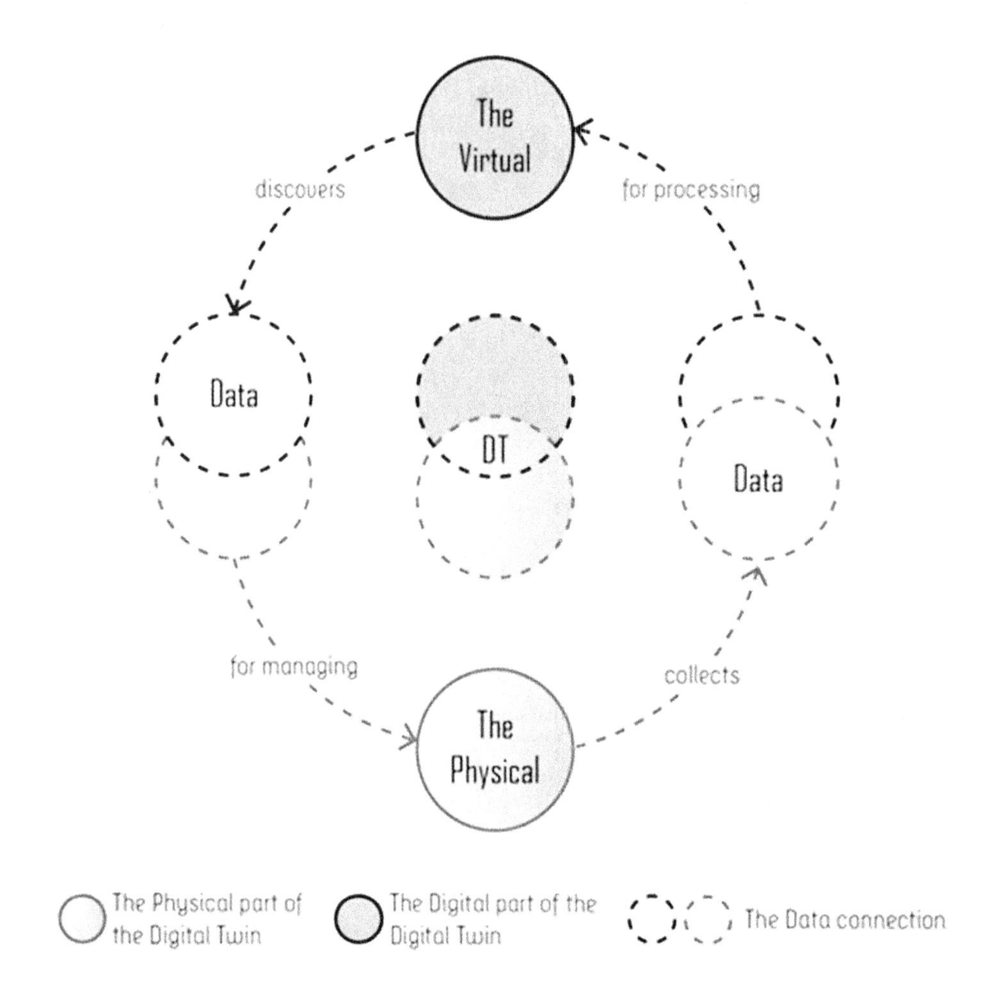

Furthermore, the increasing use of digital information models in the Construction industry allowed the integration of Computer Science, Machine Learning and Artificial Intelligence for various purposes. Certainly, these technologies dialogue better with a digital representation of the built heritage rather than the "physical" one and give the possibility to use data on buildings/cities/infrastructures included in the models to develop simulations and prefigure scenarios and related consequences. Another example of innovation producing real benefits in the AECO sector is related to the production of machine-readable data-documents, which bring automation from the design phase to the construction.

On the other hand, the other side of the coin of digital transformation is the necessity to consider that automation put at risk the labor-market. Within this context, COVID-19 pandemic did not help at all; indeed, the pandemic had a great impact on boosting productivity through automation, digitalization and reorganization of activities and operations, including a fast shift from office to at-home work, but some issues are still unsolved. Furthermore, this trend of growing productivity should spread from the large firms to those small and medium-size enterprises that have been reluctant to increase their investments in automating or digitalizing processes and operations. Indeed, such a refusal increased even more the productivity gap between large companies and smaller competitors, "exacerbating the trend towards greater inequality in economic performance across firms and regions and more market concentration" (Tyson & Mischke, 2021). To conclude, technology alone will not solve poor productivity and decline of AECO sector, because fundamental culture change is needed alongside adequate systems and processes to embrace the new opportunities previously described. However, such a culture change is possible by making the new generations of professionals understand the importance to promote continuous research towards innovation and efficiency. For this reason, education assumes a key-role in such a radical awareness.

EDUCATION AND DIGITAL INNOVATION

Considering digital transformation and its own dynamics, the education sector should better manage the shift towards new competencies and the need for innovation presented by the urge of implementing digital technologies. To resolve these pitfalls, the higher educational sector should improve both its methods and tools. In the last years, the application and use of digital technologies within the educational sector increased a lot, starting from the use of interactive multimedia whiteboards, experimenting with tablets and electronic registers to gamification and collaborative learning environments. However, digital innovation is extremely fast, and education must be able to give young people the tools they need to be safe in using technologies, such as social media, virtual reality etc. For this reason, the application of such technologies should be part of education at different levels.

Indeed, it is undeniable that the Information and Communication Technologies (ICTs) generate added and new values, approaching interaction and social behavior in a different way (Madriaga & Rubio, 2012). However, it is not just a question of technology; for instance, digital games provide entertainment by encouraging a new learning approach, peer interaction and social engagement. To achieve this objective, it is necessary to work in an interdisciplinary way in order to address important questions related to inclusive education, game design and gameplay, human-computer interaction, targeted public, teach-learning strategies, among others (Alves, Schmidt, Carthcat, & Hostins, 2015). Furthermore, educators have the role to remain aware of the student demographic; applying the historical assessment methods to digital native population, who have intimate familiarity with technology, is indeed anachronistic (Smith & Peck, 2010). As outlined by (Prensky, 2001), the digital native is attracted more to 'games' than to work; for this reason, gamification approach and serious games have experienced great development in the past years.

The next step of such procedure is represented by metaverses, considered the new interfaces that are expected to be used to carry out any type of human-computer interaction (Prieto, Lacasa, & Martínez-Borda, 2022). Such paradigm is based on Augmented Reality (AR), mirror worlds, virtual worlds, and lifelogging, as the technique of recording personal life with the help of smart devices. Within this context, is the idea of metaverse applicable to education too? Metaverses represent environments under construction, its origin comes from science fiction and is currently more like a sum of technologies (Lee, et al., 2021) such as edge computing, blockchain and Artificial Intelligence (AI). When considering the interest in youth practices, metaverses are conceptualized from the point of view of user experience, with emphasis on their dimension as narrative interfaces. Such narrative dimension could be exploited within education, to make students more directed towards learning by "experiencing" rather than learning to meet the target. Through the exploitation of digital innovation and enabling technologies the role of education could be defined, so education models could adapt and reconnect its procedures to the realities and needs of global economies and societies to create a more inclusive, cohesive, and productive world.

The experience of a global pandemic had a great effect on the educational sector, as treated in several scientific papers (Gözüm, Metin, Uzun, & Karaca, 2022) (Childs & Taylor, 2022).

Indeed, the online learning method became popular because of the situation even if it addresses several issues. First of all, the difficulty of interaction among learners and teachers was one of the biggest pitfalls of online learning (Aboagye, Yawson, & Appiah, 2020), in addition to the fact that students reported a lack of support during online learning (Kibuku, Ochieng, & Wausi, 2020). Of course, the reason of such issues is linked to the poor research and study when it comes to think of and design innovative learning systems and learning materials (Colchester, Hagras, & Alghazzawi, 2017). However, integrating digital innovation within education is more than online learning. It involves the creativity of the teacher/instructor, who should be able to design new educational approaches, based on its experience but also on the desire to engage students as much as possible.

Implementing the use of metaverse within teaching approaches becomes an opportunity, because students are in a space where avatars do everything on their behalf, thus, they need to apply their knowledge in the virtual situation. The use of metaverse system in higher education has been deepened by (Akour, Al-Maroof, Alfaisal, & Salloum, 2022) who developed a conceptual model to investigate students' perceptions towards metaverse system for educational purposes, analysing users' satisfaction and perceived trialability, observability, compatibility, and complexity. The results suggested that students' perceptions to use metaverse were significantly associated with their innovativeness, which is, in turn, influenced by perceived ease of use and perceived usefulness. For this reason, students who prefer to use an innovational technology have positive feelings towards uncertain situations and can develop more positive intentions toward it (Akour, Al-Maroof, Alfaisal, & Salloum, 2022). This study is interesting because it proves that experience and practice of metaverse could produce significant effects in terms of education, such as having the chance to travel and move around the world to learn from it. To conclude, using digital tools could help both teachers in offering a more interactive way of learning and on the other hand, students would be more impressed and facilitated in remembering facts learnt using systems such as the metaverse.

EDUCATIONAL APPROACHES INTEGRATING SEVERAL TECHNOLOGIES

As previously enounced, the idea of integrating technologies in order to produce innovative learning experiences is well founded. For this reason, after having introduced digital innovation as a phenomenon and given examples of possible application of new learning approaches for students using tools currently under implementation in the AECO sector, the goal of the present chapter is to present two examples of educational approaches slightly different from usual ones. The courses presented are currently held at the faculty of Architecture of the Politecnico di Torino. Such courses are based on Building Information Modelling (BIM), parametric algorithmic modelling and building and construction management. Within this context, advanced didactics experimentation using, for instance, Virtual Reality enabled the presentation and discussion of students' projects to be delivered for the final exam. In this case, digital tools and technologies in general are used also to overcome physical barriers, which represents one of the greatest pitfalls of online learning, to establish an effective interaction even at a distance.

The first course taken into consideration is the "Parametric Modelling" one, which is included in the MSc degree program in Architecture Construction and City at the Politecnico di Torino.

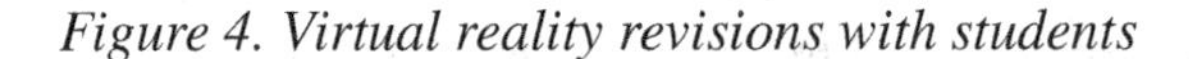
Figure 4. Virtual reality revisions with students

The course was totally held during the pandemic moment due to COVID-19, when all students attended the course remotely. In this course, tools such as VR technologies were used to better explore Building Information Models (BIM) developed by students attending, to better understand the compositional aesthetics and the architectural layout (Ugliotti, De Luca, Fonsati, Del Giudice, & Osello, 2021). The procedure for the revision can be summarized in the following steps:

1. After having downloaded the software used both by students and tutors to set the digital environment in which it is possible to meet the others using avatars, the tutor launched the functionality to visualize the model;
2. At this point, one of the students was assigned the role of master avatar, i.e. leads the review by guiding the teacher through the model with regard to the specific features of the project. In this course students used the desktop mode to navigate the model, while tutors used the HTC Vive viewers to move inside the model and check both architectural components and information related to them.
3. The project revision forced interaction among students and the tutor. However, experiencing such a way of validating and monitoring the progress of the model was strongly encouraged also during internal revision of the groups, in order to check that what was developed was correct and to discuss ideas before meeting the teacher.

Indeed, being able to navigate the models in real-time, from the inside as avatars, enhanced the collaboration and involvement of the students towards the project. However, such procedures committed students more in the development of the model itself, rather than on graphic outputs, because all necessary information must be contained inside the model. For instance, the correct visibility of objects and projects views must be set to make the revision time more effective. In addition to the geometric component, the tool is used to access file and project information. Using the controller of the HTC Vive set, the tutor was able to select and inspect parametric objects in order to validate the associated attributes as well as their geometry. Discussion became interactive because students and tutors virtually reproduced a real interaction, also thanks to the possibility to track the comments and the activities to improve the model.

Feedback from students was fundamental to define strengths and weaknesses of the experience, such as analyzing the room for improvement for the methodology. Firstly, amongst the main appreciated aspects there was the chance for a "better understanding of the model and possible modelling mistakes" due to a more interactive approach to design, which also helped to detect mistakes and uncorrected elements. Indeed, some students stated that "the result is an extremely dynamic reality where the user can interact freely and naturally and that changes according to these interactions" (Ugliotti, De Luca, Fonsati, Del Giudice, & Osello, 2021). In addition, the use of VR tools made the interaction with tutors more impactful and interesting than the traditional one, giving the chance to "meet the interlocutor as an avatar to overcome the difficulties caused by the online didactics". On the other side, the weaknesses of such procedures revealed the necessity to have high performance laptops or PCs for the use of such technologies, which means that a great effort should be put in providing the right digital devices needed by students. To conclude, such an experience represented an answer to the conditions of physical distancing due to the health emergency situation of the last years, which fostered the need to innovate the approach towards teaching activity, in the era of digitalization. The results obtained showed high potential of improvements of the adopted technologies to support teaching and their impact on students' learning processes based on critical thinking also.

The other experience reported as example of a different approach towards education in the AECO sector is the course of "Parametric and Algorithmic Modeling" part of the MSc degree program in Architecture for Sustainability at the Politecnico di Torino. The course was held after the pandemic moment due to COVID-19, after two years of remote classes, which had pros and cons as previously stated. The course was held in English and a lot of students were temporary because of the exchange programs such as the Erasmus one. The primary aim of the course was to provide the cultural, critical and operational tools necessary to introduce students to the topic of parametric modeling, investigated in its various meanings: first of all, the BIM applied to the process of design, construction and management and maintenance of buildings. Such a knowledge was integrated with VPL (Visual Programming Language) applications consistent with algorithmic approaches. The VPL, which is defined as "object-oriented logical programming language, uses algorithms by manipulating codes graphically rather than textually" (Rapetti, 2019). The theoretical developments, supported by an overview of reference legislation and the most advanced levels of international research, were supported by application activities in order to develop students' ability to use BIM in a critical and conscious way, starting from the essential concept of interoperability between software for the optimal management of information The teamwork and the application of the different concepts to real case studies will prepare students for professional practice, in line with cutting-edge experiences at international level. The teaching allows to acquire new methodologies for the control of complexity, not only in the purely formal aspects but also in the integration of different systems (spatial, structural, coating, etc.) that make up the architectural object and that allow an effective development of the project in an integrated way. Therefore, the course was built to be useful for students, also in consideration of professional practice, because the use of Parametric modeling is currently very diffuse in different sectors, both within the architectural and infrastructural design sectors for several reasons, such as the implementation of information within digital models, but also for modelling purposes, even if it is not just working with a software, but there is more to it than that. The focus of the course was on modelling with the aim not to just develop digital models using VPL, but also using it to implement information on several aspects of an architectural project. The course started from theoretical lectures not only focused on the use of parametric modelling, but treating topics such as IoT, Big Data, Digital Twin, presenting practical examples of applications of such concepts in the professional daily practice. The aim was to make students understand the real possible benefits resulting from the integrations among such disciplines. As practical part of the exam, students were asked to develop a case study not only in terms of modelling using parametric modelling but giving real possible approaches that could enhance the use of digital tools, laying the foundation for optimized and more efficient processes.

The results of the briefly described educational approaches showed that students better approached the modelling phase knowing the parameters and the details they had to consider for further developments. Fig. 5 and Fig. 6 show some graphic outputs from the groups of students who worked on this project.

Figure 5. Workflow developed by the following students: Marta Ottin Bocat, Melisa Domanic, Lorenzo Mulatero, Alp Arda

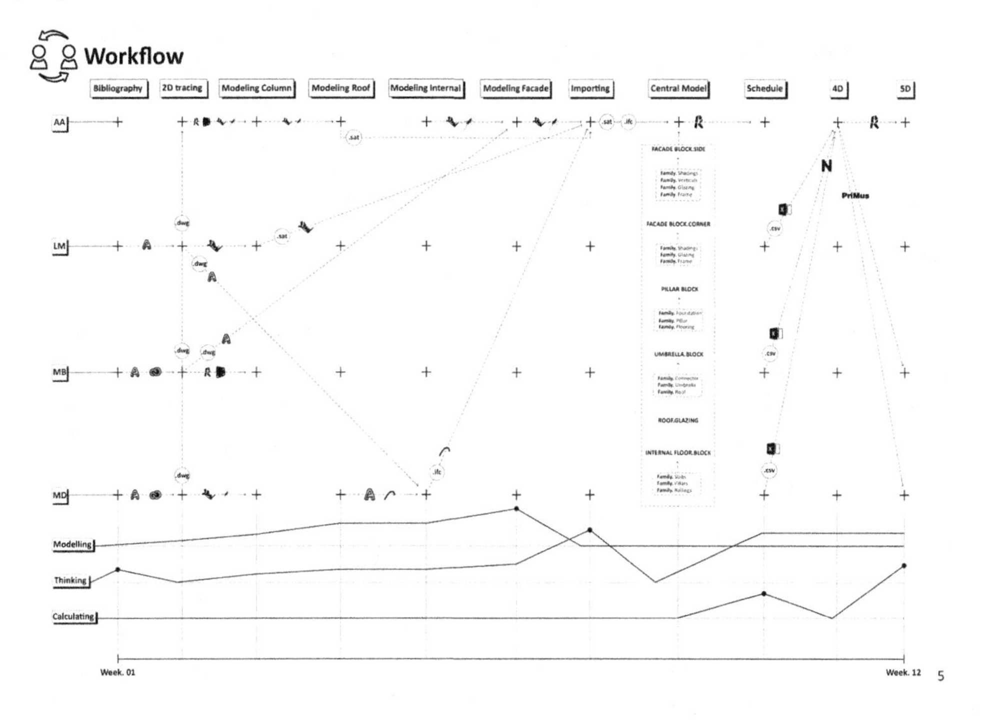

One of the most interesting considerations on these works is the fact that students tested different workflows in order to have the chance to evaluate the most efficient ones. Indeed, students showed great ability to analyze which process was the most effective according to the objective, developing their critical thinking and approaching the world of process approach (ISO 9001 series), both useful for their careers.

Figure 6. Different approaches and modelling algorithms. Developed by the following students: Francesco Maria Massetti, Mayra Bogdanova, Arda Özker Çinçin, Haihong Li, Manli Wang, Punhan Karimli, Nasrin Ghasemi

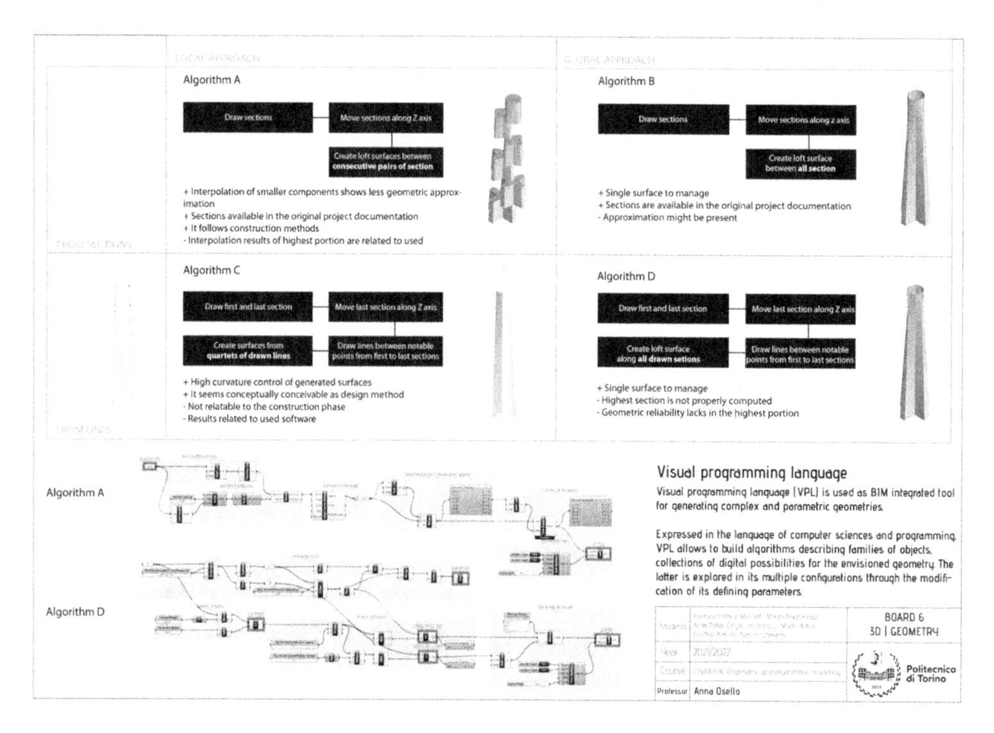

CONCLUSION

Digital innovation impact increased in the past years thanks to the introduction of progressively high-performance technologies and tools. The educational sector is moving towards this direction, even if it is highly fond of traditional teaching strategies. Some of the "enabling technologies" that could produce radical shifts in such strategies and more widely in society are cloud-computing, IoT, Big Data and Virtual and Augmented Reality (VAR). Such enabling technologies and tools are useful for educational purposes too, to train highly competitive professionals within the professional market. Indeed, digital transformation affected the society and also the educational sector from a "macro-level" point of view, which refers to the ability of digital transformation to influence industries organization and companies' interconnection, considering contextual conditions which are in turn related to social, economic, political environments. In a broader perspective, the innovation in learning techniques for higher education within construction subjects could encourage innovation in the AECO sector too, where digital innovation is still limited to the difficulties related to the change of perspectives and daily procedures. This way, engineers and architects approaching the professional environment would lead even more innovation, also towards the approach of problem-solving. In order to improve its methods, the educational sector could integrate new technologies and new ways of working, teaching since the very first years of university how to make the most out of the use of such technologies.

The chapter presented two examples of master courses focused on different ways of implementing enabling technologies within the teaching strategies. On the one side, it focuses on the use of visualization tools to better analyze projects and digital models, checking not only geometrical components but also information linked to it. On the other side, the chapter analyses the use of parametric modelling, not just as simple tool for modelling, putting together algorithms, VPL and systems for information sharing, but also the implementation workflows and its evaluation to be guided towards the most efficient application of digital tools.

To conclude, the analysis of digital innovation and the use of technologies seem to be a great way to attract students towards education. Furthermore, digital tools could help both teachers in offering a more interactive way of learning and students in being more impressed and facilitated in remembering facts learned using the previously enounced technologies. Indeed, students are inevitably linked to the digital environment and the best approach is to fully exploit all the possibilities that the digital world itself could offer.

REFERENCES

Aboagye, E., Yawson, J., & Appiah, K. (2020). COVID-19 and E-learning: The challenges of students in tertiary institutions. *Social Education Research*, 109-115.

Aheleroff, S., Xu, X., Zhong, R., & Lu, Y. (2021). Digital Twin as a Service (DTaas) in Industry 4.0: An Architecture Reference Model. *Advanced Engineering Informatics*.

Akour, I., Al-Maroof, R., Alfaisal, R., & Salloum, S. (2022). A conceptual framework for determining metaverse adoption in higher institutions of gulf area: An empirical study using hybrid SEM-ANN approach. *Computers and Education: Artificial Intelligence, 3*.

Alves, A., Schmidt, A., Carthcat, K., & Hostins, R. (2015). Exploring technological innovation towards inclusive education: Building digital games - an interdisciplinary challenge. *Procedia: Social and Behavioral Sciences*, *174*, 3081–3086. doi:10.1016/j.sbspro.2015.01.1043

Appio, F., Frattini, F., Petruzzelli, A., & Neirotti, P. (2021). Digital Transformation and Innovation Management: A Synthesis of Existing Research and an Agenda for Future Studies. *Journal of Product Innovation Management*, *38*(1), 4–20. doi:10.1111/jpim.12562

Boje, C., Guerriero, A., Kubicki, S., & Rezgui, Y. (2020). Towards a semantic Construction Digital Twin: Directions for future research. *Automation in Construction*, *114*, 103179. doi:10.1016/j.autcon.2020.103179

Childs, J., & Taylor, Z. (2022). The Internet and K-12 Education: Capturing Digital Metrics During the COVID-19 Era. Technology, Knowledge and Learning.

Colchester, K., Hagras, H., Alghazzawi, D., & Aldabbagh, G. (2017). A survey of artificial intelligence techniques employed for adaptive educational systems within E-learning platforms. *Journal of Artificial Intelligence and Soft Computing Research*, *7*(1), 47–64. doi:10.1515/jaiscr-2017-0004

Coupry, C., Noblecourt, S., Richard, P., Baudry, D., & Bigaud, D. (2021). BIM-based Digital Twin and XR Devices to Improve Maintenance Procedures in Smart Buildings: A Literature Review. *Applied Sciences (Basel, Switzerland)*, *11*(15), 6810. Advance online publication. doi:10.3390/app11156810

Dilberoglu, U., Gharehpapagh, B., Yaman, U., & Dolen, M. (2017). The Role of Additive Manufacturing in the Era of Industry 4.0. *Procedia Manufacturing*, *11*, 545–554. doi:10.1016/j.promfg.2017.07.148

European Commission. (2018). *Digital Economy and Society Index Report 2018. Integration of Digital Technologies*. European Commission.

European Commission. (2020). *Digital Economy and Society Index (DESI) 2020. Integration of digital technology*. European Commission.

Glaessgen, E., & Stargel, D. (2012). The Digital Twin Paradigm for Future NASA and U.S. Air Force Vehicles. *Structure and Dynamics*.

Governo Italiano. (2021). *Piano Nazionale di Ripresa e Resilienza*. National Recovery and Resilience Plan.

Gözüm, A., Metin, S., Uzun, H., & Karaca, N. (2022). Developing the Teacher Self-Efficacy Scale in the Use of ICT at Home for Pre-school Distance Education During Covid-19. In D. Ifenthaler (Ed.), *Technology, Knowledge and Learning*. Springer.

Grieves, M., & Vickers, J. (2017). Digital Twin: Mitigating Unpredictable, Undesirable Emergent Behavior in Complex Systems. In *Transdisciplinary Perspectives on Complex Systems* (pp. 85–113). Springer International Publishing. doi:10.1007/978-3-319-38756-7_4

Gupta, M. (2020). *What is Digitization, Digitalization, and Digital Transformation?* ARC Advisory Group. https://www.arcweb.com/blog/what-digitization-digitalization-digital-transformation

Haleema, A., Javaid, M., Qadri, M., & Suman, R. (2022). Understanding the role of digital technologies in education: A review. *Sustainable Operations and Computers*, 275-285.

Kibuku, R., Ochieng, D., & Wausi, A. (2020). E-learning challenges faced by universities in Kenya: A literature review. *Electronic Journal of E-Learning*, 150-161.

Kodratoff, Y. (1989). *Introduction to Machine Learning*. Elsevier.

Kritzinger, W., Karner, M., Traar, G., Henjes, J., & Sihn, W. (2018). Digital Twin in Manufacturing: A Categorical Literature Review and Classification. *IFAC-Pap*, 1016-1022.

Lee, L., Braud, T., Zhou, P., Wang, L., Xu, D., Lin, Z., . . . Hui, P. (2021). *From Internet and Extended Reality to Metaverse: Technology Survey, Ecosystem and Future Directions*. arXiv preprint arXiv:2110.05352.

Li, F. (2018). The digital transformation of business models in the creative industries: A holistic framework and emerging trends. *Technovation*.

Madriaga, A., & Rubio, I. (2012). *Videojuegos y discapacidad. El reto de la inclusion* [Video games and Disability. The challenge of inclusion]. Ministerio de Sanidad, Servicios Sociales e Igualdad. Secretarìa de Estado de Servicios Sociales e Igualdad. Instituto de Mayores y Servicios Sociales.

Mahroum, S., Ferchachi, N., & Gomes, A. (2018). Inside the black box. Journey mapping digital innovation in government. INSEAD. The Business School for the World.

McKinsey & Company. (2017). *Reinventing construction: A route to higher productivity*. McKinsey & Company.

Microsoft. (2021). *What is cloud computing?* Retrieved from Azure Microsoft: https://azure.microsoft.com/en-gb/overview/what-is-cloud-computing/#uses

Mitchell, T. (1997). *Machine Learning*. McGraw Hill.

Niewohner, N., Asmar, L., Wortmann, F., Roltgen, D., Kuhn, A., & Dumitrescu, R. (2019). Design fields of agile innovation management in small and medium sized enterprises. *Procedia CIRP*, *84*, 826–831. doi:10.1016/j.procir.2019.04.295

Oiha, A., Jebelli, H., & Sharifironizi, M. (2023). Understanding Students' Engagement in Learning Emerging Technologies of Construction Sector: Feasibility of Wearable Physiological Sensing Systems-Based Monitoring. In *Proceedings of the Canadian Society of Civil Engineering Annual Conference 2021*. Springer Nature Singapore.

Pienimäki, M., Kinnula, M., & Livari, N. (2021). Finding fun in non-formal technology education. *International Journal of Child-Computer Interaction*.

Prensky. (2001). Digital natives, digital immigrants. *On the Horizon*, 1-10.

Prieto, J., Lacasa, P., & Martínez-Borda, R. (2022). *Approaching metaverses: Mixed reality interfaces in youth media platforms*. New Techno-Humanities.

Rapetti, N. (2019). Strumenti e metodi per la progettazione InfraBIM. In A. Osello, A. Fonsati, N. Rapetti, & F. Semeraro (Eds.), InfraBIM. Il BIM per le infrastrutture (pp. 45-63). Roma: Gangemi Editore.

Rindfleisch, A., O'Hern, M., & Sachdev, V. (2017). The digital revolution, 3D printing, and innovation as data. *Journal of Product Innovation Management*, *34*(5), 681–690. doi:10.1111/jpim.12402

Saarikko, T., Westergren, U., & Blomquist, T. (2020). Digital transformation: Five recommendations for the digitally conscious firm. *Business Horizons*, *63*(6), 825–839. doi:10.1016/j.bushor.2020.07.005

Sejnowski, T. J. (2018). *The Deep Learning Revolution*. The MIT Press. doi:10.7551/mitpress/11474.001.0001

Smith, A., & Peck, B. (2010). The teacher as the 'digital perpetrator': Implementing web 2.0 technology activity as assessment practice for higher education Innovation or Imposition? *Procedia: Social and Behavioral Sciences*, *2*(2), 4800–4804. doi:10.1016/j.sbspro.2010.03.773

Tyson, L., & Mischke, J. (2021). *Project Syndicate*. Retrieved from Productivity After the Pandemic: https://www.project-syndicate.org/commentary/productivity-after-the-pandemic-by-laura-tyson-and-jan-mischke-2021-04

Ugliotti, M., De Luca, D., Fonsati, A., Del Giudice, M., & Osello, A. (2021). Students and teachers turn into avatars for online education. In *15th International Technology, Education and Development Conference* (pp. 4556-4565). Valencia: IATED Academy.

KEY TERMS AND DEFINITIONS

Digital Transformation: It is actual business transformation enabled or forced by digitalization technologies.

Digital Twin (DT): It is a multi-scale representation of a whole consisting of a potential or existing system (physical product, user, and activity) in the real environment, its virtual reflection in the digital space and the processes of automated exchange of data and information in real-time and using simulation algorithms and historical data or collected from smart sensors to predict the system's future state or its response to a given situation.

Digitalization: It refers to enabling or improving processes by leveraging digital technologies and digitized data, so digitalization presumes digitization.

Digitization: It is about converting something non-digital into a digital representation or artifact.

DESI: Digital Economy and Society Index.

Macro-Level of the Debate on Digital Transformation and Innovation Management: It is the ability of digital transformation to influence industries organization and companies interconnection, considering contextual conditions which are in turn related to social, economic, political environments.

Meso-Level of the Debate on Digital Transformation and Innovation Management: It focuses on the way companies structure their processes and capabilities to embrace digital transformation.

Metaverse: Is a collective virtual open space, created by the convergence of virtually enhanced physical and digital reality. It is physically persistent and provides enhanced immersive experiences.

Micro-Level of the Debate on Digital Transformation and Innovation Management: It refers to the changes in micro-foundations of companies, investigating the way digital transformation influences routines and work practices underlying innovation processes.

Section 2

Smart Environments and Systems

Chapter 8
Augmented Co-Design Methods for Climate Smart Environments:
A Critical Discourse and Historical Reflection

Pia Fricker
Aalto University, Finland

ABSTRACT

The research is based on the hypothesis that integrating site-specific and global data into the design process requires a methodological design approach, which connects local to global systems and extends the application of available predefined algorithmic scripts and singular solutions. These tools allow the designer to apprehend and simulate possible future scenarios with unparalleled precision and speed. Computational design thinking will help us master increasingly complex design challenges as well as build a profound theoretical knowledge base to meaningfully integrate current and future technologies. After re-evaluating the principles of the computational pioneers, computationally driven methods for pressing urban challenges through data-informed design speculations are discussed. Cutting-edge design speculations aim to open up new immersive design simulation and participatory processes in environmental design and urban development and give sustainable answers to societal and environmental challenges, ultimately shaping our future world.

INTRODUCTION AND HISTORICAL REFLECTION

Oversaturated with the diversity and arbitrariness of digital and social media and rapidly evolving automized design possibilities, a critically re-thinking of the future of computational design processes is needed. The ongoing process of global urbanization is affecting not only our condition of living, the social, the economic, the political, the cultural, but also the environmental beyond the Anthropocene.

DOI: 10.4018/978-1-6684-4854-0.ch008

In contrary to the ongoing discussion in the area of smart cities, we are asked to rethink our relationship towards digital and automated innovations, in order to steer towards a "technologically-mediated relationship with space" (Sayegh, 2020).

If one observes the changing of form language and complexity through technical advancements from an historical perspective, one recognizes the repetitive pattern of direct correlation. Thereby, one concludes that all relevant technical and mathematical achievements, especially with respect to the development from mechanical to electronical technologies, and even more recently the development from the electronic to the digital era, have accelerated the articulation of already present trends in the field of architecture (Oxman 2008).

The interaction of technology with society has given rise to incredible impacts on the field of architecture throughout human history. Marco Frascari describes in his article "An Age of Paper" the importance of the medium of paper and its relationship to the architecture of the 14th century (2017). Here, the medium of paper is compared with the invention of the Internet. The use of paper and its multifaceted usage in architecture, allowed for the elimination of boundaries of time and distance. This analogue element of architecture, which can be understood as the source of data, articulated in different aspects (text and numbers, sketches, images, fiction, instructions…) was almost completely eliminated by the dawn of the digital era, which allowed for the increased capacity of use of digital communication- and visualization possibilities.

The wide-reaching impact of technical developments on society can be especially seen when one compares the length of time of the past four industrial revolutions. While the first Industrial Revolution took 200 years for the invention of steam engines to replace historical agricultural societies, the second Industrial Revolution or 'Technological Revolution' where technical inventions became more widespread took only 100 years. The Third Industrial Revolution, the so-called 'Electronic Era', with a duration of ca. 50 years focused on the invention of computers and electronics. Since around 2010 we are to be found in the Fourth Industrial Revolution, which has accelerated even more rapidly into the beginning of the Fifth Industrial Revolution at the present time (Van Eerden, 2020). All these developments influenced greatly the interaction of technology with society and had wide-reaching impacts on the built environment.

The emergence of steam engines, led to the construction of massive factories to replace menial hand labor, becoming the workplace for hundreds of people, and leading the increased growth of cities. Walter Benjamin was fascinated by the Paris of the 1930s. In his uncompleted book *Passagen Werk*, or *Arcades Project*, he describes the influence of emerging technologies on the development of the city, with focus on the notion of "aesthetic affect" (Benjamin & Tiedemann, 2015). According to Prof. Dr Arie Graafland, "the notion of aesthetic affect" were so important for the philosophers of the Frankfurt School. The Frankfurt School was the first to give serious attention to mass culture, today known as 'Cultural Studies' (Graafland 2012, 14).

This time was marked by a new understanding of time and distance. Connections and communication over longer distances could be made more simply and rapidly through the development of the combustion engine and the telephone. The notion of time and spatial relationships and their interaction were thereby newly defined.

The invention of the computer and other electronic devices quickly led to a quantum leap in our handling of technology, described by van Eerden as a phase of "miniaturizing technology and personal computing" (Van Eerden 2020). This trend, the Digital Era, rose to the stage of being 'hyper-connected' through smart small-scale devices distributed over the whole globe. According to van Eerden, the Fifth

Industrial Evolution will close existing connection gaps and will make devices disappear, as "brain-computer interfaces will replace them" (Van Eerden 2020).

Creativity Must Meet Technology

Already in 2016, Klaus Schwab described our current time as being at the beginning of a revolution that is fundamentally changing the way we live, work, and relate to one another: "A time characterized by new technologies fusing the physical, digital and biological worlds" (Schwab 2016).

'Computational Design', with special focus on 'Data-Informed Design Methodologies', is a broad theme that first requires a reflection on the history of the early computational design protocols (Hensel, 2015). The integration of information technological developments in the field of architecture accelerated especially during the 60s and 70s through an intensive exchange between cybernetics and its influence on architecture (Menges, 2011). The marriage of these two areas lay in new questions related to the rise of global ecological challenges, which also changed our relationship to data and our interaction with the information it contained (Fuller, 1969; Meadows et al., 1972). At a time, pushed by technological innovations and early accessibility of computers, Nicholas Negroponte and Leon Groisser founded 1967 MIT's Architecture Machine Group (Arch MAC) – precursor of the MIT Media Lab. Outstandingly forward looking, the Arch MAC united knowledge out of the field of architecture, engineering, and computer science in order to formulate novel avenues for teaching and research (Negroponte 1998).

According to Anderson, the increasingly broad ecological consciousness, and related complex issues that it crystallized required new methods for designing architecture, "supported by memory and retrieval systems and manipulative possibilities of the computer" (Steenson n.d.). These radical expectations could be answered by the future-oriented technological infrastructure and visionary conceptional orientation of Arch MAC.

Negroponte placed great value on the multidisciplinary composition of his lab; this mixture of competencies, experience, and manners of thinking also showed in the composition of his students. Undergraduates, graduate students, PhD students and post-docs worked together on research projects in different functions. Teaching and research worked together hand-in-hand and the methodology of the projects were the pioneers of today's multidisciplinary research-led studios. Most significantly, Negroponte recognized the necessity for a new pedagogic concept for teaching architecture students how to program. Negroponte experimented with the potential of formal descriptions of architectural solutions, implemented through a program and deployed as 'Computer Aided Participatory Design'. Thus, he laid the foundation for current methods in the field of Artificial Intelligence (AI) and emphasized, "However, remember that these systems assume the driver to be an architect" (Negroponte, 1975, p. 365). Basic programming skills, as well as the recognition of workflows and tools from the area of computer science became an essential component of an architect's training and gained the same value as traditional analogue design methods. Negroponte conceived a new didactic framework in order the translate these abstract fields of expertise into the language of architecture. The archetypical architecture student, according to Negroponte in Appendix 3 of his Aspects of Teaching and Research "...is accustomed not only to working with his hands but also to physical and graphical manifestations; and he is accustomed to playing with those" (Negroponte, 1975, p. 191).

For Negroponte, the manner of gathering information as well as the flowing transition between scales and spatial sequences stood contrary to the handling of symbolic notion. The innovative use of the digital within architecture was to the team of Negroponte the only possibility to put new light on complex problems.

One source of his inspiration in particular was Moshe Safdie's design for Habitat 67[1] (Figure 1), which demonstrates a radical new typology for dense housing. Even though designed in a classical, pre-digital manner, the conceptual thinking behind the project unveils a systemized approach, arranging prefabricated housing elements in relation to their specific needs, e.g., in relation to their distance from main access points, according to a set of defined rules.

Figure 1. Habitat 67, aerial view, image by Tim Hursley (copyright Safdie Architects).

This formal rule-based articulation of a complex arrangement of diverse articulations of variations of the same base unit formation has influenced several the Arch MAC`s computational projects in the field of automized arrangement of building blocks and floor plans.

Negroponte discussed in depth the difference between 'talent' and 'competence' as a main reason, why the 'talent' should be in the hand of the future owner of a house. Meaning, the machine can support the designing of the 'perfect design', according to the owners needs and preferences (Figure 2). As the 'competence' lies in the hands of the architect, it is his task to define the rules, logic, and structure of the program, which is executed by the machine. This was interpretated often as working in close cooperation with the machine/computer: "They won't help us design; instead, we will live in them" (Negroponte, 1975, p. 5).

Figure 2. URBAN 5, developed in 1966 was one of the first projects of Negroponte and was digitally translating the complex architectural vision of Habitat 67. The user was interaction with a light pen, as well as special keypad to assign the attributes of each unit. In addition, the program contained an in-build dialogue, through which the system was able to learn and to enlarge its design understanding. It was understood as an experimental urban design partner (Negroponte, 1975). (Image courtesy of Nicholas Negroponte and Guy Weinzapfel)

With the advancements of machine learning, the importance of data is increasingly rising. Furthermore, ongoing discussion about novel methods for automated data-informed design strategies, outline fundamental shortcomings in large-scale data-handling versus small-scale data-handling and result often in green-washing solutions (Fricker, 2021).

Information Revolution – A Discussion on Data and Information

Historically viewed, the interaction with large data sets and the current discussion in the field of AI and machine learning enhanced workflows to gather firstly information and meaning out of the abundance of data, in order to secondly transfer this knowledge into a novel design method, is not a novel phenomenon and, like the development of the digital in architecture and landscape architecture.

The availability of 'Big Data 'and direct accessibility to a diverse range of datasets has a great influence on the field of architecture. Before automated workflows out of the field of data-mining, already in the

1990s. In their book Metacity/Datatown MVRDV postulated a design strategy that is mainly generated through the interpretation of abstract datasets (Maas, W. 1999). Historically, the integration and usage of data has started with statistical charts and maps[2] *displaying different data structure, which have aimed to communicate complex ideas in a clear, exact and efficient manner (Klanten, 2010). (Fricker & Munkel, 2015)*

The second industrial revolution was a decisive milestone in history, which made it possible to achieve a new way of dealing with large data sets by integrating different technical achievements. Picon also refers to the second industrial revolution as the "Information Revolution", in which society began to create new processes to archive and process large amounts of data, which were initially mostly used for administrative and economic purposes.

This approach and interest in archiving complex data sets generated completely new business models and the establishment of new professions. Since the handling of large data sets also produced physical, large amounts of paper, this component had a direct impact on the design of internal spatial structures. The definition of office building typologies was adapted to the production of large data sets in paper format and their processing processes.

The definition of office building typologies was adapted to the production of large data sets in paper format and their processing processes. Here, too, a pattern can be recognized in which form, medium and content are closely linked. If one looks at this in connection with those of McLuhan's formulated thoughts, discussed in the essay "The Medium is the Message", an interesting parallel to the articulated new networks of relationships of technology, media and society can be seen in relation to "dynamics and effects of acceleration" (Hassan, 2004; McLuhan & Baltes, 2001). McLuhan argues that even the content represents a new, abstract level of interaction using a (communication) medium and thus redefines the social network of relationships that is influenced by technology. McLuhan's research is often reduced to the fact that for him the medium has a greater meaning than information, but this is only a limited interpretation, since McLuhan's reflections aim at a deeper, more complex dimension – the influence of the medium on the development of society, and thus also for architecture of great relevance.

For the "message" of any medium or technology is the change of scale or pace or pattern that it introduces into human affairs. The railway did not introduce movement or transportation or wheel or road into human society, but it accelerated and enlarged the scale of previous human functions, creating totally new kinds of cities and new kinds of work and leisure. (Marshall McLuhan, 1964, 203)

Influenced by McLuhan's theoretical considerations, which focus heavily on the social aspect and the element of communication, Martin analyzes this system of relationships in relation to architectural typologies of the 1950s and 60s. This epoch was strongly influenced by technical achievements and a new approach to everyday technology. For Martin, the question arises as to the abstract meaning of, for example, structural elements such as the 'curtain wall' and other organizational elements as a receiving and processing instance of information (Martin, 2003). In his elaboration, based on the theories of Gyorgy Kepes, he decomposes the layers of a curtain wall into layers of communication, information and according to scale as an element of interaction with the urban, or macro, space, as well as with the private, or micro, interior [3] (Kim, 2004).

According to Picon, "the rise of the notion of pattern at the intersection of deep information structures and spatial information" illustrated by Martin and the influence of the field of cybernetics in connection with social patterns had a direct influence on well-known architects of the 50s and 60s[4] (Picon, 2017, p. 87).

New Possibilities for AI-enhanced Co-Design

In cybernetics, the discussed approaches in relation to the area of communication between individuals, groups and networks led to the Information Age and the development and integration of computers and digital communication platforms such as the Internet.

While the fact that people collect data is in itself a major step toward greater democracy in data-informed design and planning processes, major importance is on the kind of data we gather and how we interpret it.

The massive introduction of the computer and other digital communication elements as a central switching point and intermediate instance between humans and the abstract digital is described by Picon as "the individual turn" (Picon, 2017). Comparable to the background for the development from 'mass production' to 'mass customization', the rapid development of a digital culture resulted in a development from the 'mass information age' to an 'individually customized information age', the precursor of the concept introduced by Negroponte as 'post information age'[5] (Negroponte, 1998). The phase of the 'individually customized information age' is characterized by a simplified access of data and the individual as an active and passive generator of data packets. At the same time, the multitude of smart technologies enables an automated search and administration process, adapted to the individual criteria and parameters (Fricker et al., 2007). Developments such as the 'Internet of Things' (IoT) and the simultaneously developing smart city concepts represented another logical step in the 'individually customized information age'.[6] The progressive development of information and communication technologies has enabled the milestone to become the smart city concepts commonly used today. In general, the term smart city includes the following components: smart infrastructure, smart transportation, smart energy, smart health care, and smart technology, which are supplemented by further technological framework conditions such as IoT and the intelligent integration of big data. Smart cities are described as being especially efficient and responsive (Mohanty, Choppali, and Kougianos 2016).

In order to fully benefit from the available amount of data and to fundamentally change the way of designing, a new kind of human-machine interaction is required, based on the vision defined by Negroponte in the 80s, in which the computer was regarded as a partner in the design – as an architectural machine.

Looking at the concept of the 'information society' in relation to the direct interaction of technology, data, users and space, this cannot be discussed separately from the topic of connectivity and relationships with a view to a new 'digital culture'. [7] Guided by these overarching trains of thought, this raises questions for Picon in relation to architecture: "... what is the subject of the city, of the new digital city and the new digital architecture?" (Picon 2017, p. 87).

DIGITAL CULTURE

Driven by the broad application and integration of digital media in architecture, the question of the built form as a symbol and expression of a 'static' computerized process is to be seen as obsolete. Based on the desired flowing formal language, the design supported by computational design thinking enables a description of the 'gene code' of a design. In his analysis of Peter Eisenman's projects, which were created under the influence of his engagement with Derrida, Picon refers to "form is no longer synonymous with stable presence. Form could be different. It's a frozen moment... as if architecture were the section of a certain moment in time of theoretically limitless geometry and technological flow" (Picon, 2017, p. 92). Looking ahead, Picon describes this new understanding of information and its integration as a digital design tool as a "performative turn in architecture". Referring to the structure of the technical and digital development phases formulated by Picon, he formulates a comparability to the development of the entire discipline during the Renaissance, which put the human being at the center by introducing the perspective.[8]

Our understanding of the word information, originating from the 14th century *information*, "act of informing, and communication of news," changed considerably through technical advancements of the

Figure 3. The 'Cybernetic Loop Diagram' abstracts the mechanism of a self-regulatory system. (According to Norbert Wiener and adopted from Baango/Wikimedia Commons).

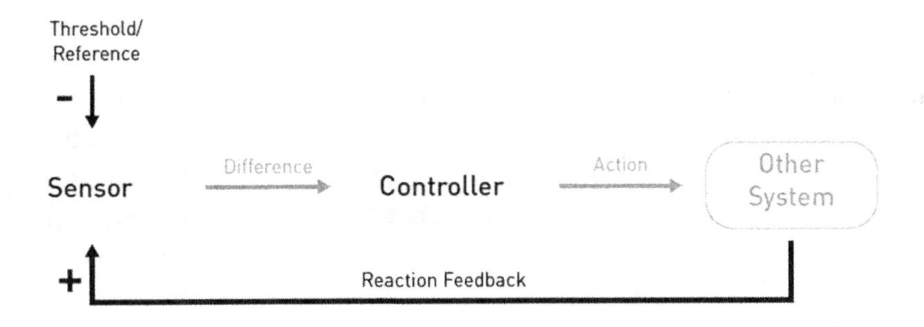

Digital Era. The diversity of the automated processes and interactions allowed for a fundamental new importance of dynamic data integration in design (Picon, 2017). As a result, the historical "act of informing", was transferred to the computational level and supplemented through the "generating of new connections" through multiple feedback loops (Figure 3).

According to Nathan Shadroff, one can conclude that "information makes data meaningful for audiences because it requires the creation of relationships and patterns between data. Transforming data into information is accomplished by organizing it into a meaningful form, presenting it in meaningful and appropriate ways, and communicating the context around it" (Diller et al, 2005). The theories developed in the area of cybernetics allowed a new computational design method to establish within architecture, which describes this complex network of relationships through the integration of System Theory and Patterns (Frazer, 1993). The discussion around AI and its potential for architecture has been very slowly moving ahead during the last decade. Revisiting the discussion of the intellectual pioneers of the 40s and 70s like, Norbert Wiener, McCulloch and Gordon Pask, especially the discourse on system theory regarding the relationship of man-machine for the field of architecture, opens a new academically founded possibility to develop the new abstract domain of AI (Menges & Ahlquist, 2011). This new perspective relates to the initial framework, which discussed the integration of the computer with the clear intent "to expand the human intellect, rather than work redundantly to the processes which form such knowledge. What arose 50 years ago, in short, was a perspective at which architecture could be both perceived and pursued as a 'system'."[9]

The Impact on the Area of Urbanism

The rise of a Society of Information has direct implications for issues related to areas of urban design and is seen by Picon as a modern cause of the development of urbanism[10] . The understanding of seeing and describing a city as an exchange of information flows laid the foundation for the development of urbanism and goes back to a period from the end of the 18th century to about the beginning of the 19th century. This development is criticized by Orit Halpern, among others, as "these architectures are performances, demonstrations of big data enacted for individuals who can neither see nor directly analyses the images" (Halpern, 2017, p. 126). Halpern considers the use of data sets, which have been used by "digitally managed systems" in different purposes, in direct connection with the development of smart cities and criticizes the increasing trend to consider "smart infrastructures as sites of value production" (Halpern, 2017).

Currently, many data-informed design projects mainly engage citizens as 'informants.' Meaningful, adaptive approaches to develop immersive data-interaction methods for sustainable design purposes, including information on atmosphere, dynamic data, and perception in relation to specific site-conditions, are to be developed (P. G. Fricker, 2013). Our time is marked by a diverse discussion on the future direction of computational technologies with respect to the increasing complex challenges we face.

Whereas previous discussions in relation to automated data-handling stressed efficiency, productivity and de-humanization, the current discussion should set its focus on quality, livability, experience and humanity.[11] These topics can be united under the slogan 'mind meets machine', regarded to be the driver of the 5th Industrial Revolution (Van Eerden 2020).

Within the introduced research, computational design thinking plays a central role in the understanding of the urban typology as an articulated 'land-scape'. It enables an integrative systemic design approach across scales and disciplines. De-coding site-specific systems in relation to complex synergies fosters the closing of a gap between otherwise separated fields of knowledge, in order to allow a transformative system design to emerge (Fricker et al. 2019).

Within the subsequent sub-chapter, a series of speculative design case studies will be introduced, which integrate data and emerging immersive design methods from an experimental and creative perspective. The profession of architecture and landscape architecture is nowadays more than ever challenged to go beyond traditional disciplinary borders, linking to phenomena in the natural sciences and computer science, in order to take full advantage of technological advancements (Girot, 2016).

The majority of these technological advancements are currently being developed and driven by a very small number of highly specialized research laboratories. Educators are missing the importance of fully integrating these advancements into the teaching of the design process and future professionals. We are currently facing the problematic situation where architects, urban designers and planners and landscape architects are being pushed to apply digital tools without any understanding of the underlying computational principles. Therefore, they are not capable of critically reviewing the quality of proposed design solutions, a reality, which is unfortunately already visible in a large number of newly built urban areas today.

A new mindset, beyond the anthropocentric point of view, will support a future-oriented reading of the city as a general action point. This experimental reading of potential points for interaction is supported by meaningful collaborations with the area of engineering, computer science, game design, robotics, human-computer interaction, neuroscience, social science as well as ecology.

Early Immersive Data-Interaction Protocols

The stimulating and innovative exploration in the field of computational design methods and early man-machine interaction at MIT, led in the 1950s and 1960s to influential research explorations in the field of virtual realities (VR)[12]. This was motivated by a general search for immersive data-interaction possibilities, in order to support the design process. The conducted research questions were in line with the development of other immersive design tools, like the MIT 3D Ball, Light-Pen, or Touch-Sensitive

Figure 4. Left: In 1950 Morten Heilig designed a hand-held display, which displayed a 140° horizontal and vertical field of view, stereo earphones, and air discharge nozzles that provide a sense of breezes at different temperatures as well as scent (Jerald, 2015). Right: 1950's visionary cinematographer Morton Heilig built a single user console called Sensorama. This enabled the user watch television in three dimensional ways. It had a 3D display and a vibrating seat to immerse the experience. (Courtesy of © Morton Heilig Legacy)

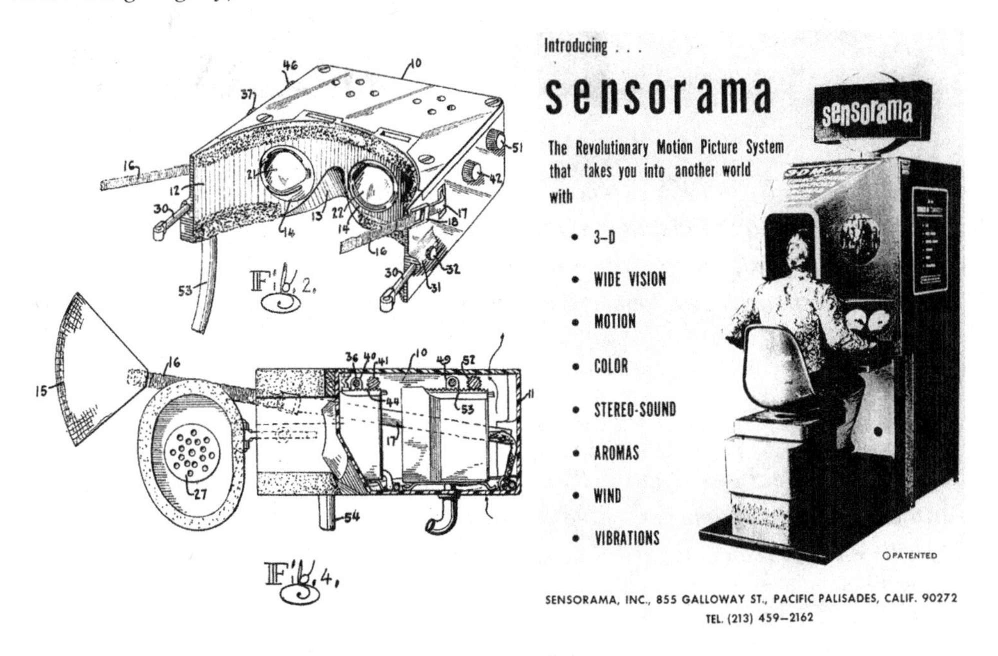

Display (Negroponte, 1998). These kinds of explorations have to be viewed in relation to Negroponte´s overall vison, to have a learning machine. "The same machine, after observing your behavior, could build a predictive model of your conversational performance. Such a machine could then reinforce the dialogue by using the predictive model to respond to you in a manner that is in rhythm with your personal behavior and conversational idiosyncrasies" (Negroponte, 1970, p. 12).

Working with data in extended realities (XR)[13], specifically referring to using digital technology to add another layer of information to the existing environment, was the next logical step to be explored (Figure 4, 5).

Driven by the rapid developments in electromechanics and computer technology, the Sensorama Machine, created by Morton Heilig in 1950, enabled the breakthrough of VR technology. If one considers the relevance of this technology for architecture, then the presentation by Ivan Sutherland and Bob Sproull of their first VR/AR14 head-mounted display, HMD (Sword of Damocles) in 1968 at MIT's Lincoln Laboratory certainly serves as a milestone. Ivan Sutherland had already described his vision

Figure 5. Ivan Sutherland's work at MIT starting in 1966 led to the first head-mounted display exhibited in 1968 (Sutherland, 1968). The display's translucent properties made it a precursor to AR (image courtesy of Ivan Sutherland).

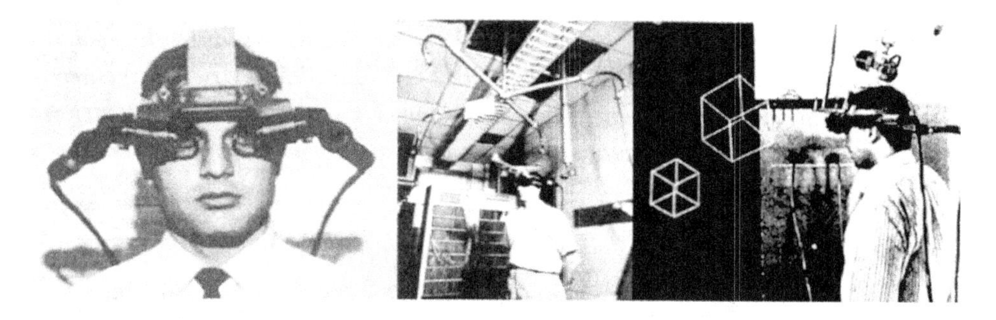

for the future of virtual reality in 1965 at a computer conference when he said that you shouldn't think of a computer screen as a way to display information, but rather "as a window into a virtual world that could eventually look real, sound real, move real, interact real, and feel real". This was a time of massive innovation in responsive and immersive technologies (Sutherland 1968). (Fricker 2018, 414)

Figure 6. Overview of the Reality-Virtuality (RV) Continuum according to Paul Milgram, 1994. Milgram sees the natural environment on one end of the continuum and the fully immersive virtual reality at the other end. The space in between is regarded as mixed reality, which integrates several sub-categories.

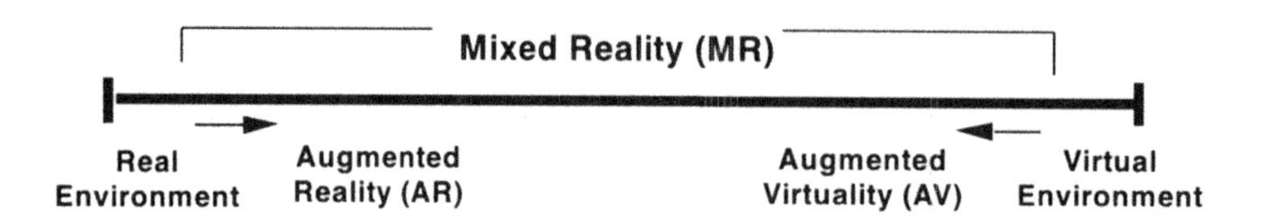

In respect to the early field of immersive data-interaction, the usage of the medium is intended to go beyond the notion of classical communication or embodied understanding of visualized data, represented through design interventions. As any form of extended reality allow the user to either fully immerse, like in the case of VR, or to interact and overlay partly with the understanding of the reality, like in the case of AR, the integration of a creative and experimental data interaction in extended realties opens up a new way for understanding hidden connections of information across scales and time dimensions (Figure 6).

After the early computational design protocols influenced mainly by Negroponte and the later development at the MIT, the introduced discussions in the field of Virtual Realty were in strong relation with the development around the late 80s until the end of the 90s. This timespan, just shortly before computers got affordable, was driven by an intellectual discussion moving towards a new theoretical knowledge domain. The articulated visions and dreams in relation to the upcoming availability of computers, was mainly influenced by imagining a cyberspace, which is marked by an immaterial stage and influenced heavily by William Gibson[15] and his thoughts on fluidity and topics like displacement and continuity (Gibson, 2002). During this time, the field of education was inspired by experimenting with a variety of media, like film, projection and interaction to overcome the static division of the real and the virtual, which originated in a parallel articulation within the practice.

RESEARCH METHODOLOGY

The introduced research is based on a newly defined methodology for computational design education, in order to explore sustainable design solutions for any specific situation or dimension, and that can react and respond to a larger continuity (P. Fricker et al., 2020). This can be accomplished by rethinking basic principles and incorporating new methods from expert systems like AI and ML (Saloheimo et al., 2021). This approach is not only capable of opening new avenues for more sustainable spatial configurations, but more importantly has the potential to overcome traditional boundaries between the fields of architecture, landscape architecture and computer science, thus allowing architects and designers to redefine their role in the shaping of a highly digitalized future by establishing a new digital material culture in built and unbuilt realms (Fricker, 2021).

While defining general strategies, the integration of the underlying hidden layers of a specific place, articulated through distinct patterns, are the local driver for a nuanced and site-specific design proposal. In order to do so, the established artificial separation between the domains of Landscape Architecture and Architecture in higher education needs to be completely re-thought. Landscape and natural processes must be duly accepted as a fundamental set of transitional concepts that operate on fields and boundary conditions. On the specific topic of urban design, reading the urban fabric as dynamic network systems with different contextual layers, allows an analysis of patterns to be linked with transitional and temporal elements of landscape systems.

Similar with the experimental structure of the computational design process, the speculative research builds on various methods from social science, computer science and educational research. The specific combination of methods chosen, pose the possibility not only to approach a theme in a linear, but rather in a circular and dynamic manner. For this, the results of each individual part were not statically grouped but rather integrated in a dynamic fabric of relationships, a strategy often found in the field of programming (Fayad et al., 2003). This meta-analysis bases on methods of pattern recognition and conceptualizes the generated conclusions, as a dynamic element of an iterative research-based design process.

CASE STUDIES

Currently, we are in a phase in which we can observe a duality of computational design development: on the one side is broad integration and access and the other the next phase of specialization and research that is initially only discussed and research in small circles of experts before becoming broadly integrated as methods for design and embedded in education.

The architectural profession needs to open up to the full implications of the computational revolution that has engulfed so many other industries. Instead of simply designing new forms in the traditional manner, architects should be designing new approaches to design, approaches based less on form per se than on informational systems, in order to harness the full potential of "Big Data" in our information age. (Leach, 2018, p. 112)

Especially due to the still ongoing discussion regarding Big Data and the potential to address the overload of information through means of machine learning (ML) (Kitchin, 2014), the current focus has shifted away from a discussion on form and towards a new understanding of the field of architecture and landscape architecture as a discipline steering, rather than controlling components to stay in an equilibrium (Ramsgard Thomsen et al., 2020). The practicability of the integration of data in large-scale discourse related to cities and territories is the theme which currently maintains the most potential in relation to computation and the global challenges we are facing. Simulating new relationships and transforming "quantitative models of building physics into qualitative sensory experiences" is currently viewed as a computational approach for the simulation of sustainable architecture (Peters & Peters, 2018, p. 1).

This current trend, the combination of algorithmic design with big data, poses challenges to the field of education, as the focus is set on the area of optimization, beyond a general discourse on theoretical principles out of the field of systems thinking (Figure 7).

Whereas the pioneers of computational design strongly emphasized the importance of translating technology, theory and design into a pedagogical framework, we are currently observing a trend to continued thinking in silos in the format of specialized research centers.

The discussed examples are based on an intellectual discourse with the field of computation and its influence on architecture. A clear conception of the possibilities and applications of digital tools and computational design thinking stands in the forefront. Theoretical discussions and sustainable architectonic design mechanisms are supplemented by the integration of coding in order to establish a unique design-specific access to the field of AI. Relating to the definition of computational thinking by the Center for Computational Thinking[16] at Carnegie Mellon University to the field of architecture and landscape architecture, it highlights the essential part of the new approach in the fact that systems thinking provides the theoretical basis connecting the individual components:

- Computational thinking is a way of solving problems, designing systems, and understanding human behavior that draws on concepts fundamental to computer science. To flourish in today's world, computational thinking has to be a fundamental part of the way people think and understand the world.
- Computational thinking means creating and making use of different levels of abstraction, to understand and solve problems more effectively.
- Computational thinking means thinking algorithmically.

Figure 7. The series of process images demonstrates the developed computational design process, which is based upon a translation of site-specific data as generative design drivers integrated into a complex system design. The design proposal is understood as a living organism, able to produce rather than to consume. This approach opens up a new way of understanding technology-driven design, combining performance, aesthetics and experience into a complex system of relationships. Project authors: Fanyi Jin Maral Alaei Tuuli Töniste – Aalto University, supervised by Prof. Dr. Pia Fricker and Prof. Dr. Toni Kotnik, Prof. Carlos Bañón (SUTD)

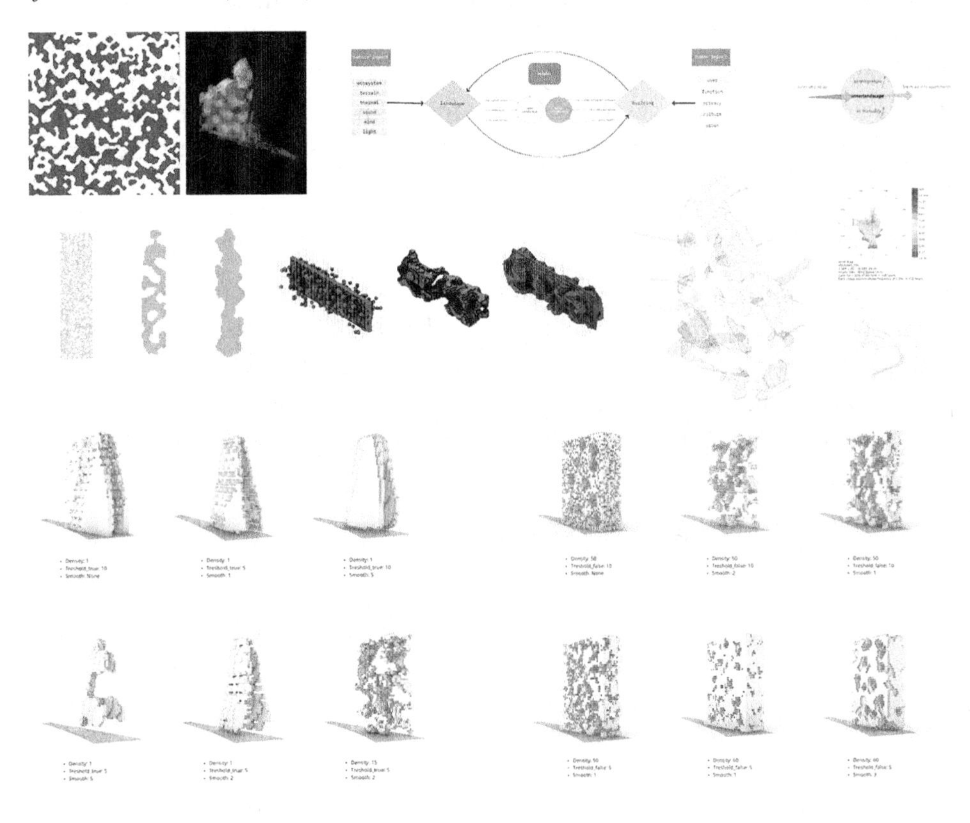

- Computational thinking means understanding the consequences of scale, not only for reasons of efficiency but also for economic and social reasons.

Within the architectural scene, the topic of Computational Thinking has been discussed broadly since over 15 years. Achim Menges and Sean Ahlquist reflected on the relevance of differentiating between computerization and computation in their publication on Computational Design Thinking:

The current transition from Computer Aided Design (CAD) to Computational Design in architecture represents a profound shift in design thinking and methods. Representation is being replaced by simulation, and the crafting of objects is moving towards the generation of integrated systems through designer-authored computational processes. While there is a particular history of such an approach in architecture, its relative newness requires the continued progression of novel modes of design thinking for the architect of the 21st century. (Menges & Ahlquist, 2011, p. 13)

From System Thinking to Computational Design Processes

With respect to data and computational processes, the integration of patterns supports an understanding of processes in nature, since they represent the interplay of material and forces across time and scale (M'Closkey & VanDerSys, 2017, p. 37). Using patterns with respect to data visualization, we gain an automatic visualization of processes, currents, and entanglements of interrelated systems (Burry, 2020). The implementation – or de-coding – of these visually represented rules, can be used to create new dependencies, as described by Alexander (1967), to be used directly as design instruments. Computation supports an understanding of complex natural behavior, which in return informs the designer of a new understanding of data. Belesky describes these informed methods as a possibility "to analyze and generate geometric information as a sequence of logic procedures, and to begin to close the gap between what constitutes a 'tool' and a 'technique'" (Tara et al., 2019).

Figure 8. Dew condensation is a process where water changes from a vapor to a liquid stat, which happens as temperature drops and objects cool down. When the surface temperature drops to the dew point, atmospheric water vapor condenses to form small droplets on the surface, as colder air is less able to hold water vapor. The dew point is affected by temperature and relative humidity. Relevant parameters for the scripting process: direction, local velocity, relative humidity, temperature, material emissivity, contact area, and crease angle.

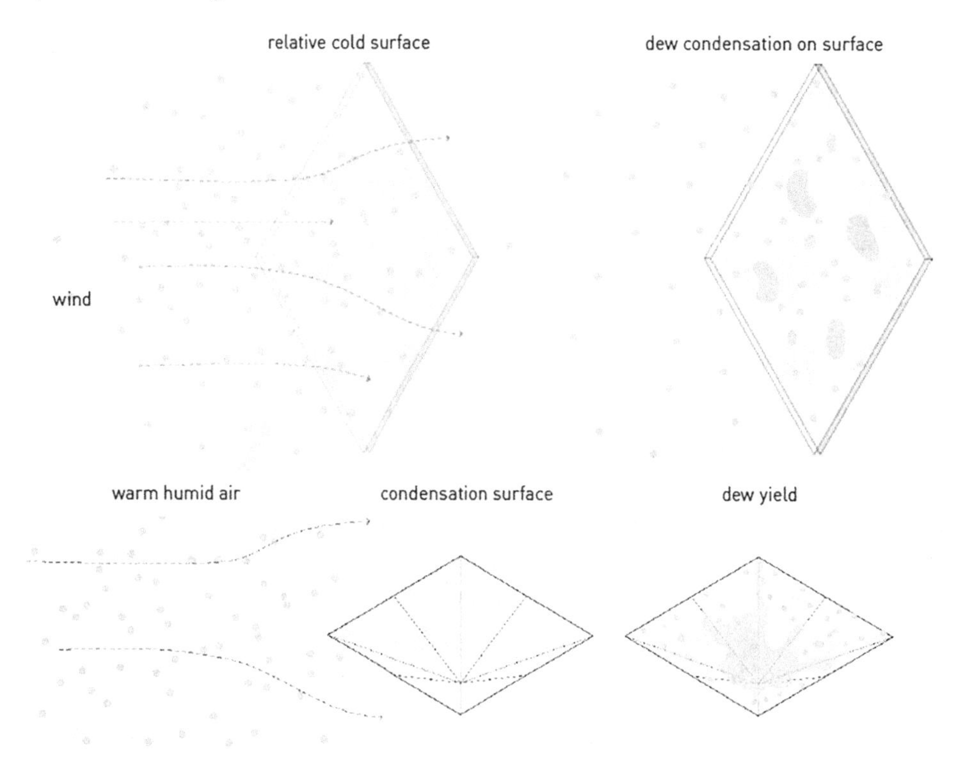

In the context of a joint speculative design studio between the Professorship of Computational Methodologies in Landscape Architecture and Urbanism – Prof. Dr. Pia Fricker and the Professorship of Design of Structures – Prof. Dr. Kotnik, students were asked to rethink fundamentals of contemporary architecture and urbanism while working on various sites in Finland and abroad. The systematic understanding of the basic physical principles of the elements, sun, wind and water, acted as a starting point for the set-up of the computational design method (Fricker et al., 2019). Within the first research phase, the physical behavior of each of the elements, sun, wind or water is explored. Each of these phenomena interacts with the field of architecture through different forms, behavior and stages of matter (Figure 8). The students asked to study at least three particular physical articulations of the assigned phenomena and to extract the most relevant knowledge onto shared knowledge templates, translating physical principles into coding language.

The explanation of the underlying physical principle is enriched through diagrams and relevant references. The findings are the starting point for further research and discussion into those conditions and the computational possibilities to modify and manipulate matter as a generative design component. This important research phase results into shared knowledge in the area of reflection behavior, thermal radiation, heat transfer, floating/buoyancy, dew condensation and Bernoulli's principle. In parallel, the computational skill building phase translates the gather systemic information into coding, expanding to Machine Learning.

This didactic framework expands from the abstract computational exploration of the chosen phenomena towards the direction of a generated design concept. As continuation of the previously developed parametric variations, the subsequent design phase focuses on the concrete definition of a possible design speculation.

Students are asked to present a schematic design concept outlining the relationship between the general design question and the possible design solution, explaining the computational operations that they develop and the effect they want to achieve (P. Fricker et al., 2020). The following questions should be addressed: What is the overarching computational design generation method you are developing (which parameters (data) and functions are you integrating…)? What possible new relationships, interactions and performative processes are you addressing between inside and outside (flows of water, sun, wind)? In which scale are you operating (building, urban, landscape…)? Which would be the optimal site for your design speculation? The schematic design concept translates the abstract previous research into visualized relational rules and interdependencies (Figure 9,10).

Project Reflection: "Aquair Oasis - A Breathable Water Machine"

From watering holes where animals and primitive men gather for resource and information to drinking fountains of Nepal or Rome, the production and acquisition of water is never merely a technical and infrastructural process, but a political one pursuing a dynamic interplay with the organization of society. Diverting from the notion of fountains sustaining water as a public good, in Antofagasta (Chile), a city lying on the fringes of the expanding Atacama Desert, the only available water resource for human consumption is the city water supply, which has been radically privatized. The scarcity of water and the large water demand of the local mining industry have resulted in the highest rates for drinking water in Latin America, impeding agricultural potential and causing severe ecological and environmental problems.

*Figure 9. Generation of Base Cell: Butterfly simulation - Dew yield for surfaces folded at different angles is estimated by the equation: D (dew yield) = -0.705+0.011*RH-0.006*N-0.01*Vi(mm), where Vi is the local wind velocity by Butterfly. With a smaller wind speed, less heat is absorbed by the surface resulting in a larger difference in surface temperature and hence more condensation. Project authors: Yikun Wang, Huiyao Fu – Aalto University, supervised by Prof. Dr. Pia Fricker and Prof. Dr. Toni Kotnik.*

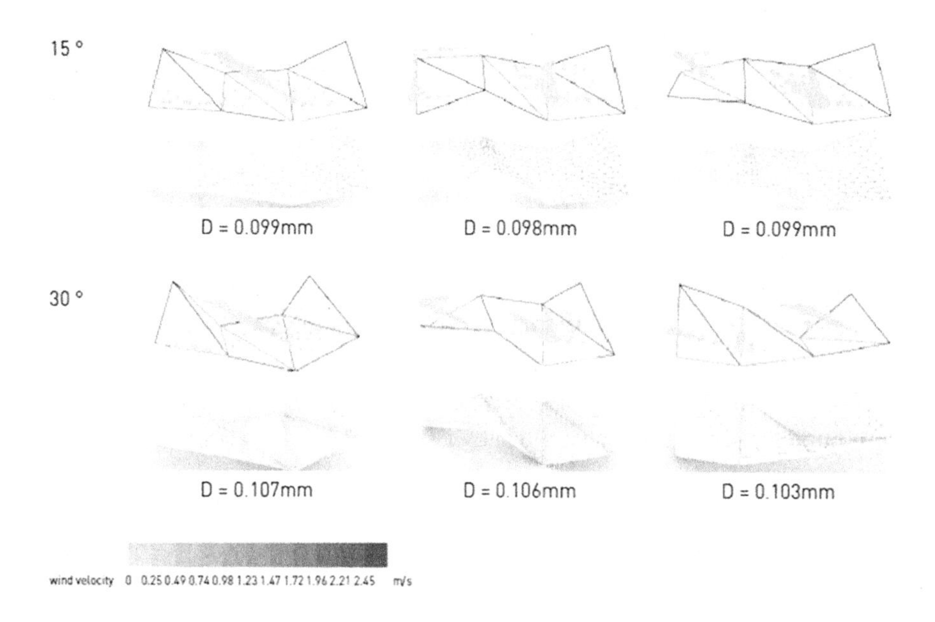

Figure 10. The simulation above displays the generation of an optimal dew condensation shape. The optimal shape produced has a concave shape with 30° crease angles, which performs best in local dew yield. Hexagonal cells are chosen for modular configuration controlled by Bezier curves and optimized by Biomorpher to maximize yield by adjusting the profile and depth of the curve. Each cell can provide a daily dew yield of 17.5 L in the chosen design location. Project authors: Yikun Wang, Huiyao Fu – Aalto University supervised by Prof. Dr. Pia Fricker and Prof. Dr. Toni Kotnik.

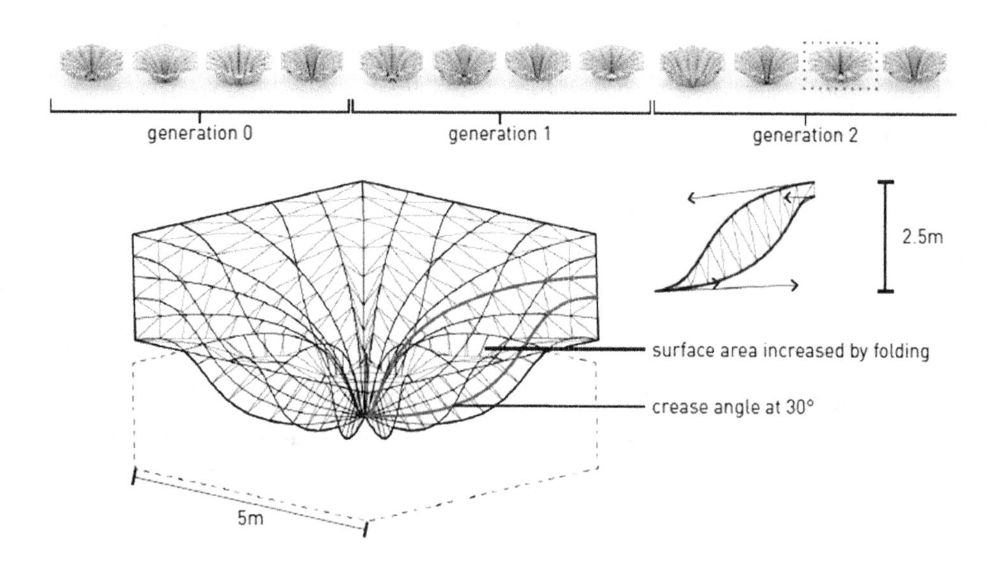

As many as 1,000 years ago, the native people of the north-built tools to gather potable water from the air by suspending animal skins from trees. Today modernized dew condensers and fog catchers are designed to harvest atmospheric water accompanied by storage and distribution systems. Unlike fog catchers which work most effectively on the high elevation of mountainous sites, dew condensers could be erected on the roof or in the yard embedded into the morphology of architecture. Starting from the simple motive of harvesting as much water as possible with the dew condensation phenomenon, the project envisions an alternative urbanization at the hydraulic and social border of the city composed of formal low-rise and informal settlements. By studying the typology of existing designs of dew condensers and extracting formalized rules, a modular canopy structure is generated which then in composed clusters forms a variety of configurations including small-scale pavilions at street corners and roof structure for public buildings accommodating an aeroponic farming lab, community library, and indoor and outdoor market spaces, adding festivity to both individual and communal life. The resultant architecture resuscitates the agential role of water treating the production of water and food as the production of social structure in pursuit of a balance of local ecological conditions, social identities and economic possibilities. The marriage of parametric design tools, advanced agriculture techniques and vernacular intelligence breeds an artificial oasis powered by atmospheric water machines, connecting the human settlement to the natural topography as a self-sustained community (by Yikun Wang, Huiyao Fu – project authors).

Project Reflection – "Water Track - City as a buffer of rainwater drainage"

With the increasingly complex process of human evolution, humans have gradually alienated themselves from the natural ecology. However, nature and the manmade should be treated as dynamic phenomena, constantly changing and interacting to create a continuously renewed ecology.

Now the buildings in the city are generally solid, for the purpose of protecting the building and creating an indoor environment, rainwater is not allowed to enter. The whole city is dense like a concrete forest. Cities and nature are like two separate systems. Thus, rainwater cannot pass through the city to the groundwater recycling system.

In four urban park design models in the United States, Galen Cranz made a concise summary of the development of park design: the first model, which began in the 19th century and was called "pleasant ground", was the product of a social movement aimed at the high proportion of diseases and high-density problems of urban people at that time, and it paid more attention to public health problems. After development, since 1965, the park design has entered the fourth mode, or'open space system': the important change in this period is that the park space no longer focuses on the construction of functional entertainment but turns to the emphasis on space experience and atmosphere. Nature is no longer regarded as a solution to urban problems. It is not a refuge in a city, or even separated from the city. Nature is the city, and the city belongs to nature. Our way to do is add some rainwater track into the city so the rainwater flows through the buildings then into water-use system of residents. Let the natural element – water – enter the building and serve as a guide for the growth of nature in the building. Various green spaces and public spaces have spontaneously penetrated into residents' lives, and residents' traditional grid like life mode has been broken. Nature and residents' lives have become an organic symbiotic system (by Yiqi Chen, Meng Xu – project authors).

Figure 11. Computational translation of the phenomena 'permeability' to generate a speculative dynamic building typology reacting to the daily rainfall in Singapore. Left: The generated residential "Eco-System" provides a variety of spaces in relation to light, water circulation and wind flow.
Right: Comparison of the performance criteria of chosen parameters with focus on the performance of the permeable building layer in relation to the velocity reduction. Project authors: Yiqi Chen, Meng Xu – Aalto University supervised by Prof. Dr. Pia Fricker and Prof. Dr. Toni Kotnik.

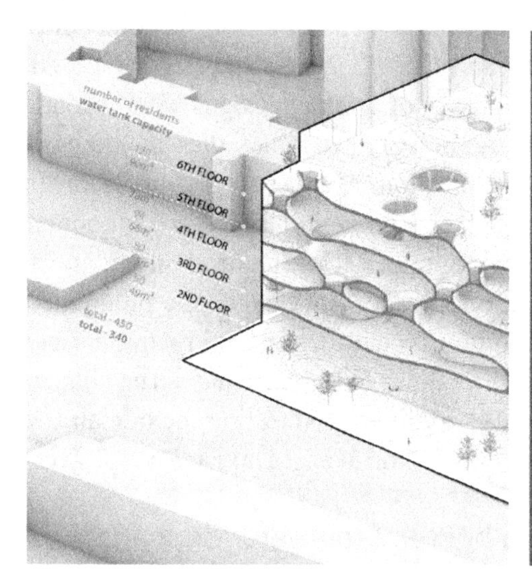

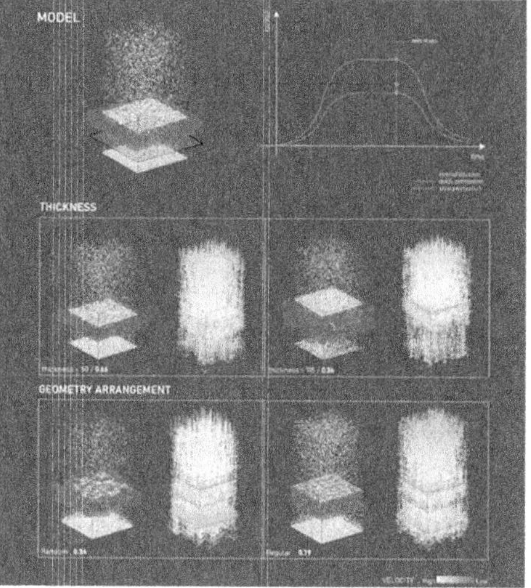

The Potential of Extended Realities

Currently explored fields in the realm of computational design show a tendency to merge the traditional areas of physical models and virtual simulations. Several research labs, like the LVML lab - Professorship of Christophe Girot at ETH Zurich or the lab of Bradley Cantrell, Professor and Chair of UVA Landscape Architecture at the University of Virginia, are developing computational tools and methods based upon exploring the availability of VR and AR tools in order to extend towards immersive design methods in the field of environmental design (Cantrell & Mekies, 2018; Hurkxkens, 2017).

At the moment, the integration of VR and other media in the field of landscape architecture seems to be rather limited in scope and focuses on topics like spatial representation. One of the main factors hindering the exploration of the full potential of the area of mixed reality in landscape architecture education is the focus on the mere solving of problems, rather than on the exploration of new strategies or approaches to their solution.

With the increasing importance of experience research, the study of user experience (UX) in VR systems has led to meaningful findings for the field of landscape architecture. As introduced by Whyte and Nikolic, the term "presence" describes the condition of being physically in a virtual environment, reaching the stage of being fully immersed in order to regard the view the virtual environment as "real" (Whyte & Nikolić, 2018). In order to reach this condition of absolute 'presence', often the focus turns to photorealistic visualizations of vegetation, surfaces, dynamic elements such as the movement of water and wind, bird song, animated groups of avatars and other romantic elements.

Only a small number of projects grapple with more complex issues, such as the integration of time within landscape architecture (Fricker 2018). The dynamic representation of vegetation spurred by the changing of the seasons, but also the visualization of vegetative growth are themes that can be developed beyond classical simulation use for design in VR and AR (Urech, 2019). User experience research in the area of VR usually centers around the user and his or her spatial experience. Technical themes of navigation, perspective and resolution often stand in the forefront (Lombard & Ditton, 2006). Whyte and Nikolić widened this traditional focus by "drawing out the implications for designing appropriate VR interfaces for information (BIM) models, in which various VR viewing perspectives, navigation modes, guides and aids can affect the illusion of an unmediated experience and support the use of VR for different applications" (Whyte & Nikolić, 2018, p. 14). These observations connect to ongoing research areas in relation to AI enhanced data integration, automized interaction through sensory-rich applications and "distributed virtual environments and teleoperations."[17]

The conducted design projects at Aalto University showcase novel computational design methods to creatively interact with global and local data. Whereas ongoing general discussions in relation to data-handling stress efficiency, productivity and often de-humanization, the exhibited projects set their focus on the exploration of immersive data-interaction design tools to enhance climate-smart solutions in the urban context. The projects showcase possibilities for making abstract data visible and tangible, to enter a new level of data-informed responsive design that has public benefit.

Augmented data-interaction is merged with the tangible sandbox environment in order to formulate sustainable design speculations for Ainonauki, an urban plaza located between Väre and Aalto Studios, located on the Aalto University Campus (Finland). The understanding of available technology able to be composed and rearranged in order to support the field specific discourse leads to innovative design speculations, based upon human-robotic interaction (Figure 12, 13). The developed design methodology emphasizes the importance of design iterations through computationally informed feedback loops.

Figure 12. The developed design methodology overcomes the distance between numeric data information and its impact on local design. The utilized 'Immersive Sandbox' environment allows for hands-on human-robotic interaction, integration AR technology to add further temporal design information (image courtesy of the author).

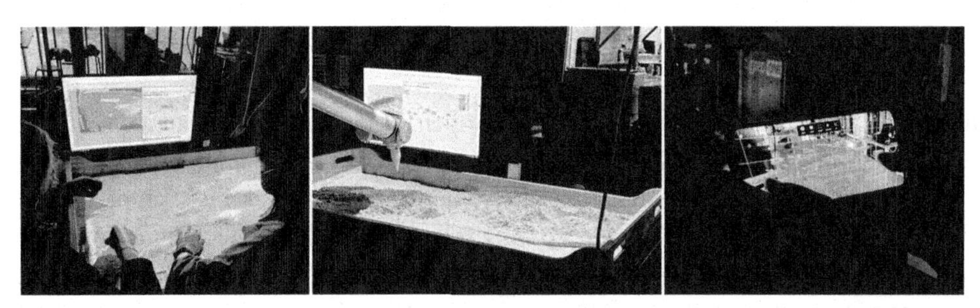

The abundance of data has created a rich pool of possible resources to support the creation of sustainable solutions to our pressing societal and environmental challenges we are facing globally. The future shapers of our environment need to be trained in computationally driven data-informed design methods, as we have already AI-informed automized design generators available to design buildings, cities and landscapes, without any architect or landscape architect involved (Yang et al., 2022). This is due the lack of knowledge in the field and affects the quality of the built environment. This is a real pressing problem of global size and major importance. The introduced projects research methods for architects and landscape architects to interact with abstract data, as a creative component in the design process and brings them back in command to interact with AI and other future technology.

Figure 13. The image series represents the computational translation of data into design. The design is conducted in the immersive 'Sandbox', allowing for intuitive human-robotic-interaction. Coding, physical representation, critical discussion and site-specific evaluation are the corner parts of the developed design process. Project authors: Antti Rantamäki, Kaisa Koskinen, Laura Tuorila, Teo Rinne, supervised by Prof. Dr. Pia Fricker, Tina Cerpnjak and Kane Borg, Aalto University

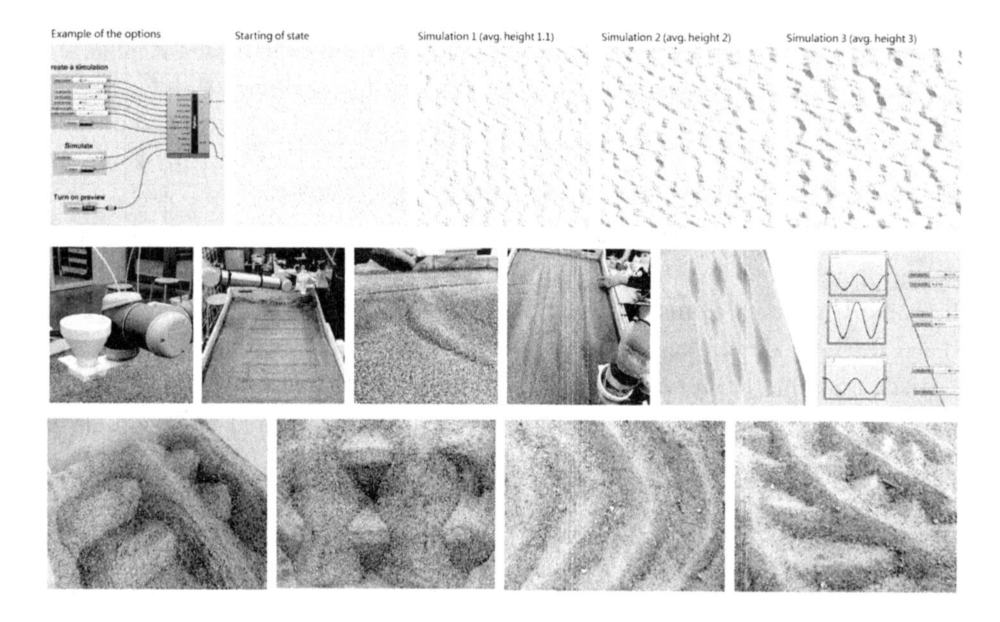

CONCLUSION

The introduced cutting-edge design speculations aim to open up new methods of immersive design simulation and participatory processes in environmental design and urban development. This series of research explorations discusses the possibilities that immersive interactive technologies can offer to reinterpret, manipulate, and interact with the flow of data coming from the acquisition techniques within the virtual landscape, both scientifically and artistically (Kowalewski & Girot, 2021). Possibilities, challenges and future visions are generated through generative design and evolutionary programming, extended by the possibilities of AI and ML and allowing for a new reading of the design process.

The research enables new future-oriented links between different disciplines (Architecture, Landscape Architecture, Urban Design, Engineering, and Media) to establish novel processes, create 'game changer' ideas, and conduct high-level interdisciplinary research.

The introduced design process generates novel computational design thinking methodology and workflows that allow for a shift in thinking towards dynamic system thinking. The case studies reflect on compositional methods, technological solutions, software, and hardware tools to address the many challenges presented. The aim is to close the current gaps between the individual design phases, which tend to lose information and take up time. The choice of the right computational methodology comes before the choice of the tool, otherwise the design will be disconnected from its surrounding system and limited by the functionality of the available 'black-box' digital tools. The generated, informed design solutions do not focus on the notion of form, but rather on the notion of process, enabling a reconnection of currently disconnected parts of our environment.

ACKNOWLEDGMENT

The introduced projects are the result of a research and teaching collaboration between Prof. Dr. Pia Fricker, Prof. Dr. Kotnik from Aalto University and Prof. Carlos Bañón from SUTD (Singapore University of Technology). The high-quality results would not have been achievable without the outstanding support from our committed teaching team: Lukas Piskorec, Kane Borg and Tina Cerpnjak. The selected design speculations represent a huge body of student work generated by the following students: Autumn Studio 2021: "Performative Patterns of High Density - Sustainable Visions between Architecture and Landscape": Ilmo Kapanen, Yanlin Liu, Yikun Wang, Huiyao Fu, Anna Li, Nikolai Fabricius, Yiqi Chen, Meng Xu, Yasmin Abdullayeva, Nuria Keeve, Yipin Wang, Qiwei Sun. Spring Studio 2022: "Mind Meets Machine: Immersive Data-Interaction": Marek Kratochvíl, Kaie Kuldkepp, Hanna-Kaisa Koskinen, Ahti Launis, Riikka Lauri, Emilia Lemmetti, Huixu Li, Loviisa Luoma, Eetu Mykkänen, Teo Rinne, Miisa-Maari Ulmanen, Eetu Mykkänen, Yanxia Qiu, Antti Rantamäki, Shenyu Sun, Laura Tuorila, Huijia Zhuang.

REFERENCES

Alexander, C. (1967). *Notes on the synthesis of form*. Harvard University Press.

Benjamin, W., & Tiedemann, R. (2015). Das Passagen-Werk: Vol. Band 5, Ed. 7, 934, Ed. 7 (7. Auflage). Suhrkamp.

Burry, M. (2020). Seeking an Urban Philosophy: Carlo Ratti and the Senseable City. *Architectural Design, 90*(3), 32–37. doi:10.1002/ad.2565

Cantrell, B., & Mekies, A. (2018). *Codify: Parametric and Computational Design in Landscape Architecture*. Routledge. doi:10.4324/9781315647791

Diller, S., Rhea, D., & Shedroff, N. (2005). *Making Meaning: How Successful Businesses Deliver Meaningful Customer Experiences*. New Riders. https://sfx.ethz.ch/sfx_nebis?url_ver=Z39.88-2004&ctx_ver=Z39.88-2004&ctx_enc=info:ofi/enc:UTF-8&rfr_id=info:sid/sfxit.com:opac_856&url_ctx_fmt=info:ofi/fmt:kev:mtx:ctx&sfx.ignore_date_threshold=1&rft.object_id=1000000000297576&svc_val_fmt=info:ofi/fmt:kev:mtx:sch_svc&

Douglas, W. Rae. (2003). *City: Urbanism and Its End*. Yale University Press. https://sfx.ethz.ch/sfx_lib4ri?url_ver=Z39.88-2004&ctx_ver=Z39.88-2004&ctx_enc=info:ofi/enc:UTF-8&rfr_id=info:sid/sfxit.com:opac_856&url_ctx_fmt=info:ofi/fmt:kev:mtx:ctx&sfx.ignore_date_threshold=1&rft.object_id=1000000000473628&svc_val_fmt=info:ofi/fmt:kev:mtx:sch_svc&

Fayad, M. E., Hamza, H., & Sanchez, H. (2003). A pattern for an effective class responsibility collaborator (CRC) cards. *Proceedings Fifth IEEE Workshop on Mobile Computing Systems and Applications*, 584–587. 10.1109/IRI.2003.1251469

Frascari, M. (2017). An Age of Paper. In A. Goodhouse (Ed.), *When is the digital in architecture?* (pp. 24–31). Canadian Center for Architecture.

Frazer, J. H. (1993). The architectural relevance of cybernetics. *Systems Research*, *10*(3), 43–48. doi:10.1002res.3850100307

Fricker, P. (2018). *The Real Virtual or the Real Real: Entering Mixed Reality*. Wichmann Verlag., doi:10.14627/537642044

Fricker, P. (2021). *The Relevance of Computational Design Thinking in Landscape Architecture. A Pedagogy of Data-Informed Design Processes across Scales* [PhD Thesis]. ETH Zurich. doi:10.3929/ethz-b-000495639

Fricker, P., Hovestadt, L., Fritz, O., Dillenburger, B., & Braach, M. (2007). Organised Complexity. *Predicting the Future: 25th ECAADe Conference Proceedings*, 695-701. http://papers.cumincad.org/cgi-bin/works/paper/ecaade2007_118

Fricker, P., Kotnik, T., & Borg, K. (2020). Computational Design Pedagogy for the Cognitive Age. *Anthropologic: Architecture and Fabrication in the Cognitive Age*, 695–692.

Fricker, P., Kotnik, T., & Piskorec, L. (2019). *Structuralism: Patterns of Interaction Computational Design Thinking across Scales*. Wichmann Verlag. doi:10.14627/537663026

Fricker, P., & Munkel, G. (2013). How to Teach New Tools in Landscape Architecture in the Digital Overload. In *Computation and Performance – Proceedings of the 31st ECAADe Conference – Volume 2*. Faculty of Architecture, Delft University of Technology. http://cumincad.scix.net/cgi-bin/works/Show?ecaade2013_028

Fricker, P., & Munkel, G. (2015). Data Mapping – Explorative Big Data Visualization in Landscape Architecture. In E. Buhmann (Ed.), *Peer Reviewed Proceedings Digital Landscape Architecture 2015* (pp. 141–150). Wichmann.

Fuller, R. B. (1969). *Operating manual for spaceship Earth*. Southern Illinois University Press.

Gibson, W. (2002). *Cyberspace*. Heyne.

Girot, C. (2016). *The course of landscape architecture: A history of our designs on the natural world, from prehistory to the present*. Thames & Hudson.

Halpern, O. (2017). Architecture as Machine: The smart city deconstructed. In When is the digital in architecture (pp. 123–175). Sternberg Press.

Hassan, R. (2004). *Media, Politics and the Network Society*. McGraw-Hill Education.

Hensel, M. U. (2015). *Grounds and envelopes reshaping architecture and the built environment*. Routledge, Taylor & Francis Group.

Hurkxkens, I. G. (2017). Robotic Landscapes: Developing Computational Design Tools Towards Autonomous Terrain Modeling. *ACADIA 2017: Disciplines & Disruption: Proceedings of the 37th Annual Conference of the Association for Computer Aided Design in Architecture (ACADIA)*, 292-297. http://papers.cumincad.org/cgi-bin/works/paper/acadia17_292

Jerald, J. (2015). *The VR Book*. Association for Computing Machinery. doi:10.1145/2792790

Kim, J. J.-Y. (2004). The Organizational Complex: Architecture, Media, and Corporate Space [review]. *Technology and Culture*, *45*(2), 468–470. doi:10.1353/tech.2004.0072

Kitchin, R. (2014). Big Data, new epistemologies and paradigm shifts. *Big Data & Society*, *1*(1), 12. doi:10.1177/2053951714528481

Klanten, R. (2010). *Data flow 2: Visualizing information in graphic design*. Gestalten.

Kowalewski, B., & Girot, C. (2021). *The Site Visit: Towards a Digital in Situ Design Tool*. Wichmann Verlag., doi:10.14627/537705022

Leach, N. (2018). Informational Cities. In P. F. Yuan & N. Leach (Eds.), *Informational Cities* (pp. 104–113). Tongji University Press Co., Ltd.

Lombard, M., & Ditton, T. (2006). At the Heart of It All: The Concept of Presence. *Journal of Computer-Mediated Communication, 3*(2). doi:10.1111/j.1083-6101.1997.tb00072.x

M'Closkey, K., & VanDerSys, K. (2017). *Dynamic Patterns: Visualizing Landscapes in a Digital Age* (1st ed.). Routledge., doi:10.4324/9781315681856

Martin, R. (2003). *The organizational complex: Architecture, media, and corporate space*. MIT.

McLuhan, M., & Baltes, M. (2001). *Das Medium ist die Botschaft = The medium is the message* (Vol. 154). Verlag der Kunst.

Meadows, D. H., Meadows, D. L., Randers, J., & Behrens, W. W. (1972). *Limits to Growth*. New American Library.

Mellon, J. G. (2008). Urbanism, Nationalism and the Politics of Place: Commemoration and Collective Memory. *Canadian Journal of Urban Research*, *17*(1), 58–77.

Menges, A. (2011). *Computational design thinking*. Wiley.

Menges, A., & Ahlquist, S. (Eds.). (2011). *Computational design thinking*. Wiley.

Mihelj, M., Novak, D., & Beguš, S. (2014). *Virtual Reality Technology and Applications* (Vol. 68). Springer Netherlands., doi:10.1007/978-94-007-6910-6

Negroponte, N. (1970). *The architecture machine: Toward a more human environment*. The Massachusetts Institute of Technology MIT.

Negroponte, N. (1975). *Soft architecture machines*. MIT Press.

Negroponte, N. (1998). Being digital. Knopf.

Oxman, R. (2008). Digital architecture as a challenge for design pedagogy: Theory, knowledge, models and medium. *Design Studies*, *29*(2), 99–120. doi:10.1016/j.destud.2007.12.003

Peters, B., & Peters, T. (2018). Introduction—Computing the Environment. In *Computing the Environment—Digital Design Toosl for Simulation and Visualisation of Sustainable Architecture* (pp. 1–13). John Wiley & Sons.

Picon, A. (2017). Histories of the digital: Information, computer and communication. In When is the Digital in Architecture? (pp. 80–99). Sternberg Press.

Prof. Dr Graafland, A. (Ed.). (2012). Architecture, technology & design. Digital Studio for Research in Design, Visualization and Communication.

Ramsgard Thomsen, M., Tamke, M., Nicholas, P., & Ayres, P. (2020). *CITA complex modelling*. Academic Press.

Saloheimo, T., Kaos, M., Fricker, P., & Hämäläinen, P. (2021). Automatic Recognition of Playful Physical Activity Opportunities of the Urban Environment. *Academic Mindtrek*, *2021*, 49–59. doi:10.1145/3464327.3464369

Sayegh, A. (2020). *Responsive Environments: Defining our Technologically-Mediated Relationship with Space*. Actar D.

StilesM. (2019, April 4). Https://medium.com/desn325-emergentdesign/reality-virtuality-continuum-868cb8121680

Sutherland, I. E. (1968). A head-mounted three dimensional display. *Proceedings of the December 9-11, 1968, Fall Joint Computer Conference, Part I on - AFIPS '68 (Fall, Part I)*, 757. 10.1145/1476589.1476686

Tara, A., Belesky, P., & Ninsalam, Y. (2019). *Towards Managing Visual Impacts on Public Spaces: A Quantitative Approach to Studying Visual Complexity and Enclosure Using Visual Bowl and Fractal Dimension*. Wichmann Verlag. doi:10.14627/537663003

Urech, P. R. W. (2019). *Point-Cloud Modeling: Exploring a Site-Specific Approach for Landscape Design*. Wichmann Verlag. doi:10.14627/537663031

Van Eerden, J. (2020, January 23). A Davos POV About a 5th Industrial Revolution. *Rea Leaders*. https://real-leaders.com/a-5th-industrial-revolution-what-it-is-and-why-it-matters/

Whyte, J., & Nikolić, D. (2018). *Virtual Reality and the Built Environment*. Routledge. doi:10.1201/9781315618500

Wiener, N. (1948). *Cybernetics, or control and communication in the animal and the machine*. Hermann.

Wu, H.-K., Lee, S. W.-Y., Chang, H.-Y., & Liang, J.-C. (2013). Current status, opportunities and challenges of augmented reality in education. *Computers & Education*, *62*, 41–49. doi:10.1016/j.compedu.2012.10.024

Yang, J., Fricker, P., & Jung, A. (2022). *From Intuition to Reasoning: Analyzing Correlative Attributes of Walkability in Urban Environments with Machine Learning*. Wichmann Verlag. doi:10.14627/537724008

KEY TERMS AND DEFINITIONS

Articulated Landscapes: An understanding to re-connect the built and the un-built, as active components of a dynamic, complex system. This viewpoint explores the city as a performative landscape, focusing on the relation and interaction between the building, its neighbourhood and the natural elements.

Climate-Smart Design: A design attitude aiming for responsible solutions focusing on environmental aspects across temporal, urban, and territorial scales.

Computational Design Thinking: Computational design thinking is driven by the idea of articulated ground that is the active use of urban and landscape strategies on a conceptual and systemic as well as operative level of design development throughout all scales of intervention.

Data-Driven Design: A strategy to integrate different datasets as creative input, feedback, and evaluation parameter into the design and monitoring process across all stages.

Human-Robotic Interaction: A design methodology allowing for co-design strategies through the integration of robotic feedback.

Immersive Co-Design: A design methodology which builds upon the integration of extended reality applications into the design process. This allows for the integration of diverse user groups into the design process.

Immersive Sandbox: An intuitive, digitally enhanced design environment allowing for real-time feedback on conducted design iterations through projected data simulation and robotic feedback.

ENDNOTES

1. Habitat 67 – a living community, designed by Moshe Safdie for the Canadian Corporation for the 1967 World Exposition in Montréal. The project developed out of architect Moshe Safdie's 1961 thesis design project and report ("A Three-Dimensional Modular Building System" and "A Case for City Living" respectively). Retrieved from https://cac.mcgill.ca/moshesafdie/habitat/concept.htm (accessed 3 November 2021).
   "The prototype for a new urban building typology, Habitat sought to mix residential, commercial, and institutional uses to create a more vital neighborhood and provide the amenities of the single-family home in a form adaptable to high densities and constrained budgets." Retrieved from https://cac.mcgill.ca/moshesafdie/fullrecord.php?ID=10816&d=1 (accessed 3 April 2022).
2. One of the most referred contribution in the field of complex data visualization is Charles Minard's map of Napoleon's disastrous Russian campaign of 1812. The graphic is notable for its representation in two dimensions of six types of data: the number of Napoleon's troops; distance; temperature; the latitude and longitude; direction of travel; and location relative to specific dates (Tufte, V., & Finley, D. (2002, August 7). MINARD'S SOURCES. Retrieved October 30, 2021, from https://www.edwardtufte.com/tufte/minard).

[3] Gyorgy Kepes (1906–2001) concentrated during his work at MIT on the establishment of a future-oriented Center for Advanced Visual Studies (CAVS), which was founded in 1967. The MIT infrastructure allowed Kepes to develop a new interdisciplinary scientific thinking for the field of design, reaching out to the field of material science and structural chemistry. Die novel combination of different disciplines attracted artists, Smithons; Kepes saw a deep importance in addressing a holistic viewpoint to the fields of Art and Ecology and was aligned with the thinking of other leading thinkers of the first ecological era, like Sir Buckminster Fuller.

[4] Reinhard Martin researches the buildings General Motors Technical Center, IBM manufacturing and training facility and the Bell Telephone Laboratories designed by Eliel Saarinen as well as Inland Steel headquarters and Chase Manhattan Bank von Skidmore, Owings & Merrill.

[5] Negroponte reflects in his book on a possible digital future and introduces in part three, "Digital Life", the vision that humanity is in the post-information age where "true personalization" is imminent. The machines will understand individuals and their preferences as humans understand humans. (Negroponte 1998, p164/165) retrieved from https://en.wikipedia.org/wiki/Being_Digital, accessed 29 October 2021)

[6] The term 'Internet of Things' (IoT), earlier defined as 'Machine-to-Machine communication', was introduced in 1999 by Kevin Ashton (cofounder of the Auto-ID Center at the Massachusetts Institute of Technology). 2013. Kramp describes the IoT as "a logical extension of the computing power in a single machine to the environment: the environment as an interface. This push-pull combination makes it very strong, unstoppable, fast and extremely disruptive."

[7] There is no uniform, scientifically accepted definition of the term "Information Society". Fritz Wiener is mostly mentioned as a prominent reference with his publication "Cybernetics or Control and Communication in the Animal and the Machine" (Wiener 1948). Wiener discussed relevant topics such as controlling the flow of information in systems with feedback loops, which are currently being discussed and applied in relation to AI.

[8] The word "information" in relation to technical and digital advancement, was used in reference to television broadcast signals from 1937; to punch-card operating systems from 1944; to DNA from 1953. *Information theory* is from 1950; information *technology* is from 1958 (coined in "Harvard Business Review"); information *revolution*, to be brought about by advances in computing, is from 1966. *Information overload* is by 1967. Retrieved November 1, 2021, from https://www.etymonline.com/word/information

[9] Ibid., page 11.

[10] In addition to the progressive consideration of the causes of the development of the phenomena urbanism, describe Mellon et al. the term Urbanism as a complex setting between a range of perimeters, the city, its density perimeter, and the hinterland/countryside, as a conclusion from a comprehensive literature research. Within their article they are outlining the diverse range of attempts to describe the term, like the definition of Rae, who *refers to an* "old urbanism" in which cities were characterized by five elements—"industrial convergence," "a dense fabric of enter-prise," "a centralized clustering of housing," "a dense civic fauna of organizations," and a "pattern of political integration"—and to the possibility of a future "new urbanism." (Mellon 2008, 60),(Douglas W. Rae, 2003).

[11] During 2020s WEF, the theme "Blockchain + AI + Humans = Magic!" was promoted. Van Eerden explained, why this equation will turn out to be correct: "AI will help increase human labor productivity. Blockchain will help give access to banking (and intangible forms of capital connected

to a person's quiddity) to the unbanked. Robots will help humans align ROI with purpose. But all this will require intentionality and moral clarity" (Van Eerden 2020).

12 Virtual Reality (VR) can be understood as "the observation of the virtual environment through a system that displays and allows interaction, thus creating virtual presence" (Mihelj et al., 2014). Morton Heilig is generally understood as the founder of Virtual Reality, with his invention of the *Sensorama* in 1957. The *Sensorama* offered a virtual bicycle experience, viewing the three-dimensional map on the display, enhanced through the sound of the city, vibrations of the seat, and the smell of scents related to the site (Mihelj et al., 2014).

13 Extended Reality (XR), "is the term referring to all real and virtual combined environments and human-machine interaction generated by computer technology and wearables. It includes representative forms such as augmented reality, augmented virtuality, virtual reality, and the areas interpolating among them. The level of virtuality ranges from partially sensory input to immersive virtual reality. XR includes the entire spectrum from the complete real to the complete virtual. In the concept of reality virtuality continuum introduced by Milgram. Still its connotation lies in the extension of human experiences, especially relating to the senses of existence represented by VR and the acquisition of cognition, represented by arcensor. With the continuous development of computer interactions, this connotation is still evolving" (Stiles, 2019).

14 Artificial Reality (AR): Different types of generated, virtual elements, such as visual, auditory, haptic, somatosensory and olfactory, are augmented as interactive experiences to the real world (Stiles, 2019). AR fulfils three conditions "combination of real and virtual worlds, real-time interaction, and accurate 3D registration of virtual and real objects" (Wu et al., 2013, p. 42).

15 William Ford Gibson (born in 1948) is an American Canadian speculative fiction writer. Gibson coined the term "cyberspace" which first appeared in print, in a short story by Gibson for the July 1982 edition of the science fiction magazine Omni.

16 What is computational thinking? – according to: https://www.cs.cmu.edu/~CompThink/

17 Ibid.

Chapter 9
Machine Learning and Artificial Intelligence for Smart Visualization, Presentation, and Study of Architecture and Engineering in the Urban Environment:
Visualizing City Progress

Andrea Giordano
University of Padova, Italy

Rachele Angela Bernardello
University of Padua, Italy

Kristin Love Huffman
Duke University, USA

Maurizio Perticarini
Università degli Studi Luigi Vanvitelli, Italy

Alessandro Basso
https://orcid.org/0000-0002-0366-975X
University of Camerino, Italy

ABSTRACT

This research experiments the theme of cultural heritage (CH) in architectural/engineering fields, located in urban space. Primary sources and new tactics for digital reconstruction allow interactive contextualization-access to often inaccessible data creating pedagogical apps for spreading. Digital efforts are central, in recent years based on new technological opportunities that emerged from big data, Semantic Web technologies, and exponential growth of data accessible through digital libraries – EUROPEANA. Also, the use of data-based BIM allowed the gaining of high-level semantic concepts. Then, interdisci-

DOI: 10.4018/978-1-6684-4854-0.ch009

plinary collaborations between ICT and humanities disciplines are crucial for the advance of workflows that allow research on CH to exploit machine learning approaches. This chapter traces the visualizing cities progress, involving Duke and Padua University. This initiative embraces the analysis of urban systems to reveal with diverse methods how documentation/understanding of cultural sites complexities is part of a multimedia process that includes digital visualization of CH.

INTRODUCTION

This article will retrace the development of Visualizing Cities, an international research initiative involving Duke University, the University of Padua and the University of Venezia IUAV, which has evolved from the fundamental strategies developed by Visualizing Venice. Visualizing Venice is a research project instituted in 2012 by Caroline Bruzelius (Duke University) and Donatella Calabi (IUAV) in which the key aspect is collaboration and teamwork (Huffman et al., 2017). This collaborative aspect includes exhibitions, reviews, publications, lectures and invited presentations in Europe, United States, Asia. The collaboration of a range of multidisciplinary skills has been fundamental for this project: History of art, History of architecture and of the city, Representation (in particular Architectural survey, Building Information Modeling - BIM, Geographic Information System - GIS, Perspective and Photographic restitutions), Structural Engineering, Restoration and Conservation of buildings, Information and Communications Technology - ICT. In particular, Visualizing Venice included multiple theoretical / operational activities through the use of interoperable models and city views, reconstructions and interactive maps, such as models that not only have an academic value but are configured as accessible to a wider audience: the aim was therefore to make open and inclusive the study of the history of a complex city like Venice. Visualizing Cities is a consequent initiative that includes examining urban systems and sites not only within Venice, but also in Padua, Carpi, Athens, Krakow, continuing to demonstrate how documentation and understanding of cultural sites and their complexities may be assumed as part of a multimedia process that includes the interpretation of digital visualizations of historical monuments (Huffman, Giordano, 2019. Giordano, 2017. Giordano et al. 2018). Moreover, precisely by using digital methods, the time has come to deepen the scientific analysis of visual and written documents that validate and / or reveal previously unknown urban circumstances. The traditional methodologies of art, architecture and city history remain the very foundation of rigorous digital approaches; in fact, the study of a city and its architectural / engineering artefacts requires the academic organization of information and visual sources to connect with a wider historical context. Therefore, the availability of new digital tools and applications, iconographic and textual sources - primary data of exceptional value not only from an historical point of view, but also for interpretative reflections - can now intertwine as a scientific practice for advanced know-how and technologies for the visualization of historic cities. There is a demand for further scientific analysis of documents that validate and / or reveal urban circumstances. Traditional methodologies of architectural and urban history must remain the fundamental core of digital approaches; the study of a city will always require the academic decoding of information and visual sources that connect them to their wider context. In this sense, the integration of multiple skills is essential: the skills of art and architecture historians with those of architects / engineers and experts in visual and media studies. With all our research projects, art and architecture historians have conducted archival research that is fundamental to our understanding. In addition, with each of our projects, the process of building a virtual model has led to new discoveries in the field. In summary, digital technologies have enabled us to improve our knowledge of how to use new tools both as a more integrated part of the research process and to visualize novel results / outcomes.

METHODOLOGY

Visualizing Cities testifies how new technologies have the ability to "revolutionize" research and teaching by implementing collaborative theory and practice in the field of Digital Humanities, to interpret, represent, teach and promote the knowledge of a city as space in time. In this sense, therefore, this research embraces one of the most innovative aspects of digital technologies: the awareness that the speculative thought of scholars / researchers does not express itself without a correct perception and treatment of 3D space (Giordano, 2019). Consequently, using various examples, the methodology of Visualizing Cities is articulated in four distinct phases for this type of investigation:

- Data acquisition: archival research, laser scans and photogrammetric surveys processed and organized through 3D modeling implemented between interoperable platforms:
- Data communication: the information collected with the methods listed above conveyed through the design of apps and interactive systems for multimedia devices and web platforms. This process involves the design and testing of augmented reality and 3D models for multimedia devices and the implementation of immersive reality.
- Integration of AI-based image management systems aimed at versatile simultaneous data acquisition and communication in the same workspace.
- Integration of models as a means of analysis in the conservation process of architectural heritage with the virtual reconstruction of architectural features.

Entering into the specifics of architectural cases, the objectives are the communication of the history of architecture and the city to a wider public with the aid of new tools, underlining how these same tools intervene in the study of the city compared to more traditional, no longer effective today. In fact, our challenge is to address issues that with the traditional tools of urban history (archive research, bibliographic, iconographic, cartographic) are difficult to narrate: an example is precisely how to deal with the dynamics of transformation that are not only historical (flows of goods and people, movements, sounds) of the life of a city. Therefore, the use of multimedia tools is configured as a moment not only for innovation but also for the development of further research. Secondly, it is a question of communicating the results of the research carried out in the field of the history of the city to a vast public consisting not only of specialists and of experts, in the belief that, with the help of new tools, it is possible to convey scientifically sophisticated acquisitions. Then, the aim is to obtain a fast and inclusive progress of studies about knowledge and sustainable development of tourism through a more direct understanding of Cultural Heritage (in our case, of the city and architecture and related transformations) through current multimedia devices.

Looking ahead, we would like to highlight the following salient points:

- How to broaden our reach to include urban studies programs engaging ongoing research related to a wide variety of cities and their distinctive phenomena.

- How partnership and collaboration could become organically more capacious (with tasks currently unforeseen in the future), as we work with complementary institutions, expertise and new research sites.
- How to visualize the progressive transformation of cities over time into pioneering virtual environments, in particular with emerging technologies.
- How to highlight the distinctive features of each city, built and natural, which mutually reinforce urban evolution as a phenomenon.
- How to articulate in a clear and transparent way the complex intersection of the social, political and economic forces involved in a virtual and documented form.
- How to animate the phenomenology of everyday life and the experiences of an urban center from past to present in a multisensory way, describe the general perspective of the chapter. End by specifically stating the objectives of the chapter.

BACKGROUND

The aim of his research is the to conception of operating digital models to convey a pioneering understanding of Urban Built Heritage (UBH), involving Artificial Intelligence (AI) and Augmented Reality (AR). Certainly, while these technologies are now extensively used, they are improving in the field of UBH, especially in the historical cities, where is still mandatory a deep exploration of their characters and mutual connections.

AI advancements - involving specifically Deep Learning (DL) - enable many applications to computer vision, thanks to Machine Learning (ML), implementing graphic computing capabilities and the availability of huge image datasets. The interaction between DL and the built environment has recently increased studies delve into the interpretation of architecture, urban space and UBH in general (Cordts et al, 2016) (Stathopoulou & Remondino, 2019). Fundamental, we need to underline the path of this connection: Visualizing Cities group developed many projects for exhibitions informative services (Palma, 2019) with the implementation of mobile apps managing multimedia contents through ML image / 3D models classification and connection. ML can also be used to process digital survey data (i.e., point clouds) by automatically generating mathematical and semantic interpretations of built elements (Andrianaivo et al, 2019). More, the implementation of AI for Cultural Heritage (CH) institutions – i.e., in cataloguing and describing digital objects, through data linking and metadata creation (Grilli & Remondino, 2020) – may help educational aspects, improving the learning setting, also suggesting the need for new models of teaching. Then the use of AR for educational purposes advantages as increased interaction between the user and the 3D model in different learning contexts, learning trail personalization, applications for university education (Akçayır et al., 2017) (Ibáñez et al, 2020). AR may be an acute apparatus for active learning, enabling real-time interaction with virtual contents, and increasing confidence and interest (Farhah Saidin et al, 2015). However, best practices and bottlenecks in understanding 3D models through AR still need to be addressed. In this case, also the link between Virtual Reality (VR) and AR may be configured as well-established components of Cultural Computing and Figuring (Greengard, 2019) (Amin & Govilkar, 2015). In fact, AR may offer instinctual / intuitive approaches to information related to the space configuration, overlaying digital substances on real-world images (Younes et al, 2017) allows the tracking of markers, images and 3D objects with AR (Panou et al, 2018), that interacts with a "more and more" huge / complex environment. More, latest research about UBH underline the integration between

AR and touristic experiences integrating itineraries and guides with the connection of digital information and real space. The same for applications of AR and VR in architecture and construction engineering (Li et al, 2018): this application effectually deciphers and resolves a variety of construction managing concerns (Ahmed, 2019) - design and data visualization, project scheduling, progress tracking, worker training, safety management, time and cost management, and quality management.

Visualizing Cities then involves the concept of Digital Twin (DT) (Kritzinger et al, 2018), reaching a realistic representation of physical assets, led to the idea a virtual reproduction that variates with the real object thanks to effective data transmission. In this sense, the large number of experimentations and studies involving AR and AI can became a type of protocol for a general approach to UBH communication and management. In this sense, Visualizing Cities will improve new approaches for the creation of an Augmented Digital Twin (ADT) for gathering information about UBH, teaching and training. Frontline AI and AR solutions will be investigated, tried and supplementary developed, networking with the interoperable information system produced. Attention will be paid to services, software, and formats to coordinate and integrate the ontologies involved in multiple topics: interoperability in CH archives (Cecchini et al, 2019), AR-document archive connection; AI and network services (Baranda et al, 2020), ontologies for CH. Furthermore, AI and AR will increase the user experience, supporting the creation of personal "narrative trails" (Mulholland et al, 2016) and developing virtual tour guides and storytelling solutions (Lim & Aylett, 2007) (Figure 1, 2).

Figure 1. Implementation of the Augmented Digital Twin (ADT) of the historical transformation (from 1500 to 2020) of the Insula of Santi Giovanni e Paolo for the immersive reality in DIVE (Duke University).

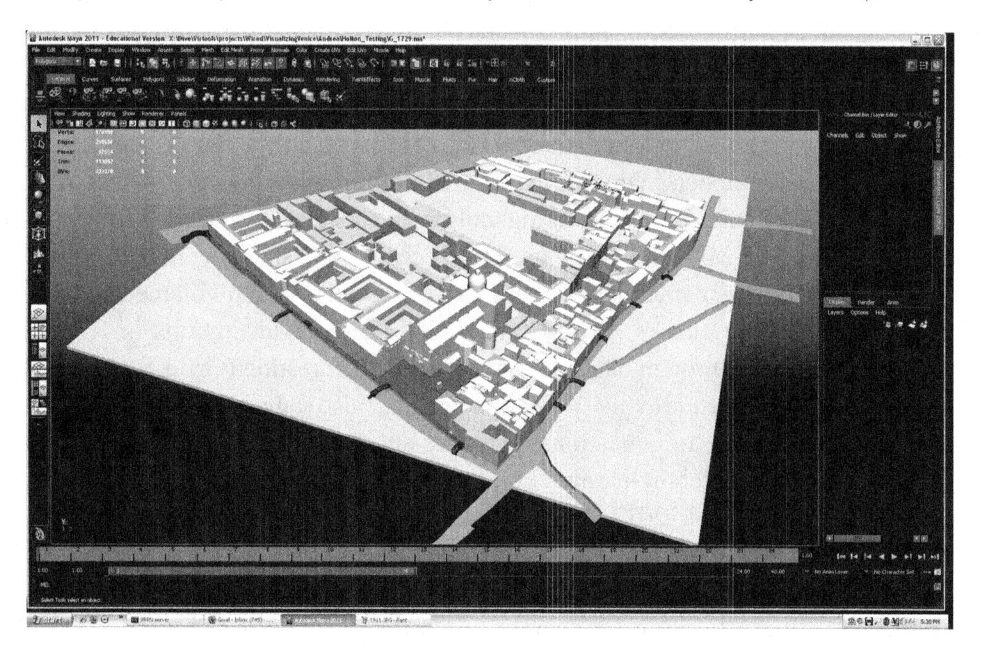

Figure 2.

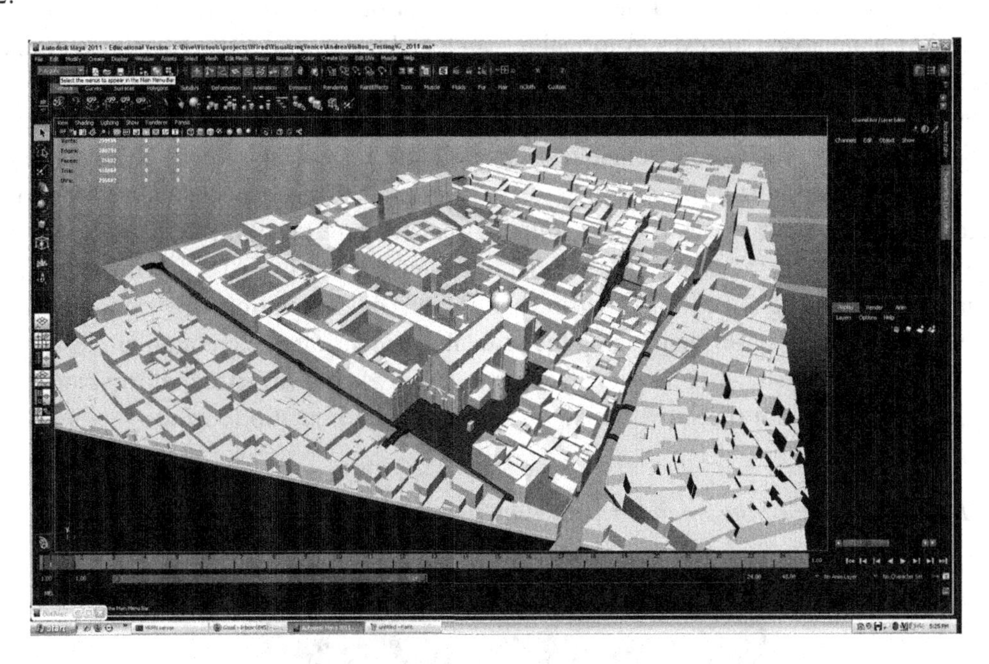

DATA ACQUISITION

Preliminary from the 2D/3D data acquisition and modeling process, Visualizing Cities organizes an integrated generation of new digital operating models: the ADT, that will allows innovative data management, moving beyond the physical surface and defining a multidimensional ontological 3D structure. Historical, geometrical, material, construction, environmental data are mixed, organized, and linked to the time evolution of the architectural/constructive topics (Szabo, 2020). This trial involves the following actions and potential results of the ADT:

- Historical research and comparison with coherent architectural/constructive phenomena;
- Integration of active and passive techniques to define an integrated visible and non-visible model;
- Point clouds segmentation and classification to identify architecture ontology;
- Construction of as-built models and multi-LOD models;
- Developing and refining the pairing of multiple raw data sources with BIM models for management qualitative control;
- Automated as-built model monitoring synchronized with the as-planned one;
- Image-based assessment to verify the structure's condition.

The 3D virtual models – ADT at different detail levels – will involve all the data referred to the built environment knowledge, also relating main geometric and material aspects. Integrated digital surveys by photogrammetry/laser scanning is planned at different scales also in relation to historical research. The BIM model will be then generated from the surveyed / collected data (Borin et al. 2021). Regarding to the urban scale, Visualizing Cities implicates methods, intervention, management procedures, to understand cultures' peculiarities (Pelliccio et al, 2017). The ADT became an integration of 3D active and passive methodologies defining a reality-based 3D model, merging various information ranks. At the architectural scale the analytical approach involves historical research and passive / active procedures to implement interoperable models, also crossing visible and invisible, investigating the geometrical configuration, the building life cycle, the historical issues, the constructive and structural features, with feedback in diagnostic and historic construction techniques. The building scale analysis will be advanced to comprehend the constructive constituents and their degradation state. The elements referred to constructive and structural features are then achieved, correlating manufacture history, predictable performances, contributing to forthcoming maintenance / safeguarding intervention. The ADT has then feedback in the training sphere: Visualizing Cities aims to achieve different 3D geometrical models to understand the built space topics (Figure 3) also challenging the students' feedback in the creation of AR applications. Relationships between 3D shape, data content, and the capacity to convey space information in different learning contexts is tested for the knowledge at the different representation scale.

Figure 3. Implementation of the Augmented Digital Twin (ADT) with BIM of San Francerso Church - Padova.

DATA COMUNICATION

The core of Visualizing Cities is to provide – with advanced ADT applications – the organization of a qualitative communication flow through simulation, predicting, formulation, and recommendation of thematic proposals in UBH. This action also adopts the UBH's inclusion, know-how transfer, and security management. An open and inclusive knowledge involves learning modules to create new competence with a collective impact at the social level. This challenge comprises the following activities and possible outcomes:

Communication with a dynamic forecast of requirements and inclusive/cross-culture dissemination;

- Adoption of open semantic interoperability and standards between BIM and ADT;
- Inclusive, interactive, and engaging open AR-based educational models;
- Expedition of AR for learning purposes.

In this sense, the ADTs can be considered research products and tools to broadcast knowledge related to UBH. Most of the communication skills intend to transfer knowledge towards different local, regional, and national realities. The commitment of social activities helps safeguard and valorization of UBH, raising social consciousness of the project's reputation among all the actors and the sense of belonging of the local communities, promoting the social and economic values. Specific connections with public institutions active in the promotion of UBH (i.e., Accademia and Palazzo Ducale in Venice, Palazzo Pio in Carpi) as well as with City and State archives (digital resources) and other local organizations (cultural associations). Communication, valorization, and knowledge sharing to the wide-ranging community is pursued through the proposal of sustainable cultural tourism, involving visual storytelling (Ippoliti & Casale, 2018). ADTs' social role is also exploited in the educational area, amplifying the ML capacities for complex data visualization. The educational system supports data communication, understanding, and transmission. Visualizing Cities is the configured as a chance to exploit scientific research and generate artifact innovation processes, according to seminar-conference / workshop of interdisciplinary research areas. Therefore, a preeminent digital communication and representation apparatus plays a primary role in the project: the ADT is usable through AR and VR. Immersive experiences are developed in situ, identifying the most suitable devices to make the AR capability useful for the project purposes. Moreover, the 3D results obtained by the research project is shown through an inclusive online platform, with data access regulated by user rights typology.

INTEGRATION OF AI FOR THE DIGITAL RECONSTRUCTION

Digital reconfigurations of urban and architectural entities attributable to conservation processes and heritage promotion objectives are linked to two progressively evolving factors, namely the structural and material survey of real space (digital cloning) and its virtual migration into immersive interactive spaces. Adherence to these new tools, with great benefit in communicating research results, allows general users not only a better understanding of architectures and complex and articulated urban and territorial structures, but also offers scholars a sophisticated analysis tool, with a better possibility of semantic-archival cataloging of 3D assets and an always possible integration of data for the definition of further research, which can be implemented with more recent data and therefore usable in the future through new tools and formats that can be used in the context of enhancing the history of city architecture. Interactivity and immersion, usable through VR / AR and XR devices, gives greater freshness and dynamism to projects for the enhancement of Heritage, with great benefit for the communication of such content that is often difficult to understand for non-professionals (Palestini & Basso, 2017). The new technologies also help digital designers to build projects that are increasingly articulated and complex and at the same time perform well from a perceptive graphic point of view, progressively tending towards extreme realism: the discipline of 3D representation and design, for example, now make extensive use of real-time raytracing - already present in various software, first and foremost Unreal Engine 5 with its Lumen algorithm for managing dynamic volumetric light - as well as the possibility of managing within a single huge virtual scene 3D assets composed of millions of polygons through the incredible Nanite proprietary algorithm - without the need to implement preliminary optimization and light baking processes. As far as innovative methodologies for cloning elements from the real world into digital format are concerned, these certainly include the new range-based survey tools (laser scanner systems, progressively more convenient and easier to use, together with hybrid Lidar systems (Perticarini et al., 2021), which today allow direct and consecutive acquisition of a complex space), and image-based survey methodologies such as photomodelling (with or without the use of drones) through the involvement of various image formats ranging from hyper HD frames to 360 equirectangular image formats. In relation to this last image-based category, especially thanks to Nvidia video cards and the new RTX architectures, systems are coming forward that are based on the use of neural networks and artificial intelligences aimed at automatically generating artificial images and 3D spatial models. Thanks to the combination of computing power and AI, visual rendering processes change, and this also strongly influences 3D acquisition methods (Xie et al., 2021) (Müller et al., 2022). Photogrammetry and laser scanning remain the most widely used methods for capturing three-dimensional objects or scenes, but another increasingly precise surveying technique based on artificial intelligence is developing: these are the Nerf (Neural Radiance fields) systems which, structured in a few different versions with specific characteristics, make it possible to generate, with the use of machine learning and very little photographic data, articulated constructed spaces, basically using the depth maps of the images to reconstruct the multilevel 3D spatial perception in order to define the density of the objects and their colorimetric properties in a 3D spatial coordinate system. The use of this new technology to support a renewed method of surveying is able, thanks to machine learning, to generate the missing frames of an image-based acquisition and to produce 3D scenes of environments or objects using inverse volumetric rendering (Figure 4).

Until now, the volume ray marching process used to take a very long time before delivering good results, but now thanks to the new Nvidia RTX GPU architecture, very detailed acquisitions can be achieved in a matter of seconds. In volumetric rendering, the processing does not stop at the data of the surfaces of the objects, but the acquisition rays penetrate inside them; it differs from traditional ray tracing (rendering that only concerns surfaces) because it uses a single ray to obtain the information about the light sources and their intensity; therefore, it does not need the secondary rays to reproduce the global illumination of an environment. The experimental workflow involves the use of the open-source algorithm through a specific compilation in which it is possible to run the code, which is based on a particular framework called Tiny CUDA. Using the Anaconda platform and the Python language, the algorithm begins a first training phase by reading all the images that are present in the survey directory and prepares the groundwork for the second phase in which the software is launched - directly from Anaconda - to process the 3D scene. Unlike traditional photogrammetry, the software does not generate a point cloud but a cloud of voxels (volumetric pixels) and the scene is presented as a cubic volume composed of a "fog" of voxels that gradually thins out to reveal the objects involved in the survey. By means of certain commands in the software interface, it is possible to optimize the final product by means of bounding boxes or certain masking operations and finally proceed to export the mesh (Figure 5).

Figure 4. Neural Radiance fields for 3d visualization and digital reconstruction of Santa Giustina in Padua

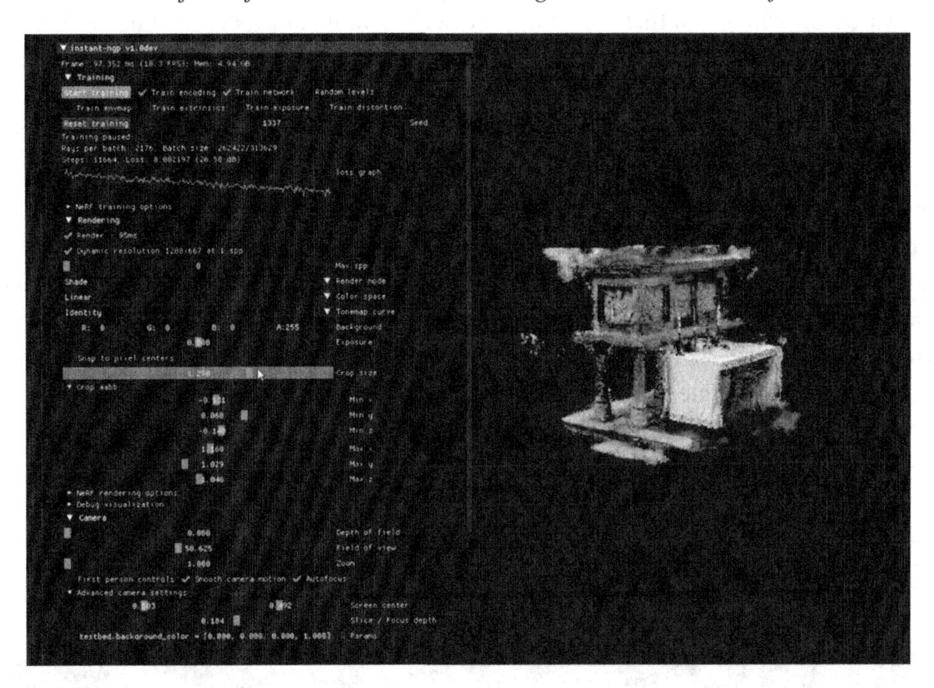

Figure 5. Altar of Santa Giustina model. Mesh result of Nerf experimentally processed with LAZPOINT and Cloud Compare

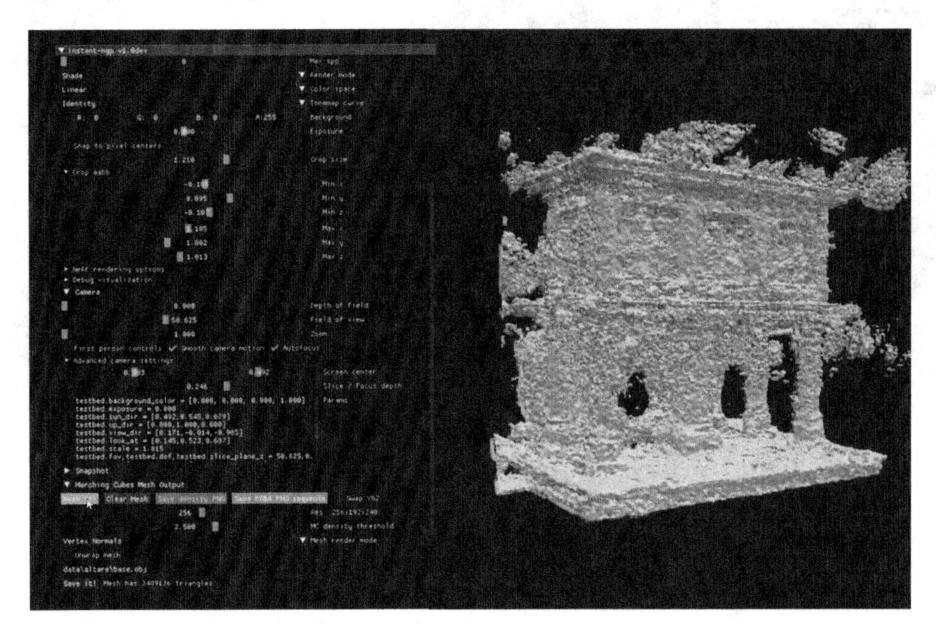

In the current version of the algorithm, the mesh is rather fragmented and, being the product of a volumetric rendering, is composed of irregular polygonal meta-balls that also invade the spaces inside the objects. These results derived from the current version of the algorithm can be improved through the use of third-party software, specific for polygonal optimization, such as Zbrush, to intervene directly on the polygonal obj with color vertex applied, or through Cinema4d (with the use of the plug in LAZ-POINT 2) which is capable of generating an additional point cloud from the polygonal model, always including vertex color, which can then be processed with canonical procedures in the Cloud Compare environment. Comparing this method with the photogrammetry technique, the big limit of Nerf is in the model export phase, which needs to be worked on and optimized later. This method is much more efficient in surveying reflective or metallic objects and is also good in experiments made with photographs with different light exposure, with obstacles that hide parts of the object and with photos that portray moving subjects. Two recent studies from the University of Washington (Park et al., 2022) and from Google Research demonstrate how it is possible to carry out a Nerf survey using selfies with your smartphone (which in photogrammetry would be impossible given that the framed subject would move and would be minus the correspondences of the homologous points between the frames) or using existing photographs taken at different times and light conditions. In the article written by Google Research researchers we talk about a reinterpretation of Nvidia's algorithm (called Nerf-W) and how it has been improved to favor the 3D acquisition from unstructured images of an artifact, most often drawn from the web and with diversified exposure and framing conditions (Figure 6).

Figure 6. Nerf-W research. A new approach for the use of generic images of architecture sourced from the web

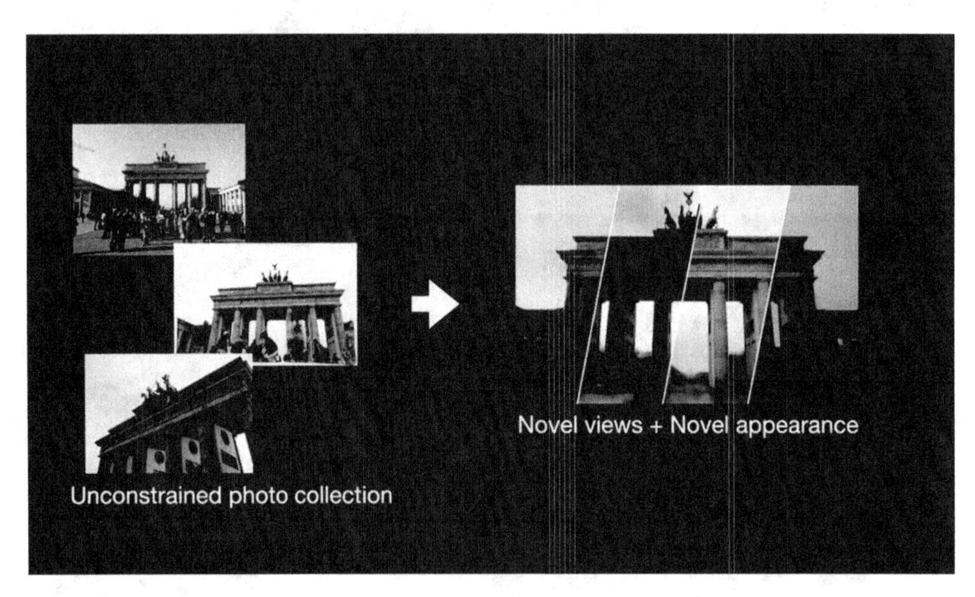

In their conclusions they assert:

We have presented NeRF-W, a novel approach for 3D scene reconstruction of complex environments from unstructured internet photo collections that builds upon NeRF. We learn a per-image latent embedding capturing photometric appearance variations often present in in-the-wild data, and we decompose the scene into image-dependent and shared components to allow our model to disentangle transient elements from the static scene. Experimental evaluation on real-world (and synthetic) data demonstrates significant qualitative and quantitative improvement over previous state-of the- art approaches. ~(Martin-Brualla et al., 2021)

The use of these methods in an integrated way allows a more expeditious approach in the correct reconfiguration of a complex entity such as that of the city, in relation to the territorial macro-scale and to a smaller scale such as the architectural one.

INTEGRATION OF MODELS AS A MEANS OF ANALYSIS IN THE CONSERVATION PROCESS

Visualizing Cities then intends the urban site as the set of elements not directly attributable to the static nature of the building but related to dynamic movements, which can constitute the very essence of the city to be represented in relation:

- To the lives that inhabit it,
- To dynamic mutations of another nature, like the passage of time that influences the very conformation of the architectural elements.

Thanks to new hardware and software technologies, digital reconfigurations thus increasingly become an excellent tool to tell the evolution of the city, showing urban configurative situations relating to different historical periods and even very distant ones in a system of parallel levels. There is the possibility of visualizing the progressive metamorphosis of cities through sophisticated digital spaces, also including hypotheses of future transformation. The quality of the obtained results and their communicability is strictly connected to the technological evolutions: the elaboration of the preliminary phase of cloning / reconfiguration through the new digital survey methods, obtaining an acquisition of the 3D data faithful to the original in order to contain – especially in Cultural Heritage area – all those recognizable signs of the artifact in its architectural evolution. In this way, the ADT will constitute the actual essence of the material and immaterial value, in relation, for example, to what seems to be lost in the past but which in the actual it has left, in the DNA of the place, a barely visible trace of its existence. In this sense, this intelligent clone of the reality may facilitate / assist the conservation process of architectural heritage with the virtual reconstruction of architectural features (Bernardello et al. 2020). On the other hand, the development of national and European directives through the extensive use of BIM has made it possible to study specific modeling practices for the analysis and conservation of buildings. In this sense, the study of buildings with differentiated construction practices have to be implemented in the interoperable 3D model: it is important to rely on the typical interrelation between architectural and structural objects. It thus becomes important to establish:

- The geometric rules for the passage of information between the BIM model and the analytical models (with Finite Element Method – FEM),
- A valid methodology for describing building pathologies of the structural/configurative surfaces.

Starting from the analysis of the analogical and digital methods to classify these conditions, in a BIM environment it is possible to integrate the information, usually incorporated into two-dimensional drawings, relating to the conservation conditions (figure 7). The ADT therefore allows to create three-dimensional semantic objects related to specific information of type of building conditions, which can help the conservation and restoration of building elements.

Figure 7. Information of the conservation conditions in the ADT of the Porte Contarine, Padova

CONCLUSIONS

The urban and architectural Digital Understanding involves conservation processes and heritage promotion objectives, interrelated to two increasingly progressing issues, specifically the survey of real space (digitally emulating) and its virtual relocation into immersive / interactive / interoperable spaces. The intelligent adoption of these new apparatuses, with great benefit in communicating research results, consents a better comprehension /conception of composite and complex urban configurations, also offering to intellectuals and researchers a sophisticated exploration apparatus, involving semantic-archival cataloging of 3D assets with the integration of data for the characterization of additional research: the opportunity is then to implement more recent data, usable in the future through new tools and formats in the context of enhancing the history of city and architecture.

Therefore, the developed ADT is configured as digital management model to achieve the previous goals transforming UBH organization/management/intervention/communication from reactive to proactive by:

- providing a digital architype to enhance intervention management and visit/exhibition organization, also involving inclusion matters. The ADT platform also enhances the situational awareness of the UBH, overcoming uncertainty in planning, physical conditions, communication organization.
- providing tools to support UBH control, including a database,
- using knowledge gathered over time through historical phases stored in the ADT system, for case-based planning.
- improving the results of three-dimensional acquisitions of the surrounding environment by reducing the acquisition time for surveying and in terms of 3D visualization, avoiding long optimization phases of the 3D models

REFERENCES

Ahmed, S. (2019). A review on using opportunities of augmented reality and virtual reality in construction. *Organization Technology and Management in Construction an International Journal, 11*(1), 1839–1852.

Akçayır, M., & Akçayır, G. (2017). Advantages and challenges associated with augmented reality for education: A systematic review of the literature. *Educational Research Review, 20*, 1–11.

Amin, D., & Govilkar, S. (2015). Comparative Study of Augmented Reality SDK's. *International Journal on Computational Science & Applications, 5*(1), 11–26.

Andrianaivo, L. N., D'Autilia, R. & Palma, V. (2019). Architecture recognition by means of convolutional neural networks. *ISPRS Arch. XLII-2-W15*, 77-84

Baranda, J., Mangues-Bafalluy, J., Zeydan, E., Vettori, L., Martínez, R., Li, X., Garcia-Saavedra, A., Chiasserini, C. F., Casetti, C., Tomakh, K., Kolodiazhnyi, O., & Bernardos, C. J. (2020). On the Integration of AI/ML-based scaling operations in the 5Growth platform. *In Proceedings IEEE Conference on Network Function Virtualization and Software Defined Networks (NFV-SDN)*, 105-109.

Bernardello, R., Borin, P., Panarotto, F., Giordano, A., & Valluzzi, M. R. (2020). BIM representation and classification of masonry pathologies using semi-automatic procedure. In J. Kubica, A.Kwiecień & L. Bednarz (Eds.). *Brick and Block Masonry - From Historical to Sustainable Masonry: Proceedings of the 17th International Brick/Block Masonry Conference (17thIB2MaC 2020)*, July 5-8, 2020, Kraków, Poland (1st ed.). CRC Press

Borin, P., Giordano, A., & Campagnolo, D. (2021). Scan-Vs-Bim Analysis for Historical Buildings. In R. P. Suárez & N. M. Dorta (Eds.), *Redibujando el futuro de la Expresión Gráfica aplicada a la edificación / Redrawing the future of Graphic Expression applied* (pp. 1257–1272). Tirant Humanidades.

Cecchini, C., Cundari, M. R., Palma, V., & Panarotto, F. (2019). Data, Models and Visualization: Connected Tools to Enhance the Fruition of the Architectural Heritage in the City of Padova. In C. L. Marcos (Ed.) Graphic Imprints. Springer, 633-646.

Cordts, M., Omran, M., Ramos, S., Rehfeld, T., Enzweiler, M., Benenson, R., Franke, U., Roth, S., & Schiele, S. (2016). The Cityscapes Dataset for Semantic Urban Scene Understanding. In *IEEE Conference on Computer Vision and Pattern Recognition* (CVPR), 3213-3223.

Farhah Saidin, N., Abd Halim, N. D., & Yahayal, N. (2015). A Review of Research on Augmented Reality in Education. *International Education Studies*, *8*(13), 1–8.

Giordano, A. (2017). Mapping Venice. From visualizing Venice to visualizing cities. In B. A. Piga & R. Salerno (Eds.), *Urban design and representation. A multidisciplinary and multisensory approach* (pp. 143–151). Springer.

Giordano, A. (2019). New Interoperable Tools to Communicate Knowledge of Historic Cities and Their Preservation and Innovation. In A. Luigini (Ed.), *Proceedings of the 1st International and Interdisciplinary Conference on Digital Environments for Education, Arts and Heritage. EARTH 2018. Advances in Intelligent Systems and Computing*, vol 919. Springer, Cham, 34-43.

Giordano, A., Friso, I., Borin, P., Monteleone, C., & Panarotto, F. (2018). Time and Space in the History of Cities. In S. Münster, K. Friedrichs, F. Niebling, & A. Seidel-Grzesińska (Eds.), *Digital Research and Education in Architectural Heritage. UHDL DECH 2017 2017. Communications in Computer and Information Science* (Vol. 817, pp. 47–62). Springer.

Greengard, S. (2019). *Virtual reality*. Mit Press.

Grilli, E., & Remondino, F. (2020). Machine learning generalisation across different 3D architectural heritage. *ISPRS International Journal of Geo-Information*, *9*(6), 1–19.

Hedman, P., Srinivasan, P. P., Mildenhall, B., Barron, J. T., & Debevec, P. (2021). Baking Neural Radiance Fields for Real-Time View Synthesis. In *Proceedings of the IEEE/CVF International Conference on Computer Vision*, 5875-5884.

Huffman, K. L., & Giordano, A. (2019). Visualizing Venice to Visualizing Cities - Advanced Technologies for Historical Cities Visualization. In F. Niebling, S. Münster & H. Messemer (Eds.), *Research and Education in Urban History in the Age of Digital Libraries: Second International Workshop, UHDL 2019*, Dresden, Germany, October 10–11, 2019. Springer, 171-187.

Huffman, K. L., Giordano, A., & Bruzelius, C. (Eds.). (2017). *Visualizing Venice: Mapping and Modeling Time and Change in a City*. Taylor & Francis.

Ibáñez, M., Uriarte Portillo, A., Zatarain Cabada, R., & Barron Estrada, M. (2020). Impact of augmented reality technology on academic achievement and motivation. *Computers & Education*, *145*, 1–9.

Ippoliti, E., & Casale, A. (2018). Rappresentare, comunicare, narrare. In A. Luigini & C. Panciroli (Eds.), Ambienti digitali per l'educazione all'arte e al patrimonio. Franco Angeli, 128-150.

Kritzinger, W., Karner, M., Traar, G., Henjes, J., & Sihn, W. (2018). Digital Twin in manufacturing. *IFAC*, *51*(11), 1016–1022.

Li, X., Yi, W., Chi, H., Wang, W., & Chan, A. P. C. (2018). A critical review of virtual and augmented reality (VR/AR) applications in construction safety. *Automation in Construction*, *86*, 150–162.

Lim, M. Y., & Aylett, R. (2007). Narrative Construction in a Mobile Tour Guide. In M. Cavazza & S. Donikian (Eds.), *Virtual Storytelling. Using Virtual Reality Technologies for Storytelling. ICVS 2007* (Vol. 4871, pp. 51–62). Springer.

Martin-Brualla, R., Radwan, N., Sajjadi, M. S. M., Barron, J. T., Dosovitskiy, A., & Duckworth, V. (2021). NeRF in the Wild: Neural Radiance Fields for Unconstrained Photo Collections. *Proceedings of the IEEE Computer Society Conference on Computer Vision and Pattern Recognition*, 7206–7215.

Mulholland, P., Wolff, A., Kilfeather, E., Maguire, M., & O'Donovan, D. (2016). Modelling Museum Narratives to Support Visitor Interpretation. In Bordoni, L., Mele, F. & Sorgente, A. (Eds.), Artificial Intelligence for Cultural Heritage. Cambridge Scholars Publishing, 3–22.

Müller, T., Evans, A., Schied, C., & Keller, A. (2022). Instant Neural Graphics Primitives with a Multiresolution Hash Encoding. *ACM Transactions on Graphics. Association for Computing Machinery*, *41*(4), 1–15.

Palestini, C. & Basso, A. (2017). The photogrammetric survey methodologies applied to low cost 3D virtual exploration in multidisciplinary field. *International Archives of the Photogrammetry, Remote Sensing and Spatial Information Sciences - ISPRS Archives*, 42(2W8), 195–202.

Palma, V. (2019). Towards deep learning for architecture: a monument recognition mobile app. *International Archives of the Photogrammetry, Remote Sensing and Spatial Information Sciences - ISPRS Archives* XLII-2/W9, 551–556.

Panou, C., Ragia, L., Dimelli, D., & Mania, K. (2018). An Architecture for Mobile Outdoors Augmented Reality for Cultural Heritage. *ISPRS International Journal of Geo-Information*, *7*(463), 2–24.

Park, K., Sinha, U., Barron, J., Bouaziz, S., Goldman, D. B., Seitz, S., & Martin-Brualla, R. (2022). Nerfies: *Deformable Neural Radiance Fields. IEEE/CVF International Conference on Computer Vision (ICCV-2021)*, 5845–5854.

Pelliccio, A., Saccucci, M., & Grande, E. (2017). HT_BIM: Parametric modelling for the assessment of risk in historic centers. *DISEGNARECON*, *10*(18), 1–12.

Perticarini, M., Marzocchella, V., & Mataloni, G. (2021). A Cycle Path for the safeguard of Cultural Heritage: Augmented reality and New LiDAR Technologies. In A. Arena, M. Arena, D. Mediati, & P. Raffa (Eds.), *CONNETTERE/CONNECTING: un disegno per annodare e tessere / drawing for weaving relationships* (pp. 2571–2579).

Stathopoulou, E. K., & Remondino, F. (2019). Semantic photogrammetry - Boosting image-based 3D reconstruction with semantic labeling. *ISPRS Int. Arch. Photogramm. Remote Sens. Spatial Inf. Sci*, *XLII-2*(W9), 685–690.

Szabo, V. (2020). Critical and Creative Approaches to Digital Cultural Heritage with Augmented Reality. In L. Hjorth, A. de Souza e Silava & S. K. Lanson (Eds.), The Routledge Companion to Mobile Media Art. Routledge, 1-14.

Tancik, M., Mildenhall, B., Wang, T., Schmidt, D., Srinivasan, P. P., Barron, J. T., & Ren, N. (2021). Learned Initializations for Optimizing Coordinate-Based Neural Representations. *Proceedings of the IEEE/CVF Conference on Computer Vision and Pattern Recognition* (CVPR), 2021, 2846-2855.

Xie, Y., Takikawa, T., Saito, S., Litany, O., Yan, S., Khan, N., Tombari, F., Tompkin, J., Sitzmann, V., & Sridhar, S. (2021). Neural Fields in Visual Computing and Beyond. *Computer Graphic Forum*, 41(2), 641-676.

Younes, G., Asmar, D., Elhajj, I., & Al-Harithy, H. (2017). Pose tracking for augmented reality applications in outdoor archaeological sites. *Journal of Electronic Imaging*, *26*(1), 1–12.

KEY TERMS AND DEFINITIONS

Augmented Digital Twin (ADT): Innovative data management, moving beyond the physical surface and defining a multidimensional ontological 3D structure.

Building Information Model (BIM): Digital place for the adoption of open semantic interoperability and standards.

Interoperability: Not only interconnection between computers, but now operative interrelation in digital dataset.

Neural Radiance Fields (Nerf): Systems structured to reconstruct the multilevel 3D spatial perception in order to define the density of the objects and their colorimetric properties.

Reconfiguration: Interpretative system of built reality, not only from a formal point of view.

Transformation: The mutation of architectural/engineering structures during time, as an organism.

Urban Built Heritage (UBH): Buildings in historical cities, where is still mandatory a deep exploration of their characters and mutual networks.

Chapter 10
Digital Transformation Stemming From a Business Assessment of Construction Industry 4.0

Edoardo Montevidoni
Politecnico di Milano, Italy

Claudio Mirarchi
https://orcid.org/0000-0002-9288-8662
Politecnico di Milano, Italy

Antonino Riccardo Parisi
Politecnico di Milano, Italy

Alberto Pavan
https://orcid.org/0000-0003-0884-4075
Politecnico di Milano, Italy

ABSTRACT

The methods, processes, and tools adopted according to the needs of the transition based on the Industry 4.0 should be based on the level of digitization of the companies, checking and monitoring their digitization over time, and considering the relation within the society. The study presented in this chapter starts from the work of the European community, directed to the assessment of the digital maturity of companies in the context of the European network of digital innovation hubs. Assessment that takes place through the compilation of questionnaires assessing the digital maturity of companies. Starting from what has been developed by the European community, the authors believe it is essential to develop specific focal points according to the peculiarities of the different sectors and in particular considering the construction one. This approach will open a new key to promote the digitalisation of the construction sector that is still lagging compared to the other industrial sectors.

DOI: 10.4018/978-1-6684-4854-0.ch010

INTRODUCTION

Construction contributes significantly to a country's socioeconomic development. The construction industry, accounts for 9% of global GDP, which indicates its significant impact on economic development. Global construction spending hit $11 trillion in 2017 and is expected to hit $14 trillion by 2025, according to McKinsey (2018). However, the construction industry is characterized by a low productivity that can be related to lack of technological advancements, computerization, and robotics.

Despite its significance, the industry has long been plagued by poor product delivery in the majority of developing countries (Ogunsemi & Jagboro, 2006). For example, South Africa, according to Emuze (Emuze, 2011), is a country where it is nearly impossible to fulfil a client's dream. This is due to a lack of project delivery within budget, time frame, and specification parameters. Although the construction industry has struggled with poor project delivery for many years, the technological innovation is demonstrating the ability to reduce the industry's long-standing issues (Aghimien et al., 2018).

When digital technologies (DTs) are used, such as Building Information Modelling (BIM), project delivery can be improved in terms of both cost and time. This can be related to the fact that clashes in designs can be more easily identified and that the possibility of faulty design and rework, as well as the associated cost and time wastage, is reduced. One of the most significant impact of digitalization, automation, and integration of processes is related to an increased productivity improving also design and construction quality. For example, technologies such as the Internet of Things (IoT) are reshaping the way data can be captured and shared during the construction process phases (Ammar et al., 2018). Big data analytics (Jin et al., 2015) promises improved predictions of future construction project delivery because patterns from previous projects can be identified and informed decisions can be made early before a project commences, allowing for better project planning and execution (Bagheri et al., 2015).

On the one hand, when it comes to adopting new technologies and automating processes, the construction industry is moving at a glacial pace. On the other hand, technological innovation is revolutionizing the construction sector, a change that requires precise strategies essential for increasing the processes efficiency in the entire sector and considering the peculiarities of each stakeholder. The objective of this chapter is to develop a process framework to assess the level of digitization of construction companies to provide a common ground of analysis that can be used to identify and develop the best strategies for the digital innovation. The phase of investigating and assessing the level of digital maturity is often neglected or discredited, bringing to the development of strategies that are not aligned with the real needs and requirements of the companies. In fact, the assessment process can work on two drivers. On the one hand it makes possible to establish which are currently the most important digital trends in the construction market with a better understanding of the advantages they can introduce. On the other hand, it highlights the digitally deficient areas of the business facilitating the identification of the existing issues and limitation for a faster and more targeted change to business needs.

The rest of the chapter is structured as follow. A background section, divided into two parts: the first one explaining the main developing technological trends about construction; the second one analyzing the main references of the scientific world for the evaluation of the digital maturity in construction sector. Then an analysis of the existing issues and open research areas is proposed followed by the presentation of the approach proposed in the research that the Authors developed around the assessment topic. Finally, future trends and conclusions are reported.

BACKGROUND

Digitization Of Construction Sector

Architecture, Engineering, and Construction (AEC) firms are increasingly using Building Information Modeling (BIM) to exchange and manage data throughout the construction project phases. However, BIM adoption in the construction industry has been slower than expected (Walasek & Barszcz, 2017). Moreover, BIM represents only the pit of the iceberg while several different drivers should be considered such as Internet of Things (IoT), 3D printing, Virtual and Augmented reality, as well as transversal topics such as cybersecurity, digital infrastructures, etc. (DigiPLACE Consortium, 2020). Construction is critical to the economy of the country, hence the need to promote the adoption of new technologies in order to better manage projects and increase productivity. In the following some of the main trends about digital innovation in the construction industry are presented to provide a common background and facilitate the understanding of the common lines then used in the development of the digital maturity assessment research proposed in this chapter.

In the Fourth Industrial Revolution, Industry 4.0 refers to a collection of future industry development trends which aim to achieve more intelligent manufacturing processes. Among several other things, these trends also include, the use of Cyber-Physical Systems (CPS), the construction of Cyber-Physical Production Systems (CPPS), the Internet of Things (IoT), and the implementation and operation of smart factories (Zhou et al., 2015). Industry 4.0 involves the increase of digitization, automation, and information and communications technology use in manufacturing in order to create a digital value chain of a product's lifecycle, from concept to development, manufacturing, use and maintenance, recycling, and disposal (Lasi et al., 2014). This results in higher-quality products at lower costs and faster time-to-market, as well as better business performance (Brettel et al., 2014). The four fundamental design principles of Industry 4.0 have been identified and have further been explained by Herman (Hermann et al., 2016).Even though the characteristics of a building project in the construction industry are vastly different from the characteristics of products in manufacturing industries, still the concept of Industry 4.0 can successfully be applied in the building construction industry. Internet of Things (IoT), Internet of Services (IOS), the Construction Planning System (CPS), Big Data, Cloud Computing, and Robotics are just a few of the Industry 4.0 components that are currently being implied in order to digitize the construction industry. However, the practical applications of Industry 4.0 are still in their infancy, as explored by Oesterreich and Teuteberg (Oesterreich & Teuteberg, 2016).

According to McKinsey & Company (McKinsey&Company, 2020), in the construction industry can be identified several driver technologies and procedures that are increasing both in terms of diffusion and impact. Above all 3D printing, modularization, prefabrication, robotics, digital-twin technology, artificial intelligence (AI), analytics and supply-chain optimization and marketplaces. For example, for De Soto (García de Soto et al., 2018), autonomous robots will soon coexist with traditional construction systems, increasing job variability, creating new roles for construction workers and modifying the existing ones. Gubbi (Gubbi et al., 2013) analyzed the application of cloud computing which provides scalable IT services over the internet to multiple external customers, and Celaschi (Celaschi, 2017) the use of augmented reality which allows users to see objects or designs in a new way. These studies also emphasized the role of digital technologies in improving construction delivery through a digital building system that can help from plot selection to project handover.

Prefabrication, Modularization, Automation

Prefabrication is defined as the use of factory automation to manufacture housing or housing components in a controlled environment (modular, prefab, panelized, precast, etc.) (Neelamkavil, 2009). Because of the industry's uniqueness, direct adaptation of technologies used in a variety of industries, such as those that support mass customization or those that support mass manufacturing, presents a number of difficulties (Shen et al., 2010). Automated Design with automatic knowledge exchange and the widespread use of BIM can support construction automation. BIM enables architects, engineers, contractors, owners, and facility managers to collaborate on the design, construction, and operation of a building throughout its lifecycle. Information such as preliminary design data, geographic information systems (GIS), financial and legal documents (such as financial and legal documents for a construction project), mechanical electrical, and plumbing layouts and specifications (including environmental and energy modelling results), and alternative data are all examples of information that can be shared. Material handling and manufacturing are the main processes that can be automated in order to reduce the risk of injury, the amount of time spent exercising, and to ensure capability, as well as boost productivity. Thanks to the automation of the manufacturing processes, production documentation will be accurate and up to date, and it will be accessible at various points throughout the manufacturing process. Artificial Intelligence (AI) can also represent a boost in this direction impaction on the automation of robots and of the construction site. In the opinion of Martinez et al.(Martinez et al., 2008), robotization and automation of the construction method have been accomplished with, for example, the following objectives in mind:

- Standard design of buildings, with robotic erection in the construction phase.
- Automatic planning and time period re-planning of offsite manufacturing, transportation and on-site assembly.
- Machine-driven transportation as well as automated assembly of prefabricated parts on-site.

The use of automation and artificial intelligence in construction covers the entire construction life cycle. This ranges from the initial design to the final activity i.e., the demolition of the building, thereby, including all stages from on-site construction to building maintenance and management. Construction project management would include the use of computer-aided design (CAD), the creation of value estimates, construction schedules, and cost accounting. Moreover, it also includes the development of software systems and ingenious machines that employ intelligent management throughout on-site operations. The advantages of automation include high accuracy, consistent quality, and the ability to do more work in a dangerous environment. Once the automated building construction system has been refined and tested on a number of occasions, it is expected to reduce construction time and costs, as well as dangers involved within.

Internet of Things and Digital Twin

Several authors (Gubbi et al., 2013) have recently discussed the significance of Radio Frequency Identification (RFID) and wireless sensor network technology in Internet of Things (IoT) applications. Because of the Internet of Things, users have been able to bring physical objects into the cyber world (IoT). Near Field Communication (NFC), RFID, and 2D barcode were some technologies used to make this possible by allowing physical objects to be assigned an identification over the internet, which enable them to be accessed conveniently. Using RIFD technology, a contractor and CI owner can benefit from a tool that allows them to gather and manage information over moveable databases. They will also benefit from the tool's assistance in communication and management directions. RFID tag-based systems are used for tracking all types of equipment as well as for creating inventory management systems. It will also be used for supply chain management. Despite the difficulties of a construction site, the Real-time Location System (RTLS) used for safety management at the job site delivered accurate and reliable results. Vehicle navigation technology such as GPS and Bluetooth can be used in the construction industry as a tracking system. In order to improve information gathering and management, an RFID-based quality review and management system (RFID-QIM) has been studied (Gubbi et al., 2013). Sensor based Monitoring Systems in the construction site's quality and testing management system is described, with specific goals for examination and design sections, data acquisition and analysis, and defect management and detection.

These examples demonstrated how the use of IoT has been extensively studied in the construction industry and more application can be mentioned considering for example the use in bridge and structures monitoring, indoor environment monitoring and quality control, building performance in real environment, construction site monitoring, etc. Moreover, the interconnection between IoT, BIM, and analytics represents the foundation for the development of Digital Twin in construction sector.

Analytics, Cloud Computing, AI

Numerous construction projects are incorporating digital sensor systems, intelligent machines, mobile devices, and new software system applications, all of which are gradually being integrated with a centralized building information modelling platform (BIM).

Big data has the potential to quickly rework each method of construction contracting and produce solutions to construction problems.

Big data is currently used, for example, to examine the construction delays, learn from Post-Project Reviews, assist with construction calls, detect structural damages to buildings, and identify staff and significant equipment actions. Also, big data is intensively being used in construction to track construction equipment. The use of big data can reveal new insights into construction management pricing, designs, and processes. The goal is to develop tools and methods for long-term construction management. These methods must be adaptable to changing user demands, technology, and other framework conditions while maintaining efficiency and productivity. For this purpose, Building Control and Monitoring Systems (BCMS) and Computer Aided Facility Management (CAFM) can all be integrated into the construction processes (BCMS).

Cloud Computing is the most common Continuous Integration (CI) technology. It allows for more efficient construction and better collaboration among all parties involved. (Armstrong et al., 2017)

Armstrong discussed the future of construction in terms of Artificial intelligence (AI) and machine learning. He emphasizes that digitalization and automation will necessitate AI and Building information modeling (BIM) integration. A new system must be designed to encourage the creativity and raise awareness regarding the potential that AI possesses. In the construction industry, huge advancement in AI has now enabled it to tool and manage several construction processes. The CI will save money and time by replacing men with virtual assistants and AI. It solves problems and completes tasks more conveniently as compared to the human beings by utilizing machine learning. Edum Fotwe et al., (2014) highlighted some critical issues related to CI, CMS, and CP in relation to striking a balance between the standardization and innovation in large projects.

Virtual, Augmented, Mixed Reality

Producing construction machinery in less time at a lower cost while maintaining high quality is possible with application of virtual reality technology within the construction machinery industry (Wu et al., 2013). The construction industry is also adopting these technologies, including Building Information Modeling (BIM), virtual reality (VR), augmented reality (AR), and mixed reality (MR). It is dependent on the application of such technologies that determines their maturity level. Compared to augmented reality and virtual reality (AR/VR), modularization and cloud computing technologies are still in their early stages due to widespread industry acceptance. A central concept of digitization is the use of seamless communication technologies to enable interconnected objects and people, through which they can share information and work together to achieve common goals (Hermann et al., 2016). Building information modeling (BIM) has been found to be the most widely used digital planning method in the construction industry, according to recent research (Oesterreich & Teuteberg, 2016). Building SMART is working toward interoperable standards for building information modelling, despite the fact that there is currently no standardized data schema in BIM and no protocol that defines who is responsible for information usage (Chen et al., 2018). Sensor data and digitalized plant models are linked to create a virtual replica of the physical world that helps Internet of Things participants make better decisions about their own operations (Hermann et al., 2016). Virtual reality (VR) and augmented reality (AR) are gaining popularity in the construction industry (Wu et al., 2013), as they have the potential to bridge the gap between the digital (e.g. 3D model, simulation model, etc.) and real world (Wu et al., 2013), (Position or condition of materials, equipment).

Despite the examples provided in the previous analysis, the construction industry hasn't yet fully embraced Industry 4.0. Construction industry must adopt digitization and automation in its workplace processes in order to improve to improve value chain, productivity, and sustainability. The first step in implementing Construction 4.0 is assessing the company's current digital maturity and setting a long-term goal. An established BIM-based project management system is required, as BIM is a digital planning method in the construction industry (Li & Yang, 2017) (Oesterreich & Teuteberg, 2016). Industry 4.0 enabled Construction Supply Chains (Dallasega et al., 2018) and a standardized document management system (Chen et al., 2018) are same measures.

The second step is to choose pilot projects to test and prove the concept. As in Industry 4.0, the company should demonstrate business value by collaborating with external digital leaders and universities to accelerate digital innovation. It is critical to learn from the pilot projects, develop strategies to improve processes by incorporating new and additive technologies like digital fabrication (García de Soto et al., 2018), robot usage (Keating et al., 2017), and train people to implement digitization. To gain a competitive advantage, the company must become a digital enterprise, where the top management and financial stakeholders are clearly committed to take the steps for digitization. Explore new technologies like construction ERP, Scrum, and eProcurement (Aguiar Costa & Grilo, 2015).

The metrics for the transformation are for improved productivity, efficiency, quality, and construction process integration as well as removing obstacles in the path of construction company digitization. Politics, economics, socio-cultural, technological, environmental, ethical, and legal factors are also likely to influence business transformation as such factors have been previously observed to influence on organizational culture and work practices while the adoption of BIM in the construction industry since the turn of the millennium.

Finally, the company should plan for a long-term system in developing customer solutions. Build a new safety culture for Construction 4.0 by using appropriate tools and standards (e.g. BIM for lifecycle management) (Woodhead et al., 2018) (Badri et al., 2018). This will necessitate more regulatory requirements and compliances.

Assessing The Digitization Level of Construction Sector Stakeholders

The ability of a company to respond and adapt towards a disruptive technological change is referred to as digital maturity (Kane. 2019). Digital maturity is a process that necessarily takes place over the course of an organization's life cycle due to the fact that technology is continuously evolving, and companies all over the world must adapt in accordance with these technologies in order to remain competitive.

By addressing a number of goals, a generic maturity model comprises dimensions and criteria that identify the areas of activity, as well as maturity stages that reflect the evolution path towards maturity (Berghaus et al., 2016). The bulk of generic maturity models may be traced back to quality management and continuous improvement activities. There are different maturity models with their own advantages and lacunae. Capability maturity models can analyze more than just a system's compliance with an ISO standard (Eadie et al. 2012).

Process-based maturity models have been criticized for overemphasizing the process perspective and the formalization of improvement initiatives, which, therefore, is accompanied by considerable bureaucracy and does not even ensure the organizational success.

While most maturity models are static and prescribe only one path of implementation and advancement, maturity model evolution should reflect those changes in contexts require changes in implementations to assure value realization. It has been shown that early in the transformation process, firms are more likely to experiment with digital innovations or react to external developments, whereas later in the process, enterprises are more systematic in their planning. Maturity models, at best, provide some insight by depicting an overview of many domains and outlining common paths (Berghaus and Back 2016).

Meng et al. (2011) offer a five-component maturity model for construction supply chain linkages based on the capability maturity approach:

1. Assessment areas (action areas, improvement range) are context-specific, based on the specific processes to be enhanced through the use of digital technology.
2. Criteria for evaluation - When determining the maturity level, describe parts of analysis, and specify contextual aspects to consider when upgrading assessment areas.
3. Maturity levels show the evolution toward maturity and are generally classified into five levels, for example, but there are variations, such as four levels (Eadie et al. 2012) or Santos et al., (2019) Industry 4.0 maturity model with six stages. Each level serves as a springboard for the next, with objectives must be met before progressing to the next (Paulk et al. 1993).
4. The framework matrix combines evaluation areas, criteria, and levels of development.

Need for descriptive, empirical investigations within operations management to produce findings and propositions useful for managers was proposed by Meredith (1993). Wernicke et al., 2021 followed the research design proposed by Meredith (1993) which included a research cycle where researchers describe, explain, and test models in their domains in order to define valid frameworks. This study is structured as follows:

1. Identification of an initial model proposed in practice.
2. Development of a preliminary maturity model based on Meng et al's study (2011) by incorporating literature and empirical findings.
3. Testing of the preliminary maturity model in its practical context to provide understanding through face validation (Brahma 2009).

The existing tools and research are usually based on the evaluation of a specific area and/or component of a company in the construction sector while the overall view and the interconnection between the different activities and areas is missing. This represents a strong limitation in the identification of digitalization paths in companies (especially construction companies) that are characterized by the interconnection of different areas. Hence the need to further analyse this research field providing a framework that is able to embrace both the different company dimensions and the different technologies that can impact on these dimensions.

MAIN FOCUS OF THE CHAPTER

Digital Maturity of Companies In Europe

Considering the crucial role of digital assessment of companies, at European level has been developed an assessment procedure as following explained. The analysis of this procedure constitutes the foundation for the development of the construction sector assessment presented in the following section according to the limitations here highlighted in comparison to the needs of construction sector companies.

Digital Europe will be the first European program dedicated to supporting the digital transformation of the European economy and society.

For the period 2021-2027, the European Commission's proposal includes 5 priority areas:

1. High Performance Computing
2. Artificial intelligence;
3. Cybersecurity and trust;
4. Advanced digital skills;
5. Implementation, optimal use of digital capacity and interoperability.

Within the proposal establishing the new Digital Europe program, the European Commission assigns a central role in the implementation of the program to the European Digital Innovation Hubs (EDIH).

With the launch of the Digital Europe program, a new generation of European digital innovation hubs will be funded to accelerate the digital transformation of SMEs, mid-cap companies and public sector organizations across the European Union. With this in mind, the progress in digital maturity of each European Digital Innovation HUB (EDIH) recipient will be measured through a new digital maturity assessment tool. This tool will be mandatory in order to ensure adequate comparability and aggregation of data at regional / national / EU level.

The Tools Monitoring Policy for The EDIH Community

The EDIH network was created with the digital program for the advancement of digital technologies for business and the public sector.

The network consists of:

- 75% from the adoption of advanced digital technologies (AI, cloud and big data)
- 90% of small and medium-sized enterprises with at least a basic level of digital intensity
- double the number of "Unicorns" in the European Union

Project performance monitoring is defined through KPIs.
Key performance indicators are reported in the EDIHs

- Number of businesses and public sector
- Amount of additional investment
- Number of collaborators

Outcomes indicators and impact indicators

- Digital maturity
- Market maturity and creation of innovation potential

Initially, a mapping of the existing methods and tools of digital maturity was carried out through a research work commissioned by the JRC to external consultants during which thirteen tools were analyzed using a SWOT analysis. This analysis provided the following main results and recommendations for the "DIGITAL MATURITY ASSESSMENT" framework for the "EUROPEAN DIGITAL INNOVATION HUB"s:

- There is no existing tool that meets the monitoring and evaluation requirements of the Digital Europe program. And you need to develop a purpose-built framework / tool (inspired by existing ones).
- It should be mainly linked to the Digital Europe program, but also fit in and help demonstrate the contribution of the "EUROPEAN DIGITAL INNOVATION HUB" initiative to the main political priorities of the EU (i.e., sustainability).
- The questionnaire or questionnaires to be developed for the evaluation of the digital maturity of the "EUROPEAN DIGITAL INNOVATION HUB" customers should cover the main sectors in which potential impacts at customer level are expected from the criteria.

The proposed method consists of two main modules (M1 and M2) which form the basis of the questionnaire. The two modules are shown in the following figure and connect to measurable digital maturity indicators in different areas:

Figure 1. Structure of the DMA questionnaire, Practical guidelines on the use of the DMA, 2022

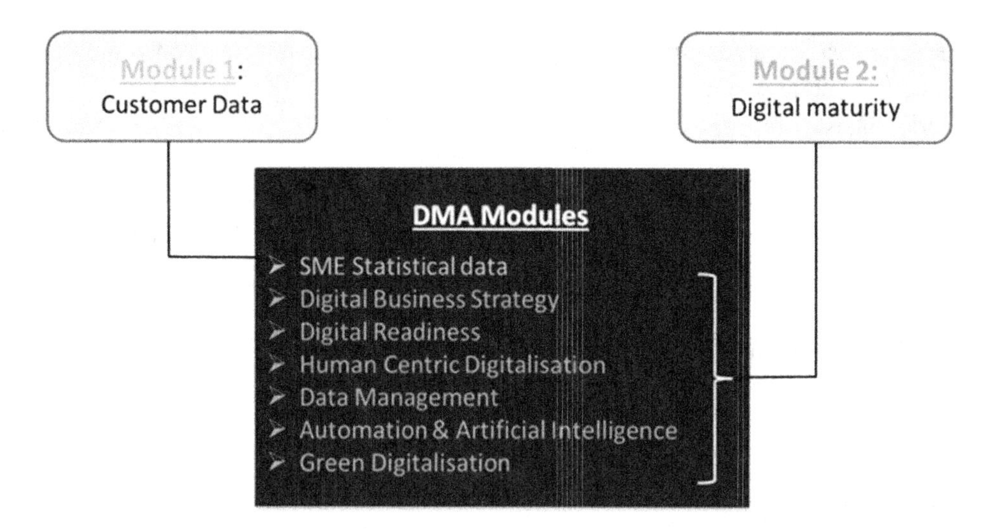

The first module collects general data on the company such as: contact, address, type and size of the organization, sector of activity and more, which will be used for statistical analysis. This information will also be used to analyze the comparison between the level of digital maturity of the company with that of others in the same sector, size or category (from micro to large), region and / or country. This, of course, will only be possible when a substantial number of data from the questionnaire is entered into the database.

The second module is the central part of the questionnaire and consists of questions for evaluating the different aspects of digital maturity within an organization, grouped into 6 areas, which contain 11 sub-dimensions and questions that all together intend to calculate the digital maturity at all. inside a business entity. Grouped into the following six dimensions:

1. Digital business strategy
2. Digital readiness
3. Human-centered digitization
4. Data management
5. Automation and Artificial Intelligence
6. Green digitization

These 6 core dimensions contain 11 sub-dimensions and questions that all together aim to capture digital maturity. This information will capture the starting point of the company's journey through digitalization and will help identify potential areas for the support of the "EUROPEAN DIGITAL INNOVATION HUB".

Figure 2. Structure of the DMA questionnaire, Practical guidelines on the use of the DMA, 2022

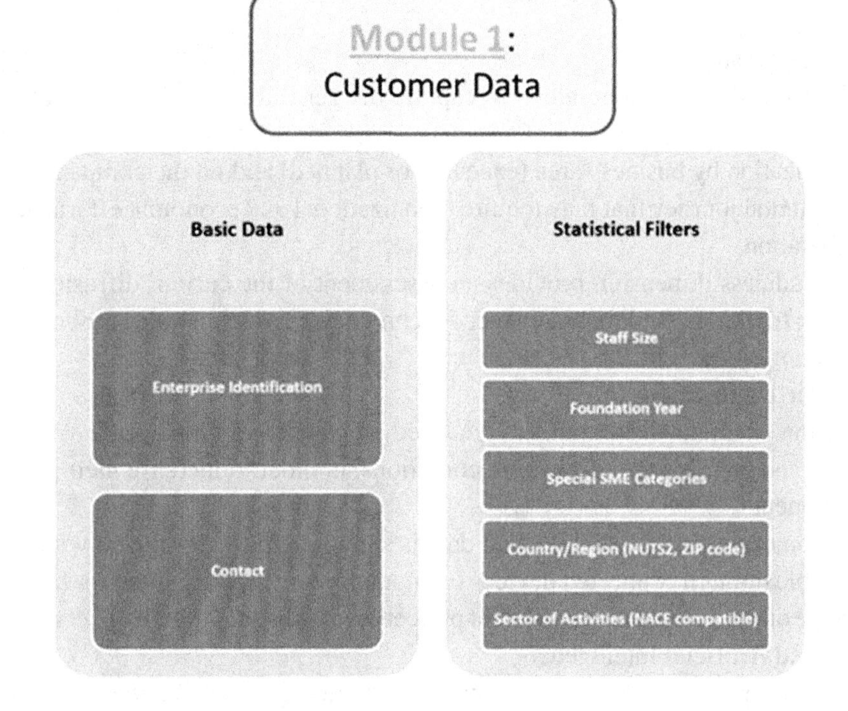

Figure 3. The six dimensions and sub-dimensions of the Digital Maturity module

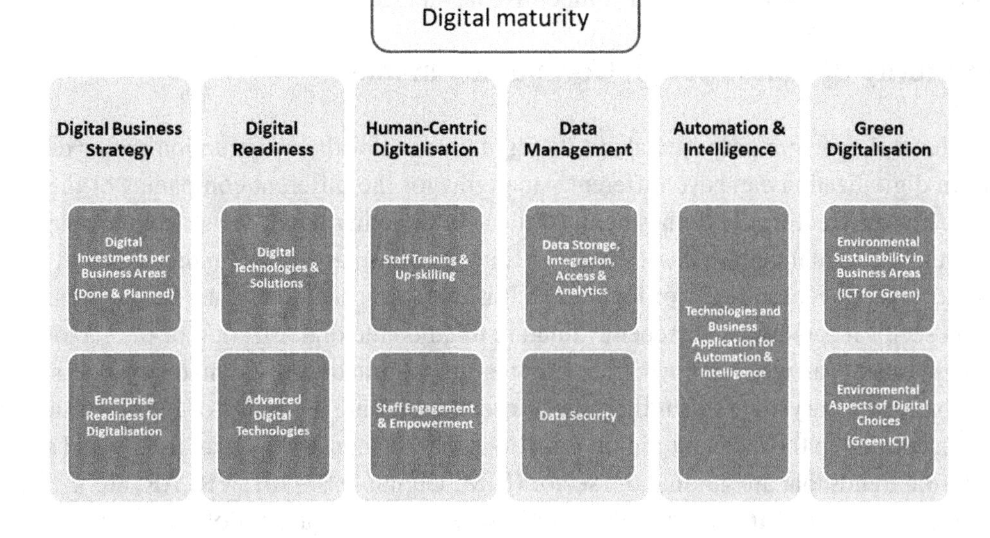

Practical guidelines on the use of the DMA, 2022
The following dimensions will be evaluated:

1. Digital business strategy
   The questions of this dimension intend to capture the general state of a digitalization strategy in the company from a business point of view. They ask for information on the company's investments in digitization by business area (executed or planned) and on the company's willingness to embark on a digital journey that may require organizational and economic efforts not yet foreseen.
2. Digital preparation
   The digital readiness dimension provides an assessment of the current diffusion of digital technologies (both traditional and more advanced technologies) which is valid for both manufacturing and service companies.
3. Human Centric Digitization
   This dimension examines how staff are qualified, engaged and empowered with and by digital technologies, and their improved working conditions, in order to increase their production being.
4. Data management
   This dimension captures the way in which data is stored digitally, organized within the company, made accessible through connected devices (computers, etc.) and exploited for business purposes, keeping an eye on the guarantee of sufficient protection of the data through IT security mechanisms.
5. Automation and Artificial Intelligence
   This dimension explores the level of automation and intelligence facilitated by digital means embedded in business processes.
6. Green digitization
   This dimension captures a firm's ability to undertake digitization with a long-term approach that takes responsibility and cares about the protection and sustainability of natural resources and the environment (ultimately building a competitive advantage out of this).

Digital Maturity of Construction Companies In Italy

On the one hand the picture proposed about the digitalization of the construction sector demonstrates that the term digitalization can have different paradigms for the different companies of the sector and according to the different needs of these companies. On the other hand, the strategies adopted for the development of the digital maturity assessment at European level are based on the possibility to capture all the sectors and are consequently of high level. It is then clear that to capture the peculiarities of the construction sector it is required a deeper evaluation to guide the digitalization of the sector itself. The European assessment can well work to create a first entry point in the digital maturity assessment, but it requires a complementary assessment that can be used to evaluate the needs of the construction sector companies according to the different digital directions that the companies may follow and considering the main digital trends that are guiding the sector (BIM, Digital Twin, IoT, VR, AR, etc.). This accordingly to the peculiarities of the construction sector such as the uniqueness of the product, the need to construct the product (building, infrastructure, etc.) in a specific physical place (the construction site), the composition of the construction sector market (mainly micro and small-medium size companies). These conditions create a difficulty in the translation of methods and approaches used in other sectors directly in the construction one highlighting the need to create tailored solutions like the one proposed in this chapter.

The tool for assessing the digital maturity of construction companies developed in this research aims to create a first point of contact between companies and the digital world and is developed according to a principle of simplicity and immediacy so as to allow its use to the widest range of subjects possible. The main expected results are following listed: an assessment of digital maturity on the different areas typically present in a construction company and a general assessment of the company according to a four-area matrix.

The entry point to the tool is a questionnaire structured in such a way as to allow an automated evaluation of the results and a consequent output according to the points listed above. The digital maturity assessment is structured according to different areas to provide a vision of the level of each company sector, as well as an overview of the general level of digitalization of the company considering, at the same time, the possible different areas of action.

1. Human resources
2. Data management and technology
3. Planning and planning of activities
4. Marketing and customers
5. Construction site
6. Corporate culture on digital transformation

In order to graphically represent the digitization level of each business area, the score is processed on a radar chart (Figure 4) and/or with a graph (Figure 5), which allows an immediate understanding of the business situation. The results from this phase allow the individual company to obtain clear answers, regarding what are its strengths and the areas that can be improved with respect to the level of digitization that is to be achieved. This is of fundamental importance for guiding the company in targeted paths aimed at developing specific skills and monitoring the evolution of skills and procedures in the company. In this way the company benefits from it, improving workflows, with less uncertainty and risks, optimizing costs and times and reducing the risk associated with the failure of innovation projects.

The first part of the questionnaire consists of the so-called "company registry", consisting of a series of questions strictly related to the specific peculiarities of the company business. This allows having an initial information and get an idea of the business category that will be analyzed. Then the questionnaire proceeds with the real digital assessment, which collects more detailed information on the level of digital maturity. In drawing up the questionnaire, aspects relating to the form of the questions to be asked to the interviewees were also considered. A questionnaire, in fact, can be made up of open questions, or questions in which the answer methods are not specified a priori by the researcher, closed questions, or questions with fixed alternatives from which the interviewee can choose. Each type of question involves a series of advantages and disadvantages, therefore we proceeded with an alternate use of the same, to probe the various aspects of the topic being researched in a coherent and reliable way. Figure 6 reports the familiarity of the interviewed companies with the enabling technologies identified for the construction sector.

Figure 4. Example of score achievable for different business functions, Radar graphic, Slim BIM, 2021

Figure 5. Functional areas of construction enterprise, Parisi A. Riccardo, 2022

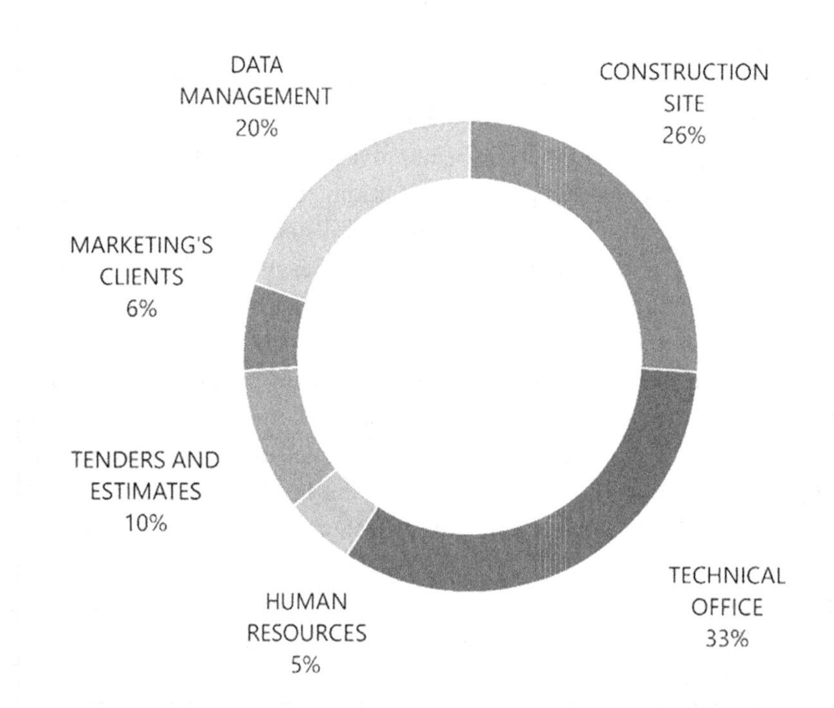

Figure 6. Familiarity with enabling technologies and Industry 4.0, Parisi A. Riccardo, 2022

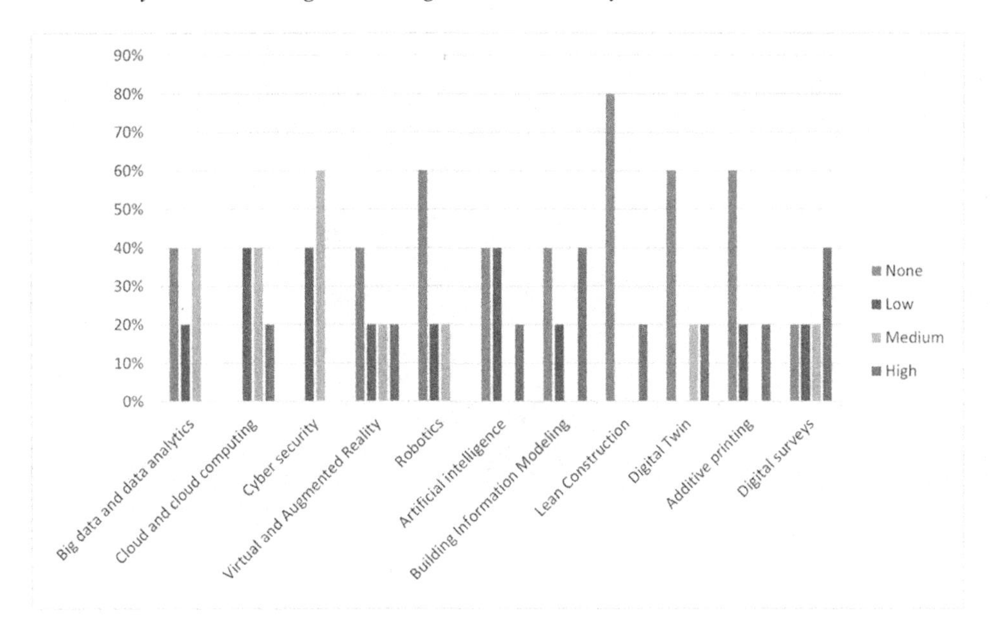

The method of returning the results is a peculiar aspect of this tool, which differentiates it from the others examined so far. The common approach, in fact, is to provide at the end of the given analysis an assessment of digital maturity, which positions the company on a certain level of digitization. This is certainly an effective method; however, it turns out to be not very informative about what the digital maturity of the company really is and what are the direction of innovation where the company may concentrate on. In fact, considering all functional areas with the same importance for every business may result in wrong decision. For these reasons, the tool does not return companies a global score (even if this may be calculated considering the weighted sum of the partial results), but rather provides an overview of the various functional areas. This allows to understand in more detail which aspects within each functional area need to be improved.

The developed tool is still being tested. The companies have been contacted to test the tool and the first results collected and processed constitute a good starting point for an analysis of the current situation of construction SMEs and how they approach industry 4.0.

Also, in this case the data and the identity of the companies must remain anonymous for confidential reasons. However, the data collected will be analyzed and the conclusions set out below. Figure 7 identify the size of the companies that participated to the questionnaire.

The companies contacted and selected as a sample for this project are active throughout the Italian territory, very diversified from each other both in terms of size and structure and in the type of activity. This means that the sample is not representative and conducting statistical analyzes would not return significant results. For this reason, the considerations that will be presented in this chapter must be understood as a qualitative analysis of the trends of the companies in the sample and a first attempt to identify common patterns of digitization.

Figure 7. Size of enterprises contacted, Parisi A. Riccardo, 2022

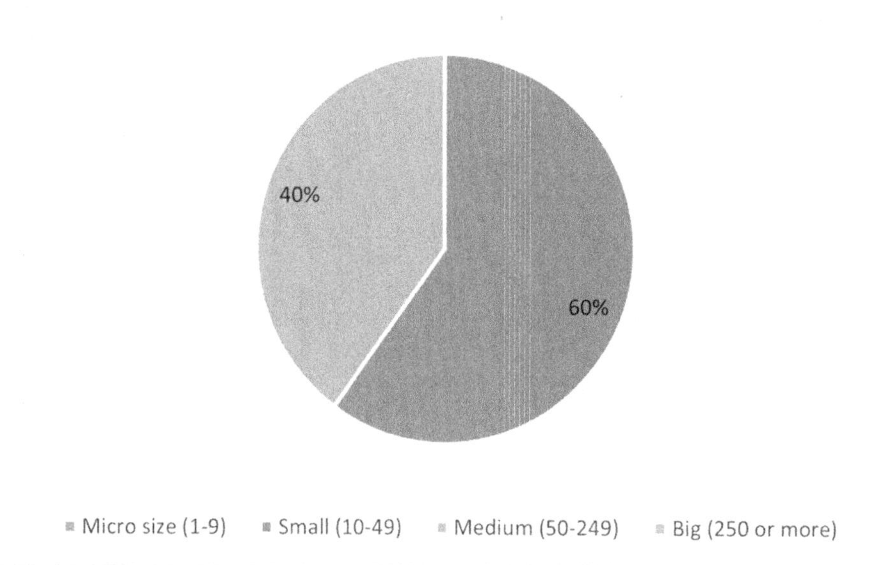

The heterogeneity of the sample also makes it impossible to create benchmarks for the analysis of the digital maturity of companies (whether they are part of this sample or hypothetical companies that will be analyzed in the future with this assessment tool). Comparing very different businesses would not have analytical validity, as each business has intrinsic characteristics that can influence various aspects related to digital maturity, such as the current presence of technologies or their need, as well as the complexity of certain processes. any constraints related to applications and many other factors.

The answers to the questionnaire were collected on an online database and are the basis on which the scores of the companies in the sample are calculated.

For the evaluation of digital maturity, the first fundamental step was to select the questions in the questionnaire that had a logical correlation with the concepts of industry 4.0 and whose answers indicated different levels of digitization.

For this reason, all those questions that are defined as contextual, necessary to identify the characteristics of the company, but which do not indicate a greater or lesser digitization of the same, have been excluded.

This category includes questions relating to company data (foundation, location, turnover, main markets, etc.) and those aimed at identifying other business factors. The first necessary step was therefore to standardize all the various answers of the questionnaire. Figure 9 depicts some of the main results obtained from the questionnaire results analysis.

After analyzing the reasons that led to obtain certain scores from the companies in the sample, it is interesting to observe how there is a correlation between the results and the performance of the companies.

It is necessary to obtain this information from the auditors as to determine the relative importance of each functional area one must have a detailed understanding of the business of each company. At the moment these parameters have not been assigned to the companies, but they will be elaborated in the near future.

Figure 8. Achieved results from Digital Questionnaire, Parisi A. Riccardo, 2022

ACHIEVED RESULTS
CONSTRUCTION SITE
40% use digital tools for automated access control
75% use digital tools to monitor machine and equipment operation
40% use digital tools for scheduling jobs (Schedule, WBS, etc.)
TECHNICAL OFFICE
60% use BIM tools for 3D visualization of the project and interference control (clash detection)
75% use project management software (eg. Microsoft Project)
DATA MANAGEMENT
60% use ERP (Enterprise Resourse Planning) software
75% use document management system (DMS)
20% use customer management (CRM) software
Data storage passes for 60% from corporate server and 40% from cloud

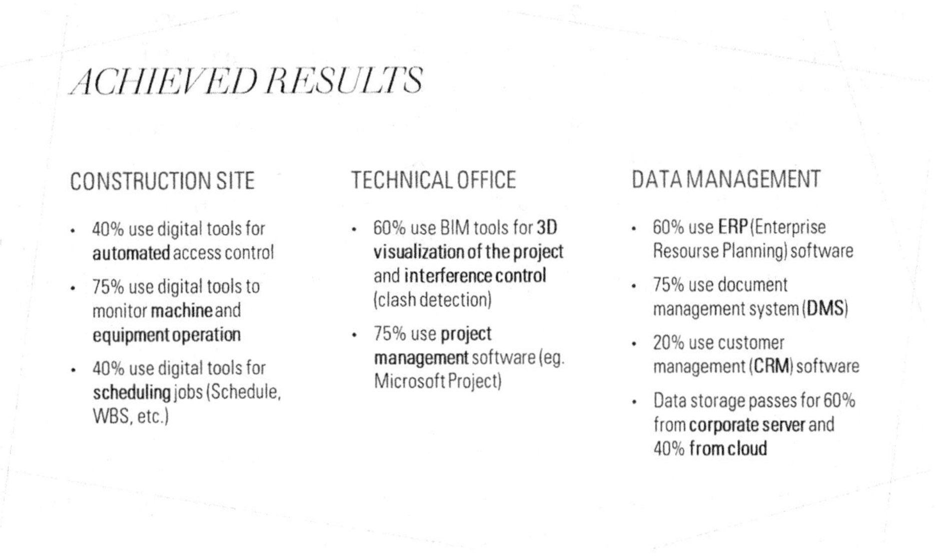

For this reason, it is difficult to state with certainty whether there is a correlation between digital maturity and corporate performance; however, the first trends that emerged from the sample show what are the "digital objectives" that companies intend to pursue in the next 5 years and are reported below (Figure 9).

Figure 9. Percentage of money spent on the digitization of the enterprise, Parisi A. Riccardo, 2022

100%
90%
80%
70%
60%
50%
40%
30%
20%
10%
0%
150,000.00 €
50,000.00 €
50,000.00 €
100,000.00 €
45,000.00 €
Enterprise 1
Enterprise 2
Enterprise 3
Enterprise 4
Enterprise 5
Costs
Percentage spent

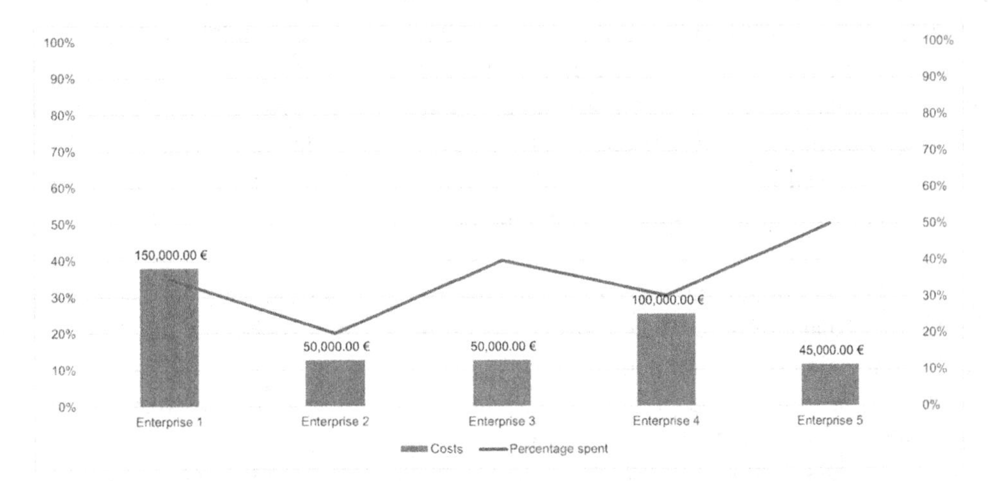

FUTURE RESEARCH DIRECTIONS

In order to be able to integrate the assessment as part of a consultancy activity for a company that decides to undertake a process of digitization, an auditing visit is necessary to allow for the collection of more detailed information. The visit requires the presence of at least two auditors and the commitment required of the company is approximately 2 man-days, which can be divided according to the availability of the company itself. During the auditing visits, the participation of some key figures of the company structure is required, who are able to provide precise and correct answers to the questions asked. The total time of data collection and processing, therefore also considering the back-office work necessary to calculate the maturity index of each company, is approximately 4/5 man-days. As regards the method of return, companies are offered a digital presentation where, in addition to the overview of the scores obtained and the relative radar charts, SWOT analyzes are provided with suggestions on potential courses of action. In addition to this presentation, it is possible, if the company requests it, to present the results through a final company visit.

CONCLUSION

The digital maturity framework enables construction companies to define their digital maturity according to a structure framework based on the main digital innovation trends in construction. Digital maturity refers to an organization's ability to analyze and adopt digital technologies, as well as the affirmation that the technology has an impact on construction site activities. It allows an organization to conduct active and goal-oriented digitally enabled short and long-term improvement processes in a standardized and structured manner while taking into account individuals, technology, organizational structure, and environment.

Understanding the benefits of digitization is a crucial first step, as it is identifying the obstacles they have encountered when integrating digital technology into their current needs and normal business operations. In order to reap the benefits of digitization, possibilities need to be explored and examined.

Using established research methods, the project examined the digital maturity of construction companies, uncover areas for improvement and future growth, and provide recommendations for further development. On the basis of pragmatic studies, it is necessary to determine the type and extent of information, data and process models and to assess the advantages of each model. The end result is a detailed study that includes recommendations on how construction organizations should evolve.

The research proposed in this chapter provides a framework for the digital maturity assessment of construction company working on three main components:

- A structured questionnaire tailored on the main technological trends in construction and the fundamental areas that composes the construction companies and can be identified as different drivers in the innovation of the company itself.
- An automated process for the analysis of the results that combines the answers providing a greed of the digital maturity on the different company's areas.
- An analysis of the possible areas of improvements dynamically produced according to the assessment results (identified as future research direction).

The assessment tool on the construction world here proposed is in line with the setting of the European assessment and can be intersected with the recent policies on the development of the DIHs for constructions. This represents a strong point for the work that have been developed, since it allows the development of a tool that can be integrated in a European context and may contribute to the creation of a shared and common criteria for the digital innovation of the construction sector at different levels.

REFERENCES

Aghimien, D., Aigbavboa, C., Oke, A., & Koloko, N. (2018). *Digitalisation in construction industry: Construction professionals perspective*. Academic Press.

Aguiar Costa, A., & Grilo, A. (2015). BIM-Based E-Procurement: An Innovative Approach to Construction E-Procurement. *TheScientificWorldJournal*, *2015*, 1–15. doi:10.1155/2015/905390 PMID:26090518

Ammar, M., Russello, G., & Crispo, B. (2018). Internet of Things: A survey on the security of IoT frameworks. *Journal of Information Security and Applications*, *38*, 8–27. doi:10.1016/j.jisa.2017.11.002

Armstrong, D., Djemame, K., & Kavanagh, R. (2017). Towards energy aware cloud computing application construction. *Journal of Cloud Computing*, *6*(1), 14. doi:10.118613677-017-0083-2

Ayodele, E. O. (2011). Abandonment of construction projects in Nigeria: Causes and effects. *Journal of Emerging Trends in Economics and Management Science*, *2*(2), 142–145. doi:10.10520/EJC133887

Badri, A., Boudreau-Trudel, B., & Souissi, A. S. (2018). Occupational health and safety in the industry 4.0 era: A cause for major concern? *Safety Science*, *109*, 403–411. doi:10.1016/j.ssci.2018.06.012

Bagheri, B., Yang, S., Kao, H.-A., & Lee, J. (2015). Cyber-physical Systems Architecture for Self-Aware Machines in Industry 4.0 Environment. *IFAC-PapersOnLine*, *48*(3), 1622–1627. doi:10.1016/j.ifacol.2015.06.318

Brettel, M., Friederichsen, N., Keller, M., & Rosenberg, M. (2014). How Virtualization, Decentralization and Network Building Change the Manufacturing Landscape: An Industry 4.0 Perspective. *International Journal of Information and Communication Engineering*, *8*(1), 37–44.

Celaschi, F. (2017). Advanced design-driven approaches for an Industry 4.0 framework: The human-centred dimension of the digital industrial revolution. *Strategic Design Research Journal*, *10*(2), 97–104.

Chen, Q., García de Soto, B., & Adey, B. T. (2018). Construction automation: Research areas, industry concerns and suggestions for advancement. *Automation in Construction*, *94*, 22–38. doi:10.1016/j.autcon.2018.05.028

Dallasega, P., Rauch, E., & Linder, C. (2018). Industry 4.0 as an enabler of proximity for construction supply chains: A systematic literature review. *Computers in Industry*, *99*, 205–225. doi:10.1016/j.compind.2018.03.039

di Milano, P. (2004). Reconciling construction innovation and standardisation on major projects. *Engineering, Construction, and Architectural Management*, *11*(5), 366–372. doi:10.1108/09699980410558566

Emuze, F. A. (2018). Productivity of digital fabrication in construction: Cost and time analysis of a robotically built wall. *Automation in Construction*, *92*, 297–311. doi:10.1016/j.autcon.2018.04.004

Gregor, S. (2013). Internet of Things (IoT): A vision, architectural elements, and future directions. *Future Generation Computer Systems*, *29*(7), 1645–1660. doi:10.1016/j.future.2013.01.010

Hermann, M., Pentek, T., & Otto, B. (2016). Design Principles for Industrie 4.0 Scenarios. *2016 49th Hawaii International Conference on System Sciences (HICSS)*, 3928–3937. 10.1109/HICSS.2016.488

Hossain, A., & Nadeem, A. (n.d.). *Towards digitizing the construction industry: State of the art of Construction 4.0*. Academic Press.

Jin, X., Wah, B. W., Cheng, X., & Wang, Y. (2015). Significance and Challenges of Big Data Research. *Big Data Research*, *2*(2), 59–64. doi:10.1016/j.bdr.2015.01.006

Keating, S. J., Leland, J. C., Cai, L., & Oxman, N. (2017). Toward site-specific and self-sufficient robotic fabrication on architectural scales. *Science Robotics*, *2*(5), eaam8986. Advance online publication. doi:10.1126cirobotics.aam8986 PMID:33157892

Lasi, H., Fettke, P., Kemper, H.-G., Feld, T., & Hoffmann, M. (2014). Industry 4.0. *Business & Information Systems Engineering*, *6*(4), 239–242. doi:10.100712599-014-0334-4

Li, J., & Yang, H. (2017). A Research on Development of Construction Industrialization Based on BIM Technology under the Background of Industry 4.0. *MATEC Web of Conferences*, *100*, 02046. 10.1051/matecconf/201710002046

Ma, X., Xiong, F., Olawumi, T. O., Dong, N., & Chan, A. P. C. (2018). Conceptual Framework and Roadmap Approach for Integrating BIM into Lifecycle Project Management. *Journal of Management Engineering*, *34*(6), 05018011. doi:10.1061/(ASCE)ME.1943-5479.0000647

Martinez, S., Jardon, A., Navarro, J. M., & Gonzalez, P. (2008). Building industrialization: Robotized assembly of modular products. *Assembly Automation*, *28*(2), 134–142. doi:10.1108/01445150810863716

McKinsey & Company. (2020). *Rise of the platform era: The next chapter in construction technology*. https://www.mckinsey.com/industries/private-equity-and-principal-investors/our-insights/rise-of-the-platform-era-the-next-chapter-in-construction-technology

Neelamkavil, J. (2009). *Automation in the Prefab and Modular Construction Industry*. Academic Press.

Oesterreich, T. D., & Teuteberg, F. (2016). Understanding the implications of digitisation and automation in the context of Industry 4.0: A triangulation approach and elements of a research agenda for the construction industry. *Computers in Industry*, *83*, 121–139. https://doi.org/10.1016/j.compind.2016.09.006

Ogunsemi, D. R., & Jagboro, G. O. (2006). Time-cost model for building projects in Nigeria. *Construction Management and Economics*, *24*(3), 253–258. https://doi.org/10.1080/01446190500521041

Parisi, A. R. (2021). *Valutazione della maturità digitale delle PMI di costruzioni a supporto della transizione verso industria 4.0*. Politecnico di Milano.

Shen, W., Hao, Q., Mak, H., Neelamkavil, J., Xie, H., Dickinson, J., Thomas, R., Pardasani, A., & Xue, H. (2010). Systems integration and collaboration in architecture, engineering, construction, and facilities management: A review. *Advanced Engineering Informatics*, *24*(2), 196–207. https://doi.org/10.1016/j.aei.2009.09.001

Walasek, D., & Barszcz, A. (2017). Analysis of the Adoption Rate of Building Information Modeling [BIM] and its Return on Investment [ROI]. *Procedia Engineering*, *172*, 1227–1234. https://doi.org/10.1016/j.proeng.2017.02.144

Woodhead, R., Stephenson, P., & Morrey, D. (2018). Digital construction: From point solutions to IoT ecosystem. *Automation in Construction*, *93*, 35–46. https://doi.org/10.1016/j.autcon.2018.05.004

Wu, Y., Zhang, Y., Shen, J., & Peng, T. (2013). The Virtual Reality Applied in Construction Machinery Industry. In *Virtual, Augmented and Mixed Reality. Systems and Applications* (Vol. 8022, pp. 340–349). Springer Berlin Heidelberg. doi:10.1007/978-3-642-39420-1_36

Zhou, K., Liu, T., & Zhou, L. (2015). Industry 4.0: Towards future industrial opportunities and challenges. *2015 12th International Conference on Fuzzy Systems and Knowledge Discovery (FSKD)*, 2147–2152. doi:10.1109/FSKD.2015.7382284

KEY TERMS AND DEFINITIONS

Big Data: The big data is data that contains greater variety, arriving in increasing volumes and with more velocity.

Building Information Modeling: Building Information Modeling is a digital representation of physical and functional characteristics of a facility. A BIM is a shared knowledge resource for information about a facility forming a reliable basis for decisions during its lifecycle; defined as existing from earliest conception to demolition.

Cloud Computing: Cloud computing is the delivery of computing services including servers, storage, databases, networking, software, analytics, and intelligence over the Internet to offer faster innovation, flexible resources, and economies of scale.

Construction Industry 4.0: Construction industry 4.0 is a variety of interdisciplinary technologies that digitize, automate, and integrate the construction process at all stages of the value chain. Industry 4.0.A belief that one's own culture is superior to other cultures.

Digital Maturity Assessment: Digital maturity assessment is a measure of an organization's ability to create value through digital is a key predictor of success for companies launching a digital transformation.

Internet of things: The internet of things, or IoT, is a system of interrelated computing devices, mechanical and digital machines, objects, animals, or people that are provided with unique identifiers and the ability to transfer data over a network without requiring human-to-human or human-to-computer interaction.

Robotics: Robotics is a branch of engineering that involves the conception, design, manufacture, and operation of robots. The objective of the robotics field is to create intelligent machines that can assist humans in a variety of ways.

Chapter 11
From the Digitization of Building Materials to Their Use in BIM Models on an Open Standard Platform:
The eBIM Project and Its Applications

Chiara Vernizzi
University of Parma, Italy

Roberto Mazzi
University of Parma, Italy

ABSTRACT

The information data that can be included in models can also relate to the different dimensional domains of BIM depending on the purpose of the model itself. On this premise, the POR-FESR eBIM project "Existing Building Information Modeling for the Management of the Intervention on the Built Environment" has developed skills, models, and solutions related to the conservation and enhancement of the built heritage using the BIM methodology implemented on dedicated IT platforms, identifying and characterizing the materials that compose it (from the shell to the structure to the covering). Among the various building materials, particular attention has been devoted to ceramic tiles and to their role and uses in the building industry for their digitization and use in BIM models on an open standard platform.

INTRODUCTION

The project "Existing Building Information Modeling for the management of intervention on existing buildings", of which the CIDEA of the University of Parma is the lead partner, is a project co-financed by the Emilia-Romagna Region within the framework of the Call for Strategic Industrial Research Projects aimed at the priority areas of the Intelligent Specialization Strategy (DGR n. 986/2018) POR_FESR 2014-2020 (ASSE 1 Research and Innovation) and has invested the Construction and ICT sectors.

DOI: 10.4018/978-1-6684-4854-0.ch011

To these two areas belong all the partners who have operated in an integrated way within the project that has seen the involvement of different sectoral supply chains, connected with the main one of construction: digital technologies applied to the acquisition and restitution of 3D data of the built environment, information technologies for the implementation of web platforms with semantic content, industries in the ceramic sector and construction companies.

The research and technology transfer activities carried out within the project concerned the conservation and valorization of the built heritage, both historical and non-historical, using the BIM methodology, by defining operational protocols for the use of the protocol on existing buildings, through the identification and characterization of materials (from the shell to the structure, to the covering), up to the transfer of information on dedicated semantic web platforms (Apollonio, Gaiani & Bertacchi, 2019).

The experience carried out, with the aim of optimizing the techniques of advanced Geomatics for the creation of three-dimensional models of historical buildings and implementing algorithms for the generation of integrated structural models in a BIM environment, has allowed to select among the possible case studies on which to test the protocol of survey and modeling in a BIM environment, a series of architectures exemplifying the diversity of construction period, construction techniques used, types and dedicated functions, up to the category of intervention to be implemented on them to achieve a reuse consistent with their characteristics.

Within the eBIM project, thanks to the parametric modeling carried out on the selected case studies, one of the main objectives was precisely the implementation of the semantic database aimed at supporting the extraction of values by the various categories of users, thanks to the wide availability of the data collected throughout the process and within shared technological environments: technologies and procedures for acquisition, integration, modelling, and representation.

BACKGROUND

As is well known, BIM is a project management methodology, in all its phases, which is optimized for collaboration and visualization during the development and realization phase of a project, but which increasingly also addresses the management and maintenance of the same during its life cycle: its purpose is therefore primarily to support the professional to develop and realize the project through a collaborative process focused on the physical, functional and user aspects of a building.

In this process, the information content within a three-dimensional model prepared according to the BIM methodology is the most important part. The information data that can be included in models can also relate to the different dimensional domains of BIM depending on the purpose of the model itself (Baiardi & Ferreira, 2020). In this sense, the issue of data interchange and the usability of data over time becomes just as important as the choice of information to be included in the models.

In the research project POR-FESR "eBIM existing Building Information Modeling for the management of the intervention on the built environment", it has been experimented how much the information can be usable in time and updatable through the creation of information models in which they are connected to external databases structuring information fields that describe building materials, making easier the updating of the information that takes place outside the BIM model (Figure 1).

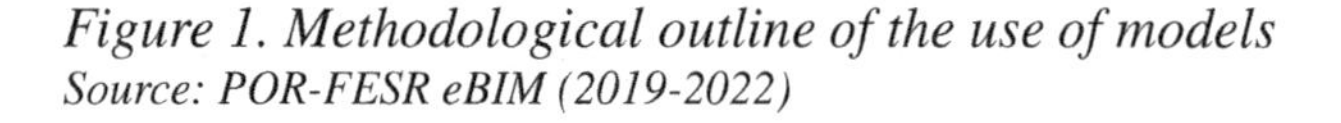

Figure 1. Methodological outline of the use of models
Source: POR-FESR eBIM (2019-2022)

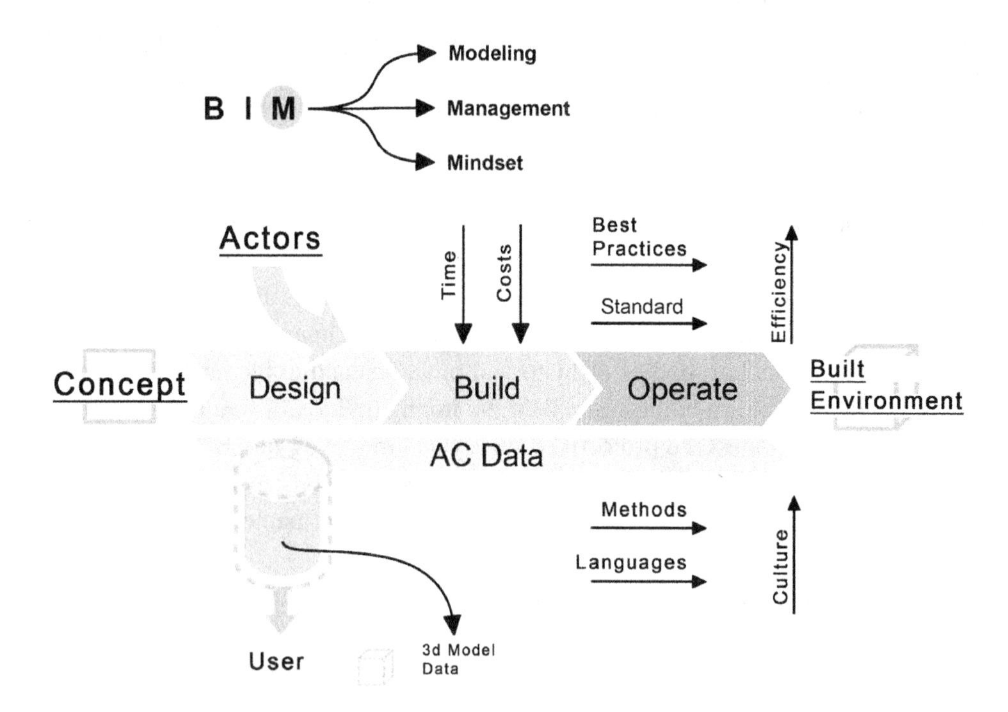

The different case studies that were used to test the information process were uploaded to the eBIM-Inception web platform, a portal with a web interface where the platform and the information it contains can be consulted without the need for specific software loaded on one's own device, whether a fixed location or a mobile device such as a tablet or smartphone.

The creation and diffusion of collaboration and sharing platforms for the management of the intervention on the built environment (Content & Business Process Management, C&BMP) based on semantic databases accessible by different categories of users (CDE, Common Data Environment or according to UNI acDat terminology) represents the response of the construction industry to overcome its characteristics of inefficiency, low competitiveness, low industrialization. The experience already gained, and the voluntary standards already adopted internationally (ISO 19650 and PAS 1192 in primis) and nationally (UNI 11337), together with the New Code of Public Contracts Legislative Decree 50 of 2016, show a rapidly changing scenario, as shown by the recent OICE data which record an increase in BIM tenders which has more than quadrupled since 2015 (Ciribini, 2013).

As far as product and service innovation is concerned, following the experiences conducted firstly in the United Kingdom, and promoted within the BIM4SME initiative, and subsequently in the German market, in the last six years an increasing number of manufacturers have been making available .ifc models of their products through dedicated sections of their web portal. Similarly, the European EU BIM Task Group is promoting, also within the Public Administration, the adoption of shared languages and collaborative technology platforms based on data environments (CDE, Common Data Environment, acDat in the UNI terminology) as a fundamental driver for the effective and widespread adoption of data-based knowledge (Knowledge Economy) (Pavan, Mirarchi & Giani, 2017).

The "eBIM" project has developed, deepened, and extended the results obtained by the partner TekneHub-UNIFE within the INCEPTION project, funded under the Horizon 2020 program (call REFLECTIVE-7-2014), regarding the structuring of an open standard web platform for the upload, access and implementation of integrated digital models by different categories of users. The results already achieved and tested within the INCEPTION project constituted the starting point of the eBIM project.

The digital model is expected to become the representation (forever, for everybody, from everywhere) and research needs to acknowledge the changing role that reconstruction, reservation and conservation now play in the representation of heritage and its analysis. INCEPTION models will be able to improve a greater understanding of European cultural assets as well as a direct reuse for innovative and creative applications. The 3D representations should go beyond current levels of visual depictions, support information integration/linking, shape- related analysis and provide the necessary semantic information for in-depth studies by researchers and users. The generation of high quality 3D models is still very time-consuming and expensive. Furthermore, the outcome of digital reconstructions is frequently provided in formats that are not interoperable, and therefore cannot be easily accessed and/or re-used by scholars, curators or those working in cultural and heritage industries.

Main aim of INCEPTION is to realise innovation in 3D modelling of cultural heritage through an inclusive approach for time-dynamic 3D reconstruction of artefacts, buildings, sites and social environments. The INCEPTION "TimeMachine" will be developed as an open standard Semantic Web platform for creation, visualisation and analysis of 3D (Heritage-) BIM models of cultural heritage over time; 3D models generated through INCEPTION methods and tools will be accessible for all types of users.

INCEPTION techniques and technologies will be tested, validated and demonstrated at various cultural heritage sites across Europe.

The outcome of the eBIM project builds on the TRL achieved by the INCEPTION project by sharing its basic approaches and extending them towards specific topics, such as those addressing the main building materials (ceramics and bricks) and structural aspects to be read in an integrated way in the BIM environment.

The acquisition protocol developed on Heritage and the structuring of the platform have in fact defined the starting point for the implementation of an ad hoc protocol, in the framework of the "eBIM" project, on the diffuse existing building, integrating the optimization of the metric-morphological acquisition procedure with data and information related to materials, structural components and technological systems (Biagini, 2007).

FINALIZED MODELING: THE DECOMPOSITION OF THE ELEMENTS

The creation of a digital model of a historical building is an articulated process already in the phase of creation of the aspect concerning the geometric components (Bianchini, 2014). Applying a logic of BIM method, that is concerning the creation of a cognitive database peculiar to aspects defined before starting the same modeling, the informative part becomes a further aspect on which to dwell and open a reflection: which informative aspects can be inserted to describe the characteristics of the model? The answer to this simple question will be various based on the scope for which the model has been created. Informative requirements vary in base to the necessities to pursue (Figure 2). The properties will be able to go to characterize single elements created in the model, like as an example wall, doors, windows, etc.

Therefore, in way tightly tied up and directly linear, the way in which such elements are created depends also on which information will have to be contained in these, in analogous way the information will be able to be inserted only if we will have elements that can contain them. This aspect, apparently simple, has influenced the workflow of the parametric modeling of the Ex Carcere-San Francesco in Parma (Figure 3, 4). The author of the parametric model, had to engage in a logic of "decomposition" of elements to pursue the purpose of the model.

Figure 2. eBIM Project Workflow
Source: POR-FESR eBIM (2019-2022)

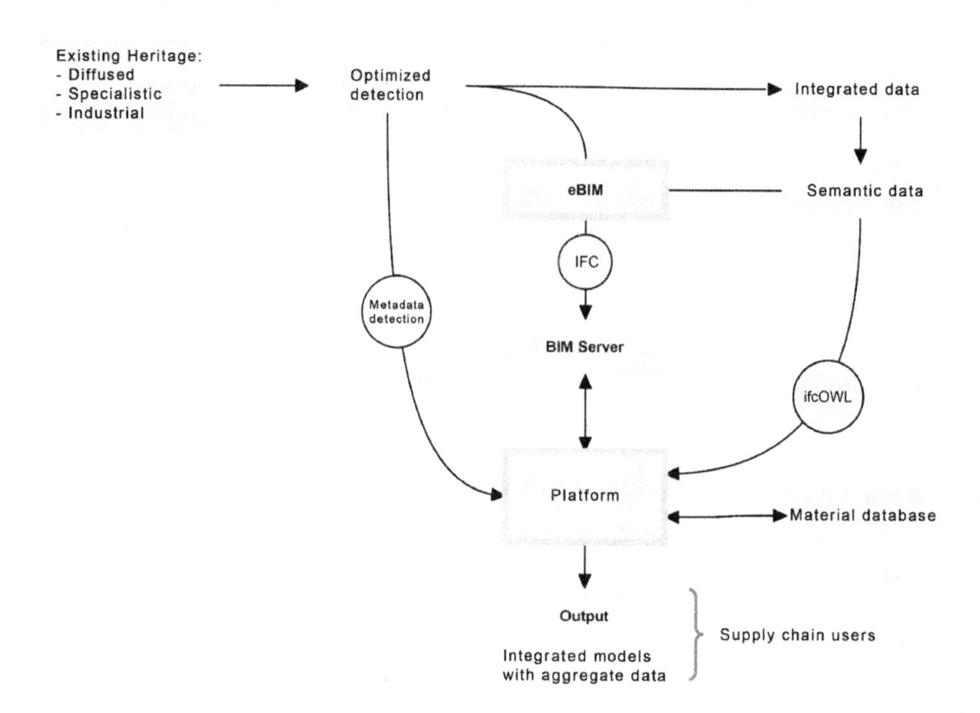

How can we separate the elements in a surveyed building? Certainly, this is another central issue. The activity of geometric survey has been the first piece in the creation of the virtualization of the building, but it is certainly not sufficient for the correct realization. Alongside the geometric survey activity, it is necessary to exploit knowledge related to the construction method of the building, the use of elements obtained from tests on materials in situ, excavations to define the depth of the foundations, a historical analysis of the work. The union of the various informative sources have given the possibility to create the elements both for what concerns the "external" geometric aspect in a precise way (Bianchini, Inglese & Ippolito, 2016), and for what concerns the internal aspect of the elements (where it has been possible to find the information), linked to the characterizing definition of the stratigraphy. Together with these aspects, the researcher had to consider the optimization of the geometries, since the parametric modeling of the elements had to be finalized for an export and upload to the Inception eBIM web platform.

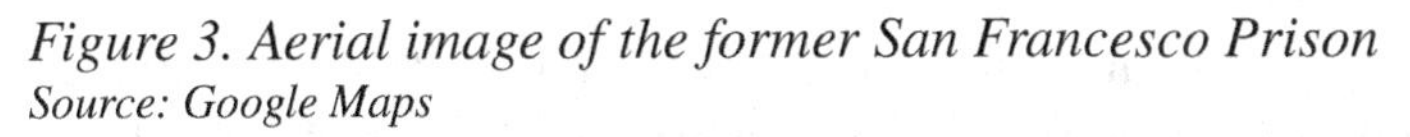

Figure 3. Aerial image of the former San Francesco Prison
Source: Google Maps

Figure 4. Orthophoto of the former San Francesco Prison complex
Source: Google Maps

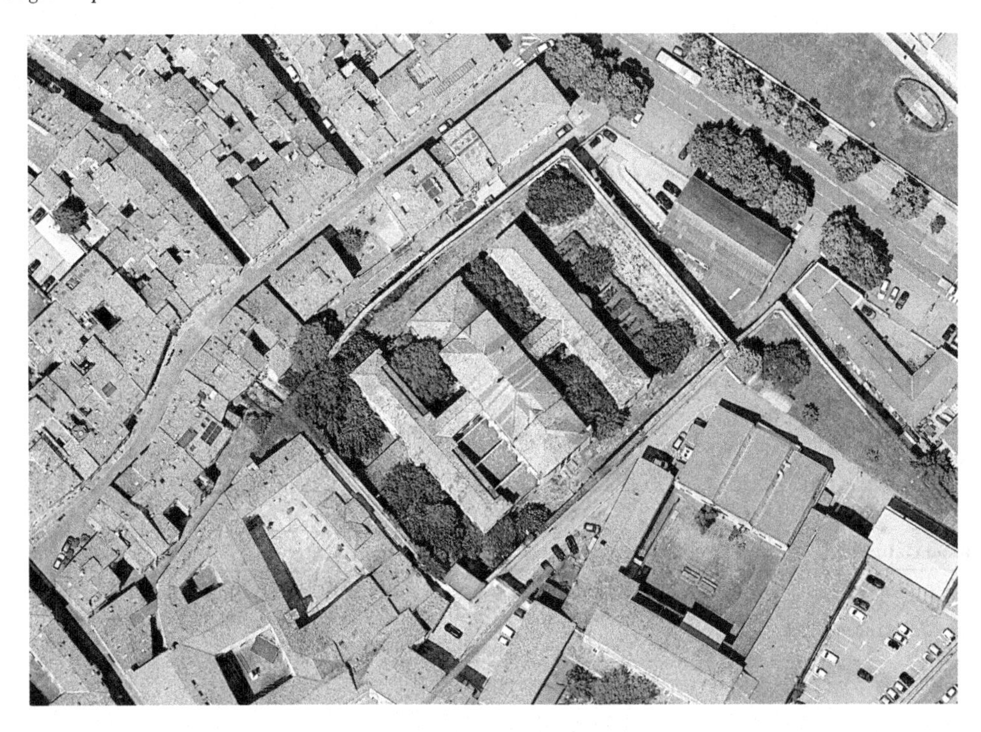

The portal gives the possibility to registered and authenticated users to upload BIM-based three-dimensional models (Pulcrano, 2020). The metadata inserted during the birth of the model in the BIM Authoring software (in the specific case Archicad was used) are characterizing the elements and building materials used in the stratigraphy (Figure 5). The platform has correctly transposed the model exported in IFC format, both for what concerns the geometric aspect and the inserted informative aspect.

Figure 5. Outline of materials used in eBIM case studies
Source: POR-FESR eBIM (2019-2022)

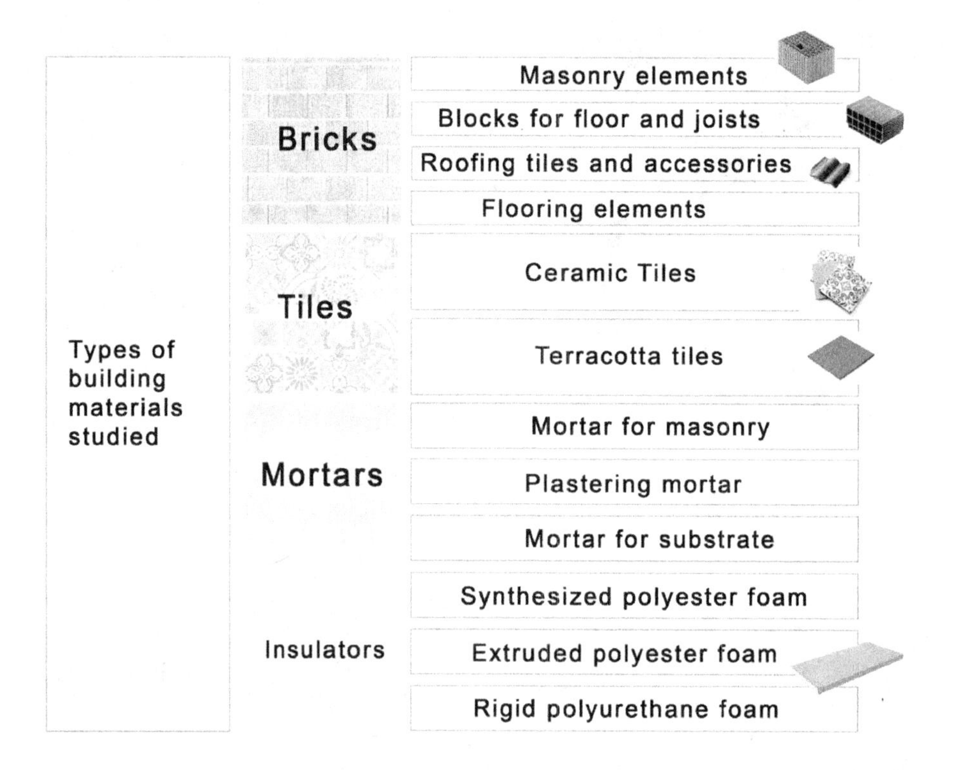

Table 1. Creation materials

Creation Existing Material			
Information Gathering	**Property Definition**	**Data Implementation**	**Result Achieved**
Retrieving information from site tests	creation of different information properties per type of material analyzed	Comparison of materials with existing databases	Reusable structure via export in dedicated proprietary format 7/10
Creation New Material			
Information Gathering	**Property Definition**	**Data Implementation**	**Result Achieved**
Retrieving information from manufacturers' data sheets	creation of different information properties per type of material analyzed	creation of data structure, creation of material textures and vector screens	Reusable structure via export in dedicated proprietary format 9/10

The modeling of the geometric elements has been faced by the author based on a precise two-dimensional restitution of the surveys, realized by the extrapolation of geometric data obtained from a cloud of points (Figure 6) and from direct measurements on site (Bolognesi & Aiello, 2020). The model is thus characterized by all those elements difficult to standardize, such as the definition of the wall elements.

In addition to having a section referable to a trapezoidal geometry, the stratigraphy of the perimeter walls is composed of materials that alternate in section both along a longitudinal axis and along a transverse axis, visible already without having to make special analysis in the elements outside the façade (Figure 7). On the external part we can in fact see ashlars of stone material alternating with bricks, giving the characteristic horizontal striped appearance to the facade.

The creation of a defined geometry could be built in a simple way, avoiding the part of creation of stratigraphic elements, but in this way, it would have been necessary to develop a customized property package for each wall element of the virtual building. Such hypothesis, even if realizable, could be suitable for models of small dimensions, in which the informative compartment, even if rich and articulated, could be directly defined with parameters and properties of the single elements, comparable to a simple description. Eventual information regarding quantities of the elements, volumes and all the relative components to eventual interrogations of the model on the quantities could not happen. The modeling is therefore directed towards the creation of the various personalized masonry walls, according to the stratigraphy and slopes detected. This multiplication of information within the masonry elements, would not have been possible if a database of building materials of the existing elements had not been developed within the research activity.

Figure 6. Point cloud image (processing by Andrea Zerbi)
Source: POR-FESR eBIM (2019-2022)

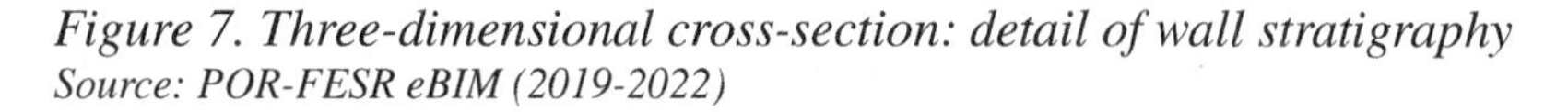

Figure 7. Three-dimensional cross-section: detail of wall stratigraphy
Source: POR-FESR eBIM (2019-2022)

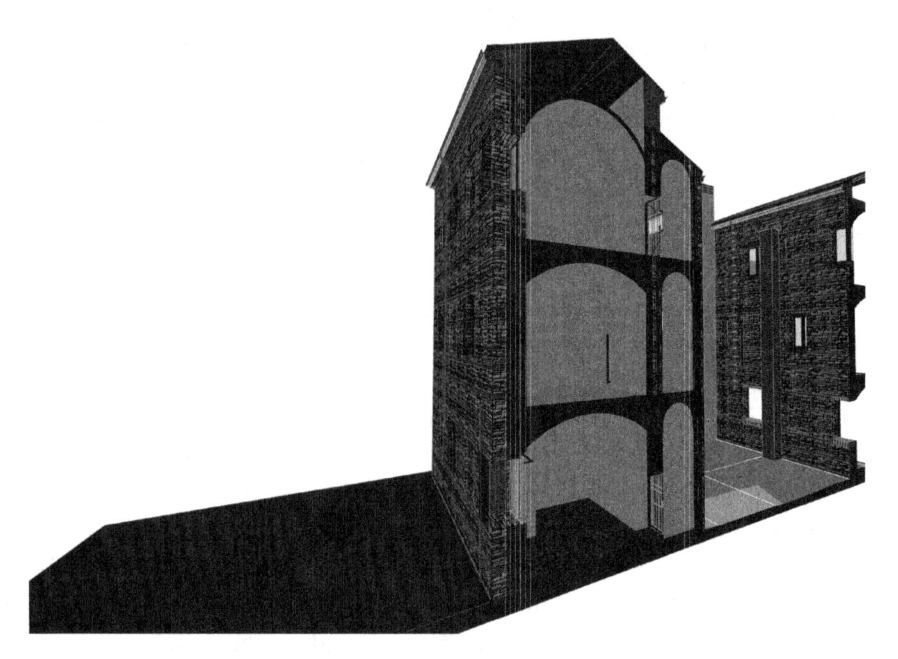

DEFINITION OF MATERIALS DATABASE

The building materials are a sort of container of information related to the single materials that compose the stratigraphy. The BIM Authoring software provides a preset set of generic materials (partially used in strata where there was no certainty or information about the data). In the modeling phase, the researchers then chose to create a specific database of building materials, characterizing the information components of individual materials, creating custom properties within the software in the materials. In addition to a characterization of the information component, the representative component was also customized. This apparently graphic aspect denotes an important advantage: it allows the visualization of stratigraphy in a planimetric view and consequently also in the third dimension of the behavior of union and intersection of the layers. A correct union, connection, and intersection of the masonry elements, consequently, produces a correct quantification of the building materials.

The rooms and corridors of the former prison are for the most part characterized by vaulted rooms. Also, the structure of the vaults (not regular lowered vaults) has been modeled adhering to the anomalies defined during the phase of survey, in which the information of the stratigraphy has been inserted. The modeling of the vaults, those of the central corridor, have been created with a decomposition and union of elements (as will be described also for the elements of the staircase bodies). The central corridor is vaulted with cruciform elements, therefore with the presence of nails near the openings facing the courtyards present between the arms.

The decomposition has been managed between the union of an irregular vaulting element and the single crossing elements of the nails. The tools used for the creation maintain the properties of the material stratigraphy. Intersection with Boolean operator tools between longitudinal and transverse elements allowed for proper three-dimensional representation. Once the vaults were created defining their stratigraphy, they were joined with the same method to the lateral wall elements. The creation of the vaults had to interface with other elements of the architecture, to maintain a correct source of information, relative to the differences in height of the floors.

In the same way the elements concerning the internal floors have been modeled (Figure 8). Also, for this type of elements, the geometric information has been used in the modeling phase by crossing the photographic documentation developed during the survey activities, together with the cores performed. The union of these three characterizing sources gave the possibility to fill the information gaps of the single aspects, being able to create precise stratigraphy of the elements. A particular aspect developed during the modeling phase was addressed by the author of the model regarding the management of the irregularities of the floor in the central separation corridor. The survey has defined a non-regular course of the floor, with irregular depressions and humpback areas. The geometric management of these elements was handled by decomposing the floor into a sum of single layers, to respect a correct IFC classification when exporting to the Inception-eBIM platform and to preserve a correct stratigraphy of the floor. The tools used for the creation of this single slab went using different tools, not specific for the creation of a slab, using this method, the researcher was able to achieve the goal.

Table 2. Creation floors

Creating floors with uneven surfaces			
Issues Addressed	**Proposed Solutions**	**Problems Solved**	**Validity of the Method**
creation of irregular stratigraphic elements	shape modelling with hybrid tools, internal to the BIM authoring program, with layer insertion	easy customization of elements	articulated method, good modelling skills needed. extended timeframe

MODELLING IRREGULAR FLOORS

How can the multiple irregularities of a floor be dealt with? The aim of the researcher was to define the geometric components of the element, including their stratigraphies. The modeling in fact has combined the potential of Mesh tools (tools typically used for the creation of land) with tools floor, combined with Shell tools (for the creation of the ceiling of the floor in which there is an irregular vault with nails), all combined with operations of subtraction and union through tools called “Boolean operators” (which allow the union, subtraction, intersections of three-dimensional geometric elements). The creation of multiple layers with the mesh tool has thus allowed the irregular management of the same stratigraphy. Tools inside the software were also used, not related to pavement modeling, but useful to achieve the purpose. The slopes detected along the corridors in the arms of the cells had a non-regular trend, they maintained a linear trend (either uphill or downhill). This characteristic could facilitate the choice in modeling, using “pitch” tools, i.e., roofing applications. In the modeling of existing buildings, it is not unusual in this type of software to take advantage of all the potential it grants, since, regardless of the name of the tool, before the final export, the element can be classified according to the IFC properties dictated by Building Smart, in an appropriate way. This is what has happened in this specific case, using tools for the creation of roofs, classifying them according to the object, as finishing elements, floors. The pitch and mesh objects in these cases, are two sides of the same coin that relate to the attic package.

These elements, which were used to create the geometric part concerning the extrados, were combined with the part of the intrados of the floor package, characterized by the shell elements of the vaults.

What could be other characterizing elements in a building of this size? This is a question the researcher asked himself during the model creation phase. The various environments, according to the plans and the typology of the room are characterized of different elements of the components of window and door frames, in particular the ex-prison of San Francesco is composed from elements to grates, placed on the external facades only in the arms of the cells that lean out towards the perimeter, excluding therefore the windows that lean out towards the inner courts and the windows of the central corridors. A second round of gratings are placed within the net span of the windows, in the middle of the wall thickness. This type of grating is directly installed with metal inserts in the wall. Unlike metal gratings, these gratings are present in the three levels where there are prison cells in each window, including those relating to the central corridors.

Figure 8. Three-dimensional cross-section of central corridor: flooring detail
Source: POR-FESR eBIM (2019-2022)

PARAMETRIC DOOR AND WINDOW MODELLING

The part related to the windows placed in the internal facade of the cells, of the central bodies and the corridors of the cells, are of different size and geometric shape (Figure 9). All these objects have been managed in a parametric way: the part concerning the external grilles has been managed through the creation of a library object in which the frame of the grill and the metal grill can be inclined in a different parametric way with respect to the three reference axes, as well as managing in the same way the dimensions of width and height of the frame.

Figure 9. Three-dimensional cross-section: detail of window frames
Source: POR-FESR eBIM (2019-2022)

The modeling of the gratings has been managed with the creation of several parametric library elements, since the survey activities have shown different steps of the elements of the grating as well as curved or inclined components. The gratings have an entirely similar typological component regarding the cell windows, while the variations are more evident within the typologies located along the central corridors.

The windows modeled by creating custom library elements. To optimize this process of creation, it has been preferred in some cases to proceed with the decomposition of the windows in a sum of separate elements, reaching in this way the desired result. The geometric definition of the windows has followed the creation of the geometric elements of the sashes, with the multiple subdivisions of the frame; in addition to this, the different lintels around the window have been customized (different for inclination, type of lintel, depth, relationship with the wall surface) and the different conditions. The researcher has also taken care to hypothesize how the elements of masonry stratigraphy could behave in the window nodes.

The difficulty in the geometric creation in the specific cases lies in the multiplicity of niches, geometric irregularities of the phorometry and frames present. For the experimental purpose the irregularities of the parapets in the windows have been deliberately not modeled, although testing the possibility of undertaking two different paths for the creation of such irregular elements: the first one going to modify the Boolean element inserted in the window tool, direct component that creates the subtraction inside the wall; the second evaluated way is to create dedicated geometric elements for the addition of the irregular component. The first way, if not parameterized, involves the creation of unique elements for each type of irregular skirting, therefore with a high expenditure of time. The second way is easier to manage, but with the limitation of an incorrect management of the stratigraphy, since this element created in addition to the wall layer can be managed only as a mono-material component, having however the advantage of being able to manage both in plan and in elevation the irregularities as represented in the surveys.

With the same method they have been modeled the elements regarding the doors: not being able to enter in the phase of modeling to the inside of the building, the creation of the typologies has been based also for this case on the photographic documentation and of the geometric surveys. The author of the model has created the different types of doors based on the information. The doors are characterized by double leaf elements, a first one characterized by a solid element with a small inspection opening towards the cell, a second one by a grated door, characterized by mixed elements between tubular and flat metal. This set of geometric elements has been managed within the parametric tools that the modeling software makes available, creating customized elements for the category. Vice versa, where the information related to these elements was not sufficient for its definition, the insertion of a generic element was chosen, keeping its geometric definition at a less detailed level, although correct regarding the main aspects such as the dimensions of the opening or the height with respect to the walking surface.

Table 3. Creation windows and doors

Creating Windows and Doors Type			
Issues Addressed	**Proposed Solutions**	**Problems Solved**	**Validity of the Method**
Creation of customized typological elements	creation of parametric library elements, in the most complex cases, deconstruction of windows into groups of elements	creation of parametric complex types	easy management and application of customized elements, significant time savings

COMPLEXITY OF GEOMETRIC DEVELOPMENT: THE MODELLING OF STAIRS ELEMENTS

The three stair bodies are an element bound in the interior and not very visible, but that had to be carefully studied for its solution in the model phase. The geometric structure is composed of elements with landings of variable height, with ramps with irregular treads and risers. The ramp is supported at the bottom by vaulted systems that connect the various landings. In addition, on the inner edge, the outline of the step is incorporated within a solid profile of masonry, which follows the slope of the steps.

The modeling of the staircase with the basic tools, without having to leave the application, had to pass from a decomposition into three elements by the researcher (Figure 10). The first element is the vaulted part, created using a tool called "shell". The characteristic is to be able to create a curved element, even irregular, inserting a defined stratigraphy. The second component used, to fit the specificity of the case, is the creation of the finishing elements (riser and tread) with the stair tool. In this case it has been used without the automated creation of the supporting structure, as this layer is fulfilled by the vault with the shell tool.

Figure 10. Three-dimensional cutaway: stairs detail
Source: POR-FESR eBIM (2019-2022)

The last element of the decomposition was the creation of the lateral wall, created by modeling an element with the "beam" tool, in which a custom profile (called complex profile) created with materials for stratigraphy was applied, and parameterizing the height of the outer layer. In this way it was possible to model with the variable slope, depending on the slopes of the ramp. The logical development of these elements by the researcher passed in a first phase through the study of the ramp through the visualization of images and surveys, in a second phase he passed to the management of the "design" of the staircase, through drawings on paper and freehand sketches. The transition through hand drawing is a very important part, which the researcher considers of equal importance to the surveying activities. Drawing an element and breaking it down into its parts means understanding how it was built, what the elements that make it up may be.

These activities seem to be far from what concerns the BIM method, but the understanding of "how it is made" is a necessary condition for a correct modeling. The three decomposed elements identified for the creation of the individual flights of stairs were used by the researcher with Boolean operations modeling. This method allows for subtraction, union, and intersection between solid elements, maintaining the characteristics of the individual materials involved, but modifying the final geometries, consequently data related to volumes and areas. The researcher subtracted the riser and tread profile of the ramp from the vault (thus maintaining the stratigraphy of the vault), the profile of the beam tool was joined to the vault component. This series of operations was repeated for the four ramps connecting one floor to the upper level, for the various levels of the building, in the three different stairwells. The cut-out vaults were then connected to the landings between the ramps.

Table 4. Creation of stairs elements

Creation of stair elements			
Issues Addressed	**Proposed Solutions**	**Problems Solved**	**Validity of the Method**
construction of non-standard stairs	deconstructing scale elements, using different tools, maintaining the stratigraphies in the final object	exceeded limits linked to standardized parametric elements	articulated method, good modelling skills needed. extended timeframe

THE COMMUNICATION OF THE DATA

The creation of the virtual model of the San Francesco Prison has been finalized trying to give answer to some simple questions: how much the complexity of a building could be limiting in its virtualization? The creation of the virtual components of the building have as limit the level of knowledge related to them, that is the more the cognitive elements deduced from geometric surveys, historical analysis, direct essays, the simpler will be the decomposition of them (Figure 11).

Comparing the modelling procedure of the building and a WBS chart, we could compare the activity of creation of the single elements to a decomposition in many small "work packages", in which the various modelling themes are deconstructed according to the aim to pursue (different stratigraphy, union between elements, personalization of irregular geometries, composition of the informative part, etc.).

Figure 11. Internal material database sheet
Source: POR-FESR eBIM (2019-2022)

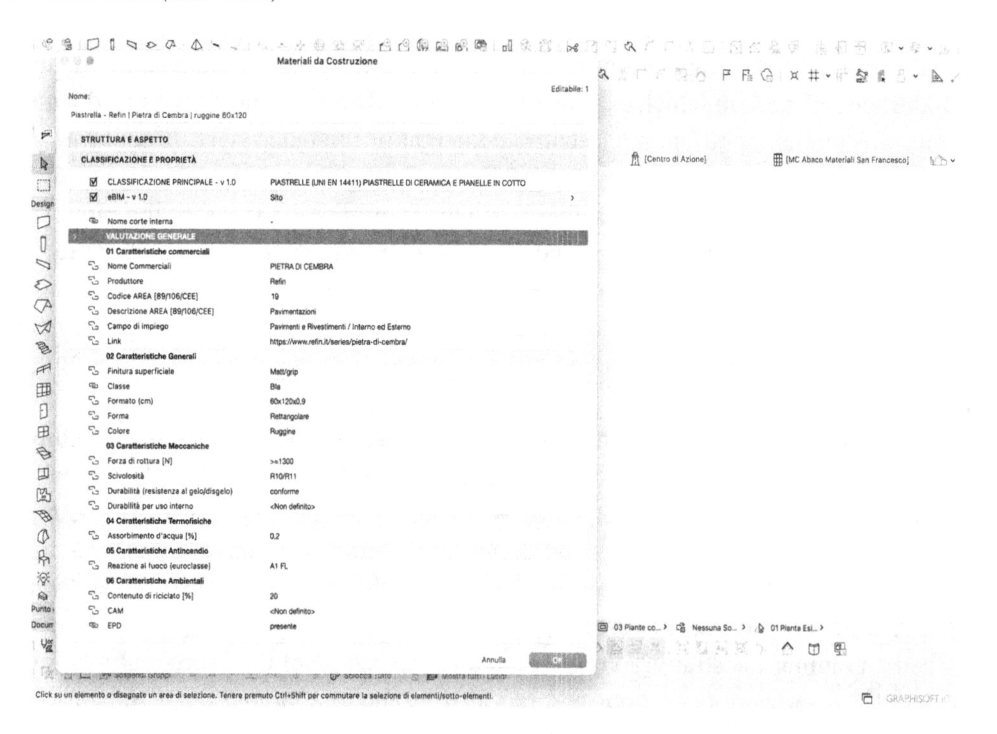

The roads that can be undertaken to reach the desired result can be created within the BIM Authoring software or with the help of third-party software that will interact with them. In fact, there are elements that do not need to be created parametrically, as they will not need to be used except for their informational components. Despite having this possibility, the researcher did not need to go down this road, the tools already internal to the software were more than comprehensive and comprehensive to address special aspects. Within a large, complex building like this one, how do you skim or research elements? Managing a building body of even modest size, with a multitude of different elements categorized in the same way, can be complicated. The solution that can facilitate this path is to implement the informative data of the single components; in the specific case, characterizing elements have been inserted in walls, floors, windows, doors, roofs, beams and joists, vaults, niches, cornices, downpipes, stairs, such as the nomenclature of the building body, as well as defining if the element is positioned inside a court between the arms of the building. A further characterization has been related to the classification of incongruent elements (easier identification purpose), comparing the different material aspects visible in the building, succeeding in this way to realize a database of such volumes.

The aspect in which the experimentation has been focused more was just the part related to the material aspect (Figure 12). The building materials created were born from the union of information sources based on tests carried out on site and weighted information obtained from the database of one of the commercial partners within the research activity. In the creation phase, the researcher created the model of the actual state and modeled the phase related to the renovation, only as far as the modifications of the first level are concerned. By modeling elements with new stratigraphy, the building materials of the new materials had to be defined accordingly.

Figure 12. Loading in web platform: textured model visualization
Source: POR-FESR eBIM (2019-2022)

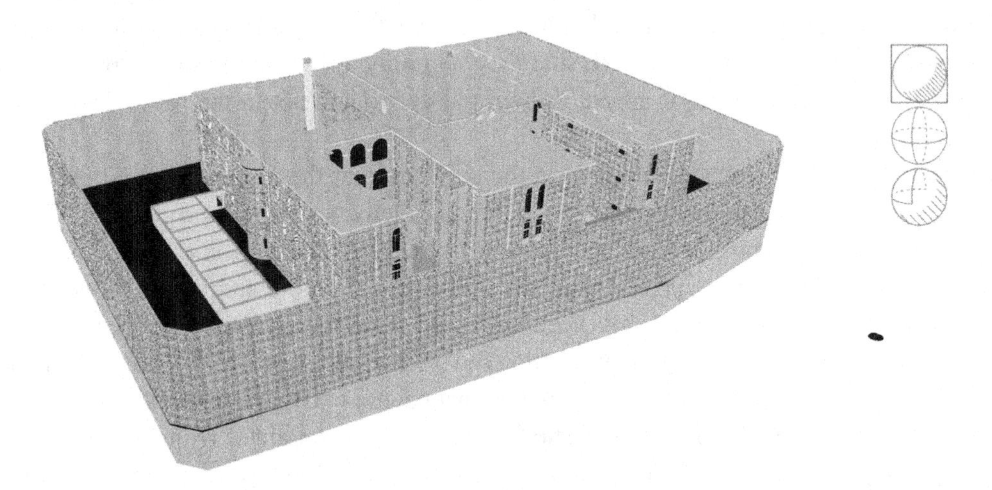

The first approach for the definition of the characteristics of existing materials was developed in collaboration with the internal business partners of the research group, creating a list of definition of information groups and material properties (definition of mechanical, physical, descriptive properties). The result produced is the creation of a coordinated database of material information. Following the same methodological track, the database of new construction materials was created, testing the various chains of materials used in the various disciplines of use. The materials have been characterized by different information fields. The parameters entered describe the main information of the materials, this implicitly determines that similar materials will have information parameters set with the same scheme.

In collaboration with the research centers within the group, databases of ceramic materials have been developed, which describe the properties of wall and floor finishes. Concurrently with the information database, the researcher developed custom mappings of individual ceramic components, along with elements for a representation of two-dimensional designs. The materials developed in the experimental phase are specific to ceramic materials, bricks, structural mortars, and materials for seismic reinforcement.

The management of the phases within the evolutionary process of the building, has been included in the modeling method. By choice of the researcher, two temporal phases were included, one related to the detected situation and one to the projected state of upcoming construction (ideally identified with the year 2022). The method by which this was handled was adapted to the standards following the needs of the platform. Remember that phases within a process (be it production, architectural, design, etc.) are sets of activities that include the creation of products created by specific work packages aimed at achieving a defined goal. The temporal decomposition put in place has the purpose of describe the evolutionary story of the real building, starting from a first phase, which is the modeling to a temporal state related to the time of the starting survey, continuing a second phase, related to the project.

The ability of the platform to detect the phases within the exported model lies in the reading of the properties of the individual elements. This implies that it does not recognize the elements classified as "existing", "to be demolished", "new construction", it recognizes the concept "phases", parallel but different. An element classified by phases will therefore express its existence in a period

specific time, that is, the elements that in a characterization of "construction/demolition" are defined with the property "state of facts" or "existing", are all those objects (masonry, doors, windows, etc.) that will not have to be demolished. These elements will have as phase property a value that includes all the inserted phases. The elements that on the other hand will be defined as objects that will be "demolished" or "constructed", must have properties such that they will only appear in phase one but not in the following phase (i.e., "demolished"), or appear only in phase two and not in the first phase (elements of "new construction").

This concept is not an automated element in the software that the researcher used. As a result, he had to vary the way he managed the software, modeling the elements with the property of "demolished" and with the property of "new construction" in dedicated layers. This concept, which may seem easy to manage, implies a different management of the properties of automated union between materials. Besides this aspect, banal elements, such as the closing of a door, the opening of an element previously modeled as a niche, have been managed by splitting the elements themselves. As it is known, an object door, window, niche are three-dimensional elements that cannot exist independently in the libraries of BIM authoring software. Such objects exist and only if they are contained within other objects (such as walls, roof pitches). Consequently, every wall that contains a new door or opening, or a plugging of an access, causes the duplication of the elements that contain them. Arriving at the solution of creating elements characterized by phases, applying this method, the researcher had to manage the automated intersection properties between these layers (which must not create incorrect intersections). When exporting to the open IFC format, the model was saved keeping all layers visible at the same time, since only active elements in the three-dimensional view could be visible in the saving. The correct visualization, separated by temporal phases, is filtered directly by the web platform, which will recognize during the loading of the IFC file, the reading of the phases, displaying in separate views the elements, choosing them through a time machine in the portal.

THE WEB PLATFORM

The modeling and loading of the information data related to the elements and materials were the basis for the next step (Maietti & Tasselli, 2020), related to the loading into the eBIM web platform (Figure 13). The portal acts as a database of the uploaded buildings, giving multiple possibilities of consultation, based on the concept of the export type, related to the MVD (Model view definition).

The mode of loading in the platform is performed directly by the user owner of the model, without having to delegate to the site administrator the task of loading. The operation takes place with the definition of several points before the closure of the procedure. The researcher performed and verified the upload of the three-dimensional model and its information apparatus. In sequence, the information for the geographical location of the building was entered, visible through an interactive map of the models uploaded in the portal. A short description of the building and its purpose has been inserted, visible in a cell below the model viewer. Still in the loading phase it is possible to insert descriptive documents, in the specific case it has been possible to go to realize photographic shots of spherical type, through a camera with double lens of type "fisheye" (Figure 14). The camera has created images already merged and georeferenced. These images of the state of the art, are particularly useful to view the state of the art of the object; exploiting the capabilities of the platform the user can thus see spherical shots taken in the rooms of the second floor, visualizing 360° of the room. These spherical images and any other document uploaded, can be linked to the geometric elements they compose. By asking the system for information related to the metadata of that single selected element, the portal will be able to view both the linked information part and the attachments (in this case the spherical images).

Figure 13. Loading in web platform: information model visualization
Source: POR-FESR eBIM (2019-2022)

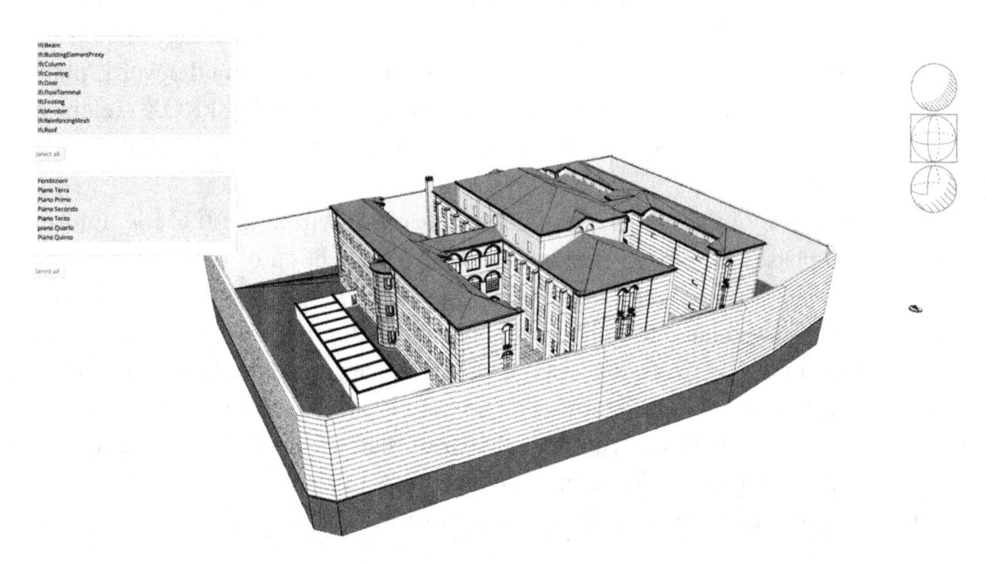

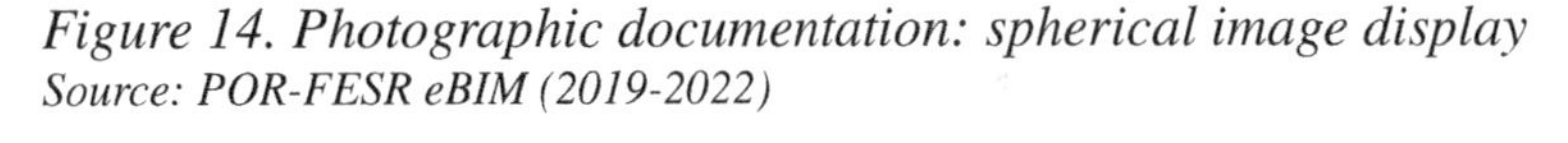

Figure 14. Photographic documentation: spherical image display
Source: POR-FESR eBIM (2019-2022)

It was not by chance that the shots were produced only for the foreground. In fact, as was mentioned earlier, the part concerning the renovation of the second floor was modeled for experimental purposes. The features created in the BIM authoring software were used to produce parallel documentation, i.e., creating spherical rendered images of second floor environments. The combination of the images taken at 360° and the new spherical images linked to objects specifically inserted in the model, can display the comparison between "before" and "after" in the environments (Figure 15). Unlike documents or information that can describe a single specific element, a spherical image is descriptive of an environment (Figure 16). To solve the problem to which element the spherical image should be attached, volumetric objects were created based on extruded text elements, representing the word "photo", placed inside the rooms where the elements were created, classifying them as IFCPROXY elements. Of the loaded documental part related to these documents, thirty-two spherical images were loaded, linked to the single proxies created.

During the loading phase of the model, the platform checks if within the IFC file, informative data related to the temporal phases have been characterized. Upon detection of this data, the platform creates a timeline. The timeline is designed to visualize the time phases of the building. Therefore, the phases do not describe a state related to the renovation states (situation, demolition, new construction, comparative), therefore, based on these concepts, it was necessary to adapt the renovation states to the time phases managed by the portal. The year in which the first phase is associated is related to the state of fact, the second phase is related to a hypothetical year of realization of the project state. In addition to the building elements, the phases placed in the export phase also characterize the proxy elements, displaying in the web page also the differences with the loaded images.

Figure 15. Rendered model documentation: spherical image display
Source: POR-FESR eBIM (2019-2022)

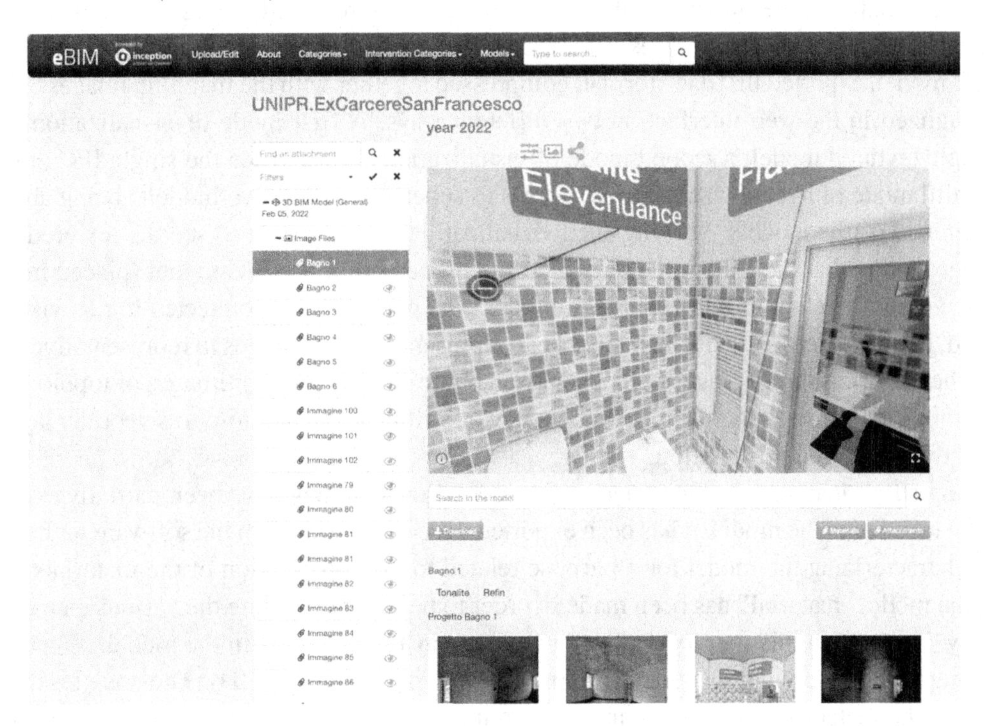

Figure 16. Rendered model documentation: spherical image display full screen opening
Source: POR-FESR eBIM (2019-2022)

During the phases of uploading documents, the portal has the further possibility to insert, in addition to the informative three-dimensional model, a further textured model. It is known that in fact, the export of the model in IFC format does not export the elements with the link to the mapping images. The second model, exported in ".dae" format, compressed together with the mapping images of its faces, can be visualized in the web interface in two different ways. A first mode of visualization allows to see the single textured model, a second mode of visualization allows to see the single IFC information model, a third mode of hybrid visualization, allows to superimpose the two models, being able to vary the transparency of the models. Varying such visualization, it is possible to see the textured elements in the foreground and at the same time to exploit the properties of the IFC model (placed in transparency), succeeding therefore to interrogate the model. The potentialities connected to this visualization are elevated, in how much the texturized model can represent relative images to representative topologic images of the materials, or to images composed from photo-rectification, or images of topological mapping. Exploiting the same principle cited previously, it will be able therefore to select the layer and to interrogate the connected information.

Following a first loading relative to a generic model, in which they have been partially exported the information inserted in the model, it has been exported a second model from the software with a different modality, characterizing the model for a purpose related to the consultation of the materials only. The export of the model "material" has been made in order to be able to separate the various elements in the stratigraphy, allowing in this way to localize in precise way the materials in the model.

The materials are individually coded by a unique identification code (ID). The code inserted in the IFC model in the materials database, connects to an internal database of the platform (Figure 17). This allows the platform to implement the information data entered in the IFC file, from an externally updatable database. This concept expands the possibilities of information exchange.

Figure 17. IFC model visualization in platform: plan filter application
Source: POR-FESR eBIM (2019-2022)

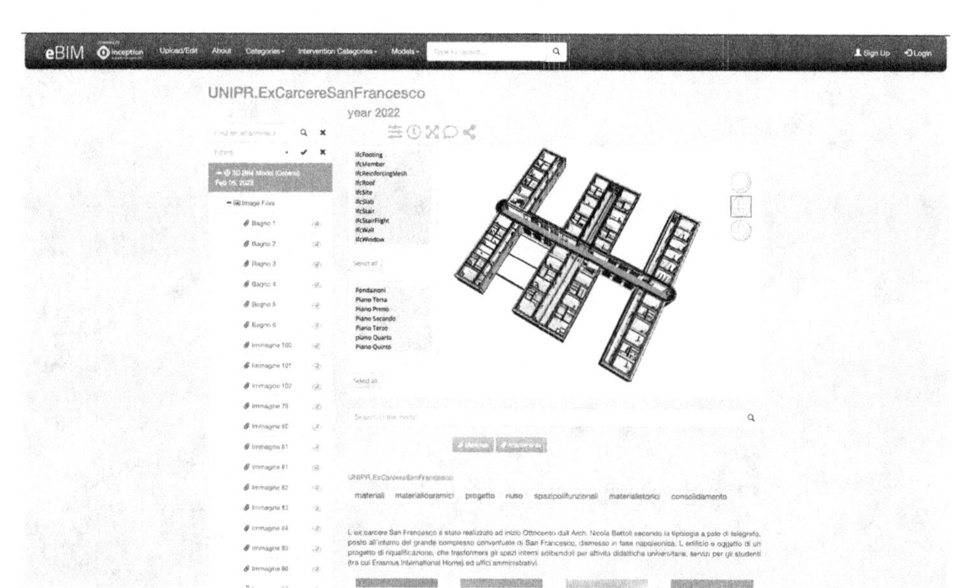

Visiting the platform, thanks to the reading of the same of the identification codes, the external user can visualize the list of the materials inserted in the IFC model and inserted in the database.

By means of this list of building materials, generated autonomously by the platform based on the intersection of information data, it is possible to visualize the univocal names of the three-dimensional elements that contain these parameters. In addition to displaying the list, three-dimensional elements can be selected from these "material abacuses" and displayed in the three-dimensional IFC model.

The IFC model can be queried in its properties related to the metadata entered, but also broken down into its three-dimensional parts (Figure 18). Applying filters directly on the web platform, exploiting the classifications given during the modeling, you can view the elements classified by the IFC classes assigned (for example ifcwall, ifcdoor, ifcslab, ifcwindows, etc..), or go to filter the elements according to the location plan, and finally you can combine these filter systems by combining the properties given by the classifications and those related to the plans (Figure 19).

Figure 18. IFC model visualization in platform: filter application for elements classification
Source: POR-FESR eBIM (2019-2022)

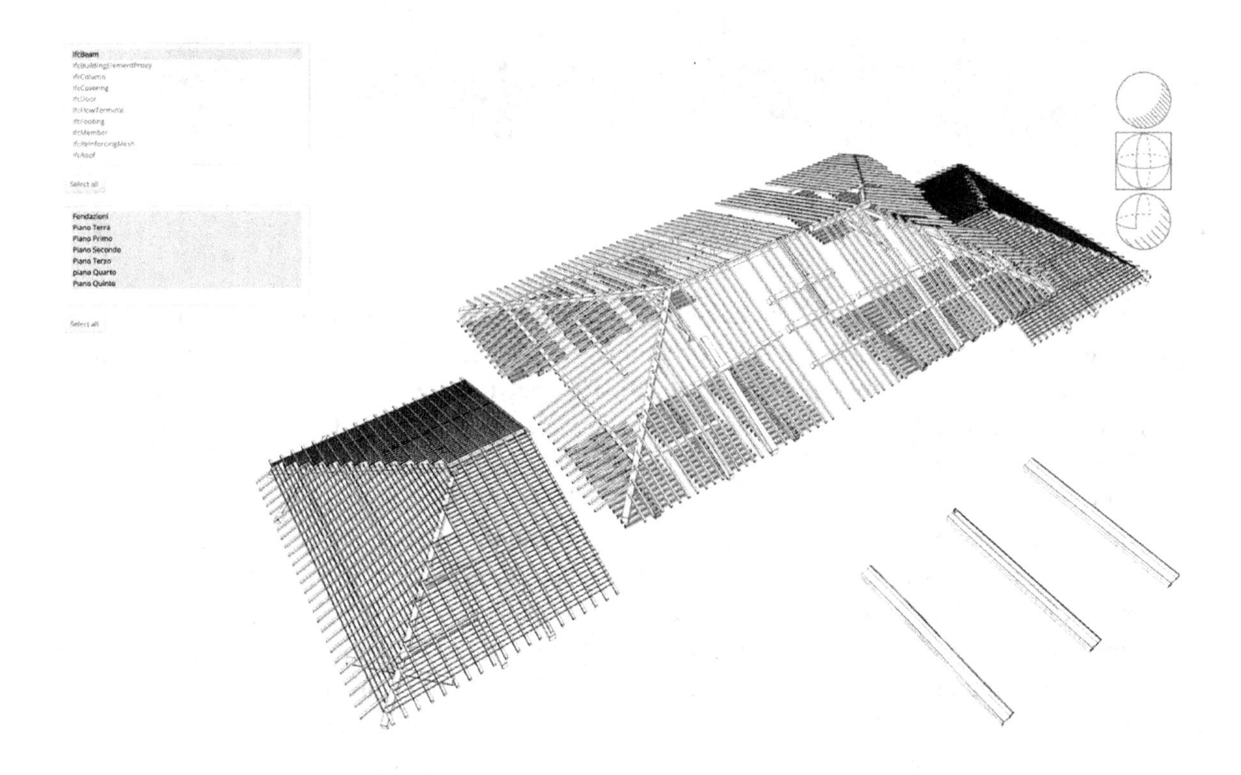

Figure 19. Display model information properties in platform
Source: POR-FESR eBIM (2019-2022)

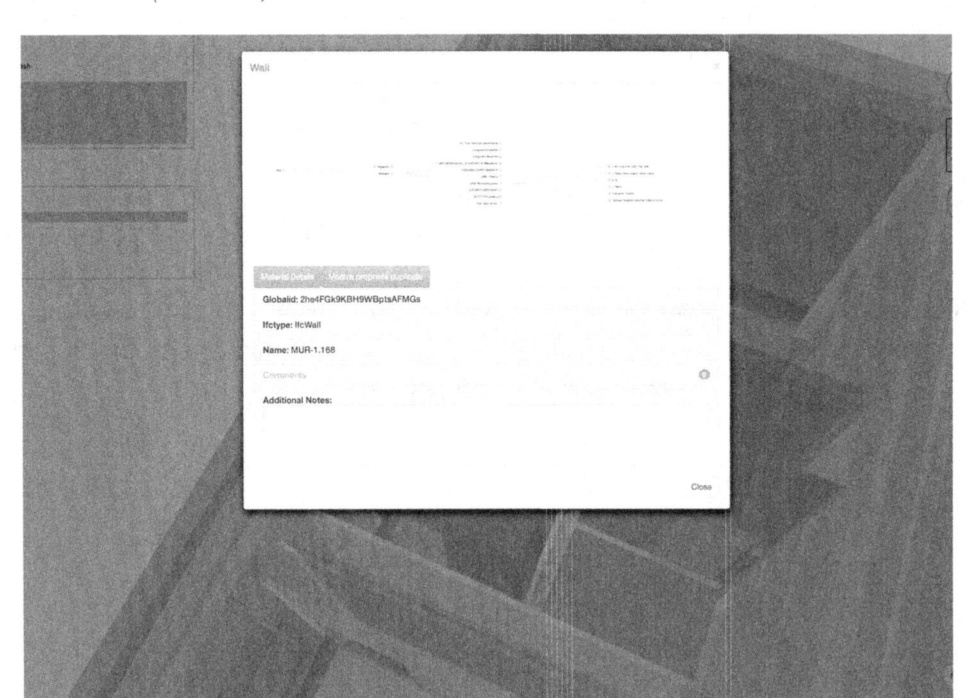

OUTCOMES

The experimentation of creating a BIM model to be imported on a web platform has highlighted the need for the creation of a methodological procedure regarding the modeling, the insertion of informative data and finally for the export phase (Figure 20). The specifications of the platform allow a correct visualization of the model and data only if different measures are followed, depending on the case and the BIM authoring software used. The ability to import even large buildings, such as the former prison of San Francesco, characterized by its geometric specificity, confirms that there are no limits dictated by the shape of the imported architecture. The fruitful collaboration between the commercial partners and the Universities has been able to achieve the result of a definition of a protocol for the standardization of the classification of material IDs, linked to an external database for easy updating.

Digital platforms are digital infrastructures capable of connecting different systems and exposing them to users through simplified and integrated interfaces, usually a mobile app or a website: the platform ensures access to contact and contextual information: it is therefore not a simple virtual shop window for choosing a product, but rather a network, a federated and cooperative mechanism in which all the operators involved can benefit and implement the availability of information or opportunities to complete or enrich with their own offer that of the other platform partners. Digital platforms enable innovative services and operate on a virtually unlimited scale by leveraging shared resources. The beating heart of the platforms are the data, and the digital technologies that allow them to be managed in real time and in an integrated manner between the different operators: the platform that is created is, therefore, an operating model that is based on the conidivision and exchange of knowledge. Through platforms it is possible to improve the customer experience, to offer products and services more in line with the needs of operators by sharing information.

The creation of information data linked to databases "bound" by an identifier, can be applied to materials, information that can change over time (e.g., maintenance status, state of preservation, etc.). Some of these examples can be easily applied, without modification of the geometric components, because they do not require precise mapping and subdivision. The same possibility of being able to upload several models with different scopes, therefore themed with information inserted in different ways, makes the platform very flexible, which stands at the midpoint between a CDE, where models or documents can be uploaded, and a web viewer.

The facility of consultation of the data is directly connected to the organization of the model and therefore of who loads in platform the elaborated ones. Too many badly organized informative data can make difficult to consult the elaborates. The various models thematic are visualized inside of a macro group, with the name of the model of the building, being able to easily choose the road to use for the consultation.

Figure 20. Ways of querying the platform by categories of intervention on the built environment, exemplified by case studies
Source: POR-FESR eBIM (2019-2022)

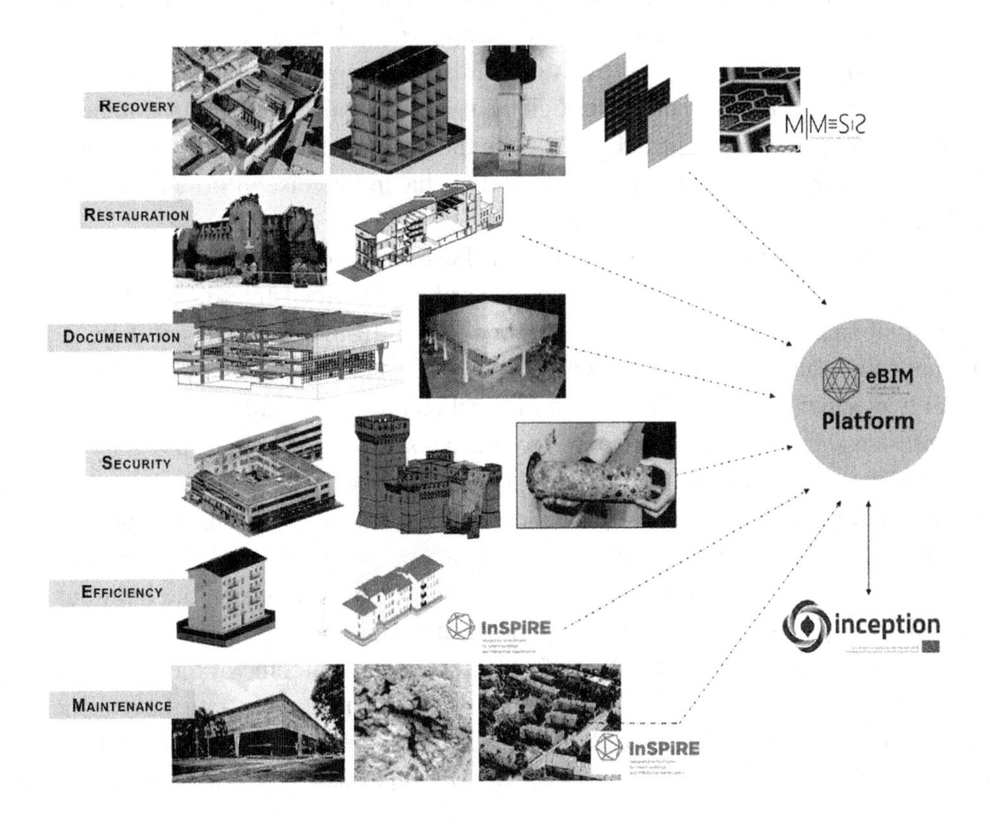

FUTURE RESEARCH DIRECTIONS

As emerged from the description of the workflow above, knowing the existing built environment implies, in addition to a careful phase of geometric survey, a research activity through historical sources, to create a model that can also collect information on the transformations that have led to the last situation detected.

Each piece of information collected, depending on the purpose of use of the model, can then be used, viewed, or implemented.

BIM systems, also in their application to the built environment, appear particularly suitable to meet the needs of designers and those who deal with aspects related to the semantic representation of data, defining a model that on a single platform allows the analytical management of visualization interfaces and assessment tools through the representation of reality; and, in addition to this, always taking into account the management of the process in terms of timing, they qualify as a key element for the proper management of any workflow on the building (Bianchini, Inglese, Ippolito, Maiorino & Senatore, 2019).

The development perspectives of this type of process can be multiple and always see the model defined in a BIM perspective at the center.

On the one hand, the materials and building elements that can expand the range of elements on which to structure and implement the information components, are numerous and customizable depending on the characteristics of the building and the categories of intervention provided on it by the regulatory instruments. The possibility of linking the information component of the BIM model to external databases also makes it possible to have data that can always be updated and increased.

On the other hand, the BIM model can evolve towards a real Digital Twin that can facilitate the subsequent design phase of each intervention through the use of sensors aimed, for example, at the thermal analysis of the building, which can be integrated with the BIM model, in order to facilitate the support to the management of the building in the use phase, thus having precise information to implement effective and targeted decision-making strategies.

As a digital replica of a physical entity, the Digital Twin can be considered as the evolution of BIM, which, thanks to an applied IoT sensor technology based on Artificial Intelligence functions, enables the development of smart buildings: the information obtained from the analysis of data, in such buildings, is automatically managed to improve their performance, enabling services, and informing users.

The digital twin should be considered as the medical record of the building, in which not only past data but also present and future data are entered for the constant monitoring of its “state of health”.

The use of the digital twin in existing buildings, to continue the design phase, can be described using the Mac Leamy curve as a basis. As it is known, the new curve will always have a much higher initial trend, as the information retrieval phase will require a greater amount of work than the one used in the same design phase for a new building and even more for the intervention on the built but continuing along the timeline, they will get closer until they merge at the end of the project.

The logic of using semantic digital web platforms, such as the outcome of the eBIM project, aligns with the research logic related to Artificial Intelligence in wanting to demonstrate the realistic possibility of using intelligent systems for common uses and in evaluating their impact on the social and economic fabric, always opening up different avenues for the realization of new types of professional figures.

CONCLUSION

The communicative capacity over time of the information data and models, together with the documentary parts of the buildings, are the fundamental aspects of the development of web platforms for BIM models. The information assets that are encapsulated in the data collection of a newly constructed building are necessary for the subsequent maintenance and management activities of the building (Osello & al., 2015).

The difficulty of collecting this information in a new building is certainly made easier when compared to the same activities in an existing building. It turns out to be considerably more valuable the creation of data package, as created through various types of survey, to be understood not only as an activity related to the measurement of geometries, but under a much broader sense, passing through a historical survey (related to the existing documentary part), material (related to the classification of specific properties of materials), typological (classifying the elements according to their function and geometry and shape).

The enormous amount of work, if managed in an uncoordinated and disjointed way, without a common space for uploading and consultation of these elements, may bring a result in the short term of rapid creation, but in the long term will break its enormous potential. The platform allows to make accessible over time the consultation of data, regardless of the proprietary software constraint, being able to upload different types of models, definitive for specific purposes of visualization and consultation. With the reading of the id associated to external databases, it is easy to assume that this feature can lead to the connection of IOT sensor systems for the query of data always updated in the web page, depending on the renewal of the connected database. The connected data could represent temperature variations, verification of local crowding, indication of access opening, etc... The amount of data can be used, depending on the phase and purpose of use, to manage planning activities for renovations, or to manage activities related to the seventh dimension (7D) of BIM, linked to the management of the building stock in use. In other words, the web tool can be used not only for archiving and cataloguing as a historical memory, but also as a tool for predictive management of activities.

ACKNOWLEDGMENT

The eBIM project was co-funded by the Emilia Romagna Region in the framework of the call for strategic industrial research projects addressing the priority areas of the Intelligent Specialization Strategy (DGR n. 986/2018) POR_FESR 2014-2020 (ASSE 1 Research and Innovation). The lead partner was CIDEA (Centro Interdipartimentale per l'Energia e l'Ambiente) of the University of Parma, under the scientific responsibility of Prof. Chiara Vernizzi. The other proposing partners were: Laboratorio Teknehub of the University of Ferrara (Scientific Responsible Marcello Balzani); Alma Mater Studiorum - University of Bologna Interdepartmental Centre for Industrial Research Building and Construction - CIRI EC (Scientific Responsible Claudio Mazzotti); Centro Ceramico (Scientific Responsible Maria Chiara Bignozzi) and CertiMaC soc. cons. a r.l. (Scientific Coordinator Luca Laghi). The following companies, linked to the fields of construction and ICT, also took part in the partnership: CMB Società Cooperativa, Carpi (MO) - POLITECNICA Ingegneria e Architettura Società Cooperativa, Modena - Buia Nereo S.r.l., Parma - Telematica Informatica S.r.l., Turin - Nemoris S.r.l., Bologna - Smart Domotics S.r.l., Faenza (RA) - Cooprogetto Società Cooperativa, (Faenza (RA) - Ceramiche Refin S.p.A., Salvaterra (RE) - Tonalite S.r.l, Modena - Monitor the Planet S.r.l., Faenza (RA) - Wienerberger S.p.A. - Fornaci Laterizi Danesi S.p.A. - Kerakoll S.p.A - SAFE LM S.r.l. - Ruregold S.r.l.

REFERENCES

Apollonio, F., Gaiani, M., & Bertacchi, S. (2019). Managing Cultural Heritage with Integrated Services Platform. *ISPRS-International Archives of the Photogrammetry, Remote Sensing and Spatial Information Sciences, 4211*, 91–98.

Baiardi, L., & Ferreira, E. A. A. (2020). The integrated project for the redevelopment of a historic building: an example of BIM and IoT integration to manage the comfort of the building. In C. M. Bolognesi & C. Santagati (Eds.), *Impact of Industry 4.0 on Architecture and Cultural Heritage* (pp. 261–282). IGI Global. doi:10.4018/978-1-7998-1234-0.ch011

Balzani, M., & Maietti, F. (2020). Data acquisition protocols and semantic modelling of the historical-architectural heritage: the INCEPTION project. Digital Strategies for Cultural Heritage, 2(1), 83-95.

Biagini, C. (2007). BIM strategies in architectural project management. In *GRAPHICA 2007. Desafio da era digital: Ensino e Tecnologia*. UFPR.

Bianchini, C. (2014). Survey, modeling, interpretation as multidisciplinary components of a Knowledge System. *SCIRES-ITSCIentific RESearch and Information Technology, 4*(1), 15–24.

Bianchini, C., Inglese C. & Ippolito, A. (2016). Il contributo della Rappresentazione nel Building Information Modeling (BIM) per la gestione del costruito [The contribution of Representation in Building Information Modeling (BIM) for building management]. *DisegnareCon, 16*(9), 10.1-10.9.

Bianchini, C., Inglese, C., Ippolito, A., Maiorino, D., & Senatore, L. J. (2019). Building Information Moldeling (BIM): Great Misunderstanding or Potential opportunities for the Design disciplines? In Architecture and Design: Breakthroughs in Research and Practice (pp. 365-386). IGI Global.

Bolognesi, C., & Aiello, D. A. A. (2020). From Digital Survey to a Virtual tale. Virtual reconstruction of the convent on Santa Maria delle Grazie in Milan. In C. M. Bolognesi & C. Santagati (Eds.), *Impact of Industry 4.0 on Architecture and Cultural Heritage* (pp. 49–75). IGI Global. doi:10.4018/978-1-7998-1234-0.ch003

Ciribini, A.L. (2013). *L' information modeling e il settore delle costruzioni* [Information modeling and the construction sector]. IIM and BIM. Santarcangelo di Romagna: Maggioli Publisher.

Ferrari, F., Maietti, F., & Balzani, M. (2019), INCEPTION – Patrimonio Culturale Inclusivo in Europa mediante la modellazione semantica 3D [INCEPTION - Inclusive Cultural Heritage in Europe through 3D semantic modelling]. In *Il Simposio UID di internazionalizzazione della ricerca. Patrimoni culturali, Architettura, Paesaggio e Design tra ricerca e sperimentazione didattica* [The UID Symposium of Internationalization of Research. Cultural Heritage, Architecture, Landscape and Design between research and didactic experimentation]. Firenze: DIDApress.

Maietti, F., & Tasselli, N. (2020). Connessioni digitali. Integrazione dati in ambiente BIM per l'intervento sul costruito esistente [Digital connections. Data integration in BIM environment for the intervention on Existing Buildings]. In *Proceedings of 42° Convegno Internazionale dei Docenti delle discipline della Rappresentazione - Congresso della Unione Italiana Disegno* [Proceedings of 42 ° International Conference of Teachers of the disciplines of Representation - Congress of the Italian Drawing Union] (pp. 585-598). FrancoAngeli.

Osello, A. (2015). BIM and Interoperability for Cultural Heritage through ICT. In *Handbook of Research on Emerging Digital Tools for Architectural Surveying, Modeling, and Representation* (pp. 281-298). IGI Global.

Pavan, A., Mirarchi, C., & Giani, M. (2017). *BIM: metodi e strumenti. Progettare, costruire e gestire nell'era digitale*. Tecniche Nuove.

Pulcrano, M. (2020). Modelli digitali interconnessi per ampliare la conoscenza e migliorare la fruizione del patrimonio costruito [Digital models interconnected to expand knowledge and improve the use of cultural heritage]. In *Proceedings of 42° Convegno Internazionale dei Docenti delle discipline della Rappresentazione - Congresso della Unione Italiana Disegno* [Proceedings of 42 ° International Conference of Teachers of the disciplines of Representation - Congress of the Italian Drawing Union] (pp. 2604-2621). FrancoAngeli.

ADDITIONAL READING

Azhar, S. (2010). BIM for sustainable design: Results of an industry survey. *Journal of Building Information Modeling*, 2728.

Bernstein H. M. Jones S. A. Russo M. A. (2010). *Green BIM: How building information modeling is contributing to green design and construction*. McGraw-Hill Construction.

Caputi, M., Odorizzi, P., & Stefani, M. (2015). *Il Building Information Modeling. BIM. Valore, gestione e soluzioni operative*. Santarcangelo di Romagna: Maggioli editore.

Di Luggo, A., Palomba, D., Pulcrano, M., & Scandurra, S. (2020). Theoretical and Methodological Implications in the Information Modelling of Architectural Heritage. In C. M. Bolognesi & C. Santagati (Eds.), *Impact of Industry 4.0 on Architecture and Cultural Heritage* (pp. 20–48). IGI Global. doi:10.4018/978-1-7998-1234-0.ch002

Gallo, G. & Rizzarda, C. (2017), *La sfida del BIM. Un percorso di adozione per progettisti e imprese* [The challenge of BIM. An adoption path for designers and companies]. AM4 Educational.

Olawumi, T. O., & Chan, D. W. M. (2019). Building information modelling and project information management framework for construction projects. *Journal of Civil Engineering and Management, 25*(1), 53–75. doi:10.3846/jcem.2019.7841

Zacchei, V. (2010). *Building information modeling. Nuove tecnologie per l'evoluzione della progettazione*. Aracne.

KEY TERMS AND DEFINITIONS

7D: Sustainability assessment of the work. References: UNI standard 11337 (internationally, 7D refers to Facility Management).

Code of Contracts PUBLIC: Legislative Decree no.50 of 18 April 2016, entitled "Code of public contracts for works, services and supplies in implementation of Directives 2014/23/EU, 2014/24/EU, 2014/25/EU" and subsequent amendments and supplements.

Database: A collection of information organized in such a way that it is easily accessible, managed and updated It is also defined as an organized collection of Datasets.

Dataset: An organized collection of observations.

Opendata: Data that are freely accessible to all (any restrictions being the obligation to cite the source or to always keep the database open).

Chapter 12
Virtual Reality for Fire Safety Engineering

Emiliano Cereda
Politecnico di Torino, Italy

Roberto Vancetti
Politecnico di Torino, Italy

ABSTRACT

The international fire safety framework defines the characteristics of an escape system that can communicate information to allow occupants to make the optimal decision to reach a safe place. Fire safety engineering is the subject that helps the designer to carry out analyses for the study of fire through the use of CFD (computational fluid dynamics) tools and escape modelling. The interaction between the escape system and the occupants is a factor that controls the effectiveness of the design solution. This factor is difficult to assess in the absence of specific tools. An analysis methodology based on numerical simulation models, aided by virtual reality tools, improves the interpretation of results. The authors set out to develop a method capable of exporting fire simulation in a virtual environment and visualising the results within a virtual reality environment. The methodology is able to improve the knowledge of the emergency dynamics within the fire scenario.

INTRODUCTION

Technological developments in the field of engineering and design have contributed to buildings with complex geometries. It is complicated to comply with the technical regulations for fire prevention when applied to very complex buildings. The development path in the field of research has also involved many innovations in fire prevention methods and approaches. There is a gradual transition from predominantly prescriptive methods to performance-based approaches. The performance approach includes a first decision-making phase regarding the fire safety measures to be adopted for the project. A second phase is the demonstration of the performance of the solution, using Fire Safety Engineering tools. The Fire

DOI: 10.4018/978-1-6684-4854-0.ch012

Safety Engineering approach is therefore the most common design path to date. Often this approach is the only one that can be applied in order to comply with architectural and functional requirements.

Fire Safety Engineering

Fire safety engineering (FSE) is a discipline whose characteristics and definitions were first presented in the technical report ISO/TR 13387 of 1999. There are various synonyms for this terminology, depending on the document in which it is discussed. The following are some of them: Engineering Approach, Performance Based Approach, Fire Safety Engineering. It can be defined as the subject that guides the designer, through scientific methods, in the choice of the most appropriate safety measures aimed at protecting people, material and the environment from the effects of fire. The technical report ISO/TR13387 provides the following statement to describe the performance methodology:

"The application of engineering principles, rules and expert judgement based on a scientific appreciation of the fire phenomena, of the effects of fire, and the reaction and behavior of people, in order to: save life, protect property and preserve the environment and heritage; quantify the hazards and risk of fire and its effects; evaluate analytically the optimum protective and preventative measures necessary to limit, within prescribed levels, the consequences of fire".

Most countries have reference legislation that includes fire safety engineering. FSE methods consider the totality of fire prevention and protection measures. The method provides a more fitting solution than traditional methods. In some cases, the approach is the only means of achieving a successful level of fire protection. International FSE standards consider the interaction between fire, buildings and occupants, and provide for the evaluation of fire scenarios. International FSE standards consider the interaction between fire, buildings and occupants, and provide for the evaluation of fire scenarios. ISO/TR 13387, provide a flexible framework in order to create a fire safety design that can be easily evaluated by the competent authorities. British Standard 7974 and the NFPA standards, together with ISO/TR 13387, provide a frame for an engineering approach to the achievement of fire safety in buildings. They contain advice and guidelines on the application of scientific and engineering principles for the protection of occupants, objects and the environment from fire. During the fire design process, a great deal of effort is required. At this stage, it is essential to identify the boundary conditions and scenarios that may interact and modify the evolution of fire and escape dynamics. Within the context, it is necessary to examine aspects such as the escape system and how it communicates with the occupants. The solution must be able to provide the right information to allow occupants to make the best decision and reach a safe location. The method is based on a careful design 'dressed' to the specific needs of the activity and the characteristics of the building. In the design phase, the fundamental requirement is to be able to predict, quantify and evaluate the elements that characterise the fire scenario. The prescriptive approach does not examine or require the in-depth examination of a number of aspects, which the fire safety engineering approach requires.

One example is the fire alarm system, whose purpose is to communicate an emergency to the occupants. The same performance required of the system can be guaranteed in different ways. Fire safety engineering requires, in this case, knowledge of the signalling and alarm times. The same performance required of the system can be guaranteed in different ways. Fire safety engineering requires, in this case, knowledge of the signalling and alarm times. These have a considerable impact on the evaluation of the system's performance, as they alter the evacuation time.

ASET/RSET CONCEPT

The philosophy of performance-based design is that a project should focus on 'performance'. This approach has generated the need to ensure that the life of the occupants is safeguarded. This requirement introduced the need for egress simulation models. Performance-based design (PBD) is based on the concept that any fire-fighting measure can be used as long as it is appropriate to allow an acceptable safety level. Fire Safety Engineering is therefore based on these principles, the main aim of which is to ensure that the occupant can evacuate the building within the necessary timeframe under acceptable conditions. The times involved are calculated using the ASET and RSET parameters. ASET describes the time during which environmental conditions remain such that the evacuation of occupants is not compromised (available safe egress time). The calculation of ASET can be performed using various methods, including mathematical approaches and by means of computational fluid-dynamic analysis using FDS calculation tools. It depends on the interactions between occupants, fire and environment. Various tools can be adopted for the evaluation of RSET, each characterised by a different level of complexity.

ASET and Tenability Conditions

The available safe egress time is defined as the instant at which the effects of the fire reach the tenability limits. The tenability threshold values are defined by international fire prevention standards for quantities considered relevant such as smoke temperature, visibility through smoke, radiation and FED.

Escape and Pre-Evacuation Times

The time required to allow the safe evacuation of the building is composed of the sum of several factors, each of which is associated with a specific egress phase. RSET (Required Safe Egress Time) represents the time interval between ignition until all occupants have left the building. The technical report ISO/TR 16738:2009 defines it as follows:

$$RSET = t_{det} + t_{a} + t_{pre} + t_{tra}$$

dove:

- t_{det}: detection time; time required for the detection system to become aware of the fire;
- t_{a}: general alarm time; time between detection and spreading of information to the occupants;
- t_{pre}: pre-movement time; time required for the occupants to carry out a number of tasks before moving to the safe location;
- t_{tra}: movement time; time taken by occupants to reach the safe location calculated from the end of pre-movement activities.

Several factors influence the RSET value, such as the extrinsic and intrinsic characteristics of the occupants, their cognitive, sensory and locomotor capacities, the fire detection and alarm system, and the planned egress design measures. For this reason, it is necessary for the designer to develop the appropriate design scenario that best matches the specific case. Fire safety engineering also introduces some attention to elements that were previously always overlooked within a design. In addition to aspects related to the best-known fire-fighting measures (compartmentalisation, fire resistance, length of escape routes, ...), there are factors (often intrinsic to each person involved) which may make the solution ineffective (Vancetti, R., & Cereda, R.,2020a). In order to understand these aspects, it is first necessary to analyse the escape of occupants during an emergency. This topic has been given increasing attention in recent years, precisely because the efficiency of the overall management system depends on it, and thus on the fulfilment of the guarantee of life safety. Indeed, in the field of fire prevention, measures to facilitate the evacuation of occupants are the most important in terms of safeguarding human life. It is important to keep in mind, how evacuation time is closely dependent on human behaviour both from an individual's point of view, but especially from a collective point of view, as discussed in the work of Cuesta et al. (2016).

To ensure effective fire safety for occupants, it is necessary to understand the factors and conditions that influence human reactions.

As Cuesta et al. (2016) describe, occupants, when exposed to an emergency situation, make decisions that differ according to a number of (random and non-random) characteristics. Some of the main aspects that can have an effect in an emergency are: gender, age, physical abilities, familiarity, social attachment, and attachment to objects.

Characteristics of the Human Being in an Emergency Situation

Any building that contains occupants constitutes an environment characterised by various peculiarities. Building-occupant interaction, precisely because of the particular type of occupants, is complex.

In addition to the ordinary alarm signalling, carried out by means of optical-acoustic systems, procedures may assign specially trained personnel a substantial role during the egress phases. Among the characteristics that influence behaviour and consequently choices during an emergency, in addition to those mentioned above, there are others that have a greater impact on egress dynamics.

The egress, therefore, is primarily influenced by the type of users.

Gender

Men are more likely to try to help the other occupants and put out the fire, while women prefer to join the family and get to safety (Tong, D., & Canter, D., 1985).

Age

Young adults have different reactions from adults when faced with hazard and emergencies in general (Fridolf, K., & Nilsson, D., 2011). Their particular behaviour is the result of their unawareness and inexperience. A child has a different perception of risk than an adult. Older people may have little resistance to the debilitating effect of smoke and heat and may therefore be more exposed to risk.

Cultural Aspects

There are other aspects that can affect the outcome of an evacuation in a preponderant way. One of these is related to the concept of culture: people with different cultures may show different behaviour in an emergency situation. This aspect is related to the different perception of risk and is more evident in those contexts characterised by populations from more than one country.

Roles/Responsibilities

The second important aspect is related to occupant roles and responsibilities. Occupants characterised by a higher hierarchical position are inclined to behave differently than occupants without responsibilities.

Moreover, an incorrect action by one of them can compromise the life of a community. This is the case with an employer or a teacher who must lead the entire class to the assembly point outside the building using the escape routes prepared and fixed in the evacuation plan. These profiles are also required to make almost instantaneous decisions when an escape route is obstructed or unavailable. This scenario represents critical issues that can only be easily overcome through the promotion of safety concepts by making all the 'actors' involved prepared and aware.

Physical Capabilities

During an emergency situation, people with physical and sensory impairments may be present, which could slow down both their own and other people's evacuation.

Familiarity

In an emergency, individuals move towards places or persons familiar to them. Familiarity in this case becomes a danger as the occupant does not really perceive risk by feeling that he is in an environment that is safe for him (Sime, J. D.,1985).

Social Attachment

Although people with social or emotional attachments can help each other during an emergency situation. Such bonds slow down evacuation and reduce the perception of danger.

Attachment to Personal Effects

Before evacuating the building, some people tend to rescue their personal belongings, even if this compromises or slows down their evacuation.

From the elements described above, one can perceive how, when faced with an emergency, each individual reacts differently. Knowing the human factor in an emergency situation is crucial to help rescue action and ensure the safety of the individuals involved.

Fire safety engineering and the tools that currently help to design our safe environments require us to develop a culture of fire safety awareness. User (and end user) information and awareness are the elements that, together with engineering tools are able to guarantee an efficient and safe system.

On the basis of what has been said, it is possible to distinguish two mainly involved contexts that are called upon to regulate and manage emergency situations:

- the public context, which is educated and regulated through laws and regulations that manage public education and culture;
- the private context, where training and company guidelines are fundamental.

Interaction Between Occupants and Fire Systems

To date, it is possible to consider the influence of many of the aspects described above, which characterise human behaviour in emergencies through management solutions. Some aspects are more aleatory and require a different approach.

In order to consider the behavioural and physical aspects stated, the relevant texts and standards [SFPE Handbook of Fire Protection Engineering (2016)] propose to categorise occupants into types. Specific velocity values and pre-evacuation times are associated with each type of occupant profile. These are based on real simulation tests. The technical report ISO/TR 16738:2009 is the document that contains this information and aims to try to rationalise the data from the different studies on the topic so that it is available by the community. It contains values in terms of time and speed to characterise occupant profiles. However, there are countless in-depth studies on the subject aimed at investigating these parameters.

EGRESS SIMULATIONS IN FIRE SCENARIOS

In the last decade, evacuation models used in fire engineering have increased their capabilities. They started out as simple computational models based on equations related only to the hydraulics laws that implemented only small spaces. Further developments have made it possible to simulate complex behaviour such as decision-making by means of the most innovative simulation techniques. A great advantage of these automatic models is that they allow the simulations to be repeated several times, enabling the best measures to be taken in the building to be evaluated. In addition, the reiteration of simulations allows a better understanding of the dynamics of evacuation and greater awareness of the phenomenon. An egress model is closer to reality than manual techniques. In order to understand the dynamics and characteristics of today's evacuation models and the principles attached to them, it is first necessary to analyse how they have evolved through the advancement of studies on human behaviour during emergency situations. The first scientific studies on egress simulations were carried out between the 1970s and 1980s. These studies represent the scientific basis on which all subsequent models were founded. The RSET calculation was at first based on calculations using simple equations in which human behaviour was not yet considered and the movement of the occupants was approximated to that of a fluid. Thus was born the hydraulic model that still forms the basis of the simulation of the movement of occupants and is based on the resolution of the Navier-Stokes fluid equations. A limitation of the RSET calculation using the model just mentioned is that it does not consider behavioural aspects. Time does not only include the time required to reach the safe place, but also an additional time in which the so-called 'decision-making process' takes place, i.e. the activity related to decision-making aspects in the phase before movement. Today, tools that were based only on hand calculations are being abandoned, and all efforts are being concentrated on the study of agent-based models. These models allow to set certain behavioural laws to regulate the interaction between occupants and between occupants and their environment. However, not all factors influencing choices during egress activities have been investigated in depth, as studies to date have not yet managed to characterise these phenomena mathematically. The limitation, to date, encountered during a fire safety engineering assessment is related to the probabilistic and irrational factors, which are difficult to evaluate. For this reason, the authors set out to explore an innovative approach that allows the effectiveness of an escape system to be assessed in first person.

General Aspects and Characteristics of Egress Models

There are different egress simulation models, and it is essential that the designer is skilled in the field and is able to select the appropriate model for the application. A distinction should be made between simulation software and the evacuation models on which the former are based.

There are scientific studies in the literature with the aim of facilitating the choice of the model that best suits the purpose. These works classify evacuation models according to their main characteristics. Kuligowski's work in 2010 analyses models and provides classifications according to them. Further research on this subject has been carried out in 'Modeling crowd evacuation of a building based on seven methodological approaches' and 'A review of optimisation models for pedestrian evacuation and design problems' in which the methodological approaches on which the models are based are classified and possible optimisations are discussed.

There are two methods for representing occupants within the model:

- macroscopic models;
- microscopic models.

Macroscopic Models

In this approach, also called "flow-based", the models represent people as one homogeneous group.

The occupants are not considered individually, but are represented as a flow of a moving fluid, thus omitting any distinction by individual. Models employing these principles consist of a network, in which nodes identify building surfaces. It uses the laws of fluid dynamics, based on a density for the fluid. These models are based on correlations between speed of movement and density. Where there are narrowings in the escape path, there will be decreases in flow. Models that are based on the analogy between flow of people and flow of continuous media are also referred to as 'Continuum models' or 'Fluiddynamic based models'. They are characterised by relatively simple equations based on fluid theory.

However, this model is less popular to date. This approach is not able to describe the population in a real and heterogeneous way respecting the intrinsic and extrinsic characteristics of each occupant.

Microscopic Models

In the microscopic model, occupants are considered as individuals with specific physical and behavioural properties. This makes it possible to represent the characteristics of the population. In the evacuation model, people can perform actions given by means of behavioural rules. Examples might be the possibility of reducing or increasing walking speed depending on the presence of other occupants. This modelling approach is called 'Agent-Based Modelling' (ABM) and its applications span a variety of fields.

Modelling of Geometries

Geometries to describe and represent the environment can be classified into three different types:

- course network;
- fine network;
- continuous.

The most widespread models represent the environment by means of a continuous space in which the occupants are free to move. It follows that this model is the one that best approximates real movement.

Movement and Behaviour Modelling: The Agent-Based Model

There are two main categories when it comes to describing the movement of people in an environment.

The first includes fluid-dynamic models, which simulate the movement of a crowd by assimilating it to a flow of a fluid by solving the mode equations by means of the Navier-Stokes equations.

These models describe through differential equations how density and speed change with time.

From the need to be able to describe human behaviour in addition to movement comes the need for the development of new models. The agent-based model is a microscopic model that allows the simulation of very complex systems. The system is modelled as a set of entities (agents) in which each of them behaves autonomously according to precise rules. Each agent is able to assess its conditions and make decisions about them following the rules that have previously been implemented within the model. Behavioural rules can be implemented within the ABM model. The agent moves within the simulation according to two steps: in the first step the individual recognises the context, after which in the second step it executes the rules predetermined in the model for that specific situation. Although from a certain point of view the ABM model may seem simple as it consists entirely of agents and their relationships, it is capable of representing complex behavioural patterns and provide information about the dynamics of the real system it simulates. One of the characteristic aspects is the bottom-up approach whereby the model considers the interactions of the individual elements of the system, after which it attempts to determine the special characteristics produced by these interactions. The interactions are repeated during the simulation with the effect that the model becomes very computationally intensive compared to its competitors. The ABM model is the most widely used in current egress simulation software, as it is very flexible and allows the heterogeneity of the population's behavioural characteristics to be faithfully described. This is impossible to achieve using purely mathematical methods that do not involve interactions. In summary, it can be said that the emergence of the ABM approach is linked to the fact that it allows the agent and agent interactions to be modelled in a flexible and optimised way. Within the world of agent-based models, there are further classifications according to certain characteristics. For instance, a first distinction can be made with reference to the level of 'intelligence' of the agents. Indeed, it is possible to obtain occupant behaviour based on a decision-making process, i.e. the behaviour depends on the decisions made by the agents themselves during the path.

THE IMPLEMENTATION OF SIGNIFICANT PARAMETERS DURING AN EVACUATION WITHIN THE MODELS

The characterisation of a fire scenario requires a complex model. The model contains information on the fire spread and the evolution of the egress. The evolution of a fire is described by the RHR - Heat Release Rate - curve and the rate at which the products of combustion are released into the environment. The effects of fire are assessed by measuring certain quantities to which occupants are subjected during their movement to a safe place. International standards suggest the quantities used to perform this test. The values of visibility, temperature, radiation, FED/FEC are usually analysed. The fire model must interface with the egress model and allow the values to which users are subjected during the fire to be extrapolated. The egress models also require the discretization of some variables, mainly related to the factors that constitute the time required for egress. These values, as mentioned, are characterised by strong randomness and make egress a complex area to study. Pre-evacuation times can be input into the models. The designer can assign the correct detection and response times to each individual. Part of the interaction between environment, fire effects and occupant can be predicted, such as the detection time and general alarm time. However, there are dynamics that are difficult to predict, such as occupant-occupant interaction, where the behaviour of one individual may influence that of others. A further limitation of egress models occurs when the study requires evaluating the effectiveness of signage, or different acoustic signalling systems. These aspects of wayfinding (which concerns a person's ability to move through an environment and reach a destination) are difficult to interpret with ordinary tools (Vancetti, R., & Cereda, R.,2019). Consequently, virtual reality tools are an optimal solution to become more aware of the designed systems.

VR FOR FIRE SAFETY ENGINEERING

The BIM (Building Information Modelling) methodology has introduced an innovative approach to design, moving the actors in the process towards a three-dimensional, multidisciplinary and shared design. The BIM process aims to follow the realisation of the building at every phase; from the design to its maintenance. The approach is a fundamental part of the entire life of the building itself. The BIM methodology fits in perfectly with the latest visualisation techniques, as it has enormous potential in this field. The applications of this approach, BIM+Fire simulation+Egress simulation+VR, represent a good method to qualitatively analyse fire scenarios. Virtual Reality refers to a tool that can faithfully reproduce a reality. Thanks to it, the user is completely immersed in a three-dimensional environment. This approach tends to involve all the senses, even those that are not usually stimulated, such as orientation, as well as sight and hearing. The user is able to explore the environment and interact with it. Virtual reality comprises two different families according to the sensory involvement of the user, namely immersive VR and non-immersive VR.

In the first family (immersive VR) the user is totally isolated from the external environment and is immersed in a virtual environment entirely modelled by computer. The immersive experience is provided by means of devices that stimulate the senses. Mainly visors placed close to the eyes are used, called HMDs (Head Mounted Displays). Visors immerse users in a 360-degree visual experience. Along with these accessories, controls can also be provided, such as joysticks that allow movements and actions to be simulated with the hands. Non-immersive reality includes technologies in which the artificially recreated environment has less sensory impact on the user. In this case, there is no possibility of interacting with the environment but only a visualisation on a monitor. In the design sector, virtual reality technologies can be used for various purposes. One of the main uses is in facility management, or to obtain information regarding the lighting perception within rooms that have not yet been constructed, or to present the concept to the owner by making him or her involved at an early phase of the design ('participatory design'). Interesting applications in the fire safety sector presented in the work of Lovreglio et al. (2020) show that an intelligent training method is safe and effective compared to traditional methods with non-interactive lessons and videos. In fact, VR tools can be used for the purpose of training specialised personnel to teach dangerous tasks (e.g. in the case of training firefighters in the use of particular fire extinguishing devices).

Based on the premises described, the authors' aim is to present an application developed in a virtual environment. This application makes it possible to evaluate the effectiveness of signalling and alarm systems in a typical office context. The experience also makes it possible to evaluate the impact these systems have on the occupants. This is achieved through the development of an application that immerses the user in a real environment (recreated using BIM modelling tools, Fire Safety Engineering and Virtual Reality). The evaluation takes place under typical fire conditions, recreating the fire scenario in terms of fire and smoke. The application allows real-time monitoring of fire effects, occupants, signage systems and fire scenarios. The method is an aid to research tools related to the field of fire safety engineering and virtual reality. Fire Safety Engineering methods involve the use of calculation models to assess the protection of occupants' lives by determining ASET and RSET values. These analyses require the use of software that allows fire simulation and building egress simulation.

The outputs returned by this type of simulation are both quantitative and qualitative. From a fire simulation, it is possible to evaluate visibility at any point in the building and visualise the spread of smoke during the fire. The outputs of escape simulations generally consist of: representative maps of crowding density, levels of service (LoS) of the escape path, evacuation times from the building or individual rooms. From the results of the simulations, the designer is able to assess the effectiveness of the fire-fighting measures planned in the design and, if necessary, optimise the system accordingly.

The aim of this paper is to show how VR represents an aid for several figures in the field of fire prevention: the designer and fire brigade. The opportunities that the applications are intended to provide to the end user are as follows:

- assist the specialist during the modelling phase, in order to verify that the fire-fighting measures adopted are the optimal ones;
- extrapolate the time needed to evacuate the building;
- verify that visibility reduced by the presence of smoke within the rooms does not compromise the safe egress of occupants.

METHOD: VISUALIZATION OF SIMULATIONS IN A VIRTUAL ENVIRONMENT

The study of wayfinding is a topic that can be explored with virtual reality applications.

This work highlights the analysis through VR of a fire scenario, focusing on the occupant/environment interaction and the impact of the egress system on occupants. Specifically, with the use of the application, only the qualitative efficiency of the signalling system configurations is evaluated, as, to date, the export of smoke is too complex. The limitation encountered does not allow quantitative evaluations as it is not possible to use the actual smoke flows in the environment, but it is necessary to recreate them qualitatively. The method recreated the smoke within the virtual environment based on data obtained from fluid-dynamic simulations: visibility isosurfaces, smoke density and obscuration. In order to achieve their objective, the authors have created an application that uses the potential of tools based on the simulation of reality and allows for feedback from the users.

The first step involves the creation of a 3D model using Revit software. The second step consists of modelling the fire using FDS (Fire Dynamics Simulator). Analyses were carried out for the study of fire simulation, through the use of CFD software. Egress modelling was also analysed through the use of numerical software. The paper proposes a method for exporting the fire simulation into virtual reality. After the import, it is possible to visualise the fire in the VR environment. FDS does not allow the direct export of 3Dplot files containing smoke information, therefore, it was necessary to carry out this step with other software. The 3Dplot files were edited on Paraview. The export of isosurfaces and slices was carried out using two different software (Blender and Deep Exploration). This was a choice related to the graphic visualisation of the components. The final result is a homogeneous isosurface. Finally, the exported elements were loaded onto Unity3D. The last step consists of modelling in Unity - by Unity Technologies - software capable of developing interactive content. Figure 1 shows the methodological process; Figure 2 describes the displayable elements within the virtual environment. Figure 3 shows an example of the output of a visibility slice.

In order to test the method, the authors chose a model with simple layout: an office building.

The environment is characterised by a single rectangular floor plan of approximately 230m^2. There is a main corridor from which each room is accessed. The aim is to immerse users in a smoke environment and collect feedback regarding the effectiveness of signage and audible alarms. Figures 4 and 5 show the layout and the main rooms.

Figure 1. Methodology

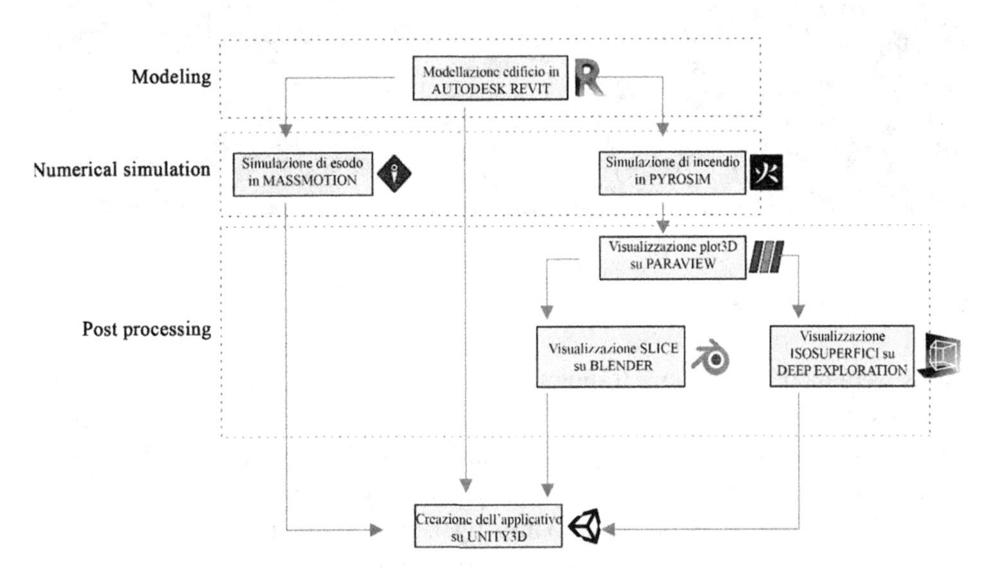

Figure 2. Visualizable elements in the application

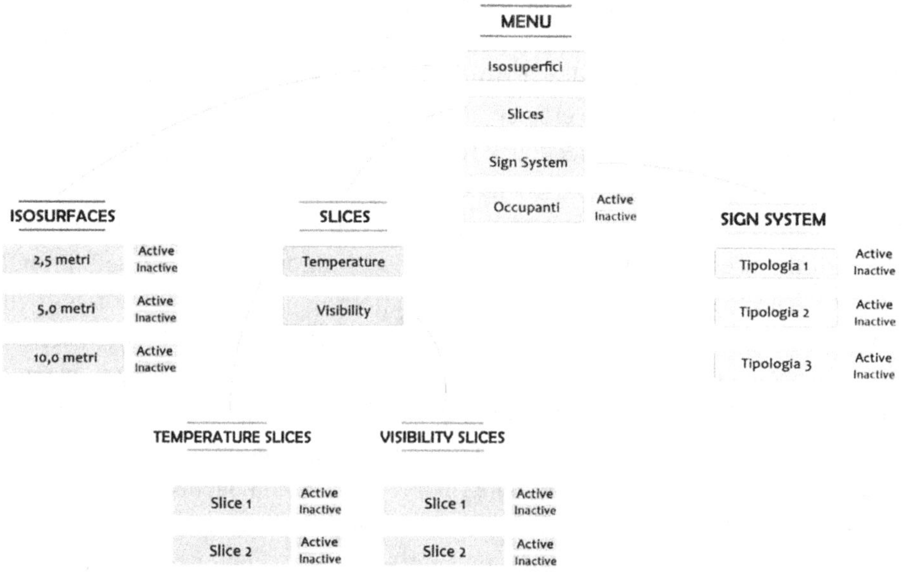

MENU
Isosuperfici
Slices
Sign System
Occupanti Active Inactive

ISOSURFACES
2,5 metri Active Inactive
5,0 metri Active Inactive
10,0 metri Active Inactive

SLICES
Temperature
Visibility

SIGN SYSTEM
Tipologia 1 Active Inactive
Tipologia 2 Active Inactive
Tipologia 3 Active Inactive

TEMPERATURE SLICES
Slice 1 Active Inactive
Slice 2 Active Inactive

VISIBILITY SLICES
Slice 1 Active Inactive
Slice 2 Active Inactive

Figure 3. Model with visibility slice

Modelled Configurations

Seven scenarios were modelled. Figure 6 shows the smoke-free environment, i.e. before the fire. The purpose of the first four scenarios is to evaluate the effectiveness of signage lighting. Italian and international regulations require minimum lighting requirements. The performance that the system must guarantee is the correct lighting of escape routes and emergency signage. There is no imposition about illuminated signage. What the authors set out to investigate, in addition to the effectiveness of ordinary lighting, is the effectiveness of illuminated signage. By comparing feedback from a sample of users, it is possible to compare the two types. The experience also makes it possible to assess the impact of signs on user decisions. The fifth scenario evaluates the effect on wayfinding of the method by which information is communicated to occupants. The comparison was made between acoustic alarm systems (alarms, sirens) and voice messages with a higher information content. The last two scenarios analysed the escape path identification system. The modelled environment includes light strips on the sides of the corridor with the function of highlighting the spaces in the environment. This type of signage was made dynamic in the last scenario. Dynamic signage is a typology that uses the signal of smoke detectors and temperature sensors. It is able to determine the location of the fire and modifies, the signage to allow occupants to reach a safe place without taking a dangerous route. Figure 7 shows the configuration with dynamic signage.

Figure 4. Model layout

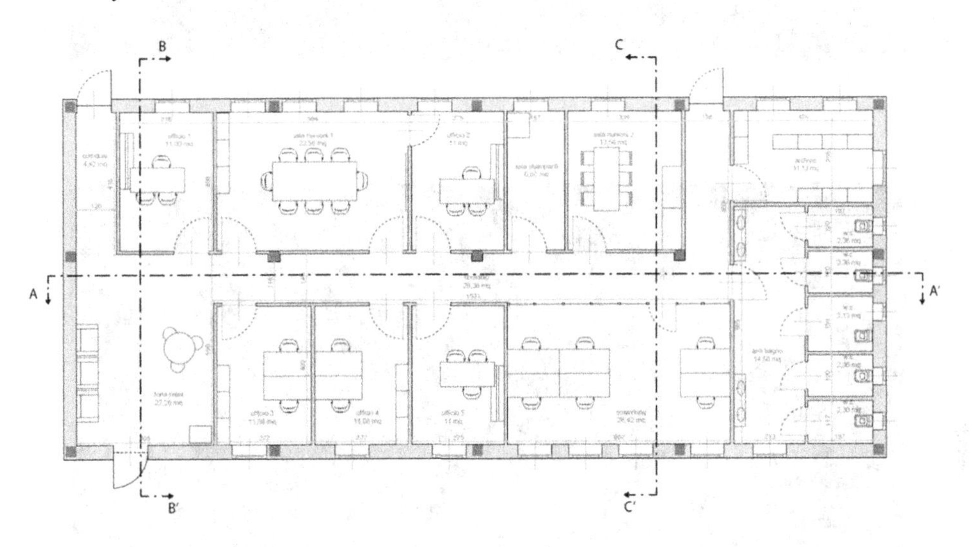

Figure 5. Interior view of modelled rooms

Figure 6. Scenario under ordinary conditions

Figure 7. Fire scenario

DATA COLLECTION

In order to obtain feedback from the users of the application, they were asked to answer a survey divided into five sections: 1. General user information, 2. Evaluation of the application, 3. Scenarios, 4. Evaluation of scenario configurations, 5. Feedback on the experience.

General user information: in this section, data such as age, gender and knowledge about fire safety were collected in order to have a general overview of the people who tried the application.

Evaluation of the application: the purpose of this section is to highlight any difficulties during the use of the application, the level of user-friendliness and the clarity of the tool's objectives.

Scenario questions: contains questions to evaluate the efficiency, strengths and weaknesses of the seven proposed scenarios.

Evaluation of scenarios: in this section, scenarios are compared in order to evaluate the effectiveness of lighting and the position of signage. Data was collected on the RSET time.

Feedback on the experience: the feedback relates to the general aspects of the application, with a focus on smoke modelling, and the achievement of the set objectives.

The survey was filled out by 80 participants.

The application was tried out by a wide range of people, diversified in terms of gender, age and pre-knowledge on fire safety.

RESULTS AND CONCLUSION

The application made it possible to collect feedback from users on the effectiveness of the type of signage adopted in the environment, but also on egress times.

This shows that similar tools are a real support to the tools currently used for these purposes.

Figure 8. Comparison of the analyzed scenarios

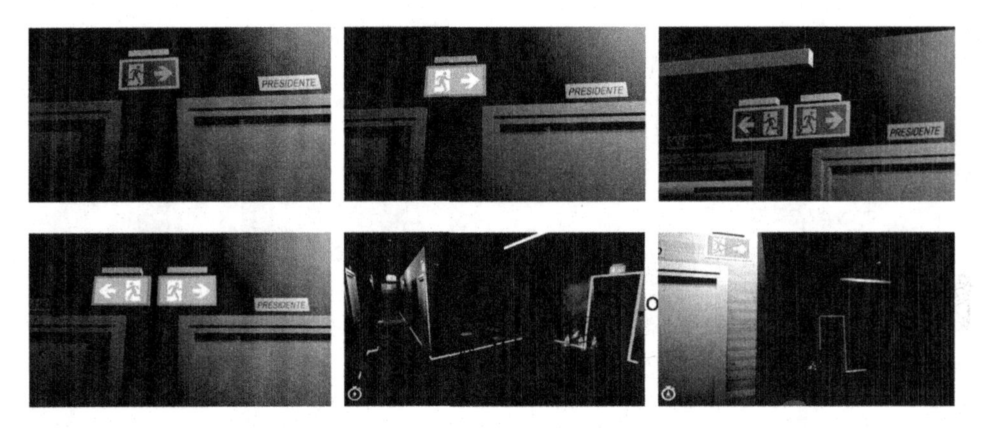

Evaluation of the Application

The responses obtained in this section show that people did not experience any problems in launching the tool and understood its objectives. A large proportion of these interviewed stated that the use of tools such as the viewer can make the experience optimal.

Scenarios

In the third section of the questionnaire, users were asked to rate the effectiveness of each scenario on a scale of 0 to 5. Here, it emerged that floor signage was very efficient, with 43.2% positive responses (rating of 5) and 52.3% responses with a rating of 4, second only to dynamic signage, which received an 81.4% response with a fully positive rating. Furthermore, of the first four scenarios, those in which internally illuminated signage was present were rated most efficient. Figure 8 shows the compared scenarios.

Scenarios Evaluation

Several results can be extrapolated from the fourth section, including:

- the characteristics of the signs (position and size), which comply with the regulations, are found to be correct in all scenarios;
- internally illuminated signs are more visible. This is in accordance with the results of the study carried out by T. Jin in 1978, in which the visibility of signs through smoke was analysed;
- Signage helped to reach the emergency exit in 82.2% of the cases. This figure confirms that proper signage inside buildings helps occupants to leave the building;
- Users rated configurations with backlit signs and with floor signs as more efficient;

Figure 9. Scenario Comparison: more effective configuration

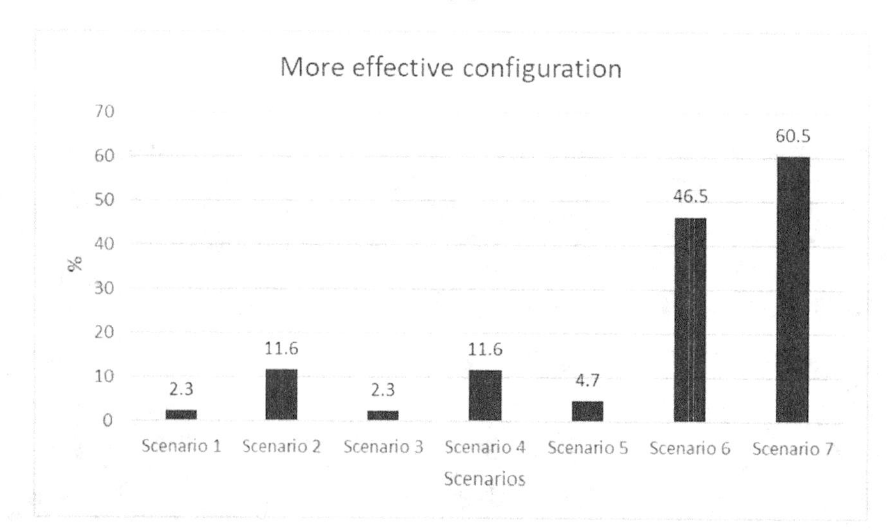

- The presence of evacuation maps, with a clear indication of 'You are here', are a main help in the evacuation phase. Almost all people, once the alarm was triggered, already knew how to get to the exit;
- A figure that allows us to make preliminary assessments of the ASET > RSET criterion, concerns the time taken to find an emergency exit. In 93.3% of the cases, it took people less than two minutes to reach the exit. The remaining 6.7% took longer only in the first scenario due to unfamiliarity with the controls.

Experience Feedback

The feedback obtained in the last section provides input on how to improve the application or what additional information can be added. In general, the application was evaluated positively for both the quality of realism and the goals it is supposed to achieve. It is important to underline that people find the use of the application educational.

REFERENCES

Cuesta, A. (2016). Evacuation Modeling Trends. Springer.

Cuesta, A., Abreu, O., & Alvear, D. (2016). *Evacuation Modeling Trends*. Springer.

Fridolf, K., & Nilsson, D. (2011). People's Subjective Estimation of Fire Growth: An Experimental Study of Young Adults. *Fire Safety Science*, *10*, 161–172.

Jin. (1978). *Visibility Through fire smoke*. Academic Press.

Kuligowski, E., Peacock, R., & Hoskins, B. (2010). *A Review of Building Evacuation Models* (2nd ed.). NIST Pubs.

Lovreglio, R., Duan, X., & Rahouti, A. (2021). Comparing the effectiveness of fire extinguisher virtual reality and video training. *Virtual Reality*, *25*, 133–145.

SFPE Handbook of Fire Protection Engineering. (2016). Academic Press.

Sime, J. D. (1985). Movement toward the Familiar: Person and Place Affiliation in a Fire Entrapment Setting. *Environment and Behavior*.

Tong, D., & Canter, D. (1985). The decision to evacuate: A study of the motivations which contribute to evacuation in the event of fire. *Fire Safety Journal*, *9*(3), 257–265.

Vancetti, R., & Cereda, R. (2019). *La Realtà Virtuale: un nuovo strumento a servizio della progettazione con la Fire Safety Engineering*. Antincendio.

Vancetti, R., & Cereda, R. (2020a). Stazioni metropolitane: la caratterizzazione degli occupanti per le verifiche di esodo ed inclusione con i metodi della fse [Metropolitan stations: characterization of occupants for exodus and inclusion checks with ESF methods]. Antincendio, 58-77.

Vancetti, R., & Cereda, R. (2020b). Progettazione e verifica del sistema di esodo con strumenti alternativi: la realtà virtuale immersiva. In Nuovi orizzonti per l'architettura sostenibile [New horizons for sustainable architecture] (pp. 1522-1530). Academic Press.

Vancetti, R., Cereda, R., & Cosi, F. (2019). Fire Safety Engineering, Universal Design, Realtà Virtuale: nuovi strumenti per una progettazione sempre più smart. Ingegno e costruzione nell'epoca della complessità - Forma urbana e individualità architettonica, 873-882.

Vermuyten, H., Beliën, J., De Boeck, L., Reniers, G., & Wauters, T. (2016). A review of optimisation models for pedestrian evacuation and design problems. *Safety Science*, *87*, 167–178.

Zheng, X., Zhong, T., & Liu, M. (2009). Modeling crowd evacuation of a building based on seven methodological approaches. *Building and Environment*, 437–445.

KEY TERMS AND DEFINITIONS

BIM: A collaborative process of creating and managing information for a built asset throughout its lifecycle from planning and design to construction and operations.

Computational Fluid Dynamics: Computational fluid dynamics (CFD) is the use of applied mathematics, physics, and computational software to visualize how a gas or liquid flows – as well as how the gas or liquid affects objects as it flows past. Computational fluid dynamics is based on the Navier-Stokes equations. These equations describe how the velocity, pressure, temperature, and density of a moving fluid are related.

Egress Simulations: A method to determine evacuation times for areas, buildings, or other spaces. It is based on the simulation of crowd dynamics and pedestrian motion. Egress simulations are used in fire safety engineering analyses.

Fire Safety Engineering: Fire Safety Engineering is the application of scientific and engineering principles based on the understanding of the effects of fire, the reaction and behavior of people to fire and how to protect people, property, and the environment.

Fire Simulations: A method of evaluating of fire scenarios to answer questions about heat, smoke, and toxic gas production.

Performance-Based Design: Approach to the design of any complexity of building. A building constructed in this way is required to meet certain measurable or predictable performance requirements, without a specific prescribed method by which to attain those requirements.

Serious Games: A game designed for a primary purpose other than pure entertainment.

Virtual Reality: The computer-generated simulation of a three-dimensional image or environment that can be interacted with in a seemingly real or physical way by a person using special electronic equipment, such as a helmet with a screen inside or gloves fitted with sensors.

Wayfinding: The process of using spatial and environmental information to navigate to a destination. Wayfinding can include physical elements such as urban design, architecture, landmarks, lighting, footpaths, landscaping, and signage. These elements work together to define paths and identify key decision points, while aiming to improve and enhance people's experiences as they move from place to place.

Chapter 13

Major Events, Big Facilities:
From FM for a Football Stadium – Tools for Augmented Experiences and Fan Engagement

Maurizio Marco Bocconcino
Politecnico di Torino, Italy

Fabio Manzone
Politecnico di Torino, Italy

ABSTRACT

Let us imagine a large sports facility and an integrated system to control its maintenance (structures, facilities, furnishings, communication systems), pre-configure temporary set-ups, procurement of goods and materials, check compliance with technical regulations concerning the safety and regularity of sports and recreational events, contracts with sponsors and suppliers, and the work of technical staff. Then, let's imagine that this mass of data is supplemented by tracking the flows of people attending events, recording their behaviour through the looks they make, the stops they make, the actions they take. This is the theme of the contribution proposed, an experimental application involving a sports facility of international importance and integrating BIM processes for design and maintenance, social and commercial information systems open to the public, marketing and usage analyses based on sensors and big data, and artificial intelligence capable of prefiguring the safest and most comfortable solutions.

INTRODUCTION

The subject of knowledge through automatic data collection requires reflection on the synthesis models that such actions produce and the construction of an analytical method for validating the qualities of metadata and its synthesis actions. In particular, a study of data concerning the interaction between humans and the environment and data more specifically involving technological infrastructures. The themes of Big Data, the Internet of things (IoT) and the smart city converge in a project that aims to bring these

DOI: 10.4018/978-1-6684-4854-0.ch013

aspects together through the innovative point of view of representation. Through models, it is possible to orient and promote the understanding of complex structures by making explicit apparently concealed relationships and mechanisms (Bocconcino, 2018). With regard to the possibility of expressing the city of tomorrow, models simulate and design the project in a dimension of physicality. In recent years, a number of instrumental and technological supports appear to have matured; they can make the design process more efficient, in particular for one relevant aspect: the integration in the cognitive framework of a series of attitudinal and behavioural data, both those concerning the individual and those relating to groups of people who move, live and use urban space.

Over time, the methods of investigation of the physical and social context in which the urban regeneration and redevelopment project is born and developed, for different reasons, have seen a progressive difficulty in the field of data collection and its processing to produce information (Lo Turco et al., 2021). The different disciplines involved in the study and the design need to search for fields of confrontation where they can express, in a language common to all, their instances, their methods, the articulation of their outcomes. This field of recomposition can be facilitated by two relevant guidelines: the integration of knowledge in the professionals in training with components that are exogenous with respect to their own field of application; the possibility of access to resources and tools for analysis and representation that are interactive, dynamic and customisable according to the user's interest (Bocconcino&Manzone, 2019).

Digital technologies have radically transformed our interaction with the built environment. Mobile devices provide tools to quickly access and share information. In the context of large event facilities, these technologies impact both capital projects and day-to-day operations. Computer-aided facilities management (CAFM) systems that support the full range of facilities management (FM) activities, both physical and IT, have become ubiquitous: information technologies that are easy to access and use open up integrated knowledge containers to professionals and workers and allow structured, organised and interrogable controls with the appropriate levels of adaptation, both in the construction and management phases. (D'Urso, 2011).

Thanks to sensor networks and IoT devices, FM teams have access to a wide range of real-time building information (Valinejadshoubi, 2022). Mobile apps and cloud-hosted file systems further enhance this functionality, providing service engineers with field access to building and equipment information and building occupants with a limited range of self-service activities, such as real-time room scheduling and problem reporting (Villa et al., 2021).

The digital information model set up in the feasibility and design phases of complex artefacts increasingly supports site monitoring and ongoing maintenance activities (Lo Turco, 2015, Lo Turco et al., 2015). From design to construction site to management, this chapter aims to define an operative frontier by illustrating a method of automatic processing and graphic representation of data that multiplies the possibilities of the construction and management model set up within information and computer systems dedicated to the maintenance process (Bocconcino, 2021).

One particular common ground is neuroarchitecture/neurourbanism; although it appears to be a new discipline, for decades its function has been to create spaces capable of arousing and ensuring well-being and improving the quality of life. A meeting point between neuroscience, architecture and urban planning, architects, engineers and neuroscientists work hand in hand within this discipline. This interdisciplinary synergy aims to design spaces and buildings focused on the functioning of the brain of those who will then live or work in them.

In this spontaneous challenge, a great influence has been directed by two main objectives in the conceptualisation of building space design: 'visual perception' and 'spatial perception'. It has been shown that the professional capacity of intervention teams cannot fully meet the needs arising from a wide range of users, who experience large performance buildings on a daily basis and develop a long-term perception that influences their habits. Therefore, there is an emerging opportunity to collect and transform citizens' perceptions into a supporting intelligence, operating as artificial intelligence, which can guide the decision and skills of professionals both by educating them at multiple levels and by producing more suitable results on building design and project transformation.

Imagine we enter a stadium to watch our favourite football team play. The mobile phone shows us a map, guides us inside the structure, shows us additional content with respect to the event we are about to attend, statistics, historical footage, interviews of past and present protagonists, and even during the match provides content that increases the sensory experience, multiplies points of view, amplifies listening, the interpretation of what is happening, allows us to interact with the facility's services, catering, gadgets, booking of present and future experiences for leisure and recreation, with the functions attached to the facility. A pocket-sized maxi-screen at our fingertips and which, if we want it, addresses only us. That recognises us and offers us routes and experiences that make us feel welcome and cared for.

Now let us imagine that all this stems from an integrated system that allows in that same stadium to control ordinary and extraordinary maintenance on structures, facilities, furnishings, communication systems, pre-configure temporary set-ups, procure goods and materials, verify compliance with technical regulations relating to safety and the regularity of the conduct of sporting and recreational events, contracts with sponsors and suppliers, and the work of technical personnel.

And finally, let us imagine that this mass of data is supplemented by tracking the flow of people, recording their behaviour through the looks, the stops, the actions they take. And that a trace of all this remains for future redesigns of spaces that take into account the ergonomic aspects that derive from reading and interpreting the recordings made.

The chapter illustrates integrated applications for the monitoring and management of a sports facility - the name of which cannot be stated at the moment - as an information base for the involvement of spectators by recording their behaviour, movements and preferences through sensors, big data, and artificial intelligence capable of prefiguring personalised and engaging solutions. The considerations that follow, after an initial definition of the state of the art, describe a meta-project and a prototype realised on an experimental basis involving a sports facility of international importance and integrating BIM and GIS processes for design and maintenance, social and commercial information systems open to the public, marketing and usage analyses based on sensors and big data, and artificial intelligences capable of prefiguring the safest and most comfortable solutions.

BACKGROUND

New high-tech systems are finding more and more areas of application in structures. Complex facilities require learning and automated systems (Roper, 2017). To achieve this, FM systems need to be more expandable and offer opportunities for data analysis (Demirdogen et al., 2020). In the literature, building automation systems (BAS) and computerised maintenance management systems (CMMS), have been used as FM systems. However, there are some problems with the use of these systems. Some of them are highly dependent on:

- sensors, implementations of new scenarios and technological extensions (Asensio et al., 2019);
- collection and recording of information in proprietary systems (Bhatt&Verma, 2015);
- predefined operational strategies, by the amount of data (Macarulla et al., 2017);
- low monitoring capacity due to dependence on process control and automation;
- missing or incorrect data due to problems with sensors;
- lack of data analysis or limited data analysis in CMMS and BAS systems, security systems and Computer-Aided Facility Management (CAFM) (Gunay et al., 2019);
- some organisational and management expectations of these technologies.

The above-mentioned restrictions for FM systems, the need for utilisation data from the design and construction phase, heterogeneous data, IoT issues and storage problems in BIM lead to the consideration of the integration of BIM, web databases, augmented reality and artificial intelligence (Fig. 1).

Figure 1. Facility Management (FM) data management vision (source: Arayici et al., 2018).

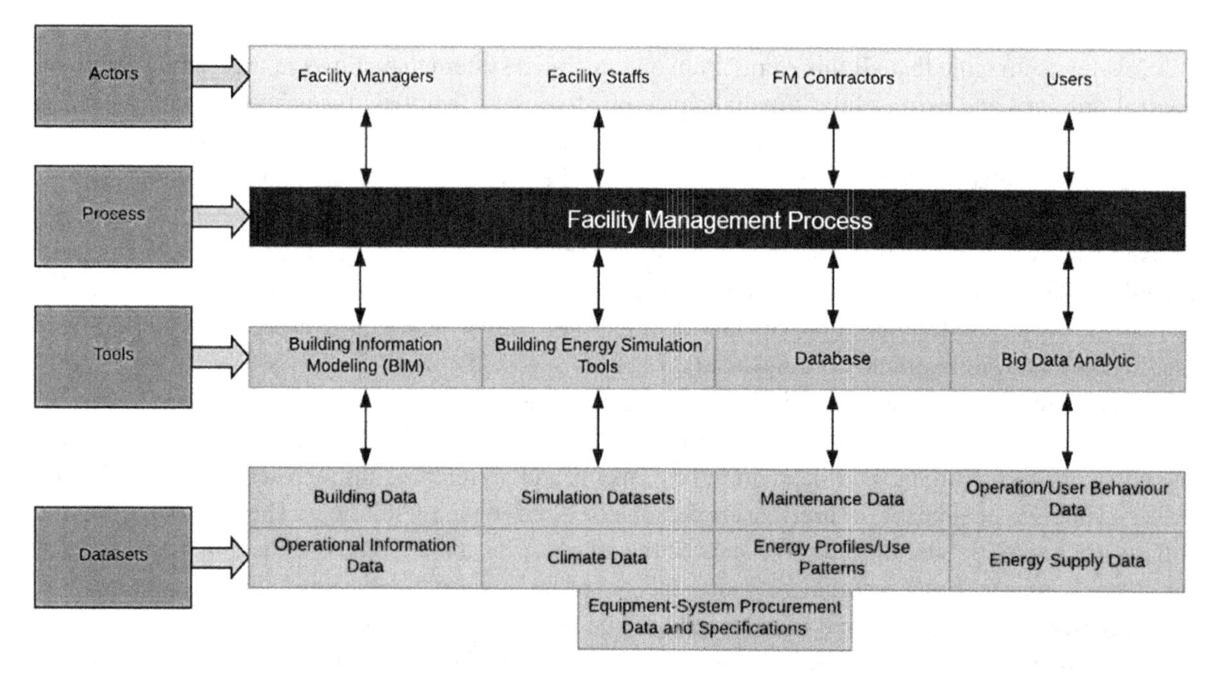

The increasing automation of site processes through BIM and AR technologies can improve decision-making and provide real-time access to information (Barbero et al., 2021). However, software products based on BIM and lean management have not yet integrated AR technology for visualising activities and related information, as well as work progress and performance. On the other hand, BIM- and AR-based applications do not manage construction processes according to lean management practices. On the contrary, they only focus on providing interactive 3D models and documents on site to assist inspections and report problems (Fig. 2). These aspects will be considered in the meta-design proposal that will be described in the following sections.

In the following, the discussion focuses on the aspects more properly related to the management of the facility in terms of user involvement. These aspects should be read in continuity with the brief framework outlined above.

The Sale, Promotion and Marketing of Sport and The Role of the Fan

The new digital technologies with their potential for dissemination and interactivity change the rules of the game, the timing, schedules, dos and don'ts of sports information, transforming sport into entertainment (Previati, 2020). At the same time, sport is considered a privileged content by the new media and also an effective promotional vehicle for industrial and service companies. For their part, sports clubs have considerable difficulty in harmonising the technical and sporting values and the passion of the public and practitioners with the different demands of businesses and communication channels. While the new media constitute an important opportunity for the dissemination and promotion of the sport seen and practised, there is a risk that sport will be overwhelmed by pure business logic, the same from which fans and sportsmen and women are fleeing in search of expression, participation, wellbeing, recreation and entertainment.

Marketing is a business function that can be expressed in the business of companies in every sector. A type of marketing is therefore also intended for the sports sector: sports marketing. Sport and therefore sports marketing must be considered as unique phenomena, given the complexity and distinctiveness of the sporting element: authors John Beech and Simon Chadwick define sports marketing as: 'an ongoing process where events with an uncertain outcome are exploited for the direct or indirect fulfilment of the needs of sporting customers, sport-related businesses and other sport-related individuals or organisations' (Beech&Chadwick, 2013).

What has transformed sport into a 'business-intensive' sector is first and foremost the new role assumed by the spectator, primarily the 'television viewer'. It is the demand of those who attend sporting events that plays a fundamental role, especially if the event is enjoyed via television. We are now in the 2000s, in which sport marketing has reached a remarkable development, and in Italy alone it turns over several billion lire a year, especially in the area of football.

The years 2010-2015, are the years in which the company is called a brand, close collaborations are called partnerships, and in which sponsorships are increasingly 'people oriented' (Kang, 2020).

Sports marketing invests in any sport, from the most popular to the minor sports, thanks to an ever-increasing number of thematic channels, magazines and media dedicated to sport, thanks to the development of increasingly advanced technologies, and thanks to the total globalisation of sport, which has reached the entire world through its most prestigious leagues, enabling investors to take advantage of communication platforms with planetary reach (Rosenthal&Eliane, 2017).

Professional football clubs have become, to all intents and purposes, businesses and their goal is no longer to achieve sporting success alone. Talking about brands is not so simple, since a univocal definition does not exist and because the brand itself dwells in the minds of customers and consumers and is identified in the global idea that customers have of that particular company or product.

Figure 2. Differences between functionalities of commercial software products for the construction management (source: Ratajczak et al., 2019, pag. 4).

Functionalities	Autodesk ® BIM 360 Plan™	Autodesk® BIM 360 Docs™	Oracle Aconex Connected BIM	Oracle Latista	Dalux TwinBIM	Trimble Vico Office (not mobile app)	VisiLean
• 3D model visualization	✓	✓	✓	✓	✓	✓	✓
• 3D object filtering	✓	✓	✓	✓	✓	✓	✓
• Visualization of 3D model in Augmented Reality (AR)	X	✓	X	X	✓	X	X[1]
• 3D model superimposed on real world in AR	X	✓	X	X	✓	X	X
• 3D interactive models (data displaying of each Building Information Modeling (BIM) object)	X	✓	✓	✓	✓	✓	✓
• Visualization of attached documents to BIM objects	X	✓	✓	X	✓	X	✓
• Notes attachment to BIM objects	✓	✓	✓	✓	✓	X	✓
• Task list related to locations	✓	X	X	X	X	✓	✓
• Instructions for the execution of construction works linked to task list	X	X	X	X	X	X	✓
• Quality Checklists linked to tasks	X	X	X	✓	X	X	✓
• Construction progress and performance tracking	✓	X	X	X	X	X	✓
• Construction progress and performance reporting (dashboard)	✓	X	X	X	X	X	✓
• Reporting of performance and progress Key Performance Indicators on BIM model in each location	X	X	X	X	X	X	X

Virtual Reality and In Store Experience

The concept of Virtual Reality is not new. What is new, rather, is its massive introduction into everyday life: a phenomenon that has been growing steadily in recent years. It now appears everywhere, in different contexts, from workplace simulations to marketing activities, from tourism to sports grounds (Fan, 2017). It has had a disruptive impact in all areas and the sports industry, in particular, has reciprocated by welcoming it with open arms. Virtual reality allows brands to generate experiences that connect them even more directly with individuals: leagues, clubs and sponsors are looking for more and more innovative ways through which to connect with audiences in a fan-engagement function (Wu et al., 2022). At first, virtual reality was one of the many 'techno fashions' of the moment, but - as times matured - it became the key to the entertainment development of the near future. Major technology brands have worked to evolve the technology to offer devices for all budgets, but also to develop sophisticated devices for professional training - just think of the race simulators used by Formula 1 drivers - or live events.

Emotional experience, and the exploration of the environment, are not simply triggered by external stimuli, but that the brain regulates emotional experience and exploratory behaviour itself, based on the perception of one's own body, and of the spatial context, put in a reciprocal relationship: we are not mere machines that merely respond to stimuli, our emotions and our behaviour are both based on the interdependent perception of ourselves and the world.

For this reason, the sensation of crossing one's physical limits, communicated by VR, has an exhilarating, entertaining and immersive effect (Kerski, 2022). Just imagine being comfortably seated at the stadium - but the same applies if we are sitting on the couch at home, in front of the TV - perhaps next to a friend, and at the same time experiencing the thrill of being close to the field or enjoying a 360° view, if desired even from above, or seeing several environments at the same time. As technology advances, experiences - both live and TV - are set to become increasingly immersive and affect the engagement generated.

Virtual shopping could soon become one of the most widely used methods of online shopping, even by clubs, especially taking advantage of highly emotional moments, such as the matches themselves: it is a much more immersive experience than usual shopping, on which companies can focus to create an emotional connection between the customer and the brand (Fathy, 2022). Here's why:

- a greater sense of ownership. People feel more comfortable with things they have established a personal connection with, they feel they own them. It is like a sense of intimacy that makes them more secure and involved. Although they do not yet own the object, they have come to relate to it as if it were already their own;
- the fan, or customer, feels at the centre of a story and, the sense of a story, is built as they go along. VR, in this sense, can be a powerful storytelling medium, telling not only the product, but also the brand, its story and its message, as well as associating them with the immense emotional imagery that a club possesses. Virtual reality allows the spectator to be in the middle of a game, it is true, but it will enable him to feel part of the story of his colours, day by day, and to choose how to become the protagonist and/or how to hear it told;
- a feeling of security and control. Everyone likes to have things under control, to decide how to live an experience, what to see and from what point of view. Conversely, the customer is inclined to distrust the brand that tries to forcibly direct him towards a service or product;

- the customer can build a personalised experience and use it at any time. Which, in reality, is not possible;
- the amount of data that the company acquires improves the knowledge of the customer and, consequently, the relationship itself, as it allows it to offer services and products tailored to the customer and increases, most importantly, the chances of monetisation.

Sport feeds on its own myths: every victory, every record, every athlete, marks the history of a team. Fans love to relive the past and virtual reality, in this respect, opens up scenarios with which other technologies cannot compete.

Augmented Experiences and Fan Involvement in Football

Often, the size and visual impact of a football stadium are the features that excite a child the first time they attend a game. Although, as one grows up, one tends to appreciate other things as well, such as the very pleasure of seeing new venues. More than eight thousand stadiums and over three hundred leagues to choose from are the best that any football fan could wish for. This is the Groundhopper App database (Futbology App, from December 2019), a very simple and intuitive mobile app that allows you to register your attendance at the stadium every time you go to watch a match. The dream of every football and travel fan, defined by the English neologism 'groundhopper', which represents a passion, but also a lifestyle. Born around the 1970s in England as a hobby, it has spread particularly in northern Europe since the 1980s, and now almost everywhere: it means travelling around the world from stadium to stadium during matches, precisely in order to see and visit as many places, teams and stadiums as possible. The database of the Futbology app is huge, with continually updated calendars in combination with the GPS function, which offers the possibility of finding a match nearby, wherever you are (even when travelling). In addition, thanks to the archive provided by the app, you can add all the matches you have seen in the past, even in stadiums that no longer exist, automatically creating your own 'history', which you can also later download in list format to your PC, and statistics broken down by stadium, team and country.

Mobile applications are becoming more and more prevalent in the lives of everyone in various fields, and football clubs could obviously not be excluded from this scenario either. In fact, more and more top clubs are deciding to make apps to allow their fans to keep up to date with the latest news or to enable personalised experiences when in or outside the stadium.

An interesting example is what Paris Saint Germain recently did, launching the 'Stadium App', an application for mobile devices that can be used both by fans in the stadium and those who decide to watch the match from home. Thanks to this service, it is possible, for example, to see the most important actions of the game in a slowed-down version, as well as to view some live content and dedicated services for fans.

Real Madrid, on the other hand, decided some time ago to offer an app for fans. Since a little over a year ago, thanks to this tool and the collaboration of Mediapro and Microsoft, it is possible to watch the matches played by the 'blancos' (Real Madrid team players) in the Champions League on mobile devices. Thanks to this initiative, the European champion club has confirmed its desire to be at the forefront of technology by offering the possibility for fans to feel close to the team even when they cannot be present at the stadium or in front of the TV. Real believes particularly in this project; the system acts as a real social network, where users can discuss the match in groups, share audio from inside the stadium or compete in virtual games for prizes such as exclusive access to content.

Barcelona has also decided to focus on digital by launching an app dedicated to its fans. This is an initiative dedicated to the fans of the blaugrana (Barcelona supporter) around the world to make them feel closer to the team. The Catalan club has thus created a multi-lingual platform to allow supporters from all over the world to get to know each other and get in touch. The service, called FCB App Studio, can also be seen as a further opportunity to increase the visibility of the club's sponsors on social networks. In fact, each user has the possibility to upload online a photo montage of him/herself in which he/she appears with some blaugrana players wearing the official uniform.

Even another top club like Manchester City did not want to be left out of this scenario. In fact, the English club recently launched a new virtual reality mobile application, where it will be possible to watch matches with specific glasses.

The Lane 360°, released by Tottenham Hotspur, is a web experience that allowed fans to explore their historic White Hart Lane stadium: for the first time ever, fans could immerse themselves in the history of a major sporting venue through 360-degree videos. The stadium - now disused - comes back to life and opens up to the fans. The pitch, the stands, the halls, the dressing rooms and familiar areas rarely accessible to supporters: everything is now part of the club's historical heritage and is shared with the fans, who are also given back the unique atmosphere of the past through a combination of 360° footage, extraordinary archive photographs and video memories of fans, players who have become Tottenham legends and players of the present. The Lane 360° is the only way to visit White Hart Lane again and, from an emotional point of view, for the fan it has a value that goes far beyond the archival one: it reminds him of his identity, reinforces his sense of belonging and links him to the positive values of the team.

But the evocative power with which virtual reality can re-present the past can also be used to show the future, to build it before the fans' very eyes: this is the case with SPVRS, another official Tottenham Hotspur app, launched last May. Back in 2018, fans of English football club Tottenham were able to take an exciting virtual tour of the new stadium, still under construction, to see what it would look like once it was finished, during a match with 60,000 spectators. SPVRS - an acronym that stands for Stadium Project Virtual Reality Suite - performs a similar function for fans as The Lane 360°, however, taking fans inside the new stadium before it is even built. As the general excitement builds, fans can follow the progress of the work as the app provides easy access to the Stadium News feed - on Tottenham's official website - ensuring up to date 'construction news'. But most importantly, fans can enjoy an extraordinary experience: thanks to a printable tracker image that can be brought to life via a smartphone or tablet, users can be transported to the centre of the pitch and, from there, enjoy a full view of the structure as it will be once completed. Not only that: the entire virtual structure is interactive, so fans can discover interesting facts and information - the number of seats, the materials used, even how many bricks will be used - about the construction of the sports facility. Premium users also have access to a special virtual area, as well as an overview from their chosen venue.

The two apps created by Tottenham are an example of how Virtual Reality - especially in its most recent developments - can push the boundaries of marketing in the quest for greater fan involvement.

In Italy, something similar, beyond apps with the latest news, is being tried by Genoa Calcio and Juventus F.C.: Genoa Virtual Reality is an engaging tool that 'dematerialises' the classic Fan Village, adapting perfectly to the characteristics of many Italian stadiums. Inside the stadium, the Rossoblù (Genoa) supporters were able to test this new technology free of charge before the match by being transported inside the pitch and experiencing the game like a footballer. Juventus F. C. focused on the Stadium, a facility that is also appreciated abroad. Before arriving at the stadium, it is in fact possible to download the 'Sosta facile' application on one's mobile device or send a simple sms to pay for parking in a simple, safe and innovative way.

PROTOTYPE STADIUM MANAGEMENT AND MAINTENANCE THROUGH WEB APPLICATION

As an information basis for fan engagement implementations, a maintenance-oriented web application is illustrated below.

Maintenance Web Management Prototype

The activity of management and maintenance of the elements present in the Stadium can be articulated in consequential phases each related to specific data organisation procedures and supporting IT products. These procedures and products must respect standards and compatibility with the most widespread technologies and, in particular, with those already adopted or being adopted by the hypothetical Customer. Furthermore, the procedures and tools must also be easily valid for future implementations on the Customer's entire estate. Below are the main development phases (Figs. 3-7).

Phase Zero: Current Status / Census and Mapping

This first phase involves the survey of all the signage elements and the corresponding location on plans of the different levels of the Stadium. The supporting computer tools consist of a series of spreadsheets that record the individual elements, univocally coded (according to criteria already in use by the Customer), with their relative types, and vector drawings of the plans with the punctual location of the signs, identified with a label bearing a unique code.

Phase One: Setting up the mySQL relational database

The set of alphanumeric data collected must be organised within a relational database so as to allow:

- insertion and updating of the data collected in phase 1 (through interfaces that can be compiled by electronic devices);
- data interrogation according to different reading filters (again through interfaces);
- production of extended or summary reports for digital or hardcopy printing.

The tool chosen for this phase, due to its compatibility characteristics and the possibility of interaction with other production environments, including future ones, is mySQL. Queries and reports are to be agreed in order to design the application to the customer's needs.

Phase One bis: DBWEB (independent of Phase 1, but co-operating with it) / Preparation of web environment interface for timely management and maintenance records

The operation and maintenance data need to be updated periodically, even at the individual sign level. This phase involves the preparation of an "agile" IT tool for recording field data that can be recorded directly (light devices) or in the back office (desktop). Recording is done by means of cards associated with each element via QRcode, or from a drop-down list to identify the code, or from Map, or from filters by category.

The records thus collected, related and linked with the permanent data referred to in the previous phases, form the history of the interventions and can allow queries regarding the planned obsolescence of the materials or the verification (check list) of the preparation of the various configurations of the Stadium or planned activities.

Phase Two: GIS / Geographical Information System Setup for Location and Signal Control

The plans set up in phase 1 must be implemented on a geographical database in order to be able to relate to the data organised in phase 2. This development of the geometric component of the data (point location) allows the relational connection with the alphanumeric attributes of the signals and thus enables:

- spatial interrogation of the data;
- production of thematic floor plans related to the queries;
- further implementation towards web management of the data.

The tool identified is OpenStreetMap, a free licence software.

Figure 3. Stadio App: Conceptual Scheme | List of areas and functionalities.

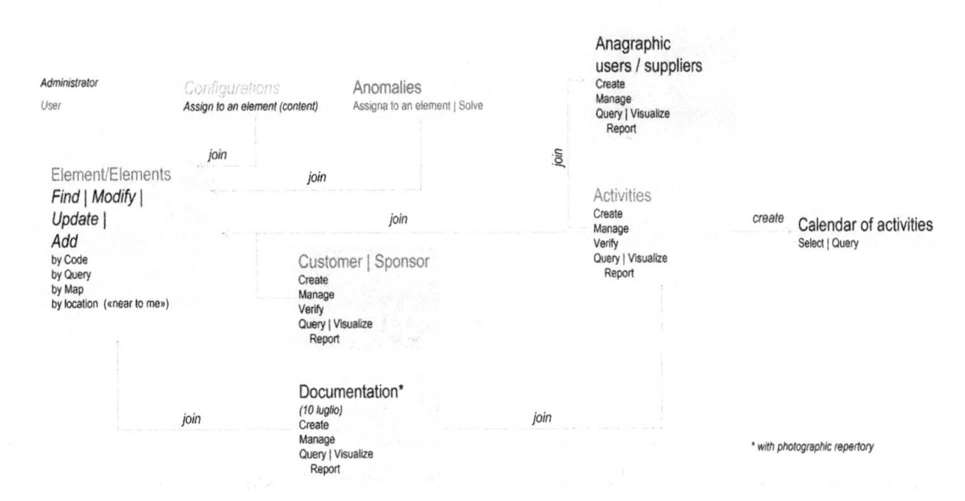

Phase Three: Integrated management environment via webGIS

The development envisages the preparation, in an environment that is fully accessible via the web and thus with common mobile devices, of all the information apparatuses (data and geolocations), so as to allow queries, updates and new entries using the plans of the various plant levels as an interface. In essence, this involves transferring the functions up to phases 2 and 3 reserved for the desktop environment also to lightweight portable devices (smartphones and tablets) (Ratajczak, 2019).

Figure 4. Stadio App: Logic Schemel Data and relationship Model.

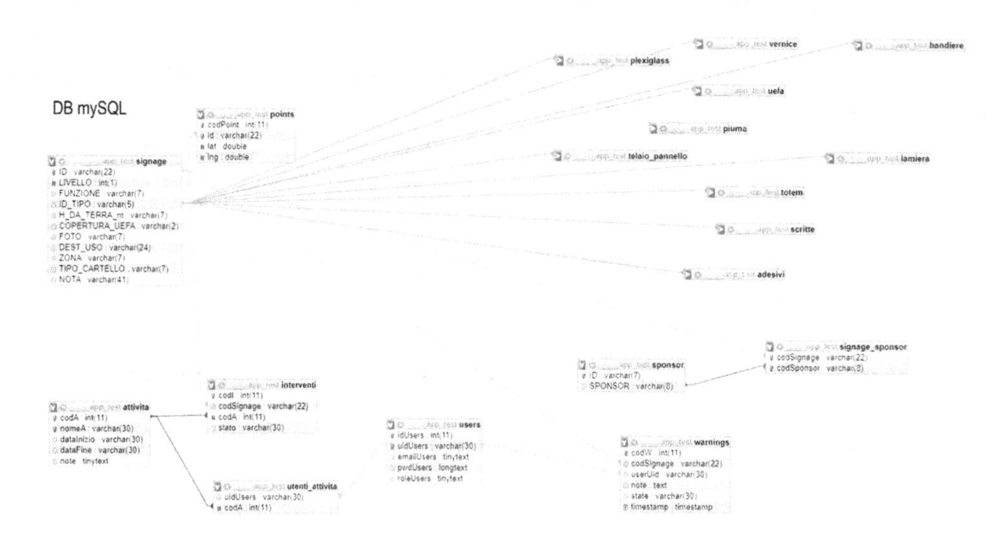

Figure 5. Stadio App: Physical Schema | Installing functions and access.

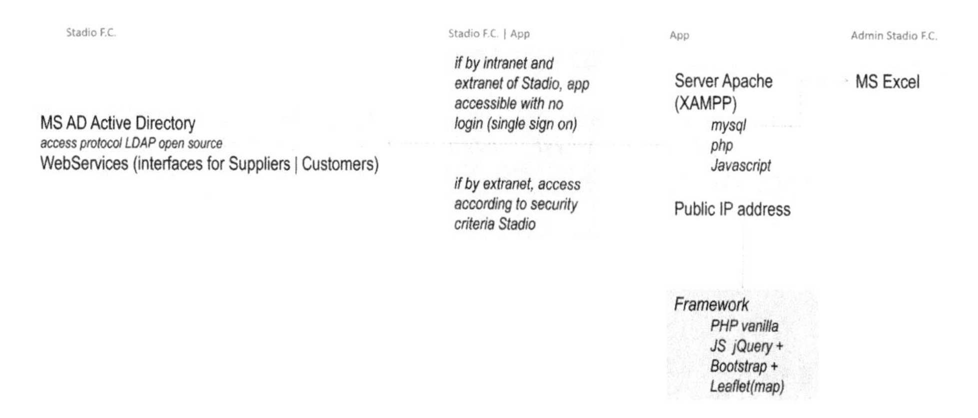

Phase Four: Coordination with existing BIM model

The previous phases are preparatory and compatible with implementation of the previously organised alphanumeric and geometric data into the existing BIM model. This phase involves the connection of the different environments developed (database, gis and web) with the stadium information model. This integration is also compatible with the maintenance system that the client is activating.

METAPROJECT SURVEY AND SHARING FOR AUGMENTED EXPERIENCES AND ENGAGEMENT

A built space valued as an environment of high perceptual quality contributes to the social and psychological health of the people and communities that inhabit it (Mehta, 2013, p. 56). Urban spaces are closely linked to the way people live and work in cities. The liveability of cities is an important issue that has an impact on the quality of life; the relief of urban form originates in the reading and interpretation of historical structuring as a consequence of a process of mutation and design in the urban, architectural, political and social spheres (Cavallari Murat, 1968). The introduction of an effective method of evaluating the emotional responses of people living in urban spaces represents a relevant step for the actions of urban design and regeneration of cities and also for those of training future professionals through the instilling of the culture of data and information. Introducing integrated support based on information systems at the urban scale and techniques for tracking behaviour, habits, and participation in social life contributes to critically assessing the relationship between public space and people's behavioural responses.

Figure 6. Stadio App: Interface functionality | Admin Profile.

Figure 7. Stadio App: Interface functionality | Operator Profile.

App | Interface functionality | Operator Profile

Stadio App

Cerca... Find
Vicino a me Near to me
Mappa Map
Intervento Intervention
Contatti Contacts
operatore operator

Research Functions of Element (for each level in Stadio)
- by proximity to User
- by Map

Visualize single Element
Complete Activity
Alert Anomaly

Assigned Activites (map)
Complete activity / Communicate Notes

Contacts
Visualize Contacts

Considerations on Perceptual Aspects Related to Meta-Design

Those who learn to design spaces and places for man and society base their knowledge process on behavioural models, demand frameworks, functional requirements, but do not in fact employ 'profiling' tools such as those used in the field of commerce and consumption, not with the same capillarity and intensity of data collection and analysis to take in information, not with the same incisiveness and not with the same level of updating and in-depth analysis (Bocconcino, 2022). It therefore seems relevant to transfer some models of analysis dedicated to profiling potential 'poor consumers' to the urban survey from its most vital component, the citizens. In some of these studies, digital urban simulation represents the starting point for the planning, design and regeneration of the built city: the use of computational methods that analyse and generate spatial configurations with iterative computational methods represents the new support tool for the definition of 'citizen journeys', a noble derivation that we could give to the concept of customer journey, here declined as citizen journey, a sort of urban marketing or neuromarketing applied to the vision (to the gaze, to observation) and to the use of the city.

The state of the art of research on human perception of the built environment is based both on questionnaires, interviews and field observations of people interacting with the environment under study, and on the use of more frontier methods, which make it possible to investigate the relationship between spatial characteristics and the senses, through the use of tools that collect and process significant masses of data, including from sensors of various kinds and from social media, often to investigate which types of places people find most comfortable (Girardin et al, 2009 and Resch et al, 2015).

The study of the city is not new to the introduction of advanced science and research fronts. Following the increasing availability of these kinds of resources, in the last decade the social sciences have introduced quantitative parameters and indicators, often labelled computational social science (Lazer et al. 2009), related to environmental perception, feeling, emotion, social connectedness, previously limited to qualitative modes of enquiry (Moretti, 2016). In particular, the emerging network science that studies complex networks has made a significant contribution to field research (Borner et al., 2007).

Network science takes an exclusively spatial perspective on urban data, focusing on the relationships and interactions between people, places, and institutions, at different scales. Manuel Castells (1996) introduced to urban studies the notion that by abstracting cities as social spatial networks of interaction, network science helps to uncover structural commonalities shared by most human systems, allowing for the construction of cognitive and predictive models of development (Batty 2013). The infrastructure of the built fabric with sensor networks has opened up real-time representations of the state and condition of places. The growth of social media is leading to new forms of participation and activism, alongside traditional forms of participation in the design of places; citizens voluntarily take on roles of monitoring and reporting, a phenomenon that has been described as the rise of the 'expert amateur' (Kuznetsov&Paulos 2010).

In recent years, more incisive approaches have been developed to understand how people perceive and feel the space and environment they experience. The analysis of the flows of citizens (human mobility), the tracking of the eye that collects the scene in glances (eye tracking), the automatic study of the movement of groups of people, who move in different physical places (body tracking), represent new scenarios - real and virtual - made available to the actors involved in the processes of management and maintenance of the real estate heritage.

In this context, the management of urban space and buildings becomes the tool for the relationship between use and wise administration of assets to adapt buildings to people's needs. Asset management as a discipline has three different and integrated purposes: i) to achieve organisational goals by balancing risks, opportunities and costs through the integration of different digital technologies; ii) to produce value through the management of the built environment; iii) to support sustainability strategies. Regarding the first objective, 'integration' is probably the most relevant topic, as a key factor of the whole process, and involves not only integration between technologies, but also between technologies and users, and between users and buildings. The second objective is closely related to the concept of heritage enhancement, and consequently to the topic of sustainable reuse, in addition to the economic dimension of the building process itself. In this challenge, two main educational objectives in the conceptualisation of building planning can have a major influence: the 'visual perception' and the 'spatial perception' of places and volumes.

Through typing on devices, we leave and collect traces. This is how our digital identity is formed, with this we can predict our behaviour or even influence it. The data market is all of us.

To expound on the objectives of the proposal, let us use a few questions: how can artificial intelligence technologies used on a commercial and consumer level be brought into the field of training and education for the design of the city? And to design spaces that are liveable and give well-being? Is it unrealistic to think of a design of the city (and of public space) that stems from the orderly and constant observation of those who breathe it, those who live it, those who see/look/observe it, those who use it, those who walk through it?

The answer to these questions should tend towards the constitution of a consultable atlas of experiences in urban space, stratified in two reading planes, lines of development of the project: perception and observation; the flow of people, behaviour, habits, use and fruition of space, production of contents, events. It must be expressed through multi-dimensional maps, digital public spaces, dynamic and interactive containers of experiences and visual stimuli that provide scenarios for the interpretations of scholars and designers.

The critical intervention of the operator in the process is required as an initial contribution to the modelling pipeline, to assess the quality of the extraction of the building form, but the process declines towards the possibility of an AI (Machine Learning) learning automatism, where a considerable support of calculation functions is provided for the adherence (fitting) of the extracted surfaces and the correspondence (compactness) of the structure of the three-dimensional entities, controlled in terms of vertices, components and parameters.

Semanticisation and Integration of the Digital Space for Digital Asset Management Responsiveness

As the databases of acquired and processed information will be composed of a large and complex extension of data, it is necessary to impose a semantic labelling approach for Repository elements, identifying 'instances' and 'annotations' in relation to parameters and query indicators that define them (Pavan et al., 2017). The methodologies adopted refer to the reading of the state of affairs in relation to heritage management, aimed at knowing:

- technological systems and management methods (type and quality of data available);
- social dimension environments (habits, movements, attendance);
- user profiling;

- methodological approach and critical review (modelling, instrumentation, environmental simulations);
- identification of representative cases and selection criteria (preliminary to measurements);
- identification of data periodicity character
- definition of supports (models from expeditious survey, to have surface models, and sensor set-up) and information system distributing info (simplified gis/bim)
- installation project, installation and data collection (automatic and questionnaires/interviews), identification of anomaly indicators that will be used for testing and verification of the practice
- identification of graphic modes and codes for representing and understanding phenomena;
- comparison with users and administrators (management savings, user education, virtuous behaviour);
- real time or asynchronous graphics and definition of minimum representation.

The main lines of development to be utilised and implemented concern the aspects highlighted below.

Visual and Perceptual Mapping for Behavioural Tracking

Eye and movement tracker systems will be applied to the case study, making it possible to obtain information on the user's behaviour in the scene, both at the level of gaze observation and movement in the scene, with measurement of dwell times and distribution trajectories. In contrast to applications focused on defined screens/frames, free and unconstrained scenes will be considered, which are functional for preliminary cognitive-behavioural considerations. Precise information will be obtained on the users' perception of architectural and building scene characterisation elements, motor sequence of scene exploration, with the potential to provide hints on the logical connections perceived by people in the building container.

Wearable eye-tracking devices, equipped with infrared cameras (to detect eye movements, gaze samples, fixations, pupil size, blinks) and an ordinary front-facing camera (to record the observed scene), record the gaze path in a totally free scene. In this case, subjects can move freely, without limiting the ocular data collected to a specific view and obtain information on how the built environment is perceived and explored, the distribution of users' visual attention, the static-dynamic elements of fixation-visual tracking, potential 'emotional states' and the cognitive load associated with performing specific functions.

Tracking Paths and Flows within the Structure

By means of head-tracking or body-tracking devices, consisting of fixed cameras installed in the scene, physical movements and pauses during the experience of the environments will be monitored, evaluating more complex behaviours, such as common exploration patterns, the permanence and distribution of users in the premises according to environmental factors, comfort, crowding, occupation of infrastructures.

In the laboratory environment, it will be possible to conduct similar experiments and investigations with virtual reality simulations, both in the non-immersive case (using a screen and a remote eye tracker) and in the immersive situation (e.g., using a wearable device such as Oculus Rift enhanced for eye tracking).

3D Modelling of the Building/Urban Container

Dynamic 3D spatial survey and scan-to-BIM modelling solutions will be applied to develop simultaneous global localisation and mapping of physical spaces with collected sensory data translated into a virtual scenario arranged for spatial machine intelligence (SMIS). Greater automation and self-management of spatial data will be investigated, for the acquisition of scene geometry and the transposition of spatial semantics. The representation of building and urban containers belonging to the study areas will define a digital background of observation and declination of models and data analysis at multiple Levels of Detail of the spatial scene, applied for the documentation of paths, volumes, open spaces and architectural/technological details (Zaker&Coloma, 2018).

The translation of different data on (sensory and visual) perception of functional space, from the building scale to the urban context of location, validates an essential methodological aspect of cognitive classification between volumes, infrastructure and open spaces within their possible building transformations (Fig. 8).

DISCUSSION

The urban and territorial relevance of large sports facilities imposes the assumption of a design and mainly recovery approach, oriented both to structural verification and to modernisation of large facilities, in relation to the performance field, which requires the use of new techniques and materials to improve their life cycle. The innovations guaranteed by the progressive development of the regulatory framework constitute the reference for design recovery and technological innovation. Integrated recovery and valorisation strategies applied to sports facilities therefore represent a new technological frontier to be explored in the light of recent developments in the field of cultural heritage and activities.

The last decade has been characterised by a progressive increase in synergy in every relational sphere. The interdependence between global and local has encouraged local cultures to rediscover the territory as a place with an identity.

The functions of human activity, such as living, producing, leisure, are linked in time and space, leading to a radical increase in the traditional representation of the urban place. The future of the urban market thus depends on the ability of cities to condition functions and forms of planning, so as to optimise their competitiveness and capacity for interaction in the context of global space. which highlights the need for sustainable and competitive economic strategies for the metropolitan context. Today, the new guidelines assume the valid principle that a sports facility can only be active if it produces a fair profit for an operator. This is to be interpreted as an incentive and stimulus aimed at a managerial nature of a sports facility and in particular of a football stadium. Otherwise, sports facilities run the risk of becoming expensive due to the degradation resulting from inactivity or the confrontation with design and safety regulations, characterised by architectural adaptation, due to facilities built with a design conception that differs from today's requirements.

Figure 8. A Framework for Integration of BIM, BEPS and Big Data Analytics in View of Lean

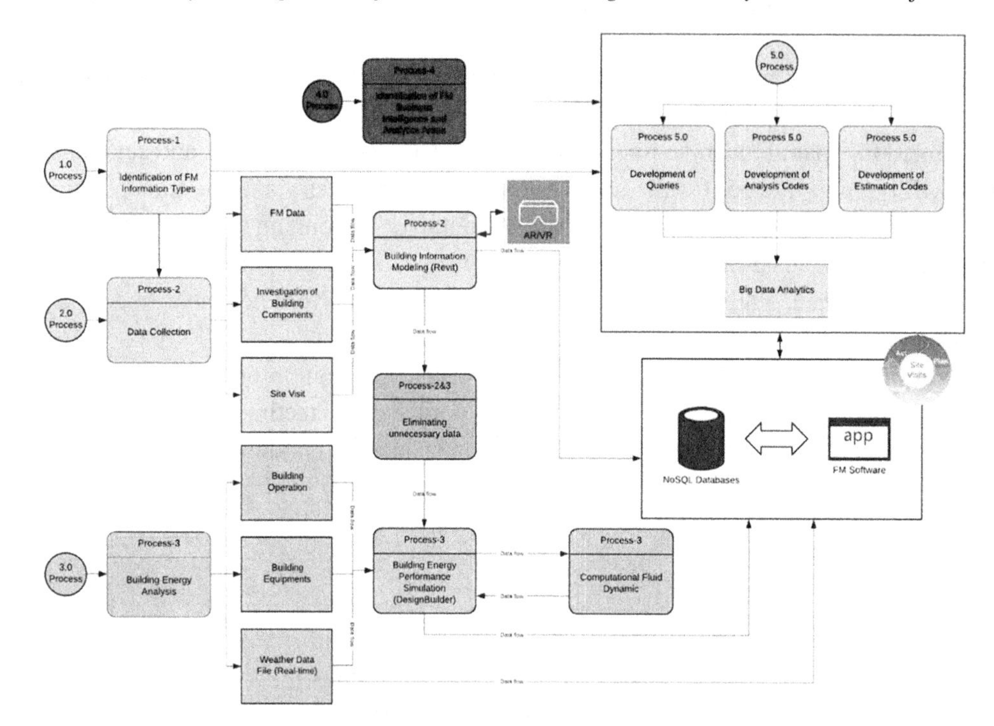

Management Philosophy (source: Demirdögen et al., 2020, pag. 20).

Richard Rogers in his autobiography states "Consistency is change (...) my goal is to adapt buildings to people's needs" (2018).

It is useful to dwell on some considerations that draw on the existing availability in terms of real-time recording of different factors. The spread and development of ICTs make possible the pervasive and ubiquitous transmission of information about people's movements and flows, fostering an interrelationship between information, humans and the environment. The traditionally static approach centred on infrastructural networks, the core of the traditional idea of smart buildings, is shifting towards a cultural vision of the smart building, where ICTs can promote the transition from "smart buildings" to "smart places", in particular through participatory applications. A new and important role is played by artificial intelligence, through which data can be collected, selected and analysed, in order to gather useful information on the behaviour of the users of spaces, and consequently be able to deduce new levels of knowledge capable of triggering the acquisition of higher level skills (critical reflection, awareness, problem solving skills). The methods of cataloguing and utilising the analysed data, through GIS and BIM information systems, dedicated to the urban environment, are important. The 2D and 3D visualisation of data processed by AI promotes the processes of 'visual computing', i.e. the analysis technique based on the visual representation of data, which favours the study of large amounts of data.

In the field of construction, the difficulty of managing large functional structures highlights the demand for tools to support the technological design and management of spaces, taking into account the activities used, behavioural practices and response mechanisms desired by users (Bocconcino, 2019). The contextual loss of quality of the common space derives from the rapid and disorderly use that is experienced, often far from the expected logical patterns of frequency, path and concentration of gaze. Through the recent limitations and alterations of the ways in which limited spaces and public compartments are experienced, the analysis of these phenomena must constantly update the knowledge behind building management, monitoring the dynamics of interaction between man and building, deciphering changes and providing solutions just as quickly.

By increasing the understanding of how individuals perceive the built environment, both inside and outside the built enclosures that make up the urban fabric, it is possible to assess which characteristics of the built space are interpreted as positive or negative on the quality of use (Donato et al., 2017).

The aim is to apply quantitative data, certified by sensory measurements of behavioural dynamics and translated with appropriate graphic codes for the representation and understanding of phenomena, to find correlations between the characteristics of the observed framework and the perception of individuals and groups of individuals within the built space. This involves defining a supervisory tool (an "explorer" database, a collector of building sensory data) as an enhanced Digital Asset Management competence.

For this requirement, the same space must be traced in parallel in its two real-digital counterparts: the real space must be the object of measurement and observation of human cognitive phenomena, and of analysis of the building containers that house them; the digital space, as a validated replica of both the spatial and technological properties reproduced, must collect and adopt such data to endow itself with responsive qualities. The direction must be twofold: management practitioners (such as energy managers, technicians) must promote analysis to develop cyclical interventions, conservation and planned management agendas; marketing practitioners will use behavioural mechanism analysis frameworks to confront users on their 'digital paradigm' and drive awareness on behavioural mechanisms of interaction with the sports facility and its functions.

Resource management is also relevant by considering the adaptive reuse approach and strategies (Lo Turco&Bocconcino, 2017). These are usually based on the assumption that buildings, areas, neighbourhoods and sites are not static entities, designed for a single use during their life cycle. On the contrary, it consists of the practice of introducing new content into an existing site, paying special attention to the needs of society and following the principle of maximum conservation and minimum transformation. Within this perspective, an integrated digital asset management strategy can integrate different tools and means related to predictive impact assessment and decision support strategies, taking real advantage of integrated and interactive models and the increase of huge amounts of data

The construction industry is rapidly pushing forward the so-called 'servitisation' process, moving from selling the physical good (the product) to the end user to selling the physical good within a set of services that can be activated after purchase. This has an enormous impact on the organisational core business, shifts much of the value generation to the utilisation phase and can be further developed through an integrated and interconnected digital approach.

It is interesting to focus on the type of interactions that can be assessed between these functions and some innovative ICT technologies, in order to evaluate the impact of the digital revolution from an asset management perspective. A short list of these disruptive ICT technologies may include: Internet of Things (IoT), communication technologies, data acquisition, Blockchain, BIM and HBIM, augmented and virtual reality (A/V R), artificial intelligence (AI), and additive manufacturing.

From a comprehensive analysis of these mutual interactions, it is possible to foresee at least two foreseeable evolutionary scenarios: i) a rapid forward shift from BIM/HBIM, IoT and AI to a widespread implementation of interactive Digital Twins; ii) an increasing integration of IoT networks with

AI techniques in order to improve connectivity and interoperability between users, digital models and buildings, with a real-time self-updating BIM/HBIM.

In the end, this scenario may configure a true 'Digital Asset Management revolution' based on two pillars: an incoming ontological mutation generated by the multidimensional and multiplatform integration and interaction between humans, things and models, both in a real and virtual context; the rise of a new 'digi-real' paradigm and ecosystem based on intelligent and self-adaptive buildings and models in the so-called 'everything everywhere' paradigm.

It will assist in design activities through data collection with quali-quantitative methodology, in order to outline and propose the profile of users interacting in specific architectural environments, their identity profiling as users of collective spaces and their ethical characterisation with respect to value systems and collective behaviour patterns. It will define the ethical-moral criteria of applications and interactions with the aim of preparing protocols and practices that outline active processes both in their latent and manifest representation, considering the environment as a characterising element for the formation and education of the subject, trying to delineate, should they emerge, counter spaces of action.

CONCLUSION

The world of sport was one of the first early adopters of VR (after the military), think of F1 drivers and simulators, now so advanced in terms of technology that they not only digitally reproduce a physical circuit but also allow athletes to train body and mind - as if they were really driving on a circuit - and engineers to acquire new data for car design or engine engineering. Other interesting examples come from cycling, golf and football. Here, too, the beginning of the use of simulators based on Virtual Reality goes back several years.

In a copy of the digital world, the sporting event itself has crossed the dimension of the real to pass - at least partially - into that of the virtual with eSports. Consumers' expectations regarding experiences - increasingly personalised, increasingly immersive - put brands under pressure, but a healthy pressure because it pushes them to find new solutions, and, new solutions mean new lifeblood for sports marketing, both for entertainment and engagement. In this context, the value of VR is invaluable because it represents a comprehensive and immersive way to communicate what the brand can offer, especially for its potential as:

- has the power to show what is not visible, 'purpose-built' situations or environments that would otherwise be impossible for a fan to reach;
- the experiences are highly customisable and not pre-set and monotonous, because they can be set and varied according to the needs, tastes and inputs of the user. This makes them even more engaging;
- the use of gamification makes the VR experience highly active and interactive. And more easily memorable;
- each user experience can be analysed to collect data and information: how much time a user has spent online, what they have seen, which products they have marked as their favourites, from which area they connect;
- millennials, and younger generations in general, were born and grow up using digital technologies as their preferred vehicle of communication.

Investing in the integration of management systems and promotion and involvement of supporters therefore means looking at the customers of today and those of the future.

Mixed Reality is making it possible to raise the quality of both virtual reality and augmented reality: much more technologically advanced than VR (because it combines the use of different technologies, sensors, wearable devices, highly advanced optics, and ever greater computing power for data analysis), Mixed Reality makes it possible to raise AR to a superior experience by allowing people to experience ever larger realistic scenarios within spaces and time that are no longer confined to the real visual experience.

It is a horizon that still needs to be deciphered but which may appear clearer if we imagine the interesting frontiers that can open up in medicine, wellness, education (training/teaching), entertainment, communication, industrial design or building construction, art and everything else. All the way to sport.

360° videos have already introduced a new way of viewing a game or a sporting event (but also a concert or a show) by allowing people to have a full view, as if they were physically present, even while sitting on the sofa in front of the TV or holding their smartphone or tablet. Soon, the experience will be live 360° and will allow fans not only to have a full view of the stadium, the dressing room, the pitch, but even to immerse themselves in the game action with a view similar to that of the athlete on the pitch.

Compared to what we are already able to touch today, therefore, and imagining what we will be able to do in the near future, the future will increasingly be a mixture of AR/VR, which, intertwined with other technologies such as IoT and wearable devices, will make Mixed Reality increasingly pervasive and allow us to experience sport at a higher level.

Mixed Reality in fan engagement represents an evolution in how existing technologies can be adapted to provide tremendous opportunities for teams to engage and entertain crowds and television audiences. The technology exists, it fits directly into existing technical and commercial structures. Most importantly, it demonstrates that with a fan-focused approach to technology implementation, we can build customised experiences from the information assets that facility management already provides.

ACKNOWLEDGMENT

We would like to thank engineers Piergiorgio Mingoia and Enrico Fatta for their cooperation during the activities.

REFERENCES

Arayici, Y., Fernando, T., Munoz, V., & Bassanino, M. (2018). Interoperability specification development for integrated BIM use in performance based design. *Automation in Construction*, *85*, 167–181. doi:10.1016/j.autcon.2017.10.018

Asensio, J. A., Criado, J., Padilla, N., & Iribarne, L. (2019). Emulating home automation installations through component-based web technology. *Future Generation Computer Systems*, *93*, 777–791. doi:10.1016/j.future.2017.09.062

Barbero, A., Vergari, R., Ugliotti, F. M., Del Giudice, M., Osello, A., & Manzone, F. (2021). *Automated semantic and syntactic BIM data validation using visual programming language*. doi:10.30682/tema0702m

Batty, M. (2013). *The New Science of Cities*. The MIT Press. doi:10.7551/mitpress/9399.001.0001

Beech, J., & Chadwick, S. (2013). *The Business of Sport Management*. Pearson.

Bhatt, J., & Verma, H. K. (2015). Design and development of wired building automation systems. *Energy and Building*, *103*, 396–413. doi:10.1016/j.enbuild.2015.02.054

Bocconcino, M. M. (2018). La tecnologia BIM per il controllo strutturale - Metodi e strumenti grafici per il monitoraggio strutturale e per la manutenzione di manufatti complessi. In *Controlli strutturali: metodologia e applicazione* (pp. 123–142). Maggioli.

Bocconcino, M. M. (2019). Graphic information and visual communication: tools for simplifying knowledge. In *Riflessioni - L'arte del disegno/Il disegno dell'arte* (pp. 1427–1434). Gangemi.

Bocconcino, M. M. (2022). *Mappe "nd" che mostrano ciò che non si vede, per un'immaginazione del concreto - Sistemi informativi e prospettive future che già sono il presente ["nD" maps that show what cannot be seen, for an imagination of the concrete - Information systems and future perspectives that are already the present]* (Vol. 3). Edifir.

Bocconcino, M. M., Lo Turco, M., Vozzola, M., & Rabbia, A. (2021). Intelligent Information Systems for the representation and management of the city. Urban survey and design for resilience. PROJECT, 5, 90-107.

Bocconcino, M. M., & Manzone, F. (2019), Sistemi informativi e strumenti grafici per la manutenzione di manufatti complessi [Information systems and graphic tools for the maintenance of complex buildings]. In Ingegno e costruzione nell'epoca della complessità. Politecnico di Torino.

Börner, K., Sanyal, S., & Vespignani, A. (2007), Network Science. Annual Review of Information Science & Technology, 41, 537-607. doi:10.1126cience.1167742

Castells, M. (1996). *The Rise of the Network Society*. Blackwell.

Cavallari Murat, A. (1968), Forma urbana e architettura nella Torino Barocca. Dalle premesse classiche alle conclusioni neoclassiche, Unione tipografico-editrice torinese.

D'Urso, C. (2011). Information Integration for Facility Management. *IT Professional*, *13*(6), 48–53. doi:10.1109/MITP.2011.100

Demirdöğen, G., Işık, Z., & Arayici, Y. (2020). Lean Management Framework for Healthcare Facilities Integrating BIM, BEPS and Big Data Analytics. *Sustainability*, *12*(17), 7061. doi:10.3390u12177061

Donato, V., Lo Turco, M., & Bocconcino, M. M. (2017). BIM-QA/QC in the architectural design process. *Architectural Engineering and Design Management*, 1-16. doi:10.1080/17452007.2017.1370995

Fan, L., Xinmin, L., Bingcheng, W., & Li, W. (2017). Interactivity, Engagement, and Technology Dependence: Understanding Users' Technology Utilisation Behaviour. *Behaviour & Information Technology*, *36*(2), 113–124. doi:10.1080/0144929X.2016.1199051

Fathy, D., Mohamed, H. E., & Ehab, A. (2022). Fans Behave as Buyers? Assimilate Fan-based and Team-based Drivers of Fan Engagement. *Journal of Research in Interactive Marketing*, *16*(3), 329–345. doi:10.1108/JRIM-04-2021-0107

Girardin, F., Vaccari, A., Gerber, A., Birderman, A., & Ratti, C. (2009). Quantifying Urban Attractiveness from the Distribution and Density of Digital Footprints. *International Journal of Spatial Data Infrastructures Research*, 4.

Gunay, H. B., Shen, W., & Newsham, G. (2019). Data analytics to improve building performance: A critical review. *Automation in Construction*, *97*, 96–109. doi:10.1016/j.autcon.2018.10.020

Kang, S. (2020). Going Beyond Just Watching: The Fan Adoption Process of Virtual Reality Spectatorship. *Journal of Broadcasting & Electronic Media*, *64*(3), 499–518. doi:10.1080/08838151.2020.1798159

Kerski, J. J. (2022). Online, Engaged Instruction in Geography and GIS Using IoT Feeds, Web Mapping Services, and Field Tools within a Spatial Thinking Framework. *Geography Teacher (Erie, Pa.)*, *19*(3), 93–101. doi:10.1080/19338341.2022.2070520

Kuznetsov, S., & Paulos, E. (2010), Rise of the Expert Amateur: DIY Projects, Communities, and Cultures. In *Proceedings of the 6th Nordic Conference on Human-Computer Interaction: Extending Boundaries*. Association for Computing Machinery.

Lazer, D., Pentland, A., Adamic, L., Aral, S., Barabási, A., Brewer, D., Christakis, N., Contractor, N., Fowler, J., Gutmann, M., Jebara, T., King, G., Macy, M., Roy, D., & Van Alstyne, M. (2009). Computational Social Science. *Science, 323*(5915), 721-723.

Lo Turco, M., & Bocconcino, M. M. (2015). La rappresentazione operativa - Operative representation. In *BIM* (pp. 131–160). Massimiliano Lo Turco.

Lo Turco, M., & Bocconcino, M. M. (2017). Esattezza, molteplicità e integrazione nell'Information Modeling & Management [Exactitude, multiplicity and integration in Information Modelling & Management]. TECHNE, 13, 267-277.

Lo Turco, M., Bocconcino, M. M., Cangialosi, G., & Serini, M. (2015). Dal disegno di progetto al modello di cantiere: le radici del FM. In *AR BIM GIS, a cura di Anna Osello* (pp. 126–139). Gangemi.

Lo Turco, M., Bocconcino, M. M., Vozzola, M., Giovannini, E., & Tomalini, A. (2021). *Il BIM per il Construction Management. Il caso della Domus Eleganza a Milano [BIM for Construction Management. The case of the Domus Eleganza in Milan]. 3D Modeling & BIM 2021 - Digital Twin.*

Macarulla, M., Casals, M., Forcada, N., & Gangolells, M. (2017). Implementation of predictive control in a commercial building energy management system using neural networks. *Energy and Building*, *151*, 511–519. doi:10.1016/j.enbuild.2017.06.027

McArthur, J. J., & Bortoluzzi, B. (2018). Lean-Agile FM-BIM: A demonstrated approach. *Facilities*, *36*(13/14), 676–695. doi:10.1108/F-04-2017-0045

Mehta, V. (2013). Evaluating Public Space. *Journal of Urban Design*, *19*(1), 53–88. doi:10.1080/13574809.2013.854698

Moretti, M. (2016). *Senso e paesaggio. Analisi percettive e cartografie tematiche in ambiente GIS* [Sense and landscape. Perceptual Analysis and Thematic cartography in a GIS environment]. Franco Angeli.

Pavan, A., Mancini, M., Lo Turco, M., Pola, A., Mirarchi, C., Rigamonti, G., & Bocconcino, M. M. (2017). *Applicazione dell'approccio INNOVance per le imprese di costruzione* [Application of the INNOVance appraoch for construction companies]. Edilstampa.

Previati, A. (2020). *The Importance of Fan Engagement and Fan Management in Sports*. Academic Press.

Ratajczak, J., Riedl, M., & Matt, D. T. (2019). BIM-based and AR Application Combined with Location-Based Management System for the Improvement of the Construction Performance. *Building (London), 9*(5), 118. doi:10.3390/buildings9050118

Rogers, R. (2018). *Un posto per tutti. Vita, architettura e società giusta* [A Place for Eevryone. Life, architecture and just society]. Johan & Levi.

Roper, K. O. (2017). Facility management maturity and research. *Journal of Facilities Management, 15*(3), 235–243. doi:10.1108/JFM-04-2016-0011

Rosenthal, B., & Eliane, P. Z. B. (2017). How Virtual Brand Community Traces May Increase Fan Engagement in Brand Pages. *Business Horizons, 60*(3), 375–384. doi:10.1016/j.bushor.2017.01.009

Valinejadshoubi, M., Osama, M., & Ashutosh, B. (2022). Integrating BIM into Sensor-based Facilities Management Operations. *Journal of Facilities Management, 20*(3), 385–400. doi:10.1108/JFM-08-2020-0055

Villa, V., Naticchia, B., Bruno, G., Aliev, K., Piantanida, P., & Antonelli, D. (2021). IoT Open-Source Architecture for the Maintenance of Building Facilities, Basel: MDPI AG. *Applied Sciences (Basel, Switzerland), 11*(12), 5374. doi:10.3390/app11125374

Wu, C., Shieh, M.-D., Lien, J.-J. J., Yang, J.-F., Chu, W.-T., Huang, T.-H., Hsieh, H.-C., Chiu, H.-T., Tu, K.-C., Chen, Y.-T., Lin, S.-Y., Hu, J.-J., Lin, C.-H., & Jheng, C.-S. (2022). Enhancing Fan Engagement in a 5G Stadium With AI-Based Technologies and Live Streaming. *IEEE Systems Journal*, 1–13. doi:10.1109/JSYST.2022.3169553

Zaker, R., & Coloma, E. (2018). Virtual reality-integrated workflow in BIM-enabled projects collaboration and design review: A case study. *Vis. in Eng., 6*(1), 4. doi:10.118640327-018-0065-6

KEY TERMS AND DEFINITIONS

Fan Engagement: Fan engagement is the watchword for any sports club. This concept is based on the consideration that real fan engagement only occurs when the team is able to build an intense and special bond with them. It represents the new frontier of the relationship between fans, brand and athletes, and the individual experience of each fan is the beating heart of this relationship. If a brand improves the way it reaches out to its fans and succeeds in providing them with memorable moments, as a direct consequence the same fans will want to experience it again and this will lead to a tangible increase in performance for the brand.

Sports Venue Management: Sports venue management is the practice of operating an establishment in which people participate in or watch sports. Football stadiums, bowling alleys, golf clubs and sports-focused entertainment establishments are examples of sports venues.

Sustainable Facility Management: Sustainable facility management seeks to ensure that each building has a greatly reduced, or even neutral, impact on the environment. Generally, this implies several changes in the daily operations, as well as changes to the structure of the building itself. One of the possible alternatives to architectural changes is to use smart sustainable technology. However, it has been shown that sustainable facility management is not just about minimising the impact of buildings. Sustainable facility management has repercussions for buildings, people, and organisations.

Venue Management Software: Venue management software can help market avenue as a major selling point for events. This can help to book more attendees, increase sales and spend less time on admin. Venue management software eliminates the busy work behind planning an event and lets users get straight to increasing awareness and sales for their company.

Section 3
Resilient Cultural Heritage

Chapter 14
Digital Technologies Towards Extended and Advanced Approaches to Heritage Knowledge and Accessibility

Federica Maietti
https://orcid.org/0000-0002-8076-0020
University of Ferrara, Italy

Marco Medici
https://orcid.org/0000-0002-9643-4721
University of Ferrara, Italy

Peter Bonsma
RDF Ltd., Bulgaria

Pedro Martin Lerones
Fundación CARTIF, Spain

Federico Ferrari
University of Ferrara, Italy

ABSTRACT

The new directions that digital reality is currently taking include an ever-greater involvement and interaction with the human being. In the field of cultural heritage, there is a need to find new ways to visit, enjoy, understand, and preserve cultural assets, also through digital fruition. The social value of cultural heritage and citizens' participation became crucial to increase quality of life, public services, creative activities, public engagement, new understanding, and education through technology development. Digital technologies can also contribute to safeguarding endangered cultural heritage preventive interventions, as well as ensuring equal and wide access to cultural assets and heritage sites. The aim is to find positive interconnections between physical and virtual spaces by applying digital systems to find additional knowledge and supporting the access to our common heritage through new technologies. The chapter explores more in detail these topics through the description of methodological approaches, applications of Semantic Web technologies, and latest projects.

DOI: 10.4018/978-1-6684-4854-0.ch014

INTRODUCTION

The new directions that Digital Reality is currently taking include and enhance an ever-greater involvement and interaction with the human being. The term "digital reality" is related to our ever-more digital societal context (Bowen & Giannini, 2021), where many aspects of everyday life are managed by digital media.

This is occurring in different fields of application, from health to urban planning, from domotics to smart cities, up to the fruition of Cultural Heritage. In particular, in this area, the emergency condition caused by the Covid-19 pandemic has highlighted the need to find new ways to visit, enjoy and understand cultural assets, including digital fruition (Europa Nostra, 2020). The social value of cultural heritage and citizens' participation became more and more crucial, considering also how to increase quality of life, public services, creative activities, public engagement, new understanding and education through technology development.

These needs are very clearly understood within the Horizon Europe Strategic Plan, released in 2021, in particular in the Cluster 2 (Culture, Creativity and Inclusive Society) where the sense of belonging through a continuous engagement with society, citizens, social partners and economic sectors is very carefully stressed. "*The full potential of cultural heritage, arts and cultural and creative sectors as a driver of sustainable innovation and a European sense of belonging is realised through a continuous engagement with society, citizens and economic sectors as well as through better protection, restoration and promotion of cultural heritage*" (European Commission, Directorate-General for Research and Innovation, 2021).

The expectation of exploiting the full potential of Cultural Heritage through research and innovation is intended to support the access to our common heritage through new technologies. In this direction, high quality digitisation and curation of digital heritage assets are of key importance in order to access and manage different information and data domains.

The key target is to find positive interconnections between physical and virtual space, by applying digital systems and virtual and augmented reality technologies to find additional knowledge levels to heritage accessing and understanding, inclusive communication and interdisciplinary collaborations.

In addition to expanding this kind of "social resilience", digital technologies can also contribute to safeguarding endangered cultural heritage from natural hazards and anthropogenic disasters by preventive interventions, as well as ensuring equal and wide access to cultural assets and heritage sites. The issue of assets at risk is indeed crucial, as well as all possible strategies for preventive documentation and the accessibility to site in critical conditions.

The UN (United Nations) 2030 Agenda for Sustainable Development (Goal 11: Make cities and human settlements inclusive, safe, resilient and sustainable) includes a focus on cities in the sustainability goals, including challenges related to inclusive and sustainable urbanization and capacity for participatory, integrated and sustainable human settlement planning and management and efforts to protect and safeguard the world's cultural and natural heritage. Under the same Goal, the vision includes a substantial increase of the number of cities and human settlements adopting and implementing integrated policies and plans towards inclusion, resource efficiency, mitigation and adaptation to climate change, resilience to disasters, and develop and implement, in line with the Sendai Framework for Disaster Risk Reduction 2015-2030, holistic disaster risk management at all levels.

In this framework, in which the strategic importance of heritage digitization had already been widely emphasized in the Horizon 2020 Programme, the economic, social and environmental impacts of Covid-19 stressed the needs to find new strategies to mitigate the crisis, accelerate the ecological and digital transition, improve training, and achieve greater gender, territorial and generational equity.

In this direction, the National Recovery and Resilience Plan (PNRR, 2021) highlights, under the section related to tourism and culture, the need to improve culture and tourist accessibility through digital investments and investments to remove physical and knowledge barriers to heritage; and regenerate small historical centres through promoting participation in culture, the enhancement of sustainable tourism. Investments to achieve these goals include the "Digital Strategy" and platforms for cultural heritage and regeneration of small cultural sites, to foster the development of new tourism/cultural experiences and balance tourist flows in a sustainable way.

According to the Next Generation EU programme, the six Missions of the Recovery and Resilience Plan are digitization, innovation, competitiveness, culture and tourism; green revolution and ecological transition; infrastructure for sustainable mobility; education and research; inclusion and cohesion; health.

The social "direction" of the programme is therefore clear, and Digital Reality is crosswise, covering several topics of interaction between people and machines to find solutions for social aspects and problems.

The Chapter includes an overview of the main issues related to new research directions to digitization and the current scenario of interconnection between physical and virtual spaces, describing extended and advanced approaches to heritage knowledge and accessibility. The objectives of this essay are mainly related to the presentation of methodological approaches and latest applications grounded on digital technologies arising from the INCEPTION project.

INCEPTION - *Inclusive Cultural Heritage in Europe through 3D semantic modeling*, has been funded by the European Commission under the Horizon 2020, Work Programme *Europe in a changing world – inclusive, innovative and reflective Societies*, started in June 2015 and has been completed in May 2019. During the project development and after the conclusion of the project, several advancements and related activities were developed, by implementing the main project' achievements and outcomes.

The main topics of INCEPTION were related to advancement in data capturing, proposing a protocol for 3D documentation tailored on Cultural Heritage: data processing within BIM environment; and models sharing, working in the direction of an open semantic web platform and related applications. The Chapter presents the main project' follow-ups, describing in particular the application of Semantic Web technologies and their potential in the 3D Cultural Heritage domain as a means for extended and advanced approaches to heritage knowledge and accessibility. Two case studies are presented in order to show application of semantics exploiting digital catalogues for enriching 3D models.

The Chapter is divided into five main sections:

- the paragraph related to the "Background" describes the research scenario starting from the outcomes of the INCEPTION project and summarizing further development and directions, including the most recent reports and studies related to heritage digitization and some considerations on the impact of Covid-19 on Digital Heritage fruition;
- a section related to 3D technologies for Cultural Heritage highlights the rapid evolution of digital tools and the need for standard use of 3D in the Cultural Heritage sector. The section includes a short overview of the ongoing EU project "4CH" and new directions in the field of semantic web technologies;
- The section "Current approaches and results exploiting semantics" includes an overview and description of some significant projects and activities developed by the authors in the field of

heritage knowledge and accessibility through digital technologies and a short overview of other relevant projects;

- "Future research directions" is a section summarizing possible scenarios in the field of heritage knowledge and accessibility through digital technologies;
- "Conclusion" is the concluding paragraph of the chapter, in which the issues presented are analyzed with a critical approach, outlining the next research steps.

BACKGROUND

New ways of ensuring better access to, understanding of and engagement with Cultural Heritage, and digitization to support heritage at risk are among the key missions of the latest research avenues developed by INCEPTION, the spin off created at the end of the European project.

The project INCEPTION - *Inclusive Cultural Heritage in Europe through 3D semantic modelling* (Maietti et al., 2020), focused on Cultural Heritage buildings advanced digitization using BIM (Building Information Modeling) to manage enriched 3D models (Bianchini et al., 2016; Empler et al., 2021) through the INCEPTION platform, toward a better knowledge sharing and enhancement of European Heritage. The project developed core technologies for managing, visualizing and archiving 3D BIM, together with all the related digital documents, based on semantic technologies (Martín-Lerones et al., 2021; Russo & De Luca, 2021). This approach was aimed at the adoption of new digital technologies from a wide audience.

The application, updating and evolution of the results achieved during the four-year project have triggered interesting synergies, such as the work within the Expert Group on Digital Cultural Heritage and Europeana.

INCEPTION results indeed contributed to the report "3D Content in Europeana" (3D content in Europeana task force, 2020), produced by a Task Force established to analyse 3D digitization of cultural artefacts at large in the perspective of their integration in Europeana and in the network of Europeana data providers. According to the report introduction: "*3D digitization of the cultural heritage has become more common in recent years. New tools and services have made it much easier to capture, model and publish. The creation of highly accurate 3D models of monuments, buildings and museum objects has become more widespread in research, conservation, management and to provide access to heritage for education, tourism and through the creative sector. Yet this is still a developing field and organisations that are commissioning 3D media need to make a series of choices on the type of content that is created, how it will be visualised and rendered online, and for which users*" (3D content in Europeana task force, 2020). The viewer developed within INCEPTION was listed as the only solution for accessing 3D BIM data together with the related semantic information.

The results provided in this report mainly focused on the fruition of the 3D content for use in education, research and in the creative industries, while the digitization itself was still out of the scope. Nevertheless, the holistic approach that is now required for documenting the Cultural Heritage is still lacking. For this reason, the European Commission (EC) tasked an Expert Group on Digital Cultural Heritage and Europeana, where again representatives from the INCEPTION project have been involved, to the development of guidelines on 3D cultural heritage assets. Thus, the Expert Group elaborated a list of 10 basic principles and a number of related tips for each of them geared toward cultural heritage professionals, institutions and regional authorities in charge of Europe's cultural heritage (European Commission, 2020).

The list of basic principles and tips is especially addressed to Cultural Heritage professionals and institutions, and other "*custodians of tangible cultural heritage, including local and regional authorities, who are in charge of cultural heritage buildings, monuments, or sites, who do not have any experience with 3D digitisation yet, neither directly nor via an external service provider*" (European Commission, 2020). The vision of this work is to support all other such professionals, institutions and authorities, who may find the document useful in achieving the best results in 3D digitisation projects.

Among the basic principles highlighted in the document, the one related to "the value of and need for 3D digitisation" considers 3D digitisation for many purposes, including conservation and preservation, reproduction, research, education, exploration, and creative or tourism-related reuses, in addition to the preservation of tangible cultural heritage at risk. Regarding interactions, social impacts and accessibility of digitized contents, the need to provide broad public access, storing and distributing 3D models via open public platforms is strongly stressed. Moreover, principle 6 "Identify the different versions and formats needed for the different use cases targeted" considers 3D raw data resulting from the digitisation process to be further processed to generate 3D models and other 3D content for diverse uses. "Digitisation may serve different purposes, which can be documentation, reconstruction, preservation, research, education, visualisation, or online discovery and access. The models and content for each purpose may include high-resolution offline models, online models, interactives, 3D printable models, augmented and virtual reality models, publications, images, videos, and panoramas".

A further development in this direction is the launch of the Commission Expert Group on the common European Data Space for Cultural Heritage (CEDCHE) as a forum for cooperation on digitisation, conservation and preservation of cultural heritage between the European Commission, Member States and UNESCO. The expert group will help monitor the implementation of the Recommendation on a common European data space for cultural heritage adopted on 10 November 2021. The recommendation sets a main focus on 3D digitization requiring all monuments and sites at risk, and at least 50% of the most visited ones, to be digitized in 3D by 2030. However, even considering the sense of urgency due to the increasing number and impact of climate change and human pressure, showing the fragility of our Cultural Heritage and the risk of loss, targets for 3D digitization seem to be too high for the current situation. If the recommendation can definitely stimulate the adoption of policies facilitating actions of digitization, there's still a significant need for tools and services that could make such targets achievable and the digitization actually accessible for a wider public.

For instance, as indicated by the EC, the European common data space for Cultural Heritage will build on the current Europeana platform and link to relevant European, national and regional initiatives and platforms. The aim is to provide interoperable access to cultural heritage databases all over Europe for providing citizens and professionals with efficient, trusted, easy-to-use and attractive access to European digital cultural content. It must be noticed that Europeana does not directly support 3D contents yet as reported in the previously mentioned report.

An additional reference in the field of Heritage digitization and the use of 3D to enhance documentation, knowledge and accessibility, is the "Study on quality in 3D digitization of tangible cultural heritage". This study was commissioned by the EU to help advance 3D digitization and support the objectives of the Recommendation on a common European data space for cultural heritage (European Commission, 2022). The study was leaded by Cyprus University of Technology and developed by nine institutions (VIGIE, 2022). The aim of the study is to map the parameters, formats, standards, benchmarks, methodologies and guidelines relating to 3D digitization of tangible cultural heritage, improving the quality of 3D digitization by enabling cultural heritage professionals, institutions, content-developers, stakeholders and academics to define and produce high-quality digitization standards for tangible cultural heritage.

In addition to mapping parameters, formats, standards, benchmarks, methodologies, and guidelines, the study analyses in particular two main issues: complexity and quality as essential topics in determining the necessary effort for a 3D digitization project to achieve the required value of the output.

Digital Heritage Fruition in the Post-Covid Era

As stated in the report developed in 2021 by the European Parliament, Policy Department for Structural and Cohesion Policies, pre-Covid-19 the Cultural and Creative Sectors were already characterised by fragile organisational structures and practices. The fragmented organisation of value chains, the project-based working and the (not well-protected) Intellectual Property (IP)-based revenue models are only a few elements contributing to this.

Since the Covid-19 pandemic hit Europe in spring 2020, the Cultural and Creative Sectors have been among the most negatively affected sectors. The containment measures that have been put in place throughout the EU have led to a chain of effects, severely impacting the economic and social situation in the Cultural and Creative Sectors. Especially the venue- and visitor-based sub-sectors such as the performing arts and heritage were most severely hit.

The global crisis had a strong impact on the Cultural Heritage sub-sector, which is in most cases venue and visitor based, due to the closing of museums and heritage sites and strict bans on internal movement and tourism, which is the first source of income for many heritage sites.

Most of the museums (80%) and religious heritage institutions (83%) increased their online presence (online communication, virtual tours, online exhibitions). Online services that increased the most were those requiring less additional financial resources and/or experience and skills (hashtags on social media or activities around an already existing online collection), while services that required time, resources and skills (podcasts, live content, online learning) increased the least. Most of the museums that increased their online presence also recorded an increasing number of online visitors, accessing especially educational material (IDEA Consult et al., 2021).

The Covid-19 crisis has impacted every dimension of the Cultural Heritage field: from research to conservation and protection, and from outreach to training and education. Implications have been categorized (Europa Nostra, 2020) as follows: Implications for personnel and security of jobs; Implications for security of heritage sites, contents and visitors; Socioeconomic implications; Cultural implications; Financial implications; Implications to ensure proper communication and keep networks alive.

The Covid-19 crisis had the advantage to foster the digital transformation of the heritage sector, but it also serves to highlight and even contribute to existing and significant inequalities. Only the largest museums and heritage organizations have the capacity to share their collections and materials online, while smaller organizations are missing digital opportunities.

The digitalization gap between smaller and bigger heritage organizations can lead to a lack of diversity in cultural and heritage content. Many heritage actors in rural areas did not have the necessary digital skills to stay active during the lockdown, which led to an important void in cultural offer. Lending and borrowing artworks among museums was totally suspended due to the closing of international borders. This was translated into a decrease in international and European exhibitions. Another important implication relates to intangible heritage. "*The impossibility for communities to gather in rites and celebrations, including religious ones, that traditionally favor intergenerational exchange, exchange between local and foreign communities, with the consequent sharing and transmission of knowledge and values, was particularly detrimental*" (Europa Nostra, 2020).

Invest in digital services and infrastructures as well as in training and capacity building for digital skills in the heritage sector is therefore crucial. Heritage organizations must be prepared to facilitate the sharing of cultural heritage assets and values through digital and on-line means.

Several digital initiatives by Museums and Cultural institutions during Covid-19 have been traced by studies and reports (Nemo, 2021; UNESCO, 2020), but the pandemic crisis highlighted also the debate on impact of mass tourism and the need to diversify tourist flows from large attractors to different contexts. The vision is to repopulate small historical centers and to enhance unknown and inaccessible heritage sites.

"The development of advanced digital technologies, such as 3D, artificial intelligence, machine learning, cloud computing, data technologies, virtual reality and augmented reality, has brought unprecedented opportunities for digitization, online access and digital preservation. Advanced digital technologies lead to more efficient processes (e.g. automated generation of metadata, knowledge extraction, automated translation, text recognition by optical character recognition systems) and higher quality content. They allow innovative forms of artistic creation, while opening up new ways of digitally engaging with and enjoying cultural content through co-curation, co-design and crowdsourcing, empowering public participation. Artificial intelligence, blockchain and other advanced technologies can also be explored for automatically identifying cultural goods that are illicitly trafficked. The uptake of such advanced technologies has a significant impact on European recovery and growth following the Covid-19 pandemic, and Member States should support it by taking appropriate measures" (Commission Recommendation, 2021).

The Commission Recommendation on a common European data space for cultural heritage stressed also the topic of the digital skills gap widened by Covid-19, also in the cultural heritage sector, "*where the digital divide leaves small institutions (e.g., museums) in particular struggling to make use of advanced technologies, such as 3D or artificial intelligence*".

In this direction, several initiatives and studies have been carried out in order to trace new research directions for heritage fruition through digital devices, discussing the emerging themes of access and human connection (King et al., 2021). The Covid-19 crisis has triggered a huge demand for social and economic recovery, also foster the resilience of cultural heritage for innovation, but concrete mechanisms concerning digitisation and cultural heritage is still lacking (Münster et al., 2021), especially when the need for digitation and data sharing comes from small museums, cultural institutions or heritage sites.

3D TECHNOLOGIES FOR CULTURAL HERITAGE: NEW DIRECTIONS

The trends highlighted above are definitely leading to a more consistent use of 3D technologies in the Cultural Heritage sector. Here indeed several factors come to a convergence. Research infrastructures and networks such as DARIAH, CLARIN, ARIADNE, CARARE and EUROPEANA, modelling approaches such as INCEPTION, 3D-COFORM, 3D-ICONS, or EPOCH, together with a direct effort by the European Commission addressed to foster the use of 3D technologies, are shaping a new approach to the use of 3D technologies. In particular, the field of buildings, monuments and sites could greatly benefit from these innovations and that's why latest European actions such as the 3D declaration (Co-operation on advancing digitisation of cultural heritage, 2019) and the already mentioned list of basic principles and tips for 3D digitization, 3D recommendation, VIGIE study, are pointing in that directions.

However, due to the rapid evolution of equipment, tools, software, applications and, nonetheless, the rise of even newer technologies (Artificial Intelligence - AI, blockchain, etc.) that can build on the adoption of a standard use of 3D in the Cultural Heritage sector, this transition can appear as a big challenge for a lot of institutions or professionals who need for re-training and up-skilling.

Towards an EU Competence Centre

That obstacle could be mitigated by some of the actions foreseen in the project 4CH - *Competence Centre for the Conservation of Cultural Heritage* (Maietti & Ferrari, 2021), funded by the European Commission and started in January 2021. The project is setting up the methodological, procedural, and organizational framework of a Competence Centre able to seamlessly work with a network of national, regional, and local Cultural Institutions, providing them with advice, support, and services focused on the preservation and conservation of historical monuments and sites.

Training, up-skilling and capacity building are also part of the strategic mission of the 4CH project and the future Competence Centre in order to make the transition to 3D technologies smoother.

On this side, several actions are already in place, such as:

- coordination and networking of training and up-skilling initiatives;
- interchanges between Universities, PhD programs, post-graduate schools or Research Institutes;
- creation and maintenance of training material for up-skilling, using different modalities, both for distance and proximate learning;
- sharing best practices and promoting the use of released standards, guidelines and tools.

In fact, on the technological side for what concerns 3D, the project is a follow up in the application of technologies developed by INCEPTION for documenting, monitoring and visualizing 3D BIM models of the built heritage. This aim is pursed by exploiting the benefits of 3D and Semantic Web technologies. In the project, indeed, the INCEPTION approach represents a pillar of the 4CH Cloud Platform, a cloud-based platform hosting a set of services for Cultural Heritage that should be made available, by the end of the project, on the EOSC (European Open Science Cloud) Portal Catalogue and Marketplace and that will power the future Competence Centre. Joining as a provider the EOSC Portal Catalogue and Marketplace allows indeed promoting the adoption of those services outside the traditional user group, managing service requests, and ensuring the compliance with European protocols for digital services. In this sense, the approach to digital technologies for Cultural Heritage begins to be outlined in a more transparent and accessible way. Together with previously listed actions aimed at capacity building, the future Competence Centre (Figure 1) will make clear the distinction between infrastructures for data storage and those for high-performance computing running services. In this way, services will run on and exploit data located in a federated infrastructure, where different institutions keep their data on their own storage (independently managing ownership and access policies). Anyway, there will a kind of one-stop-shop for accessing services released directly from the Competence Centre or being addressed to external ones. Among the services directly offered by the Competence Centre, there will be for sure tools that can bridge the gap between 3D and Semantic Web technologies.

Figure 1. Outline of the main objectives of the 4CH project.

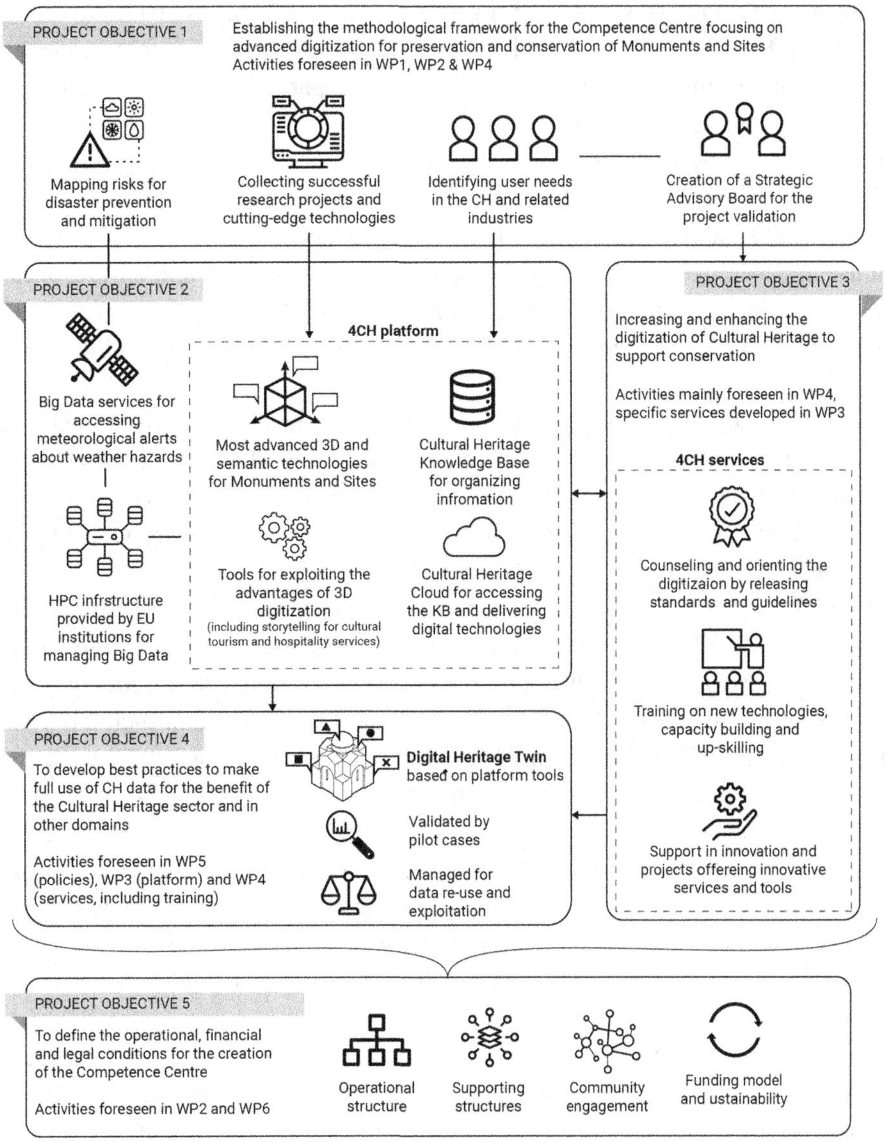

Joining 3D and Semantic Web Technologies

In past years, one of the main drawbacks of Semantic Web technologies in big production environments was the use of 3D. This while the power and expressiveness of Semantic Web potentially adds a lot of value to the 3D world. In fact, the realization of this approach makes possible to involve a widely distributed available knowledge of Cultural Heritage, exploiting the advantages offered by the Semantic Web technologies (Bonsma et al., 2018). One of the core issues is that 3D tends to grow very rapidly in size, especially on individual components / instances like vertices, lines or triangles. On the other hand, the key benefits of Semantic Web suffer from scalability issues when the content grows to millions of instances as is typical for geometrical representations.

Modern mature open exchange standards for geometry like IFC - Industry Foundation Classes (ISO 16739) and STEP (ISO 10303 AP 242) are able to grasp the original design intent and with this describe geometry in a much more semantically rich manner. This compared to pure visualization exchange formats like glTF, FBX, Collada, U3D, X3D etc.

Therefore, ontologies based on ISO 16739 and ISO 10303 AP 242 like IfcOWL (for the Industry Foundation Classes – IFC standard for BIM data) and GEOM (Geometry) are in essence the practical approach to combine 3D and Semantic Web. The crucial point of this interoperability among data relies on the correct use and extension of existing ontologies or the implementation of new ones. Ontology is indeed as a formal depiction of a shared conceptualisation, and therefore the issue in semantic search engine is actually how to exploit them to collect meanings based on their formal description (Colucci et al., 2021).

However, even here scalability issues arise on real world data: especially as within the Cultural Heritage domain. This issue becomes even more clear when 3D BIM models are translated in semantic triples since we don't have to deal only with 3D geometries but even with related information, attributes and metadata regarding state of conservation, history, events, or anything connected to a building, a monument or a site.

Within INCEPTION, several solutions have been applied to overcome these scalability issues and allow the integration of 3D and Semantic Web on real world project sizes for the Cultural Heritage domain (Iadanza et al., 2019). The ifcOWL and the GEOM ontologies are definitely the starting point in this process, but part of that knowledge has to be mapped also on others such CIDOC CRM (Conceptual Reference Model), PROV (Provenance) Ontology or Time Ontology or Getty Architecture&Art Thesaurus (AAT) (Lopez at al., 2018) just to mention few of the most used ones.

The project 4CH is improving the important steps made already in INCEPTION to increase the benefit from Semantic Web into the 3D domain in general and for Cultural Heritage specifically (Ziri et al., 2019). Then, the models can be furthermore enriched by the mapping the knowledge also on ontologies vertically developed on specific domains such as Europeana Data Model (EDM) for content available in Europeana, ArCO for the Italian Catalogue of Cultural Heritage of BIMerr for building renovation processes, as reported in the following paragraph.

In this scenario, current approaches to semantic modeling of the Digital Twin are emerging, considering the potential of these digital representations to support interventions such as restorations, and to aggregate data and interdisciplinary information (Bolognesi & Signorini, 2021).

Moreover, current research are working in the direction of designing the semantic infrastructure of the data space for cultural heritage as envisaged by the European Commission (Niccolucci, Felicetti & Hermon, 2022). Several applications on architectural heritage are at the centre of frontier research connecting virtual 3D models with several applications related to reconstruction, virtual reality, 3D interoperable BIM models and semantic-aware 3D digitisation (Bevilacqua et al., 2022; De Luca, 2020).

Beside an effective use of semantic technologies, it is of utmost importance also the development of search engines capable of querying and retrieving all available resources on Cultural Heritage at European level that cannot otherwise efficiently be accessed and analysed.

CURRENT APPROACHES AND RESULTS EXPLOITING SEMANTICS

The use of Building Information Modelling (BIM) and Historical/Heritage BIM (HBIM) is becoming more common for management and conservation of historic buildings, and in some cases for archaeological monuments. BIM is a mature system, developed for the architecture, engineering and construction sectors over twenty years, which allows for great information management and collaboration between multi-disciplinary teams. It typically involves the use of high-end professional software and hardware systems. Researchers have been exploring methodologies for exporting HBIM models to more accessible applications, open standard platform and open formats with the aim of making HBIM more accessible. However, a standard data scheme in which include all the available pieces of information is still under investigation by several projects.

Metadata for 3D objects is indeed another important area that needs further standardisation. The Europeana Network Association 3D content task force, amongst others, considered the state of development of metadata standards for the description, discovery, preservation and re-use of 3D objects. The task force evaluated the following metadata models: LIDO (Lightweight Information Describing Objects), CARARE, CIDOC CRM (Conceptual Reference Model) and Europeana's EDM (Europeana Data Model) schema. These are established schemas in community use with different strengths, some placing more emphasis on capturing information about the digitisation process and the provenance of the 3D object. EDM's strength lies in its ability to integrate information from a wide range of different cultural institutions. It provides for a full description of the conceptual thing that is being provided and binds this together with information about the digital representation and where these may be found online. However, EDM does not currently readily allow for information about the provenance or the technical characteristics of a 3D dataset to be included.

The INCEPTION Core Engine

The Inception Core Engine (ICE) is the solution, based on a semantic approach, for easily exploring BIM and HBIM models and the web and accessing other contents developed within the INCEPTION project (Bonsma et. Al, 2018; Iadanza et al. 2019; Iadanza et al. 2020). ICE interprets each element of a 3D model as a single entity that can be connected to a specific knowledge. The approach consists of transforming all the geometries of a 3D model of a monument or a site into semantic triples that connect one element to another using specific predicates, defined in a dedicated semantic ontology. The 3D models have to be provided in the form of a Building Information Model (BIM) as an IFC (Industry Foundation Classes) standard file. Once the models and related information are transformed into triples, all of these are stored in a semantic triple store that is accessed via HTTP through a dedicated Apache Fuseki SPARQL server. ICE technologies can power web applications that allow users to enrich their models with new semantic metadata. Indeed, the web client allows to enrich the models with new data (e.g., a date, a value, some textual remarks) as well as attachments (e.g., pictures, thermographic images, 3D models of specific details, videos, etc.), all of which are related to the cultural heritage site or a specific geometrical element.

Similarly, the Core Engine allows not only enriching the model but also easily navigating it together with semantic metadata and attachments. Three different modes are natively included in the ICE: IFC, texture and hybrid views (Figure 2). The IFC view mode allows to select geometric elements, filter them by levels or classifications and query their metadata, while the texture mode does not present selectable elements but offers a visualization capable of offering an intuitive material understanding. The hybrid mode superimposes the previous ones by means of an editable transparency layer and allows you to select IFC geometric elements enriched this time by their real appearance. It is also possible, for each element, to access its metadata: it is possible to read the unique Global-ID, the name, the IFCType, notes and comments. Through a graphical schema, it is also possible to navigate between categories, parameters and values attributed during the modeling phase. The values, which may contain links to internal references (other models in the platform) or external (in the web).

Figure 2. 3D visualization of the H-BIM model processed by ICE services in IFC (left) and hybrid (centre) visualization modes. Through the selection by elements (right) it is also possible to access the detailed information of each object.

The generality of this approach allows representing both tangible and intangible information. To give an example, a single element (e.g., a brick) can be linked to a wall, as well as to one or more documents, or to some metadata, or even to external information on the web, using nothing but semantic triples. There are also other numerous benefits of switching to a semantic solution. For example, using standard and open protocols makes the system intrinsically interoperable. It is worth recalling, here, that defining a dedicated ontology is much more than just defining a taxonomy (i.e., a tree of categories and subcategories). A semantic ontology, indeed, allows you to define properties and inference rules that link one category to another. Hence, the unique chance of enabling a real semantic reasoner, or AI "reasoning engine", based on these rules, able to extract new knowledge.

Exploiting Digital Catalogues for Enriching 3D Models: Two Case Studies

As stated, among the benefits of the semantic approach there is interoperability with other sources sharing the same standard. For instance, under a framework agreement between the Department of Architecture of the University of Ferrara and the Italian Central Institute for Catalogue and Documentation (ICCD), a connection between ICE technologies and the New Italian Digital Catalogue of Cultural Heritage (Catalogo Generale dei Beni Culturali, 2022) has been recently developed. The aim was to exploit every single component of the H-BIM model in the Linked Open Data network for automatically querying and making inferences to other data sources such as the New Digital Catalogue of Cultural Heritage. The approach has been tested on two case studies: the Church of Santa Maria delle Vergini in Macerata and the archaeological area of the Gran Carro of Bolsena of the Early Iron Age.

The church of Santa Maria delle Vergini is a majestic building with a Greek cross plan, surmounted by an octagonal dome erected on a drum, supported by four imposing pillars with a quadrangular base; the arms of the Greek cross end in semicircular apses, each with two cross-vaulted "scarsella" chapels

(Canullo, 2016). A first plant of the church dates back to 1355 but the consecration of the current configuration of building took place in 1577. The facade, developed on two horizontal registers, does not correspond to the interior space, because it was later completed. The interior of the church has eleven chapels decorated between the late sixteenth and late eighteenth centuries. The church has also undergone the action of various seismic events: among the most recent and important, we can mention those of 1997 and 2016, with particular injuries to the drum supporting the dome, still under restoration (Figure 3).

Figure 3. Aerial photo of the church of Santa Maria delle Vergini in Macerata before the 2016 earthquake (on the left) and the dome damaged in 2016 (on the right).

The site of the Gran Carro of Bolsena is unique for its state of conservation among the pile-dwelling settlements in Italy, and it certainly represents one of the most important discoveries that took place at the end of the 1950s. It is currently submerged halfway along the eastern coast of the Bolsena Lake and it is the first protohistoric deposit identified in the inland waters of peninsular Italy (Figure 4).

Figure 4. Aerial photo of the Bolsena Lake with the indication of the excavation area (on the left) and a picture of underwater excavation activities (on the right).

In both cases, the starting point was to elaborate a BIM model that could operate as the main reference and federator of information about the architectural or archaeological asset. Since the purpose of the modelling is different from the logic of intervention on buildings, it was decided to operate a modelling that was adequate to accommodate the information of material nature and state of preservation, as well as of historical-documentary data from the ICCD Catalogue. 3D-BIM geometry modelling was performed with the use of free-form techniques since the majority of elements (in particular for the archaeological case study) were out of the scope of use of the BIM software used (Autodesk Revit).

On the other hand, from the information point of view, we proceeded to browse the New Catalogue of Cultural Heritage where it is possible to search by elements, authors, cultural places, sectors, locations, possibly filtering the search by specific parameters (Carriero et al. 2019). In this phase, it was decided to extrapolate all the data of the RDF (Resource Description Framework) resource related to the assets, which will be subsequently queryable through SPARQL endpoint by inserting the URI of reference of the element. In this case, the URI of the element has been then used for the informative enrichment of the BIM model, representing a univocal identification according to the standardization of the catalogue.

The data thus elaborated were processed by the INCEPTION Core Engine for accessing the benefits of the Semantic Web, to visualize and aggregate heterogeneous sources of data. In fact, based on the identification code previously identified and annotated on the element during modeling, a function of the ICE services was developed in order to automatically operate a query of the RDF data available online through an SPRQL Query on the Virtuoso EndPoint located at http://dati.beniculturali.it/sparql. The data retrieved from a SPARQL endpoint conforms to the Linked Data approach for publishing data in the Semantic Web context. RDF-formatted data is normally represented as a list of semantic triples but, through a specially developed function, these triples are displayed in the graph as if they were data directly entered during modelling. This approach allows the end user to read them seamlessly and, in the same way, to read data that is always up-to-date and aligned with the catalogue (Figure 5).

Figure 5. Every element of the building can be used to perform a live SPARQL query that returns all the details for that element, according to the HBIM ontology. Each value can be updated via web, thanks to the SPARQL 1.1 Update functionalities.

This approach, in addition to having been developed for the entire federated H-BIM model, has also been extended to detail models, creating a true multiscale model. In the case of Santa Maria delle Vergini, for some of the eleven chapels, in fact, the 3D attachments referring to minor cultural assets (paintings, decorations and sculptures) have also been loaded, creating a complete navigation path able to give a spatial location to the cards of the Digital Catalog (Figure 6). Conversely, on the Gran Carro di Bolsena has been created a hierarchy of 3D models that replicate the same organization deriving from the Catalogue, moving from the overall site to the excavation sector to the finds where everything is documented in 3D and enriched automatically by the appropriate RDF resource.

Figure 6. 3D textured visualization of the overall H-BIM model processed by the ICE services (left) and visualization of the high resolution photo of one of the chapels (right) retrieved from the ICCD Digital Catalogue.

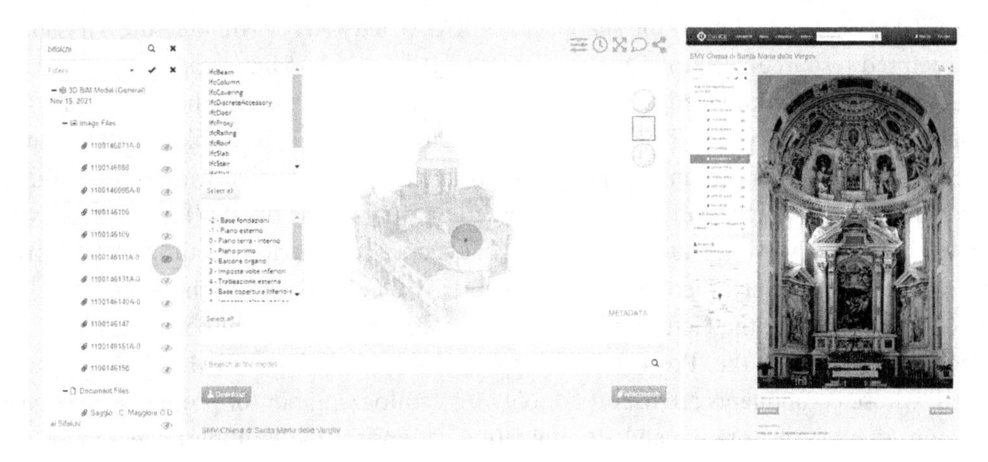

Related Projects Scenario

Analysing the most innovative European RTD projects, CORDIS Results Pack features sixteen projects (twelve from the original Pack published in 2020 and four new projects added in 2021) that are making important contributions to digital cultural heritage efforts. In particular, the ARCHES (Garcia Carrizosa et al., 2019), DigiArt (Anastasovitis et al., 2017) and EU-LAC-MUSEUMS (Repetto Málaga and Brown, 2019) projects have been harnessing technologies such as 3D modelling and Augmented Reality to increase the accessibility and enjoyment of museums as a key institution of cultural curation and preservation.

Other projects have focused on using technology to increase social awareness and interest in Cultural Heritage and preservation. For instance, PLUGGY (Lim et al., 2018) developed the first-ever social network dedicated to promoting European cultural heritage, whilst the I-Media-Cities project (Loos and Weigel, 2018) launched a revolutionary platform that uses audio-visual material to allow anyone to discover the rich cultural heritage of nine European cities. Meanwhile, EMOTIVE (Katifori et al., 2018) tapped into the raw power of storytelling by offering tools to heritage professionals allowing them to create interactive storytelling experiences to engage, inform and provoke the interest of audiences.

Finally, the ArchAIDE project (Gualandi et al., 2016) developed innovative software to identify fragments of pottery found during excavations and to store them in a dedicated database, thus helping the vital work of archaeologists and other professionals in the cultural heritage field. Time Machine (Münster et al., 2019) has been developing large-scale digitisation and computing infrastructure using AI and Big Data mining in order to extract and analyse the vast amount of data generated when digitising archives from museums and libraries.

Of the four new projects that have added in 2021 (those of 2022 are pending on decision), the NewsEye project (Jean-Caurant and Doucet, 2020) tries to help better preserve newspapers and allow for easier search options when using historical newspapers for historical research. In a similar way, TROMPA is pioneering its own digital tools to enrich and democratise publicly available classical music heritage through a user-centred co-creation set-up (Weigl et al., 2019). Technology created by the V4Design project (Avgerinakis et al., 2018) will allow architects and video game designers to integrate existing digital content into their designs in stunning 3D. Lastly, SILKNOW took on the challenge to help preserve silk textile heritage (Portalés et al., 2018).

This context reinforces INCEPTION's innovation in graphical-semantic interoperability. Specifically, López et al. (2018) details the link of the spatial data and the Getty Art & Architecture Thesaurus (Getty AAT) as structured resource used to improve access to such a kind of information, released as Linked Open Data (LOD) to make knowledge freely available and published under the Open Data Commons Attribution License (ODC-By) 1.0. Thus, Getty AAT vocabulary is uniquely linked to 3D parametrically modelled elements (architectural units) upon a multilingual approach. The way to make this association depends on the BIM package in which it is worked on. In the case of REVIT (the most widespread worldwide for how closely it integrates with the BIM concept), it is carried out through the option 'Type of properties' tool. Conversely, the Getty AAT vocabulary can be linked to the modelled elements in two ways: (i) by associating the Getty AAT url with the 'Identity data' parameter within the 'Property type; (ii) generating a link in the 'Properties'. This way, tailored Heritage-BIM (HBIM) libraries could be generated, whose components are useful not only for cataloguing but for proper performance on the protection, conservation, restoration and dissemination actions for the built heritage.

FUTURE RESEARCH DIRECTIONS

There are several current important issues into which framing close research actions.

EU endorsements and initiatives are fostering pan-European collaboration between heritage institutions, and promoting interoperability and open access, while respecting copyright. In addition to the initiatives mentioned above, the Commission recommendation on digitisation and online accessibility and digital preservation of cultural material (2011/711/EU) is oriented to direct Member States to develop strategies and improve conditions for the entire digitisation lifecycle. The Current Council Work Plan for Culture (2019-2022) cares about six priorities: (i) sustainability, (ii) cohesion and wellbeing, (iii) creative industries; (iv) gender equality and (v) culture, as drivers for sustainable development; while Europeana embodies the continuous effort of the European Commission and the Members States to democratise access to cultural heritage.

The ecological and digital transitions are, in fact, the pillars of the Recovery Plan for Europe (European Commission-2021a). EU Member States have agreed on the need to invest more in improving connectivity and related technologies to strengthen the digital transition and emerge stronger from the Covid-19 pandemic, transforming the economy and creating opportunities and jobs for that Europe into which citizens want to live (European Commission-2021b). During the confinement, society has shown that Heritage in digital format indeed is a true social balm, with buildings, museums and collections open online 24 hours a day.

Moreover, recent advancements in 3D digitization and data management are focused on 3D point cloud segmentation by hierarchically organize data through semantic classification based on Machine Learning and Deep Learning (Matrone et al., 2020; Pierdicca et al., 2020) as the basis for different levels of data knowledge and interpretation (Grilli and Remondino, 2020). These research directions will involve the recognition of several features, in addition to shapes and geometries.

The impact of new technologies and virtual dimensions such as blockchain and the metaverse will involve also the field of Cultural Heritage, considering also the growing applications of Virtual Reality, Augmented Reality and Mixed Reality to explore heritage sites in an immersive way.

According to the already mentioned VIGIE report - Study on quality in 3D digitisation of tangible cultural heritage, the development of these systems will have a direct impact on the Cultural Heritage industry (i.e., Virtual Museum, Virtual Sites, Smart Cities, 3D digital libraries, fabrication and eArchiving). The potential of these technologies together with open data, Heritage Building Information Modelling (HBIM), Holistic BIM (HHBIM) and Digital Twins will impact more and more on the documentation, conservation and dissemination of Cultural Heritage. This will create the need of new skills in order to improve the knowledge, conservation and management of a such extensive and articulated heritage. Deepening the semantic interrogation of digital models of historical-architectural buildings can be an effective research direction, considering the increasing digitisation potential and the need for new forms of sharing and accessibility. An additional requirement in the field of digital data processing is the "thematic" management of digital data, considering different levels of knowledge. As a matter of fact, one of the main issues is the capability to find optimized data management and sharing processes able to combine critical-interpretative skills with automatic procedures to hierarchize data according to different cognitive levels. BIM-based digitisation processes need an increasing data integration towards an actual interdisciplinary and the implement of open standard semantic web platforms for monitoring, maintenance and management of cultural heritage through semantically enriched HBIM models, facilitating accessibility and use of complex information models by different users.

CONCLUSION

As stated by the Executive Vice-President for A Europe Fit for the Digital Age, Margrethe Vestager, during the launch of a common European data space for cultural heritage, "*The tragic burning of Notre Dame Cathedral in Paris showed the importance of digitally preserving culture and the lockdowns highlighted the need for virtually accessible cultural heritage. A robust data infrastructure coupled to easy data pooling and sharing are the necessary ingredients of a common European data space for cultural heritage*" (Common European data space for cultural heritage, 2021).

The research framework and activities here presented highlight the need to find new paths for digital technologies applications towards extended and advanced approaches to heritage knowledge and accessibility. These new directions do not just have to be driven by the advancement of new technologies, but should be consistent with the values and significances of Cultural Heritage, which is characterized by uniqueness.

Starting from the main outcomes of the INCEPTION projects, and considering the changes occurred in the overall scenario (the Covid-19 pandemic, but also the initiatives launched by the European Commission), the contribution explores in particular the new research directions opened by the 4CH project, and the potential of Semantic Web technologies.

The 4CH project has been applied under the Work Programme *Europe in a changing world – inclusive, innovative and reflective Societies* (Call - *Socioeconomic and Cultural Transformations in the Context of the Fourth Industrial Revolution* - H2020-SC6-Transformations-2018-2019-2020). The application addressed the requirement by the European Commission to set up a European Competence Centre aiming at the preservation and conservation of European Cultural Heritage using new state-of-the-art ICT technologies. Concerning Heritage 3D models, 4CH further develops the main outcome of INCEPTION for documenting, monitoring and visualizing 3D BIM models of the built heritage, by applying advantages of 3D and Semantic Web technologies, fostering data connections and knowledge sharing and accessibility.

In this direction, the Inception Core Engine (ICE), based on a semantic approach, is presented as a valuable tool for easily exploring BIM and HBIM models and the web and accessing other contents.

In fact, BIM-based digitisations are recently growing in numbers thanks to a variety of factors; faster data capturing technologies where a higher number of points comes together with a significantly reduced time for data processing (i.e., pre-registration or registration on field), improved procedures for point cloud segmentation and shape recognition, increasing demand of BIM models in Cultural Heritage for both documentation and restoration activities. However, due to the long, complex and articulated data pipeline of BIM models where the three-dimensional geometries are accompanied by pieces of information, it's even more important tracking each step of the processing workflow for making data accessible and actually reusable. The transformation of 3D geometries into semantic triples already allows enriching models with semantic metadata, as previously demonstrated. Since this approach can connect intangible information, future research activities will exploit the possibility of correlating on field operations, instrument specifications and settings, and data processing operations with the BIM model, realising a consistent source of knowledge.

This research direction will be driven by the development of the European Competence Centre, towards new abilities to acquire knowledge and interact with Cultural Heritage. The vision is to merge together digital evolutions, sustainability in a broad sense and human development.

ACKNOWLEDGMENT

The case studies of the Church of Santa Maria delle Vergini in Macerata and the archaeological area of Gran Carro of Bolsena have been developed by INCEPTION Srl, Spin off incubated at the University of Ferrara, under a framework agreement between the Department of Architecture of the University of Ferrara and the Italian Central Institute for Catalogue and Documentation (ICCD).

The project 4CH - *Competence Centre for the Conservation of Cultural Heritage* is being developed by a Consortium led by INFN (Coordinator), INCEPTION Srl (Scientific coordinator), PIN Scrl (Technical Coordinator): Fundacion Tecnalia, Spain; Visual Dimension, Belgium; RDF, Bulgaria; Iron Will, Moldavia; KNAW, Netherlands; University of Bologna, Italy; Athena Research Centre, Greece; Laboratorio Nacional de Engenharia Civil, Portugal; The Cyprus Institute, Cyprus; FORTH, Greece; ICCU, Italy; CARARE, Ireland; Michael Culture Association, Belgium; Institutul National al Patrimoniului, Romania; Universite de Tours, France; Leica Geosystems, Switzerland. 4CH is a Horizon 2020 project funded by the European Commission under Grant Agreement n.101004468-4CH.

The INCEPTION project has been developed (2015-2019) by the Department of Architecture of the University of Ferrara as project coordinator and the University of Ljubljana (Slovenia), the National Technical University of Athens (Greece), the Cyprus University of Technology (Cyprus), the University of Zagreb (Croatia), the research centers Consorzio Futuro in Ricerca (Italy) and Cartif (Spain), DEMO Consultants BV (The Netherlands), 3L Architects (Germany), Nemoris (Italy), RDF (Bulgaria), 13BIS Consulting (France), Z + F (Germany), Vision and Business Consultants (Greece). The research project has received funding from the European Union's H2020 Framework Programme for research and innovation under Grant agreement no 665220.

REFERENCES

Amann, S., & Heinsius, J. (2021). Research for CULT Committee – Cultural and creative sectors in post-Covid-19 Europe: crisis effects and policy recommendations. European Parliament, Policy Department for Structural and Cohesion Policies.

Anastasovitis, E., Ververidis, D., Nikolopoulos, S., & Kompatsiaris, I. (2017). Digiart: Building new 3D cultural heritage worlds. In *2017 3DTV Conference: The True Vision-Capture, Transmission and Display of 3D Video (3DTV-CON)* (pp. 1-4). IEEE.

ArCO Ontology. Available online: http://wit.istc.cnr.it/arco

Avgerinakis, K. (2018). V4design for enhancing architecture and video game creation. In *2018 IEEE International Symposium on Mixed and Augmented Reality Adjunct* (pp. 305-309). IEEE.

Bevilacqua, M. G., Russo, M., Giordano, A., & Spallone, R. (2022, March). 3D Reconstruction, Digital Twinning, and Virtual Reality: Architectural Heritage Applications. In *2022 IEEE Conference on Virtual Reality and 3D User Interfaces Abstracts and Workshops* (pp. 92-96). IEEE.

Bianchini, C., Inglese, C., & Ippolito, A. (2016). The role of BIM (Building Information Modeling) for representation and managing of built and historic artifacts. *Disegnarecon*, *9*(16), 10–11.

BIMerr Ontology. (n.d.). Available online: https://bimerr.iot.linkeddata.es/

Bolognesi, C. M., & Signorini, M. (2021). Digital Twins: combined surveying praxis for modelling. In *ARQUEOLÓGICA 2.0–9th International Congress on Archaeology, Computer Graphics, Cultural Heritage and Innovation. GEORES–3rd GEOmatics and pREServation*. (pp. 275-280). Editorial Universitat Politècnica de València.

Bonsma, P., Bonsma, I., Ziri, A. E., Iadanza, E., Maietti, F., Medici, M., Ferrari, F., Sebastian, R., Bruinenberg, S., & Lerones, P. M. (2018), Handling huge and complex 3D geometries with Semantic Web technology, Florence Heri-Tech – The Future of Heritage Science and Technologies. *IOP Conf. Series: Materials Science and Engineering, 364*.

Bowen, J. P., & Giannini, T. (2021). Digitality: A reality check. *Proceedings of EVA London, 2021*, 12–19.

Canullo, G. (2016). I Bifolchi e l'eucarestia. La cappella maggiore della chiesa di Santa Maria delle Vergini a Macerata. *Journal of the Section of Cultural Heritage*.

Catalogo Generale dei Beni Culturali. (n.d.). https://catalogo.beniculturali.it

CIDOC-CRM Ontology. (n.d.). Available online: http://www.cidoc-crm.org/

Colucci, E., Xing, X., Kokla, M., Mostafavi, M. A., Noardo, F., & Spanò, A. (2021). Ontology-Based Semantic Conceptualisation of Historical Built Heritage to Generate Parametric Structured Models from Point Clouds. *AppliedSciences*, *11*(6), 2813.

Commission recommendation on a common European data space for cultural heritage. (2021). https://digital-strategy.ec.europa.eu/en/news/commission-proposes-common-european-data-space-cultural-heritage

3D content in Europeana task force. (2020). https://pro.europeana.eu/project/3d-content-in-europeana

Cooperation on advancing digitisation of cultural heritage. (2019). *Digital Day 2019*. Retrieved from https://ec.europa.eu/digital-single-market/en/news/eu-member-states-sign-cooperate-digitising-cultural-heritage

De Luca, L. (2020, October). Towards the Semantic-aware 3D Digitisation of Architectural Heritage: The" Notre-Dame de Paris" Digital Twin Project. In *Proceedings of the 2nd Workshop on Structuring and Understanding of Multimedia heritAge Contents* (pp. 3-4). Academic Press.

Empler, T., Caldarone, A., & Rossi, M. L. (2021). BIM Survey. Critical Reflections on the Built Heritage's Survey. In C. Bolognesi & D. Villa (Eds.), *From Building Information Modelling to Mixed Reality* (pp. 109–122). Springer.

Europa Nostra. (2020). *Covid-19 & Beyond. Challenges and Opportunities for Cultural Heritage*. https://www.europanostra.org/wp-content/uploads/2020/10/20201014_COVID19_Consultation-Paper_EN.pdf

European Commission. (2020). *Basic principles and tips for 3D digitisation of cultural heritage*. https://digital-strategy.ec.europa.eu/en/library/basic-principles-and-tips-3d-digitisation-cultural-heritage

European Commission. (2021a). *Recovery plan for Europe*. https://ec.europa.eu/info/strategy/recovery-plan-europe_en

European Commission. (2021b). *Europe's Digital Decade: digital targets for 2030*. https://ec.europa.eu/info/strategy/priorities-2019-2024/europe-fit-digital-age/europes-digital-decade-digital-targets-2030_en

European Commission. (2022). *Study on quality in 3D digitisation of tangible cultural heritage*. https://digital-strategy.ec.europa.eu/en/library/study-quality-3d-digitisation-tangible-cultural-heritage

European Commission, Directorate-General for Research and Innovation. (2021). *Horizon Europe: strategic plan 2021-2024*. https://data.europa.eu/doi/10.2777/083753

Europeana Data Model Ontology. (n.d.). Available online: https://pro.europeana.eu/page/edm-documentation

Expert Group on Digital Cultural Heritage and Europeana. (n.d.). https://digital-strategy.ec.europa.eu/en/policies/europeana-digital-heritage-expert-group

Garcia Carrizosa, H., Diaz, J., Krall, R., & Sisinni Ganly, F. (2019). Cultural differences in ARCHES: A European participatory research project—working with mixed access preferences in different cultural heritage sites. *The International Journal of the Inclusive Museum, 12*(3), 33–50.

Getty Architecture & Art Thesaurus. (n.d.). Available online: https://www.getty.edu/research/tools/vocabularies/aat/

Grilli, E., & Remondino, F. (2020). Machine Learning Generalisation across Different 3D Architectural Heritage. ISPRS International Journal of Geo–Information, 9(6).

Gualandi, M. L., Scopigno, R., Wolf, L., Richards, J., Heinzelmann, M., Hervas, M. A., Vila, L., & Zallocco, M. (2016). ArchAIDE-archaeological automatic interpretation and documentation of cEramics. In C. E. Catalano, & L. De Luca (Eds.), *EUROGRAPHICS Workshop on Graphics and Cultural Heritage* (pp. 1-4). The Eurographics Association.

Horizon Europe strategic plan 2021-2024. (n.d.). *European Commission.* https://ec.europa.eu/commission/presscorner/detail/en/ip_21_1122

Iadanza, E., Maietti, F., Medici, M., Ferrari, F., Turillazzi, B., & Di Giulio, R. (2020). Bridging the Gap between 3D Navigation and Semantic Search. The INCEPTION platform. *IOP Conference Series. Materials Science and Engineering, 949*(1), 012079.

Iadanza, E., Maietti, F., Ziri, A. E., Di Giulio, R., Medici, M., Ferrari, F., Bonsma, P., & Turillazzi, B. (2019). Semantic Web Technologies Meet BIM for Accessing and Understanding Cultural Heritage. *The International Archives of the Photogrammetry, Remote Sensing and Spatial Information Sciences, 42*(W9), 381–388.

Jean-Caurant, A., & Doucet, A. (2020). Accessing and investigating large collections of historical newspapers with the NewsEye platform. In *Proceedings of the ACM/IEEE Joint Conference on Digital Libraries in 2020* (pp. 531-532). ACM.

Katifori, A., Roussou, M., Perry, S., Palma, G., Drettakis, G., Vizcay, S., & Philip, J. (2018). The EMOTIVE Project - Emotive Virtual Cultural Experiences through Personalized Storytelling. CIRA@EuroMed.

King, E., Smith, M. P., Wilson, P. F., & Williams, M. A. (2021). Digital Responses of UK Museum Exhibitions to the COVID-19 Crisis. *Curator, 64*(3), 487–504.

Lim, V., Frangakis, N., Tanco, L. M., & Picinali, L. (2018). PLUGGY: A pluggable social platform for cultural heritage awareness and participation. In *Advances in Digital Cultural Heritage* (pp. 117–129). Springer.

Loos, A., & Weigel, C. (2018). I-MEDIA-CITIES: Automatic Metadata Enrichment of Historic Media Content. In Metrology for Archaeology and Cultural Heritage (MetroArchaeo) (pp. 351-356). IEEE.

López, F. J., Lerones, P. M., Llamas, J., Gómez-García-Bermejo, J., & Zalama, E. (2018). Linking HBIM graphical and semantic information through the Getty AAT: Practical application to the Castle of Torrelobatón. *IOP Conf. Ser.: Mater. Sci. Eng., 364*.

Maietti, F., Di Giulio, R., Medici, M., Ferrari, F., Ziri, A. E., Turillazzi, B., & Bonsma, P. (2020). Documentation, Processing, and Representation of Architectural Heritage Through 3D Semantic Modelling: The INCEPTION Project. In C. Bolognesi & C. Santagati (Eds.), *Impact of Industry 4.0 on Architecture and Cultural Heritage* (pp. 202–238). IGI Global.

Maietti, F., & Ferrari, F. (2021). Un Competence Centre europeo per la conservazione, il restauro e la valorizzazione del patrimonio culturale. Il progetto 4CH. In R. A. Genovese (Ed.), *Il patrimonio culturale tra la transizione digitale, la sostenibilità ambientale e lo sviluppo umano* [Cultural heritage between the digital transition, environmental sustainability and human development] (pp. 203–216). Giannini Editore.

Martín-Lerones, P., Olmedo, D., López-Vidal, A., Gómez-García-Bermejo, J., & Zalama, E. (2021). BIM Supported Surveying and Imaging Combination for Heritage Conservation. *Remote Sensing*, *13*(84), 1584.

Matrone, F., Grilli, E., Martini, M., Paolanti, M., Pierdicca, R., & Remondino, F. (2020). Comparing Machine and Deep Learning Methods for Large 3D Heritage Semantic Segmentation. ISPRS International Journal of Geo–Information, 9(9), 1-22.

Münster, S., Apollonio, F. I., Bell, P., Kuroczynski, P., Di Lenardo, I., Rinaudo, F., & Tamborrino, R. (2019). Digital cultural heritage meets digital humanities. *The International Archives of the Photogrammetry, Remote Sensing and Spatial Information Sciences*, *42*(W15), 813–820. https://doi.org/10.5194/isprs-archives-XLII-2-W15-813-2019

Münster, S., Utescher, R., & Ulutas Aydogan, S. (2021). Digital topics on cultural heritage investigated: How can data-driven and data-guided methods support to identify current topics and trends in digital heritage? *Built Heritage*, *5*(1), 1–13.

NEMO - Network of European Museum Organizations. (2021). *Follow-up survey on the impact of the COVID-19 pandemic on museums in Europe*. https://www.ne-mo.org/fileadmin/Dateien/public/NEMO_documents/NEMO_COVID19_FollowUpReport_11.1.2021.pdf

Niccolucci, F., Felicetti, A., & Hermon, S. (2022). Populating the Data Space for Cultural Heritage with Heritage Digital Twins. *Data*, *7*(8), 105.

Piano Nazionale di Ripresa e Resilienza [National Recovery and Resilience Plan]. (2021). Available online: https://www.mise.gov.it/index.php/it/68-incentivi/2042324-piano-nazionale-di-ripresa-e-resilienza-i-progetti-del-mise

Pierdicca, R., Paolanti, M., Matrone, F., Martini, M., Morbidoni, C., Malinverni, E. S., Frontoni, E., & Lingua, A. M. (2020). Point Cloud Semantic Segmentation Using a Deep Learning Framework for Cultural Heritage. Remote Sensing, 12(6), 1-23.

Portalés, C., Sebastián, J., Alba, E., Sevilla, J., Gaitán, M., Ruiz, P., & Fernández, M. (2018). Interactive tools for the preservation, dissemination and study of silk heritage—An introduction to the silknow project. *Multimodal Technologies and Interaction*, *2*(2), 28.

PROV-O. (n.d.). *The PROV Ontology*. Available online: https://www.w3.org/TR/prov-o/

Repetto Málaga, L., & Brown, K. (2019). Museums as Tools for Sustainable Community Development: Four Archaeological Museums in Northern Peru. *Museum International*, *71*(3-4), 60–75.

Russo, M., & De Luca, L. (2021). Semantic-driven analysis and classification in architectural heritage. *Disegnarecon, 14*(26), 1–6.

Time Ontology in OWL. (n.d.). Available online: https://www.w3.org/TR/owl-time/

UNESCO. (2020). *Museums around the world in the face of COVID-19.* United Nations Educational, Scientific and Cultural Organization. https://unesdoc.unesco.org/ark:/48223/pf0000373530

VIGIE. (2022). *Final report.* https://op.europa.eu/en/publication-detail/-/publication/dc1c4098-b551-11ec-b6f4-01aa75ed71a1/language-en/format-PDF/source-255964403

Weigl, D. M. (2019). Interweaving and enriching digital music collections for scholarship, performance, and enjoyment. 6th International Conference on Digital Libraries for Musicology. doi.org/10.1145/3358664.3358666.

Ziri, A. E., Bonsma, P., Bonsma, I., Iadanza, E., Maietti, F., Medici, M., Ferrari, F., & Lerones, P. M. (2019). Cultural Heritage sites holistic documentation through Semantic Web technologies. In A. Moropoulou, M. Korres, A. Georgopoulos, C. Spyrakos, & C. Mouzakis (Eds.), *Transdisciplinary Multispectral Modelling and Cooperation for the Preservation of Cultural Heritage* (pp. 347–358). Springer.

KEY TERMS AND DEFINITIONS

Holistic Documentation: Overall retrieving of critical information regarding a heritage's main attributes and characteristics, able to define it as a whole and to identify its significance and main needs. It includes morphometric survey, historical documentation, features, state of conservations, etc.

INCEPTION Core Engine (ICE): Solution, based on a semantic approach, for easily exploring BIM and HBIM models and the web and accessing other contents. It consists of a framework of software tools and a set of APIs capable of transforming each element of an IFC BIM model into semantic triples, described according to an RDF data model.

INCEPTION Platform: Semantic-based BIM platform for Cultural Heritage sites grounded on semantic web technologies using WebGL and RESTful APIs to enrich heritage 3D models by using Semantic Web standards.

Semantic Approach: Integration and connection of semantic attributes hierarchically and mutually aggregated to 3D geometric models to manage heritage information.

Semantic Triple: or RDF triple or simply triple, is the atomic data entity in the Resource Description Framework (RDF) data model. It is a set of three entities that codifies a statement about semantic data in the form of subject–predicate–object expressions.

Semantic Web: Extension of the current Web into an environment where documents are associated with information and data (metadata) specifying their semantic context in a format suitable for querying and interpretation and, more generally, automatic processing.

Chapter 15
Towards a Smart Cultural Heritage in a Post-Pandemic Era:
Enhancing Resilience Through the Implementation of Digital Technologies in Italian Heritage

Riccardo Florio
University of Naples Federico II, Italy

Raffaele Catuogno
University of Naples Federico II, Italy

Victoria Andrea Cotella
https://orcid.org/0000-0002-3160-7601
University of Naples Federico II, Italy

ABSTRACT

Preservation and dissemination of cultural heritage symbolizes a problem already present before the pandemic period and amplified during the COVID-19 crisis. As a result, the dematerialisation of architecture by digital technologies is the approach to connect Society 5.0 and architecture in cyberspace. The ambition of this chapter is to achieve an approach aimed to explain the impact of ICT during the pandemic and post-pandemic period, using HBIM technology, an essential tool for the approximation of Society 5.0 to the tangible smart heritage. On the other hand, the creation of a virtual tour breaks down architectural barriers (physical and spatial) allowing access to all users as a benefit of the dematerialisation of the asset. The work represents the use of technologies to create new knowledge and values, generating connections between people and tangible and non-tangible things.

DOI: 10.4018/978-1-6684-4854-0.ch015

INTRODUCTION

Contemporary challenges of (post)modern societies are multiple and strongly related to the significant progress of Information and Communication Technologies (ICT), the Internet of Things (IoT), Artificial Intelligence (AI), competitiveness, productivity, connectivity, and welfare. This society is moving towards a resilient approach, especially in a post-pandemic era immersed in an environment of increasing risk and uncertainty. In consideration of these statements, it is reasonable to assume that contemporary societies require a "broad" or "general" concept of resilience that is transdisciplinary and applicable to multiple contexts.

With these assumptions, after a comprehensive analysis of the state of the art, the ambition of this paper is to clarify how in a context of crisis and instability, societies look to adapt to the actual situation by using the success of ICT during the pandemic and post-pandemic period moving towards the concept of the society 5.0, explicitly or implicitly, by the government policies and standards implemented. Technologies should be used to create new knowledge and values, generating connections between people and tangible and non-tangible things. Also, ICT become an essential tool for the approximation of the society 5.0 to the tangible Smart Heritage.

The case study under analysis is The Castle of Baia, a landmark of southern Italy where the immersion of the user in cyberspace is made possible thanks to the use of Heritage Building Information Modelling (HBIM) and the creation of Virtual Tours (VT), being fundamental components for the knowledge and dissemination of the historical heritage in a period where physical tours are impossible. With this aim, an accurate methodology was established: a main digital survey of the entire building was developed, resulting in a high-resolution three-dimensional point cloud that will become the geometrical basis for the two paths to be followed.

In the first instance, the present research aims to provide an informative and geometrical integration of the HBIM model of the Cavaliere Pavilion previously developed. The main goal is to respond to the problem of fragmentation and lack of information, improving the tools provided by HBIM technology in addition to the results of the integral survey: archiving, digitization and systematization of historical data, interoperability between disciplines in the implementation of interventions, management of costs, time and resources. This final model will be mainly oriented towards the AECO (Architectural, Engineering, Construction, and Operations) sector. Also, to this integration, a library of BIM families will be developed, to standardise the complex morphology of historical architecture and be able to support heritage conservation planning and restoration and enhancement interventions.

On the other hand, the creation of a Virtual Tour, directly linked to the model, breaks down architectural barriers (physical and spatial) allowing access to all users as a benefit of the dematerialisation of the object, which however will never replace the physical experience but will allow its optimal dissemination in the cyberspace.

The preservation and dissemination of cultural heritage represents a problem already present before the pandemic period and amplified during the COVID-19 pandemic crisis. As a result, the dematerialisation of architecture achieved by digital technologies is the resource to link society 5.0 and architecture digitally, articulated in the concept of Smart Heritage. The implementation of ICTs allows the fruition of the asset by all users, not only the technical sector, even at a distance with the possibility of creating virtual tours to make the spaces accessible for everyone, anytime, anywhere (Trillo, et al., 2021). Ethically, these virtual tours should never be considered as an alternative way to experience the architecture, but simply as promotional tools to strengthen and widen the accessibility of Cultural Heritage exploring the role of digital tools during and mainly after the COVID-19 crisis finding an alternative way to enjoy it even in times of impossibility. It is crucial to highlight the importance of high-quality surveys, otherwise, an incorrect "Digital Twin" is created, generating misinformation and uncertainty in the database created for management and dissemination of Cultural Heritage.

TOWARDS A SOCIETY 5.0 AND SMART HERITAGE IN A POST-PANDEMIC ERA BY IMPLEMENTING ICT: A LITERATURE REVIEW

Nowadays, with the sophistication of ICT, the impact of globalization will spread very quickly due to the rapid and massive circulation of information, becoming Big Data a key basis of competition, underpinning new waves of productivity growth, innovation, and consumer surplus (McKinsey Report 2011). This impact puts pressure on the progress of Smart Cities (Chui, Lytras, & Visvizi, 2018), residents, and developers to constantly adjust to global trends, but various innovations continue to be created without the assistance of cultural awareness and promotion policies provided by the local government.

In a world where the propensity of industrial automation to incorporate some new production technologies is dominating, arises a global consensus on the importance of creating a super-intellectual social model as a backbone between Industry 4.0 and sustainable policies. As a reaction, in 2019, the Japanese researchers introduced the concept of Society 5.0 (Gurjanov, Zakoldaev, Shukalov, & Zharino, 2020) to create an anthropocentric society that strongly integrates cyberspace and physical space to balance economic and technological progress with the resolution of human social problems (Žižek, Mulej, & Potocnik, 2021). The concept of Society 5.0 addresses the economy and citizens, thus promoting the idea of a Smart Society, in which the ICTs will be the starting point of a super-smart society in line with the future sustainable strategies developed with the 17 UN goals and therefore of the 2030 Agenda for Sustainable Development (Israilidis, Odusanya, & Mazhar, 2019).

This UN 2030 Agenda further promotes technological innovation applied to infrastructures, encouraging the field of research into the interplay of heritage conservation and digital technologies. A further emphasis on the broader role of cultural heritage in society is demonstrated through an explicit target for heritage, Target 11.4, in the Sustainable Development Goals which commits countries to make efforts to protect and safeguard the world's cultural and natural heritage. In this context, the concept of Smart Heritage has been introduced: a relatively new system to respond to the challenge of Cultural Heritage Conservation in the Era of Smart Cities using the advances of Information and Communication Technologies (ICT). Examples of IT tools include the use of Geographic Information systems (GIS), Global Positioning System (GPS), digital camera, digital survey techniques (Florio, Catuogno, & Della Corte, 2019), (Barba, Di Filippo, Cotella, Ferreyra, & Amalfitano, 2021), virtual and augmented reality (VR and AR) (Chui, Lytras, & Visvizi, 2018), Building Information Modelling (BIM), and artificial intelligence.

In the light of these challenges, the European Commission published the "2030 Digital Compass: The European way for the Digital Decade" to help advance EU ambitions for a digital transformation by 2030. Point three identifies four cardinal points for mapping the EU's trajectory: 3.1 Digitally skilled citizens and highly skilled digital professionals; 3.2 Secure, performant and sustainable digital infrastructures; 3.3 Digital transformation of businesses; 3.4 Digitalisation of public services. According to 3.2, the Commission encourages Member States to digitise by 2030 all monuments and sites that are at risk and half of the most physically visited monuments and sites. The goal is to preserve the European

Cultural Heritage, exploiting technological knowledge, artificial intelligence, data, and extended reality (Promoter, 2021).

In the building sector, digital technologies are playing an important role in systematising the architectural and social languages of diverse local cultures and optimising information management processes. In these circumstances, the concept of Smart Heritage comes into play: a management system and method developed to respond to the challenge of Cultural Heritage Conservation in the Era of Smart Cities by using the progress of technologies.

Smart Cultural Heritage involves tangible and intangible values supporting the diversity of cultural societies and allowing holistic and multiple perspectives of the diverse infrastructures that the smart city contains. Furthermore, it considers the field of cultural heritage as a vital organism where its various dimensions have dynamic properties allowing it to cover a whole range of perspectives in the socio-cultural timeframe.

As a response to the social and economic crisis resulting from the COVID-19 pandemic, the European Commission has released the Recovery and Resilience Facility to mitigate the economic and social impact of the coronavirus pandemic and make European economies and societies more sustainable, resilient, and better prepared for the challenges and opportunities of the green and digital transitions. Additionally in a national context, the Italian government has promoted the National Recovery and Resilience Plan (NRP) where task 1 called "digitisation, innovation, competitiveness, culture, and tourism", involves the building sector for the implementation of the Recovery and Resilience Facility.

CONSERVATION AND DISSEMINATION OF CULTURAL HERITAGE: IMPACTS OF THE COVID-19 EMERGENCY

Conservation and dissemination of cultural heritage always have been a problem, also before the pandemic period and amplified during the COVID-19 emergency.

Regarding the building sector, the pandemic crisis was particularly strongly felt in Italy, where construction activities fell by 71.5%, between February and April 2020 (Figure 1). The recovery between April 2020 and January 2022 was quite strong (303.2%) – mainly as a result of the extremely low index level of April 2020 (Eurostat, 2022). One of the biggest factors behind this strong recovery was the implementation of the "Relaunch Decree" by the government in May 2020. The key measure of this decree was the so-called 'Superbonus 110' and consists of two types of interventions: the 'Super Ecobonus' stimulates energy efficiency works and the 'Super Seismabonus' encourages anti-seismic measures. This measure creates a virtuous market mechanism that offers benefits to all stakeholders: citizens can renovate their homes for free, reduce the cost of their utility bills and enhance the value of their property; companies can increase their turnover thanks to the increased volume of work; the State can make homes more efficient and safer and support increased employment and incomes.

In terms of Cultural Heritage, the buildings continued their natural degradation process and the works in progress were interrupted. Digital databases with sensors implementation would have been the optimal solution for remote monitoring and control of the asset.

On the other hand, the tourism sector could be among the sectors most damaged in Italy. The pandemic meant that tourism stakeholders were faced with a completely new and unpredictable situation, in which any certainty and knowledge gained in the past was no longer certain. (Solazzo, Maruccia, Ndou, & Del Vecchio, 2022). Beyond the economic aspect, a weak point was the incapability to visit an architectural monument in place: the lack of physical accessibility to tangible heritage, begins to produce a disconnection with the human and culture. In this perspective, the dematerialisation of Cultural Heritage adopting and exploiting smart technologies can help to minimize the aforementioned knowledge gaps while limiting the consequences of the spread of Covid-19 (Sigala, 2020). The principal results were Virtual Tours, allowing the fruition of the asset even at a distance with the possibility of creating virtual tours to make the spaces accessible for everyone, anytime, anywhere.

Figure 1. EU development of construction production January 2020-January 2022.
Source: Eurostat 2022.

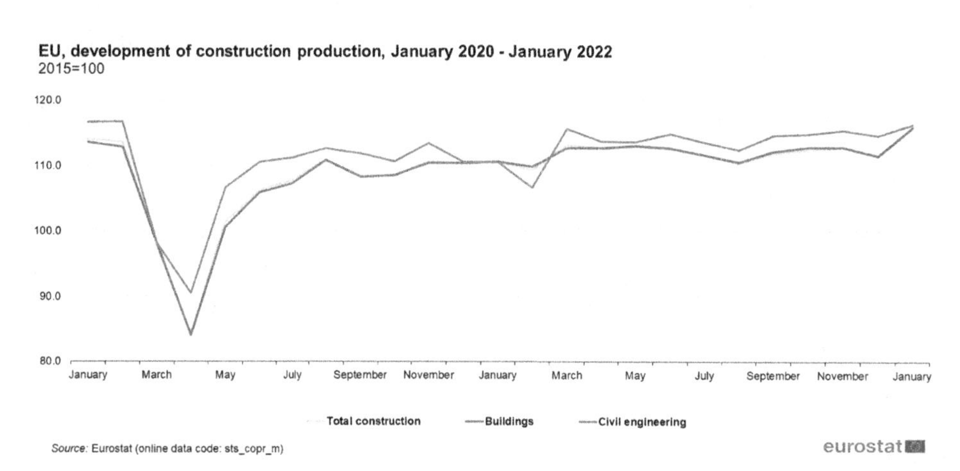

RESEARCH METHODOLOGY

The present research concerns the intensive use of digital technologies as a main tool to dematerialise the tangible Cultural Heritage. Furthermore, it is foreseen to implement innovative technologies that facilitate access to services and provide faster information exchange, respond to the needs of stakeholders (architects, engineers, archaeologists, tourists, tourism institutions and organisations, government, citizens, etc.) and to minimise the use of resources and pursue a sustainable society (Duran & Uygur, 2022).

This vision anticipates the use of ICT-based tools and other digital technologies (such as GIS, BIM, VR, AR) for data collection, processing and analysis, in order to ensure a more efficient and effective use of resources in the post-pandemic society. In particular, Big Data and Analytics techniques face unprecedented challenges and opportunities after COVID-19 to take advantage of the huge amount of data and a new data sharing model that is more sustainable, smarter and safer than the ones implemented before (Aguirre Montero & López-Sánchez, 2021).

Practically, the research study aims to highlight the interaction between digital technologies, a resilient post-pandemic society and the conservation and dissemination of tangible heritage, clarifying the dynamics between the three fields and their implications in the practice. The discussion offered in this paper is a body of knowledge accumulated as a result of an interdisciplinary work for the restoration and valorisation of the 'Padiglione Cavaliere' at Baia Castle, in the framework of the 'Programma Operativo Nazionale (PON) Cultura e Sviluppo' 2014-2020. This program contributes to the implementation of the European Union's Cohesion Policy by targeting the "development lagging regions" (Campania is one of them) and by setting as a priority the enhancement of the Cultural Heritage, a potentially decisive asset for the country's development, through interventions for the preservation of cultural buildings, the strengthening of the tourism services system and the support to the entrepreneurial chain linked to the sector.

The research methodology consisted in the graphic and informative restitution of the case study through the application of digital technologies. As a consequence, in a first step a digital survey was carried out with terrestrial laser scanning techniques (TLS) and aerial photogrammetry with drone (UAV). Then, with the processed data, a point cloud of the whole complex was obtained, which was the starting point for the subsequent restitutions (Figure 2). Among the most important outputs, the HBIM model as first pilot case for the management and monitoring of the complex is highlighted. On the other hand, the importance of developing a virtual tour was fundamental not only to use it as a visual diagnostic tool for the technical sector, but also to promote the dissemination of the heritage by allowing virtual "visits" in a period when the tourist activity had not yet awakened. Both outputs have required extensive data collection, both in terms of data acquisition and stakeholder involvement on an informative level.

Figure 2. Workflow process of Scan to BIM for Baia Castle.
Source: Univ. of Naples Federico II, Department of Architecture. Research group coordinated by Prof. R. Florio.

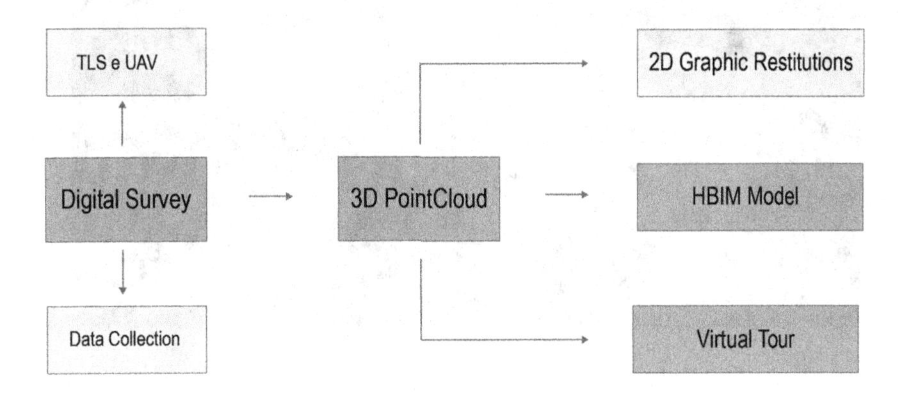

The following sections structure the methodological process described above: a brief description of the case study, the integrated digital survey and consequently the two main outputs embedded in a socio-cultural framework: HBIM Model and Virtual Tour. Finally, an analysis of the results is presented, as well as future research perspectives and a conclusion.

ENHANCING CULTURAL RESILIENCE THROUGH THE IMPLEMENTATION OF DIGITAL TECHNOLOGIES IN ITALIAN HERITAGE

Case study: the 'Padiglione Cavaliere' at castle Baia

The Baia fortress covers an area of 45,000 square meters and reaches a height of about 94 meters above sea level (Figure 3). It appears today as a set of architectural overlays built over the centuries, the most important of which are those carried out by Don Pedro Alvarez de Toledo, Don Emanuele Fonseca and Ferdinand IV. The building dominates the entire bay of Pozzuoli and represented a limit for anyone attempting to land along its coast: The most remote traces relating to the construction of the Castle of Baia date back to 1490, when, by order of King Alfonso II of Aragon, a castle was built on the remains of the Roman villa of the Caesars to defend the Phlegrean coast from Saracen incursions (Piccone, 1978).

After the unification of Italy, the castle went through a period of slow decline and inevitable decadence. Considered no longer useful for military purposes, in 1887 the castle came under the administration of various ministries: first the Navy, then the Interior, and finally the Defence. Finally in 1984 it was definitively handed over to the 'Soprintendenza Archeologica di Napoli e Caserta' to become the seat of the 'Campi Flegrei' Archaeological Museum.

Figure 3. General overview of the whole building as a result of the photogrammetric survey.
Source: Univ. of Naples Federico II, Department of Architecture. Research group coordinated by Prof. R. Florio.

Digital Survey as a Key Tool for tangible Heritage Dematerialisation

Data acquisition has always been of crucial relevance in the documentation of Cultural Heritage. In fact, architectural survey is an evolving field that has changed significantly during the past decades by the technological advancements in the field of 3D data acquisition (Porras, Carrasco, Carrasco, González-Aguilera, & Lopez Guijarro, 2021), (Balletti, Bertellini, Gottardi, & Guerra, 2021). Currently, the use of digital technologies such as LiDAR (Light Detection and Ranging) with the Terrestrial Laser Scanner (TLS), Aerial photogrammetry with the Unamend Aerial Vehicle (UAV) as main instrument, and finally the Wearable Mobile Laser System (WMLS).

The presented systems mentioned above, from a philosophical point of view, can be considered as the first step to "dematerialise" the tangible cultural heritage and bring it into cyberspace. The result of each digital survey, after a careful phase of data elaboration and processing, is a 3D point cloud: a reliable representation of the asset under study represented by an infinity of points located on the XYZ axes. Therefore, the use of three-dimensional models in the assessment and research of cultural heritage allows to improve the comparison of research results making available a tool that enables the analyses, the management and the development of data collected, also provided by several disciplines involved in the process of knowledge (Marra, 2017).

In order to carry out the digital survey of Baia Castle, the fieldwork was divided into two phases in which suitable methods of interaction between the innovative aspects of the range-based modelling method and the image-based modelling method were experimented (Figure 4). In the first phase, the terrestrial laser scanner (TLS) Faro Focus 3D x330 was used. Laser scanners can have a wide range of applications in the documentation of cultural heritage, from small objects to large complex buildings, due to real-time data acquisition on a real scale, high accuracy and high speed and the production of large number of points (Hassani, 2015).

Figure 4. Cross section of the whole complex with Crystal Clear View technology by PointCab Software.
Source: Univ. of Naples Federico II, Department of Architecture. Research group coordinated by Prof. R. Florio.

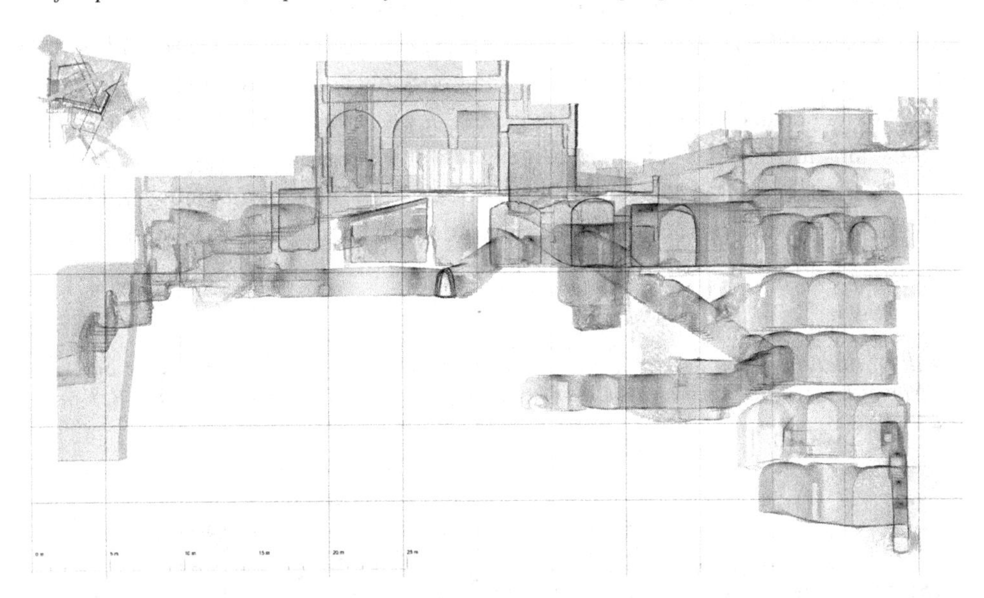

Considering the large dimensions of the complex, approximately 480 scans were taken inside and outside the building. With a resolution of 6,136 mm on 10 meters and a duration of approximately 6 minutes per each scan, the TLS survey campaign lasted a total of 47 hours.

On the other hand, regarding the image-based modelling method, an aerial photogrammetric survey was carried out. In addition to the ground photos taken with a 24.2-megapixel SLR camera, two different UAV systems were used: a DJI Phantom 4 and a DJI Mavic 2 Pro flying at a height of 18 m and 40 m respectively. To control the metric error, 16 GCP are detected on the arena floor by a Geomax Zenith 25 used in nRTK mode. The accuracy of planimetry is below 1 cm and 2.5 cm for altimetry. For the acquisition of the frames, two flight which each UAV were prepared, both automatic and with double grid: a first one for the acquisition of nadir photogrammetric images and a second one, with the optical axis tilted about 45°, to survey the vertical walls and any shadow cones. Parallel flights lines are programmed using the DJI Terra software package to have an image overlap of 60% as well as sidelap of 60%, setting the proper camera parameters (dimensions of the sensor, focal length, and flight height). The image acquisition is planned bearing in mind the project requirements - a Ground Sampling Distance

(GSD) of about 1 cm - and, at the same time, with the aim of guaranteeing a high level of automation in the following phases.

Photogrammetry consists of techniques for interpreting, measuring, and modelling the objects based on their acquired images. This method has indeed proved its efficacy in the documentation of cultural heritage and archaeological sites (Florio, Catuogno, & Della Corte, 2019), (Remondino, Barazzetti, Nex, Scaioni, & Sarazzi, 2011). Documentation and 3D modeling and surveying of historic sites and structures can be performed using low-altitude flight. In this case, with the UAVs data output it is possible to produce panoramic images, Digital Surface Model (DSM), ortho-photo, and three-dimensional models with high accuracy of the surveyed objects.

In the data processing phase, a high-level, point-based data fusion approach was chosen, in which all raw data streams are kept separate and processed independently as partial point clouds. Only at the end are the resulting point clouds merged to obtain a final 3D point cloud. In TLS Scan Registration phase, the clouds are characterized by a high degree of overlap and for this reason are registered employing a global bundle adjustment procedure, accomplished after a top view-based and cloud-to-cloud pre-registration. Given the set of scans, the algorithm searches for all the possible connections between the pairs of point clouds with overlap. For each connection, a pairwise ICP is performed and the best matching point pairs between the two scans are saved. A final non-linear minimization is run only among these matching point pairs of all the connections. The global registration error of these point pairs is minimized, having as unknown variables the scan poses (Santamaría,, Cordón, & Damas, 2011). The resulting maximum value of the RMSE on all the registration pairs is about 2.34 cm.

Figure 5. Result of the aerial photogrammetry data processing in Agisoft Metashape Software.
Source: Univ. of Naples Federico II, Department of Architecture. Research group coordinated by Prof. R. Florio.

The photogrammetric process data treatment is performed by Agisoft Metashape (Figure 5). Its workflow is based on four steps: Align Photos, Build Dense Cloud, Build Mesh and Build Texture. At the first step an algorithm evaluates the camera internal parameters (Focal Length, position of the principal point, radial and tangential distortions), the camera positions for each photo and the Sparse Cloud. In the next phase, a greater pixel number is re-projected for each aligned camera, creating the Dense Cloud. In the Build Mesh step, it is possible to generate a polygonal mesh model based on the dense cloud data. Finally, the polygon model is textured in the Build Texture step. The outputs of the photogrammetric model, necessary for further documentation studies and data integration with active sensors, are a nadir orthophoto of the entire villa and the dense point cloud (Barba, Di Filippo, Cotella, Ferreyra, & Amalfitano, 2021). The extracted point cloud has more than 72 million points.

The data collected through the techniques explained above (UAV and TLS) determine the first step to dematerialise the tangible cultural heritage (Fig.6). On the other hand, thanks to the information obtained through research and fieldwork, it is possible to enrich "cyberspace" built with the intangible values of each culture. In a post-pandemic context where the world is rising from a crisis, the aim is to help solve more precisely every social and economic problem of the population, so that Society 5.0 is constantly being optimised thanks to the use of digital technologies.

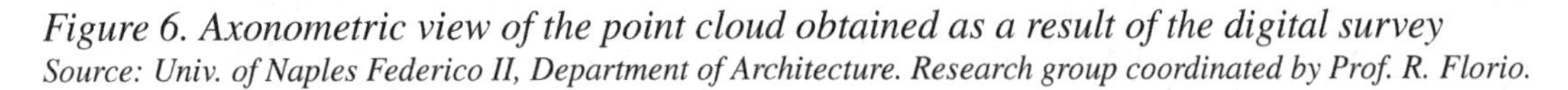

Figure 6. Axonometric view of the point cloud obtained as a result of the digital survey
Source: Univ. of Naples Federico II, Department of Architecture. Research group coordinated by Prof. R. Florio.

HBIM Systems for enhancement Cultural Heritage management

Nowadays digital technologies offer a unique opportunity in order to respond to the challenge of managing different sources of information. Research is moving towards the preparation of interdisciplinary databases that allow the documentation produced throughout the life of the Cultural Heritage asset to be structured in such a way that future interventions can be properly prepared. In facts, they offer the possibility to integrate multiple layers of information and to link across industry, community and higher education with a flexibility and timeliness that traditional techniques such as paper-based drawings did not allow at such an extent (Udeaja, et al., 2020).

Building Information Modelling (BIM) offers the opportunity to link a variety of information concerning heritage assets and convey them across multi-disciplinary professionals (Pocobelli, Boehm, Bryan, Still, & Grau-Bové, 2018). In this context, the international scientific community has been dealing for some years with Heritage Building Information Modelling (HBIM): the application of BIM systems to historic buildings (Murphy, McGovern, & Pavia, 2011), investigating and experimenting with approaches suitable for the digitisation of the existing heritage. The most recent experiences of HBIM are addressing with particular attention the issue of Facility Management, to effectively guarantee permanent maintenance of the building (Hull & Ewart, 2020), (Patacas, Dawood, & Kassem, 2020).

It has been demonstrated that HBIM can improve the efficiency of heritage information management (Parisi, Turco, & Giovanni, 2019) as it allows the geometric, semantic and documentary information on heritage properties from all disciplines involved to be centralised in a common repository and facilitates collaborative work and the coordinated exchange of information among multidisciplinary teams (Hawas & Marzouk, 2017).

However, regarding the geometrical aspect, BIM of existing or historic buildings remains an ongoing field of research and also a challenge, due to the complexity of geometric data and its interpretation within a strongly standardized BIM software as Autodesk Revit or Graphisoft Archicad, that allows 3D modelling based on predefined architectural families or very simplified objects. A first solution could be to work with NURBS (NonUniform Rational B-Splines) in specifics softwares like Rhinoceros for example (Diara & Rinaudo, 2019). The problem is that most of these programs are not parametric and therefore does not allow to modify the construction parameters associated to a 3D object, although these operations are possible by externally editing the objects or by implementing several plug-ins but slowing down the modelling process.

The HBIM process includes several steps to get the maximum benefit from BIM technology, in addition to the geometric information obtained from the digital survey, manual data import is still essential to include additional information that can help architects, engineers and restorers. The physical characteristics of materials is one of the main requirements for restoration and maintenance work. Such data can be generated through specific surveys and added to the HBIM model. While geometry can be derived from the point cloud, other data, such as thermal characteristics of materials or strength, can be generated and added manually (Figure 7).

Figure 7. Workflow process of HBIM model creation.
Source: Univ. of Naples Federico II, Department of Architecture. Research group coordinated by Prof. R. Florio.

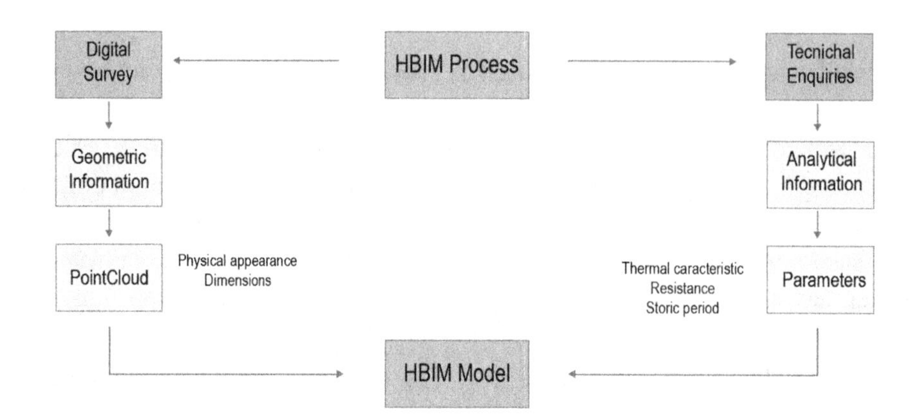

In the present case study, the methodology chosen is the Scan-to-BIM approach, a reverse modelling technique that uses digital sensing technologies to obtain point clouds that become the basis for BIM modelling. On this basis, there are several approaches based on Scan-to-BIM applied to Italian Heritage: (Spettu, Teruggi, Canali, Achille, & Fassi, 2021) conducted a preliminary test on the possibility of using the digital tools belonging to the manufacturing industry field for the digitisation of the Milan Cathedral; (Barazzetti, Brumana, & della Torre, 2018) presents a workflow for finite elements model generation from point clouds; (Bruno, et al., 2020) studies the requirements of the CH for the realisation of a historical BIM-GIS system, and (Banfi, Bolognesi, Bonini, & Mandelli, 2021) intend to support the transmissibility of the historical and cultural background by developing Virtual Visual Storytelling (VVS) with an emphasis on data acquisition phase.

In detail a four-step workflow is applied: decision-making, data acquisition, integration and analysis, redesign. Starting with the decision-making aspect, it is important to remember that the development of this BIM model is not linked to specific purposes or contingencies, which may be specific restoration works, structural consolidation, energy analysis or others. It is therefore a challenge to generate a first pilot model that, precisely because of its generality, is flexible and can become the container for all the heterogeneous information collected up to now, thus demonstrating the basis for the development of Smart Heritage to realise the sense of belonging to the community. Regarding to the Italian UNI 11337-4 regulation (Pavan, Mirachi, & Giani, 2017) a Level of Development (LOD) for the project should be defined at this point, a topic that will be discussed in more detail in the following sections as well as the Level of Accuracy (LOA) (Graham, Chow, & Fai, 2018).

About data acquisition and integration, details have already been provided in precedent sections. Turning to the analysis, automatic data segmentation is not employed in the proposed application. The structure investigated, characterised by numerous stratifications and modifications over the centuries, required a manual cataloguing of information, relying on the skills of qualified personnel capable of differentiating the elements studied by interweaving numerous parameters, which are difficult to manage simultaneously by an algorithm.

The result is a first 3D unique geometrical dataset aimed to optimise decision making on the use and management by automatically sharing several data with building managers in real time and adopting more participatory and collaborative approaches, making cultural data freely accessible and, consequently, increasing opportunities for interpretation, digital curation and innovation. The use of Building Information Modelling allows an accurate management of both the timing of implementation (evaluating any discrepancies and potential anomalies with respect to what has been estimated) and the economic aspect (based on the results of analysis and comparisons between different possible project scenarios). These evaluations, applied to the field of the built heritage, make BIM technology suitable to Cultural Heritage for evaluating the different hypotheses at the time of implementation.

It is becoming necessary in the construction industry to achieve higher standards of efficiency, which means increased productivity, quality, reduced timeframes and greater cost-effectiveness with less impact on resource consumption. Based on these assumptions, the digital reconstruction of the Cavaliere Pavilion is the starting point of a long journey; the stratified urban fabric of the Phlegraean Fields, with its history and vibrant heritage, positions the Castle of Baia as an ideal reference for sustainable management of Smart Heritage conservation.

Virtual Tours' Role in Tourism Recovery Post COVID-19 emergency

Virtual tours (VT) can play a crucial role in preventing the total suspension of many tourism activities during emergency periods by creating new business models and providing various opportunities for different entities in the tourism ecosystem. Examples of these opportunities include enabling tourists to experience and learn about different sites and destinations during times of travel restrictions and bans while staying safe at home, enabling museums and different touristic sites to remain engaged with their public and providing job opportunities for employees such as tour guides through the provision of Virtual-based tour guiding (El-Said & Aziz, 2022).

Some studies even suggest that virtual tourism during the pandemic has become a source of revenue for various attractions and these remote visits also have a positive impact on the physiological well-being of the visitors (Itani & Hollebeek, 2021). On the other hand, regarding the issues related to the praxis of conservation, the use of the VT related to condition assessment, is currently being developed by a variety of studies and applications for tangible heritage (Lee, Kim, Ahn, & Woo, 2019), (Trizio, et al., 2019) and refers to the remote collection and exchange between different stakeholders - technical, authorities, operators - of data related to deterioration patterns and anomalies/performance defects of building components, in order to schedule monitoring, maintenance and repair activities. To this purpose, several solutions are proposed, including: virtual tours of 360° panoramic scenes; 3D reality-based models from terrestrial laser scanners (TLS) and/or digital photogrammetry surveys; and 3D computer-based models from Computer-Aided Design (CAD), Historical Building Information Modelling (HBIM) and digital rendering graphics as videogames or cinema 4D (De Fino, Bruno, & Fatiguso, 2022).

In order to develop the Virtual Tour of the Cavaliere Pavilion the first step was to select the 360° spherical images produced for the laser (Faro Focus X330), in coincidence with the location where the scanning was performed, making visible both the interior and the exterior of the asset. The software used was SCENE2Go, an extension from FARO, where in a previous step the point cloud processing was performed.

SCENE 2go is a facility to share read-only scanning projects with the stakeholders involved. It consists of an application for (Windows of macOS) and a viewer: the application is installed together with the main software and is used to transfer the viewer and the scanning projects to the respective storage media. The viewer is portable and does not require an internet connection for most functions. Figure 8 represents the general menu showing a roof plan, where each selected scan is represented by a coloured dot with the name and number. Each point includes 3 options: a panoramic view, a 3D view and the respective properties.

ANALYSIS OF THE RESULTS

As explained in the previous sections, the methodological process in the field was structured in three phases taking as a pilot case the Cavaliere Pavilion at Baia Castle. First, an integrated digital survey phase applying TLS and UAV methods was carried out. This was the main step to start with the process of "dematerialising" the architecture. Then, a second phase of data processing in which the point cloud of the asset is obtained as a result. Finally, in a third step, the point cloud is taken as a reference to carry out the production of the main outputs: the 3D Virtual Tour and the HBIM Model.

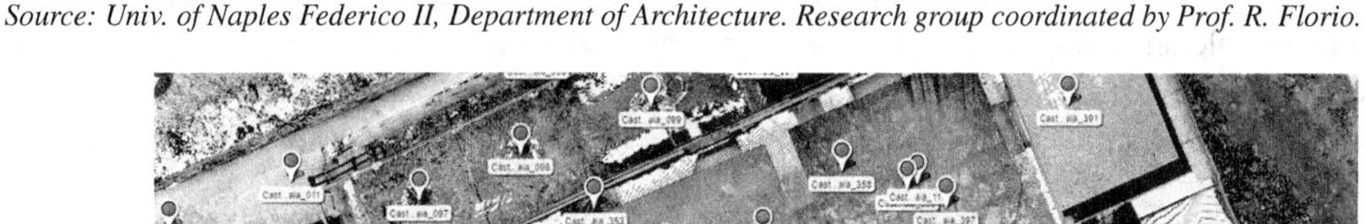

Figure 8. Superior view with the different spots of the virtual tour in Scene2Go Software.
Source: Univ. of Naples Federico II, Department of Architecture. Research group coordinated by Prof. R. Florio.

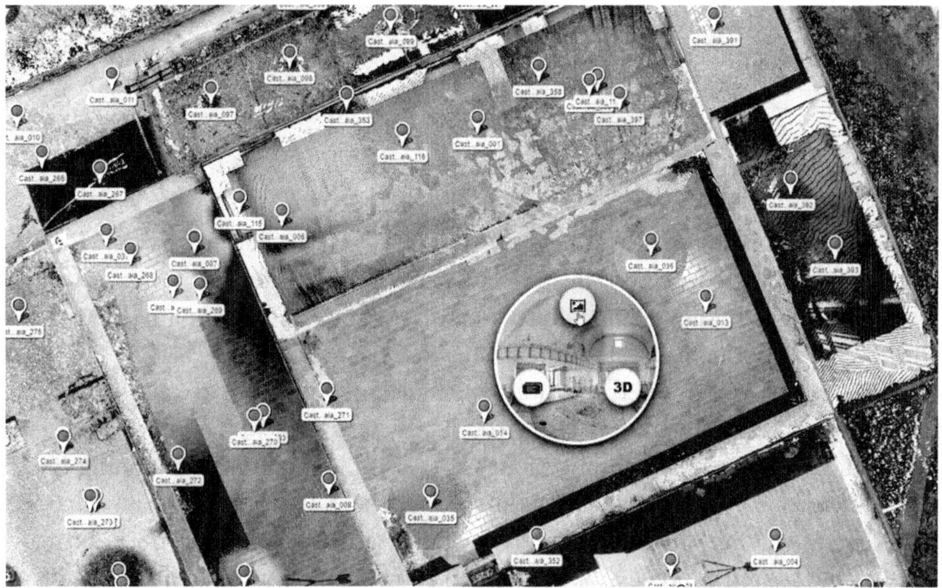

In this section, we proceed to analyse the survey quality and the two main outputs mentioned in the previous paragraph, finally focusing on the sharing of the data produced.

ANALYSIS OF GEOMETRICAL AND SURVEY ACCURACY

To ensure the reusability of survey and BIM data, it would be advisable to provide an indicator of the quality of the results. The definition of the required Level of Accuracy should be related to the objective. For the present case study, reference is made to the framework proposed by the USIBD (U.S Institute of Building Documentation), which is structured in two parts. The first part is called Measured Accuracy and the second part is called Represented Accuracy (Graham, Chow, & Fai, 2018). The Level of Accuracy (LOA) is structured in five incremental intervals from 0 to 5 cm, plus one that can be user-defined according to the project. Each of these can be applied within the same project because the LOA can be applied to individual building elements and not necessarily to the whole project. The framework is intended to be flexible enough to work on both small and large projects. It is also designed to assist the specifier by offering suggestions for the most used LOA, differentiated for recent and historic buildings. In the former case, wider tolerances are accepted, in the latter more limited.

The LOA intervals are used to perform an in-process check of the modelled parts. Using the As-Built plug-in for Revit, able to calculate the distance between objects and the reference cloud, a visualization profile is structured where a specific colour corresponds to a certain LOA. This profile is then applied to the model surface with a 5 cm sampling mesh. This approach allows to promptly correct any inaccuracies and to meet the pre-set accuracy, corresponding to LOA 20 for this case study. As a result, a maximum deviation of 7.6 cm was obtained in the most geometrically complex areas.

Concerning the accuracy of the survey, two different aspects can be identified: For photogrammetry, a statistical upper tolerance range of (0 cm, 4.94 cm) is obtained for the residuals at the control points, compatible with a LOA 20 (1.5 cm - 5cm). For the TLS cloud, the upper range is (0 cm - 2.34 cm), which is also compatible with an LOA 20.

HBIM MODEL FOR DATA STORAGE

As mentioned above, a crucial step in the decision-making phase is the definition of the LOD, in this case according to Italian standards. However, the first difficulties arise. Standard levels are conceived with reference to a forward engineering methodology, where geometric and informative contents increase as one moves from the idea to the real system (Brumana, Stanga, & Banfi, 2022).

Based on these observations, referring to an existing building surveyed and then modelled, one could be led to attribute the product to a LOD F (built structure), where the digital objects express the verified virtualisation of the single building elements (As-Built situation), containing the trace of the management, maintenance, repairs, carried out during the whole life cycle of the building. If this direct correspondence can be valid for the geometrical aspect, the same is not true for the information content, which depends on the cognitive process of acquisition and elaboration of the information according to one's own level of behaviour. A solution to the problem could be to decouple the two aspects, identified as Level of Geometry (LOG) and Level of Information (LOI) which, however, cannot be treated separately.

A main problem concerns geometry: the geometric and compositional complexity of Cultural Heritage, together with the phenomena of ageing and deterioration that can alter the material, make modelling particularly complex, especially if reference is made to the pre-modelled objects present in commercial software, designed for standardised elements typical of new buildings.

Returning to the case study, based on the above considerations, we could indicatively attribute it a LOD C, while reiterating the need to specifically define the contents necessary to describe the heritage and differentiating the geometric and material characterisation of the surfaces, attributable to a LOD F.

The result of the applied methodology is a digital volumetric model (Figure 9) developed in Autodesk Revit 2020, where the object is decomposed into sub-elements described by several quantitative and qualitative parameters. This model then becomes the database of all existing information for the Castle of Baia in the BIM environment, becoming a digital support for future conservation, restoration and dissemination of the archaeological heritage.

The identification of the families necessary for the project, such as walls, ceilings, doors, etc., is an operation that precedes the actual modelling. As this is an architecture of considerable antiquity, this identification process is not very complex: the architectural asset is made of walls and vaulted systems of original stone, excluding the presence of particular elements such as doors, windows, stratigraphy in floors and ceilings, etc. Only in the sectors where interventions were carried out or subsequently built (e.g., the upper area of the museum) are there a greater level of detail.

These historical phases and constructive interventions find their expression in the BIM, configured according to an increasing chronological order. In addition, a brief description of each is added for better understanding and differentiation. This information is associated with visualisation filters that allow easier management of the model by means of the chronological selection of the elements that compose it.

Figure 9. Axonometric view with the intervention areas of the executive project highlighted in blue.
Source: Univ. of Naples Federico II, Department of Architecture. Research group coordinated by Prof. R. Florio.

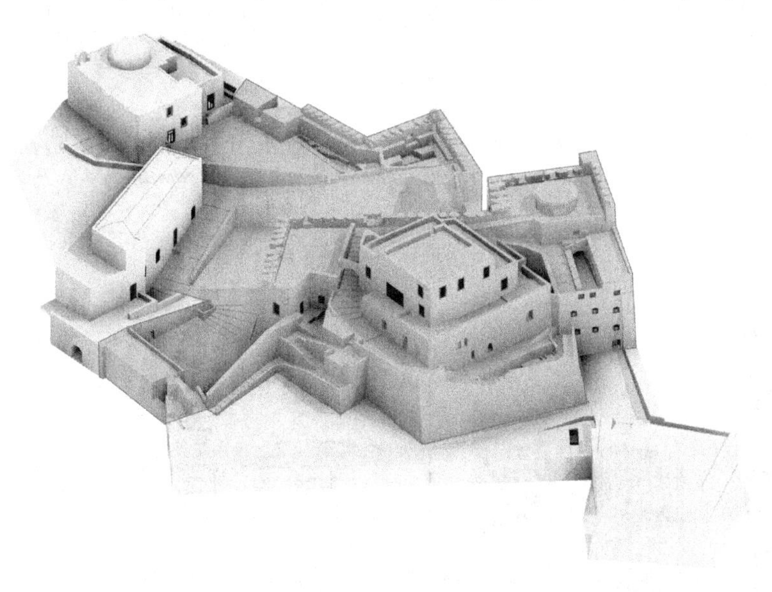

A library of BIM families (Figure 10) has been developed with the aim of standardising the complex morphology of historical architecture and supporting heritage conservation planning, restoration and enhancement. The main step was to organise the architectural elements and sub-elements to model, the method described in the UNI 10838:1999 and UNI 8290:1981 standards was used, where the building is divided into classes of technological elements (foundations, vertical structures, horizontal structures, etc.) and sub-classes (plinth, column, masonry, etc.). After this first classification, according to the amount of information coming from external sources (structural tests, historical sources) to be used as a parameter, the different architectural elements could vary between LOI C and LOI D. while on a geometrical level, the standardisation remains stable at LOD C.

Modelling parametric smart objects is one of the essential components of BIM since this consist of modelling information in the form of objects; it is, therefore, necessary to assign to all the objects constituting a particular geometric model of a building all the detailed information like the materials used, the constructive process, historical details and any other kind of information that can be useful to create accessible information management, any intervention or just maintenance. This type of information (photos, texts, state details) cannot be coded as Revit project parameters and are highly necessary in the field of heritage. For this reason, the possible mechanisms of connection between the graphic entities that constituted the model, together with their attributes, and the alphanumeric information included in an external database and coming from historical, cultural, constructive and structural notes, have been studied.

The solution has been found in the creation of Revit Shared Parameters, which can be used in multiple families or projects and by multiple entities: Revit creates a .txt file independent of any Revit family or project file, allowing us to access the file from different families or projects and even different devices. Figure 10 represents the parameters used, among which the following are highlighted: Construction Period; First, Second and Third period of renovation (which allows to visualise the model historically thanks to Revit filters and understand the different interventions in the building); Current Use, Sector Name (in which the family is located), Family Name and Number, among others. In this way, it is possible to store into one place all the intangible information related to that particular family and the building to which it belongs.

Figure 10. View of the loadable family of the museum main door with all parameters defined.
Source: Univ. of Naples Federico II, Department of Architecture. Research group coordinated by Prof. R. Florio.

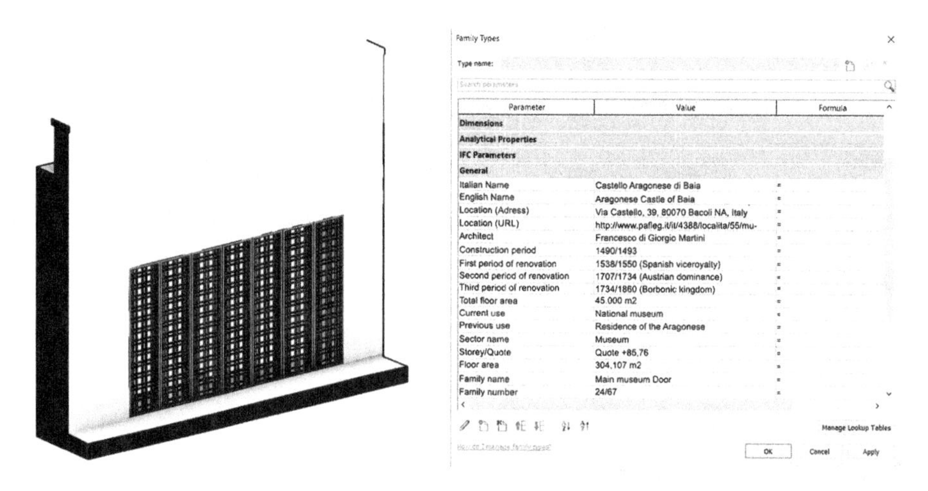

In addition, after the geometrical and therefore visual survey, another absolutely important information parameter was the state of conservation. This parameter has been created to be applied to heritage maintenance and conservation work, especially in the field of restoration. For this purpose, the RICS (Royal Institution of Chartered Surveyors of the United Kingdom) condition classification was taken as a reference scale, by means of which the element was evaluated and assigned a condition classification ranging from 3 to 1, where 1 means need no repairs and has no area of concern, level 2 highlights non-urgent areas with defect that need repairing, and finally level 3 highlights defects that are in need of urgent or series repairs. As a result of the general assessment, a majority of elements can be considered to be in state 2 where interventions are not urgent. this allows restorers, structuralists and architects to carry out the appropriate analyses.

VIRTUAL TOUR FOR A NON-IMMERSIVE VIRTUAL REALITY

The VT development allows the visualisation and querying of data and the virtual use of the space identified, through navigation in a digital space made it by Panoramic Views and Overview Maps (Figure 11).

From a technological point of view, VT is built on the basis of the Virtual Reality. VR consists of a series of digital tools capable of obtaining information about the user's actions (input tools), which are integrated and updated in real time by the computer to build a cyberspace. In the present case of study, a Non-Immersive VR was developed: this type is determined by a monitor that acts as a "window" through which the user sees the world in 3D; interaction with the virtual world can be done through the mouse, the joystick or other peripherals such as gloves (Banfi & Previtali, 2021).

The tour was developed according to the guided exploration logic; therefore, the user can visit the building thanks to a teleportation system that can be activated by selecting special hotspots. During the visit, it is possible to interact with the objects of the scenes activating the multimedia contents (audio, texts and images), some of these can be also downloaded, such as the data forms that can be acquired in pdf format. Furthermore, the tour is able to be used in offline mode.

In addition, the user can view a menu with additional technical contents: to verification actions and control procedures, it is also possible to carry out measurement operations, making an archive of information available for subsequent analysis. Therefore, it is possible to access the results of the research: the architectural survey, the 2D designs developed, the analysis on the conservation state, and also a link to the Autodesk platform to consult the BIM model in order to get additional information.

Regarding the display options, one of the menu items allows to view digital models, both the point cloud and the photogrammetric model with photorealistic traces.

Figure 11. Screenshot of the virtual tour in the museum sector, highlighting the spots for moving around the building and the project options on the right side.
Source: Univ. of Naples Federico II, Department of Architecture. Research group coordinated by Prof. R. Florio.

DATA SHARING

Concerning the data sharing, a first issue to face is the lack of compatibility of the virtual tour platform. In order to visualise it, first of all the user needs to download the SCENE2go application and the virtual tour created is shared as any file to be opened from the desktop. This type of data sharing prevents public access by all types of users, limiting it only to the technical sector that owns the file. The problem arises when the dimensions of the project increase (mainly due to the number of scans, quality of the images, etc.). In addition, a large file can only be used and processed by high-level computers and is not accessible to the whole community. A viable solution is to provide an online platform where the Virtual Tour can be uploaded, making it available online and accessible to all, avoiding heavy transfers of large files on computers accessible only to a sector of the community.

On the other hand, regarding the HBIM Model this was uploaded to the Autodesk Construction Cloud. This online platform allows to share the geometrical and informative data with all the professionals that could be involved in a future management process, allowing an interdisciplinary workflow. In this case, it was considered a future intervention of professionals such as architects, engineers and archaeologists who would have to consult the database at different times and stages of progress of the work, so a single database updated in real time is crucial.

From an operational point of view, it should be considered that territorial entities in charge of heritage assets, such as local governments, tourism associations and private stakeholders, could greatly benefit from a coherent and comprehensive compilation of their cultural, architectural, technical and functional characteristics, in order to address compatible and effective protection and enhancement measures.

FUTURE RESEARCH DIRECTIONS

Moving towards Smart City by an Open platform for Data-Sharing and Management

Currently, there is a large number of studies on the development of collaborative BIM data platforms that allows to ensure the real time access and control of IFC files on a 3D viewer. In a first instance (Logothetis, Karachaliou, Valari, & Styli, 2018) proposed a Cloud-based system for providing a web-based service for managing BIM combining different technologies such as Cloud computing technology with the Nextcloud software, 3D display technology with BIMserver equipped by plugins "BIM Surfer" and BIMviews. These kinds of platforms are free web applications or Software-as-a-Service (SaaS) solutions, especially designed for multiple device access to a common cloud system. Although the actual panorama of cloud-BIM is strongly affected by the presence of fee-based solutions (e.g., Autodesk Construction Cloud), the number of free and open-source web applications are growing, also thanks to the increase of WebGL apps and the success of the ICT especially Artificial Intelligence (AI) and Programming Languages (PL) such as JavaScript and Phyton.

Consistent solutions include BIMServer.center (https://bimserver.center/en/) a system that, besides managing, sharing and updating projects in the cloud, allows one to access the BIM project in Virtual Reality (VR) and Augmented Reality (AR) environments, unlocking a new level of accessibility and immersion; and BIMServer (https://github.com/opensourceBIM/BIMserver) an open-source web platform (via browsers) reachable through a JavaScript server app that allows the management and sharing of BIM projects.

(Diara & Rinaudo, 2021) develops ARK-BIM Platform (https://ark-bim.github.io/) with the guidelines of BIMData (https://bimdata.io/). This platform is principally based on VueJS (design system), JavaScript and XEOKIT representing the perfect match between features, design stability, a user-friendly interface and source code accessibility. This web platform (SaaS license) is probably the most interesting and flexible cloud solution for managing BIM projects in a smart environment. It includes three suites (free, professional, enterprise) where the differences principally concern cloud storage, the IFC editor, email support and Application Programming Interface (API) access.

With these assumptions, future perspectives for research will include the development and implementation of a digital platform dedicated to the study and dissemination of Cultural Heritage that will fully implement the potential of the Digital Twin to prepare 3D-based models of CH tangible assets, accessible, interoperable and dynamic. The main goal is to become an intangible infrastructure of the historical, architectural and archaeological heritage that improves the usability of culture and promotes the tourism by launching processes of regeneration of historical contexts and territorial areas, also in order to increase the levels of cultural and tourist activity in the country after the pandemic period. The platform will be able, through geospatial modelling, to generate HBIM models containing information on urban inclusivity. This information can be used for the automatic construction of traditional and inclusive tourist routes. Involvement of the general public in the construction of memory repositories on digital technologies and cultural heritage in general through various communication channels.

Essentially, the purpose is to connect the Italian cultural heritage by starting to development an integrated network of assets that, both physical and virtual, increasing over time the number of entities and relationships identified in the pilot case The Baia Castle, and prospecting its structure towards a more inclusive "virtual museum" for the South of Italy. The main step is to start connecting the historical assets of the 'Campi Flegrei' area.

A detailed 3D geospatial model (thanks to the GIS system) integrated into the platform that can be adapted for various purposes to benefit both citizens and the government sector is a key target. This 3D data repository could be made available to visitors to the city, via a web interface, to facilitate virtual tours and tourism. This would benefit tourism, increase the visibility of some of the museums and enhance the value of areas of archaeological and architectural interest that are disconnected from the city center and not easily accessible. It would also fulfil the UNESCO mandate to improve access to areas and assets of important cultural heritage.

ENHANCING IMMERSION BY USING VR

In order to improve the accessibility and reach of the virtual tour, a future line of research could focus on the implementation of Immersive Virtual Reality (IVR) to the VT previously generated from 360° images for the Castle of Baia in order to increase the levels of interactivity between the user and the asset. The incorporation of technological devices capable of displaying digital objects in an immersive environment with a 3D depth perspective, such as headsets, visualization, movement and tactile devices (VR helmets, gloves and sensory trackers) that isolate the subject's perceptive channels, fully immersing them, on a sensory level, in the virtual experience they are going to carry out, will be considered.

It is crucial to have the support and incentive of both the government and private associations that support the use of new devices, especially in public educational institutions and universities in order to start generating a sense of community between the students. In addition to educational centers, it is essential to generate accessibility for users who cannot reach the physical visit and the use of Virtual Reality is the only option.

CONCLUSION

The emerging Society 5.0 requires the development of highly elaborated systems that offer all participants in society the optimal solutions in which the human being is at the focus of the transformations, together with technological development and sustainability. Society 5.0 is based on related advanced technologies that merge physical spaces with cyberspace. This contributes to the creation of structures that can solve the social problems of the population with the resources of the digital field.

In this framework, also the recent COVID-19 pandemic has shown how the possibility of visiting a complex (archaeological, architectural, museum) remotely, without being able to replace an in-place visit, constitutes a valid reference for direct experience. But it has also shown the need to reshape and rethink virtual cultural experiences in an innovative way, in order to enhance their experiential values and meanings, instead of proposing them as a mere substitute for the physical reality of the assets. Moreover, the connection between pandemics and tourism are the most important elements to a deeper understanding of health safety and global transformation.

In a first instance, HBIM models integrated in a visualisation and management digital platform can combine historical, cultural and technical data that can be used both for the general public and for the technical one. In the technical field, the HBIM as a data repository register and store all the geometric and non-geometric data as well as the different interventions that have been done along the heritage building life. These data are essential for future works of heritage preservation, conservation and management.

For this reason, the case study of the Castel of Baia made it possible to define a digital method capable of going beyond the Scan-to-BIM process, HBIM models and Virtual Tours. Thanks to the 3D survey, aerial photogrammetry and laser scanner data acquisition, it was possible to set the basis for the generation of a digital representation with a high level of detail. In addition, the quality of the survey was essential for the development of the Virtual Tour where the 3D mapping and texturing of the objects were enhanced by the post-production of orthophotos and 360 images, capable of expressing the material characteristics of the architectural and structural elements surveyed.

On the other hand, the impacts of technology can also be found in the quality of life of the citizen. Low-income people and people who live in isolated and with low population density areas, still face different opportunities to preserve the local culture and develop competences that permit them to have a more sustainable life. So, HBIM was developed to explore different kinds of social engagement by the effective participation of people in the CH management, conservation by smart and participative solutions for a dynamic and sustainable fruition of Cultural Heritage.

Smart Heritage has potential applications for enhancing the valorisation, knowledge diffusion, fruition, and growth of a positive and engaging heritage experience through active interaction with museum visitors. Institutions play a key role in leveraging technology by acting as the backbone among the interplay between digital technologies, a resilient post-pandemic society and the preservation and dissemination of Cultural Heritage.

An essential step towards the achievement of Society 5.0 is social dialogue, the application of technologies and participation of all levels of society. The involvement of stakeholders (government, tourism bodies, private companies) is essential to achieve the enhancement of data processing technologies in order to achieve sustainable development goals. Another fundamental point to highlight is the importance of the participation of the state and private organizations to apply policies and standards of dissemination and training for the use of technologies and digital tools by marginalized communities to bring them closer to their own culture under the technological challenges of nowadays.

REFERENCES

Aguirre Montero, A., & López-Sánchez, J. (2021). Intersection of Data Science and Smart Destinations: A Systematic Review. *Frontiers in Psychology, 12*, 12. doi:10.3389/fpsyg.2021.712610 PMID:34393952

Balletti, C., Bertellini, B., Gottardi, C., & Guerra, F. (2021). Geomatics techniques for the enhancement and preservation of Cultural Heritage. *The International Archives of the Photogrammetry, Remote Sensing and Spatial Information Sciences, 42*, 133–140.

Banfi, F., Bolognesi, C., Bonini, J., & Mandelli, A. (2021). The virtual historical reconstruction of the cerchia dei navigli of Milan: From historical archives, 3D survey and HBIM to the virtual visual storytelling. *The International Archives of the Photogrammetry, Remote Sensing and Spatial Information Sciences, XLVI-M-1-2021*, 39–46. doi:10.5194/isprs-archives-XLVI-M-1-2021-39-2021

Banfi, F., & Previtali, M. (2021). Human–Computer Interaction Based on Scan-to-BIM Models,Digital Photogrammetry, Visual Programming Language andeXtended Reality (XR). *Applied Sciences (Basel, Switzerland), 11*(13), 6109. doi:10.3390/app11136109

Barazzetti, L., Brumana, R., della Torre, S., Gusmeroli, G., & Schiantarelli, G. (2018). Point clouds turned into finite elements: The umbrella vault of Castel Masegra. *IOP Conference Series. Materials Science and Engineering, 364*, 364. doi:10.1088/1757-899X/364/1/012087

Barba, S., Di Filippo, A., Cotella, V., Ferreyra, C., & Amalfitano, S. (2021). A SLAM Integrated Approach for Digital Heritage Documentation. *Culture and Computing. Interactive Cultural Heritage and Arts. HCII, 2021*, 27–39.

BIMserver. (n.d.). *BIMserver*. Retrieved from BIMserver: https://github.com/opensourceBIM/BIMserver

BIMServer.Center. (n.d.). *BIMServer.Center*. Retrieved from BIMServer.Center: https://bimserver.center/en

Brumana, R., Stanga, C., & Banfi, F. (2022). Models and scales for quality control: Toward the definition of specifications (GOA-LOG) for the generation and re-use of HBIM object libraries in a Common Data Environment. *Applied Geomatics, 14*(S1), 151–179. doi:10.100712518-020-00351-2

Bruno, N., Rechichi, F., Achille, C., Zerbi, A., Roncella, R., & Fassi, F. (2020). Integration of historical GIS data in a HBIM system. *Int. Arch. Photogramm. Remote Sens. Spatial Inf. Sci.*, 427–434.

Chui, K., Lytras, M., & Visvizi, A. (2018). Energy Sustainability in Smart Cities: Artificial Intelligence, Smart Monitoring, andOptimization of Energy Consumption. *Energies, 11*(11), 2869. doi:10.3390/en11112869

De Fino, M., Bruno, S., & Fatiguso, F. (2022). Dissemination, assessment and managementof historic buildings by thematic virtual tours and 3D models. *Virtual Archaeology Review, 13*(26), 88–102. doi:10.4995/var.2022.15426

Diara, F., & Rinaudo, F. (2019). From reality to parametric models of Cultural Heritage assets for HBIM. *The International Archives of the Photogrammetry, Remote Sensing and Spatial Information Sciences, XLII-2*(W15), 413–419. doi:10.5194/isprs-archives-XLII-2-W15-413-2019

Diara, F., & Rinaudo, F. (2021). ARK-BIM: Open-Source Cloud-Based HBIM Platform. *Applied Sciences (Basel, Switzerland)*, *11*(18), 8770. doi:10.3390/app11188770

Duran, G., & Uygur, S. (2022). A Comprehensive Systematic Literature Review About Smartness in Tourism. In Handbook of Research on Digital Communications, Internet of Things, and the Future of Cultural Tourism. IGI Global. doi:10.4018/978-1-7998-8528-3.ch011

El-Said, O., & Aziz, H. (2022). Virtual Tours a Means to an End: An Analysis of Virtual Tours' Role in Tourism Recovery Post COVID-19. *Journal of Travel Research*, *61*(3), 528–548. doi:10.1177/0047287521997567

Eurostat. (2022). *Impact of Covid-19 crisis on construction*. Obtenido de https://ec.europa.eu/eurostat/statistics-explained/index.php?title=Impact_of_Covid-19_crisis_on_construction

Florio, R., Catuogno, R., & Della Corte, T. (2019). Integrated methodologies for the knowledge and regeneration of the Paestum site. The role of the nature between the temples and the sea. *Sustainable Mediterranean Construction*, 93-101.

Florio, R., Catuogno, R., & Della Corte, T. (2019). The interaction of knowledge as though field experimentation of the integrated survey. The case of Sacristy of Francesco Solimena in the church of San Paolo Maggiore in Naples. *SCIRES. SCIentific RESearch and Information Technology.*, *9*(2), 69–84.

Graham, K., Chow, L., & Fai, S. (2018). Level of detail, information and accuracy in Building Information Modelling of existing and heritage buildings. *Journal of Cultural Heritage Management and Sustainable Development*, *8*(4), 495–507. doi:10.1108/JCHMSD-09-2018-0067

Gurjanov, A., Zakoldaev, A., Shukalov, A., & Zharino, I. (2020). The smart city technology in the super-intellectual Society 5.0. *Journal of Physics: Conference Series*, *1679*(3), 032029. doi:10.1088/1742-6596/1679/3/032029

Hassani, F. (2015). Documentation of Cultural Heritage. Techniques, potentials and constraints. *The International Archives of the Photogrammetry, Remote Sensing and Spatial Information Sciences*, *40*(5), 207–214. doi:10.5194/isprsarchives-XL-5-W7-207-2015

Hawas, S., & Marzouk, M. (2017). In Y. Arayaci, J. Counsell, L. Mahdjoubi, G. Nagy, & K. Dewidar (Eds.), *Integrating Value Map with Building Information Modelling Approach for Documenting Historic Buildings in Egypt* (pp. 62–72). Heritage Building Information Modelling.

Hull, J., & Ewart, I. (2020). Conservation data parameters for BIM-enabled heritage asset management. *Automation in Construction, 119.*

Israilidis, J., Odusanya, K., & Mazhar, M. (2019). Exploring knowledge management perspectives in smart city research: A review and future research agenda. *International Journal of Information Management.*

Itani, O., & Hollebeek, L. (2021). Light at the end of the tunnel: Visitors' virtual reality (versus in-person) attraction site tour-related behavioral intentions during and post-COVID-19. *Tourism Management*, 84.

Lee, J., Kim, J., Ahn, J., & Woo, W. (2019). Context-aware risk management for Architectural Heritage using Historic Building Information Modeling and Virtual Reality. *Journal of Cultural Heritage,* 242–252.

Logothetis, S., Karachaliou, E., Valari, E., & Styli, E. (2018). Open source cloud-based technologies for BIM. *The International Archives of the Photogrammetry, Remote Sensing and Spatial Information Sciences, 42*, 607–614. doi:10.5194/isprs-archives-XLII-2-607-2018

Marra, A. (2017). Il complesso monumentale di Santa Chiara a Napoli: un modello innovativo per la conoscenza e la. *Conoscere, conservare, valorizzare. Il patrimonio religioso culturale, 3*, 141-146.

Murphy, M., McGovern, E., & Pavia, S. (2011). Historic Building Information Modelling - Adding Intelligence to laser and image-based surveys. *International Archives of the Photogrammetry, Remote Sensing and Spatial Information Sciences, 38*(5).

Parisi, P., Turco, M., & Giovanni, E. (2019). The value of knowledge through H-BIM models: Historic documentation with a semantic approach. *The International Archives of the Photogrammetry, Remote Sensing and Spatial Information Sciences, 42*(W9), 581–588. doi:10.5194/isprs-archives-XLII-2-W9-581-2019

Patacas, J., Dawood, N., & Kassem, M. (2020). BIM for facilities management: A framework and a common data environment using open standards. *Automation in Construction, 120*.

Pavan, A., Mirachi, C., & Giani, M. (2017). *BIM: metodi e strumenti. Progettare, costruire e gestire nell'era digitale* [BIM: methods and tools. Design, build and manage in the digital age]. Tecniche Nuove.

Petti, L., Trillo, C., & Makore, C. (2019). Towards a Shared Understanding of the Concept of Heritage in the European Context. *Heritage*, 2531-2544.

Piccone, G. (1978). *The Castle of Baia - History, Legend and Poetry*. Società Editrice Napoletana.

Pocobelli, D., Boehm, J., Bryan, P., Still, J., & Grau-Bové, J. (2018). BIM for Heritage Science: A review. *Heriage Science, 6*(1).

Porras, D., Carrasco, J., Carrasco, P., González-Aguilera, D., & Lopez Guijarro, R. (2021). Drone Magnetometry in Mining Research. An Application in the Study of Triassic Cu–Co–Ni Mineralizations in the Estancias Mountain Range, Almería (Spain). *Drones (Basel), 5*(4), 151. doi:10.3390/drones5040151

Promoter. (2021). *Digitalmeetsculture*. Obtenido de Digitalmeetsculture: https://www.digitalmeetsculture.net/article/eu-commission-recommendation-to-accelerate-the-digitisation-of-cultural-heritage-assets/

Remondino, F., Barazzetti, L., Nex, F., Scaioni, M., & Sarazzi, D. (2011). UAV photogrammetry for mapping and 3D modelling– current status and future perspectives. *The International Archives of the Photogrammetry, Remote Sensing and Spatial Information Sciences, 38*(1).

Santamaría, J., Cordón, O., & Damas, S. (2011). A comparative study of state-of-the-art evolutionary image registration methods for 3D modeling. *Computer Vision and Image Understanding, 115*(9), 1340–1354. doi:10.1016/j.cviu.2011.05.006

Sigala, M. (2020). Tourism and COVID-19: Impacts and implications for advancing and resetting industry and research. *Journal of Business Research, 117*, 312–321. doi:10.1016/j.jbusres.2020.06.015 PMID:32546875

Solazzo, G., Maruccia, Y., Ndou, V., & Del Vecchio, P. (2022). How to exploit Big Social Data in the Covid-19 pandemic: The case of the Italian tourism industry. *Service Business*. Advance online publication. doi:10.100711628-022-00487-8

Spettu, F., Teruggi, S., Canali, F., Achille, C., & Fassi, F. (2021). A hybrid model for the reverse engineering of the Milan Cathedral. Challenges and lesson learnt. *ARQUEOLÓGICA 2.0 - 9th International Congress & 3rd GEORES - GEOmatics and pREServation.*

Trillo, C., Aburamadan, R., Makore, C., Udeaja, C., Moustaka, A., Gyau, K., . . . Mansouri, L. (2021). Towards smart planning conservation of heritage cities: Digital technologies and heritage conservation planning. *Culture and Computing. Interactive Cultural Heritage and Arts. HCII 2021.*

Trillo, C., Aburamadan, R., Mubaideen, S., Salameen, D., & Makore, C. (2020). *Towards a Systematic Approach to Digital Technologies for Heritage Conservation. Insights from Jordan.* Preservation, Digital Technology & Culture. doi:10.1515/pdtc-2020-0023

Trizio, I., Savini, F., Giannangeli, A., Fiore, S., Marra, A., Fabbrocino, G., & Ruggieri, A. (2019). Versatil tools: Digital survey and virtual reality for documentation, analysis and fruition of cultural heritage in seismic areas. *The International Archives of the Photogrammetry, Remote Sensing and Spatial Information Sciences*, *52*(2-3), 377–384. doi:10.5194/isprs-archives-XLII-2-W17-377-2019

Udeaja, C., Trillo, C., Awuah, K., Makore, C., Patel, D., Mansuri, L., & Jha, K. (2020). Urban Heritage Conservation and Rapid Urbanization: Insights from Surat, India. *Sustainability*, *12*(6), 2172. doi:10.3390u12062172

ADDITIONAL READING

Territoriale, A. p. (n.d.). *Next Generation EU e il Piano Nazionale di Ripresa e Resilienza* [EU and the National Recovery and Resilience Plan]. https://www.agenziacoesione.gov.it/dossier_tematici/nextgenerationeu-e-pnrr/

Zharinov, I. (2020). The smart city technology in the super-intellectual Society 5.0. *Journal of Physics: Conference Series*, 1679.

Žižek, S., Mulej, M., & Potocnik, A. (2021). The sustainable socially responsible society: Well-being society 6.0. *Sustainability*.

KEY TERMS AND DEFINITIONS

Building Information Modelling (BIM): Approach for optimising the management of a built asset in its entire lifecycle with the help of software that allows different professionals to work interdisciplinarity on the same database.

Historic Building Information Modelling (HBIM): Reverse engineering process that, starting from data obtained with appropriate survey and investigation techniques from the existing building, allows its digital representation integrated with all significant information for its interdisciplinary management.

Intangible Cultural Heritage: Voices, values, traditions, oral history, dress, forms of shelter, traditional skills and technologies, religious ceremonies, performing arts, that define and characterise a culture in a determined time and space.

Resilience: Process and outcome of successfully adapting to a new reality determined by a single traumatic event or challenging experiences for the human being, especially through mental, emotional, and behavioural flexibility.

Smart Heritage: Management system aimed at responding to the challenge of cultural heritage conservation through the use of advances in information and communication technologies (ICT) by generating new immersive experiences for interacting with digitised historical sites, works of art and objects in museums, galleries and public spaces.

Society 5.0: Anthropocentric society that strongly integrates cyberspace and physical space to balance economic and technological progress with the resolution of human social problems by promoting the idea of a Smart Society based on the use of ICTs, in line with the future sustainable strategies developed in the framework of the 2030 Agenda for Sustainable Development.

Tangible Cultural Heritage: Artefacts or physical assets produced, maintained and intergenerationally transmitted in a society and which determine the sense of identity and belonging to a community, region, or nation.

Chapter 16
Experiences of Digital Survey Data Applied for the Involvement of Societal Smart-Users in Cultural Heritage Awareness

Sandro Parrinello
University of Pavia, Italy

Raffaella De Marco
https://orcid.org/0000-0002-4857-3196
University of Pavia, Italy

ABSTRACT

Heritage accessibility has been highlighted as a fundamental condition to convey multi-sphere values (social, artistic, economic, territorial), necessary for assigning the label of cultural heritage. Similarly, it permits to include new frontiers of educational processes for smart communities within digital data and VR systems developed from 3D survey actions. In this way, digital technologies can convey the societal challenge to evaluate the efficacy of cultural heritage communication beyond the in-situ physical experience, assessing the learning impact of virtual heritage environments. The scientific research on the production of effective heritage learning objects, from the EU project PROMETHEUS, is presented, enhancing opportunities of communication and virtual smart-fruition for sites along cultural heritage routes. Sites' virtual models are joined to physical prototypes to increase awareness and sustainable knowledge from the users' interactions with digital heritage.

DOI: 10.4018/978-1-6684-4854-0.ch016

INTRODUCTION

Accessibility has insistently emerged as one of the fundamental conditions to convey societal values and assign the nature of Cultural Heritage to communities' users (Welch, 2014; Chong et al., 2021).

The possibility of accessing a direct experience of Cultural Heritage has always been a function of factors surrounding the social and socio-geographical situation affecting the specific heritage site. Many influencing conditions can be considered: from environmental climatic conditions, which can limit the visit to short or seasonal periods (as for sites in the Nordic regions, Finland, Russia, or the tropics similarly), to political conditions of territorial interest (common in areas of the Middle East and, recently, in the geographically European territories of Ukraine). Regarding ordinary conditions, also gender, religious and safety conservation limits can be assumed.

Considering the worldwide experience of the COVID pandemic, the assurance of heritage accessibility has emerged as one of the main conditions of deprivation caused by the need for social distancing (Ginzarly, Jordan Srour, 2022). The reference is not only for museum spaces (Miłosz et al., 2022), classified as delimited premises with a strict requirement for crowd control but also for outdoor sites, such as monumental areas and parks, where the need for access regulation has emerged more clearly due to health risks.

Concerning the significant reduction in accessibility and visit opportunities suffered by many sites, it has been emphasised how Cultural Heritage acts as a fundamental resource both for territorial valorisation and societal education. Its intellectual enjoyment is strictly connected to the sustainable assets developed by a community. This issue does not only concern the touristic-economic framework, which is often of fundamental sustenance for many regional realities, but it also concerns a broader panorama regarding the impact of the education of tangible culture on the development of local communities, on behaviours of societal users, and the recognition and characterisation of society concerning the identity of its territory.

The relevance of the accessibility of Cultural Heritage by international citizens and visitors does not only concern symbolic heritage and monumental sites of higher visibility, often supported by a higher global awareness due to international awards (e.g. UNESCO sites). It permeably characterises the entire ecosystem of sites spread over the territory, which are often linked together in networks of relationships and mutual historical influence. Their cultural relevance needs to be traced and enhanced as a character of resilience within the multiple actions that have undergone the transformation of territory and the local communities (Korro Bañuelos et al., 2021).

In particular, the "widespread architectural heritage" is a natural phenomenon that concerns networks of heritage sites linked by historical and cultural connections, which create a relationship between their tangible and intangible elements through more than physical and geographical existence. Widespread heritage concerns a system of relations and interconnections of communication models disseminated at the territorial level, where the recognition of a basin for its development is related to the modalities of cultural exchange and development of a common identity applied by communities, from the past until the present. It crosses physical and administrative boundaries that clarify the formal cultural value of the homogeneous heritage system (Oppio, Dell'Ovo, 2021).

This topic is more evident in the case of tangible heritage scales with characteristics of greater immovability, as in the case of Architectural Heritage, and of sites involved in territorial transformations that divide and fragment their original arrangement and foundation network (e.g., geo-political reorganisations, societal restraints, territorial infrastructural planning). In these cases, the various entities in charge of the administration and management of heritage sites through the territorial system are not even able to overcome social barriers and geo-political events that decontextualise their accessibility and the transmission of their cultural significance.

Architectural heritage enriches the territory with the anthropological value of the construction tradition, providing the territorial space with human meanings. Opposing the natural dispersion of assets, the opportunities for connection and joint accessibility between widespread architectures can thematize and exalt some aspects of their heritage, for the communication of didactic contents regarding historical, artistic, and economic sustainability (Soler et al., 2013).

In this way, the advancement of 3D digital representation technologies has pursued the concept of "Digital Twins" also applicable to the field of architectural sites. It regards digital reproductions that are developed from the documentation datasets of a physical site, processing a first reliable archive on the asset and state of preservation of a Cultural Heritage object. From this step, multiple formats of information can be integrated, coming from historical archives or collective memories, and the replicas can be directed towards different technical uses, such as the mapping of territory in GIS (Geographic Information Systems) tools or the support to intervention and maintenance goals in BIM (Building Information Modelling) tools.

In both cases, digital translation of Cultural Heritage can overcome the spatial dimension of the distance between sites, as well as other barriers of geography, gender, and user mobility, bringing physically distant sites closer together in the virtual space, to find common features and cultural associations guaranteed from the reliability of the digital asset. This grouping procedure, on the one hand, increases the asset value of Cultural Heritage sites from a perspective of physical identification and conservation; on the other hand, it rediscovers and reinforces bonds and memory collections between people, involving multiple dimensions of society ranging from the local-regional sphere to the global involvement of citizens and users (Buhalis et al., 2022).

The topic of "Accessibility to Cultural Heritage" can thus be translated into "Accessibility to Digital Heritage", and in this sense, it brings into consideration the ability to engage and educate communities through reliable products and services based on digital development of Cultural Heritage assets. It is focused on the possibility to build a digital capacity of users' involvement and awareness of Cultural Heritage, educating a new generation of societal smart-users in interacting with the digital replica and in accepting it as a means for the accessibility to sites' knowledge and its transfer (Chung et al., 2015; Han et al., 2017).

However, the terms of the digital translation of Cultural Heritage are not without dangers and criticalities that may undermine their educational and awareness relationship with society. If unregulated, Digital Heritage can misrepresent and distort the truth of real physical heritage, undermining its authority, or it can undermine the participation of locals and tourists in its learning experience, representing inaccurate and inauthentic information. This treat has not to be overestimated even within the adhesion to supporting initiatives in digital transformation promoted by Cultural Agendas and European funding programs, focused on addressing Digital Heritage to lead cohesion of a local and even international community, for raising people's inclusion and societal awareness.

The term "reliability" of information conveyed by Digital Heritage, as associated with its "accessibility", is considered: Digital Heritage must be designed and verified not only at the source phase of development and enrichment of the digital replica (Doerr, 2003) but also in the methods of achievement by its users (Jan, 2018). It must be referred to the knowledge levels of the users' classes and to their confidence with languages of interaction with the world of ICT that are promoted within the daily dynamics of society (Meyer et al., 2007). It is central to affirm the role of Digital Technology and Digital Replicas as tools, which need to organise their educational offers in functions and contents related to the typology of a user, even adapting the modalities of their enjoyment in physical or mixed forms, to improve communication.

Considering the availability of databases and documented information on physical Cultural Heritage sites, Digital Technologies are assigned to the role of certifying not only the correspondence of Digital Survey Data of Cultural Heritage assets to their physical versions (in terms of resolution, accuracy, completeness, and reliability of the survey) but also to convey the experience modalities for its enjoyment (Parrinello, Dell'Amico, 2019). It regards a societal challenge in evaluating the efficacy of Cultural Heritage communication beyond the in-situ physical experience, concerning the educational impact of its digital reproduction until considering the behavioural implications conceived by the fruition of Virtual Heritage Environments in the societal public.

The following chapter explores the communication basis related to the concept of Digital Replica for Architectural Heritage, within the last guidelines given by "digital transformation" programmes for public and community-oriented education to Cultural Heritage. While outlining the key features concerning the identification and involvement of societal smart-users in Cultural Heritage awareness, the methods and products for communicating Cultural Heritage in terms of digital-real interaction will be presented, showing as a reference the experience of engagement between citizens and the scientific sector in 3 editions of the European Research Night held in Pavia, between 2019 and 2021, which adopted data developed within the Horizon 2020 project "PROMETHEUS" on the topic of Cultural Heritage Routes and the network of heritage sites in Upper Kama (Russia).

The chapter aims to outline how the theme of accessibility can be considered an integral part of the nature of Digital Heritage and its means of communication and to discuss the elements of awareness and their impacts conveyed by the interaction with Virtual Heritage, towards assuming new frontiers of education processes to Cultural Heritage for smart communities.

BACKGROUND: A CRITICAL REVIEW ON THE DIGITAL TRANSITION FROM CULTURAL HERITAGE TO VIRTUAL HERITAGE

Cultural Heritage is inevitably implicated in the creation of a sense of community (Howard, 2003; Byrne, 2008), and in this way, it can be addressed as a tool to increase the sense of societal cohesion (Arenghi et al., 2016). Also, it can be influenced by the sensibility of its educational practices in a specific social and territorial context. Furthermore, the community itself constitutes an integral component of the societal ecosystem of education to which the Cultural Heritage field belongs, and it is from the community scale that the concept of Heritage resilience comes to force (Longstaff et al., 2010), in particular when considered in its physical declination of Architectural Heritage.

Accessibility to Cultural Heritage is therefore intended as a means of equalisation of the enjoyment experience for members of the community (Puyuelo et al., 2013). It links the concept of "heritage democratisation" (Rodéhn, 2015), and it assumes the form of Heritage (despite the physical or digital format used for its communication) as an exceptional vehicle for spreading knowledge and learning mechanisms toward society. The aim is to improve the facilitation of access and awareness of Cultural Heritage among citizens, to trigger behavioural and cognitive mechanisms of identity between people and their heritage, and encourage its extended protection (Chiapparini, 2012).

Even if most commonly replaced by the suffix "digital", the concept of "smart" Cultural Heritage (Borda and Bowen, 2017) can be used to promote the goal of accessibility, since it has the potential to play a significant role in the development of both smart visualisation technologies and smart users (Ciurea et al., 2020), and to relate learning practices coming from Cultural Heritage experience.

The modalities of experience able to replicate direct contact with Cultural Heritage (even in remote ways) remain the key issue for the learning enrichment of users (Treccani et al., 2016). They especially regard Architectural Heritage, where the user is directly involved in the 360° perception of physical spaces, morphological qualities and sensory conditions (such as illumination, acoustics, orientation within the context, etc.).

The condition of cultural learning as expressed by the in-situ experience of architecture has to fit in the educational challenge introduced by the use of digital replicas. It can be addressed by the application of Information and Communication Technologies (ICT) to provide expanded access to Cultural Heritage Education while introducing significant changes in the visual perception of Digital Heritage (Ott, Pozzi, 2010). These changes can regard the enrichment of information and multi-perspective views of Digital Heritage objects, and their relation to innovative teaching and learning methods.

The Digital Transformation, also advanced by the Education Action Plan 2021-2027 (Yanli, Danni, 2021), inevitably reflects the direction of European educational initiatives regarding Digital Strategies for Cultural Heritage. Their key role is to establish data collection, knowledge and smart-oriented application of digital data, involving users in background training to interact and share the received learning stimuli within the digital dimension of Architecture and Cultural Heritage. Furthermore, it also represents a transversal means of progress in the diffusion of common digital languages and know-how practices in advanced digitalization for tangible heritage, supporting replicability across different sectors and functions (both social and technological).

To consider a critical review of the topics connected to the research experience of processing digital survey data for communication and knowledge awareness of Cultural Heritage through virtual means, the following points are considered:

- The centrality assumed by "Digital Competences" in influencing citizens' and communities' expectations to relate with societal and educational services.
- The increasing assumption of digital products and replicas as a basis for the construction of interactive learning processes.
- The application of virtual fruition methods to smart-communication channels to share Cultural Heritage environments and contents for a broader valorisation.

The Centrality assumed by "Digital Competences" in influencing Citizens' and Communities' expectations to relate to Societal and Educational Services

The Digital Competence Framework 2.0 (Mattar et al., 2022) has outlined five main areas interested in the development of key components which can be related to Cultural Heritage and Digital Heritage. Alongside the skills more related to developers of digital methods and virtual products (1. Information and data literacy, 3. Digital content creation) and their stakeholders (4. Safety, 5. Problem-solving), it is possible to assign precisely to societal users a key component: 2. Communication and collaboration. For this way, it focuses on "*interact, communicate and collaborate through digital technologies while being aware of cultural and generational diversity; to participate in society through [...] participatory citizenship"*, this competence exactly relates to the conditions expected by the user-citizen following the contact with Digital Heritage in an educational perspective. It relates to the reception of the cultural content communicated through digital practice, and the user's ability to contribute in cultural terms to a "participatory" response, starting from influencing the same modalities of virtual interaction with the digital object.

If transposed into the terms of Cultural Heritage and its digital translation, the DigComp 2.0 reference model succeeds in describing a relationship between Digital Heritage and the user based on the key actions of "Interacting, Sharing, Engaging, Collaborating", exactly following the steps of an educational process. From the interaction with the digital object and the cognitive learning of its information content, becoming part of both the individual and collective identity of the community, the process of virtual experience can trigger behaviours of coordination and collaboration of users on Cultural Heritage topics, to apply the acquired critical "digital" knowledge also in the real field.

The developer of the digital product remains in charge of the core competence of being aware of behavioural norms and know-how while using digital technologies, and of their design following educational targets and the settings of digital environments.

This goal finds practical realisation in the adaptation of communication strategies to the specific audience, with awareness of the cultural and generational diversity of Cultural Heritage perception that can be established through digital environments. The result can be the creation and management of a "digital identity" for Cultural Heritage also suitable for educational activities, ensured by its reputation of reliability to the real physical object.

The increasing assumption of Digital Products and replicas as Primary Data for the Construction of Interactive Learning Processes within Smart Cultural Heritage

The State of the Art in experimental processes for remote learning established on digital replicas of Cultural Heritage objects includes many references (Pescarin, 2016; Champion, Rahaman, 2020; Kaldeli et al., 2022; van Ruymbeke et al., 2022). International projects are considered among this topic about the type of digital information (e.g., EUROPEANA digital library) (Macrì, Cristofaro, 2021; Giannakoulopoulos et al., 2021), the standards established for digital data processing (e.g., CIDOC CRM reference models) (Nicolucci, Felicetti, 2018; Faraj, Micsik, 2021), the sector of interest (e.g., ARIADNE and ARIADNE-plus project for archaeological collections) (Nicolucci, 2017) or the historical timeline reference (e.g., INCEPTION ontologies) (Di Giulio et al., 2017). Through these experiences, different case studies and scales of Built Heritage are investigated, conceiving preliminary criticalities and optimal proposed strategies compatible with many varieties of Cultural Heritage objects.

The difficulty in coherently representing a complex physical system of Architectural Heritage digitally, lies mainly in the management process of a large amount of data, coming from the 3D survey and scanning procedures adopted for its documentation (Picchio et al., 2020).

Even with the development of new automatizations and technological processes, the development of reliable narrative systems for Architectural Heritage remains dependent on 3D complex databases (Georgopoulos, 2017). The reality-based condition of these data, certified as corresponding to the real physical conformation of an architectural site, motivates the higher complexity of these elements. Spatial data, usually acquired in the format of discontinuous 3D point clouds and processed following different methodological approaches, constitutes the background information to be managed (within sampling, semantic, information, or modelling actions) for designing the communication approach of "digital" Architectural Heritage to users.

The increased studies on Digital Databases and Information Systems have introduced how digital practice can offer a common platform to integrate this different sectoral knowledge and to set a common basis of technical language for 2D/3D products. Digital Environments and Virtual Spaces have been developed and slowly increased with even more realistic reproductions of contents from the object to the urban scale. 3D motion-capture, VR animations and real-time immersion systems have been tested and supported by research-oriented institutions to extend from the ICT sector to other scientific purposes, and in the last 2 years, the COVID pandemic crisis has enhanced their application to the widening and sharing of cultural and creative contents from Digital Twins to Digital Humanities, Museums, Creative Industries and Educational channels (Cardozo, Papadopoulos, 2021).

Based on the development of this kind of product, it is the translation from Cultural Heritage to Digital Heritage that is placed at the centre of the explication of models and modalities of involvement and educational cohesion instituted directly by citizens and social users. This process must take the form of a participatory dynamism, where users are not only observers of Digital Heritage but are made capable (in tools and skills) of establishing interactions and stimuli for exploration in an alternative but parallel process to the knowledge and visit of the real site.

The centrality of integration of 3D technologies, considering both sensors and techniques, is at the basis of the collection of primary data for smart digital replicas of Cultural Heritage. It includes the macro-groups of laser scanning and photogrammetry applications, combined with 3D tools and modelling methods for further representations and analysis of the object data in the virtual space (Balletti, Ballarin, 2019; Choromański et al., 2019).

In this way, the adoption of Virtual Models and 3D printing from Digital Replicas, and their communication into contexts already disposed of high learning involvement (e.g., museums), find a transversal application through scientific sectors of research, also considering Geomatics and Metrology experiences suitable for educational purposes. It is related to the purposes of digital models to display and share, in possible tangible and intangible ways, information that is related to the "usability" of the object (Ballarin et al., 2018). Metric data, as well as the visual/perceptual choice for its representation, involves the opportunity of a user to increase "clearness" and "immediateness" in the various opportunities and levels of perception of the Cultural Heritage object (Rossetti et al., 2018), even in cases of virtual reconstruction, inaccessibility, or destruction of the real object.

Furthermore, the issue of compatibility and integration between data formats needs to be addressed. The sectorisation in the adoption of end-user platforms mainly regards BIM (Building Information Modelling) or GIS (Geographic Information Modelling) solutions. It includes the solving of compatibility issues regarding formats, level of details and multi-scale structuring of information (Colucci et al., 2020), which assumes an influence on the quality of data both considering the acquisition phase (Bitelli et al., 2017; Barba et al, 2019; Mohammadi et al., 2021; Moyano et al., 2020; Balado et al., 2022) and the processing tools phase (Ahmad Yusri et at., 2022; Ramm et al., 2022). The development of accurate 3D models (Tucci et al., 2017) represents the final step of this control and validation process of digital replicas, concluding with the verification of standards and quality requirements for the application of tools (Gireesh Kumar, 2021).

As the process of modelling and programming visual experiences on Cultural Heritage is creative, the heritage content created in the virtual reproduction can be viewed as a new form of "authenticity" (Di Giuseppantonio Di Franco et al., 2018) rather than as a substitute for real heritage. This issue relates to a double verification: the assurance of the validity of information, considered the main "authenticity" feature, and the "accuracy" in the representation of information, influencing the reliability of perception and immersive experience. "Authenticity" and "accuracy" define the main expectation of the public from digital replicas, indifferently coming from the research results of archaeologists, architects, historians, and museum curators.

The process of "Building digital capacity" (Europeana, 2021) can also be made explicit concerning Architectural Heritage, declining its key themes. To overcome the concept of "Digital divides" existing between physical Heritage and audiences, the developers of Digital Replicas are called to act as an "Agency for change", triggering the human dynamics and 'soft skills' of empathy between users and Digital Heritage. The resulting "Collaboration" in declining educational tools and defining Virtual Heritage Learning Environments is foreseen also to develop and adopt shared standards, and to share infrastructures for the sustainability of the action.

The development objectives towards the transfer of learning purposes on Digital Heritage emerge in (i) Presenting and monitoring learning materials and needs, prioritising the capacity-building needs strictly related to Cultural Heritage; (ii) Offering and facilitating the development of open and community-managed digital heritage use systems, enriched with an information apparatus, to provide learning guidance and practical tools; (iii) Driving the increased quality and diversity of learning resources, foreseeing their increase, extent and modality of engagement with users, contexts and languages.

The Application of Virtual Fruition Methods to Smart-Communication Channels to share Cultural Heritage environments and Contents for a Broader Valorisation

A key opportunity relies on the availability of novel platforms and cloud services to set up a virtual interacting solution for users directly from source datasets, saving time and formats' adaptability (Burkey, 2022). The user must be enabled to navigate within the virtual system that reproduces the Cultural Heritage object and to interact directly with the place and its elements following the implementation of both visual and information contents. In this way, through the development of supports based on Virtual Reality (VR) or Augmented Reality (AR), the architectural information acquired and catalogued during the documentation activity can become accessible and usable for educational purposes. The idea of sites' accessibility can be no longer limited to the territorial configuration, but it can be duplicated in the digital concept of the virtual dimension that hosts the replica (Chong et al, 2021). Similarly, the interconnections of the Digital Heritage with other sources of data, such as documents, multimedia, and hyperlinks, can integrate and expand the real experience of the visitor.

This way of fruition for Cultural Heritage through Virtual Heritage Environments acts on the "accessibility" and "inclusion" impact that Digital Heritage can arise, converting every kind of interaction to a communication mechanism "with" and "from" the virtual prototype, which can be dynamically manipulated also with reduced risks for the sustainable conservation of the real one, and with increased capabilities and multi-perspectives of interactions by the users.

Following the rapid advancements offered by emerging technologies, heritage information is increasingly represented via more dynamic and interactive formats (King et al, 2016). Their functionality varies in terms of modality, immersion, and integration of virtual environments to solutions for their physical use by users. Website platforms, smartphone apps, and VR/AR systems suitable for personal devices, tablets, or within visors for immersive fruition, represent some examples of smart devices able to communicate with societal users, as smart technology integrated within the daily digital transition of communication services. Adopting their technical languages and modality of interaction, it is possible to imagine a more sustainable acceptance of immersive reality technologies, and the adoption of their exploitation impact for educational, explorative, and exhibition enhancement purposes on Cultural Heritage.

To extend Virtual Heritage Environments out of research labs and into the public agenda (Champion, Rahaman, 2019), the educational, enjoyment and even recreational aspects must be emphasised. The key issue is not only to visually represent, in an immersive manner, Cultural Heritage objects which exist or may be reconstructed in time (both in past and in future). It deals with the communication strategy of enhancing meaningful and engaging opportunities, able to transform the "visual representation" into "experience".

The evaluation of communication channels suitable for Digital Heritage Objects is, therefore, basic for the structuring of Virtual Heritage Environments able to convey learning purposes (Ibrahim, Nazlena, 2018; Ghani et al., 2020). It is the virtual amplification to guarantee the spreading of knowledge of the existence of heritage, which competes to inspire new but real emotional experiences from the simulated visit.

In this sense, it is important to give visibility to widespread heritage and monuments reserved in secondary evidence, not only physically but also culturally overshadowed, by the great notoriety of a few others. It is a mechanism that acts through any learning and educational purpose, to redirect also touristic flows on the territory in influencing the structuring of territorial management systems, embracing the well-known heritage also the less visible one. This opportunity restores the "valuable" characterization of sites' notoriety, concerning knowledge, services, and resource impacts, and even consolidates historical routes and networks between people and culture.

THE SOCIETAL EDUCATION CHALLENGE RELATED TO CULTURAL HERITAGE ROUTES

The study of the connections between Architectural Heritage and territory is based on the analysis of the cultural imprint of mutual influence between sites, landscapes, and communities' networks where it belongs. Between settlements and stylistic flows based on "basin" connections, classification of images and information on a type of "widespread heritage" is recognized with increasing emphasis, generated by the intertwining of geographical and cultural crossroads.

The recognition of extensive networks of cultural contamination, in particular between Europe and closer continents, classifies Cultural Heritage contexts as learning scenarios capable of affirming and reinforcing a societal knowledge linked to the identity of a community. The understanding and structuring of related educational contents are necessary not only to operate their management but also to conform to Cultural Heritage as a unique learning apparatus able to convey best practices and cultural sustainability between the memory of people and places, sensitizing citizens and the public on the recognition of a specific "cultural landscape" (Salerno, Casonato, 2008). Their experience is assumed as a meaning of both geographic (cross-border cultures) and timing (sites' transformation) results of decisions and societal policies adopted by communities about their cultural sensibility and education.

The recognition of a Cultural Heritage Route is intended as a "*sum of elements referring to a whole [...] within a joint system which enhances their significance*" bringing to a "*dynamic conformation of cultural landscape*" (ICOMOS, "Charter on Cultural Routes", 2008). The recent European attention on Cultural Routes has brought the Council of Europe to recognise numerous examples: Hansa 1991, El Legado Andalusì 1997, the industrial heritage of Rhine Valley 2002, the Pyrenees Route 2003, the Cluniac Sites 2005, the Transromanica 2007, the Cistercian abbeys 2010, the historical thermal towns 2010, the fortresses of Charles V 2015, the fortified cities of Grande Région 2016. Currently, they are at the centre of communitarian programs for their enhancement and management, and they have sensitised on the phenomena of cultural exchange and dialogue in centuries between peoples and regions with the contemporary frameworks of Cultural Heritage education of communities.

Cultural continuity through space and time thus becomes the key to classify Cultural Heritage along Cultural Heritage Routes, defined as an inclusive category between material heritage and territorial memory capable of understanding tangible but dispersed values of communities in time. The connective qualities testified by the visual and constructive characteristics of architecture and place are researched and classified, linked to the understanding of construction practices, traditional forms and communities' practices that dynamically conform society in cultural landscapes. They entrust societal connection, collective memories, and considerations to Heritage sustainability.

In this way, their documentation represents an opportunity to testify historical and physical knowledge of Cultural Heritage, to convey more engagingly an educational stimulus to its presence and permeable influence on the societal dimension of citizens. Architectural Heritage of Cultural Heritage Routes should not be interpreted just as the existence of physical constructions but as the tangible evidence of meanings and values forming a territorial identity.

In this way, Architectural Heritage therefore can be interpreted as a communication process (Kepczynska-Walczak, Walczak, 2015), in which different types of learning purposes can be perceived, understood and appreciated by a wide range of visitors. (Figure 1, Figure 2, Figure 3).

INVOLVING USERS IN CULTURAL HERITAGE AWARENESS: IDENTIFYING RECEPTORS AND STRUCTURING MEANS OF COMMUNICATION

The communication of Cultural Heritage emphasises at its basis the need to identify information relevant to the pursued educational process, in an interconnected relation between Digital Heritage and Digital Learning Capacities, and to design the structure of this relation to optimise the process of sharing and knowledge between educational source and program receptors.

To understand the mechanisms of familiarity with Digital Heritage, it is necessary to consider it at the same time as a tool for the action of cultural preservation, a possible way of open accessibility for a wider public, and a means of establishing interaction between different and otherwise unconnected users, to foster their ever-increasing participation both as recipients of the educational service and as agents of development and involvement for further initiatives.

If the action is focused on communicating how Digital Heritage can improve citizens' well-being within society in terms of cultural consumption possibilities, then the same citizen will engage himself in sensitising his reception mechanisms to overcome sensory, operational and cultural filters due to the different cultural experience of fruition in the virtual environment.

Figure 1. Cultural Heritage Route's order: temporal order scheme. Source: Parrinello S., 2022

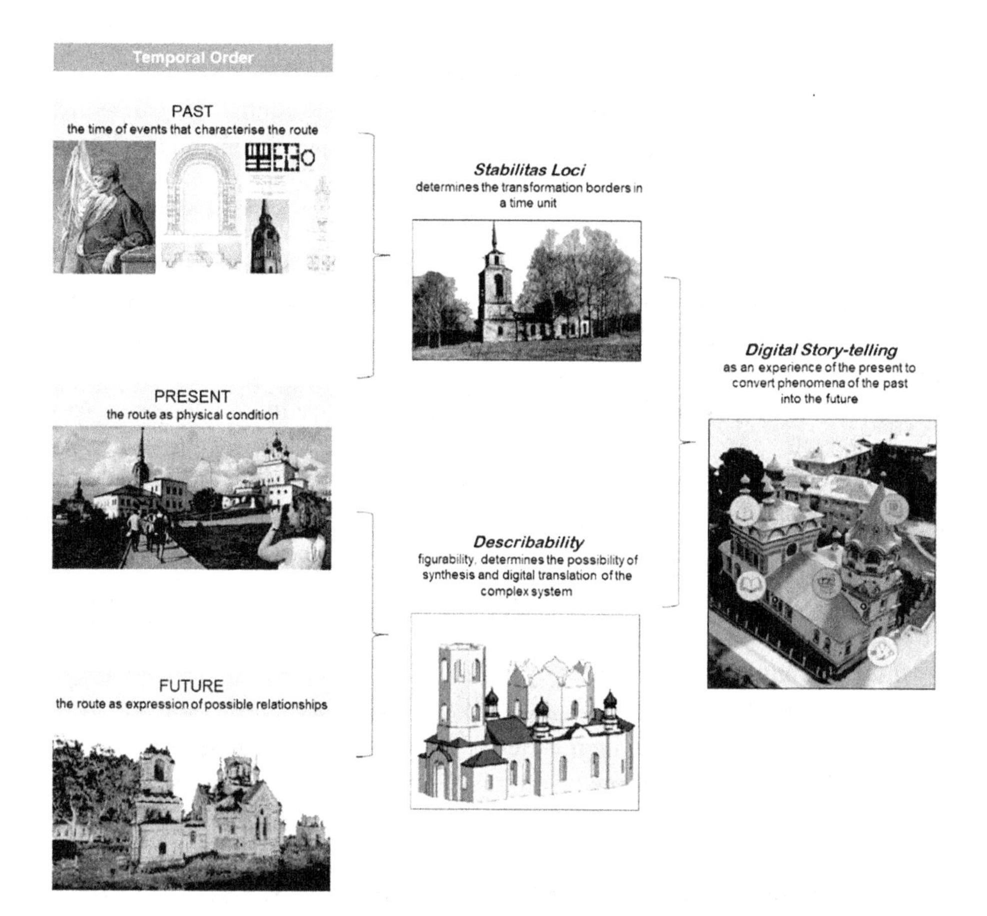

Regarding the disturbances and noise caused by the physical environment in which the enjoyment of Digital Heritage necessarily takes place, where the user interacts in a multi-sensory manner involving predominantly sight, hearing and touch, these may or may not be intentionally relevant to the impact of the educational purpose, considering the depth of perception in the user's experience of the digital replica. The placement of the virtual experience platform in a physical environment functionally dedicated to the educational process (e.g, a museum, a school or a cultural space), can be translated into a learning design system not only of the content and the virtual container, but also of the physical container itself in which this process takes place, and from whose physical participation the user cannot be disregarded.

On the contrary, the freedom left by the fruition of the virtual platform through a personal remote device (e.g., screen, tablet or smartphone), in a domestic or in any case unguided environment, does not allow to control the nature and entity of the disturbance influencing the recipient's reaction, and in general the impact of the entire system. The possibility is to search for solutions to enhance the degree of immersiveness of the experience, focusing the user's attention on the context with interactive content or through the connection of technological devices that can implement a filter of isolation from the physical context.

Figure 2. Cultural Heritage Route's order: spatial and relational order scheme. Source: Parrinello S., 2022

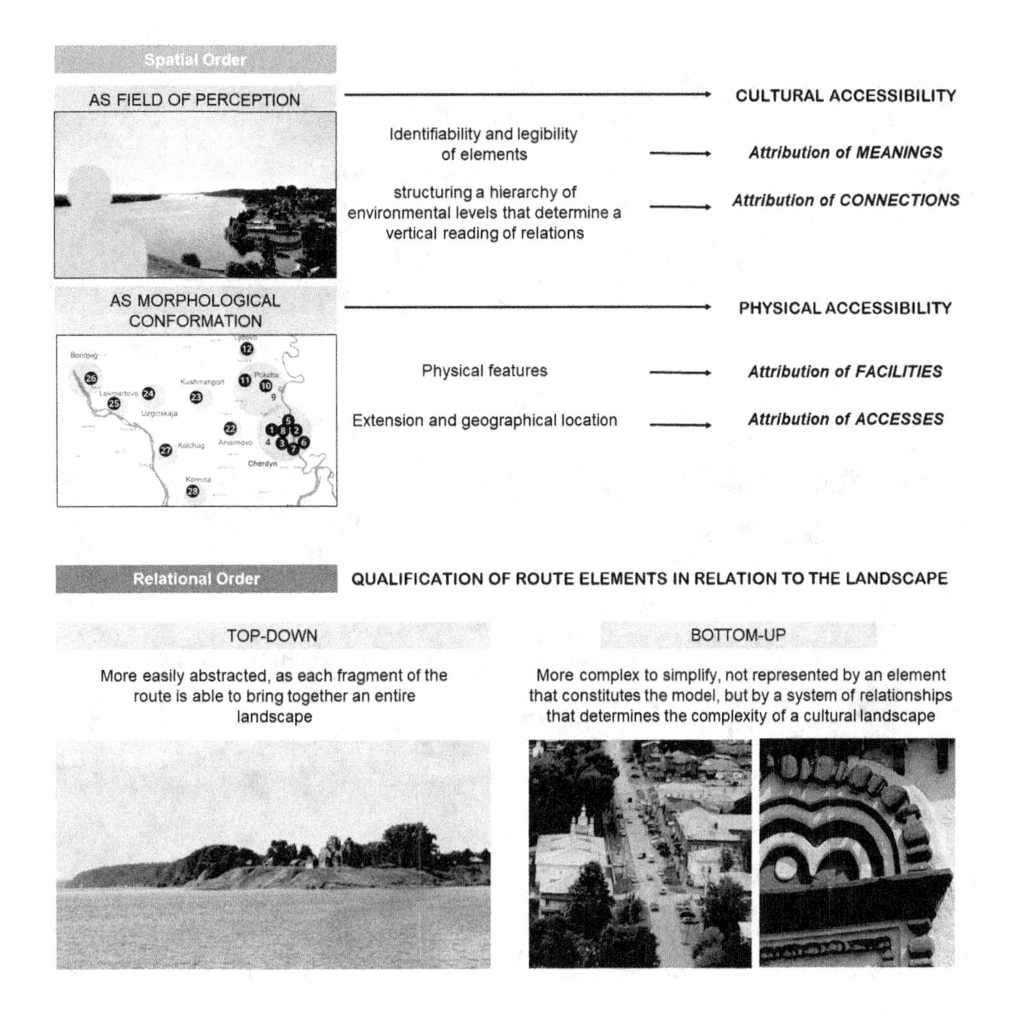

In this case, the use of visors as an interaction device with the virtual environment can be both an instrumental limitation but also a guarantee of control over how the reception process takes place. The use of a visor limits the sensitivity of interaction between user and context, hindering the factors of physical perception with an absolute involvement in the virtual environment: by occupying sight (ocular visors), hearing (integrated headphones) and touch (controllers/joysticks), it replaces the real experience with a perceptual alternative linked to the language and contents of the virtual platform, whose control is dependent on the immersive representation implemented by the developer.

At the basis of the design of sensitised or forced perception features for the immersive enjoyment of Digital Heritage products developed for the educational process, which can be defined as Heritage Learning Objects (HLO), are two fundamental analysis characteristics:

- The perceptual qualities of reception assignable to the medium of learning communication.
- The stimulable cognitive mechanisms and factors according to the profile of users.

Figure 3. Views of widespread heritage sites along Upper Kama Route. Source: Parrinello S., 2022

The Perceptual Qualities of Reception Assignable to the Medium of Learning Communication

"Visual features" on Built Heritage can be condensed and conveyed by graphic and representational content of different types (Cardone, 2015):

- explicit, such as geometric and spatial characteristics of 2D drawings and 3D models, which can be interpreted through measuring comprehension processes and direct observation.
- encoded, as degradation or elevation maps, where colour and texture features are encoded as info-graphic elements, associated with physical conservation or revealing other conditions.

However, more "intangible meanings" and values, memories, and experiences about heritage, can be more challenging to be communicated to the public, both in terms of unreleased perceptions and as knowledge contexts of greater media complexity and even educational richness. Also, in this case, they are associated with visual perception characteristics, linking different meanings to the degree of abstraction of the scene, the enrichment of textures and renderings or the discovery of educational content associated with the elements. Furthermore, "motion sensitivity characteristics", replicated in the virtual environment, influence the reliability of immersion in the experience, such as the design of visit barriers, suggested paths, camera settings (height, angle), avatar movements and speed modes or direct visualisation of the scene.

Considering also 3D printed prototype models, the tactile perception of the details of the solid form, as well as the perception of the 3D texture on the surface, due to the extrusion of the filament and its thickness, are also relevant in the process of transposition and communication of Digital Heritage.

Furthermore, "target groups" are very important when designing the features of a virtual environment, assuming end-users goals that come from interests calibrated on gender, social role, cultural expertise, and digital and even physical abilities. Above the citizens' users, within different age groups, capabilities, also researchers, professionals, policymakers and public authorities need to be considered, according to the different structures of information that they expect to collect from Cultural Heritage communication. Some examples can regard traditional constructive shapes, the configuration of the volumetric architecture, the disposition of the heritage element regarding the surrounding land.

The Stimulable Cognitive Mechanisms and Factors According to the Profile of Users

The study of the cognitive mechanisms involved in the user experience of digital-virtual products defines some guiding factors to optimise the impact of communication:

- to establish a contextual relationship between users, virtual content, and cultural context.
- to allow collaboration between users, to improve the educational impact or the contribution to content creation.
- to enable engagement with the cultural context "through" and "with" the virtual environment itself.

The orbiting visualisation of 3D digital representations, as well as the virtual immersive perception, instinctively develop participation and awareness towards the elements of the scene. The user is prompted to investigate the architectural object, approaching where necessary to distinguish details or to solicit information hotspots, thus making the learning process towards the object dynamic. The possibility of overcoming physical constraints of movement, for example, raising the observation altitude concerning the ground until an aerial observation experience, contributes to the cognitive originality allowed by Virtual Heritage Environments for a real physical experience.

In the case of vision, the process also relies on the declination of visual variables of control and signic/mimetic processing of the 2D/3D image to the discontinuous source data (Bertin, 1960), on the visualisation and immersive experience features (Kersten et al., 2018), on the type of multimedia information to which the user interacts during the experience. For 3D printed models, this is the case of tactile sensitivity of perception and interpretation of the formal resolution defined for the prototype (Peinado-Santana et al., 2021), including the colour of the printing filament and the light/shadow pattern emphasised.

In the case of the physical medium, the individual "portability" of the model also becomes relevant for a static observation limited to the moment of learning, leaving the user the possibility of regenerating and verifying the learning mechanisms of the heritage experience also at home and with the necessary time.

The integrated design of Heritage Learning Objects from Virtual Heritage Environments, as elements or entities in digital format and can be reused as content in WEB-based learning environments (Meegan et al., 2020), is thus assumed as a preliminary phase. It is necessary to assume values and benchmarks to adapt the processes of representation and manipulation of the virtual object according to the demands of communication and awareness of heritage culture. The management of immersion modes, with simultaneous visual stimulation of virtual or prototyped Cultural Heritage, aims to better contextualise the communication of the intrinsic value of Heritage, as more effective in supporting learning activities.

Physical-Digital, Real-Virtual: the "Phygital Heritage" opportunity as a Structuring Concept

As a synthesis of the integration of the perceptual qualities and cognitive mechanisms associated with Virtual Heritage for educational purposes, it emerges the purpose to identify guidelines for the combination of virtual products with physical media of communication. In this way, the concept of "Phygital Heritage" (Nofal, 2019) emerges between the opportunities of communication of Architectural Heritage by digitised means, considering the integration of digital technology versions of virtual replicas to possible physical solutions for supporting their perception, to be activated in a simultaneous communication manner.

The aim is to address the modes of transfer of knowledge on Cultural Heritage to users in a more engaging, educational and meaningful way, adopting simultaneous and integrated physical and digital means to facilitate the integration of users within its educational offer. This concept can act not only on several channels of experience (sight, touch), but it can modulate for each of them different levels of complexity in the representation of Cultural Heritage.

Such considerations, applicable through the representation strategy of immersive virtual environments, can also be extended to the perceptual qualities of solid prototypes and 3D prints. In such products, it is possible to control the description of detail on an architectural apparatus, such as the proportion of the printed model and tactile perception sensitivity of its surface texture and component system. By acting on decimation operations of digital geometries, processed as virtual 3D models that are then conformed after the printing process into physical models, it is possible to define the type of detail required for the virtual model as a function of the mechanism of information perception to be triggered.

The design of "phygital" representation formats for the development of Heritage Learning Objects related to PROMETHUES project was based on:

- "phygital affordance", considering which features were able to establish a perception interface between users and Virtual Heritage objects, facilitating the immersion and learning task.
- "phygital situatedness", considering the modalities of integration between the technological solutions, the contents and the fruition devices into the physical reality where users conduct their experience (the organised location of the Research Night), to improve the impact of the communication.
- "phygital engagement", considering the response of users to the learning stimulus triggered by the phygital communication of Heritage Learning Objects from Upper Kama route, to their memory from the experience, their novel sensibility through the Heritage object, their feedback and storytelling expression with their groups.

Databases and Information Applied for Cultural Heritage Social Communication: the PROMETHEUS Project Background

The pilot case of Upper Kama, a basin in the Russian territory west of the Urals mountains, has been interested from many years of international research experience on the digital documentation of some historical complexes (Parrinello, Cioli, 2018). The conformity of constructive elements and traditional significance in Upper Kama architectural sites highlights a critical analysis not limited to the specific territorial reality of the pilot area, but observable in the architectural heritage of all European Cultural Routes.

The complexity of relationships that qualifies the structure of the identity of the place, consisting of the monuments and their relationship with the landscape, attempts to be transposed into a new "virtual" conformation. In this way, the act of digitization becomes a methodological input associated with a deeper cognitive process, with the decomposition and critical reconstruction of the architectural-historical-territorial site through a semantic classification of its sub-components that can be linked to other formats and contents of information, characterising a learning purpose for users (Parrinello et al., 2019).

The project PROMETHEUS, funded by the European Union through the Horizon 2020 program, has provided an international partnership to develop both a "digitization", scanning and creation of digital copies in integrated 3D databases (e.g., laser scanning, photogrammetry, UAVs Structure from Motion processing), and a "digitalization", implementation of these databases within Information and Communication Technologies (ICTs), actions on the territorial widespread heritage sites.

The project has developed activities of documentation of the architectural and urban sites along the Upper Kama route, with close-range digital survey campaigns integrated with a census activity. Following a preliminary identification and cataloguing, the monumental sites have undergone operations of the integrated survey including multi-instrumental acquisition methodologies.

Spatial metric data was collected through the structuring of Terrestrial Laser Scanning (TLS) survey activities (adopting a FARO S150 Focus model), to reach the documentation of the architectural and constructive detail of each site. At the same time, a photographic campaign for Structure-from-Motion photogrammetry (adopting a Canon 3300D model) was conducted, aimed both at the acquisition of the territorial macrosystem and at the description of the single architectural site on the scale of decorative architectural elements. For the territorial system, wide-range acquisition tools have been applied, with a Mobile Laser Scanner (MLS) (Stencil KAARTA model) for the identification and structuring of territorial and urban interconnection routes, and photogrammetric flight plans from UAVs (adopting DJI Phantom Pro and DJI Spark models) for the restitution of a territorial photogrammetric model to integrate the metric data acquired at the architectural scale.

In an overall scale of analysis, digital sparse databases, in the format of point clouds, have been collected with a resolution of 1mm from TLS, 1cm from UAVs and 5-10 cm from MLS. Regarding the accuracy of data referencing, it has been provided for 3 mm at the architectural/constructive scale, 2-5 cm for architectural features at the urban/territorial scale, and 10-20 cm for the territorial localisation scale of sites (also considering the GPS coordinates provided from the UAV photogrammetric datasets).

The optimization and data collection for the structuring of PROMETHEUS digital library was followed by the information implementation on construction technologies and census archives relating to architectural, historical and territorial planning practices, collected within the compiling of technical cards designed on descriptive criteria necessary for the structuring of a unique digital modelling language, considering the architectural, accessibility, landscape, construction and conservation fields.

The objective of simulating a cognitive and educational action similar to the on-site learning experience is crucial due to the geographical extension and accessibility conditions of the sites, which are located in the Perm region far from the main tourist flows of the cities, and often through the forest or river navigation. Today, more than ever, the recent political events and conflict related to the occupation of Ukraine (2022) make the issue of providing a communication and knowledge strategy of regional Architectural Heritage that overcomes socio-political barriers, extending a unique educational sensibility to an increasingly European basin of influence, even more relevant.

The development of digital products from the PROMETHEUS database proved not to be sufficient on its own to enable a broad communication of heritage. The technical data format and information structures were developed by researchers for technological documentation and conservation practices and mainly transposed into BIM architectural models and GIS mapping. These products immediately revealed a complexity of info-graphical languages and ways of interacting with databases that were designed for the professional or academic sector, and a difficulty to be accessed by social users and communities public.

The virtual processing of Digital Heritage is intended to configure the production of Virtual Heritage Environments which will be characterised, in terms of content, mode of interaction and awareness, into Heritage Learning Objects. The path started by the open access databases, mainly cartographic and photographic ones (e.g., Sobory.ru for the Upper Kama region), is already heading towards the implementation of 3D contents coming from the processing of 3D documentation and survey sources (e.g., referring to the wide platforms of EUROPEANA and INCEPTION projects). The opportunities that PROMETHEUS aims to improve and advance concerning current projects concern the enhancement of Virtual Learning associated with Architectural Heritage, to integrate missing factors in the impact of communication to societal users for the on-site experience:

- Immersive/Tactile experience with guidance on the relevant characters of Heritage.
- Portability and smart-device fruition of learning contents by societal users.

SOLUTIONS AND RECOMMENDATIONS: TOWARDS A SMART-USER COMMUNICATION OF CULTURAL HERITAGE

The challenge of accessibility of Digital Heritage to smart societal users has been structured by comparing the spectrum of PROMETHEUS products and results and understanding their usability for the implementation of interaction methods able to predetermine and assist the relevance of Virtual Heritage Environments in their application as Learning Objects. Technically, it consisted in identifying the best approach for integrating specific forms of immersive reality and the interaction with prototypes to enable cultural learning in a specific Virtual Heritage Environment.

The digital databases obtained from the 3D survey of Architectural Heritage sites have been analysed for their potential of establishing a versatile 3D VR gallery, considering data formats and structures. In this way, it has been possible to dispose of a basic overall repository for different comparisons of platforms and software, to develop immersive realities and digital interactions in an expeditious manner adapted to the evaluation of digital skills achieved by smart users.

The design of a PROMETHEUS apparatus suitable for smart communication benefits actions for education and awareness of Architectural Heritage and its related topics, such as:

- Visualise and provide accessibility to abandoned places, ruins, unreachable sites, or sites that even may no longer exist, due to time, distance, scale, safety, and protection issues.
- Arise interest in relevant features and stimulate cultural connection, historical awareness and collective memory between cultural characters and local identities that are more familiar.
- Provide strong motivational impact on recommendable best practices for young citizens.
- Trigger asynchronous and personalised learning approaches, to maintain interaction, collaboration and responsive fruition of Digital Heritage.

Considering the "phygital" heritage model and its balance of physical affordance and level of situatedness (Nofal, 2019), three categories of products for Virtual Heritage communication have been considered and experimented with mixed configurations: 3D virtual environments, website contents, and 3D printed artefacts.

Regarding their categorisation (Nofal, 2019), the classification as Augmented phygital heritage products defines the opportunity of a "*continuous interaction between heritage objects or assets (physical) and electronic devices (digital)*". Furthermore, the possibility to enjoy these products from portable devices and the provision of open-access and open-format content makes it possible to amplify the accessibility of Cultural Heritage also in methodological terms of data and formats.

The development of Heritage Learning Objects (HLO) included reality-based products with a digital structure, enhanced by both the virtual 3D format and enrichment meta-tags concerning meanings, historical notes, and multimedia social information. HLO can be singular or provided in combined solutions, specifying their semantics in addition to the level of detail of visual and information links.

The key quality of HLOs concerned the conceptual structure to convey cognitive attention and the aptitude of the users to interact with the Digital Heritage medium. This structure concerned the info-graphic qualities attributed to the developed 3D virtual models, such as geometry, textures, interface design, the setting of the fruition modalities and the physical supports of interaction between the user and the digital content. The research experimentation concerned the case of deciding what type/combination of digital resources can be arranged in smart processing of 3D survey data, and which level of detail and scale of perception arranged for the Virtual Heritage Environments can correspond to the participants' learning impact.

Considering the involved features for the design of Virtual Learning Environments (Meegan et al., 2020), technical requirements of 3D survey data have been applied to organise both synchronous and asynchronous mechanisms for the remote accessibility to HLO. Data and file formats have been processed to avoid possible system requirements, increasingly adapting to the most common smart users' devices. In the same way, the security of accessible information and the scalability of Virtual Environments have been provided within the digital design of contents and models.

Products, Fruition Mechanisms and Feedback from the European Research Night Experience

The dissemination and evaluation of the functionality of Heritage Learning Objects required the possibility of implementation through experiences of direct involvement of a broad, not specifically professional, audience. Considering the purpose of Cultural Heritage awareness set by the project PROMETHEUS, the opportunity of social sharing and communication offered by the Research Nights actions has been chosen to host the launch of PROMETHEUS HLOs.

European Research Nights are promoted by the European Commission as part of the Marie Skłodowska-Curie Actions program, to encourage enhanced opportunities to interact with citizens and improve communication skills, competencies and learning experiences to interact with a non-research audience. The purpose is to combine education with entertainment, especially with young audiences as stated in the Horizon 2020 WorkProgramme. The University of Pavia is involved in the SHARPER Project, aimed at enhancing the contribution of researchers and the role of scientific research in sharing and educating citizens on issues of interest to society.

Pursuing this mission, the team of researchers at DAda-LAB, the coordinating team of PROMETHEUS project, has designed different solutions of interactive virtual and phygital prototypes to be proposed yearly to the citizens, to observe their reaction and confidence in interaction related to the different modalities of HLOs digital communication.

Participation in the SHARPER event took place in 2019, 2020 (in remote mode due to the pandemic situation) and 2021. As part of the programme, a day of extensive contact with citizens was planned, with the organisation of over 30 University stands in the courtyards of the Visconti's Castle in Pavia, with free access for citizens. Each stand was able to organise activities aimed at the direct involvement of visitors to scientific topics, considering both guided groups from primary, secondary and high schools and the visit of families, children and professionals from various sectors. In 2020, due to the lockdown conditions caused by the pandemic situation, the event was transposed in telematic mode, with live presentation slots of activities and interactive contents shared from the research groups to the users.

During the three years, three different products and platforms solutions for involvement and communication on the topic of awareness of Cultural Heritage were tested. The evolution of the solutions, from the same available 3D documentation database, was improved as a result of on-site observations of users' involvement and awareness, both to the Cultural Heritage theme and to the digital-phygital communication medium itself. The observations made it possible to conduct targeted and substantiated reflections on the relationship between citizenship, Architectural Heritage and Digital Heritage. It was developed by observing how users spontaneously interact with HLOs and, concerning the answers and clarifications requested, assessing their mechanisms of reception and processing of information on Cultural Heritage. The process was assumed towards a 360° knowledge experience given by both visual-tactile perception and the knowledge of digital data and information on Cultural Heritage integrated with HLOs.

2019 – Immersive Fruition of VR Visits within Point Clouds Acquired on Upper Kama sites

The first participation in the European Research Night included, as a phygital heritage communication solution, the setting up of a stand for VR enjoyment of Upper Kama's virtual environments (digital component), through the adoption of Oculus Rift visor and joysticks for immersive interaction (physical component).

As the purpose of the communication strategy, it was chosen to provide users with versions of Digital Heritage from the original databases acquired on site on the architectural complexes, avoiding processed versions of 3D models. The aim was to expedite the construction of the platforms and to provide a reliable experience from the on-site visit. The production of 3D models in the context of research regardless of the method implemented (GIS or BIM among the most common) implies a critical reinterpretation of the modelling, which depends on the intended functionality of the product (for conservation, restoration, technical intervention, or spatial mapping) provides for a criticization of the available information.

Such an interpretation may inevitably lead to a discretisation of the visual information, depending on the associated functional meaning, with the consequent geometric, texture and setting characterisation associated with the object design.

In the case of the HLOs of the PROMETHEUS project, the choice was to adopt 3D point clouds, acquired by on-site laser scanning procedure with RGB metric and colourimetric information, as direct import data in the VR environment. At the same time, a software platform capable of hosting this data format, critical both in terms of file size (extended to tens of Gigabytes) and spatial data extension and density (more than 500.000 points/sqm for each dataset) was envisaged, taking advantage of expeditious supplementation and minimal processing.

In this case, the solution was adopted through the *Veesus Arena4D* platform. *Veesus* technology allows interaction between external data and its applications enabled by the conversion to the proprietary *.vpc* format, conducted with the *VPC Creator* tool. The automatic conversion preserves the topological, colorimetric and georeferencing features that characterise the source point cloud, and it self-interacts with the *Oculus* VR navigation technology. The user involved in the VR experience finds himself in an immersive environment where he interacts as an avatar with a first-person navigation view of the spatial dataset. The size of the avatar is scaled and adjusted in view's height about the proportions of an observer concerning the real monument (Figure 4).

Figure 4. Virtual environments from 3D laser scanning point clouds, as fruited in an immersive mode in Arena4D platform through Oculus Rift visors. Source: De Marco R., 2019.

In the interaction with the HLO, structured as an *Arena4D* project enjoyed together with the narrative guidance of the operator, the user was able to move freely (but with a fixed ground level) within the scene, without the insertion of "invisible walls" or "semi-transparent fences" limiting the motion simulation. In this condition of freedom, the user was interested in two cognitive perception mechanisms:

1. by approaching the virtual architectural surfaces without limitations, he became aware of the "discontinuous" topology of the domains of point clouds, made by a non-continuous grid of points that, despite the dimension associated with each point, reveals at a certain distance its "virtuality" and evanescence about the solidity of real Heritage objects.
2. The absence of "virtual solidity" of the architectural replica also permitted the user to freely move not only in the identified spaces of the virtual visit but also through them and in the portions without data, corresponding to wall interstices, shadow cones or inaccessible rooms.

This "phantom mode" navigation, voluntarily chosen by users or shown within their involuntary motion approach to details, has solicited playful and discovery behaviours for the environments, as well as curiosity and learning stimuli concerning the technology issues of the virtual system.

However, it also resulted in a reduction of the immersive component for users, who never lost the cognitive awareness of being in a virtual replica of the system, often with conditions of disorientation and psycho-motor imbalance linked to the relationship between perceptual simulation and the physical conditions maintained by the user in the real context of the stand (e.g., crowded and noisy environment, presence of the guiding operator, accidental contact motion outside the stand).

The prevailing response of the involved citizens included feelings of curiosity towards the technological and innovation aspect provided by the digital technologies adopted for HLO, and the playful-exploratory involvement in the modes of interaction offered by the visor and remote controls. The learning impact was greater on the technological and digital smart solutions themes than on the themes of discovery and awareness of the conservation of Cultural Heritage sites. (Figure 5)

2020 – Remote Access to 3D and Multimedia Information on Upper Kama sites through a Web Dashboard

The second participation in the European Research Night included, as a phygital heritage communication solution, the design of a web-accessible dissemination portal (physical component), structured as a dashboard, enriched with information content and virtual 3D results (digital component) on the sites documented in Upper Kama.

The communication strategy was updated from the previous year, keeping fixed the intention to engage users in an immersive approach concerning the replica of Digital Heritage, but implementing the information apparatus to allow the learning process to take place also remotely, without the assistance of an operator. In addition, the organisation of the European Research Night in a virtual mode during the Covid-19 pandemic period has forced the need for improvement of independence and personalisation opportunities for participation in the educational process.

The information and digital data were structured on a web-hosting platform, managed through *WordPress*, designing a layout format useful to provide both an introduction to the context and physical reality of the site and to function as a graphic hyperlink to access virtual 3D web-sharing contents. The graphic layout, developed for each site, was structured on three areas of information (Figure 6):

Figure 5. Stand and users' interaction moments at the Research Night in 2019. Source: De Marco R., 2019.

- an introductory General Information section, identifying the monument, the geographical area, and a description of the main historical and architectural features. Contents were developed for the general public and users. A credits section, relating to the scientific project and the researchers involved, was added.
- a Documentation Methodology section, with a brief description of the methodological development of data acquisition and processing, applied to the production of the Digital Heritage system. The section was enriched with smart buttons linked to slide galleries of images, hyperlinks and videos from web platforms (e.g., Youtube), to enrich the learning content with graphic and multimedia material, related to historical images, survey images, and digital output graphics. Contents were developed for the general public, with increased multimedia involvement in communication, and for a non-specialised technical/professional audience regarding ICT application information. A supporting bibliography of scientific publications on the subject was provided.
- an interactive Virtual Switch section. The section mainly includes three possible content implementation solutions: (a) a descriptive photo gallery of the physical and digital version of Cultural Heritage, developed with *MetaSlider* tool for *WordPress*; (b) a virtual tour, developed with *WP VR* tool for *WordPress*, including spherical or panoramic photos of the site; (c) a viewer of 3D

datasets, virtual models or point clouds from *SketchFab* platform. All contents can be navigated and orbited directly from the *Dashboard*, but they also link to the hosting source.

The virtual datasets were selected from the digital survey database available for each site and followed processing aimed at optimising both the info-graphic components of VR representation and the web visualisation performance related to their size and technical format.

Figure 6. Heritage objects cards available on the Dashboard: general structure and main VR contents. Source: De Marco R., 2020.

For the 3D point clouds by Laser Scanner source, they were extracted from the overall source datasets and underwent a process of decimation from the original densities (1 point/1 mm) to lighter versions (1 point/10-50mm), however capable of providing a constitutive perception of the elements and surfaces present in the scene and on the architectural structures. For the 3D models, mainly from UAV sources, were processed with polygonal meshes from 3D point clouds reconstructed with SfM photogrammetry in *Agisoft Metashape* (Figure 7).

Figure 7. Digital databases with a high density of 3D spatial data, collected from the on-site 3D digital survey: above, 3D laser scanning point cloud, below, 3D Structure from Motion UAV point cloud. Source: De Marco R., 2019.

The 3D meshes had double processing:

- the optimisation of the polygonal mesh decimated over the entire dataset with multiple intensities. In particular, in the case of single monuments, a uniform decimation was applied to 100 poly/sqm. While in the case of larger sites, consisting of a landscape component and several monumental elements, a higher decimation was conducted (10 poly/sqm) on the landscape context, and a minor one (50 points/sqm) was applied to the architectural elements, aimed at maintaining a higher geometric resolution on the monuments.
- the optimisation of the texture, processed from the photo gallery underlying the photogrammetric reconstruction. The texture was checked for both resolution (8,192-pixel grid) and blending mode, conducted in Average mode (meaning that it uses the weighted average value of all pixels from individual photos, in selecting the way how colour values of pixels from different cameras are combined).

Through the *SketchFab* interface, it was possible to calibrate the environment conditions (lighting, background) and VR navigation sets, setting the scale of the viewer's avatar, the field of view (angle), the background, and limiting or not limiting the directions of movement and the distance of navigation or approach to the 3D model, even in vertical elevation. (Figure 8, Figure 9).

The layout of the Dashboard was designed on a more traditional HLO approach, where a static interface of textual content and images also includes viewer containers. Within them, the user finds a type of interactive content, which dynamically engages him/her to interact with the virtual object and unconditionally discover its details. The use of the system from a personal device, in a free remote environment, on the one hand, leaves the context of the development of the educational process uncontrolled. On the other hand, it allows the user to be more familiar with the technical languages and control modes of the device, adopting those already experienced with the personal device.

The prevailing response of the involved citizens included extensive navigation of the content and hyperlinks associated with the Dashboard, both for users and professionals, who were monitored in their accesses and the cascading structure of engagement and visitation of the associated content.

Figure 8. Polymesh model optimised with different densities of poly-grid concerning the morphological detail on heritage monuments and territorial context. Source: De Marco R., 2020.

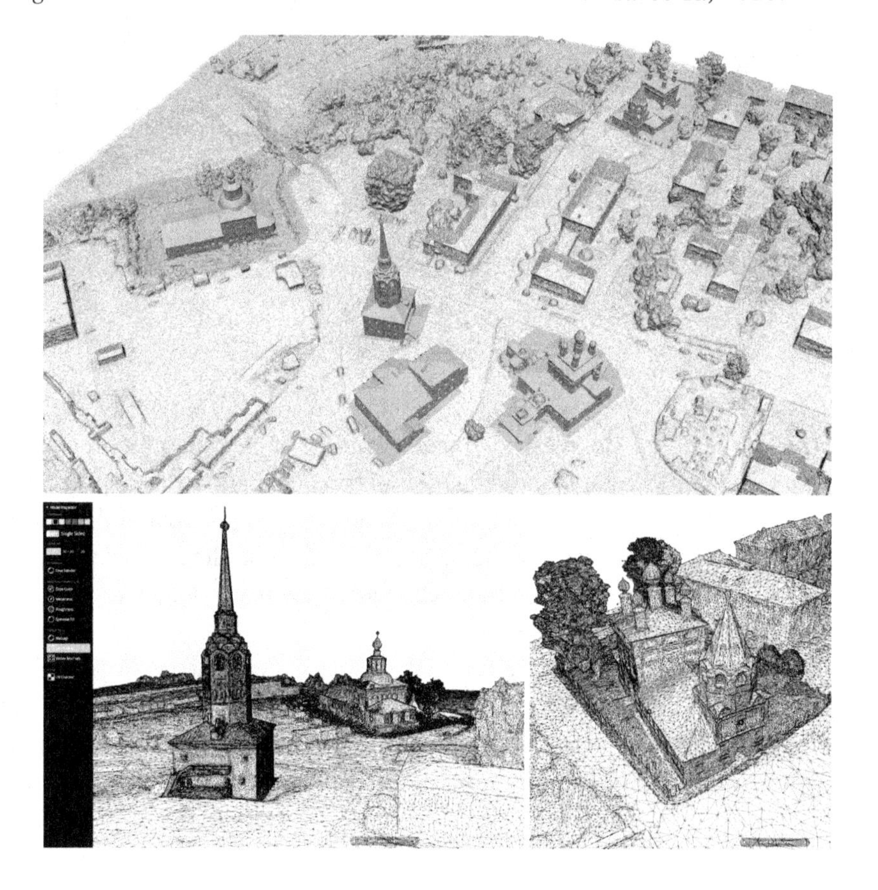

Figure 9. Smart 3D VR model on SketchFab in solid, textured and rendered view. Source: De Marco R., 2020.

2021 – Delivery of take-away learning sets on Upper Kama sites, integrated within Information Cards, Web-Based 3D Virtual Models and 3D Scaled Physical Prototypes of Monuments

The third participation in the European Research Night has improved the solution of the previous year, reinforcing the scientific integration of digital and real 3D models as communication support for Cultural Heritage. Web-access 3D virtual models (digital component) were extended and enriched with interactive information, associated with 3D print models and explanatory cards (physical component) for each site.

The communication strategy was aimed at detailing and optimising the virtual 3D models communicated through the *SketchFab* platform in the previous edition, while also providing educational gadgets that, distributed at the event stand, could constitute a tangible orientation and learning support component to the educational process on Cultural Heritage. Compared to the previous editions, this solution introduced a "tangible portability" and a fully customised version of HLO for the user, who has been untied from the condition of presence on the site or his experience in web browsing, triggering a more shared social involvement among the users themselves.

Similarly to the Dashboard, paper cards (A5 format) were structured, describing the general information on the site, a brief historical introduction and a photographic overview of the monument. Each card was associated with a 3D printed plastic model of the architectural object, reduced to a scale of 1:1000.

The 3D prototypes were developed from the 3D mesh models achieved with SfM aerial photogrammetry acquired on-site, adopting the same datasets already applied for the 3D VR models on the *SketchFab* platform. The mesh models, already optimised in the polygonal surface, were transformed into manifold solids with the addition of a parallelepiped stabilisation basement, processed in an *Autodesk MeshMixer* reverse-modelling environment.

Subsequently, the final meshes were processed with *CuraSlicer* software to simulate the 3D printing process with the technical characteristics of the desired prototype, enabling the setting of parameters. Some parameters were related to the technical operations of the printer (filament flow, printing speed and temperature), while others were calibrated due to the portability and robustness of the printed 3D model, to guarantee an adequate use and tactile experience even for thin architectural shapes, such as onion domes, bell towers, or ruined portions. Among these parameters, the Layer height (0,4 mm, as the extrusion diameter), the Fill density (20%, corresponding to the internal filling percentage of the prototype) and the Shell thickness (0,8 mm) were evaluated. They have been calibrated to ensure both the tactile perception sensitivity of the geometric components of the architecture and the resistance of the prototyped model to the touch and handling even by young users.

In the final solid 3D models, printed with monochrome plastic filament, users could appreciate the architectural configuration of each monument, reproducing the traditional forms of architecture and allowing them to appreciate the articulation of elements, blocks, openings, and roofs that conform to each building, linked to the architectural and geometric lexicon of their cultural belonging. The sensitivity of detail in the final printed models for delivery to users was 50cm. Other models, in scale 1:300, were presented at the stand, accompanied by the operators' narration to appreciate the features of Cultural Heritage also at the construction-technological scale. The printing scale was chosen for the optimisation of printing time/cost and the proportion of the architectural elements in the final prototype, allowing the printing of more than 1,000 pieces per monument for the event. The printing processes, limited to up to 1.5 hours of printing per piece, were conducted by the simultaneous work of 10 printers (Figure 10, Figure 11).

To supplement the learning components not represented by the material 3D prototype, a QR Code was placed on the HLO card, accompanying each model as a hyperlink to the VR model version available on *SketchFab*. These versions have maintained a high level of texture optimisation, integrating information on both the material data and the more particular details of the architectural apparatus, such as the decorative system and artistic-sculptural details. (Figure 12, Figure 13).

In the design of the immersive virtual visit process, *SketchFab* 3D models were completed by the addition of annotations, containing both textual and photographic information about historical events, transformations, interventions and curiosities about the conservation and management of monuments. The annotations, directly visible and interactive in the VR model as *balloon buttons*, made it possible to critically integrate notions and contents characterising the educational process towards Cultural Heritage, strengthening the awareness-raising impact of users and citizens on the topic of Heritage valorisation (Figure 14).

Figure 10. 3D printed prototypes from 3D models of Heritage sites, with double dimensional scale (target size of 6 cm): above, portable models for users in scale 1:1000, below, model in scale 1:300. Source: De Marco R., 2021.

Figure 11. The difference in tactile and visible detail between the 3D prototyped models printed in scale 1:300 (left) and 1:1000 (right). Source: De Marco R., 2021.

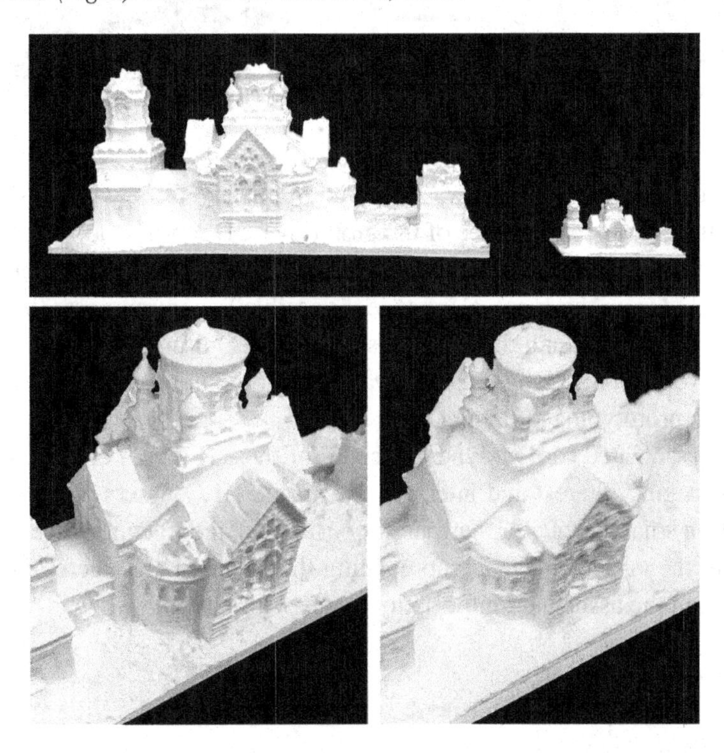

Figure 12. Heritage Learning Object card, with QR code for the VR view of the model through a smart device. Source: De Marco R., 2021.

The reception by users of this solution was reported high. The users, while visiting the stand, were directly involved by the tangible experience of the prototyped 3D models, adopted by the operators as a means of support to trigger their interest in the architectural object. The presence of a real-time working 3D printer at the stand, during the event, also made it possible to communicate the scientific relevance of research in the documentation and digital representation of Cultural Heritage, highlighting the new technologies of creative manufacturing, raising awareness of the issue of access and reproducibility of Cultural Heritage with prototyping systems.

Subsequently, the associated card and the QR code for the VR model constituted key materials for maintaining and prolonging interest and interaction with the HLO. Users were enabled to personally scan the QR code from smart-devices (smartphones or tablets) and to experience the VR model on the web, during and after the in-person event, appreciating the different architectural characters in greater detail and learning from the associated information notes within them.

Figure 13. Smart connection and fruition of 3D VR models on SketchFab through QR code on the card. Source: De Marco R., 2021.

In the case of young users and children without a personal device, the availability of the VR object's link on the HLO card made it possible to extend the interaction process with Digital Heritage to parents and tutors, involving them in the digital connection and virtual fruition of 3D products. From the point of view of these users, the specific strategy of HLOs' sharing made it possible to activate spontaneous training on the digital skills envisaged for Digital Heritage communication, and children acted as enthusiastic teachers towards the sharing of manoeuvrability and confidence skills with smart-devices to older users (Figure 15).

DISCUSSION OF ACHIEVED RESULTS AND FUTURE RESEARCH DIRECTIONS

The presented research experience has highlighted the opportunities for HLOs to increase the interaction of users within Digital Heritage and the means for the virtual enjoyment of PROMETHEUS sites, acting on the improvement of immersive experience to assign educational targets to the perception of Heritage qualities, as replicated from the real object.

Figure 14. 3D VR models on SketchFab are enriched with annotations of multimedia contents for historical, conservation and management information on Cultural Heritage. Source: De Marco R., 2021.

However, technologies and interaction methods for education on Cultural Heritage are still not supported by guidelines, to assist in predetermining its relevance to attaining the intended objectives of Virtual Heritage applications. In this way, their reapplication must consider different implications in building a sustainable educational action through tools for digital tourism:

- the supporting of smart citizens and cultural institutions in adopting digital strategies and tools for an educational purpose coming in the virtual accessibility of architectural heritage sites.
- the encouragement of professionals and public institutions in promoting and expanding digitization initiatives on Cultural Heritage, adopting fast survey and remote control opportunities of data acquisition for sustainable documentation also of remote and inaccessible sites.

In this way, the indications from fast survey experiences and the development of smart models for web-fruition or 3D printing can be replicated from PROMETHEUS results to other examples of Cultural Heritage sites, in particular considering Cultural Heritage Routes typology. However, besides the production of digital replicas, the considerations on users' expectations of involvement with the HLO need to be analysed and stated, based both on the users and Heritage characterization within "phygital" communication features.

Figure 15. Stand and users' interaction moments at the Research Night in 2021. Source: De Marco R., 2021.

The comparison between the guidelines for the design of Virtual Learning Environments and ICT advancements highlights multiple opportunities for interaction between communication modes and digital content on Cultural Heritage. The choice for the most appropriate development directions lies in the evaluation of which main features characterising synchronous learning mechanisms (from on-site accessibility to Heritage) can be reliably adapted in the virtual version, to ensure asynchronous smart accessibility for users, even despite the participation in physical events of communities' engagement.

The solution for communication of Cultural Heritage performed by the association between 3D virtual models and their 3D prototyped version for PROMETHEUS sites represents the result of a pyramidal experimental procedure performed to adapt the digital survey dataset to phygital fruition means. It has been affirmed as a synchronous result of smart-learning devices to trigger the asynchronous development of learning mechanisms from virtual contents, and it has methodologically contributed to the extension of open-access galleries of HLOs from reality-based documentation of Cultural Heritage.

The evaluation of users' experience (such as the time of VR interaction, the contextualization into the site's context, and the recognition of Cultural Heritage features) and its impact on the educational purpose has been central in the research process to understand and calibrate the VR parameters linked to the simulation of PROMETHEUS Heritage sites and their virtual accessibility. In the same way, users' confidence in digital languages has been calibrated following the adoption of foreseen smart devices to access the products and their contents, improving usability, user-friendliness and reducing assistance in the asynchronous fruition of HLOs.

The choice of the VR mode of fruition enables the extension of accessibility to Cultural Heritage by allowing an interaction free of geographical contextualisation. While Augmented Reality would impose a physical co-presence with the site, VR breaks down the limitations of users' degrees of freedom: its representation remains confined to the computer-generated virtual environment, but within this confinement it is possible to freely condition the user's simulation of experience, breaking down the sensory barriers and standard motion mechanisms of physical vision.

It is in the transition of the learning-storytelling modalities from "static" to "dynamic" that the evaluation of the impact of digital media on the educational receptors (users and citizens) takes place. This transition is constituted by a double design scale: the "identification of the educational receptors", as behavioural mechanisms that are triggered in users, and the "structuring of information", as the design of the communication means that convey the digital translation of the Cultural Heritage physical replica and its related contents.

The adaptation of VR environments into HLOs can give the sense of belonging of the learner to the Cultural Heritage Site: it brings the result of Immersive Learning, bringing the experience of the user to further enhance his creative attitude based on the learned information. Since the imagery displayed in VR is not predetermined or pre-recorded but generated in real-time, users can continuously adapt their behavioural learning to the stimulus perceived from the scene. In this way, the comparison of immersive realities and interaction educational modalities must be evaluated on three factors:

1. contextual relationship between object and contents
2. collaboration between digital contents and the fruition platform.
3. engagement between digital content and learning receptors.

In this way, two perspectives of products and medium must be discussed:

1. focus on forms/categories of Virtual Learning Environments
2. focus on enabling technologies and virtual applications for the communication and accessibility to Cultural Heritage.

In this way, the Cultural Heritage communication experience through the results of the PROMETHEUS project is to be delivered in an unlimited and unconditional manner, through access sessions to the platform even structured for school classes, with the freedom to navigate in the virtual environment and to learn from the contents arranged. The platform was also disseminated among master's students and professionals at an international level, forming part of the panorama of virtual 3D content web galleries set up for the digitisation of heritage.

It is central to consider the impact of smart culture and smart accessibility to Cultural Heritage on three levels: personal, industrial, and societal. The positive effects considered ranging from the overcoming of time, cost, and distance barriers, to the inclusion of vulnerable categories of citizens. Furthermore, the standpoint of industries on creative reproduction joins the citizens' purpose of engagement with the consumption of cultural learnings and services. In this way, influential factors arise from the fruition of Virtual Learning Environments and HLOs, regarding the learner's digital media literacy and changing attitudes on the traditional experience of Cultural Heritage visits and assumptions into daily societal mechanisms.

Within PROMETHEUS advances of research on communication strategies for Cultural Heritage, it is foreseen the development of further experimentation and progresses on:

- the engagement of learners on deeper levels of education enabled by the Virtual Learning Environments, introducing learning contents also with a project purpose of civic responsibility and territorial sustainability towards the administration and conservation of Cultural Heritage.
- the fostering of the authenticity of Cultural Heritage values for learning mechanisms, including the attitude to problem-based learning and solving suggestions to the critical issues encountered by the user concerning the virtual scene and the perceptions developed, for example on conservation conditions of sites and the possibilities of the community to intervene, integrating intentions of active contribution to the learning contents of the platform.
- the expansion of the virtual dimension of access to the context of territorial location and networking of Cultural Heritage sites, including challenging issues that belong to the system of services, infrastructures and cultural extension characterising the scale of the route.
- the adaptation of digital infrastructure to the more dynamic interactions developed by users with smart devices, considering also inter-personal connection and the materialisation of the user inside the "metaverse" of the virtual platform, to replicate the experience with the site not only in a stand-alone manner but through the engagement between groups.

CONCLUSION

Multiple modalities can be established to foster accessibility to Cultural Heritage. Pursuing this aim, it is necessary to foresee, exactly in the exchange of data between the proposing system and the receiver of information, a possibility for the receiver to state himself within the information.

Accessibility corresponds to the opportunity to become part of a place, to be able to experience it and establish relationships within it. That is the reason why, in digital transposition, it is necessary to attempt the re-creation of narrative forms of Storytelling able to project the user into the dimension of knowledge communication. It regards the capacity to develop learning experiences, which corresponds to the achievement of a circuit of digital products in which different ways of enjoyment, playing and deep learning can be faced.

Through digital practice and its related products, it must be possible to develop a sense of belonging to Cultural Heritage and to define a common sense for Heritage as a background source of knowledge developed from real experiences. It is necessary to link the concept of Heritage to the dimension of "experiences" precisely because it is in the translation of signs and forms, achieved as an action of "drawing", that a virtual experience of Heritage must be reconstructed and personalised, preferably in an autonomous way, by each user. In this way, multiple versions of Heritages will be achieved due to the multiplicity of meanings that users will associate, generating in this way many other natures of Heritage.

It is considered the deeper sense of interpreting the nature of Cultural Heritage Routes: their features, transformations and frameworks originate from a historical phenomenon linked to multiple reasons, and in this way, the identity of their complex system has to consider even more complex temporal dimension.

Historical traces are joined to the present societal mechanisms, and the digital processes become the means to connect this double dimension of knowledge through the future. Regarding digital translation, objects that cannot find an appropriate form of rewriting, interconnection and dialogue, are destined to disappear from memory, defining a new type of "prehistory" compared to the present form of knowledge. With the advance of digital practice, the need to interconnect its forms and models and to increase synergies between data preludes to a scenario where each community must attempt to assert meaningful syntheses to its Cultural Heritage, to face the process of globalisation and commercialisation of its memory. The task of Cultural Heritage Routes is therefore very delicate: it is a matter of experimenting with narrative forms for the communication of Heritage through society, trying to understand also the most "convenient" modes to sustainably commercialise its meanings, in a process never completely defined due to its dynamic changing.

Drawing and the science of Representation have the central role of defining appropriate strategies and methodologies to interpret, simplify (not in the sense of "reduction" but in the way of achieving "order") and reshape the image of Heritage, its elements and sites. This aim corresponds to the construction of experiences and dialogues on Digital Heritage, between its drawing and the public. It defines the construction of a core message to be completed by many narrative forms, that find complex expressions between movement, music, art, dance, languages, and geometries. Thus, the aim deals with the "drawing of experience".

To be communicated, Heritage must adopt the network dimension of a Cultural Heritage Route, and the Cultural Heritage Route must be definable even in its dynamic, fluid reality. The lack of borders and boundaries, to define its extension, characterises the assumption of meaning and identity to the route of communication of Heritage knowledge. However, due to the same knowledge dynamism, it will never be possible to affirm, with the drawing of models or any digital medium, a communication limit.

The storytelling of Cultural Heritage collides with the narration of a topic as extended as the history and development of communities. It assumes the reflection proposed by T. Gilliam in "*The Imaginarium of Doctor Parnassus*": it is not a single story that funds the universe, but it is the self-perception of each person to structure his narration. Each user communicates himself, as the protagonist of his imagery, and he actively becomes part of the structure of digital projections that complete the narration of a Cultural Heritage Route, contemplating all forms and values linked to society, community, and Heritage. In the film, the narrative modality of storytelling and play is animated by choice, by the possibilities of interaction with the story itself, explicating an important lesson about the truth of storytelling: the co-presence of several narrative dimensions that constantly intertwine with the users. But if "*you can't stop a story being told*", it also assumes the risk of reaching a condition when "*no one will listen to Parnassus' stories anymore*".

The definition of a Cultural Heritage Route necessarily means defining a simplified vision of a more complex system, developing syntheses and translating them into narrative expressions. The extent of this simplification, and reduction of narration defines the task of the drawer. The choice of deepening the communication, reception and memory contribution to the dimension of Cultural Heritage for continuing its narration remains in charge of everyone.

ACKNOWLEDGMENT

The presented activities, data processing and results have been developed under the international project "PROMETHEUS". PROMETHEUS project is funded by the EU program Horizon 2020-R&I-RISE-Research & Innovation Staff Exchange Marie Skłodowska-Curie.

This project has received funding from the European Union's Horizon 2020 research and innovation programme under the Marie Skłodowska-Curie grant agreement No 821870.

The authors would like to thank Francesca Galasso, Elisabetta Doria, Hangjun Fu, for their strong support in the management of activities and products presented during the Research Nights 2019-2021 in Pavia. Gratitude is also expressed toward the international research group of the PROMETHEUS project, from Universitat Politècnica de València and Perm National Research Polytechnic University, for the implementation of information and databases on the Cultural Heritage sites in Upper Kama.

SHARPER is a European Researchers' Night project funded by the European Commission under the Marie Skłodowska-Curie actions. GA 101061553.

The editorial responsibility is due to Sandro Parrinello for the paragraphs "Introduction", "Background: a critical review on the digital transition from Cultural Heritage to Virtual Heritage", "The societal education challenge related to Cultural Heritage Routes", "Conclusions"; to Raffaella De Marco for the paragraphs "Involving users in Cultural Heritage awareness: identifying receptors and structuring means of communication", "Solutions and recommendations: towards a smart-user communication of Cultural Heritage", "Discussion of achieved results and future research directions".

REFERENCES

Ahmad Yusri, M. H., Johan, M. A., Khusaini, N. S., & Ramli, M. H. M. (2022). Preservation of cultural heritage: A comparison study of 3D modelling between laser scanning, depth image, and photogrammetry methods. *Jixie Gongcheng Xuebao, 19*(2), 125–146.

Arenghi, A., Garofolo, I., & Sormoen, O. (Eds.). (2016). Accessibility As A Key Enabling Knowledge for Enhancement of Cultural Heritage. Franco Angeli.

Balado, J., Frías, E., González-Collazo, S. M., & Díaz-Vilariño, L. (2022). New Trends in Laser Scanning for Cultural Heritage. In D. Bienvenido-Huertas & J. Moyano-Campos (Eds.), *New Technologies in Building and Construction* (pp. 167–186). Springer. doi:10.1007/978-981-19-1894-0_10

Ballarin, M., Balletti, C., & Vernier, P. (2018). Replicas in cultural heritage: 3D printing and the museum experience. *The International Archives of the Photogrammetry, Remote Sensing and Spatial Information Sciences, 2018*(42), 55–62. doi:10.5194/isprs-archives-XLII-2-55-2018

Balletti, C., & Ballarin, M. (2019). An Application of Integrated 3D Technologies for Replicas in Cultural Heritage. *ISPRS International Journal of Geo-Information, 2019*(8), 285. doi:10.3390/ijgi8060285

Barba, S., Barbarella, M., Di Benedetto, A., Fiani, M., Gujski, L., & Limongiello, M. (2019). Accuracy Assessment of 3D Photogrammetric Models from an Unmanned Aerial Vehicle. *Drones (Basel), 2019*(3), 79. doi:10.3390/drones3040079

Bertin, J. (1970). Le graphique. *Communications (Englewood)*, 15.

Bitelli, G., Balletti, C., Brumana, R., Barazzetti, L., D'Urso, M. G., Rinaudo, F., & Tucci, G. (2017). Metric documentation of Cultural Heritage: Research directions from the Italian GAMHER project. *The International Archives of the Photogrammetry, Remote Sensing and Spatial Information Sciences, XLII-2*(W5), 83–90. doi:10.5194/isprs-archives-XLII-2-W5-83-2017

Borda, A., & Bowen, J. (2017). Smart Cities and Cultural Heritage: A Review of Developments and Future Opportunities. In *Proceedings of EVA London* (pp.9-18). London, UK: BCS London. 10.14236/ewic/EVA2017.2

Buhalis, D., & Karatay, N. (2022). Mixed Reality (MR) for Generation Z in Cultural Heritage Tourism Towards Metaverse. In J. L. Stienmetz, B. Ferrer-Rosell, & D. Massimo (Eds.), *Information and Communication Technologies in Tourism 2022. ENTER 2022* (pp. 16–27). Springer. doi:10.1007/978-3-030-94751-4_2

Burkey, B. (2022). From Bricks to Clicks: How Digital Heritage Initiatives Create a New Ecosystem for Cultural Heritage and Collective Remembering. *The Journal of Communication Inquiry, 46*(2), 185–205. doi:10.1177/01968599211041112

Byrne, D. (2008). Heritage as social action. In G. Fairclough, R. Harrison, J. H. Jameson, & J. Schofield (Eds.), *The heritage reader* (pp. 149–173). Routledge.

Cardone, V. (2008). Modelli grafici dell'architettura e del territorio [Graphic Models of architecure and the territory]. CUES.

Cardozo, T. M., & Papadopoulos, C. (2021). Heritage artefacts in the COVID-19 Era: The aura and authenticity of 3D models. *Open Archaeology, 7*(1), 519–539. doi:10.1515/opar-2020-0147

Champion, E., & Rahaman, H. (2019). 3D digital heritage models as sustainable scholarly resources. *Sustainability, 11*(8), 2425. doi:10.3390u11082425

Champion, E., & Rahaman, H. (2020). Survey of 3D digital heritage repositories and platforms. *Virtual Archaeology Review, 11*(23), 1–15. doi:10.4995/var.2020.13226

Chiapparini, A. (2012). *Communication and Cultural Heritage: Communication as Effective Tool for Heritage Conservation and Enhancement* (Unpublished doctoral dissertation). Politecnico di Milano University.

Chong, H. T., Lim, C. K., Ahmed, M. F., Tan, K. L., & Mokhtar, M. B. (2021). Virtual reality usability and accessibility for cultural heritage practices: Challenges mapping and recommendations. *Electronics (Basel), 10*(12), 1430. doi:10.3390/electronics10121430

Choromański, K., Łobodecki, J., Puchała, K., & Ostrowski, W. (2019). Development of virtual reality application for cultural heritage visualization from multi-source 3D data. *The International Archives of the Photogrammetry, Remote Sensing and Spatial Information Sciences, XLII–2*(W9), 261–267. doi:10.5194/isprs-archives-XLII-2-W9-261-2019

Chung, N., Han, H., & Joun, Y. (2015). Tourists' intention to visit a destination: The role of augmented reality (AR) application for a heritage site. *Computers in Human Behavior, 50*, 588–599. doi:10.1016/j.chb.2015.02.068

Ciurea, C., Pocatilu, L., & Gheorghe Filip, F. (2020). Using Modern Information and Communication Technologies to Support the Access to Cultural Values. *Journal of System and Management Sciences, 10*(2), 1–20.

Colucci, E., De Ruvo, V., Lingua, A., Matrone, F., & Rizzo, G. (2020). HBIM-GIS Integration: From IFC to CityGML Standard for Damaged Cultural Heritage in a Multiscale 3D GIS. *Applied Sciences (Basel, Switzerland), 2020*(10), 1356. doi:10.3390/app10041356

De Marco, R., & Dell'Amico, A. (2020) Connecting the Territory between Heritage and Information: Databases and Models for the Cultural Heritage Routes. In *Proceedings UID 2020 CONNECTING | drawing for weaving relationships* (pp. 258-277). Franco Angeli.

De Marco, R., & Pettineo, A. (2021). The recognition of Heritage qualities from feature-based digital procedures in the analysis of historical urban contexts. *Int. Arch. Photogramm. Remote Sens. Spatial Inf. Sci., XLVI-2/W1-2022*, 175–182.

Di Giulio, R., Maietti, F., Piaia, E., Medici, M., Ferrari, F., & Turillazzi, B. (2017). Integrated data capturing requirements for 3D semantic modelling of Cultural Heritage: The INCEPTION protocol. *The International Archives of the Photogrammetry, Remote Sensing and Spatial Information Sciences, XLII-2*(W3), 251–257. doi:10.5194/isprs-archives-XLII-2-W3-251-2017

Di Giuseppantonio Di Franco, P., Galeazzi, F., & Vassallo, V. (2018). *Authenticity and cultural heritage in the age of 3D digital reproductions*. McDonald Institute for Archaeological Research.

Doerr, M. (2003). The CIDOC CRM- an ontological approach to semantic interoperability of metadata. *AI Magazine, 24*(3), 75.

Fanea-Ivanovici, M., & Pan, M. C. (2020). From Culture to Smart Culture. How Digital Transformations Enhance Citizens' Well-Being Through Better Cultural Accessibility and Inclusion. *IEEE Access: Practical Innovations, Open Solutions, 8*, 37988–38000. doi:10.1109/ACCESS.2020.2975542

Faraj, G., & Micsik, A. (2021). Representing and Validating Cultural Heritage Knowledge Graphs in CIDOC-CRM Ontology. *Future Internet, 13*(11), 277. doi:10.3390/fi13110277

Georgopoulos, A. (2017). Data Acquisition for geometric Documentation of Cultural Heritage. In M. Ioannides & N. Magnenat-Thalmann (Eds.), *Mixed Reality and Gamification for Cultural Heritage* (pp. 29–73). Springer. doi:10.1007/978-3-319-49607-8_2

Georgopoulos, A., & Stathopoulou, E. K. (2017). Data acquisition for 3D geometric recording: state of the art and recent innovations. In M. Vincent, V. López-Menchero Bendicho, M. Ioannides, & T. Levy (Eds.), *Heritage and Archaeology in the Digital Age. Quantitative Methods in the Humanities and Social Sciences* (pp. 1–26). Springer. doi:10.1007/978-3-319-65370-9_1

Ghani, I., Rafi, A., & Woods, P. (2020). The effect of immersion towards place presence in virtual heritage environments. *Personal and Ubiquitous Computing, 24*(6), 861–872. doi:10.100700779-019-01352-8

Giannakoulopoulos, A., Pergantis, M., Poulimenou, S. M., & Deliyannis, I. (2021). Good Practices for Web-Based Cultural Heritage Information Management for Europeana. *Information (Basel), 12*(5), 179. doi:10.3390/info12050179

Ginzarly, M., & Jordan Srour, F. (2022). Cultural heritage through the lens of COVID-19. *Poetics*, *92*(A), 101622.

Gireesh Kumar, T. K. (2021). Designing a Comprehensive Information System for Safeguarding the Cultural Heritage: Need for Adopting Architectural Models and Quality Standards. *Library Philosophy and Practice*, *2021*(5392), 1-19.

Han, D. I., Claudia tom Dieck, M., & Jung, T. (2018). User experience model for augmented reality applications in urban heritage tourism. *Journal of Heritage Tourism*, *13*(1), 46–61. doi:10.1080/1743873X.2016.1251931

Howard, P. (2003). *Heritage: management, interpretation, identity*. Continuum. doi:10.5040/9781350933941

Ibrahim, N., & Ali, N. M. (2018). A conceptual framework for designing virtual heritage environment for cultural learning. *Journal on Computing and Cultural Heritage*, *11*(2), 1–27. doi:10.1145/3117801

Jan, J. F. (2018). Application of open-source software in community heritage resources management. *International Journal of Geo-Information*, *7*, 426.

Kaldeli, E., García-Martínez, M., Isaac, A., Scalia, P. S., Stabenau, A., Almor, I. L., & Herranz, M. (2022). Europeana Translate: Providing multilingual access to digital cultural heritage. In *Proceedings of the 23rd Annual Conference of the European Association for Machine Translation* (pp. 297-298). European Association for Machine Translation.

Kepczynska-Walczak, A. &Walczak, B.M. (2015). Built heritage perception through representation of its atmosphere. *Experiential Simulation*, *1*.

Kersten, T. P., Tschirschwitz, F., Deggim, S., & Lindstaedt, M. (2018). Virtual reality for cultural heritage monuments–from 3D data recording to immersive visualisation. In *Euro-Mediterranean Conference* (pp. 74-83). Springer. 10.1007/978-3-030-01765-1_9

King, L., Stark, J. F., & Cooke, P. (2016). Experiencing the Digital World: The Cultural Value of Digital Engagement with Heritage. *Heritage & Society*, *9*(1), 76–101. doi:10.1080/2159032X.2016.1246156

Kontogianni, G., Koutsaftis, C., Skamantzari, M., Chrysanthopoulou, C., & Georgopoulos, A. (2017). Utilising 3D realistic models in serious games for cultural heritage. *International Journal of Computational Methods in Heritage Science*, *1*(2), 21–46. doi:10.4018/IJCMHS.2017070102

Korro Bañuelos, J., Rodríguez Miranda, Á., Valle-Melón, J. M., Zornoza-Indart, A., Castellano-Román, M., Angulo-Fornos, R., Pinto-Puerto, F., Acosta Ibáñez, P., & Ferreira-Lopes, P. (2021). The Role of Information Management for the Sustainable Conservation of Cultural Heritage. *Sustainability*, *13*(8), 4325. doi:10.3390u13084325

Longstaff, P. H., Armstrong, N. J., Perrin, K., Parker, W. M., & Hidek, M. A. (2010). Building resilient communities: A preliminary framework for assessment. *Homeland Security Affairs*, *6*(3).

Macrì, E., & Cristofaro, C. L. (2021). The Digitalisation of Cultural Heritage for Sustainable Development: The Impact of Europeana. In P. Demartini, L. Marchegiani, M. Marchiori, & G. Schiuma (Eds.), *Cultural Initiatives for Sustainable Development. Contributions to Management Science* (pp. 373–400). Springer.

Mattar, J., Ramos, D. K., & Lucas, M. R. (2022). DigComp-Based Digital competence Assessment Tools: Literature Review and Instrument Analysis. *Education and Information Technologies*, 1–25.

Meegan, E., Murphy, M., Keenaghan, G., Corns, A., Shaw, R., Fai, S., Scandura, S., & Chenaux, A. (2021). Virtual Heritage Learning Environments. In Proceedings Digital Heritage. Progress in Cultural Heritage: Documentation, Preservation, and Protection. EuroMed 2020 (pp. 427–437). Academic Press.

Meyer, E., Grussenmeyer, P., Perrin, J. P., Durand, A., & Drap, P. (2007). A web information system for the management and the dissemination of Cultural Heritage data. *Journal of Cultural Heritage*, *8*(4), 396–411.

Miłosz, M., Montusiewicz, J., Kęsik, J., Żyła, K., Miłosz, E., Kayumov, R., & Anvarov, N. (2022). Virtual scientific expedition for 3D scanning of museum artifacts in the COVID-19 period – The methodology and case study. *Digital Applications in Archaeology and Cultural Heritage*, *26*, e00230.

Mohammadi, M., Rashidi, M., Mousavi, V., Karami, A., Yu, Y., & Samali, B. (2021). Case study on accuracy comparison of digital twins developed for a heritage bridge via UAV photogrammetry and terrestrial laser scanning. *Proceedings of the 10th International Conference on Structural Health Monitoring of Intelligent Infrastructure*, *10*, 1-8.

Moyano, J., Nieto-Julián, J. E., Bienvenido-Huertas, D., & Marín-García, D. (2020). Validation of close-range photogrammetry for architectural and archaeological heritage: Analysis of point density and 3D mesh geometry. *Remote Sensing*, *12*(21), 3571.

Niccolucci, F. (2017). Documenting archaeological science with CIDOC CRM. *International Journal on Digital Libraries*, *18*, 223–231.

Niccolucci, F., & Felicetti, A. (2018). A CIDOC CRM-based Model for the Documentation of Heritage Sciences. *2018 3rd Digital Heritage International Congress (DigitalHERITAGE) held jointly with 2018 24th International Conference on Virtual Systems & Multimedia (VSMM 2018),* 1-6.

Nofal, E. (2019). *Phygital Heritage. Communicating Built Heritage Information through the Integration of Digital Technology into Physical Reality* (Unpublished doctoral dissertation). KU Leuven, Arenberg Doctoral School.

Nofal, E., Reffat, R., Boschloos, V., Hameeuw, H., & Vande Moere, A. (2018). Evaluating the role of tangible interaction to communicate tacit knowledge of built heritage. *Heritage*, *1*(2), 414–436.

Nofal, E., Reffat, R., & Vande Moere, A. (2017) Phygital heritage: An approach for heritage communication Immersive Learning. In *Proceedings of Research Network Conference* (pp. 220-229). Verlag der Technischen Universität Graz.

Oppio, A., & Dell'Ovo, M. (2021). Cultural Heritage Preservation and Territorial Attractiveness: A Spatial Multidimensional Evaluation Approach. In P. Pileri & R. Moscarelli (Eds.), *Cycling & Walking for Regional Development. Research for Development* (pp. 105–125). Springer.

Ott, M., & Pozzi, F. (2010). Towards a new era for Cultural Heritage Education: Discussing the role of ICT. *Computers in Human Behavior.*

Paladini, A., Dhanda, A., Reina Ortiz, M., Weigert, A., Nofal, E., Min, A., Gyi, M., Su, S., Van Balen, K., & Santana Quintero, M. (2019). Impact Of Virtual Reality Experience On Accessibility Of Cultural Heritage. *The International Archives of the Photogrammetry, Remote Sensing and Spatial Information Sciences*, *XLII-2*(W11), 929–936.

Parrinello, S., & Cioli, F. (2018). Un progetto di recupero per il complesso monumentale di Usolye nella regione della Kama Superiore. *Restauro Archeologico*, *26*(1), 92-111.

Parrinello, S., & Dell'Amico, A. (2019). Experience of Documentation for the Accessibility of Widespread Cultural Heritage. *Heritage*, *2*(1), 1032–1044.

Parrinello, S., Picchio, F., De Marco, R., & Dell'Amico, A. (2019). Documenting The Cultural Heritage Routes. The Creation Of Informative Models Of Historical Russian Churches On Upper Kama Region. *The International Archives of the Photogrammetry, Remote Sensing and Spatial Information Sciences*, *XLII-2*(W15), 887–894.

Peinado-Santana, S., Hernández-Lamas, P., Bernabéu-Larena, J., Cabau-Anchuelo, B., & Martín-Caro, J. A. (2021). Public works heritage 3D model digitisation, optimisation and dissemination with free and open-source software and platforms and low-cost tools. *Sustainability*, *13*(23), 13020.

Pescarin, S. (2016). Digital heritage into practice. *SCIRES-IT-SCIentific RESearch and Information Technology*, *6*(1), 1–4.

Picchio, F., Bercigli, M., & De Marco, R. (2018). Digital Scenarios and Virtual Environments for the Representation of Middle Eastern Architecture. In C. L. Marcos (Ed.), *Graphic Imprints The Influence of Representation and Ideation Tools in Architecture* (pp. 541–556). Springer.

Picchio, F., De Marco, R., Dell'Amico, A., Doria, E., Galasso, F., La Placa, S., Miceli, A. & Parrinello, S. (2020). Urban analysis and modelling procedures for the management of historic centres. Bethlehem, Solikamsk, Kotor and Santo Domingo. *Paesaggio urbano, 2020*(2), 103-115.

Puyuelo, M., Higón, J. L., Merino, L., & Contero, M. (2013). Experiencing Augmented Reality as an Accessibility Resource in the UNESCO Heritage Site Called "La Lonja". *Procedia Computer Science*, *25*, 171–178.

Ramm, R., Heinze, M., Kühmstedt, P., Christoph, A., Heist, S., & Notni, G. (2022). Portable solution for high-resolution 3D and color texture on-site digitization of cultural heritage objects. *Journal of Cultural Heritage*, *53*, 165–175.

Rodéhn, C. (2015). Democratization: The performance of academic discourses on democratizing museums. In K. L. Samuels & T. Rico (Eds.), *Heritage Keywords: Rhetoric and Redescription in Cultural Heritage* (pp. 95–110). University Press of Colorado.

Rossetti, V., Furfari, F., Leporini, B., Pelegatti, S., & Quarta, A. (2018). Enabling access to cultural heritage for the visually impaired: An interactive 3D model of a cultural site. *Procedia Computer Science*, *130*, 383–391.

Salerno, R., & Casonato, C. (2008). Paesaggi culturali. Rappresentazioni, esperienze, prospettive [Cultural landscapes. Representations, experiences, perspectives]. Gangemi.

Schweibenz, W. (1991). The Virtual Museum: New Perspectives for Museums to Present Objects and Information Using the Internet as a Knowledge Base and Communication System. In H. Zimmermann & H. Schramm (Eds.), Knowledge Management und Kommunikationssysteme, Workflow Management, Multimedia, Knowledge Transfer (pp. 185-200). UKV.

Soler, F., Torres, J. C., Leon, A. J., & Luz. (2013). Design of cultural heritage information systems based on information layers. *ACM Journal of Computing and Cultural Heritage, 6*(4), 1-17.

Tucci, G., & Bonora, V. (2016). Documenting Syrian Built Heritage to Increase Awareness in the Public Conscience. In M. Silver (Ed.), *Challenges, Strategies and High-Tech Applications for Saving the Cultural Heritage of Syria* (pp. 83–94). Austrian Academy of Sciences Press.

Tucci, G., Bonora, V., Conti, A., & Fiorini, L. (2017). High-quality 3d models and their use in a cultural heritage conservation project. *The International Archives of the Photogrammetry, Remote Sensing and Spatial Information Sciences*, *42*, 687–693.

van Ruymbeke, M., Nofal, E., & Billen, R. (2022). 3D Digital Heritage and Historical Storytelling: Outcomes from the Interreg EMR Terra Mosana Project. In *International Conference on Human-Computer Interactio*n (pp. 262-276). Springer.

Welch, J. (2014). *Cultural Heritage, What is it? Why is it important?* Retrieved July 28, 2022 from https://summit.sfu.ca/item/16150

Yanli, X., & Danni, L. (2021). Prospect of Vocational Education under the Background of Digital Age: Analysis of European Union's "Digital Education Action Plan (2021-2027)". In *2021 International Conference on Internet, Education and Information Technology (IEIT)* (pp. 164-167). IEEE.

KEY TERMS AND DEFINITIONS

Accessibility: It refers to Cultural Heritage fruition and open access through "basin" connections, linked to the character of diffusivity of Heritage on the territorial dimension. It considers the classification of images and information generated by the intertwining of geographical and cultural crossroads, where a deeper relationship between sites and monuments belonging to the scale of the territory still emerges from the form of anthropic intervention on the landscape.

Digital Learning Capacities: Educational and behavioural capacities related to the use of digital tools effectively, by an extended part of citizens and society users, for critical learning, digital communication. They are necessary to be addressed for increasing engagement and preparation for higher education learning environments based on digital data and fruition modalities.

Heritage Learning Objects: Products from object-oriented programming practice in computer science applied on Cultural Heritage elements, for educational purposes. It means an application in technology-supported learnings, centered in the field of Cultural Heritage, with potential in supporting the design and reuse of technology-based educational materials and individuals contributes from involving wider public in the communication process.

Phygital: The term derives from the fusion of the words "physical" and "digital". It means an incorporation of digital functionality within the physical customer experience, and it also includes the reverse process, introducing new path of hybrid experience proposition for digital data.

Smart-Users: Users who either use most of the smart technologies (or work in this developing sector) providing skills of integration with digital contents already disposed from daily experience and basic education. Smart technologies are defined as computers or services that perform digital contents and qualities guiding the human experience and decision-making within them.

Virtual Learning Environment: It refers to a virtual space for remote learning, with remote interaction for consultation of multimedia learning contents, exchange of information and sharing of materials. It does not only replicate the mechanisms of a "physical" teaching environment, but it allows for the creation and management of a more effective learning ecosystem based on digital tools and interactions for a personalised management of educational time and users' attitudes.

Widespread Architectural Heritage: Heritage system consisting of building monuments and minor built assets, identified as discrete, punctual elements that are fundamental in understanding the complex and interrelated systemic cultural relationships to be recognised in the territory. It is defined as "widespread" in reference also to the collective and identity value on the territory.

Chapter 17
Phygital Heritage Experiences for a Smart Society:
A Case Study for the City of L'Aquila

Luca Vespasiano
https://orcid.org/0000-0003-1332-6475
University of L'Aquila, Italy

Stefano Brusaporci
https://orcid.org/0000-0002-8505-7895
University of L'Aquila, Italy

Fabio Franchi
University of L'Aquila, Italy

Claudia Rinaldi
University of L'Aquila, Italy

ABSTRACT

Starting from a recognition of the progressive settlement of the conception of cultural heritage through years, and the role that digital technologies have played, the contribution analyses how ICT (information communication technology) solutions, altogether intended, could provide a new human centrality in interpretation and presentation of cultural heritage. This opportunity is provided from the experience of INCIPICT project (INnovating CIty Planning through Information and Communications Technology), developed in L'Aquila since 2012. Within its framework, several reflections and applications on the field of cultural heritage have been developed to achieve results in terms of theory and praxis on the route toward a culture-based smart society.

DOI: 10.4018/978-1-6684-4854-0.ch017

INTRODUCTION

Over the last decades the international debate has progressively extended and deepened the idea of cultural heritage to make it more inclusive and respectful of the different cultures of the world and their manifestations. At the same time, technological development, particularly in the ICT field, has invested our society, generating substantial and irreversible changes. The scenario of a society 5.0 that is emerging, with its intent to promote human-centred services, bringing the society back to the center of development (Deguchi 2020, De Felice 2021), calls for a rethink of the relationship between the sphere of cultural heritage and the sphere of technological innovation aiming to build a greater integration for the benefit of all.

Indeed, also the safeguard, and the conservation of Cultural Heritage can benefit greatly from the relationship with technology.

So, what is the perimeter of cultural heritage nowadays? What contribution can it make and what role can it play in in strengthening society and furthering its well-being? What challenges and what opportunities do new technologies present in this field?

This contribution seeks answers to these questions.

The first section, Cultural Heritage in the becoming of society 5.0, aims to provide a framework of the progressive development of the concept of cultural heritage. A chronological exploration of the main international documents in the field of cultural heritage will allow deepening various theoretical aspects necessary for providing a comprehensive understanding of the subject. The progressive extension of the cultural heritage perimeter, the inclusion of intangible aspects and the dynamics of community involvement in the interpretation process are still overlooked and remain peripheral to the mainstream discourse, but essential factors in outlining future developments. By delving deeper into these questions, we will be able to actualize the extraordinary opportunity offered by technology to mediate between the tangible and intangible aspects of cultural heritage, facilitating accessibility and the preservation of authenticity. From this point of view, we can observe how the continous digital transcoding of the heritage, its gradual digitalization, gives rise to a phygital convergence, in which we arrive at a synthesis between the tangible and intangible instances of the heritage, functional to its presentation.

In any case, structural integration of applications and services from the infrastructure level of communication networks is essential if these opportunities are to become real. In this sense, the integration of the fields of content development related to the interpretation of cultural heritage and research related to the ICT field must aim at an Open Innovation model, to ensure at the same time the sustainability of the system, its effectiveness, and the necessary support for the full development of society and communities.

The second part of the contribution presents a series of experiences gained within the project INCIPICT - Innovating City Planning through Information and Communication Technologies and related to the case study of the historic center of L'Aquila, in Italy.

These experiences, ranging from digitalization to the development of interpretative and restitutive models and presentation methodologies, are united by a constant reference to the user experience. This reference to the user experience starts first of all from the need to identify the platforms that guarantee greater accessibility and consequently a greater diffusion of content. The proposed experiences aim to define a workflow that, starting from digitalization, guides the operators to the phase of presentation and maintenance of the contents.

Taking into account that many aspects of this scenario are still evolving, starting from hardware for Augmented Reality and Mixed Reality, to networking technologies, and platforms and software for presentation, the general intention is to identify strategies that allow rapid adaptation to the innovations introduced by the research, based on a stable theoretical and epistemological system.

CULTURAL HERITAGE IN THE BECOMING OF SOCIETY 5.0

From Monuments to Contexts

In order to fully understand the role played by cultural heritage in the scenario that is taking shape of a society 5.0, it is necessary to retrace the process of its evolution in light of its theoretical definition, that is to say, in relation to the theoretical reflections and innovations emerged as responses to the technological transformations that have progressively and always more pervasively spread in this as in all areas of our social life.

Throughout the years, the concept of cultural heritage evolved significantly: by Acknowledging and promoting theoretical discussions and experimental practices, internationals institutions such as UNESCO and ICOMOS, have registered and disseminated this continuous transformation through their documents. This evolution has led to a progressive expansion of the concept of heritage, from the object, a single element, to its context, intended as a whole environment.

Starting from the *Convention for the Protection of Cultural Property in the Event of Armed Conflict* (1954), which is the first international document with a global and worldwide ambition, we can see how the focus is placed on material and tangible objects:

"For the purposes of the present Convention, the term `cultural property' shall cover: movable or immovable property of great importance to the cultural heritage of every people, such as monuments of architecture, art or history, whether religious or secular; archaeological sites; groups of buildings which, as a whole, are of historical or artistic interest; works of art; manuscripts, books and other objects of artistic, historical or archaeological interest; as well as scientific collections and important collections of books or archives or of reproductions of the property defined above" (UNESCO, 1954, Art.1).

Having been adopted shortly after the Second World War, this document, in a certain way, reacts to the terrible destructions caused by the war and in this sense, it is clear that the first objective is to oppose the material destruction of the heritage. It is also interesting to note the use of the words: the expression "cultural property" clearly refers to the French "bien culturel", which implies a relationship with economic value and with possession, with property. After this expression, the one of "cultural heritage" will be preferred, with the double intention to remove the economic implications and to underline the importance of its transmission to the future.

With the UNESCO *Convention Concerning the Protection of the World Cultural and Natural Heritage* (1972, aka World Heritage Convention), natural heritage comes into play, presented as distinct and autonomous from cultural heritage. This distinction is very clear and made explicit in the first two articles:

"Art. 1 For the purpose of this convention, the following shall be considered as 'cultural heritage':

- monuments: architectural works, works of monumental sculpture and painting, elements or structures of an archaeological nature, inscriptions, cave dwellings and combinations of features, which are of outstanding universal value from the point of view of history, art or science;
- groups of buildings: groups of separate or connected buildings which, because of their architecture, their homogeneity or their place in the landscape, are of outstanding universal value from the point of view of history, art or science;
- sites: works of man or the combined works of nature and man, and areas including archaeological sites which are of outstanding universal value from the historical, aesthetic, ethnological or anthropological point of view."

Art. 2 For the purposes of this convention, the following shall be considered as 'natural heritage':

- natural features consisting of physical and biological formations or groups of such formations, which are of outstanding universal value from the aesthetic or scientific point of view;
- geological and physiographical formations and precisely delineated areas which constitute the habitat of threatened species of animals and plants of outstanding universal value from the point of view of science or conservation;
- natural sites or precisely delineated natural areas of outstanding universal value from the point of view of science, conservation, or natural beauty."(UNESCO, 1972, Art.1-2).

The clear distinction between cultural heritage and natural heritage becomes more nuanced when dealing with landscape. Over the decades, a profound reflection has evolved around the idea of landscape that led to overcome the stringent definitions of the third quarter of the twentieth century, towards broader, more inclusive, and more effective concepts in terms of safeguarding.

Probably the first sign of this extension of attention from the single object to the context can be found in *The Chart of Venice* (ICOMOS, 1964). This document, specific to the architectural field, expresses awareness of the intrinsic and unavoidable link between the building and the "context", the latter intended not only as a physical and spatial context, but in fact as a cultural surrounding determined by social, economic, political, historical, and environmental factors too.

Building on this awareness a new consideration was developed, a consideration not only on the heritage, but above all on conservation strategies, starting from the re-evaluation of the role of those who live the "site" and of the activities that take place in it. In parallel, as in the ecological field the idea of safeguarding expanded to include in addition to the safeguard of the biotype or endangered species, the protection of the whole ecosystem, in the cultural field, the concept of heritage expanded from that of monument to a wider notion encompassing the context, the landscape, and the widening of the concept of site to the environment, intended as a whole.

Reflection on the Landscape as a Space for Discussion on the Idea of Heritage

The landscape has been the subject of numerous UNESCO documents and has long been the focus of debate on cultural heritage. Its promiscuous essence, suspended between natural and anthropic, consisting of punctual and extensive elements, its changeability in line with the seasons, time, interaction with traumatic phenomena, have led to a series of very stimulating reflections on the consideration of heritage. For example, *The Landscape Convention* (Council of Europe, 2000) defines landscape as *an area, as perceived by people, whose character is the result of the action and interaction of natural and/or human factors (Council of Europe, 2000, p.9).*

While the interaction between human and natural factors is explicit, the task of establishing a limit, of defining a perimeter, is left to people and their perception: it is the perception of people that makes an area or a territory a landscape, without any objective element being able to definitively determine it.

Following in this direction The *World Heritage Cultural Landscape* (UNESCO, 2009) specifies:

The very notion of landscape is highly cultural", where the "term "cultural" has been added to express the human interaction with the environment and the presence of tangible and intangible cultural values in the landscape (UNESCO, 2009, p.17).

Moreover,

Landscape interpretation and cultural landscape go together, for both are about ideas and meanings, concepts and interpretations, dynamics and dialogues (UNESCO, 2009, p.22).

This establishes a dialogical relationship between the different components of the landscape, whose integration and correlation is such that it is impossible to distinguish their limits. It is not just about people's perception, but the long-lasting relationship between people and their environment, and the reciprocal influence that is exerted between territories and communities, and that leads both to being in constant transformation. In this dynamic, the first transcendent reference is that of memory:

Landscapes also exist in people's memories and imaginations and are linked to place names, myths, rituals, and folklore. In people's minds there is rarely a clear distinction between the visible and invisible – or tangible and intangible – components of the landscapes. Stories and myths endow landscapes with meanings transcending the directly observable and thereby help to create people's 'mental maps', or awareness of place [...] Cultural landscapes can be seen as the repository of collective memory (UNESCO, 2009, p.22).

From this last passage it is clearly understood how the work around the concept of landscape has provided the opportunity to lead to a significant extension of the concept of cultural heritage, making evident the presence of an intangible component. As said, the first transcendent reference is memory. But it is only the first step in the definition of the "mental map", in the collective process of attributing a transcendent meaning to the material component of the environment that surrounds us.

The Immaterial Dimension of Heritage

The *Convention for the Safeguard of Intangible Cultural Heritage* (UNESCO, 2003b) represents a decisive step forward in establishing a culture of protection and conservation, as it Sanctions the definitive overcoming of the cultural/natural dualism. These two instances of heritage are absorbed and reread in function of the distinction between tangible and intangible heritage. This distinction, however, is not a dichotomy, and opens to the possibility of intersections between these two spheres which are always in dialogue with each other. Let's see the definition that is given:

The "intangible cultural heritage" means the practices, representations, expressions, knowledge, skills – as well as the instruments, objects, artefacts, and cultural spaces associated therewith – that communities, groups and, in some cases, individuals recognize as part of their cultural heritage (UNESCO, 2003b, Art.2).

As already seen with regard to the landscape, the definition gives great importance to the perception of people. In this case the concept of "people" is more specific in communities, groups and, in some cases, individuals. Also in this instance, the attention is placed on memory, understood in this case as a tool of transmission between generations and therefore as a process that enables people to appropriate places and build their own identity:

This intangible cultural heritage, transmitted from generation to generation, is constantly recreated by communities and groups in response to their environment, their interaction with nature and their history, and provides them with a sense of identity and continuity, thus promoting respect for cultural diversity and human creativity. (UNESCO, 2003b, Art.2).

It undermines the traditional concept of heritage as a monolithic object, unchanging over time, and it suggests that we have to take into account that the heritage is continuously is something alive and that continuously changes.

The crucial aspect presented by the convention is the paradigm shift in terms of time and people, the latter term intended both as a community or, in some cases, individuals. In fact, time, considered as an enemy to conservation for its ability to corrupt and transform, becomes the means through which heritage acquires its semantic value, which is developed, transmitted, and continuously re-elaborated by individuals and communities, in a system in constant evolution and in a dialogic dynamic that involves the heritage and the people who carry it over time. In fact, the transmission process of intangible heritage is emphasized as something that involves different generations through time in a continuous way. This ensures that the intangible assets are continuously recreated in relation to the context and the community.

The basis of this evolution of the concept of heritage can be found in germ already in the *Nara Document on Authenticity* (UNESCO, 1994), which in its ambition to adapt historically Eurocentric or in any case Western principles to the global and globalized heritage, finds the opportunity to reflect on the role of the "intangible" as deep source of value. The preamble, at Art. 4, defines the situation:

In a world that is increasingly subject to the forces of globalization and homogenization, and in a world in which the search for cultural identity is sometimes pursued through aggressive nationalism and the suppression of the cultures of minorities, the essential contribution made by the consideration of authenticity in conservation practice is to clarify and illuminate the collective memory of humanity (UNESCO, 1994, Art.4).

and at Art. 5 expresses as follows

The diversity of cultures and heritage in our world is an irreplaceable source of spiritual and intellectual richness for all humankind. The protection and enhancement of cultural and heritage diversity in our world should be actively promoted as an essential aspect of human development (UNESCO, 1994, Art.5).

For this reason:

All cultures and societies are rooted in the particular forms and means of tangible and intangible expression which constitute their heritage, and these should be respected (UNESCO, 1994, Art.7).

This call for the 'intangible' was made to include non-European cultures, but today it is renewed by participative culture challenges.

THE CONVERGENCE BETWEEN TANGIBLE AND INTANGIBLE: A NEW PARADIGM

This drive towards greater inclusiveness in the way heritage is defined has also led to greater attention to inclusiveness in the use and perception of heritage. The *Faro Convention* (Council of Europe, 2005) is decisive in this respect, locating at its core the notion of community. Particularly interesting is the definition of cultural heritage that is provided in these terms:

a group of resources inherited from the past which people identify, independently of ownership, as a reflection and expression of their constantly evolving values, beliefs, knowledge, and traditions. It includes all aspects of the environment resulting from the interaction between people and places through time (Council of Europe, 2005, Art.2, lit. a).

the Heritage community is defined as follows:

people who value specific aspects of cultural heritage which they wish, within the framework of public action, to sustain and transmit to future generations Council of Europe, 2005, Art.2, lit. b).

The most innovative aspect is the acknowledgment of the fundamental role of people and communities, which become absolutely protagonists. Indeed their active role is crucial in both the preservation and interpretation of cultural heritage. Heritage-making becomes a social activity of interpretation, and in fact, according to a postmodern perspective, it is the discourse on interpretation that defines the content. From this point of view, heritage looks like a "performance".

The *Faro Convention* is an extraordinarily innovative instrument, both for the theories that inspire it, and which are therefore spread and disseminated through it to institutions and communities, and for the operational possibilities to which it preludes. The first articles alone could shed light on these possibilities:

Article 1 – Aims of the Convention

The Parties to this Convention agree to:

a) recognize that rights relating to cultural heritage are inherent in the right to participate in cultural life, as defined in the Universal Declaration of Human Rights;
b) recognize individual and collective responsibility towards cultural heritage;
c) emphasize that the conservation of cultural heritage and its sustainable use have human development and quality of life as their goal; [...]

Article 4 – Rights and responsibilities relating to cultural heritage
The Parties recognize that:

d) everyone, alone or collectively, has the right to benefit from the cultural heritage and to contribute towards its enrichment;
e) everyone, alone or collectively, has the responsibility to respect the cultural heritage of others as much as their own heritage, and consequently the common heritage of Europe;
f) exercise of the right to cultural heritage may be subject only to those restrictions which are necessary in a democratic society for the protection of the public interest and the rights and freedoms of others." (Council of Europe, 2005, Art.1, 4).

Other general aspects, such as the relationship between economic activities and cultural heritage (Art.10) or the relationship between cultural heritage and participation (Art.12), are of great relevance, but for our purposes, it is enough to consider the profound integration that can be found throughout the document between tangible and intangible aspects, and the role of communities in the continuous re-elaboration of the meaning of heritage.

This theorical approach is confirmed also by The Burra Charter (ICOMOS, 2013), which, delving deeper in this topic, considers at the same time tangible characteristics – such as elements, objects, spaces and views – and intangible dimensions, taking both into account as essential aspects of places.

An innovative contribution is the introduction of the expression "cultural significance". In particular, the document clarifies what this notion stands for by stating it

means aesthetic, historic, scientific, social or spiritual value for past, present or future generations. Cultural significance is embodied in the place itself, its fabric, setting, use, associations, meanings, records, related places and related objects (ICOMOS, 2013, p.2).

Thus, it roots the intangible in the tangible. Furthermore, a Place

may have a range of values for different individuals or groups (ICOMOS, 2013, p.2)

An extensive and inclusive consideration of heritage is also inherent in the expression "Urban landscape" that in the *Recommendation on the Historic Urban Landscape* (UNESCO, 2011) is defined as:

the urban area understood as the result of a historic layering of cultural and natural values and attributes, extending beyond the notion of "historic centre" or "ensemble" to include the broader urban context and its geographical setting (UNESCO, 2011, Art.8).

In particular, the Art.9 says:

This wider context includes notably the site's topography, geomorphology, hydrology and natural features, its built environment, both historic and contemporary, its infrastructures above and below ground, its open spaces and gardens, its land use patterns and spatial organization, perceptions and visual relationships, as well as all other elements of the urban structure. [...] It also includes social and cultural practices and values, economic processes and the intangible dimensions of heritage as related to diversity and identity (UNESCO, 2011, Art.9).

From this conception emerges clearly the idea of a deep integration between natural, anthropic and cultural aspects, independently in their tangible and intangible manifestations. This integration results from a complex relational system based o environmental logic in which each aspect manifests its repercussions on the others, and in which therefore each element is simultaneously conditioned by and conditions the others. In this sense, there is a substantial convergence between the tangible and intangible aspects of the place, which anyway do not appear to everyone in the same way. Ultimately, we should remark how communities play a fundamental role, and the greatest effort must be made to make heritage inclusive and central to society.

MEDIATING BETWEEN TANGIBLE AND INTANGIBLE CULTURAL HERITAGE

Heritage as participation

In the previous paragraph we have seen how the concept of cultural heritage has progressively been extended, assuming ever more inclusive connotations. Above all, we have seen how the paradigm shift that took place in recent decades has led to the conception of heritage as something changing, dynamic, evolving, closely linked to people, intended both as a community and as individuals, and capable of becoming part of their ordinary life. However, this dynamic requires specific attention in order to be effectively implemented. To enter this order of ideas, it is necessary to consider the ICOMOS *Charter for the Interpretation and Presentation of Cultural Heritage Sites* (ICOMOS 2008, aka Ename Charter), which defines "the basic principles of interpretation and presentation as essential components of heritage conservation efforts and as a means of enhancing public appreciation and understanding of cultural heritage sites"; In relation to its principles digital technologies can play a decisive role mediating between tangible and intangible cultural heritage. The added value is that every effort in the construction of this transcoding system between the two instances of heritage contributes to the development of the digital heritage, increasing its overall value.

The *Ename Charter* arises from the needing of

stress the importance of public communication as an essential part of the larger conservation process (variously describing it as "dissemination," "popularization," "presentation," and "interpretation"). They implicitly acknowledge that every act of heritage conservation – within all the world's cultural traditions – is by its nature a communicative act (ICOMOS, 2008, p.2).

Therefore, the Charter presents the following definitions:

- Interpretation: "refers to the full range of potential activities intended to heighten public awareness and enhance understanding of cultural heritage site. These can include print and electronic publications, public lectures, on-site and directly related off-site installations, educational programmes, community activities, and ongoing research, training, and evaluation of the interpretation process itself. (ICOMOS, 2008, p.4).

- Presentation: "more specifically denotes the carefully planned communication of interpretive content through the arrangement of interpretive information, physical access, and interpretive infrastructure at a cultural heritage site. It can be conveyed through a variety of technical means, including, yet not requiring, such elements as informational panels, museum-type displays, formalized walking tours, lectures and guided tours, and multimedia applications and websites. (ICOMOS, 2008, p.4).

A very important role is played by "Interpretive infrastructures": They

refer to physical installations, facilities, and areas at, or connected with a cultural heritage site that may be specifically utilised for the purposes of interpretation and presentation including those supporting interpretation via new and existing technologies (ICOMOS, 2008, p.4).

The Charter offers seven cardinal principles on which Interpretation and Presentation should be based:

Principle 1: access and understanding: interpretation and presentation programmes should facilitate physical and intellectual access by the public to cultural heritage sites

Principle 2: information sources: interpretation and presentation should be based on evidence gathered through accepted scientific and scholarly methods as well as from living cultural traditions

Principle 3: attention to setting and context: the interpretation and presentation of cultural heritage sites should relate to their wider social, cultural, historical, and natural contexts and settings

Principle 4: preservation of authenticity: the interpretation and presentation of cultural heritage sites must respect the basic tenets of authenticity in the spirit of the Nara Document (1994).

Principle 5: planning for sustainability: the interpretation plan for a cultural heritage site must be sensitive to its natural and cultural environment, with social, financial, and environmental sustainability among its central goals

Principle 6: concern for inclusiveness: the interpretation and presentation of cultural heritage sites must be the result of meaningful collaboration between heritage professionals, host and associated communities, and other stakeholders

Principle 7: importance of research, training, and evaluation: continuing research, training, and evaluation are essential components of the interpretation of a cultural heritage site

The Australia ICOMOS *Charter for Places of Cultural Significance* (aka Burra Charter, 2013), specifies and clarifies some aspects. It says:

Interpretation means all the ways of presenting the cultural significance of a Place (ICOMOS, 2013, Art.1.17).

In addition, it explains:

Interpretation may be a combination of the treatment of the fabric (e.g., maintenance, restoration, reconstruction); the use of and activities at the place; and the use of introduced explanatory material (ICOMOS, 2013, Art.1.17).

The "Practice Note_Interpretation" of The Burra Charter (ICOMOS Australia, 2013) is expressly dedicate to interpretation. It underlines the importance of: making inventories, defining the audience; developing interpretation policy; defining key interpretive themes and stories; establish interpretation methods and techniques; systematic implementation (p.4-6). The presented "Key principles" are noteworthy:

1. "Facilitate understanding and appreciation of cultural heritage sites and foster public awareness and engagement in the need for their protection and conservation.
2. Communicate the meaning of cultural heritage sites to a range of audiences through careful, documented recognition of significance, through accepted scientific and scholarly methods as well as from living cultural traditions.
3. Safeguard the tangible and intangible values of cultural heritage sites in their natural and cultural settings and social contexts.
4. Respect the authenticity of cultural heritage sites, by communicating the significance of their historic fabric and cultural values and protecting them from the adverse impact of intrusive interpretive infrastructure, visitor pressure, inaccurate or inappropriate interpretation.
5. Contribute to the sustainable conservation of cultural heritage sites, through promoting public understanding of, and participation in, ongoing conservation efforts, ensuring long-term maintenance of the interpretive infrastructure and regular review of its interpretive contents.
6. Encourage inclusiveness in the interpretation of cultural heritage sites, by facilitating the involvement of stakeholders and associated communities in the development and implementation of interpretive programs.
7. Develop technical and professional guidelines for heritage interpretation and presentation, including technologies, research, and training. Such guidelines must be appropriate and sustainable in their social contexts." (ICOMOS Australia, 2013, p.2)

Interpretation is strictly related to knowledge and understanding. Therefore, conservation – and consequently presentation, communication, engagement, maintenance and enhancement – is founded on interpretation.

Interpretation and presentation are not issues to be faced after the overall process of knowledge, but they are strategies for conservation and for knowledge and understanding. Moreover, when we recur to participative approaches, where the authoritative role of the scholar collaborate with the one of inhabitants – i.e. with people who constitute, build, transform, and give meaning to the heritage itself –. Interpretation and presentation acquire a wider meaning and purpose:

From an educational to a cultural function oriented to everybody involved in the heritage conservation, both for knowledge and planning.

Digital Transcoding of Cultural Heritage

In order to draw a comprehensive framework for understanding cultural heritage, it is necessary to consider that, nowadays, this notion has expanded to integrate a new, digital dimension: the *Charter for the Preservation of Digital Heritage* (UNESCO 2003) defines digital heritage as "unique resources of human knowledge and expression"; this kind of heritage should not be understood as opposed to tangible or intangible "natural" ones but as new manifestations that can play an important role in the processes of interpretation, communication and conservation of the heritage, with a view to accessibility and participation.

Digital technologies create brand new scenarios relating to cultural heritage, capable of transforming, through their mediation in a virtual environment, the relationship that is established between material and immaterial instances of heritage, putting in place a new system of transcoding. The opportunities offered by technology are not simply to be intended as technical expedients capable of creating suggestive or pedagogical experiences, but as a versatile tool, at the service of human creativity, capable of generating value and facilitating exchanges and cultural relations.

The development and diffusion of digital technologies for surveying, modeling and visualization have actually achieved a new level of reality. The digitization of the heritage, seen as the process of creating digital content starting from tangible heritage, and related to it, can be seen as a transcoding system that brings material instances of physical reality to exist also in a virtual dimension. But this transcoding process is not only limited to this: in fact it allows to enrich the real, by associating to the material and tangible data an interpretative and communicative apparatus that also takes into account the immaterial aspects. It is therefore possible to attribute a double relevance to the adoption of digital models and contents: first of all they are embedded with a testimonial value in that, as documents, they are characterized by the direct relationship between the physical object and its digital image, which in this sense is autonomous and usable for various purposes, even if not predictable at the start. On the other hand, they constitute an interpretative mediation tool, which can be used to communicate complex values from a presentation perspective.

In consideration of this certainly interpretative, but also creative nature, the model is not limited to documenting the architectural object, but becomes itself a document: Centofanti (2010) writes:

The representative and restitutive model of an architectural heritage is an integral part of historical knowledge and an autonomous text that can be subjected to further analysis and interpretations. Indeed the model provides a series of useful information on the architectural significant [...]. In turn, the model lends itself to be historicized and studied no more, and not only for its relationship with the presented object, but as a document itself in relation to the historical and cultural context that produced it (Centofanti, 2010, p.47).

Taking into account the principles of the *Ename Charter*, digital media can undoubtedly have a strong impact in terms of the effectiveness of the process of both interpretation and presentation; however, they also present a series of risks and problems. Precisely in this sense, it is useful to refer once again to documents of international institutions, to clarify the limits and factors to be considered in order for this transcoding process to be carried out in the best way. The most important concept in this sense is transparency. On 'transparency', cornerstones are *The London Charter* and *The Principles of Seville* that deal with issues related to "Research Sources", "Documentation", "Authenticity", and "Historical Rigor" of computer-based visualization of cultural and archaeological heritage. And they find in paradata an important tool to reach transparency.

The London Charter (2009) defines paradata for cultural heritage as:

Information about human processes of understanding and interpretation of data objects. Examples of paradata include descriptions stored within a structured database of how evidence was used to interpret an artefact, or a comment on methodological premises within a research publication. It is closely related, but somewhat different in emphasis, to "contextual metadata", which tend to communicate interpretations of an artefact or collection, rather than the process through which one or more artefacts were processed or interpreted (The London Charter, 2009, p.13).

Paradata configures as a sort of metadata useful to the philological reconstruction of the modelling process (Bentkowska-Kafel, Denard, Baker, 2012). In this way *The London Charter* defines "Intellectual transparency" as:

The provision of information, presented in any medium or format, to allow users to understand the nature and scope of "knowledge claim" made by a computer-based visualization outcome (The London Charter, 2009, p.12).

The concept of paradata accompanies the one of "scientific transparency", following the ambition to shun the seduction of some photorealistic computer-based visualizations, and to declare the scientific fundaments, reasons, and reflections from which the digital representation derives.

The Principles of Seville (2012) say:

All computer-based visualization must be essentially transparent, i.e. testable by other researchers or professionals, since the validity, and therefore the scope, of the conclusions produced by such visualization will depend largely on the ability of others to confirm or refute the results obtained (The Principles of Seville, 2012, p.8).

The principle of transparency has been widely developed in the archaeological field, where the digital reconstruction of artifacts (in their ancient configuration) is largely based on indirect information, comparative analysis, interpretative hypothesis.

In 2012 a publication titled "Paradata and Trasparency in Virtual Heritage" renovates the debate on transparency (Bentkowska-Kafel, Denard, Baker 2012). This is a necessary requisite because:

Digital technologies offer flexible analytical tools, both sensory and semantic, for the study and representation of the past, but the digital techniques – it is argued here – are only useful and valid if interpretative frameworks and processes are published≫, therefore the need to ≪emphasize the importance of reliably, documenting the process of interpretation of historical materials and hypotheses that arise in the course of the research". From here descend the use of paradata: "[...] the term borrowed from other disciplines that rely on recording information processes. Paradata document the process of interpretation so that the aims, contexts and reliability of visualization methods and their outcomes can be properly understood [...] Paradata may be seen as a digital equivalent to "scholia", as well as an addition to the traditional critical apparatus for describing the process of reasoning in scholarly research (Bentkowska-Kafel, Denard, Baker 2012, p.1-2).

In this way paradata is a sort of "comments" and "explanations" that refer to the choices made in visualizing raw-data (i.e. the "documents") and describe the critical interpretations. Although the archaeological experience rises from the necessity of computer-based visualization of ancient heritage, it is an important lesson in our online state. A claim for Transparency is very important and urgent now, in our constant dimension of "digitality" melt into reality (Brusaporci, 2017b).

Phygital Convergence

The term phygital is the fusion of the words physical and digital. It expresses the idea of a deep integration between the digital and physical spheres, in a single environment. This integration provides, on a perceptual level, that there is no distinction, but rather a nuance between these two instances of reality.

In the context of marketing, the term pyhigital refers to the integration between the e-commerce sector and physical stores, in particular with regard to the possibility of having physical consequences for actions carried out in the digital environment and *vice versa*. This integration is strongly related to the diffusion of connected interfaces and in general with the Internet of Things (IoT). From a broader perspective, on the idea of interaction and integration between physical and digital, systems for collecting, sharing and communicating information in a participatory manner can be effectively implemented. Referring to the field of cultural heritage, the phygital concept allows for the integration of the expressive, interpretative and communicative potentials of the digital with those of the physical dimension, opening up new perspectives in the presentation and interpretation of heritage.

In fact, retracing the list of principles proposed by *The Ename Charter*, we can realize what impact this concept can have on the issue of accessibility, for example, being able to adapt the cultural level of communication according to the public, but also overcoming physical or perceptual obstacles depending on any motor or sensory disabilities. Similarly, with regard to the second principle, the digital extension of reality allows the direct and detailed use of the sources from which the presented contents derive, facilitating a deeper, more transparent and more open interpretation.

Then there is a complex and still only partially developed theme, relating to authenticity: certainly, the possibility of making sources more accessible and therefore to achieve greater transparency regarding the interpretation of cultural heritage allows a more solid guarantee of the principle of authenticity as well as it is intended in the *Nara Document*. However, there are important implications in the field of restoration as well as conservation that still need to be fully explored: if the heritage is no longer only the tangible one, and if a digital instance is also part of its ontological consistency, our way of acting on the tangible component may not suffer any repercussions? If the value of cultural heritage can be expanded to its intangible aspects through a phygital conception, our relationship with the material, with which restoration and conservation are concerned, must move along and realize a convergence with the digital world, in an integrated system, in a way that is not limited to how the public or scholars perceptually relate to the heritage, but also operational.

In short, the scenario that is currently taking shape sees the overcoming of the tangible / intangible dualism and physical/digital dichotomy, rather moving towards a profound integration. The approach to the phygital heritage focuses on a new interpretation of the ontological reality, considering its physical and digital aspects to be coextensive. This approach allows a more inclusive conception, and consequently, a more inclusive perception of the heritage, which is configured as a single landscape, with an environmental sense, made up of physical and digital elements.

The exploration of the landscape thus conceived, necessarily passes through the creation and implementation of specific assets, capable of maximizing the accessibility of contents and their usability both at the infrastructural level and at the level of services.

ICT INFRASTRUCTURE AS CATALYST FOR COMMUNITIES EMPOWERMENT

Building on the above considerations, the phygital paradigm may be the key for achieving and improving inclusivity, accessibility and usability of the overall experience of the cultural heritage.

From an ICT point of view, the phygital paradigm can be realized only by guaranteeing the creation of an infrastructure capable of providing innovative services based on the performance that is required for the realization of the listed principles. In this way the INCIPICT project (INnovating CIty Planning through Information and Communications Technology), described in this contribution, has to be intended. Its main focus is the development, in the city of L'Aquila, of an innovative optical network, which surrounds the city center and connects the main public sites of interest. The optical network represents an open testbed for researchers in optical transmission and networking as well as the basis for innovative wireless technologies and smart city applications, especially since the project is integrated with the 5G experimentation carried on the city of L'Aquila. This peculiar condition makes of L'Aquila a "living lab" where the paradigm of Smart City can be experimented, and to be employed to define the end point of the reconstruction process after the earthquake of 2009 (Antonelli 2018).

The distinctive aspect of this experimentation is the opportunity to implement a model of Open Innovation (OI), where research and development as part of the infrastructural layer and the design of services layer are interrelated and industries and research are able to share knowledge and experiences as well. The success of the Fifth Generation (5G) mobile network in Europe, in fact, is strictly dependent on the creation of a community able to design and develop applications that properly exploit the 5G network potential. Network slicing, following this direction, allows a network operator to provide dedicated virtual networks with functionalities that are specific to the service or customer over a common network infrastructure. The next generation mobile network should inherently address requirements of such a hybrid network. This new approach of integration between the infrastructure and the services provided through it, is fundamental not only to increase the quality of service, but also to ensure the sustainability of the system.

The new 5G network will be also able to operate in different application contexts, i.e., low power solutions and/or high-reliability, low latency solutions, providing multilevel network architectures. In this scenario, classic macro-cell structures coexist and integrate, in a functional manner, with different network types, such as small-cell.The latter are able to provide different communications modes, as relay and device-to-device (D2D), to various heterogeneous devices (smart objects, cyber physical systems, connected vehicles) that are also characterized by different requirements in terms of Quality of Service (QoS) (Franchi 2022).

Creating this kind of fluid hierarchy is the only practicable way in future scenarios for ensuring a full and integrated development of technologies and communities. Precisely to validate the research process, attention was paid to experimenting with Mixed Reality (MR) technologies, unanimously considered among the killer applications of 5G.

The experience carried out in L'Aquila on MR refers to many on field trials aimed at exploiting the capabilities of the 5G network to support CH applications.

All the activities mainly refer to the creation of services for the safeguard, preservation, promotion, and enjoyment of the regional cultural heritage (tangible and intangible, historical, architectural, artistic, and naturalistic) and the promotion of innovative politics for tourism support. The initiative is concerned with the sites' heritage which is composed by cultural assets under preservation, restoration, promotion and safeguard and cultural activities contributing to the cultural industry.

The 5G network under deployment will be an enabler for experiential and sustainable models of tourism aimed at the enjoyment of cultural heritage with the possibility of experiencing increasingly engaging and personalized travel experiences directly from tourists' mobile devices (smartphones, tablets, headsets), without latency and availability issues, thanks to applications specifically developed in Augmented Reality (AR). Furthermore, the 5G network would represent a formidable tool in the hands of conservators, restorers and art historians allowing the use of diagnostic tools and their results for the proper conservation of cultural heritage in real time and in-situ.

5G technology will indeed provide the computational and storage resources with very low latency: tourists will be able to "immerse" totally in the reality discovering cultural and artistic events and information, as well as commercial and promotional information; the same will happen for conservators, who will be able to compare results of different diagnostic techniques and/or carried out at different times, as well as other useful data (e.g. historical reconstructions, documents, drawings and so on).

The applications designed within the project allow the users to enrich the visit experience, simply targeting with the proper device, the observed asset also receiving augmented audio descriptions (Rinaldi 2021). Thanks to the AR the viewport will be enriched by information and images, in addition to audio content, which will greatly increase the quality of the visit experience and provide elements otherwise not available "on site". The 5G component will not be limited to the data transfer but will be used in its whole levels using the storage and computational resources made available in the distributed nodes of the network. The idea is to use the different levels of location accuracy to identify the contents to be moved from the cloud, passing through Multiaccess Edge Computing (MEC) nodes up to the user devices from which they will be consumed.

USER EXPERIENCE CENTERED APPLICATIONS

The scenario, that we want to realize, about a smart and phygital heritage, supported by the infrastructure described in the previous paragraph, provides for the development and publication of specific and innovative content. With a view to sustainability, understood also in the sense foreseen by the fifth principle of The *Ename Charter*, and applying the seventh principle, the idea is to develop content in the field of research work, to make the results usable first through experimentation, and then with the systematic use of MR technologies. It is intended to implement a cyclical workflow that allows for each new research, for each area and object of study, to lead and guide the process, from the realization of the contents to their publication and maintenance, guaranteeing their usability and updating over time.

The case study of these experimentations is the historic center of L'Aquila. The city, founded in the thirteenth century, is the result of a process of urbanization absolutely peculiar. During its eight centuries of history, the city has undergone significant transformations due to both the violent and recurrent earthquakes that have affected it, as well as social and political transformations. Following the last destructive earthquake of 2009, a renewed interest in its architectural and urban history has led to a series of new and extensive studies, in particular aimed at rereading the signs of past eras that, hidden by the various repair and transformation interventions, have resurfaced precisely as a result of the earthquake, both by means of the damage and by means of the subsequent repair and restoration works. The image that comes out from these studies is that of a palimpsest defined by the urban structure and the system of hierarchical polarities, such as the market square, the square of civil power, the system of large convents and monumental streets, which underwent very few transformations until the nineteenth-twentieth century. Of this palimpsest, each epoch has elaborated its own spatial and architectural interpretation, with specific formal characteristics. A peculiar aspect of great interest is that this process of continuous spatial and architectural re-elaboration has not been pursued through the realization of brand new architectures, but through the reworking of existing architectures establishing a dynamic of continuous dialogue between conservation, reinterpretation and innovation. This dynamic dialogue corresponds to a deep stratification of the architectural heritage in which elements from different periods are integrated not only from a constructive point of view, but also often, from a figurative and formal point of view. In this regard, reading the architectural stratification does not only mean moving in a synchronic perspective to reconstruct the spatial configurations of a specific epoch, nor does it mean moving only in a diachronic way to reconstruct the chronological sequence of transformations, but enter into a deeper dynamic, going in search of the correspondences between the different architectural languages and the progressive settling of the collective identity: the architectural heritage, in this sense becomes a reservoir of the will to representation, of mentality, aspirations, social tensions, euergetism, spirituality and crisis of the generations that have built, transformed and preserved those architecture and those space systems. Consequently, a phygital conception of heritage has been placed at the center of the work to be carried out: the peculiarity of material immanence and digital correlation provide the opportunity to build complex and dynamic information systems. On this basis is possible to structure an offer of personalized and smart services able to guarantee the interpretation and presentation of the cultural heritage.

The creation of this environment comes first of all from digitalization, and then from the identification of the possibilities that allows to mediate this complex system of information in view of its presentation. The examples that will be presented in the following paragraph, are part of this system, which is inevitably still under development, but help to define replicable and integrable methodologies according to the general project.

Digitize Cultural Heritage

From an operational point of view, the first phase of work is the digitization of cultural heritage. This section will mainly refer to the architectural heritage, which for its peculiarities of spatial extension, relational heterogeneity and territorial diffusion is a paradigmatic example of the complexity of the type of work required.

The development and dissemination of digital technologies in the architectural field, has concerned not only the representation and digital modeling of architecture, but also the instrumental field at the service of architectural survey. Tools such as terrestrial laser scanner (TLS), SfM-based photogrammetric technologies (structure from motion), or Simultaneous Localization And Mapping (SLAM) technology allow you to capture a large amount of data in a very short time. The spread of software, often opensource, and the progressive lowering of the costs of many tools is making these techniques increasingly accessible and widespread. In particular, photogrammetry, with the use of SfM algorithms, makes it possible to build 3D models in a short time, even using only a smartphone and freeware programs, reducing costs and obtaining useful results.

From this point of view, we can consider that the border between the physical world and the digital world is always easier to cross. The typical result of these data acquisition processes are point clouds (Figure 1).

Figure 1. Point cloud of the memorial of Beatrice and Maria Pereyra Camponeschi, in the apse of the Basilica of San Bernardino in L'Aquila. On the right is visible part of the major altar with a statue of St. Francis. The survey was carried out using a drone.

These point clouds have a singular nature: they consist of discrete elements, but as a whole they return a continuous image of the object; generally the information that can be obtained does not concern only the geometric and dimensional aspects, but since each point that constitutes the cloud brings with it information about the color of the point, and approximating the continuity of the physical surface, information can also be obtained about the material and its state of preservation for example. Ultimately, what is obtained from this type of processing is an image, in a mathematical sense, of the physical object, transcoded into a virtual dimension. This "image", this cloud of points, is not yet a critical result, except for having chosen that object, that portion of physical reality, and not another and having determined spatial limits, beyond which no data have been acquired.

The critical choices made at this point in the elaboration process were taken, in general, only at a technical level, in order to obtain good quality data, but they are choices related, precisely, to a purely technical area. This type of choices, which also determine the subsequent processing of refinement and interpretation of the acquired data, should not be underestimated, because it determines the quality of the work as a whole, but it is important to clarify that these are not interpretative choices, which, as we shall see, are translated to a later step.

It is clear the difference with the process of survey with traditional techniques, in which, the acquisition of measures takes place after a series of preliminary operations. In that case, the process has a critical nature, directed to the choice of which and how many measures to acquire, depending on the characteristics of the architectural object. We can then consider that in a certain sense, the work of analysis and acquisition of knowledge, typical of the architectural survey, no longer takes place in the physical environment, but mainly in the digital environment, or rather, between these two environments, where a relationship of deep integration is created that can be defined already phygitale and represents, in this environment, a sort of reversal of the traditional methods for the critical reading activity takes place after, and not before the measurement.

The result obtained from this first digitization step can already be considered useful in itself to be introduced in the field of presentation, opening to issues that will be analyzed in the next paragraph. Undoubtedly, point clouds are instruments of extraordinary scientific value, which allow, also because of the absence of critical determinant connotations, further independent elaborations. So it is certainly useful to think of being able to share them in a perspective of open access for academic purposes.

However, usually we do not stop at this level, but it is on the basis of the data acquired, that is, as mentioned, of the point cloud and its subsequent processing that the vector models are made in a reverse modelling dynamic. This modelling process is not necessarily aimed at reproposing the current configuration of the building, but also configurations from the past, which can be inferred from historical research in conjunction with the analysis of the current state. Precisely in order to give rise to such reconstructions it is necessary to produce an apparatus of elaborates, drawings, diagrams, based both on the survey and on archival documents, or on other examples of architectures, which together will be the ideal environment in which reconstruction is developed. Those just mentioned, can be considered intermediate products, which also have their own value and are able to justify, explain and enrich the perception of the model. Essentially, in addition to a model, the work of interpreting the architectural heritage produces an ecosystem of different contents, which has a constant reference in physical reality, but extends indistinctly in the phygital environment. It is of crucial importance in terms of transparency and respect for authenticity (principle 4 of The *Ename Charter*) to keep track of both the sources to which reference has been made and of the operating methods adopted in the various elaborations, since it must always be borne in mind that every interpretation, every result is partial and transitory, and must be presented in this light, as a consequence of the fact that, as we have seen before, it is in the specifical nature of the cultural heritage that it is not stable and unchangeable, but become an entity in ideal continuity with the community that claims it.

Experiences of Phygital Integration

We have seen the impact of new digitalization tools, particularly point clouds, in the survey and documentation of cultural heritage. But how can these products be used for the presentation of cultural heritage? The first problem concerns the hardware and its portability: point clouds, complex models such as h-BIM models, are heavy objects, requiring high-performance components like CPUs and GPUs to be better managed. In addition, they need specific software, which is not always possible to have on mobile devices, both for the incompatibility of operating systems, and because mobile devices often have performance characteristics below the minimum requirements of the software. Therefore, there is a need to find an exchange interface able to make the models produced on mobile devices available. The need to use mobile devices derives from two fundamental considerations: first of all, in order to give rise to MR applications, in this context, it is necessary to allow the application to take place in the specific place for which it was developed; on the other hand, in order to allow greater accessibility, it is necessary to ensure that applications are usable on common devices, within the reach of the widest possible audience.

It is certainly to be expected that both on the hardware side and on the ICT infrastructure side in the future there will be evolutions that will facilitate this step, but in order to be able to create the conditions for this development, it is necessary to operate, as already mentioned in the previous paragraph, in an Open Innovation mode, while developing the tools and their applications. The first experimentation in this field concerned as a case study a series of models of particular architectural elements that is capitals of columns. The occasion was offered by a study on the diffusion within the architectural heritage of the city of L'Aquila of a specifical type of columns with a circular base, which spread especially between the late XV and early XVI century (Figure 2). The study, still in progress, has for the moment covered 14 courtyards and 67 capitals.

Figure 2. Models obtained from SfM of some of the round column capitals included in the study. Each model can be viewed in virtual environment or physical space via the AR function.

Figure 3. The online map used to collect and georeference 3D models

The particular situation of the historic center of L'Aquila, due to the reconstruction process following the earthquake of 2009, makes the accessibility of the courtyards, in which this type of columns is located, rather complex. In addition, these are mostly found in private courtyards and can only be visited occasionally. Thus, the possibility of having digital instances on which to conduct comparative investigations has proved to be strictly necessary in order to conduct the study. The models were made through photogrammetry, and then processed with the AgSoft Metashape software obtaining 3D meshes with photorealistic textures.

These models, once processed to obtain 3D meshes, did not present any particular management problems, being rather limited in size and relatively simple. They have therefore lent themselves to experience their usability on mobile devices, in particular on smartphones.

Several Android environment applications available on the Google Play Store have been tested for this purpose. However, the results were unsatisfactory in some cases due to the difficulty of handling the texture, in others due to the impossibility of changing the interface and the conditions of viewing the objects. The best option turned out to be the SketchFab.com platform, which also provides a good interface for viewing in AR. The platform's ability to embed the display frame into websites and create direct links, allowed all models to be collected in a gallery on a web page and on a map where you can geolocate your device (Figure 3). In this way, both remotely and in situ, it is possible to view and navigate the models, with the possibility of making comparisons between the different capitals through the AR interface. Obviously, in addition to the detailed study on the capitals of which the 3D models have been elaborated, the map shows a large amount of information: bibliographical information on buildings and churches, historical photographs, current photographs from the ground, normal and spherical and aerial drone photos. To give an idea of the scale of the work in progress, more than 1700 punctual bibliographical references have been collected, 189 historical photographs have been uploaded georeferenced in the original point of capture, and 216 aerial photographs. The general aim is to systematically digitize the entire architectural heritage of the historic center. At the moment the greatest attention has been placed on the architectural elements. 897 portals and 1254 windows with 2D photographs have been mapped. Of the most relevant cases 3D models has been made, as in the case of the lodges. Also, in this case the models were obtained by photogrammetry, but using not a point of view from the ground for the realization of the photographs at the base of the process, but taken using a drone, because the items of interest are located on the top floor, just below the coverage line (Figure 4).

In this case the intention, regarding the presentation, is to facilitate the vision from the ground. The loggias under study, in fact, are placed in private buildings, and being at the top, overlooking narrow alleys, are very difficult to observe. In this case, the visualization on mobile device is intended to support the physical experience and integrated by a series of schemes and explanations that highlight the technical and cultural aspects of this formal expression within Renaissance architecture (Figure 5).

The case of architectural elements is useful to show how a phygital conception of heritage opens up new working scenarios for researchers: in the purely physical reality, it is obviously not possible to conduct a comparison activity like the one described, since the architectural elements are by nature immovable from their context. Even in the presentation the pyhigital conception of the heritage opens new perspectives, in fact a virtual museum is being developed. It will be experimented through Oculus visors, and the users will have the opportunity to set up in the virtual room the objects they want to view, being guided in their exploration.

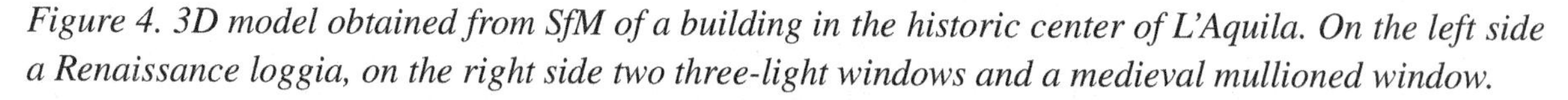

Figure 4. 3D model obtained from SfM of a building in the historic center of L'Aquila. On the left side a Renaissance loggia, on the right side two three-light windows and a medieval mullioned window.

This idea of bringing in the same virtual space elements and objects physically and visually isolated, and that can therefore not be compared directly or altogether, is also the basis of the virtual collection of works by Silvestro Aquilano. This artist, active with his workshop in L'Aquila between the XV and XVI centuries, in addition to some mobile works exhibited at the National Museum of Abruzzo, left a series of works scattered in churches and palaces of the city: from the Memorial of Beatrice and Maria Pereyra Camponeschi (Figure 6), to the Mausoleum of San Bernardino, in the Basilica of San Bernardino, to some bas-reliefs and architectural elements, attributed not certain, but related to its cultural environment. The virtual museum can then become a place of research and a tool that promotes a more conscious exploration of the urban landscape in search of the signs left by the artist.

Thus far, this study has considered examples of architectural elements or works of art limited in size and in any case included in larger scale contexts. Turning now to the scale of the entire buildings, it should be noted, first of all, that a greater difficulty comes into play when managing these models because of their size. This not only results in an increase in the amount of calculation needed, and therefore requires more performing devices, but above all it puts in crisis all those restitutive systems thought and optimized for the exploration of objects with an external point of view. An architecture is designed to contain users, and therefore the point of view cannot be always and solely an external one, otherwise the concept of architectural scale would be put into crisis. It is precisely the possibility of space exploration and the overcoming of the human scale that constitutes the essential prerogative of architecture, and that distinguishes it from the other arts. Moreover, an architecture is generally a complex and articulated system, defined in its spatial and constructive conformation by the relationship with the human point of view. To convey these characteristics in a virtual experience, the model must necessarily contain the viewer, who must be completely immersed in the architecture. It is also necessary to provide the possibility to move freely within the space, choosing any point of view. It is indeed the need for free space exploration, rather than performance limits, that determines the change of scenery between the scale of the architectural element or sculpture and the scale of the entire building.

Figure 5. Example of analysis drawings in support of 3D models of architectural elements.

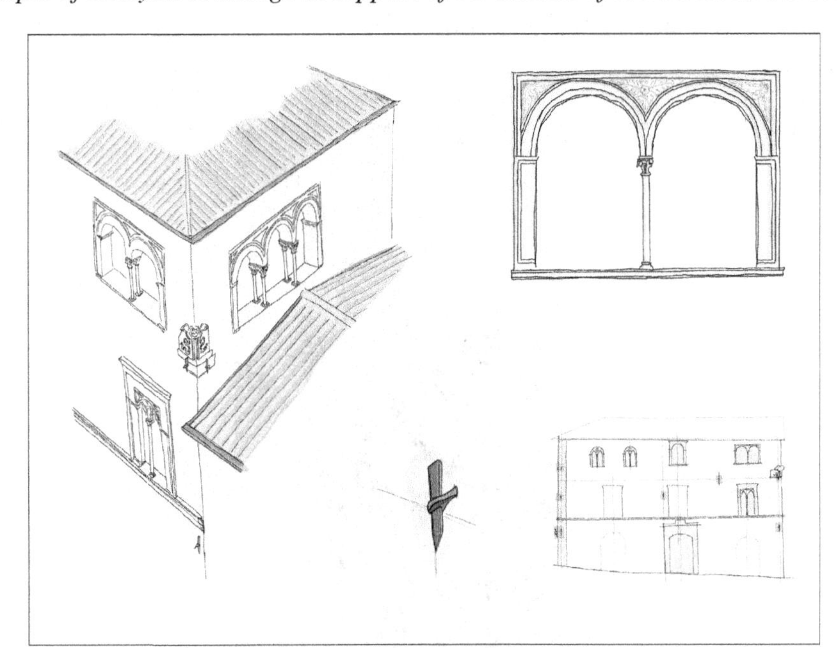

With this goal to be achieved, none of the applications tested in the previous scope meets the conditions requirements. The trial then examined the possibilities offered by the rendering engine Unity. The possibilities offered by this tool are considerable, in fact, first of all, it allows importing a wide range of different formats of models, objects and textures, so it has a certain versatility in entry; there is also the possibility to program the behavior and possible animations of objects and create complex scenes for lighting and types of materials. Finally, there is the possibility to export the results to all major operating systems, thus ensuring the conditions for a release on different platforms. The general idea is to validate a workflow that allows, from any type of input model, to implement an application in VR that can return to the user the experience of space exploration on any device, from PC and Smartphone to a model in Oculus' space.

The experimentation in this area concerned Palazzo Carli, in the historic center of L'Aquila, one of the main buildings of the University of L'Aquila until the earthquake of 2009. This palace, whose origins date back to the most ancient portions in the fifteenth century, underwent a first reconfiguration during the sixteenth century and an overall restoration in the eighteenth century following the earthquake of 1703. The most interesting space is the main courtyard, which combines the late Renaissance layout with rich 18th century ornamentation. Just in this courtyard there was, following the earthquake of 2009, a major collapse that destroyed one of the four interior facades. The model returns the spatial configuration of the eighteenth, as can be inferred from the interpretation of a series of archival documents and historical investigations. This configuration is different both from the present state after the collapse, and from the state before the earthquake, since a series of changes and transformations had been carried out over the centuries.

Figure 6. Axonometric view of a cross section of memorial of Beatrice and Maria Pereyra Camponeschi, in the apse of the Basilica of San Bernardino in L'Aquila. The model is obtained from SfM.

The model, made on the basis of a survey with laser scanner, re-reads, on the basis of the still existing elements, the documentary traces and proposes a hypothesis on the ancient spatial articulation. The model was imported in. fbx format and the materials were processed in Unity (Figure 7). A First Person Controller was then added, designed to return the point of view of a visitor of average height (1.75m). The scene was exported as an application .exe for the Windows environment and .apk for the Android environment (Figure 8). Since the building is still unusable, it is not yet possible to experience the effectiveness of the model in AR applications, but the development of VR applications is intended to make up for its current inaccessibility.

Figure 7. Screenshots of the Unity interface: in the frame the model of Palazzo Carli in L'Aquila with the First Person Controller in the foreground.

Figure 8. Screenshot of the .apk application during a test on smartphone.

CONCLUSION

Certainly, there are still a number of obstacles in the pervasive application of AR and VR technologies in the creation of a fully accessible and usable phygital landscape. To overcome these obstacles, it is necessary to operate on different layers, from infrastructure, to the development of services, to the definition and compliance of standards compatible with the technologies available today. Above all, it is necessary to change approach, not to consider these layers separately, but on the contrary, to examine them deeply interrelated and to operate on them in a strategically integrated way.

Although experimentation is still ongoing, and some technical solutions still need to be developed further, the overall model in its essential components is outlined. Much as all processes relating to the application in a broad field of innovative technologies, research needs to go hand in hand with the development and testing of new methodologies, progressively building the necessary tools to identify and validate innovative strategies, always paying due attention on the theoretical and conceptual implications of what is operationally realized.

The improvement in the field of ICT has paved the way for a confrontation between those disciplines related to cultural heritage and new and in some way unexpected fields. This confrontation has certainly helped to extend the conceptual boundaries of such disciplines, making them more accessible and capable of adapting to the changing society. It was noted in the first paragraphs that the evolution of the concept of cultural heritage has been positively affected by globalization, making use of the wealth deriving from the comparison between cultural heritage and communities extremely different from each other and how they have all benefited. An attempt has been made to clarify the evolving scenario in the specific field of ICT with regard to cultural heritage, stressing the importance of sharing an open innovation model, not only to optimize the integration between the various parties involved, but also to ensure greater sustainability of the system and make more accessible to communities the impact of the innovations that are carried out.

Finally, a series of trials were presented. These help to define a scenario of development and application of technologies in the field of digitalization and visualization. Certainly, it is necessary to develop these experiences further considering also how to improve them; it is furthermore crucial to build and agree on a direction to follow, one that takes into account possible users in terms of dissemination and acceptance. In order to achieve the objective of greater inclusiveness in the interpretation and presentation of cultural heritage, increasing attention to the public and society is essential from the very first stages of work.

However, what needs to be done, the core around which to build the future development, is based on the idea, proposed by the *Faro Convention*, that cultural heritage shall be considered structural to society, a right that belongs to everyone and that as such, compels individuals to be engaged not only in its protection and conservation, but also in its valorization and mediation. Technology is an instrument of extraordinary value in this challenge for it offers the opportunity to overcome social, cultural, perceptive and sensory disabilities, and therefore to achieve inclusiveness and accessibility. At the same time, the phygital concept of heritage, which allows transcending its physical boundaries and to project it into to the digital dimension while always acknowledging and respecting its physical consistency, opens very interesting perspectives regarding a more appropriate and complete communication of the cultural significance of places and contexts, allowing greater participation in their definition and a deeper and more solid communication of their values: deeper because more effective through the use of immersive visualizations, clearer because more understandable than traditional means, as well as more persuasive and pervasive; more solid, because able to show directly, through the use of multimedia platforms, the complex system of sources, knowledge, analysis and interpretations that build the foundation of historical-artistic knowledge, with a view to transparency and authenticity.

In conclusion, even with reference to the field of cultural heritage, we cannot imagine the paradigm of the smart city, disconnected from a contextual adaptation of society in synergy with it: centering the development of services and contents on the user experience and doing so within a framework of open innovation, defines a sustainable scenario for a cultural-based smart society.

ACKNOWLEDGMENT

This research was supported by the the Italian Government under CIPE resolution n.135 (Dec. 21. 2012) within the project INCIPICT - INnovating City Planning through Information and Communication Technologies and the project "SICURA – CASA INTELLIGENTE DELLE TECNOLOGIE PER LA SICUREZZA CUP C19C20000520004 - Piano di investimenti per la diffusione della banda ultra larga FSC 2014-2020."

REFERENCES

Antonelli, C., Cassioli, D., Franchi, F., Graziosi, F., Marotta, A., Pratesi, M., . . . Santucci, F. (2018, July). The city of l'aquila as a living lab: the incipict project and the 5g trial. In *2018 IEEE 5G World Forum (5GWF)* (pp. 410-415). IEEE.

Bentkowska-Kafel, A., Denard, H., & Baker, D. (Eds.). (2012). *Paradata and transparency in virtual heritage*. Ashgate Publishing, Ltd.

Brusaporci, S. (Ed.). (2017a). *Digital Innovations in Architectural Heritage Conservation: Emerging Research and Opportunities: Emerging Research and Opportunities*. IGI Global. doi:10.4018/978-1-5225-2434-2

Brusaporci, S. (2017b). *The importance of being honest. 3D Printing: Breakthroughs in Research and Practice*. Hershey, PA: IGI Global.

Brusaporci, S. (2020). Toward smart heritage: Cultural challenges in digital built heritage. In *Applying Innovative Technologies in Heritage Science* (pp. 271–296). IGI Global. doi:10.4018/978-1-7998-2871-6.ch013

Brusaporci, S., Graziosi, F., Franchi, F., & Maiezza, P. (2018). Remediating the historical city. Ubiquitous augmented reality for cultural heritage enhancement. In *International and Interdisciplinary Conference on Digital Environments for Education, Arts and Heritage* (pp. 305-313). Springer.

Brusaporci, S., Graziosi, F., Franchi, F., Maiezza, P., & Tata, A. (2021). Mixed reality experiences for the historical storytelling of cultural heritage. In *From Building Information Modelling to Mixed Reality* (pp. 33–46). Springer. doi:10.1007/978-3-030-49278-6_3

Centofanti, M. (2010). Della natura del modello architettonico. In S. Brusaporci (Ed.), *Sistemi informativi integrati per la tutela, la conservazione e la valorizzazione del patrimonio architettonico e urbano* [Integrated information systems for the protection, conservation and enhancement of the architectural and urban heritage] (pp. 43–54). Gangemi.

Coluccelli, G., Loffredo, V., Monti, L., Spada, M. R., Franchi, F., & Graziosi, F. (2018). 5G Italian MISE Trial: Synergies Among Different Actors to Create a "5G Road". In *2018 IEEE 4th International Forum on Research and Technology for Society and Industry (RTSI)* (pp. 1-4). IEEE.

Council of Europe. (2000). *European Landscape Convention*. Retrieved on April 27, 2022 from https://rm.coe.int/1680080621

Council of Europe. (2005). *Framework Convention on the Value of Cultural Heritage for Society*. Retrieved on April 27, 2022 from https://rm.coe.int/1680083746

De Felice, F., Travaglioni, M., & Petrillo, A. (2021). Innovation Trajectories for a Society 5.0. *Data*, *6*(11), 115. doi:10.3390/data6110115

Deguchi, A., Hirai, C., Matsuoka, H., Nakano, T., Oshima, K., Tai, M., & Tani, S. (2020). What is society 5.0. *Society*, *5*, 1–23.

Franchi, F., Marotta, A., Rinaldi, C., Graziosi, F., Fratocchi, L., & Parisse, M. (2022). What Can 5G Do for Public Safety? Structural Health Monitoring and Earthquake Early Warning Scenarios. *Sensors (Basel)*, *22*(8), 3020. doi:10.339022083020 PMID:35459005

ICOMOS. (1964). *International charter for the conservation and restoration of monuments and sites*. Retrieved on April 27, 2022 from https://www.icomos.org/charters/venice_e.pdf

ICOMOS. (2008). *Charter for the Interpretation and Presentation of Cultural Heritage Sites*. Retrieved on April 27, 2022 from http://icip.icomos.org/downloads/ICOMOS_Interpretation_Charter_ENG_04_10_08.pdf

ICOMOS. (2013). *The Burra Charter*. Retrieved on April 27, 2022 from http://portal.iphan.gov.br/uploads/ckfinder/arquivos/The-Burra-Charter-2013-Adopted-31_10_2013.pdf

ICOMOS Australia. (2013). *The Burra Charter. Practice Note_Interpretation*. Retrieved on April 27, 2022 from https://australia.icomos.org/wp-content/uploads/Practice-Note_Interpretation.pdf

Nofal, E. (2019). *Phygital Heritage: Communicating Built Heritage Information through the Integration of Digital Technology into Physical Reality* [PhD Thesis]. KU Leuven.

Principles of Seville. (2012). Retrieved on April 27, 2022 from http://sevilleprinciples.com/

Rinaldi, C., Franchi, F., Marotta, A., Graziosi, F., & Centofanti, C. (2021). On the Exploitation of 5G Multi-Access Edge Computing for Spatial Audio in Cultural Heritage Applications. *IEEE Access: Practical Innovations, Open Solutions*, *9*, 155197–155206. doi:10.1109/ACCESS.2021.3128786

The London Charter. (2009). Retrieved on April 27, 2022 from http://www.londoncharter.org/

UNESCO. (1954). *Convention for the Protection of Cultural Property in the Event of Armed Conflict*. Retrieved on April 27, 2022 from http://portal.unesco.org/en/ev.php-URL_ID=13637&URL_DO=DO_TOPIC&URL_SECTION=201.html

UNESCO. (1976). *Recommendation concerning the Safeguard and Contemporary Role of Historic Areas*. Retrieved on April 27, 2022 from http://portal.unesco.org/en/ev.php-URL_ID=13133&URL_DO=DO_TOPIC&URL_SECTION=201.html

UNESCO. (1994). *The Nara Document on Authenticity*. Retrieved on April 27, 2022 from whc.unesco.org/document/9379

UNESCO. (2003a). *Charter on the Preservation of the Digital Heritage*. Retrieved on April 27, 2022 from http://portal.unesco.org/en/ev.php-URL_ID=17721&URL_DO=DO_TOPIC&URL_SECTION=201.html

UNESCO. (2003b). *Convention for the Safeguarding of Intangible Cultural Heritage*. Retrieved on April 27, 2022 from https://ich.unesco.org/en/convention

UNESCO. (2009). *World Heritage Cultural Landscape*. Retrieved on April 27, 2022 from https://whc.unesco.org/en/culturallandscape/

UNESCO. (2011). *Recommendation on the Historic Urban Landscape*. Retrieved on April 27, 2022 from https://whc.unesco.org/en/activities/638

UNESCO. (2015). *Operational Guidelines for the Implementation of the World Heritage Convention*. Retrieved on April 27, 2022 from https://whc.unesco.org/en/guidelines/

KEY TERMS AND DEFINITIONS

Heritage Community: An expression introduced by the *Faro Convention* that identifies the community of people who take public action to support, transmit and disseminate the values of cultural heritage. The idea of the community that takes care of its heritage, and collaborates in its interpretation is the basis of the idea of participation in the sphere of protection and conservation of cultural heritage.

Interpretation (of Cultural Heritage): Process of knowledge and understanding of cultural heritage in the different levels of its complexity and in its relational system of values.

Open Innovation: Development model in which the innovation process is carried out in a collaborative way with the society, sharing knowledge and tools.

Phygital: Fusion of "physical" and "digital" concepts. The term defines a scenario in which the physical and digital instances fade into each other, ultimately resulting coextensive from a perceptive point of view.

Phygital Heritage: Application to the sphere of cultural heritage of the idea of phygitality. The integration between physical and digital results in a greater possibility of interaction with tangible heritage, and a more effective possibility of mediation of intangible heritage.

Point Clouds: Intermediate product of the survey process, extremely useful for storing or sharing large amounts of information about the tangible elements of cultural heritage. They can be obtained with different techniques (SfM, TLS, SLAM) and can be used for different purposes, even those not initially planned.

Presentation (of Cultural Heritage): Set of activities and actions planned to mediate the communication of interpretive content of cultural heritage.

Chapter 18
Digitization of Cultural Heritage:
The Farnese Theatre in Parma

Andrea Zerbi
University of Parma, Italy

Sandra Mikolajewska
University of Parma, Italy

ABSTRACT

Cultural heritage represents the identity of people and, as such, is a fundamental element of our lives. The numerous projects carried out in recent years in the field of CH digitization have shown that the operation of dematerialization may be considered an essential tool for its preservation, conservation, and enhancement. Since advanced technology allows to valorize artifacts and bring a positive impact on the people's life to whom they belong, in the context of Society 5.0 it can be considered as a key tool. Starting from the analysis of the state of the art in the field of digitization, the main goal of the present study is to investigate the role that this process can take on within the complex process of valorization of monuments. To this aim, a research carried out on the Farnese Theatre will be illustrated. Particular attention will be paid to the methodological choices made for the creation of an extremely versatile three-dimensional model and for its possible uses.

INTRODUCTION

The numerous projects carried out in recent years in the field of digitization of Cultural Heritage have shown that the operation of dematerialization may be considered an essential tool for its preservation, conservation and valorization. The digital technologies that are used with this aim can be considered as a key tool within the "Society 5.0". As well known, this concept was introduced in the 5th Science and Technology Basic Plan adopted by the Japanese Cabinet in 2016. This term refers to a human-centered

DOI: 10.4018/978-1-6684-4854-0.ch018

society, in which the goal is to improve people's quality of life by using advanced technology in different fields (healthcare, environmental, transport, production and so on).

In recent years, in fact, a rapid and continuous evolution of digital and information technologies has taken place, and today they able to produce extraordinary potentialities. Among the fields interested, particular attention should be given to the cultural one. In this field, new technologies can really produce extremely positive impacts, both from a social and economic point of view, both on an individual and collective level.

A recent research funded by the European Spatial Planning Observation Network (ESPON) called "HERIWELL - Cultural Heritage as a Source of Societal Well-Being in European Regions" has clearly highlighted the positive impacts that Cultural Heritage can have on the society. In the final report drafted at the end of the research (ESPON, 2021), a link between social welfare and Cultural Heritage was identified. It is interesting to note that the short- and long-term outcomes regard all types of heritage: tangible, intangible or digital.

The impacts identified concern first of all people's quality of life: "growth in happiness and life satisfaction, improvements in eudaimonic conditions and health rates, improvements in education levels and empowerment in adults' capacities, including digital skills, higher levels of knowledge and research, improved quality and sustainability of environment" (ESPON, 2021, p. 20). This report also identifies additional positive impacts, mainly related to material and social factors. For the latter, the most interesting effects are related to increasing community awareness, greater sense of belonging and social inclusion.

It is also true that, during the lock-down period, caused by the pandemic situation related to the spread of the SARS-COV-2 virus that began in 2020, the benefits just mentioned were inevitably denied to society. A survey conducted as part of this European research, specifically aimed at assessing the individual perceptions of citizens on the relationship between Cultural Heritage and their well-being, showed clearly that the restrictions adopted for health reasons created a negative impact on the population. All citizens of the eight European countries involved in this survey (Belgium, Czech Republic, Germany, Ireland, Italy, Norway, Poland and Spain), despite coming from very different cultural backgrounds, declared a strong sense of exclusion related to the fact that they no longer had access to Cultural Heritage.

On the other hand, in order to deal with the temporary restrictions adopted, new ways of alternative use of heritage were widely experimented. These included also virtual solutions. In that period, the increasingly pressing need to share knowledge virtually was favored by the fact that new technologies make it possible to overcome the normal space-time limits, offering the possibility of extending the spectrum of users of Cultural Heritage (Agostino, Arnaboldi & Lampis, 2020).

However, the current pandemic situation is not the only problem. In fact, we live in a historical moment in which, in the name of "globalization" that affects any field today, national boundaries are crossed and the historical memories of a single population, from heritage of a specific territory, become heritage of the whole humanity. It is important to stress that all the assets are vulnerable, often in danger. Natural disasters, armed conflicts, neglect, vandalism and so on, have always caused (and still cause) the more or less significant destruction of unique and unrepeatable artifacts (Figure 1). Despite their intrinsic value is universally recognized, even today we still witness too often the destruction of documents, buildings and even entire cities. It is clear that these events destroy the memory and identity of the populations (Bevan, 2016).

The events currently taking place in Ukraine demonstrate once again the fragility of heritage. After the Russian invasion on February 28, 2022, UNESCO adopted immediately the procedures established by the 1954 "Hauge Convention for the Protection of Cultural Property in the Event of Armed Conflict".

In fact, one of the first operations carried out regarded marking all sites and monuments with the "Blue Shield" emblem, used to prevent intentional or accidental damage. But even the sites under protection are constantly monitored, there is no guarantee that they will not be attacked by the military or that the artifacts will not be stolen and sold on the illegal market. As claimed by the UNESCO World Heritage Center, more than a month after the beginning of the Russian invasion on Ukrainian territory, there are almost one hundred cultural and religious sites damaged.

In many cases, when historical sources were available, in order to preserve the memory of citizens, the damaged heritage was chosen to be reconstructed. There are countless cases of artifacts destroyed over time (by intentional or accidental events) and then reconstructed in order to preserve the memory of history: the Bell Tower of San Marco in Venice, the Bridge Kriva Cuprija in Mostar, the Frauenkirche in Dresden, the Theatre La Fenice in Venice, the Cathedral Notre-Dame in Paris, and so on. With great probability, once the Russo-Ukrainian conflict will be over, the damaged buildings will be at the center of debates for their reconstruction aimed at the recovery of the identity of Ukrainian citizens.

The growing awareness of the environmental vulnerability in which each community lives and works, led to the necessity of documentation of Cultural Heritage. In fact, its digitization process is becoming increasingly widespread. The virtual reconstruction of heritage offers many benefits: it allows an accurate documentation of the artifacts, it contributes to their preservation, it provides also an important tool for their study, analysis and dissemination (CORDIS, 2021).

It is clear that thanks to the recent evolution of digital technologies, today we have a wide range of possibilities, previously unthinkable. Since they allow the virtual reconstruction of any kind of artifact, which through the process of digitization can be valorized, in the context of "Society 5.0", advanced technology can be considered as a key tool. As said before, Cultural Heritage represents the identity of people and, as such, is a fundamental element of their lives. If it is enhanced, it can really have a positive impact on the people's life to whom it belongs.

Figure 1. Examples of tragic events that caused damage of well-known architectures. From the left: the Bell Tower of San Marco in Venice, the Frauenkirche in Dresden, the Cathedral Notre-Dame in Paris
Sources: Boni, G. (1912). Sostruzioni e macerie. In Fradeletto, A. (Ed.), Il campanile di San Marco riedificato: studi, ricerche, relazioni (pp. 27-65), Officine Grafiche Carlo Ferrari di Venezia; Ambrogio, K. (2013). Restoration faces the gap. Ottagono, 257, 02/2013, 126-149; Dalla Negra, R. (2019). Notre-Dame de Paris: the Restoration, the unknown one. Paesaggio urbano, 2.2019, 8-17

On the basis of the issues mentioned above, strongly related to the use of digital technologies for Cultural Heritage valorization, a research on a unique historical Italian architecture was carried out (Mikolajewska, 2021). As it is characterized by some typical elements of cultural artifacts, but at the same time by its own uniqueness, the Farnese Theatre in Parma was chosen as a case study. This monument is particularly significant to investigate the role that the digitization process can take on within the complex process of valorization and dissemination of Cultural Heritage, and specifically within the "Society 5.0". As the theatre represents an extremely large and complex architecture, characterized by numerous decorative elements and frescoes, it requires an integrated multiscale survey and particular procedures for the management of different datasets. Such characteristics, common to many other artifacts, make the monument an excellent example for dealing with the issues related to the process of historic architecture digitization (Chiabrando, Sammartano, Spanò & Spreafico, 2019). Another reason that makes the monument an interesting case study is related to the fact that it was completely reconstructed after the 1944 destruction. The survey of the theatre allows to verify the support of digital technologies in the artifact knowledge process.

Starting from the analysis of the state of the art in the field of digitization of Cultural Heritage, the main goal of the present study is to illustrate and critically analyze the research carried out on the Farnese Theatre. Particular attention will be paid to the methodological choices made for the creation of an extremely versatile three-dimensional model and for its possible uses.

Cultural Heritage Digitization

In recent years, the need of Cultural Heritage documentation has become evident. In fact, it represents one of the main topics treated within the plans promoted by the European Union. This activity, in addition to contributing to the preservation of historical memory, is inevitably linked to the safeguarding and protection of the assets. In many cases, it also became fundamental for the transmission of the artifacts to future generations.

It is important to stress, that like any other activity related to Cultural Heritage, digitization operations still face some problems. One of the main issues refers to the fact that dematerialization process involves extremely different artifacts, such as documents, sculptures, architectures, and so on. It is easy to understand that depending on the object, the operations to be carried out for its dematerialization need to be different. The main problem is that despite the fact that digitization is now an operation universally recognized, there are still no clear procedures. The same object can be digitized using different techniques and workflows, and the procedures adopted are not always clear. Still today there are many problems related to the integration of different datasets. The management of data provided from survey campaigns performed on the complex architectures present a low level of automation.

With the aim to standardize procedures, in the last years the most important institutions that manage Cultural Heritage published a series of reference able to ensure rigor and scientific validity to this operation. Of particular importance is the "London Charter for the Digital Visualization of Cultural Heritage" published in 2009 and updated two years later in the "Seville Charter", which extends some concepts to archaeological heritage. The document establishes six principles for the use of digital visualization, upon which the intellectual integrity of the methods and results themselves depend. In this context, a significant step forward is the publication in 2020 of the "Guidelines for 3D Digitization of Tangible Cultural Heritage". The document, addressed to professionals, institutions and regional authorities involved in the activities of 3D documentation of Cultural Heritage, contains ten basic principles aimed at improving these projects (European Commission, 2020).

There are numerous experiences of digitization aimed at the preservation of assets, both in the field of architectural and documentary heritage. The documentary apparatus is extremely fragile and constantly exposed to inevitable processes of deterioration, often also linked to continuous manipulation by users. Digital copies of documents can make consultation faster and more accessible to a greater number of users, preserving the physical integrity of the sources. These materials, once digitized, could be also subjected to traditional restoration operations in order to ensure their preservation over time.

As far as the written documentary apparatus is concerned, in the last two decades, a large number of projects aimed at the digital preservation of archival material was carried out, both at national and international level. Among these, the "Codex Sinaiticus" project can be considered particularly interesting. In 2007, the pages of the Codex Sinaiticus (a fourth-century manuscript of the Christian Bible) kept in four institutions (the British Library in London, the Leipzig University Library, the Monastery of St. Catherine in Egypt and the National Library of Russia in St. Petersburg) were digitized. Thanks to this international collaboration it was possible to obtain the digital version of the entire manuscript and make it accessible to a global audience on-line. The benefits of this kind of operation were immediately recognized by many institutions. For example, numerous Italian projects aimed at the digitization of documentary material are collected on the "Biblioteca Digitale Italiana" which hosts heritage coming from libraries of the Ministry for Cultural Heritage and Activities (MiBAC), the Ministry of University and Research (MIUR), local institutions, foundations and other cultural institutions.

In the context of the documentary digitization projects, particular attention should be paid to those focused on cartographic and iconographic apparatus. The characteristics of these drawings, often made on large paper supports, almost always folded repeatedly, makes them deteriorate quickly. The digitization of this material, not only represents the first step towards its preservation, but offers multiple possibilities to users. At the same time, it simplifies their consulting and promote activities of study and research. From this point of view, particularly interesting are the projects carried out in the field of Historical Geographic Information Systems (HGIS) (Bruno, Bianchi, Roncella & Zerbi, 2015) or those related to the iconographic heritage inherent to the specific field of architecture. For example, in the context of this research, which is also linked to the iconography of the Farnese Theatre in Parma, the digital library of the Institut National d'Histoire de l'Art in Paris is particularly significant.

Architectural Heritage Digitization

The digitization operation is particularly useful not only for the documental apparatus but also in the field of monumental heritage, especially the architectural one. The possibilities offered by the dematerialization of this kind of artifacts are countless.

First of all, the rapid evolution of surveying techniques (laser scanner and photogrammetry) allows today to obtain accurate point clouds to be used as a base for the creation of digital models of existing heritage. The 3D models that we are able to create using digital technologies, in many cases obtained by integrating data surveyed with different techniques (Balletti, Bertellini, Gottardi, & Guerra, 2019), document the state of the artefacts in a specific historical moment, providing useful information for their conservation, maintenance and management.

At the same time, the digital models represent an extraordinary tool for the communication and dissemination of Cultural Heritage. As well know, historical architectures are inevitably characterized by more or less transformations taken place in different periods. In these cases, their virtual copy can be used for the creation of audiovisual supports, useful to facilitate understanding of the monument and its transformations (Giordano, Borin, Cundari & Panarotto, 2014).

One of the most important filed in which an accurate heritage digitization is extremely useful is the restoration one. When dealing with historical buildings, it is increasingly common to use 3D models for the structural analysis (Tucci, Bonora, Conti, Fiorini & Riemma, 2014), consolidation activities (Ottoni, 2008), anastylosis and restoration operations. From this standpoint, particularly interesting project is the one related to the Basilica of San Marco in Venice, where accurate surveys have been carried out for many years for restoration purposes (Fassi, Fregonese, Adami & Rechichi, 2017).

Another interesting project related to the restoration purposes, but of the frescoes, is the one carried out within the museum itinerary of the Camera di San Paolo in Parma. Thanks to an accurate integrated survey, it was possible to document all the painted surfaces, including the well-known frescoes made by Correggio in the "Camera della Badessa" (Bruno, Mikolajewska, Roncella & Zerbi, 2022). The drawings obtained were fundamental for restoration activities which, as previously mentioned, are essential for the transmission of this heritage to future generations.

The architectural heritage digitization offers many possibilities in the specific field related to the management purposes. The 3D models can be integrated with qualitative data and be used within BIM environments (Building Information Modeling) (Osello, 2012). Despite the fact that this approach initially was adopted for the design and management of new buildings, in recent years there has been a particular interest in its application to existing ones, including those historical (Historical Building Information Modeling). In fact, numerous BIM processes were applied to important cultural artifacts (Figure 2): the Cathedral of Milan (Fassi, Achille, Mandelli, Rechichi & Parri, 2015), the Basilica of San Marco in Venice (Fregonese et al., 2017), the Basilica of Santa Maria di Collemaggio (Brumana et al., 2020), etc.

Even more ambitious seems the be the project by the French laboratory "Modèles et simulations pour l'Architecture et le Patrimoine" in relation to the Notre-Dame Cathedral in Paris, affected by a serious fire on April 15, 2019. It is intended to create a unique database, based on the integration of all the scientific and technical data on the cathedral, to use for the management of the entire monument.

Figure 2. Examples of digitization projects for restoration, conservation and management purposes. From the left: the Basilica of Santa Maria di Collemaggio, the Basilica of San Marco in Venice
Sources: Brumana et al. (2020); Fassi, Fregonese, Adami & Rechichi (2017)

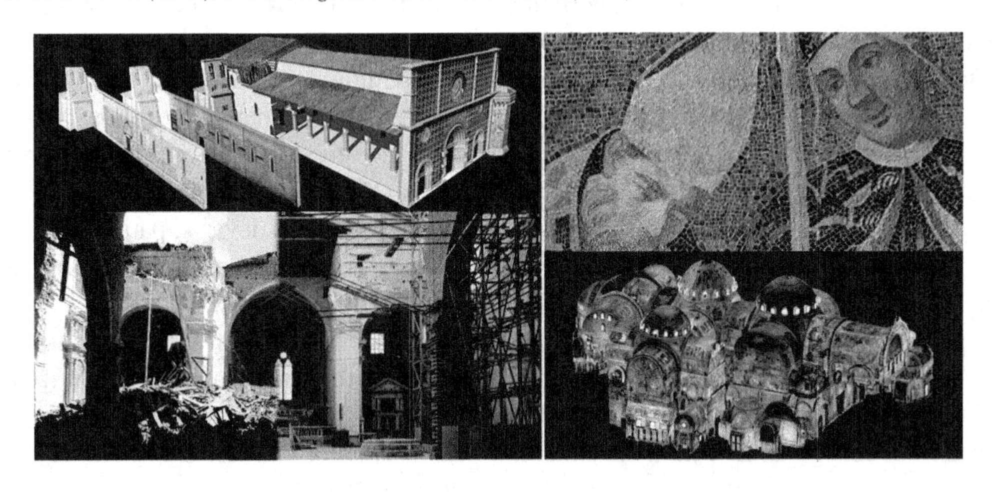

Regarding the dissemination of knowledge and the desire to spread it without any spatial-temporal limits, 3D models can be used for the development of virtual tours of particularly valuable artistic environments. Several world-renowned museums offer the possibility of virtual tours of their collections: Louvre Museum in Paris, National Gallery in London, Uffizi Gallery in Florence, Egyptian Museum in Turin, and so on. These virtual reconstructions are also important for the public unable to physically go to the site, due to various reasons (distance, economic difficulties, physical limitations of the users, etc.). One of the most emblematic experiences conducted in this sense, aimed at the knowledge dissemination, concerns the virtual tours of the Sistine Chapel and the Vatican Museums (Belardi & Bori, 2018).

Since a few years, the repertory of assets that can be consulted virtually has been extended to archaeological sites and architectural artifacts. From this point of view, particularly significant is the activity carried out by a non-profit organization CyArk, which, thanks to the "Open Heritage" project developed in collaboration with Google Arts & Culture, makes accessible digital models of many historical monuments located in different parts of the world.

The digitization of the architectural heritage offers extraordinary potentialities also in the field of alternative and multisensory use of the artifacts. When their accessibility is limited or even denied to the public, due to the particular morphological conformation of the places, to the space-time limits or to the motor or visual difficulties of the visitors, the search for alternative solutions becomes fundamental (Figure 3). Even the national and international plans (for example "Declaration of Cooperation to Promote the Digitization of Cultural Heritage"-2019, "Three-Year Plan for the Digitization and Innovation of Museums"-2019) promote the use of augmented reality, virtual reality, gaming experiences, tactile museums, etc.

Virtual models are extremely important in contexts characterized by particular morphological conditions or architectural barriers that prevent full accessibility to people with motor difficulties. It is the case of the Mausoleum of Theodoric in Ravenna, in which the upper floor is accessible only by a staircase. Thanks to the creation of the virtual model of the structure, it was possible to overcome the physical limits imposed from the Mausoleum (Incerti, D'Amico, Giannetti, Lavoratti & Velo, 2018).

The alternative use of heritage is also fundamental for the users who are lacking in one or more of the five senses, such as visually impaired people. In recent years, the increasing diffusion of projects aimed at the creation of tactile museum itineraries has led to the foundation of the first Italian museum entirely dedicated to this topic, State Tactile Museo Omero in Ancona (Riavis, 2019). Taking inspiration of this museum, several tactile itineraries related to the architectural heritage are being created (Sdegno, 2018). For example, the project called "A portata di mano" includes a physical model of the UNESCO Modena Site composed by the Square with the Duomo and the Ghirlandina Tower.

An interesting application linked to the digitization of Cultural Heritage and to its multisensory fruition has also involved the medical sector. Particular attention to such a delicate issue can be found in the project called "SAM4Care". Among the many topics investigated in a broader research aimed at combining BIM (Building Information Modeling) and VAR (Virtual and Augmented Reality), the use of digital reconstructions is experimented. Thanks to the use of various devices, patients have the possibility to virtually visit monuments without having to make any movement and, above all, to relieve, at least emotionally, a moment of their lives so difficult and painful (Osello, 2018).

It is important to stress that the progress achieved in the field of information and communication technologies, allow to establish new strategies capable of promoting the dissemination of heritage (Empler, 2018). At the same time, with the introduction of digital technologies applied to Cultural Heritage, there was a significant change in the communication and transmission of knowledge (Buono & Giugliano,

2021). Today, traditional learning methods are combined with more interactive and engaging solutions, which can also be useful tools for the study activities (Lo Turco et al., 2019).

In parallel with the progress achieved in the digitization of Cultural Heritage, the evolution of multimedia technologies has led to the use, for its enhancement, of new media traditionally used in other sectors. Among these, of extreme interest is a particular form of augmented reality, known as video mapping. Several projects carried out in different parts of the world have shown how this technique is able to bring the attention of a wide audience to an artifact. In this way it becomes the potential tool for its valorization (Maniello, 2014). The video projection, non-invasive and completely reversible, can become an extraordinary tool for the dissemination and communication of knowledge. Several performances carried out to celebrate cultural events, such as the opening of the National Library of Sarajevo (2014), or the celebration of the bicentennial of the Prado Museum in Madrid (2018), clearly highlight how a conscious use of video mapping can contribute to the valorization of historical architectural heritage. In addition, in contexts characterized by the lack of decorative apparatus, it can also be useful to strengthen the collective memory of citizens (Figure 4); (Peter, 2018); (Font Sentias, 2020); (Giannetti, Lodovisi, Incerti, Grassivaro & Sardo, 2019); (Maniello, 2020).

Figure 3. Examples of digitization projects for alternative use and knowledge dissemination. From the left: the physical model of the Duomo in Modena, the virtual model of the Mausoleum of Theodoric in Ravenna, the virtual reconstruction of the Church of the Eremitani in Padua

Sources: Mendoza, H.R. (2017, December 7). 3D Printing Brings Modena Cathedral to Fingertips of Visually Impaired (https://3dprint.com/196322/3d-printing-modena-cathedral/); Incerti, D'Amico, Giannetti, Lavoratti & Velo (2018); http://eremitani.beniculturali.unipd.it/video (July 15, 2022)

THE CASE STUDY: THE FARNESE THEATRE IN PARMA

The Farnese Theatre in Parma is worldwide known for being one of the few seventeenth-century wooden theatres still existing. The monument was designed by Giovanni Battista Aleotti and was built from the end of 1617 in about one year. Although it was commissioned by Ranuccio I (Duke of Parma and Piacenza) to celebrate the 1618 visit to Parma by Cosimo II de' Medici (Grand Duke of Toscany), the official inauguration of the structure took place ten years after the end of its construction. In fact, the theatre was used for the first time in 1628, to celebrate the wedding between Duke Odoardo I Farnese and Princess Margherita de' Medici.

Despite the undoubted uniqueness of the structure, before the nineteenth century the Farnese Theatre hosted less than ten events. The reasons that caused the limited use of the monument in that period are related to the significant number of resources needed in such a vast structure. The last performance held within the theatre was celebrated in 1732. This event signed the end of the first part of the active life of the monument.

Figure 4. Examples of digitization projects for knowledge communication through video mapping. From the left: the Temple of Dendur (USA), the Church of Sant Climent de Taüll (Spain), the Rocca in Vignola (Italy)

Sources: Nofal, E., Stevens, R., Coomans, T., & Moere, A.V. (2018). Communicating the spatiotemporal transformation of architectural heritage via an in-situ projection mapping installation. Digital Applications in Archaeology and Cultural Heritage, Vol. 11; Maniello (2020); Giannetti, Lodovisi, Incerti, Grassivaro & Sardo (2019)

Since then, the theatre gained fame throughout Europe for more than a century. This led the Farnese Theatre to become a place of travel study for various artists. Numerous visitors came to Parma with the aim of seeing, admiring and analyzing it. At the same time, the state of abandonment led inevitably to an increase in decay. In fact, in that period, more and more numerous were the records, both written and iconographic, aimed at denouncing the state of degradation in which the theatre was.

In order to stop the state of decay, from the second half of the nineteenth century began to be carried out some restoration works. One of the most significant activities regarded the complete renovation of the roof in 1867. About fifteen years later the entire proscenium was restored, including the painted decorations on the wooden structure. Other restoration operations were carried out in subsequent years. They were mainly related to the equestrian monuments placed in the triumphal arches and the structure of the bleachers, entirely rebuilt in 1913.

In the twentieth century the monument was used for few occasions. Among the most significant events worth mentioning are: the first Italian Congress of Sciences (1907), the fiftieth anniversary of the plebiscite of Parma (1909), the celebration of the centenary of Giuseppe Verdi's birth (1913), the National Conference dedicated to Correggio (1935). Despite these few uses of the theatre, with the beginning of the Second World War, the monument was closed again.

In fact, as well known, in 1944 many Italian cities were bombed by the Allies. As far as concern the city of Parma, between April and May, large portions of the urban fabric were destroyed. Regarding the monumental buildings, the biggest damages were registered on May 13, 1944, when the Pilotta Palace was bombed. In that occasion, the State Archives, the Palatina Library, the Picture Gallery and the Farnese Theatre were strongly affected by the bombing. In fact, almost all of Aleotti's structure was damaged in an irreparable way. Only part of the proscenium, some of the structures leaning against the perimeter walls and the frescoes painted on the latter (which were protected by the loggias), were saved. The sculptural apparatus, originally present in the entire theatre, was almost totally destroyed (Figure 5).

Figure 5. The Farnese Theatre (from the left: first half of the twentieth century, 1944 destruction, 1950s reconstruction)
Source: Photographs from Pinazzi private collection

After years of debates concerning the future of the ancient structure, starting from the end of 1956, the theatre was completely rebuilt. In this process, the original configuration of the monument was respected. Only the rich sculptural apparatus and the pictorial decorations that once completely covered the wooden structures were not reconstructed. So, what we see today is a twentieth-century reconstruction (Figure 6). In particular, the theatre is characterized by a U-shaped cavea with two orders of loggias that contain 17 Serlian windows on each level. In front of the cavea there is a proscenium surrounded by two triumphal arches placed on the side walls with the equestrian statues of Alessandro and Ottavio Farnese. The hall that contains the theatre is covered by a roof system composed of 20 trusses able to cover a distance of 32 meters.

While the historical events of the theatre are well known and have been studied by numerous scientific researches, the events related to the reconstruction of the monument are less investigated.

For these reasons, in this research, the historical period of the monument's reconstruction was mainly examined. To this aim, the main tools available in the digitization field were used. In particular, a workflow based on the use of digital technologies for the valorization of Cultural Heritage was experimented.

Figure 6. The Farnese Theatre in Parma (on the left: view of the proscenium, on the right: view of the bleachers)
Source: Author photographs

Data Acquisition Phase

The data acquisition phase of the theatre was divided into two parts: the first involved the collection, digitization, cataloguing and analysis of the documentary and iconographic material, the second focused on an accurate integrated survey of the monument.

It is important to stress that, in order to understand the transformations that have characterized the theatre from the seventeenth century, it is necessary to have a complete overview of its most significant graphic representations produced over time (project drawings, survey drawings, elaborates made for the purpose of documentation, pictorial views and representations, and so on). Since the monument has always raised interest among architects and artists, the available documentation is extremely rich. However, as always happens when dealing with historical heritage, the structure representations provided by the various authors are not always homogeneous and, in some cases, are completely different.

Most of the historical iconographic documentation regarding the Farnese Theatre that has survived over time and was at least partially used for the post-war reconstruction, is today preserved in some national and international archives. In this phase of the research, the drawings kept in these archives were examined and digitized. Among the main iconographic documents collected, particular importance was given to: the drawings of Giovanni Battista Aleotti (first half of the seventeenth century), the drawings attributed to Francesco and Ferdinando Bibiena (second half of the seventeenth century), the drawings of Pier Paolo Coccetti (first half of the eighteenth century), the anonymous drawing of the floor plan (probably after 1728), the survey drawings by Louis Auguste Feneulle (seventies of the eighteenth century), the survey drawings by Louis-Hippolyte Lebas (first decade of the nineteenth century), the reconstruction project drawings by Arrigo Stanzani (1953).

In parallel, a research aimed at the collection of the photographic documentation of the monument was carried out. In this specific context, the analysis of these sources was particularly significant in acquiring a more complete knowledge of the artifact. Thanks to the photographs collected, it was possible to reconstruct the transformations of the theatre that took place between the end of the nineteenth century and the beginning of the twentieth century.

In this phase, the first reference was made to the bibliographic sources. Particular attention was also paid to the photographs published in newspapers. In addition, further research was conducted in national archives and on-line digital collections. Finally, the most significant sources were found within two private collections: the first, of the family of Italo Pinazzi (who contributed to the reconstruction of the theatre) and the second, of the historian Gianni Capelli, donated to the University of Parma.

The photographic corpus collected in this phase was properly digitized and catalogued. In this way, a database to use as a support for the study activities was created. The comparison of these photos with the present state of the monument has made it possible to identify some transformations to which the theatre has been subjected over the years, to recognize some elements of dissimilarity between the reconstructed structure and the one prior to the destruction and to reconstruct digitally the decoration that once characterized the wooden surfaces of the artifact.

The second phase of the data acquisition process focused on an accurate survey campaign of the theatre. The choice of the most suitable survey methods was based on the analysis of geometric and material characteristics of the artifact and was influenced by some logistical problems related to the accessibility conditions of the structure.

Given the complexity of the monument and the purposes of this research, it was chosen to use mainly indirect survey methods. The aim was to integrate different methodologies in order to obtain maximum performance from each of them.

First of all, in the planning phase of the survey campaign, it was decided to carry out a topographical survey to use as a reference for subsequent measurement operations. In particular, it would serve for the Terrestrial Laser Scanner (TLS) survey and for the photogrammetric survey. Finally, for some detailed elements and for those that cannot be documented through the previously mentioned methods, it was decided to use direct survey method.

Before starting the survey campaign, all the points in which to place targets and stations were identified, as well as the resolution to be used for each scan. These considerations were made in order to optimize survey times and to obtain a density of points as uniform as possible on all surfaces.

Since the survey operations could only be carried out on the one day of the week when the theatre was closed to the public, it was necessary to pay particular attention where to place stations and targets. In general, the least invasive locations were selected. Regarding the targets, two were placed on the walls of the stage and four on the bleachers. As for the station points, it was verified that from each of them it was actually possible to measure at least three of the six targets placed inside the theatre and at least three different stations. Only the stations under the bleachers were an exception to this workflow. At the level of the loggias (inaccessible to the public), the points were materialized on the ground, while at the floor level (open to the public) were chosen characteristic points, of which monographs were made. A total of 35 station points were identified.

After placing the targets and materializing the station points, the topographic survey campaign began. In order to ensure the correctness of the measurement operations, a closed network with ramifications was created. A Topcon Image Station IS2 total station was used. The topographic survey performed concerned only the measurement of targets and station points positioned at floor level. For the survey of the other station points, the Leica ScanStation C10 laser scanner was used, taking advantage of the possibility to use it as a total station.

Then, the laser scanner was used to perform 35 scans inside the structure, distributed as follows: six scans at the floor level, six in the space under the bleachers, two on the stage, nine at the level of the first loggia, nine at the level of the second loggia and three at the level of the latter's roof (Figure 7). Most of the scans were carried out at high resolution (1 point/5cm at 100m), while only in the points where the device was close to the structures of the theatre, and therefore would have had to measure points placed at a very short distance, a medium resolution scans were made (1 point/10cm at 100m).

At the moment, the only part of the monument that was documented with less detail is the one related to the stage. Only two scans were carried out in this area: one in the initial part and one between the two arches that separate the theatre from the Hall of Triumph. The area below the stage and its lateral parts were not surveyed. However, due to the characteristics of the survey carried out, these parts can be documented and integrated into existing database in the future.

Figure 7. Laser scanner survey of the Farnese Theatre (on the right: identification of station points and targets limited to the area of the bleachers)
Source: Author photographs

The registration of the 35 scans took place keeping as a reference the point cloud obtained from the topographic survey and was performed within the software Cyclone 9.2. A unique point cloud was generated. The latter was then optimized (reducing its size but maintaining data quality) and was used as a basis for the creation of the three-dimensional model of the monument.

In addition to the TLS survey, a photogrammetric survey of the pictorial apparatus was also performed. For the moment, the methodology adopted was experimented on the fresco placed on the back wall of the hall. This choice was based in order to verify the relationship between the fresco and the architectural proscenium, always considered a specular copy of each other (Gandolfi, 1980).

The main challenge regarded the light conditions and the geometric characteristics of the decoration (Figure 8). Since the fresco is painted on a 32-meter-long wall, close to the wooden structures that divide it into seven unconnected areas, it was necessary to carry out a survey of single parts, then connected to each other. In addition, because of the strong contrasts between light and shadow generated by the windows placed on the east side of the hall and the spotlights pointing to the wooden structure leaning against the fresco, it was necessary to take frames with different exposures. In particular, from each point at least two frames with different exposure were acquired.

Figure 8. Photogrammetric survey of the fresco placed on the back wall of the hall with some particulars of the wooden structures leaning against it
Source: Author photographs

A Nikon D7200 with a resolution of 6000x4000 pixels and 24 mm optics (36 mm equivalent focal length) was used. The camera was mounted on a tripod and the images were acquired using the scheme with the optical axis of the camera as much as possible perpendicular to the wall. The different parts of the fresco do not always have the same height: at the level of the first loggia the fresco is height approximately 5.80 m, at the level of the second loggia, 4.20 m, and at the last level, 3.50 m. So, in relation to the height of the decoration, one to three stripes of photographs were necessary. In the parts characterized by the high closeness of the loggias to the wall (for example near the stairs), other photos were taken. An overlap between frames of at least 60% was kept. A total of about 360 photographs were acquired using this scheme.

The data processing was performed using the software Agisoft Photoscan Professional 1.2.0.

Multifunctional Three-Dimensional Model of the Theatre

At the end of the data acquisition phase, two different point clouds were obtained. The first one (from the laser scanner survey) was related to the entire structure, the second one (from the photogrammetric survey) to the back wall of the theatre. Both point clouds were then used for the creation of the three-dimensional model of the theatre.

Given the complexity of the monument and the desire to create a 3D model that could be used for different purposes, it was necessary to carefully choose the best procedures to be adopted. The aim was to find the correct balance between the level of detail, the time needed for digital reconstruction and the management of the model. These issues are at the center of the scientific debate related to three-dimensional modeling of architectural heritage (Tommasi, Achille & Fassi, 2016). Another important issue is that the field of modeling of historical Cultural Heritage is still characterized by the lack of common standards.

The Farnese Theatre can be considered an emblematic case of historical architecture: it is extremely complex, is characterized by an articulated plastic apparatus, is subject to deformation phenomena and has few regular elements. Those characteristics and its unicity make it hardly suitable for parametric or mesh modeling. The first is based on the use of standard and regular elements and the second, applied to such a complex structure, would inevitably lead to management problems. An interesting solution seemed to be the free-form modeling.

As said before, the model had to be multifunctional. In particular, its purposes included: support to ordinary management operations, to restoration activities, to study research and to dissemination of knowledge about the monument. In the case of large and complex buildings like the Farnese Theatre, the good solution is to divide the digital model in several parts, allowing to optimize the management of the large amount of data. Therefore, it was decided to create a model composed by elements characterized by several parts, with different levels of adherence to reality. Each part was then modeled using different procedures.

With the aim of achieving the above-mentioned needs (regarding level of detail, time and management issues) 5 fundamental parts of the theatre were identified. These were denominated as "classes" and were selected according to their function within the structure. The identified parts are: masonry perimeter walls, wooden elements characterizing the artifact (loggias, bleachers, proscenium), repetitive decorative elements, bleachers system, roofing system.

It is important to note that, at the moment, some elements such as complex plastic decorations were not modeled (the coat of arms, the equestrian statues and the medallions in relief on the loggias). However, when detailed data of these elements will be collected, it will be possible to integrate them into the general theatre model.

Definition of the Level of Accuracy

Before the modeling operations it was necessary to establish the levels of adherence to the point cloud to be used.

An interesting reference to take inspiration from is the USIBD Level of Accuracy Specification Guide (USIBD, 2019). In the guidelines, a classification of Levels of Accuracy (LOA), determined by comparing different datasets or independent measurements, is proposed in terms of standard deviation (Table 1). The most interesting point suggested in this guide is that LOA have to be applied to each single element and not to the entire project. This implies that, within the same model, there may be elements with different LOA. These levels depend on how the element was surveyed and how it was modelled. As the document is mainly addressed to new constructions, it provides a subdivision of a building into different categories (such as Substructure, Shell, Interiors, etc.).

Table 1. The LOA classification provided by USIBD

Level	Upper range	Lower range
LOA10	Defined by user	5 cm
LOA20	5 cm	1,5 cm
LOA30	1,5 cm	0,5 cm
LOA40	0,5 cm	0,1 cm
LOA50	0,1 cm	0

Source: USIBD (2019)

It is easy to understand that in a structure like the Farnese Theatre, it is not possible to achieve equal levels of accuracy for the entire model. Therefore, 5 different levels of adherence to reality for the classes of elements previously mentioned were identified. For each level, the maximum value of the allowed standard deviation was established (the standard deviation value is obtained by comparing the point cloud with the 3D model). Its minimum value was always assumed to be 0 (Table 2).

Table 2. Standard deviation levels adopted for the modeling of 5 classes of elements identified in the Farnese Theatre

Level	Class	Max value
Level 1	Roofing system	Not defined
Level 2	Bleachers system	10 cm
Level 3	Repetitive decorative elements	4 cm
Level 4	Wooden elements characterizing the artifact	1,5 cm
Level 5	Masonry perimeter walls	0,5 cm

Source: Graphic elaboration by the authors

It is important to note that the standard deviation represents an average value. This means that in some points the deviation between the data can be even higher than the defined value. For this reason, it is essential to ensure that the deviation between the data always remains within acceptable values. During the modeling phase, a continuously control was performed, verifying that the punctual deviation was no more than twice the defined maximum value.

The highest levels of accuracy (Level 5 and Level 4) were assigned to two parts of the theatre: masonry perimeter walls (L5) and the wooden elements characterizing the artifact, such as loggias, bleachers, proscenium (L4). High accuracy was maintained for these two classes, in order to ensure the use of the model for the management and restoration purposes.

The medium level accuracy (Level 3) was assigned to repetitive decorative elements. Apparently, the maximum value of the standard deviation decided may seem rather high for elements of small dimensions. However, a faithful 3D reproduction of these would have to be done separately, it would be too time-consuming and would make the model too heavy.

The lowest levels of accuracy (Level 2 and Level 1) were assigned to the last two parts of the theatre: bleacher support elements (L2) and roofing system (L1). The faithful reproduction of these elements would have required an accurate survey which, at the current state of this research, was not possible.

Processing Laser Scanner and Photogrammetric Data

The processing of laser scanner data was performed using a commercial free-form modeling software, Rhinoceros 6.0. According to their characteristics, the previously mentioned "classes" of elements were modeled adopting slightly different procedures (Figure 9). It is important to note that, due to the complexity of the theatre, all the procedures adopted present a low level of automation.

The class of perimeter walls, characterized by numerous irregularities, imperfections, out of plumb, plaster blisters, was modeled with a high level of adherence to reality. To this aim, it was decided to operate directly on the point cloud, which was used as a base for creating mesh surfaces. Some optimization operations were necessary: filling holes, eliminating self-intersections and finishing edges, decimation operations, etc. The last operations on the mesh were aimed at obtaining easy manageable surfaces. The number of triangles was decreased using different percentages of reduction (higher reduction percentages were adopted in the flat parts). The class of wooden structures was also modeled with a high level of adherence to reality. In this case, NURBS surfaces were created. In particular, three different procedures were adopted: the sliding of vertical sections along one or more paths (for complex details, such as entablature), the extrusion of two-dimensional profiles along the related paths (for flat surfaces, such as basement) and the generation of surfaces from edge curves (for non-planar surfaces, such as columns and half-columns).

The class of repetitive decorative elements (such as capitals or balusters) was modeled by regularizing the geometry of a typical element, which was then systematically repeated in every point where it was present. For each element, a block model was created. Then, it was copied several times. This approach was fundamental to not overloading the 3D model.

Figure 9. Photographs and digital reconstruction of the five classes of elements identified in the Farnese Theatre (from the left: masonry perimeter walls, wooden elements characterizing the monument, repetitive decorative elements, bleachers system and roofing system)
Source: Graphic elaboration by the authors

In the theatre there are three types of capitals: one type of the Ionic capital and two versions of the Corinthian capital (with circular and square plan). For the Ionic capitals a photogrammetric survey was performed. The mesh model obtained was subsequently processed and transformed into NURBS surface. The latter was then placed in all 68 points where the capital is actually present. As the Corinthian capitals are not directly accessible to operators because of their too high position, it was not possible to perform a photogrammetric survey for them. Therefore, a different methodology was adopted. Starting from section curves obtained from the point cloud, two capitals were modeled. Then, these elements were copied in different parts of the model. As for the balusters (present in 3 versions for a total of 730 elements), starting from the data acquired from the point cloud, revolution surfaces were created. In particular, they were obtained through the rotation of the vertical section profile around an axis of rotation.

The class of bleachers system was also regularized. As it is extremely complex (it is composed of 56 beams, each of which is made of at least 7 elements properly interlocked and connected between them), it was chosen to accurately model one truss. The modeling workflow was based on the extrusion of two-dimensional profiles along the related paths or on the creation of the surfaces from edge curves. Each truss was then copied, adapting the beam type to the variation of the length of the covered span.

The same approach was adopted for theatre roofing system. A typical truss was reconstructed using the already mentioned methodologies for the construction of NURBS surfaces (Figure 10).

It is important to note that all surfaces were continuously compared with the point cloud, both in the modeling phase and at the end of it. This operation allowed to verify the correctness of the accuracy levels decided for the different classes of elements previously identified. To this aim, a commercial software, Polyworks Inspector 10.0, was used. For example, the comparison performed on the proscenium, showed a standard deviation value (StdDev) between the model and the point cloud equal to 6.78 mm (in order not to falsify the data the model was considered without capitals).

As previously mentioned, the processing of photogrammetric data was performed using software Agisoft Photoscan Professional 1.2.0. At first, it was necessary to operate separately on the single parts of the fresco.

First of all, the seven parts of the decoration were correctly scaled and oriented within the same reference system used in the laser scanner survey campaign. To this aim, for each part, three to nine easily distinguishable points were identified (both in the laser scanner point cloud and in the frames). This operation was fundamental to tie all parts together and for the reconstruction of the entire decoration.

Figure 10. Digital model of the Farnese Theatre (on the left: view of the bleachers, on the right: views from the bleachers)
Source: Graphic elaboration by the authors

The workflow of each part of the fresco included the correct alignment of the frames, the generation of the dense point cloud and its cleaning through the elimination of elements not related to the wall (such as wooden structures).

Starting from the point cloud obtained, it was possible to create mesh models of the fresco portions, to use as a reference for the generation of the orthophotos. During this phase, for each single frame, particular attention was paid to the selection of the parts considered qualitatively better. For example, areas characterized by strong contrasts between shaded and illuminated parts and those in which there were disturbing elements such as loggias were eliminated.

A vertical reference plane, parallel to the wall and the plastic proscenium, was used for the generation of the orthophotos. The single orthophotos were extrapolated with a resolution of 5mm/pix. Then, some optimization operations were carried out (related to brightness and contrast correction). These activities were performed with the aim of making orthophotos as homogeneous and faithful as possible from the colorimetric point of view, with reference to the original fresco.

Finally, the seven orthophotos produced were integrated within the vertical-section of the theatre. The section was created near the wall, in scale 1:50. In this way it was possible for the first time to admire the painted composition in its entirety (Figure 11).

Figure 11. Orthophoto of the fresco placed on the back wall of the theatre, original scale 1:50 (on the right: detail of the orthophoto)
Source: Graphic elaboration by the authors

Digital Reconstruction of the Original Decoration of the Proscenium

Once the 3D model of the theatre was created, it was possible to proceed with the digital reconstruction of the monument's original decoration.

As already mentioned, before its 1944 destruction, the entire Farnese Theatre was originally covered by a rich pictorial decoration. However, the sources available today do not allow to provide its complete description. Since not all parts of the monument are documented in the same way, a methodology for reconstruction of the proscenium decoration was tested. This part is one of the most documented areas of the theatre by painters and photographers (Figure 12).

First of all, a critical analysis of all available sources, continuously compared with each other, was necessary. In particular, the following types of sources were analyzed: written sources (the 1629 book by Marcello Buttigli, in which a detailed description of the architectural structure and its pictorial decoration is present) (Buttigli, 1629), iconographic and photographic sources (historical drawings of the theatre, pictorial representations and photographs), plastic and pictorial models (19th century model of the theatre attributed to Fanti and Rousseau), and fresco located on the back wall of the hall.

From a methodological standpoint, the critical source analysis phase required two different steps. Initially, by comparing elements in all available sources, it was possible to identify them from a figurative point of view. Then, a chromatic analysis of the elements was performed (Mikolajewska & Zerbi, 2021).

Once all the elements were identified, it was possible to proceed with the digital reconstruction of the original decoration of the proscenium. This operation was divided into two phases: the first one regarded the reconstruction of the elements in raster format, the second on their mapping on the 3D model.

In the first phase, dedicated to the reconstruction of the decoration elements, a commercial software, Adobe Photoshop CC 2017, was used. Four different workflows were adopted, depending on the type of elements.

Figure 12. Details of the original proscenium decoration
Source: Photographs from Pinazzi private collection

The first workflow regarded elements clearly documented in the historical monochromatic photographs, such as grotesques, figures represented on the pillars or armors. The pipeline adopted was based on the following steps: photo scanning, vector drawing of contour lines of the elements, their rectification and scaling, definition of chromatic aspect. Particular attention was paid to make this aspect easily editable in the future, by keeping the monochrome image level separate from the color one.

The second workflow was adopted for the reconstruction of inadequately documented elements, for example present in low-quality photographs, not completely legible. In particular, this operation involved festoons and some decorations of moldings. The reconstruction of these elements was performed taking as a reference the orthophoto of the fresco placed on the back wall of the hall, which, as previously mentioned, represents a mirror copy of the wooden proscenium. The pipeline adopted in this case was based on the following steps: definition on the orthophoto of contour lines of the element, scaling of element (according on where it would be placed), optimization operations and homogenization from the chromatic point of view. To this aim, parameters related to the brightness and contrast of the images were modified. To make always possible to modify the images, new separated adjustment levels were added.

The third workflow was used for the reconstruction of the principal materials of the decoration, such as porphyry, gold and marble. Also in this case, the fresco placed on the back wall of the hall was used as a reference. In particular, sufficiently large portions in which these materials were present were identified and cut out on the orthophoto. The three images thus obtained were used to generate as many patterns (images that can be copied repeatedly to complete specific areas). To make this operation more reliable, accurate comparisons with the decorations present in the church of San Paolo in Ferrara (painted in part by the same artists of the Farnese Theatre) were also made.

The last workflow regarded repetitive geometric elements, such as pedestal decorations, cornices and meanders. The pipeline adopted included the followings steps: photographs scaling, definition of the contour lines of the elements, their vector drawing and coloration using the previously mentioned patterns.

It is important to note that the illustrated workflows do not allow to reproduce the totality of the decoration in an always reliable manner. In fact, due to the lack of adequate photographs, in some cases it is hard to reconstruct the original form of the decorations. Elements such as frieze, decorations of the plinth, two pilasters placed in correspondence of the perimeter walls of the hall and masks painted on the pedestals, are not exhaustively documented. For these parts of the decoration, at the moment, evocative integrations seemed to be the best solution. Obviously, if more reliable documents emerge, it will always be possible to modify each element.

For example, as for the frieze and masks, the previously mentioned orthophoto was used as a reference. Although the painted composition of the frieze is non perfectly identical to that documented in the photographs, it reproduces the same type of figures, painted with the same colors. Regarding the masks, the one present in the fresco is very similar to those documented in the photos, so the best solution seemed to be to repeat the one painted in the fresco. For the two perimeter pilasters and for some of the figures present in the other four similar surfaces, the decorations already reconstructed for the corner pilasters were used. The less documented area regards the plinth decorations. In this case, only the contour lines were defined, without specifying further formal elements of detail. Those areas were then colored using methods described above.

Figure 13. Types of decorative elements identified for the reconstruction of the original decoration of the proscenium
Source: Graphic elaboration by the authors

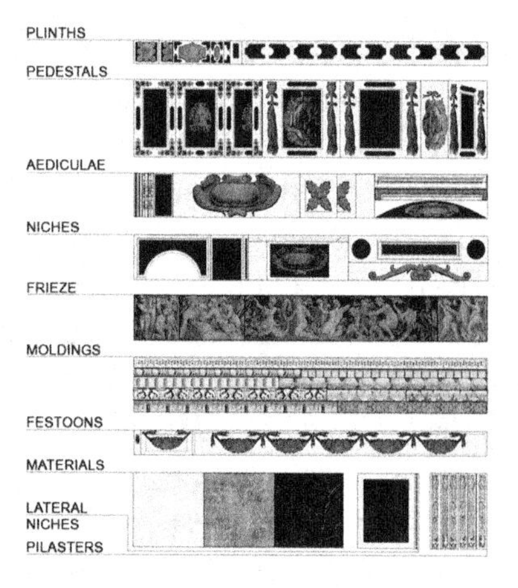

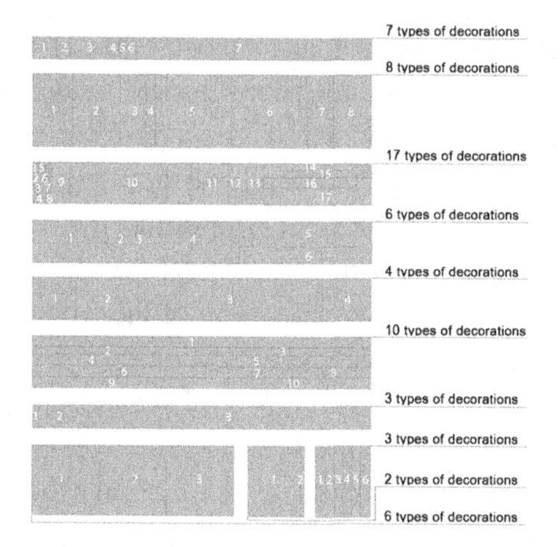

It is important to note that in most cases, for each face of the 3D model ad hoc files were created. Thc aim was to faithfully reproduce even the shading. As the artists painted simulating a central light source, placed in front of the proscenium, all decorations result different. For example, observing pedestals, it is possible to note that each face is different, depending on its position with respect to the imaginary light source.

At the end of the first phase of decorations' digital reconstruction, approximately sixty types of different decorative elements were identified and more than two hundred images were produced (Figure 13). Particular attention was paid to their quality. All images were generated in a .jpg format with a resolution of 300 dpi and an average weight of about 1 MB.

The second phase necessary to define the hypothetical original state of the proscenium regarded mapping the obtained raster images on the digital model. To this aim, the same software adopted in the modeling phase, Rhinoceros 6.0, was used.

First of all, for each previously obtained image it was necessary to create a new "material". In particular, more than 200 custom "materials" were generated and placed in the "Material Library". For materials such as porphyry, gold and marble, the above mentioned patterns were used. Then, it was possible to map the decorations on the 3D model. To this aim, three different methods were adopted.

The first method was used to map all the flat surfaces of the model (for example pedestals, pilasters, friezes, etc.). In this case, ad hoc images were created. Using planar mapping, each image was applied to the surface of interest at a 1:1 scale.

The second method involved all the complex surfaces (for example moldings). In this case, custom UV mapping was used. First of all, the interested NURBS surfaces were transformed into mesh surfaces and were projected onto a 2D plane, on which "materials" textures were placed. According to the shape of the textures, the position and scale of the meshes were then optimized.

The third method regarded the parts of the proscenium, which were painted in a fashion that simulated precious materials, such as gold, porphyry and marble. In this case, the scale parameters of the three materials were modified, according to the area to be mapped. The "porphyry material" was mapped on the columns, the "gold material" on their bases and capitals and the "marble material" was applied to all other surfaces.

Figure 14. Free-form model of the proscenium of the Farnese Theatre with the digital reconstruction of its original decoration (on the right: detail of the textured model)
Source: Graphic elaboration by the authors

The digital reconstruction of the original decoration of the proscenium obtained following the above-mentioned methodology allowed to create a support tool for study, communication and dissemination purposes (Figure 14).

Uses of the Farnese Theatre 3D model

As the 3D model of the theatre was created from the beginning with the intent to make it extremely versatile, it is characterized by numerous areas of application.

First of all, the model created provide an accurate documentation of the monument which, at present, lacks reliable surveys. In fact, not even the traditional two-dimensional drawings of the theatre are available. Today, it is possible to extrapolate accurate drawings from the model created, in any point of the structure and at any time. This is clearly a significant advantage for the monument's management.

The 3D model is also an important support tool for research activities. For example, starting from the survey drawings obtained directly from the digital model, it was possible to verify the accuracy of the twentieth-century reconstruction of the monument. To this aim, the elaborates that document the current state of the structure were compared with some historical representations dating before the 1944 destruction of the artifact.

Among the numerous existing drawings of the theatre, it was considered particularly significant to make a comparison with those made in two different moments: on the occasion of the eighteenth-century survey campaign (drawings by L.A. Feneulle) and of the twentieth-century reconstruction of the artifact (drawings by A. Stanzani). From the one hand, this operation aimed to verify the correctness of the Feneulle drawings, and from the other, to examine the correspondence of the reconstruction project with the current state of the structure.

The comparison process was based on the overlapping of the drawings, once they were rectified and scaled (Figure 15). As far as concern the drawings of the French architect, it was possible to verify the validity of the reconstruction of the monument. In particular, the comparison highlighted that the current configuration of the theatre reflects almost perfectly the structure surveyed by Feneulle in the eighteenth century. The evaluation of these drawings was fundamental to provide scientific support for some considerations about the transformations of the artifact that have taken place over the centuries.

Regarding the drawings made by Stanzani, the comparison operations allowed to identify some differences between the initial reconstruction project and the solutions actually adopted on site.

The digital model of the theatre was also used as a support to verify some theses elaborated by architectural historians regarding the painted decoration hidden by the loggias. In particular, it made possible to carry out a first metric analysis of the relationship between the painted fresco and the wooden proscenium. The aim was to verify whether the two elements, so distant from each other, and so different from the materic point of view, are perfectly specular. The relationship between the plastic proscenium and the fresco is recognized by several authors, but there was no adequate documentation that demonstrate it from a metric-dimensional point of view.

Despite the fact that the distance separating the fresco from the proscenium is almost fifty meters, the two elements are placed on almost perfectly parallel planes. Thanks to this particular position, it was possible to project the digital model of the fresco (obtained with TLS techniques) and those of the proscenium (obtained with photogrammetric techniques) on the same vertical plane. In doing so, the two motifs overlapped, although mirroring each other.

Figure 15. Comparison between the vertical section obtained from the current survey and the one elaborated by L.A. Feneulle (seventies of the eighteenth century)
Source: Graphic elaboration by the authors

The comparison showed immediately the strong relationship between the painted fresco and the plastic proscenium (Figure 16). Except for a few localized differences, the two elements coincide almost perfectly (Mikolajewska & Zerbi, 2019). Considering also all the alternations that may have interested both motifs over the last years, there is no doubt that the two are the exact mirror copy of each other.

Thanks to the accurate digitization of the theatre, it was possible to provide a solid and scientific base to the theses elaborated in recent years by many historians, confirming the surprising precision adopted by seventeenth-century artists in the elaboration of the frescoed composition.

Another application of the 3D model created concerned its use as a support tool for the management and preservation of the monument. This study was conducted in the specific field of BIM. The goal was to verify if the free-form model could be used for the informative modeling.

Figure 16. Projection of the proscenium model and the mesh model of the fresco on the same vertical plane (on the right: detail of the overlapping between the two elements)
Source: Graphic elaboration by the authors

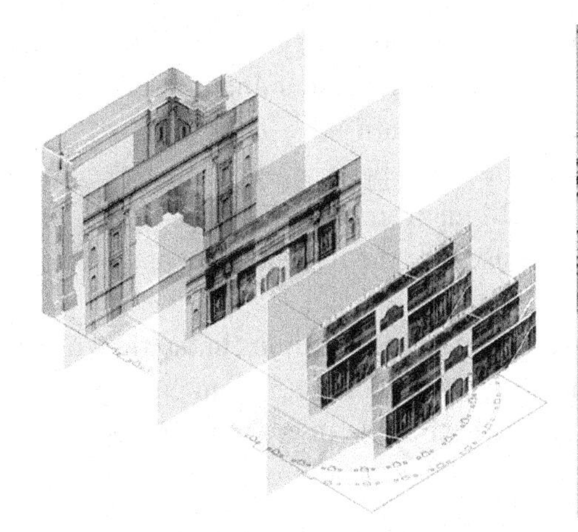

To test the validity of the procedure from a methodological standpoint, the experimentation was conducted only on a part of the model. As it can be considered a significant area of the model, a portion of the bleachers was chosen (Figure 17). In this phase a commercial software, Archicad 24, was chosen.

As well known, in the BIM context, a correct organization of the model is fundamental. Therefore, before importing the selected part of the bleachers in Archicad it was necessary to make some integrations. The goal was to define a single, solid 3D entity for each element.

Figure 17. Photographs of the bleachers system and transformation of the free-form model into the informative model used within the BIM platform
Source: Graphic elaboration by the authors

The numerous solids thus defined were then imported and placed in the Archicad Object Library. Each element was then classified according to the constructive principles of architecture. In particular, three-dimensional components of building were identified: wall, pillar, beam, etc. Then, for each type of element, descriptive attributes were added. It is important to stress that BIM systems were developed mainly for the management of new constructions, so the attributes that can be associated to the various elements are related to the geometry and characteristics of materials. In the context of historical heritage, it is easy to understand that they are not sufficient for the complete description of the monuments.

Considering the uniqueness of the Farnese Theatre, it was necessary to customize all attributes. The aim was to provide information that was functional to its management and conservation. To this end, for each type of element, new descriptive parameters were identified.

The experimentation carried out, deliberately limited to a portion of the model, allowed to verify that the model obtained can be effectively used in BIM environment. Through simple procedures and integrations, the model of the Farnese Theatre can be used for the management and conservation of the monument.

Finally, the digital model of the theatre can have numerous applications in the field of dissemination of knowledge. For example, it may be used for the visualization of the monument original decorations. From this point of view, the digital model played an essential role in the video mapping project aimed at the projection of the decorations on the proscenium wooden structure.

Since only a few signs of the original decorations remain today, it is really difficult to understand what the theatre might have looked like before its 1944 destruction. The few items exposed today under the bleachers (historical photographs, painted representations or plastic models) are not sufficient to allow the visitors to fully understand the extraordinariness of the original decorative apparatus of the monument.

For this reason, it was decided to project a video mapping performance. The aim was to overcome the limitations imposed by the space reserved for the exhibition and to visualize the decoration directly on the wooden proscenium. There is no doubt that this operation contributes to improve readability and understanding of this historical artifact and allows the communication of knowledge in an unconventional way.

On the basis of the accurate 3D model of the monument and of the previously mentioned digital reconstruction of the pictorial apparatus of the proscenium, it was possible to develop two different performances: the first project concerned a static performance and the second a dynamic one (Figure 18).

Each solution required an accurate study of the light conditions, morphology, material and color of the surfaces involved in the projection and so on. At the same time, the number and type of devices, their placement inside the monument and the digital content were analyzed.

Unfortunately, due to the pandemic situation and to the limited availability of funds, it was not yet possible to conduct the final on-site testing operations.

Figure 18. Textured digital model of the theatre used for the static projection of the pictorial apparatus on the wooden proscenium
Source: Graphic elaboration by the authors

CONCLUSION

The presented research investigates how the use of digital technologies can contribute to the development of a wide range of activities in the specific field of Cultural Heritage, and which effects they can produce within the “Society 5.0”. To this aim, the Farnese Theatre in Parma was chosen as a case study. In particular, the procedures typical of the field of Representation were applied in order to provide scientific support to the activities of valorization of historical monumental architecture.

In this research, the digital reconstruction of the architectural space took a fundamental part. It was considered from the beginning both a valid tool for the documentation, preservation and analysis of the artifact, as well as essential support for any operation of maintenance and restoration, or any activity of valorization and dissemination of the theatre. Since the areas of application are numerous, it became important to create an extremely versatile three-dimensional model of the monument. In order to obtain an accurate and easily manageable model, the definition of the operative procedures in the modeling phase was fundamental.

In terms of the needs of the future society, the digital model can be used to several purposes. As it accurately documents the monument and allows to obtain detailed drawings, in any point of the structure and at any time, it optimizes its preservation and management process, which is crucial for its transmission to future generations. Since the virtual model can be disseminated on-line, it overcomes the concept of inaccessibility of Cultural Heritage. In addition, it can be used for study activities by researchers from around the world, allowing better information connection. Finally, the 3D model can be used for the video mapping project aimed at the visualization of decorations that no longer exist. This operation can improve the readability and understanding of the artifact. At the same time, it can strengthen the collective memory of the citizens. The many purposes to which the digital model can serve demonstrate once again how technology can produce positive impacts on the people's well-being and consequently also within "Society 5.0".

The future developments of the research can be numerous. On the one hand, they are determined by the vastness of the themes involved, on the other, by the importance and complexity of the architecture analyzed.

First of all, the *modus operandi* proposed in this study could be tested in other similar contexts. Since there are many historical architectures not yet accurately digitized, they could be valorized using the methodology tested in this research. In many cases, even the most complex ones, the multiscale approach in function of the different parts of the model and its division in several parts would allow to optimize the handling of large amount of data. At the same time, it would allow to preserve data quality. The method adopted for the digital reconstruction of decorations that no longer exist could also be used in other similar contexts. Since historical architectures do not always have exhaustive documentation, it became essential to make the whole process implementable at any moment. Choosing to adopt different workflows, depending on the type of elements and the source used, became essential to make the reconstruction rigorous.

In regard to the Farnese Theatre, it would be useful to verify the validity of the method proposed for the virtual reconstruction of the decorations. The video mapping performance would allow to highlight the potential of this technique in the specific field of architectures that have been deeply transformed over time. At the same time, it would prove that the methodology proposed can be replicated in many other contexts.

As previously mentioned, originally all the structure was completely painted. The workflow adopted for the digital reconstruction of the decoration that once covered the proscenium could be extended to the entire monument. To this aim, it would be necessary to continue research on iconographic documents. This would also allow to extend the video projection performance to the entire theatre.

It is important to note that the digital model created allows to accomplish all the activities described here. However, since some parts of the artifact have not been documented exhaustively, both the survey and the model could be completed. The data collection operations could be extended to the roof structure, the area of the stage, the entrance portal, the decorative elements not adequately documented, the frescoes on the longitudinal walls and the spaces adjacent to the theatre.

Another future development of the research concerns the BIM field. The experimentation conducted on a single part of the theatre could be extended to the entire structure. This operation would allow to create a fundamental tool for the management and planned conservation of the monument. At the same time, it would allow to deeply investigate the use of a free-from model in a parametric environment.

Another issue that deserves further development is the digitization and systematization of the entire documental corpus related to the monument. At the moment, the documents are collected in different archives, mainly located in Italy. It would be necessary to adopt common operating procedures that would make available to researchers high-quality digitized documents. This operation would facilitate the further research, dissemination and restoration activities.

Finally, the research field related to the alternative use of Cultural Heritage could be further investigated. Starting from the digital model of the artifact, the relationship between an integrated survey, three-dimensional modeling and the physical print of the monument could be more investigated. For example, a physical model of the theatre could be placed within the exhibition itinerary of the structure. The latter could also be combined with multimedia content, offering the possibility of alternative and multisensory use.

REFERENCES

Agostino, D., Arnaboldi, M., & Lampis, A. (2020). Italian state museums during the COVID-19 crisis: From onsite closure to online openness. *Museum Management and Curatorship*, *35*(4), 362–372. doi: 10.1080/09647775.2020.1790029

Balletti, C., Bertellini, B., Gottardi, C., & Guerra, F. (2019). Geomatics techniques for the enhancement and preservation of cultural heritage. *The International Archives of the Photogrammetry, Remote Sensing and Spatial Information Sciences*, *XLII-2*(W11), 133–140. doi:10.5194/isprs-archives-XLII-2-W11-133-2019

Belardi, P., & Bori, S. (2018) Sistina Experience. *Paesaggio urbano*, *4*, 41-49.

Bevan, R. (2016). *The Destruction of Memory: Architecture at War*. Reaktion Books Ltd.

Brumana, R., Oreni, D., Barazzetti, L., Cuca, B., Previtali, M., & Banfi, F. (2020). Survey and Scan to BIM Model for the Knowledge of Built Heritage and the Management of Conservation Activities. In B. Daniotti, M. Gianinetto, & S. Della Torre (Eds.), *Digital Transformation of the Design, Construction and Management Processes of the Built Environment. Research for Development* (pp. 391–400). Springer. doi:10.1007/978-3-030-33570-0_35

Bruno, N., Bianchi, G., Roncella, R., & Zerbi, A. (2015). An open-HGIS project for the city of Parma: database structure and map registration. In M. A. Brovelli, M. Minghini, & M. Negretti (Eds.), Geomatics Workbooks (Vol. 12, pp. 189–203). Academic Press.

Bruno, N., Mikolajewska, S., Roncella, R., & Zerbi, A. (2022). Integrated processing of photogrammetric and laser scanning data for frescoes restoration. *International Archives of Photogrammetry, Remote Sensing and Spatial Information Sciences*, *XLVI-2/W1-2022*, 105-112.

Buono, M., & Giugliano, G. (2021). Systems and models of intelligent connection and interaction for society 5.0. *Revista Iberoamericana Académico-Científica Universitaria de Humanidades. Arte y Cultura, 9*, 195–208.

Buttigli, M. (1629). *Descrittione dell'apparato fatto, per honorare la prima, e solenne entrata in Parma della serenissima prencipessa, Margherita di Toscana, duchessa di Parma*. Piacenza, & c., Appresso Seth, & Erasmo Viotti.

Chiabrando, F., Sammartano, G., Spanò, A., & Spreafico, A. (2019). Hybrid 3D Models: When Geomatics Innovations Meet Extensive Built Heritage Complexes. *ISPRS International Journal of Geo-Information, 8*(3), 124. doi:10.3390/ijgi8030124

CORDIS. (2021). *How digital technologies can play a vital role for the preservation of Europe's cultural heritage.* https://cordis.europa.eu/article/id/413473-how-digital-technologies-can-play-a-vital-role-for-the-preservation-of-cultural-heritage

Empler, T. (2018). Musei tradizionali, musei virtuali. La funzione divulgativa delle ICT [Traditional museums, virtual museums. The divulgation role of ICT]. *DisegnareCon, 11*(21), 13.1-13.9.

ESPON. (2021). *HERIWELL – Cultural Heritage as a Source of Societal Well-being in European Regions Draft findings on the linkages between cultural heritage and societal well-being.* https://www.espon.eu/HERIWELL

European Commission. (2020). *Basic principles and tips for 3D digitisation of tangible cultural heritage for cultural heritage professionals and institutions and other custodians of cultural heritage.* https://digital-strategy.ec.europa.eu/en/library/basic-principles-and-tips-3d-digitisation-cultural-heritage

Fassi, F., Achille, C., Mandelli, A., Rechichi, F., & Parri, S. (2015). A New idea of BIM system for visualization, web sharing and using huge complex 3d models for facility management. *The International Archives of the Photogrammetry, Remote Sensing and Spatial Information Sciences, XL-5*(W4), 359–366. doi:10.5194/isprsarchives-XL-5-W4-359-2015

Fassi, F., Fregonese, L., Adami, A., & Rechichi, F. (2017). BIM systems for the conservation and preservation of the mosaics of San Marco in Venice. *The International Archives of the Photogrammetry, Remote Sensing and Spatial Information Sciences, XLII-2*(W5), 229–236. doi:10.5194/isprs-archives-XLII-2-W5-229-2017

Font Sentias, J. (2020). El mapping de Sant Climent de Taüll [The mapping of the Sant Climent de Taüll]. *Mnenòsine*, 10.

Fregonese, L., Taffurelli, L., Adami, A., Chiarini, S., Cremonesi, S., Helder, J., & Spezzoni, A. (2017). Survey and modelling for the BIM of Basilica of San Marco in Venice. *The International Archives of the Photogrammetry, Remote Sensing and Spatial Information Sciences, XLII-2*(W3), 303–310. doi:10.5194/isprs-archives-XLII-2-W3-303-2017

Gandolfi, V. (1980). *Il teatro Farnese di Parma* [The Farnese Theatre in Parma]. Luigi Battei.

Giannetti, S., Lodovisi, A., Incerti, M., Grassivaro, A., & Sardo, A. (2019). Esperienze di projection mapping per la valorizzazione delle facciate dipinte nei territori estensi. In P. Belardi (Ed.), *Riflessioni. L'arte del disegno e disegno dell'arte, Atti del 41° Convegno Internazionale dei Docenti della Rappresentazione* [Reflections. The art of drawing and art design, Proceedings of the 41st International Conference of Representation Teachers] (pp. 1621-1628). Gangemi Editore.

Giordano, A., Borin, P., Cundari, M. R., & Panarotto, F. (2014). La chiesa degli Eremitani a Padova: rilievo, documentazione, storia [The church of the Eremitani in Padua: Survey, documentation, history]. In P. Giandebiaggi & C. Vernizzi (Eds.), *Italian Survey & International Experience, Atti del 36° Convegno Internazionale dei Docenti della Rappresentazione* (pp. 869–876). Gangemi Editore.

Incerti, M., D'Amico, S., Giannetti, S., Lavoratti, G., & Velo, U. (2018). Le digital humanities per lo studio e la comunicazione di beni culturali architettonici: Il caso dei mausolei di Teodorico e Galla Placidia in Ravenna [Digital humanities for the study and communication of architectural cultural heritage: the case of the mausoleums of Theodoric and Galla Placidia in Ravenna]. *Archeologia e Calcolatori, 29*, 297–316.

Lo Turco, M., Piumatti, P., Calvano, M., Giovannini, E. C., Mafrici, N., Tomalini, A., & Fanini, B. (2019). Interactive Digital Environments for Cultural Heritage and Museums. Building a digital ecosystem to display hidden collections. *Disegnarecon, 12*(23), 7.1-7.11.

Maniello, D. (2014). *Realtà aumentata in spazi pubblici. Tecniche base di video mapping* [Augmented reality in public spacers. Basic techniques for video mapping]. Le Penseur.

Maniello, D. (2020). Digital anastylosis for digital augmented spaces: spatial Augmented reality applied to Cultural Heritage. In M. Lo Turco, E. C. Giovannini, & N. Mafrici (Eds.), *Digital & Documentation. Digital strategies for Cultural Heritage* (Vol. 2, pp. 141–151). Pavia University Press.

Mikolajewska, S., & Zerbi, A. (2019). Uno specchio dell'arte: il proscenio e l'affresco sulla parete di fondo del teatro Farnese di Parma [Mirror of the art: the proscenium and the fresco on the back wall of the Farnese Theatre in Parma]. In P. Belardi (Ed.), *Riflessioni: l'arte del disegno, il disegno dell'arte, Atti del 41° Convegno Internazionale dei Docenti della Rappresentazione* (pp. 1027–1034). Gangemi Editore.

Mikolajewska. (2021). *Tecnologie digitali integrate per la conoscenza, la conservazione e la valorizzazione del patrimonio culturale storico. Il teatro Farnese di Parma* [Integrated digital technologies for the knowledge, conservation and enhancement of historical cultural heritage. The Farnese Theatre of Parma] [Doctoral dissertation]. University of Parma, Parma, Italy.

Osello, A. (2012). *Il futuro del disegno con il BIM per ingegneri e architetti* [The future of drawing with BIM for engineers and architects]. Dario Flaccovio editore s.r.l.

Osello, A. (2018). BIM, Virtual and Augmented Reality in the health field between technical and therapeutic use. In R. Salerno (Ed.), *Rappresentazione/Materiale/Immateriale, Atti del 40° Convegno Internazionale dei Docenti della Rappresentazione* (pp. 1535–1538). Gangemi Editore.

Ottoni, F. (2008). From geometrical and crack survey to static analysis method: the case study of Santa Maria del Quartiere dome in Parma (Italy). In D. D'Ayala & E. Fodde (Eds.), *Structural Analysis of Historical Construction* (Vol. I, pp. 697–704). Taylor & Francis Group. doi:10.1201/9781439828229.ch79

Peters, E. A. (2018). Coloring the Temple of Dendur. *Metropolitan Museum Journal, 53*, 8–23. doi:10.1086/701737

Riavis, V. (2019). Discovering Architectural Artistic Heritage Through the Experience of Tactile Representation: State of the Art and New Development. *DisegnareCon, 12*(23), 10.1-10.9.

Sdegno, A. (2018). Rappresentare l'opera d'arte con tecnologie digitali: dalla realtà aumentata alle esperienze tattily [Representing the artwork with digital technologies: from augmented reality to tactile experiences]. In Ambienti digitali per l'educazione all'arte e al patrimonio (pp. 256-271). FrancoAngeli.

Tommasi, C., Achille, C., & Fassi, F. (2016). From point cloud to BIM: A modelling challenge in the Cultural Heritage field. *International Archives of the Photogrammetry, Remote Sensing and Spatial Information Sciences. XLI, B5*, 429–436.

Tucci, G., Bonora, V., Conti, A., Fiorini, L., & Riemma, M. (2014). Il rilievo digitale del Battistero: dati 3D per nuove riflessioni critiche [Digital survey of the Baptistery: 3D data for new critical reflections]. In F. Gurrieri (Ed.), *Il Battistero di San Giovanni. Conoscenza, diagnostica, conservazione* (pp. 105–117). Mandragora.

USIBD Level of Accuracy (LOA) Specification Guide v. 3.0-2019. (2019). Academic Press.

Zerbi, A., & Mikolajewska, S. (2021). Digital technologies for the virtual reconstruction and projection of lost decorations: the case of the proscenium of the Farnese Theatre in Parma. *DisegnareCon, 14*(27), 5.1-5.11.

ADDITIONAL READING

Adorni, B. (1997). Il teatro Farnese a Parma. Vita, miracoli e morte del teatro [The Farnese Theatre in Parma. Life, miracles and death of the theatre]. *Casabella, 650*, 60–77.

Adorni, B. (2003). Il teatro Farnese a Parma. [The Farnese Theatre in Parma] In C. Cavicchi, F. Ceccarelli, & R. Torlontano (Eds.), *Giovan Battista Aleotti e l'architettura* (pp. 205–226). Diabasis.

Adorni, B. (2008). Il teatro Farnese. [The Farnese Theatre] In B. Adorni (Ed.), *L'architettura a Parma sotto i primi Farnese 1545-1630* (pp. 99–123). Diabasis.

Capelli, G. (1990). Il Teatro Farnese di Parma. Architettura, scene, spettacoli [The Farnese Theatre in Parma. Architecture, scenes, shows]. Public Promo Service.

Cavicchi, A. (1974). Il teatro Farnese di Parma [The Farnese Theatre in Parma]. *Bollettino del Centro Internazionale di Studi di Architettura Andrea Palladio di Vicenza, XVI*, 333–351.

Dall'Acqua, M. (1992). Il teatro Farnese di Parma. [The Farnese Theatre in Parma] In *Lo spettacolo e la meraviglia. Il Teatro Farnese di Parma e la festa barocca* (pp. 17–149). Nuova Eri.

KEY TERMS AND DEFINITIONS

3D Model: A digital model of the physical object. Depending on the type of modeling procedures adopted, defined according to the specific purposes of the research, different models can be obtained: mesh models, NURBS (Non-Uniform Rational Basis-Splines) models, parametric models, etc.

BIM: Integrated process for information modeling and information management on a construction project or an existing building throughout its life cycle.

Cultural Heritage Digitization: A process based on the use of different techniques aimed at digital documentation of tangible assets (architectural or documentary heritage, artworks, etc.).

Farnese Theatre: Wooden theatre built in Parma (Italy), completed in 1619 by Ranuccio I Farnese and designed by Giovanni Battista Aleotti. Almost completely destroyed during an Allied air raid in 1944, starting from the 1950s, the theatre was reconstructed following a philological approach.

Laser Scanning: Non-contact survey technique that digitally captures the shape of physical objects using a laser beam. The laser scanner provides a 3D point cloud of the surveyed object, from which metrical and chromatic information can be obtained.

Painting Reconstruction: A philological process based on the collection and critical analysis of historical sources, aimed at virtual reconstruction of decorations that no longer exist.

Photogrammetry: Image-based survey technique that allows to obtain metrical and chromatic information of an object and its 3D virtual reconstruction starting from suitably taken photographs.

Video Mapping: A projection of two-dimensional visual contents (static or dynamic) on any type of surface, in order to achieve a perfect correspondence between them and to modify the visual perception of the objects.

Chapter 19
Interactive Virtual Participation for Opera and Theatre Using New Digitization Information Systems

Daniela De Luca
Politecnico di Torino, Italy

ABSTRACT

Society 5.0 has implemented the use of new digital technologies, overcoming traditional active learning systems with means and methodologies that extend the involvement of the digitized user. This trend has revolutionized how organizations and companies deliver their services through interconnected and interoperable platforms. The prevalence of new media has led to the adoption of applications that exploit gamification techniques and serious games to transfer reality into new virtuality. The contribution analyses procedures and methodologies that can be adapted to digitalize cultural heritage, focusing on the theatrical and musical entertainment sector (i.e., opera and theatre). During the COVID-19 pandemic, cultural organizations received significant containment measures to cancel events and openings. Therefore, investing inaccessible and reality-like digital applications through advanced participatory techniques reduced financial and target losses. In this way, the shift from the digital model to the interactive service model for sensory experiences skills the Citizen 5.0.

INTRODUCTION

New digital technologies in recent years have overtaken traditional methods of knowledge and sharing with innovative tools that involve a broad and heterogeneous audience. In this sense, the increasing use of digital objects such as Smartphones, fast Web Connections, Media in all forms, Virtual Reality (VR), and Augmented Reality (AR) have enabled new forms of interaction. In a society centered on human needs and behavior, users need to manage tools and information in the best possible way to increase their cultural knowledge, diversifying how they are delivered. Experiences become engaging only when service

DOI: 10.4018/978-1-6684-4854-0.ch019

delivery methods are innovative and accessible. This is reflected within the Society 5.0 for economic and social spheres where digitization and technologies overcome intellectual and physical barriers in safe and efficiently controlled virtual environments. Therefore, differentiation of experiences leads to sustainable and inclusive Information and Communications Technology (ICT) solutions and virtuality complements and supports the real world.

Every year, the top strategic tech trends are identified in a report (Burke, 2020), highlighting the main strategies that companies and the 5.0 society will adopt in the following years to streamline activities with medium- and long-term benefits. Based on these strategies, it is possible to identify Artificial Intelligence as a valuable tool to improve the relationship between users, machines, and realistic artefacts or to define original and easy-to-disseminate contents. Business models and integration between heterogeneous data sources are implemented thanks to an integrated and transversal approach. Within this technological evolution, another trend is emerging to unite the technological world with Gamification and Serious Game, namely the creation of a Total Experience (TX). This is understood as the all-around involvement of the user. The interweaving of multiple experiences and relationships through a multi-experience system accelerates the growth of companies and increases trust and end-user satisfaction through scalable and customizable applications.

During the health emergency linked to the Covid-19 pandemic, this trend became increasingly solid and necessary. The web world saw strong growth in digital sharing environments that would allow people to continue with their daily activities. According to some surveys, the market for devices, software and hardware increased by 20.5% in 2020 compared to pre-pandemic years (NetConsulting cube, 2020). Consequently, each product introduced into the digital market requires adequate means for its distribution on each device and optimized management of the systems in use with continuous monitoring of the data transmitted. According to technology trends, the effects on the stability of the technological choices made are expected to take five to ten years. However, many initiatives still need adequate means for more significant deployment before settling down. For example, multi-experience is a helpful approach to define the best way a user interacts with a digital product to take metadata management into account (TechRepublic, 2020). There are many advantages to this trend, which may be confirmed in the years to come (i) scalable actions at several levels with repercussions also on business and social, economic models, (ii) simplification of the construction processes of digital applications, (iii) expansion of the target audience of end-users thanks to intelligent and interconnected systems (Gartner, 2022).

Based on these technological trends, also in Italy, the digital sector has seen a series of investments (about 50 billion euros) at a national level through the National Recovery and Resilience Plan (PNNR). Many organizations and sectors have undertaken initiatives based on Total Experience to improve services to people and support the growth of new departments to drive the development of innovative products usable with new visualization and interaction technologies.

Among the sectors that the PNNR foresees a strong development, we can find the cultural and tourism sectors strongly affected by the pandemic (Anitec-Assinform, 2021). The plan aims to implement the industry through actions and applications that act on attractiveness and social inclusion levels. To fulfil these guidelines, methods and tools of Gamification and Serious Games can be combined, bringing the public closer and making any cultural artefact accessible.

The concept of Total and Digital Experience highlighted above is a crucial element in creating game applications, allowing the correlation of pleasurable experiences with information flow to facilitate concentration, challenge, skill, control, the goal, immersion and social interaction through feedback. These activities are made possible through the technological transfer of virtual data into an accurate model with advanced three-dimensional modelling techniques and visualizations intended as educational games.

In this way, active learning of digital content promotes a greater awareness of the cultural assets that characterize the society of the future. In this type of interaction, the player is supported through specific inputs to achieve learning objectives adapted to their abilities (Mariotti, 2021). The creation of applications that follow the principles of Serious Game transforms settings into simulation spaces that guide the user in exploring places with a full historical and artistic value, interacting with artefacts and representations typical of the cultural and entertainment sector. Enabling through game mechanics users to achieve individual and group goals with different paths of advancement is necessary to test the capabilities of the digital spectator. In addition, collective gamification aspects enable social comparison and support to work together towards a common goal (Krath et al., 2021).

This paper investigates the issues related to gamification and educational games in theatre and culture through innovative digital and visualization technologies to engage the user and analyze the responses regarding fun, goal achievement and emotional involvement. Several evaluation metrics are adopted to define the best strategy for transforming a simple navigation model into an educational game. Digital modelling, the Building Information Modelling (BIM) methodology, is used to create the three-dimensional model of a theatre building. It can transfer new knowledge and accompany the user towards a new sensorially, sharing actions and emotions in a dynamic way for optimized control of human behavior. The cultural and theatrical entertainment environment is characterized by many aspects that require an adequate visualization according to the audience's preferences, improving their communication. Therefore, the contribution proposes adopting new digitization information systems that enable interactive virtual participation through non-invasive Virtual Reality technologies.

BACKGROUND

Thanks to new communication technologies, the way we use spaces, and the built environment are subject to continuous change. For this reason, making our cultural heritage usable and comprehensible requires the adoption of information and technology systems based on knowledge approaches suitable for all types of users and ages. To facilitate information flows through digital tools, precise rules must be analyzed to create new social models. Studies carried out during the pandemic, Covid-19, show that digital transformation required adequate investment in technological infrastructure capable of providing communication and general entertainment services with even low-cost smart devices (Hennig-Thurau et al., 2021). Many organizations have redesigned their business models to adapt to these new requirements, investing economic and human resources in digitization. The cultural sector has also seen the emergence or downsizing of its business models with significant market growth through the creation of digital content and innovative visualization technologies. To increase and improve access to such data, web platforms during the periods of closure due to the pandemic have upgraded their ICT infrastructures to meet the increased demand.

The new ways of sharing cultural heritage activate the personalization of content by making it available to the viewer at any time of the day. The container, i.e., the historical building, has also benefited from this transformation by extending the reception of the built heritage with virtual visits.

Adopting technological means should expand cultural content and implement the user-consumer base through educational games. However, this change requires implementing new digital models related to entertainment called Entertainment 4.0 and gamification methods. They develop dynamic and easily accessible communication scenarios with high-medium-low visualization tools. Content automation is closely linked to innovative learning systems using specific devices to analyze the asset being visited near. The constant advancement of media uses cyber-physical systems and Digital Twin (DT), where information is interconnected within virtual scenarios. Underlying these models is a strong need to develop applications that reflect the characteristics of video games to accommodate a young and evolving audience. The game mechanics are introduced to increase the user's involvement to such an extent that new active participation strategies are defined even in non-gaming contexts. In fact, during the pandemic, the video game sector in Italy saw remarkable growth, about 51.6% more between 2018 and 2023. If we add augmented reality to these figures, the expected growth is about 23% more in the same period (Censis, 2021).

From this substantial increase, all areas of cultural entertainment and enjoyment of the architectural heritage have developed solutions/applications that could encompass multiple involvement scenarios, expanding their cultural offerings. In this way, the 5.0 society is more and more resilient towards humanity in which culture and entertainment become meeting places even in virtuality to share and develop thoughts and visions with new forms of participation.

Therefore, high-performance computer graphics solutions must transfer reality into virtuality akin to everyday life (Khan et al., 2020). Photorealism effects, challenges and levels are intertwined with well-structured information data for another value of the experience. As a result of this change, Virtual and Augmented Reality technologies facilitate the creation of experiences that add new deals and sensorially to the 5.0 citizen.

These technologies make it possible to bring the public into the theatre even if they cannot create a 24-hour opening of the cultural heritage without affecting the authenticity and economy of the asset. Moreover, it is possible to relate actions of the territory with needs and users even with fragility through universal languages and techniques that give a concrete answer to the needs of the digitized user.

Digital Experience For Society 5.0 - Game-Based Learning (GBL) and Serious Game (SG) in a Cultural heritage

In the era of digitization, many cultural organizations and museums have transformed the use of artefacts with promotional tools that combine virtual and actual content. Virtual environments have moved from simple story-telling applications to interactive three-dimensional digital scenarios that create experiences based on educational learning. This trend has changed the user's cognitive style so much that video game companies must invest resources in the cultural heritage sector to interpret information better and use new communication technologies as an integral part of everyday life (Cosovic et al., 2020). In the 5.0 society where man and his behaviors are highlighted, Game-Based Learning (GBL), understood as the use of digital games for educational purposes is the basis of the new generations called Digital Natives. The use of these systems allows the adoption of ludic measures that, based on the involvement and interactivity of the users, guarantee an increase in people's participation. Underlying principles and mechanisms that determine the application's success can be identified as promotional and marketing aspects of reaching a broad and heterogeneous target audience. The main characteristics on which GBL is based are motivational aspects and involvement through fun actions and contextualization of the objectives, finally reaching or not reaching the phases proposed by the game. To facilitate the achievement of the objectives, simple and clear rules must guide the decision-making process, increasing the difficulties progressively through direct control or feedback that guarantees the sharing of experience and forming of bonds. In this way, a new section of the public is involved, and existing relationships are stabilized (Al Fatta et al., 2018).

The logical evolution of digital games is the Serious Game (SG), which are associated with cultural heritage and can have a dual value, i.e. applied in formal or informal environments depending on the technological solutions adopted and the context. Several techniques for the realization of serious games must be considered: (i) simplicity of the game according to the final location, (ii) the reduction of the personnel in charge of explaining the game, (iii) single-player or multiplayer choice according to the time available or the location (Mortara et al., 2014). Following these procedures and methods, learning allows the implementation of cognitive, psychomotor and affective aspects, i.e. the user at the end of the game can acquire new knowledge, skills, and behavior (Bloom et al., 1956).

For this reason, the design of serious games is not trivial; on the contrary, it requires sound design that considers usability, graphics, good information content and sharing between players. The only effort required of the user is to pay close attention to what he sees and hears and reason about the actions to be taken. In this way, complex concepts can be assimilated through investigative and motivational attitudes. Therefore, the rise of these techniques and the increasing use of digital systems have given rise to many projects using educational approaches to cultural heritage. Here are some examples:

- **Feidias Workshop:** is a serious game that uses gamification techniques to provide a virtual learning and interactive experience. The game reproduces the construction site of the 15-metre-high golden ivory statue of Zeus. Once wearing the VR visors, the user finds himself inside the construction site, where he has to reconstruct the unfinished version of the sculpture. They can interact with the sculptor's tools, move between scaffolding benches, define the materials, and finally prepare the molds. In this way, the active role of the visitor takes on several faces, first as assistants and finally as actual sculptors using virtual tools to complete the work. They can grasp materials and virtual equipment to work the plates and apply them to the wooden core of the statue, shaping each part. Thanks to the continuous guides and information, the user can learn the sculpting techniques, materials, and procedures that the past greats used for their works (Anderson et al., 2009).
- **Grand Master Challenge:** is a game designed to facilitate the knowledge of the medieval city of Rhodes, a UNESCO heritage site. The game's objective is to retrieve information from different sources and customize it according to one's interests. Among the documentary sources, the game is supported by an iBook to understand the plot of the game and the necessary clues, while Google Maps facilitates navigation. As many as 32 monuments of the city were analyzed, divided into game levels in which an audiovisual narration increases the cognitive value. The significant aspect is transforming play into multisensory experiences of fun and relaxation. Thus, interactions aim to define new social contexts and learn about history. The project also links to web pages that refer to services, geolocation, transport and accessibility of each place with the ability to identify the main elements (Papathanasiou-Zuhrt et al., 2017).
- **ArkaeVision project**: the cultural heritage becomes accessible through interactive thematic virtual models (ITVE) where the visitor is immersed within a platform in which monuments, works of art and ancient artefacts can be visualized with real-time information (Bozzelli et al, 2019). Data are stored by an external database and transmitted during the experience in the form of gamification. Real-life simulations and audio descriptions accompany the monuments reproduced in their original context. For this reason, the project proposes new communication paradigms where

information is provided according to levels of experience based on AR and VR technologies. The user is stimulated through the progressive discovery of the historical artefact and a system of rewards based on the actions and interactions carried out between user-user and user-virtual character. The technology includes a cybernetic glove that recognizes gestures and allows interaction with the virtual space. Each user has a personalized activity with profiling within the platform that calibrates the type of experience. In addition, it is possible to share everything in multi-user mode (Pagano et al., 2020).

METHODOLOGY

Case Study

Visualization technologies and digital models defining the development of serious games are considered the new communication strategies to entertain and educate the visitor. In this context, the cultural entertainment sector turns into a real museum to ensure the accessibility of the container and the contents of the present and the past. The main aspects to be considered for the digitization and visualization of historical heritage are: (i) flexibility of strategies to render geometries and alphanumeric properties, (ii) the ability to engage the user and ensure the possibility of actual visits, (iii) updatability of technological systems according to the cultural evolution of the audience and devices. During the Covid-19 pandemic, this area suffered considerable limitations in the actual accessibility of the heritage, so the way of delivering services had to be revised. Some relied on streaming some content, and others implemented web services with images and virtual tours of some building parts. Still, adapting to the new modes was problematic for several reasons. Difficulties in adopting stable platforms for viewing works, numerous archives that cannot be digitized, inadequate means and resources.

Therefore, the choice of the case study to become a use case focused on a strategic building for the Piedmont region with a solid cultural and entertainment value. It can represent multiple national realities and become a reference point for adopting innovative technologies applicable during the pandemic and replicable in the future with user-friendly applications. In this research, the case study selected is the Teatro Regio Torino in Figure 1. The theatre was first entrusted to Juvarra as court architect. Still, after his death, the architect Benedetto Alfieri built it at the behest of King Carlo Emanuele III of Savoy in 1738. In 1740, when it was inaugurated, it had a horseshoe-shaped hall with a frescoed vault and a total capacity of 2500 spectators distributed between stalls and five tiers of boxes. It is located in Piazza Castello in Turin.

Unfortunately, in 1936 the historic baroque building was devastated by a fire, which destroyed the entire theatre, leaving the city with only the landmark wing overlooking the main square. For several years, a succession of projects was presented to the town, but it wasn't until 1967 that architect Carlo Mollino assumed the role of designer and director of works for the new theatre.

The design features a main entrance with twelve double glass doors separated by granite walls (diaphragms). Continuing into the interior, solid volumes and curved lines dictate the forms of every component: from the stairs to the balconies, doorknobs and the bar counter, arranged symmetrically concerning the main entrance.

The stalls have an ellipsoidal floor plan with 1398 seats and 31 boxes with 194 spectators. The lighting system of the entire hall is entrusted to an enormous chandelier composed of 1762 aluminum tubes with a final light point and 1900 reflecting Perspex stems.

An analysis of the compositional elements of the building reveals specific reflections which characterize and reflect the cultural evolution of society and the desire to emphasize the role of music and theatre. The inclusion of double doors, all identical, expresses the determination to eliminate the distinction of social class by giving everyone the possibility of attending a performance. Similarly, the entire floor plan of the new theatre is reminiscent of musical elements, i.e., it takes the form of a cello case. At the same time, the stalls are half-open oysters to implement sound and achieve high acoustic values (Bardelli, 2010).

Figure 1. Case Study

Workflow

Starting with Society 5.0, in which visualization and communication technologies are extended to everyday actions, it is necessary to focus on user needs to innovate flows and processes. Each person can thus change their lifestyles thanks to digitalization, using tools that become new business and development models. Based on these models, Figure 2 identifies strategies applicable to different use cases, prototyping applications that follow gamification rules to engage the public. Finally, through dissemination actions and tests with specially analyzed metrics, it is possible to determine the success of a strategy and apply it with precise rules. The basis of this workflow is a digital model that evolves thanks to specific algorithms that manage and optimize the information and relational interactions between user and machine.

The methodological approach proposed for this contribution in Figure 3 outlines the development of a virtual model understood as an interactive digital service model (IDSM). The basis of a dynamic process is the knowledge of the building, the evolutionary phases, the management and the staging of the works combined with the needs of the citizens in the 5.0 society. Technical elaborations of data mining, gamification, and Augmented and Virtual Reality (VAR) require defining algorithms to process cultural data for interactive dissemination. The duality of content and container allows the public to enjoy the best possible personalized experience according to their interest. The modelling, for this reason, must be both three-dimensional, photographic, and alphanumeric restitution to connect external information databases. Different sources collaborate within the model to best render the services required by the user.

Figure 2. Identification of needs and processes

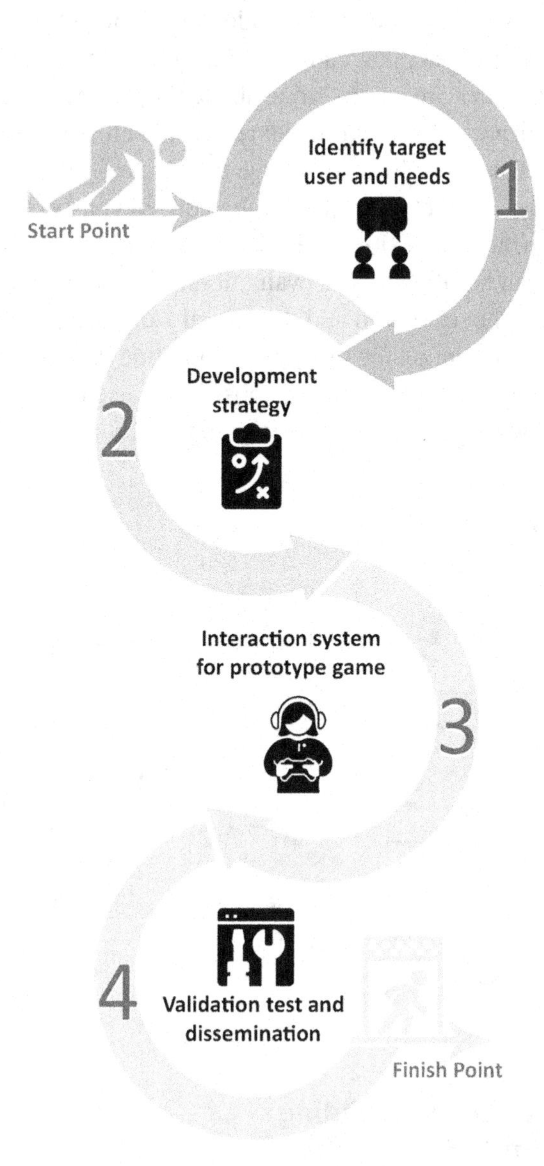

Data cataloguing and visualization methods are intertwined with metrics related to effectiveness for learning, Playability, Engagement, Immersion, Presence, Flow, Absorption. Based on the data collected and measured, various scenarios are modified, expanding the perception of real objects within virtuality. Only in this way it is possible to upgrade into a sensory journey where every perceptual stimulus is overcome thanks to the dynamic involvement of the user. Every interaction defines a level of play through precise tasks where every piece of information is assimilated and passed on to the next generation. At the end of the methodological process, verifying learning and enjoyment through precise standards can lead to the definition of decision-making strategies. The implementation of services is possible with a mechanism of facilities and bonuses that facilitate, on the one hand, the use of serious games for knowledge and, on the other hand, increase the number of real visitors. In this way, the circular management of the asset is optimized by taking full advantage of technologies that entertain, involve, and create curiosity in new generations. If we add to this competition techniques between users and the time variable, the amount of data that can assimilate could increase, involving more and more "players". Pedagogical learning rules are applied to evaluate the design and graphical aspects to return a version ready for commercialization. Several procedures adopt before making available an application that follows these indications. This contribution will use the Game User Experience Satisfaction Scale (GUESS) system to evaluate the satisfaction of the game in different aspects based on effectiveness for learning, Playability, Engagement, Immersion, Presence, Flow, and Absorption. The most interesting aspect that this methodology can highlight is the evaluation of the social connection, reducing, on the one hand, the distances between the real world and the virtual world and, on the other hand, making the user participate in the activities present on the territory. In addition, the tourist aspect can also evaluate since, during waiting times or periods of closure, it is possible to use these applications to increase the cultural background with the awareness of what one is going to see next.

Figure 3. General methodology

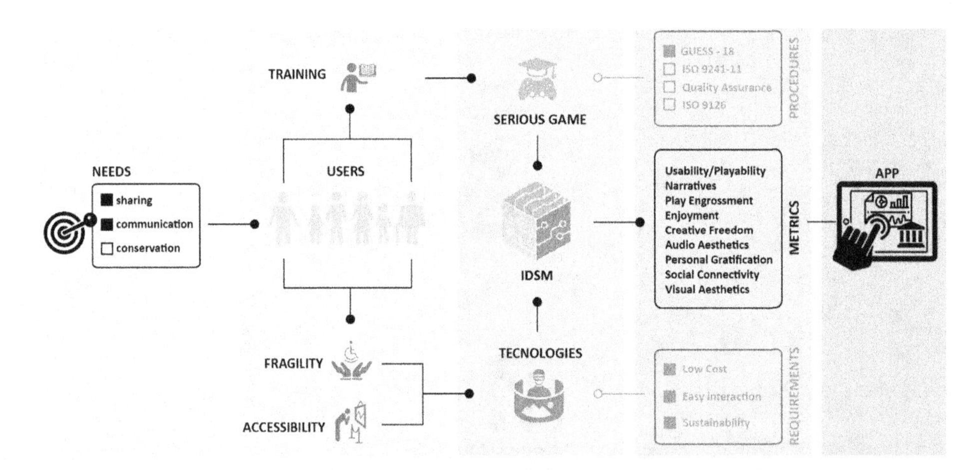

Digitalization Procedures Using Building Information Modelling for IDSM System

The real innovation brought in the era of digitization is the need to adopt digital processes where virtual models learn actual behaviors by acquiring data and cataloguing them in different domains to analyze them with other assets. This system governs the so-called Digital Twins or virtual replicas of the real twin, where each object corresponds to its virtual copy equipped with sensors and algorithms that interpret the data (Jones et al., 2020). The construction of a Digital Twin involves a layered conformation where multiple elements interact to determine the completeness of an object. It is equipped with: (i) connectivity thanks to a network of intelligent sensors that become "bridge" tools between reality and virtuality, (ii) modelling and simulation techniques, (iii) visualization and communication tools between users.

Thanks to the introduction of DT, parametric modelling methodologies (BIM) have also undergone enormous developments to the extent that they have become true digital twins in which interoperability and information exchange regulate the continuous transformations in increasingly interconnected and responsive systems. Through its cognitive phases, the BIM model allows the definition of a virtual component that includes data collection, data restitution, data integration, and data visualization. For each of these phases, it is possible to adopt adequate levels of detail and development to create an IDSM.

For this reason, the virtual model developed in this contribution is based on the BIM methodology, considering the reference standard for the national context on the geometric and informative definition of the fundamental properties that an object must contain in virtuality. The measure considered is the UNI 11337 part 4, in which the levels of detail and development are catalogued through literary codes and according to the amount of information required, there will be more or less technical specifications associated (UNI 11337, 2017). The case study presented in Figure 4 adopted several levels.

Figure 4. BIM Model

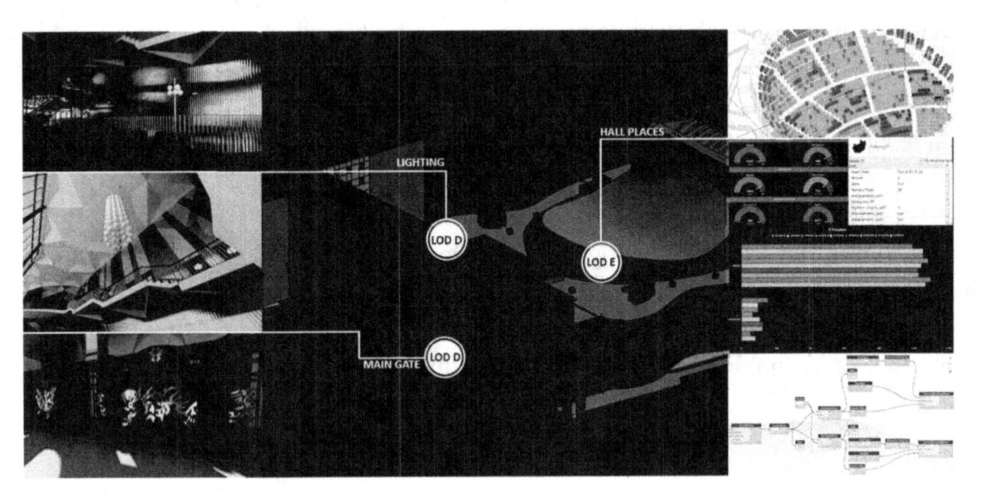

The first defines the container, and the second regulates the geometric and informative detail of the individual components. In particular, the whole building follows a LOD C where the objects represent in a defined way. The qualitative and quantitative characteristics describe generically, respecting the technical standards' references for similar things. At the same time, the individual objects present a further subdivision of the LOD. Some elements adopt a LOD D where each part is represented geometrically in great detail. Qualitative and quantitative characteristics are well specified with typological, spatial and maintenance data. Within the model, this LOD has been associated with elements: (i) the lighting system, the central chandelier of the foyer and the under-lighting of the different floors and finally, the large cloud of the stalls; (ii) the main entrance gate. Other objects follow LOD E, where they are graphically virtualized with a specific geometric system linked to product-specific quantitative and qualitative information. This LOD has been associated with the hall places with the entire database of tickets sold, identification codes and sectors to carry out employment and revenue scenarios. In addition to the manufacturing, assembly and installation method, the entered data concerns certain event management aspects.

Visual programming and data mining techniques were adopted to create an interactive service model that retrieved information via the web and virtual visualization interfaces to carry out detailed analyses of the information contained within the model. This phase adds to the levels that make up the DT, the concept of interoperability to cross the border between reality and virtuality with application solutions adapted to the behavior of the digitized user. It is precisely here that the BIM model, in Figure 5, is transformed into IDSM, i.e., using educational feedback, the data received from the outside world modified and improved scenarios are defined for reliable replication of reality. The needs to be met by the IDSM concern: (i) predefined system templates to satisfy user responses; (ii) integration of multiple document sources to implement user and system knowledge; (iii) dynamic selection of searchable data. This translates into customizing the application with specific interfaces managed by sub-modules involving aspects of gamification metrics and processes.

That is, defining clear objectives achievable through: (i) sub-modules; (ii) systematizing the final objective to express specific needs; (iii) mapping of objectives with advanced graphic and multimedia qualities; (iv) ease of access to information with identification of flows; (v) speed of execution of assigned tasks; (vi) quick and straightforward commands to terminate the service request. In this way, the customization of a model through flexible processes provides high decision-making capabilities to the user, thanks to new functionalities to improve the interaction between user-user and service provider (Cao et al., 2006).

The main features developed in this contribution for the definition of an IDSM are:

- Multilevel interaction between a real object and virtual model
- Data processing through visual programming algorithms to implement an information system approach
- Connection for visualization of web services
- Non-immersive Virtual Reality techniques for defining the game platform

Figure 5. IDSM model

Thanks to this type of model, it is possible to communicate and transmit the emotions of cultural heritage with easily accessible cognitive paths within everyone's reach. In addition, the personalization of services ensures that even those with disabilities can fully enjoy art and culture, overcoming physical and social barriers (De Luca et al., 2021). This process initiates sustainable growth actions on the part of each user to broaden their cultural background with technological investments appropriate to social inclusion.

Evaluation Metrics for Virtual Experience

The validation of the effectiveness of the serious game for the education and dissemination of cultural content passes through metrics based on business concepts making an application marketable. In fact, in recent years, the video game market has defined through sales strategies the main characteristics to evaluate in the analysis and research of game requirements (Gupta et al., 2015). Usually, the scope of evaluation falls into three main areas: technical, functional, and content testing. For the technical testing part, the response of the game platform and the associated hardware can be analyzed through compliance, compatibility, and responsiveness tests. Functional tests look for issues related to the game and network stability, the integrity of resources that can be used to control game actions and the user-machine interface.

The graphics, the design, the structuring of the storyline, the general presentation and the objectives are of fundamental importance to understanding the user perception for Quality Assurance or User Experience.

The content tests, therefore, have a double value, i.e., they determine the quality of the video game but at the same time relate the experience perceived by the player to adoptable tools (Drachen et al., 2009). Other metrics that can measure quality include game design, user satisfaction, usability, usefulness, understandability, motivation, performance, playability, pedagogical aspects, achievements, engagement, user experience, effectiveness, social impact, cognitive behavior, enjoyment, acceptance, and user interface. Game design evaluates the correct graphic and visual design to make the graphical interface appealing and creative. User satisfaction to influence by perceived usefulness, ease of use and flexibility. It is helpful to consider age, gender, and pre-and post-activity digital skills acquired when assessing this property.

Usability values are standardized by ISO 9241-11 (ISO, 2018), where specific parameters are used to determine the goal's achievement with effectiveness, efficiency, and satisfaction. These three aspects, combined with other attributes such as knowledge, memory, and error identification, can facilitate tasks, concentration, and productive user during the gaming experience.

The utility is the measure for establishing the success of the learning and dissemination system, as it reflects the efficiency and appropriateness of the application of knowledge. It defines an application's ability to adapt to the user's levels of understanding through specific tasks and conditions of use. Understandability is also standardized through ISO 9126 (ISO/IEC9126, 2002).

Motivation is another parameter used to understand how much the game impacts and influences the user's choices, bringing high levels of creativity and curiosity. Adding elements of competition and challenge helps to understand the positive or negative impact on the motivational level.

The performance parameter is related to the functionality and effectiveness of testing the main components.

On the other hand, playability focuses on the totality of functionality with the integration of tools that best tolerate interaction with the game. This concept is the main characteristic of quality and effectiveness.

The pedagogical aspects associated with the other elements focus on the educational evaluation of the cultural content transmitted, which must be well structured and calibrated according to the user and the levels. The latter determines the success of the knowledge. Through the results obtained during the experience, the game can be considered an accurate educational tool. The categories analyzed to validate learning outcomes are related to cognitive, motor, affective and communicative skills.

Involvement in the user's engagement with the game is based on the degree of immersiveness, direct control, challenges faced, and purpose achieved through high-interest levels. Effectiveness is connected to employment, attention, involvement, enjoyment, and difficulty in reaching the end of the game. The ability of the game can improve the knowledge of the participants. The user experience also follows the rules established by the ISO 9241 standard, in which the perceptions and responses received derive from the use of a system (ISO 9241, 2010).

Social impact is the measure for defining the expected results in a sustainable society where different users cooperate. On the other hand, cognitive behavior understands as the ability to create effects and changes on the cognitive behavior of the user. Fun is a crucial factor in marketing serious gaming, focusing on enjoyment, competition, and motivational aspects.

In conclusion, the user interface aspect is fundamental to developing user-friendly graphic layouts with maximum accessibility and usability. The element of acceptance is associated with perceived usefulness and ease of use. If the user is hostile to accepting the game mechanisms, his engagement is limited, and he will not learn properly (Abdellatif et al., 2018).

These parameters are the starting point for constructing an application based on gamification techniques that consider fun the fundamental element for learning with a high level of involvement in a specific time frame. At the end of the experience, the user must be able to establish lasting relationships with what they are analyzing, i.e., the cultural heritage in question, to implement the information exchange. Therefore, the construction of history through layers allows for the transmission of educational and cultural messages to many generations.

To be effective, the serious game must best identify: (i) the business objectives, i.e. outline the concrete results; (ii) monitor the expected user behavior through validation metrics; (iii) reflect the expectations of the audience with motivational aspects; (iv) design activities that can be integrated with incremental steps of user experience; (v) include fun aspects and allow for the possibility of adding levels with new learning tools (Werbach et al., 2015).

Based on these considerations, the contribution adopts the Game User Experience Satisfaction Scale (GUESS) as a metric for validating the application. This scale considers 55 items divided into sub-categories (Usability/Playability, Narratives, Play Engrossment, Enjoyment, Creative Freedom, Audio Aesthetics, Personal Gratification, Social Connectivity, and Visual Aesthetics. The analysis conducted using this method guarantees a good validation of the video game as every single aspect is analyzed. It covers: (i) the ease of play in which clear objectives allow overcoming obstacles with an easily controllable interface; (ii) the ability to engage the user's interest, modifying his emotions through adequate narrative aspects; (iii) the level of attention related to the degree of perceived pleasantness as final feedback; (iv) the ability to express creativity and curiosity, leaving the user free to tell about himself even if he is involved in the game; (v) the inclusion of audio effects through multiple sensory aspects; (vi) the desire to achieve goals through engaging challenges and levels; (vii) the social connection through social channels and sharing the experience with other players. Finally, the attractiveness of the graphic aspect is assessed, with icons, care for the three-dimensional model and texts (Phan et al., 2016). For each category, the user is asked questions with a rating (from 1 = strongly disagree to 7 = strongly agree). At the end of the test, each answer gives a score. The average sum of each answer determines a value. If it is high, it decrees the possibility of marketing the video game by improving the categories that received a slightly lower score. If the final sum is a low average value, the individual aspects of each part must be modified, and the application retested. This system is advantageous for understanding where to focus more on developing the video game. From the analysis of various scientific studies, 55 items could become time-consuming, and multiple questions focused on similar aspects, making differentiation difficult. For this reason, leaner versions of this method to develop in which 18 items or two/three questions per sub-category are considered to differentiate the topics to be tested (Keebler et al., 2020).

In detail, in this contribution, the matrix of questions administered to the users for the developed application follows the indications of the GUESS scale with 18 items with the addition of other questions. They frame the target users, the familiarity with video games and finally, the game's effectiveness in transmitting the curiosity to really go and visit the cultural heritage even if one has some weaknesses.

According to the classification by subcategories, the developed matrix considers: (i) how simple the navigation commands are, and the matured interface facilitates the exploration for the Usability/Playability category; (ii) the degree of the pleasantness of the contents and attention - Narratives; (iii) how much the application engages us to the point of not considering what happens in the real world - Play Engrossment; (iv) the fun or not aroused - Enjoyment; (v) the creativity and imagination - Creative Freedom; (vi) whether the audio-visual effects with their interactions enhance the experience - Audio Aesthetics; (vii) whether the challenges proposed engage the user to want to do better - Personal Gratification; (viii) whether the inclusion of weblinks and multiplayer may or may not support social interaction; (ix) whether the graphics adopted and in general the whole application is visually attractive - Visual Aesthetics. In addition, another metric introduces to understand whether certain cultural content delivered through a video game can be better understood and thus entice a visit by expanding the user base.

The spread of this rating scale is not only limited to the field of video games but to use in different domains, such as the health sector, mixed or virtual reality, and social interaction. Thanks to its standard method, it can adapt health simulation to measure the effectiveness of specific exercises combined with new digital technologies.

Figure 6. Interaction and metrics of IDSM

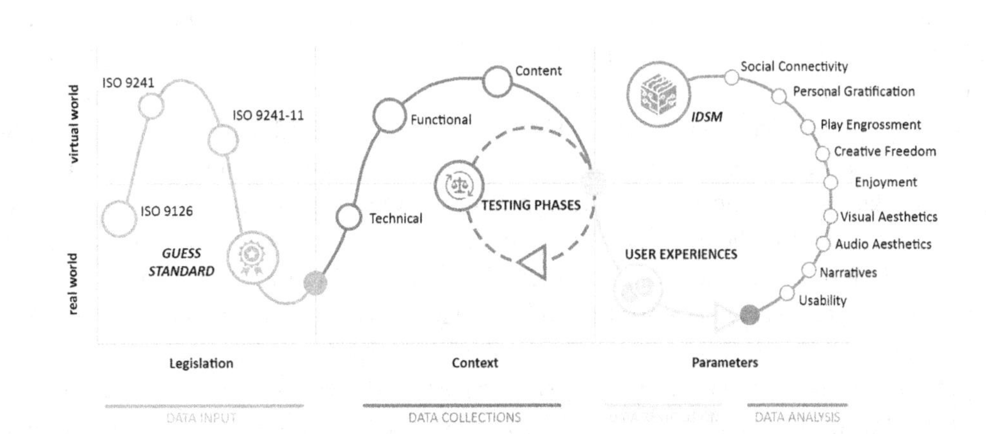

Therefore, the choice of this method, combined with the usability and utility values from ISO 9241-11 and 9126 in Figure 6, allows the designer to hit the target and the client, in this case, the cultural institution, to expand the cultural offer with dynamic, entertaining, and competitive virtual tours.

In this direction, the virtual environment doesn't refer to as static models but as interactive multilevel services systems, i.e., IDSM.

RESULTS

Use of Game Platform for Virtual Experiences

Multilevel interactive service systems can be adopted as the basis for serious gaming applications, thanks to data exchange platforms that connect three-dimensional digital models and the associative database. To build an IDSM compliant with gamification rules, gaming software is suitable, using high performance for customizable interactions. Once the use and purpose are defined, users to provide applications based on the geometric and alphanumeric interoperability of the digital model. In this direction, Unity 3D software using AR, VR and MXR procedures connected to the digital model can validate the measurements and methods adopted. The data exchange process, in Figure 7, can be activated thanks to standard formats that communicate the virtual model developed on Revit and Unity. The format used for this use case is the DAE. Using this format links to the ease with which materials can be associated. Unfortunately, from the BIM model, it is not possible to export the DAE directly, but only Filmbox formats (.FBX) which, when opened with other software such as 3Ds Max, allows you to obtain different formats. Several tests have been carried out to achieve more significant geometric and material interoperability, while the data transmission doesn't optimize for the alphanumeric part. Most of the information is lost, and it is necessary to create external codes to call up the database. However, basic knowledge of programming languages is necessary to set up small interactions and visualize data lost during the information exchange process.

The first test was to export the DWG and FBX format of the RVT model and import it onto a bridge software called SimLab Composer. Through this platform, intuitively, every single object keeps the associated texture and allows to convert the model directly in the DAE format used for the video game. Once imported into Unity, the model retains its material folder, and the software can use the textures associated with the object directly. Although this process is lengthy and requires additional software to pass through, it reduces the time it takes to associate materials and does not require any further changes to the game platform. The second test is to use the model's FBX format on Unity directly. Still, it cannot get the folder with all the textures, so it is necessary to use 3Ds Max to export the single textures and then manually associate the material to the object. This step is not optimized and requires a lot of time and human resources to associate the material correctly. Therefore, sometimes using more than one "bridge" software reduces data acquisition time and human errors as the association automatically occurs. The only problem encountered is the heaviness of the initial file, which carries with it every texture. The meshes are divided into dense triangles to best match the material to the object's geometry. Once the model was imported into Unity, we moved on to building the video game with characters, animations, lighting, and interactions that additional information. The application is based on scripts that activate elements and data whenever the user explores salient parts and needs to acquire helpful information. The user will be able to interface directly with the contents using simple keyboard commands, explained even before starting the experience. The main activities included in the application to involve the user and transform it into a sensory experience are listed below. Starting from the gate, the user will meet virtual characters who, through questions, will invite him to discover the individual pieces of information. In this case, approaching the totem pole will reveal historical and technical information about the author and the design of the entrance.

Once through the entrance, various information about the season and the booklets of the operas in the theatre. Thanks to virtual hostesses positioned in the foyer and the waiting room, images, videos, and avatars entertain the user to learn and get to know better what they will see first in virtual and then in reality. After passing the entrance, the user can go to the conference and reception room (Sala del Toro) on one side and view the room with tables for different events. Or enter the hall directly and interact with additional virtual characters to learn about the seat highlighted in the ticket and watch the show. In this way, the user will be able to see the view. In addition, it is possible to interact with the orchestra and the lighting in the auditorium and visit the backstage area. Once moved backstage, the spectator can learn about the scenic mechanisms with the movement of the bridges and the latticework. At the end of the experience, to earn a bonus or not according to the score obtained from each activity carried out, and you will be able to keep up to date with live events by connecting to social pages. In addition, throughout the experience, you will be able to chat with other users thanks to the inclusion of a virtual chat system that, based on the login made, connects all users connected to the application at the same time. A plugin, Photon, integrated into Unity, transforms the game from single user to multiplayer. This type of application can be made available through the institution's website. In addition to proposing the virtual tour by traditional means, one can download the game onto PCs or mobile devices. The individual user or the entire family can interact with innovative and engaging virtualization systems. For this reason, each interaction is designed to involve the primary senses, from sight to hearing, and ensure that each activity adds new meanings to everyday life.

Figure 7. Interoperability process for game platform using IDSM

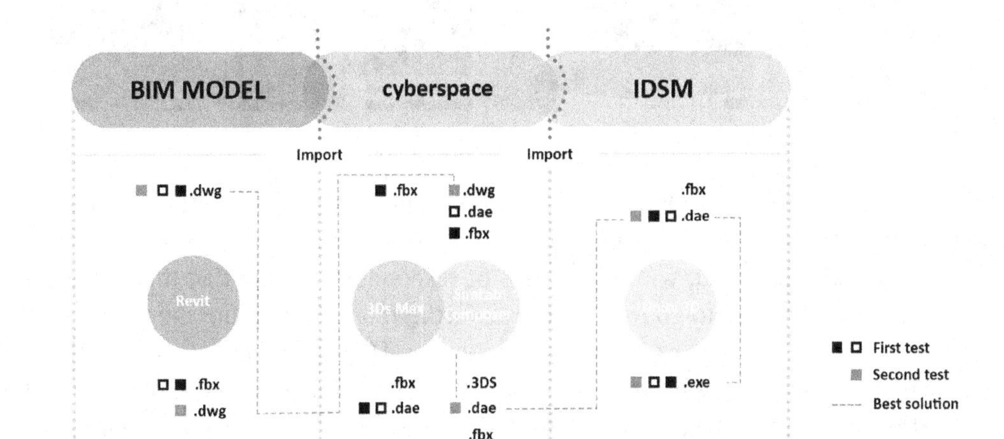

Virtual Experiences Activities

The application was tested at several events during the year, in Figure 8. The first phase involved Researchers' Night with excellent results in single-user mode. About 60 people took part in the experience by filling in the questionnaire described in the previous paragraph. From the analysis of the data entered in general, it is possible to state that a heterogeneous target of users, with a prevalence of familiarity with video games, aspects related to the simplicity of navigation and an easy interface obtained positive values for 50%. Similarly, the contents in the way they present, and the type of data included showed positive values for 44% of the respondents. The degree of enjoyment was rated as excellent by 57%. On the other hand, the detachment from the real world during the experience was complex for more than 35% of the respondents to assess since the general event did not allow them to concentrate only on their activity. Although not strongly developed within the application, creativity and imagination aroused strong interest, thanks to the music and video aspects that recalled intrinsic elements. The interactions and audio-visual effects included had an excellent response with 40%, this stimulating even more to achieve the objectives set by the game.

Figure 8. Dissemination experiences

The link to social channels was also very much appreciated, and some users started to follow the organization through the new media. Finally, in Figure 9, the degree of understanding of the cultural context provided was also assessed, with about 90% of respondents saying that this application helps dissemination and entices people to go to the theatre. Furthermore, comparing the target users in both experiences shows that the prevailing age is relatively young. Hence, aspects of usability and involvement through chat and multiplayer systems allow cultural content to be transmitted to the best effect. By activating gamification mechanisms, it is easier to attract a young audience as they are used to learning through video games that give the message in a clear and defined manner. Therefore, the results show that the correct content design can meet the different needs of an inclusive and heterogeneous society.

During the second phase, in multiplayer mode, the application was made available for the Orientation days held by the DrawingTOthefuture Laboratory and VR@polito to high schools. On this occasion, the target users were mainly young people between 16 and 18 years old, experts in video games. Also, in this case, the questionnaire disclosed follows the GUESS-18 items matrix as described above, and for each category, the user answered two questions specifying their interest and liking. Specifically, they found positive values (80%) for simple controls during navigation and the interface (83%). The cultural content provided caught their attention and was 91% pleasant. Taking place in a secure and controlled zone, the application can involve the best without any interference with positive values (62%). Overall, the fun aspect involved about 85%, while only 15% were bored during the experience. Creativity and imagination related to the graphic elements and audio-visual interactions reached 67% approval. The choice of sound and music effects enhanced the experience for around 80% of users. This makes it easier to achieve the goal in the best possible way. Above all, the addition of multiplayer with chat communication facilitated sharing and involved the user more concerning the challenges and the content provided. It can also be seen in the results on graphics, as about 85% found the interface attractive and appealing. Finally, through this application, most of them approached the world of the Opera House and thought of buying a ticket.

In conclusion, in Figure 10, they were also asked to evaluate if this kind of application could help someone with frailties easily enjoy cultural content and if it could be helpful even if one cannot physically visit the place. 86% of the users responded positively, so it can say that based on the value achieved by the GUESS matrix, the multiplayer application can be marketed and become the starting point to increase the pool of users and make up for any periods of closure by facilitating virtual visits. If we compare the individual items, it can be seen that in predominantly young users, aspects of usability, narrative involvement and multimedia characteristics are particularly appreciated as they recognise elements common to other video games they frequently use. Therefore, it is necessary to adopt these mechanisms to involve the audience of the future, i.e. the new generations who are not familiar with the theatre world, to visit it to keep alive the cultural identity of a place and a society.

Figure 9. Results metrics first and second phase

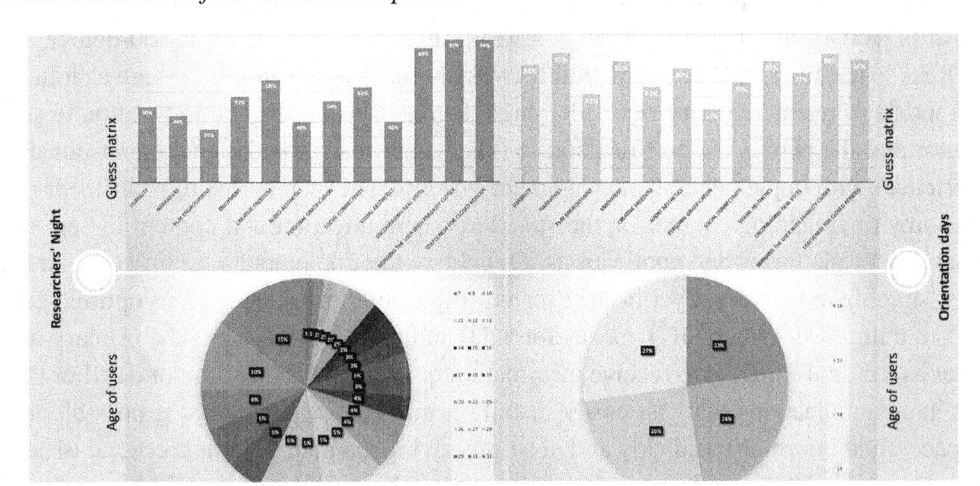

Figure 10. Comparison of metrics between the first and second phase

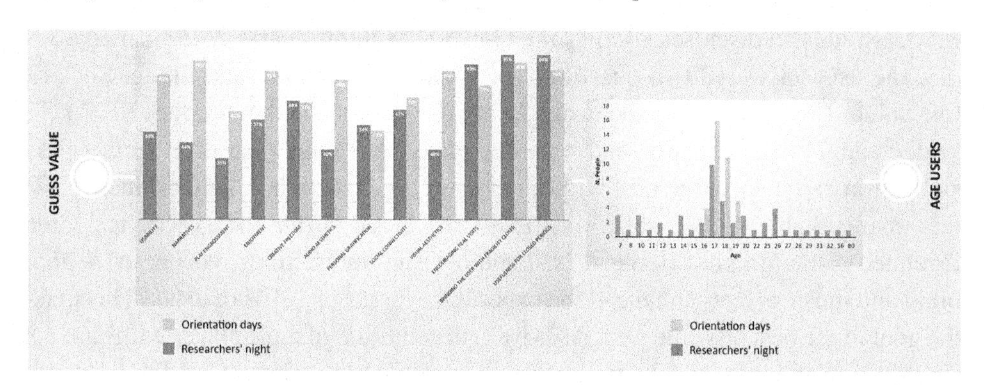

CONCLUSION

In the era of digitization and the advent of increasingly innovative technologies, the performing arts and culture sector can reap the benefits of related applications, as highlighted by the research presented in this contribution. The growing demand for technological tools has led to an increase in the services provided. Thanks to gamification techniques and serious games, the desire to learn about virtual places has grown to such an extent that they have been transformed into real educational experiences with innovative technological systems for a sensory visit. Their use, availability and integrity increase viewer involvement when associated with Virtual and/or Augmented Reality tools. In this regard, virtuality can help this transition through a technological system of preserving the historical artefact. It reflects reality without damaging it or applying the digitization of innovative services. The analysis of the feedback received during the different public activities shows that the other appropriately designed applications can establish objectives and playful activities as a promotional element for a 360-degree sharing of scenic art. Real space, time and the interaction between humans and technologies are as follows to create digital models of interactive services with a high degree of technological innovation to support the cultural sector and the public. To best respond to this evolutionary trend, generate sustainable educational experiences with implementable management systems. The factors that most influence this trend are the economy of technological means, the speed of communication and content dissemination and the sustainability of platforms and applications. Of course, the implementation process and the search for the most suitable tools are not yet perfect for turning virtual experiences into optimal educational moments. Yet, cultural sharing by ICT means for a 5.0 audience is necessary. The primary need of the 5.0 spectator as a digital expert is to receive information quickly and at any time of day. For this reason, developing game applications that can easily install on mobile devices makes it possible to promote services, spaces, and information quickly and in a time-saving manner. The most critical issue is how to best manage the continuous updating of information within the cultural games. While, on the one hand, there is a reduction in user time and research, on the other hand, the economic cost for organizations to provide human resources to build the upgrade of the proposed solution is significant. For this reason, while recognizing the direction that needs to take, human and economic resources need to invest in continuing to make the cultural and entertainment sector usable with new visual technologies.

FUTURE RESEARCH DIRECTIONS

Digitization within cultural heritage organizations is changing the way audiences access content through web services that connect users from different parts of the world. Therefore, adopting these systems requires a significant effort on the part of citizens, associations, and society to improve the accessibility of information without losing the collaboration and experience-sharing aspects. In this direction, Serious Games applications, starting from the typing of cultural heritage, are growing thanks rapidly to learning 2.0, where training develops in games, challenges, levels, and achievement of objectives with creativity and joy. This can be developed as cultural entertainment 5.0 in which digital models and media tools intertwine with AR and VR technologies to a neural network where specific Artificial Intelligence algorithms can deliver the desired experience in a personalized way. In this direction, the metaverse is preparing to provide a solid basis to give space to the virtual world and make the everyday life and history of places the digital world. The user will be able to choose where to live and with whom to share experiences in any place and by any technological means. It is hoped that in the future, human-hardware and community interaction will be increasingly integrated and benefit sociality and human behavior aimed at inclusion and preservation of heritage memory.

ACKNOWLEDGMENT

The author would like to thank the Teatro Regio Torino for their valuable collaboration in defining the demanding framework underlying the experiments and for the material used for the case study. In addition, the author would like to thank the DrawingTOtheFuture laboratory and VR@Polito for their technical support and availability during the various events where the validation phase of the methodological process was carried out.

REFERENCES

Abdellatif, A. J., McCollum, B., & McMullan, P. (2018, March). Serious games: Quality characteristics evaluation framework and case study. In *2018 IEEE Integrated STEM Education Conference (ISEC)* (pp. 112-119). IEEE. 10.1109/ISECon.2018.8340460

Al Fatta, H., Maksom, Z., & Zakaria, M. H. (2018). *Game-based learning and gamification: Searching for definitions. International Journal of Simulation: Systems, Science and Technology.* doi:10.5013/IJSSST.a.19.06.41

Anderson, E. F., McLoughlin, L., Liarokapis, F., Peters, C., Petridis, C., & de Freitas, S. (2009). Serious games in cultural heritage. *The 10th International Symposium on Virtual Reality, Archaeology and Cultural Heritage VAST - State of the Art Reports.*

Anitec-Assinform. (2021). *Il Digitale in Italia 2021, Mercati, Dinamiche* (Vol. 1). Policy.

Bardelli, P. G., Garda, E., Mangosio, M., Mele, C., & Ostorero, C. (2010). *Il teatro Regio di Torino da Carlo Mollino ad oggi*. Dario Flaccovio Editore.

Bloom, B. S., Engelhart, M. D., Furst, E. J., Hill, W. H., & Krathwohl, D. R. (1956). *Taxonomy of educational objectives: the classification of educational goals. In Handbook I. Cognitive Domain*. Longmans, Green.

Bozzelli, G., Raia, A., Ricciardi, S., De Nino, M., Barile, N., Perrella, M., & Palombini, A. (2019). An integrated VR/AR framework for user-centric interactive experience of cultural heritage: The ArkaeVision project. *Digital Applications in Archaeology and Cultural Heritage*, *15*, e00124. doi:10.1016/j.daach.2019.e00124

Burke, B. (2020). *Gartner Top Strategic Technology Trends for 2022*. Academic Press.

Censis. (2021). *Il valore economico e sociale dei videogiochi in Italia* [The Economic and Social Value of video games in Italy]. Primo Rapporto IIDEA-CENSIS.

Cosovic, M., & Ramic-Brkic, B. (2020). Application of Game-Based Learning in Cultural Heritage. *Proceedings of the 2nd International Workshop on Visual Pattern Extraction and Recognition for Cultural Heritage Understanding*.

De Luca, D., Ugliotti, F.M., (2021). *Digital meets Opera: an interactive service model towards accessibility and sustainability*. DN.

Drachen, A., & Canossa, A. (2009, September). Towards gameplay analysis via gameplay metrics. *Proceedings of the 13th international MindTrek conference: Everyday life in the ubiquitous era,* 202-209. 10.1145/1621841.1621878

Gartner. (2022). *Top Strategic Technology Trends for 2022*. Gartner.

Gupta, S., Kaushik, P., & Suma, D. (2015). Game Play Evaluation Metrics. *International Journal of Computers and Applications*, *117*(3), 32–35. doi:10.5120/20538-2902

Hennig-Thurau, T., Ravid, S. A., & Sorenson, O. (2021). The Economics of Filmed Entertainment in the Digital Era. *Journal of Cultural Economics*, *45*(2), 1–14. doi:10.100710824-021-09407-6

ISO. (2010). *ISO 9241-210:2010. Ergonomics of human-system interaction — Part 210: Human-centred design for interactive systems*.

Jones, D., Snider, C., Nassehi, A., Yon, J., & Hicks, B. (2020). Characterising the Digital Twin: A systematic literature review". *CIRP Journal of Manufacturing Science and Technology*, *29*, 36–52. doi:10.1016/j.cirpj.2020.02.002

Khan, I., Melro, A., Carla, A., & Oliveira, L. (2020). Systematic review on gamification and cultural heritage dissemination. *Journal of Digital Media & Interaction*, *3*(8), 19–41.

Krath, J., Schürmann, L., & Von Korflesch, H. F. (2021). Revealing the theoretical basis of gamification: A systematic review and analysis of theory in research on gamification, serious games and game-based learning. *Computers in Human Behavior, Vol.*, *125*(106963), 2–33. doi:10.1016/j.chb.2021.106963

Mariotti S. (2021). The Use of Serious Games as an Educational and Dissemination Tool for Archaeological Heritage Potential and Challenges for the Future. *Consolidation, 1.* . doi:10.30687/mag/2724-3923/2021/03/005

Mortara, M., Catalano, C. E., Bellotti, F., Fiucci, G., Houry-Panchetti, M., & Petridis, P. (2014). Learning cultural heritage by serious games. *Journal of Cultural Heritage, 15*(3), 318–325. doi:10.1016/j.culher.2013.04.004

NetConsulting Cube. (2020). *Mercato Digital Workspace In Italia, 2017-2022E.* Author.

Pagano, A., Palombini, A., Bozzelli, G., De Nino, M., Cerato, I., & Ricciardi, S. (2020). ArkaeVision VR Game: User Experience Research between Real and Virtual Paestum. *Applied Sciences (Basel, Switzerland), 2020*(10), 3182. doi:10.3390/app10093182

Papathanasiou-Zuhrt, D., Weiss-Ibanez, D. F., & Di Russo, A. (2017). The gamification of heritage in the unesco enlisted medieval town of Rhodes. GamiFIN, 60-70.

TechRepublic. (2020). *Gartner's top tech predictions for 2021.* CBS Interactive Inc.

UNI. (2017). *UNI 11337-4:2017. Edilizia e opere di ingegneria civile - Gestione digitale dei processi informativi delle costruzioni - Parte 4: Evoluzione e sviluppo informativo di modelli, elaborati e oggetti. Consturction and civil engineering works – Digital Management of Construction Information Processes. Part 4: Evolution and Information of Model.* Designs, and Objects.

UNI EN ISO. (2018). *ISO 9241-11:2018. Ergonomia dell'interazione uomo-sistema - Parte 11: Usabilità: Definizioni e concetti* [Ergonomics of Human-System Interaction – Part 11: Usability: Defintions and Concepts].

Werbach, K., & Hunter, D. (2015). *The gamification toolkit: dynamics, mechanics, and components for the win.* University of Pennsylvania Press.

KEY TERMS AND DEFINITIONS

Building Information Modelling: Is the graphic and alphanumeric transposition of the information that defines a building. The digital model can be related to other data domains and visualization techniques through three-dimensional and informative exportation.

Cyber Physical System: Is a computer system that processes data using a computer to communicate with the real world dynamically and directly the information system.

Digital Model: Is a computerized digital reproduction of an information model that describes the properties and shape of existing objects within simulated environments.

Gamification: Consists of game techniques to implement user involvement during the experience. Everything that does not start as a game application is transformed into a formative learning activity.

Interactive Digital System Model: Is the three-dimensional digitization of an object/component in which visual programming techniques intervene to define the interactions that the user can perform during the virtual experience. In this way, customizable services can be called up via web services and be easily accessible.

Interoperability: Is a process in which the three-dimensional and alphanumeric restitution is reproduced digitally. Once the characteristics of digital modelling have been defined, it is possible to define data exchange formats and procedures.

Serious Games: Are games designed for educational purposes, not primarily based on entertainment, but based on dynamic and fun education.

Virtual Reality: Is the reproduction of a real object within computerized environments where visors and devices connect the user in virtual environments with personalized interaction systems.

Chapter 20
Visualization and Fruition of Cultural Heritage in the Knowledge-Intensive Society:
New Paradigms of Interaction With Digital Replicas of Museum Objects, Drawings, and Manuscripts

Fabrizio Ivan Apollonio
Alma Mater Studiorum – University of Bologna, Italy

Marco Gaiani
Alma Mater Studiorum – University of Bologna, Italy

Simone Garagnani
https://orcid.org/0000-0002-9509-6564
Alma Mater Studiorum – University of Bologna, Italy

ABSTRACT

The knowledge-intensive society paradigm fosters relationships between technology and human actors with data, values, and knowledge that become mutual drivers for social innovation. The cultural heritage sector is naturally influenced by this vision, and museums and cultural institutions have a prominent role in dissemination of cultural values. This chapter focuses on a method developed to combine the power of the computer visualization technology with the cultural elements spread across collections, introducing some notes and remarks on how digital replicas of drawings, manuscripts, and museum objects can be successfully employed to spread knowledge. Through a custom application called ISLe, aimed at visualizing 3D models that accurately replicate the original items, some experiences in the production of digital replicas are introduced, highlighting opportunities and criticalities to be considered in the adoption of technology that can be potentially shared and exploited by many possible figures involved in cultural heritage.

DOI: 10.4018/978-1-6684-4854-0.ch020

INTRODUCTION

Over the last decades many digital technologies bound to the cultural dissemination improved, leading to brand new ways to make information accessible at different levels, enhancing how it can be shared among interested people. Key concepts like dynamic web applications (i.e., Web 2.0), eXtended and Augmented Reality (XR and AR), Artificial Intelligence (AI), virtual space and metaverse represent today intertwined terms, often combined to explain how technical expertise can be useful to share cross-sectional knowledge (Tenenbaum, 2006). However, this scenario is frequently too technology-centered, with most of the relevance dedicated to the Information and Communication Technologies (ICT) rather than to the cultural values they can convey. The many possible implications of themes belonging to tangible and intangible Cultural Heritage (CH), for example, have been more and more influenced by recent digital platforms and data schemes, often developed to shift the attention of users from the computer application *per se* to the extended fruition of the knowledge it can trigger. What was still not enough pervasive in the information society (the so-called *Society 4.0*), becomes now largely considered in a context where, in the words of Deguchi et al. (2020), relationships between technology and society contribute to describe a knowledge-intensive society, in which data, information, and knowledge are mutual drivers for social innovation, as the *Society 5.0* paradigm suggests (Fukuyama, 2018).

Following these premises, this chapter introduces the visualization outcomes of a more general and complete workflow, meant to produce a system able to replace, investigate, describe and communicate Cultural Heritage objects; these goals are reached through the generation of digital communicative artefacts, dedicated to fine drawings, objects belonging to museum collections and ancient manuscripts, following a research path that started more than ten years ago. The working pipeline that will be presented, travelling through the Society 5.0 model, follows a basic scheme made of data collected from the real world and processed by computer applications; it leads to output results applied back to the real world for an extended fruition by many different human users, though a custom visualization interface at the core of this chapter discussion. These processes are rooted on the assumption that all the elements that were investigated, even with apparently flat geometry, are actually 3D objects with specific behaviors in terms of color, light reflectance and shape.

With particular care to visitors in museums, art historians, scholars, conservators and restorers, who rely more and more on digital applications meant to visually explore and understand the characteristics of surfaces and materials belonging to many kinds of objects, a novel process to replicate and exhibit CH artefacts is detailed here with a focus on visualization purposes, whose main fruition directives are *realism*, requiring accurate shapes and surfaces representation, and *responsiveness*, since models have to change their appearance when directly manipulated by observers in museums or laboratories. The visualization framework evolved to display artefacts including a wide set of different features, from diverse materials to complex fabrication or drawing techniques and tools usages. Beginning from some experimentation on ancient drawings, in which the third dimension was reasonably less perceptible than the whole surface, the proposed system later proved its versatility in managing digital replicas originated from many input sources, surrogating the user experience on real 3D objects. More in general, this contribution focuses on this interactive manipulation, illustrating the paradigms and technical features behind the platform authored by the research team who introduced it, its methodological approach based on a consolidated scientific foundation and the final interface targeted to users with heterogeneous ergonomic needs.

After a proper section, referred to as *Introduction*, where main themes and cultural contexts are presented, the *Background* section expresses a wide state-of-the-art in scientific literature related to these topics, introducing known technologies and previous works on CH digitalization and cultural dissemination based on applications recalling the knowledge-intensive society precepts. The section titled *A comprehensive solution for the dissemination of CH* delves into the proposed approach, illustrating in detail modalities and mathematical references to properly replicate the appearance of CH artefacts. This section is split into two paragraphs: one titled *From digital acquisition to 3D models: the pipeline in ISLe (InSight Leonardo)*, in which the custom software replication framework is introduced, and *The visualization technique*, in which the final visualization stage is described, focusing on the adopted render techniques dedicated to exhibits in museums through an accurate and customized Real-Time Rendering (RTR) pipeline that was expressly created. The different steps to digitally replicate ancient drawings, manuscripts and 3D complex items are then culminating in the development of a visual environment inferred by the direct observation of many relevant case studies, as presented in the section titled *Some real applications*.

Five among the most well-known Leonardo da Vinci's drawings, covering in a comprehensive way his entire activity as a draftsman, some pages of an original manuscripts dedicated to Dante's Divina Commedia, handwritten in the XIV century, and a set of museum objects belonging to the Sistema Museale di Ateneo at the University of Bologna are mentioned to assess the outcomes of the visualization technique. These case studies where chosen according to specific criteria, since they individually bring very different features that guarantee the versatility of the techniques and the platform developed. These case studies are expression of an accurate analysis of drawing and writing techniques together with geometry, shape and materials of articulated objects, in which their peculiarities are digitally reproduced at the various scales, exploiting solutions that favor the accuracy of perceived reproduction instead of the numerical fidelity to the physical model.

Then, the *Future research* section opens to possible developments all over the proposed pipeline, while the *Conclusion* paragraph wraps all the illustrated topics and strategies up, in order to clarify once more the chapter contents.

BACKGROUND

In 2016, the *Japan Business Federation Keidanren*, a comprehensive economic organization whose mission is to support corporate activities, which contribute to the sustainable development of the Japanese economy and improvement in the quality of life (Keidanren, 2016), published the declaration "*Toward realization of the new economy and society - Reform of the economy and society*". The document established a new vision for further responsible development, introducing the term Society 5.0 as "*A human-centered society that balances economic advancement with the resolution of social problems by a system that highly integrates cyberspace and physical space*" (Keidanren, 2016, p. 5).

While Society 5.0 was born as a Japan's own growth strategy, it is not limited to this country, since its goals are the same as those expressed by the United Nations 2030 Agenda for the Sustainable Development Goals (SDGs), in order to face a number of challenges both as individuals and as communities. Together with the definition of new economic models, these challenges are related to a novel approach to CH since cultural contexts and capitals based on heritage have a significant impact on the social, economic, cultural and technological development of modern societies at individual, collective or institutional levels.

Furthermore, the documented growing number of museums in the whole world during the last 50 years (Negri and Marini, 2020) highlighted the importance of the cultural legacy they preserve. About 35,000 museums are today estimated to be operative all over the United States of America, while 40,000 are in Europe with almost 6,000 of them in Italy. Most of these institutions are basically small-to-medium museums (with less than 50,000 visitors per year) hosting mainly physical artefacts (such as works of art, sculptures, furnishing, building's decorative parts, stuffed animals, etc.). About the 84% of the European Museums, in fact, owns 3D man-made movable objects (Europeana, 2020) representing large part of the European Art Heritage.

Additionally, the COVID-19 pandemic also impacted the fruition of CH too. Lockdowns, which forced the closure of museums as well as many other activities, did not prevent the world of culture from finding alternative ways to guarantee the use of its assets, exploiting the possibilities offered by digital tools, following again the Society 5.0 route.

According to Burke et al. (Burke et al., 2020), the COVID-19 pandemic affected the cultural offers by museums forcing to try new ways to replicate the visit experience, from replacements making use of online platforms, to initiatives that envision radically different digital relationships to audiences. As Kist pointed out (Kist, 2020), the necessary isolation that took place suggested a novel way to understand the legacy of the past as shared in museums, especially those related to emotional historic contents who need a comprehensive strategy to disseminate also ethic messages through clear media.

These issues were nevertheless already a priority, as investigated by some funded research projects such as *Digital Media for Heritage: Refocusing Design from the Technology to the Visitor Experience* (DIME4HERITAGE, which was completed in 2015), Material EncounterS with digital Cultural Heritage (MESCH, Petrelli et al., 2014) or the still running Digital incubatOr fOR muSeums (DOORS, 2021), initiatives that investigated how to question and rethink digital media design practices for better experiences of visitors.

According to the foreword by the Directorate-General for Research and Innovation of the European Commission in "*Innovation in Cultural Heritage Research For an Integrated European Research Policy*" (European Commission Directorate, 2018), the "*Cultural Heritage is our bond with the past come to life in the present. It shapes our thinking and identity, our environment, and the places we live in. European cultural heritage is unique and diverse*". However, artistic or historical physical artefacts are not the only one expression of heritage, since museums evolved during the last decades in more articulated institutions that foster research and dissemination of values as well as the preservation of sites and regions with intangible cultural diversities. Some issues and criticalities affect this precious work of legacy conservation, spanning from the lack of consolidated standards proposed to collect cultural information, to the costly digital applications developed to manage collections from a commercial perspective, which are usually not affordable by small to medium institutions.

From the technical perspective, in scientific literature several papers have been written over the years about digital methods applied to the protection, documentation and understanding of humanity's shared CH, particularly in the architectural and archaeological fields, which historically count the majority of case studies related to the application of hardware and software, as they are commercially released. Since computers have become powerful enough to edit and process huge quantities of geometric and hyper textual data, the digital domain has been investigated in order to find more and more effective systems aimed at the knowledge management. Faithful digital surrogates are also a dynamic support today for the complex activities that imply the preservation and restoration of CH assets. Unfortunately, to be effective, they entail the expertise of many professional figures (restorers, architects/engineers, art historians, chemists, or photographers), who individually produce a massive amount of documentation for each activity in which they are involved. The introduction of 3D forms of representation improves significantly their workflows, since 3D digital models are a solid basis for future implementation and integration of data collected into vast, cognitive, spatial information system, which adopts a comprehensive and holistic perspective.

Nevertheless, the digitization of cultural assets is still a marginal practice, mainly due to economic and technical knowledge issues. In fact, after more than 40 years of dedicated campaigns, the 35% only of the European CH hosted by museums has been already digitized, while 27% of it is barely archived in Europeana, the web portal who provides cultural heritage enthusiasts, professionals, teachers, and researchers with digital access to European CH materials in form of artworks, books, music, and videos on art, newspapers, archaeology, fashion, science, sport, and much more (Europeana, 2020). In addition to these critical issues, a standard and shared policy for the digitization and classification of museum stocks and a common strategy aimed at the introduction of virtual 3D repositories for the CH life-cycle management are far from being consolidated.

However, over the last ten years, a wide number of studies were published on the application and use of Virtual Reality and Augmented Reality (VR/AR) in museums (Petrelli, 2019), the development of virtual exhibitions (Pierdicca et al., 2015) and interactive experiences through 3D reconstructions (Berthelot et al., 2015; Fassi et al., 2016), the introduction of low-cost solutions for 3D interactive museum events (Pescarin et al., 2018) and their analysis (Agus et al., 2012). But many of these studies cover specific features and casual experiences, spanning from laser scanning techniques to more invasive ones. For instance, in analyzing drawings and manuscripts, colors and artists' techniques are usually captured, as for paintings, using multispectral 2D images such as X-rays, UV fluorescence, and IR reflectography (Remondino et al., 2011; Elias et al., 2011). These methods, however, do not match the features belonging to the artifacts. On the one hand they are not able to capture properly important aspects of the drawings or manuscript, on the other hand, they return redundant data such as repeated sampling of the number of shades, where the range of colors in the drawings is limited compared to paintings. Among those methods, the ones dealing with object-safe digitization processes, employing photographic technical solutions from the acquisition to the spatial visualization of massive quantities of artifacts, are the most promising. Also, hybrid techniques were introduced to reproduce accurately CH 3D artefacts, such as the *Fraunhofer Cultlab3D* (CultLab3D - Fraunhofer IGD) or *Witikon* (Witikon), both exploiting laser scanning and/or photogrammetric workflows, even though they seem to be impractical for small to medium museums, who need huge economic investments in human specialization, spaces and dedicated equipment. Low-cost open-source solutions to acquire, visualize and promote CH objects, have been developed as well (Menna et al., 2017), but they are usually too specific and they require either very specialized hardware or a huge customization effort covering the peculiarities of the objects to document or the peculiar museum policies. Due to these reasons, it is necessary to foster an integrated approach to the problem and, as underlined by previous research, supplying non-experts working in museum area with robust, easy-to-use workflows based on low-cost widespread devices for the study, preservation, dissemination and restoration (Gaiani et al., 2019).

This chapter focuses mostly on this strategy, concentrating on acquisition and visualization of ancient fine drawings, manuscripts and physical items using digital media, to address both the needs of scholars and museums visitors. This wide range of users requires the understanding and perception of the object as it is in its real domain, following a common generalization related to the perception of lightness and color in scenes approximating the three-dimensional environment and materials surface properties, including glossiness or roughness. Some efficient answers to the lightness and color digitization problem for CH objects come from the single-camera, multi-light techniques. Among these, the most common is surely the Reflectance Transformation Imaging (RTI), a per-pixel function fitting technique that interactively displays objects under variable light conditions, as originally introduced by Tom Malzbender in form of Polynomial Texture Maps (PTM), an extension of the conventional texture map with greater control over the rendered appearance (Malzbender et al., 2008). Pledging photo-realistic objects relighting, RTI also recovers surface characteristics, considering orthographic cameras and light sources at infinite distance. The original RTI solution, developed by *Cultural Heritage Imaging* (Mudge et al., 2008), and later improved by ISTI CNR (Ponchio et al., 2018), consists of flexibility, low-cost application and ease of use: in fact, a point light source, a camera, and a reflective sphere are enough to capture RTI images. There are downsides, however, often generated by a process that is generally slow, time consuming and which needs complex devices to output accurate results, such as light domes or similar. Thus, RTI may result in inaccurate color reproduction due to the difficulties arising in evaluating light sources while it usually allows observation from a single predefined viewpoint only preventing the observer to explore in the same viewer the front and back of drawings and manuscripts. RTI does not follow physical rules to replicate surface material characteristics overlooking effects like self-shadowing, interreflections, and complex light-matter interactions.

A totally different class of solutions to improve the perceived quality of lightness and colors of digital replicas exploits ultra-high-resolution techniques, such as the web-based viewer developed by the *Centre de Recherche et de Restauration des Musées de France* based on the open-source software *IIPImage* (Pitzalis et al., 2009), or the viewer by *Google Arts & Culture*. The principle behind these systems is to exceed the maximum resolution perceived by the human eye, which is known to isolate 0.2-0.3 mm at 20-25 cm, showing details focused on accurate qualitative visual assessment and quantitative measures on works of art. However, the lack of tridimensional geometry limits the visualization of all those elements still visible with the naked eye but not responsible for the global shape definition of the model, often stored in large files hard to manage on consumer devices by unskilled operators. Moreover, surfaces are assumed as matte with a uniform albedo, preventing the observation of the apparent color, one of the main features to correctly classify a material (e.g., paper or parchment; chalk or charcoal). The second generalization is meant to store and visualize the whole optical properties of materials to be reproduced (color, surface texture, translucency, gloss). This implies the acquisition, representation, and visualization of the *Bidirectional Reflectance Distribution Functions* (BRDF, Nicodemus, 1965), a mathematical function that describes the real light reflection in a quantitative way, considering the entire hemisphere surrounding the object. A more detailed introduction of the BRDF theory and its applications can be found in Guarnera et al. (2016).

Many attempts to author correct BRDF models in the CH field were promoted in the last twenty years, introducing a highly accurate radiometric reproduction, or a series of photo textures taken from different angles. Today these solutions were framed in mature systems, such as the already mentioned *Fraunhofer Cultlab3D*, which applies complete 3D workflows from acquisition to accurate visualization of CH artefacts; nevertheless, they are not completely safe for drawings and manuscripts in the digitization phase and sometimes impractical to use where objects to be digitized are preserved. Also, many scientific approaches have been proposed to capture the angular appearance variation of fine art surfaces (Tominaga and Tanaka, 2008), usually represented by a *Spatially Varying Bidirectional Reflection Distribution Function* (SVBRDF): this mathematical model describes the existing relationship between incoming irradiance and the reflected radiance, considering every point on a surface. Several approaches employ sparse sampling using point light sources (Hasegawa et al., 2010) however, the specular reflectance is not separately modeled or becomes noisy. As a solution, surfaces are sorted into material groups, and the angular measurements are combined within each group and shared across the spatial domain (Wang et al., 2008). An efficient solution to estimate SVBRDF (including translucency), suitable for drawings and manuscripts, is illustrated in Gardner et al. (2003).

The *Portable Light Dome* system (PLD), developed at KU Leuven (Watteeuw et al., 2016), is another possible solution successfully experimented for archival documents (Vandermeulen et al., 2018), designed to support scholars and restorers in the identification of materials to monitor their changes and decay. The system captures and models the BRDF of non-Lambertian surfaces, using reflectance maps as input to model lower-dimensional analytic or tabular BRDFs, and it aims to show in *Real-Time Rendering* (RTR) virtual relighting and enhancements. However, the software solution associated with the PLD shows different spectra from IR to UV, extracting surface properties and reflectance distribution (reflectance maps) using per pixel photometric stereo techniques (Woodham, 1980). The PLD, as the RTI, is a single-camera, multi-light technique.

The *Four Light Total Appearance Imaging of Paintings* (Cox and Berns, 2015; Berns and Tongbo, 2012) is another interesting application that uses photometric stereo and CG techniques to virtually reproduce the shape, the color, the macrostructure, and the microstructure of the artwork. This solution needs only the typical equipment made of a digital camera and some strobes to capture the diffuse color and macrostructure of paintings (estimating normal vectors to surfaces). Finally, the *Tangible Imaging System* (Ferwenda and Darling, 2013) is worth to be mentioned, since it is based on a couple of fundamental assumptions: digitized objects must be realistic, representing the shapes and material properties of surfaces with accurate color and gloss reproduction, and the visualization system must be responsive, letting the user change point of views with direct manipulation.

A COMPREHENSIVE SOLUTION FOR THE DISSEMINATION OF CH

From Digital Acquisition to 3D Models: The Pipeline in *ISLe* (InSight Leonardo)

Beginning from the premises expressed in the previous sections of this chapter, a comprehensive solution targeted to the digitization first and visualization later of manuscripts, ancient drawings and museum objects was developed in order to preserve original artefacts, whose nature is often delicate and fragile. The complete system introduced was initially developed to replace, investigate, describe and communicate ancient fine drawings authored by Leonardo da Vinci, so it was named *ISLe* (*InSightLeonardo*). The system was later applied to different case studies such as medieval manuscripts and objects, as it will be later introduced. *ISLe* consists of five intertwined sub-systems (Figure 1):

1. a LED-based lighting system that avoids typical problems in fluorescent illuminators that prevent the acquisition of reflected information at certain light wavelengths (Gaiani at al., 2020);
2. an accurate and safe on-site 48-bit color IMAGES capture, supported by a precise fully automated Color Correction (CC) from RAW images based on SHAFT (*SAT & HUE Adaptive Fine Tuning*) software (Gaiani and Ballabeni, 2018);
3. a solution developed to replicate the original surface with micro and macroscopic fidelity (Gaiani et al, 2020);

4. a solution to accurately visualize the communication artifact using a low-cost rendering engine;
5. a kiosk visualization interface for museums visitors based on usual touch gestures.

Figure 1. The five different sub-systems that define the ISLe solution

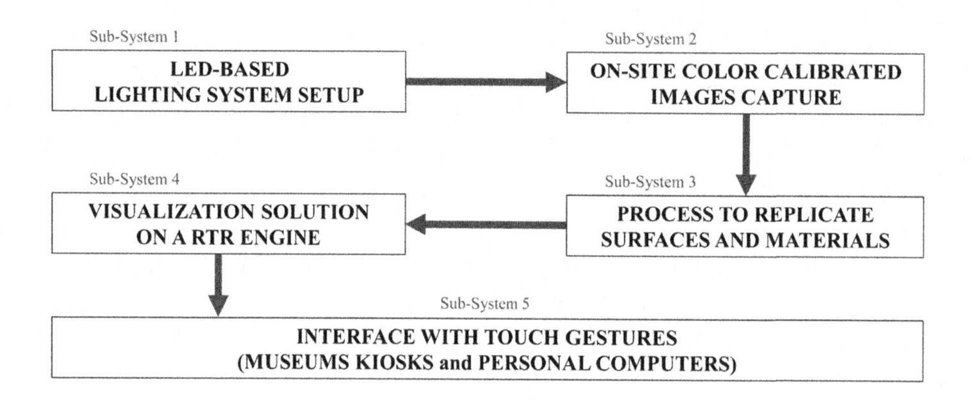

While RAW images were privileged to minimize the artifacts generation by proprietary software embedded into commercial cameras, particular LED lights were chosen to replicate mostly all the frequencies in natural light (Apollonio et al., 2020). The illuminant was supplied by a series of Relio² devices (Relio), a very small lamp (35×35×35 mm) emitting continuous spectrum light at 4000°K Correlated Color Temperature (CCT), 40000 lux at 0.25 m in brightness and a Color Rendering Index (CRI) > 95% with high color reliability on all wavelengths. These LED lights also avoid excessive emission of heat and harmful UV and IR radiation. The first sub-system, in fact, is meant to be less harmful as possible to CH objects, and it relies on four groups of independent Relio² LED lights positioned at a proper 45° angle. Ishii et al. (2008) compared the color degradation due to white LEDs and fluorescent lamps, discovering that the former LEDs generate more limited saturation losses compared to the latter. Piccablotto et al. (2015) evaluated the discoloration generated by white LEDs too at different CCTs compared to a traditional halogen lamp. The results obtained highlighted that, in general, white LEDs are less harmful than traditional halogen lamps.

The second sub-system requires five pictures of the original to reconstruct and digitally restore the three-dimensionality of objects by reproducing their formal quality and surface reflectance. The CC technique, applied with pre-measured spectral color targets (Wandell et Farrell, 1993), translates the camera's sensor responses into a device-independent colorimetric representation, to get also an accurate color representation. The reference target, as a common solution, usually consists in the X-Rite ColorChecker Classic (McCamy et al., 1976), which shows 24 standardized patches with known reflectance. The patches are organized in a 4 by 6 array, with 18 familiar colors that include the representation of true natural colors (such as skin, foliage and sky), additive and subtractive primary colors, and six grayscale levels with optical densities from 0.05 to 1.50 and a range of 4.8 f-stops (EV), at a gamma of 2.2.

The third sub-system aims at modeling scale-dependent light phenomena on the 3D morphology generated before, following a classification by Westin et al. (1992), in which geometric structures are divided into three different levels (Figure 2):

- a *macrostructure*, which describes the shape and the geometry of the object;
- a *mesostructure*, where all elements still visible with the naked eye are represented but not responsible for the global shape definition of the model (for example small bumps);
- a *microstructure*, considering the microscopical structure not visible to the human eye that contributes to the final aspect of the object occluding or deviating light and projecting shadows and highlights.

Figure 2. Leonardo da Vinci, "Study of proportions of the human body known as The Vitruvian Man" (around 1490) and the different structures described in the 3D model: the global shape and color (a), the 3D geometry (b) and details invisible to the naked eye, in form of normal map (c)

The fourth subsystem is based on the faithful visualization of the 3D model into a dedicated virtual scene, in which materials and light behaviors of the original CH object are replicated in the digital domain, taking advantage of Real Time Rendering (RTR) software environment usually adopted in video game design, embedding replicated textures with color and reflectance with imperceptible difference to the expert observer, at a resolution of 50 μm.

The last sub-system, the fifth, consists of an adaptation of the traditional multitouch interaction paradigm to fit the exploration of 2D or 3D contents to minimize uncommon gestures and let user explore the virtual object from different points of view, under different light conditions (Figure 3) and on different devices.

Figure 3. The ISLe 3D model of the drawing "Two mortars launching explosives", by Leonardo da Vinci (around 1485), many details can be explored this way, under different lights: at a 45° angle (a), and at a 20° angle (b)

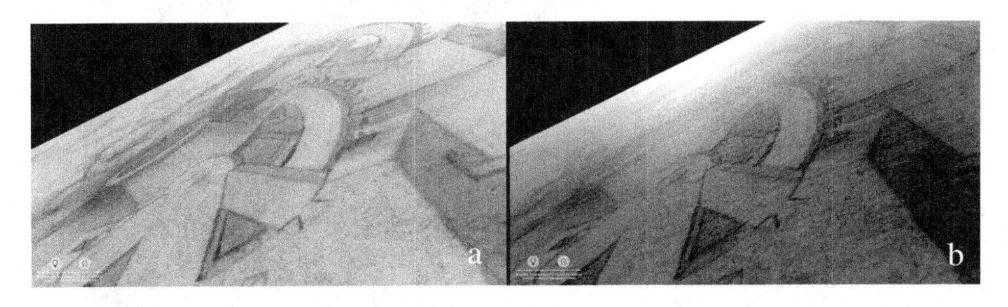

Since this chapter is mainly focused on new paradigms of interaction with digital replicas of relevant museum objects, drawings and manuscripts, details concerning the acquisition stages, the photographic set, and the later processing techniques are referenced in already published previous works (Apollonio et al., 2021). Thus, in the next section, some deepening on the visualization technique is presented.

The Visualization Technique

The reproduction of visually realistic 3D models that can credibly replicate interactions between surfaces and the whole light spectrum, as implemented in *ISLe*, needs at least three different steps to be performed in ordered sequence:

Step 1. A fine reproduction of the appearance for the materials involved;
Step 2. An RTR visualization with accurate color reproduction on a full sRGB capable display, adopting an RTR engine running on multiple devices;
Step 3. The authoring of a visualization and navigation graphic interface based on common gestures.

Details on these steps are reported as follows.

The Appearance of Reproduced Materials

The first step is mainly dedicated to the definition of *macrostructure* and *mesostructure* features for the objects to be replicated, which are reproduced embedding the low-frequency geometric details into the final mesh and using a rendering approach for the high-frequency ones based on the application of normal maps.

Figure 4. The representation of paper's translucency and gall-ink's reflectivity in ISLe: a portion of the 3D model of the Fortress with a square plan, with very high scarp wall and concentric layout, with corner towers and grandiose ravelin in front, by Leonardo da Vinci (1507 or later)

Splitting high-frequency from the low-frequency geometric details, a better geometry estimation including features such as self-shadowing or parallax deformation at macrostructure is possible. The behavior proper of microstructure is addressed developing a superset of many shaders able to correctly reproduce the light-matter interaction of a lot of materials which the museum objects are made of. This way, microstructure behavior is usually addressed at shader level.

Among the many existing solutions, the Torrance-Sparrow theory for off-specular reflections on micro-faceted surfaces (Torrance and Sparrow, 1967) allows to model the BSDF of a very large range of materials and its improved version by Burley allows better quality and simplest implementation in the RTR engines (Burley, 2015). These differentiations proved to be extremely effective in the definition of visually accurate models (Figure 4).

Over the last decade, there have been significant progresses in real-time visualization software design (Akenine-Möller, 2018), and many applications were developed upon existing graphic game engines (Eberly, 2015; Salama and ElSayed, 2018, Zarrad, 2018). *ISLe* followed this path, after extensive investigations on existing software implementations; a custom-made render template was developed into the *High Definition Render Pipeline* (HDRP), a rendering framework tailored on C# scripts and easily customizable on top of Unity 3D graphic engine (Unity3D).

Unity is a videogame development platform, originally released in 2005, which consists of a graphical user interface (GUI) and an advanced game render engine. A large community of developers uses this piece of software to author interactive simulations, ranging from small mobile and browser-based applications to high-budget console games and AR/VR experiences.

An RTR Visualization Running on Multiple Devices

The second step is related to the fine tuning of a dedicated RTR virtual environment that can successfully display 3D models, their shading and the complexity of materials associated to them. In fact, the proper modulation of descriptive maps for diffusive colors, normals, height and glossiness, can be implemented into a multi-texture script defined as *shader*. It is indeed an algorithm that estimates the appropriate levels of light, darkness, reflectivity and color when applied to 3D surfaces of a rendered scene. Many software frameworks support multitexture shaders, but those supporting cross-platform deployment and visualization are preferred, due to their use also by non-expert operators on common devices such as smartphones or tablets. Other features like low-cost deployment, high render quality and the ability to allow an easy, customized exploration of artefacts at many scales are relevant, since these are the typical requirements by visitors in museums, art scholars and restorers in the field of CH.

In order to improve the output quality with the multi-texture approach some custom shaders in form of specific code plugs can be written to mimic the behaviors to light scattering of parchments and inks. These shaders must however ensure a precise conformity to:

- *light*, with its actual spectral composition in rendered drawings, to replicate it in an accurate light distribution model,
- *color*, with an accurate simulation to mimic material color appearance under different light directions,
- *surface*, with precise replica at different scales of the paper's surface with custom implementation of algorithms to simulate the physical behaviors of superficial scattering light.

ISLe is not limited to a simple rendering view, but it provides the ability to freely visualize the fine details of the surface at any observation angle and under different kind of lights instead. Users can zoom in on high-resolution images to evaluate characteristics of a view, edit the light behavior, explore various parts of the drawings or manuscripts with different shaders or lighting techniques, to isolate details, notes or remarks (Figure 5).

Figure 5. A deep exploration in ISLe of the drawing "Landscape" (1473), by Leonardo da Vinci, in which small details can be analyzed zooming close to the 3D replica

As introduced in the *Background* section of this chapter, color, surface texture, translucency, and gloss are essentially a result of a proper acquisition, representation, and visualization of the *Bidirectional Reflectance Distribution Function.* A detailed dissertation on the BRDF theory and its applications can be found again in Guarnera et al. (2016). The *ISLe* visualization system extended this paradigm, since the materials developed for the 3D models were based on two assumptions:

- the effect of the subsurface dispersion, which is modeled extending the BRDF to the *Bidirectional Scattering Distribution Function* (BSDF), a quantity that consists of the sum of the BRDF and the *Bidirectional Transmittance Distribution Function* (BTDF). The latter function expresses how light passes through a (semi)transparent surface;
- the interactions between lights and materials are considered as scale-dependent phenomena, to better fit the features and behavior of the RTR engine.

The formulation of the BSDF model, used for example in the paper or parchment reproduction, is a superset of the BSDF formulation adapted for other materials with minimal errors. Papas et al. (2004) introduced an accurate model of paper's BRDF, with the formulation of a BSDF where transmission and diffusion of light are modeled from observations on dispersity (Bartell et al., 1980). The paper is an optically thick material that entails light behaviors given by the combination of many effects: subsurface dispersion, mirror reflection, retro-reflection, surface glossiness, and transmission. This formulation proved to be effective also to reproduce the paper used by Leonardo, as presented in the next section. This way, paper behavior in *ISLe* is simulated using a BSDF representation exploiting a reduced version of the multi/layered model of Donner and Jensen (2005), the single dispersion theory model (Hanrahan and Krueger, 1993) and the microfacets model for surface reflections and refractions, as in Walter et al. (2007). These features are embedded into the material's shader code.

More in detail, if superficial reflections scatter from the light incoming vector l and the outgoing direction view v, as the microfacet model postulates, a small portion of the surface (a *microfacet*) with an equiangular normal direction between l and v must exist. This 'half-vector' is the micro-surface normal and it is expressed as $h = l + v$.

The general mathematical formulation of a micro-faceted model for isotropic materials is represented by:

$$f(l,v) = \frac{c_{diff}}{\pi} + \frac{F(l,h)G(l,v,h)D(h)}{4(n \cdot l)(n \cdot v)}$$

where:

- c_{diff} = light portion diffused by albedo;
- $F(l,h)$= Fresnel reflectance; a term that calculates the fraction of light reflected by an optically flat surface. Its value depends on two factors: the angle between the light vector and the normal surface and the refractive index of the material. Since the refractive index can vary over the entire visible spectrum, the Fresnel reflectance is a spectral quantity;
- $G(l,v,h)$ = statistical amount of microfacets occluded or shaded by the surface shape when hit by light along the l direction toward the v view vector, considering the h normal for every facet. Our approach to mimic paper's micro facets considers also inter-reflections;
- $D(h)$ = normal distribution function for every microfacet; $D(h)$ expresses the number of microfacets with normal equal to h and determines the size, brightness, and shape of the specular illumination;
- $4(n \cdot l)(n \cdot v)$ = correction factor that considers quantities that must be transformed between the local space of the microfacets and the global one of the entire surface.

However, as proved again by Papas et al. (2014) the Fresnel equation does not seem to be an appropriate solution for rough surfaces like matte paper. Starting from measurements, they found that the transmission is correlated to the direction of the incident light, to the refractive index and to the roughness. To consider this phenomenon, an attenuation function was introduced. To simplify the computational process, a BSDF implementation provided by the Walt Disney Animation Studios (Burley, 2012) can be successfully adopted, as it was tested in *ISLe*. The Fresnel refraction calculation is then done twice, both for incoming and outgoing light that scatters from the paper's surface, to preserve Helmholtz's principle of reciprocity. Then, the Schlick approximation (Schlick, 1994) is introduced, to expresses the grazing retro-reflection response, using a value determined by the material roughness. The error introduced by this approximation is significantly smaller than that due to other factors, improving the overall original non-natural response to grazing light for opaque paper. To properly model the subsurface scattering in *ISLe*, the diffuse lobe was refactored evaluating directional microsurface effects and non-directional subsurface effects, and the Lambertian portion of the diffuse lobe was replaced with a diffusion model combined with a volumetric dispersion model. This technique preserves the effects of the microsurface, letting the diffusion model converge to the same result as the diffused BRDF when its dispersion distance is small. The diffuse reflection model was then refactored as in the solution proposed again by Burley (2016), to allow the light exchange of the Lambertian diffusion with the subsurface scattering, while keeping the Fresnel factors for rough surfaces and retro-reflective gains. Along the *ISLe* visualization pipeline $G(l,v,h)$ is evaluated as in the Smith's solution (again in Walter et al. 2007), while the multi-dispersion anisotropic GGX written in Heitz's (2014) formulation is used to compute $D(h)$.

This rigorous mathematical scheme frames the features of the basic shader that was customized to visualize materials in *ISLe*, applying principles inferred by *Physically Based Rendering* (PBR), to support both an accurate light transport and the Torrance-Sparrow theory for off-specular reflections on micro-faceted surfaces (Torrance and Sparrow, 1967) in the modified version described before.

Thus, the key for credibility of CH objects resides in the proper definition of these set of diversified features, which were carefully processed following again the less invasive photographic approach. To get diffuse and normal maps for the general shader, for example, a custom application was specifically written, called *nLights*.

nLights exploits the simple stationery zenithal acquisition of 9 images with 4 LED lights placed at 45° and 4 LED lights placed at 20°. This allows a correct evaluation of the specular component. The application is actually designed to output several maps for the visualization shader, starting from the analysis of the incident light direction, usually calculated from pictures representing a black reflective sphere in the same light conditions of the manuscript or drawing to be acquired. *nLights* this way can produce autonomously:

- an *albedo map*, estimating lighting matrix by solving a nonlinear least squares problem;
- a *normal map*, inferred by the different light directions to define the materials microstructure;
- a *depth map* recovered from estimated normal vector field;
- a *reflection map*, important for metallic materials such as gold leaf on manuscripts;
- a *geometric 3D representation* of the manuscript shape, in form of STL or OBJ file, to define the macrostructure.

A Visualization GUI Based on Common Gestures.

The general-purpose approach illustrated so far needs a final third step dedicated to the visualization on a variety of software platforms, making again the Unity3D engine an ideal development platform for CH visualization purposes, since an effective software need to be scalable and versatile in order to meet both museums and scholars' requirements. The custom HDRP pipeline collects shaders including the replication features, then translates them into runtimes to be run on different 3D graphics APIs (D3D11, D3D12, Metal, and Vulkan) letting *ISLe* work on Microsoft Windows, Apple Os X, and mobile devices with Google Android or Apple iOS.

However, this multitude of platforms is bound to a central *Graphical User Interface* (GUI) that was designed with versatility as a major feature in mind. The GUI was developed in Unity 3D in form of a digital canvas overlaid on 3D model views. Specific touch buttons were layered on this transparent canvas to avoid placing them too distant from the bottom side of the screen, close to their effective position on the 55 inches touchscreen, final destination of the GUI framework.

These buttons let the users navigate the model moving back-and-forth to specific positions on the drawings, manuscripts or objects, where details are framed by a dynamic camera with commentaries that pop up on the screen to describe in detail the peculiarities in those views. By touching lightly other buttons, users may rotate the drawings to see their back face (*verso*), for instance, or they can dynamically change the light sources, switching from diffuse lights to raked ones, so that the model's shaders can perform in a realistic way.

Another touch button activates an automatic guided tour, during which the most important details in the models are selected and automatically visualized in sequence. A dedicated button turns the free navigation environment on, so visitors can freely rotate, pan, enlarge and switch lights on the 3D model, as if it were in their hands, swiping fingers on it. Some care was also dedicated to the robustness of the GUI, guaranteed by some customized C# scripts written to avoid multiple buttons selections or to program the application to restart from the beginning if no activity is detected within a timed interval. But the versatility of the application is also related to its ability to be targeted to scholars: in fact, some specific commands may be executed also via traditional devices. For example, pressing the space bar of a connected keyboard, the interface will change again getting out from the kiosk mode to let the drawings or manuscripts be explored with the use of a common pointer device, i.e. a mouse, connected to a laptop computer or a desktop one. The video resolution, usually set to 4K during exhibitions, is also automatically set to the maximum values that can be guaranteed by the hardware in desktop mode. This way, the application conforms to many possible uses.

SOME REAL APPLICATIONS

Since 2015, the visualization system equipping the *ISLe* system was continuously improved; its visualization system in particular, was also tested on many case studies to estimate its versatility on different objects belonging to diverse collections in many museums and cultural institutions. In detail, the ancient drawings and manuscripts investigated and visualized during exhibitions were:

1. Leonardo da Vinci, *Landscape*, 1473, recto: pen and iron-gall inks, lead point, blind point on paper; verso: pen, iron-gall inks, black, and red chalk, lead point, blind point on paper, 194 x 285 mm, Florence, Le Gallerie degli Uffizi, GDSU, inv. 8P.
2. Leonardo da Vinci, *Study of various buildings in perspective* (study for the background of the *Adoration of the Magi*), around 1481, metal point, reworked with Pen and iron-gall ink, diluted iron-gall brush and ink, partially oxidized white gouache highlights (basic lead carbonate), stylus and compass on light brown prepared paper, 164 x 290 mm, Florence, Le Gallerie degli Uffizi, GDSU, inv. 436 E.
3. Leonardo da Vinci, *Study of proportions of the human body known as The Vitruvian Man*, around 1490, metal point, pen, and iron-gall ink, watercolor ink touches, stylus on white paper, 345 x 246 mm, Venice, Gallerie dell'Accademia, Gabinetto dei Disegni e delle Stampe, inv. 228.
4. Leonardo da Vinci, *Two mortars launching explosives*, around 1485 (or shortly after), traces of black pencil (?), stylus tip, pen, and iron-gall ink, diluted ink and watercolor with reworking on the right side, 219 x 410 mm, Milan, Veneranda Biblioteca Ambrosiana, Codex Atlanticus, f. 33.

5. Leonardo da Vinci, *Fortress with a square plan, with very high scarp wall and concentric layout, with corner towers and grandiose ravelin in front*, 1507 or later, pen and iron-gall ink on black pencil, 131-207 x 436 mm, Milan, Veneranda Biblioteca Ambrosiana, Codex Atlanticus, f. 117.
6. Manuscript n.589, *Divina Commedia con chiose in volgare* (Dante Lambertino), Biblioteca Universitaria di Bologna, Alma Mater Studiorum - University of Bologna, page 1.
7. Manuscript n.589, *Divina Commedia con chiose in volgare* (Dante Lambertino), Biblioteca Universitaria di Bologna, Alma Mater Studiorum - University of Bologna, page 69.
8. Manuscript n.589, *Divina Commedia con chiose in volgare* (Dante Lambertino), Biblioteca Universitaria di Bologna, Alma Mater Studiorum - University of Bologna, page 137.

Coming to drawings by Leonardo, *ISLe* aims to spread knowledge and culture by reproducing and analytically showing the three-dimensionality and features of the Master's graphic signs, starting from originals defined in analog form on paper. *ISLe* helped in generating 3D digital reconstructions and visualize them in high-quality RTR with an imperceptible difference compared to the originals. The whole system proved to be efficient following two paradigms: '*drawings explored as they were in your hands*', and '*models showing what you cannot see'*. The visualization platform customized to reach these goals was tested indeed during many real exhibitions such as "*Leonardo in Vinci, at the origin of the genius*", open at Museo Leonardiano in Vinci (Italy) and visited by more than 140.000 people, "*Leonardo, drawing anatomy*", at Museum of Palazzo Poggi in Bologna (Italy), with more than 12.000 people attending and more than 100.000 when the same event lately moved to Vinci. This remarkable amount of museum visitors and their collected feedbacks on the application interface proved how the platform was successful in narrating at different levels which details are behind these noticeable and famous artefacts. In fact, to collect any suggestions for further developments, an evaluation by exhibitions visitors, together with an evaluation of the *ISLe* quality by professional operators (art historians, conservators, restorers, museum services) were requested through questionnaires. The feedback indicated a high level of satisfaction regarding the qualitative and perceptive characteristics of the digital artifact reproduced, with great benefits from the comparative analysis, in the study of drawings up to the conservation and documentation of artifacts.

These important exhibitions were systematically considered in improving the Graphical User Interface that was developed for the visualization stage of the platform, which had to be robust enough in order to guarantee a constant level of interaction for all visitors and users. Since this platform can also be used by scholars, restorers or simple museum visitors, it required to be efficiently implemented over a standard GPU-accelerated RTR pipeline. This requirement, together with the challenges and the adopted solutions to develop a user-friendly, multi-purpose environment, lately ported to different final fruition platforms, from wide 4K touch screens to VR mobile goggles or more sophisticated devices such as Oculus Rift and Oculus Quest (Figure 6).

In these drawings, the custom shader affecting the 3D models was customized to reproduce iron-gall inks, paper and different signs left by silver points, black and red chalks and white gouache.

During the exhibit "*Dall'Alma Mater al Mondo. Dante at the University of Bologna*", held in 2021 at the Biblioteca Universitaria di Bologna (BUB), the visualization kiosk application in *ISLe* was used again to exhibit the manuscript no. 589, *Divina Commedia con chiose in volgare (Dante Lambertino),* written in the XIV century and originally collected in 1755 by the Library of the Institute of Sciences together with the entire library belonged to Pope Benedict XIV, born Prospero Lambertini (1675-1758), hence the name of Codex Lambertino (Figure 7).

Figure 6. The ISLe visualization system running on a 55 inches wide touch screen during the exhibition "Dall'Alma Mater al mondo, Dante all'Università di Bologna": the 3D model is representing the Manuscript n.589

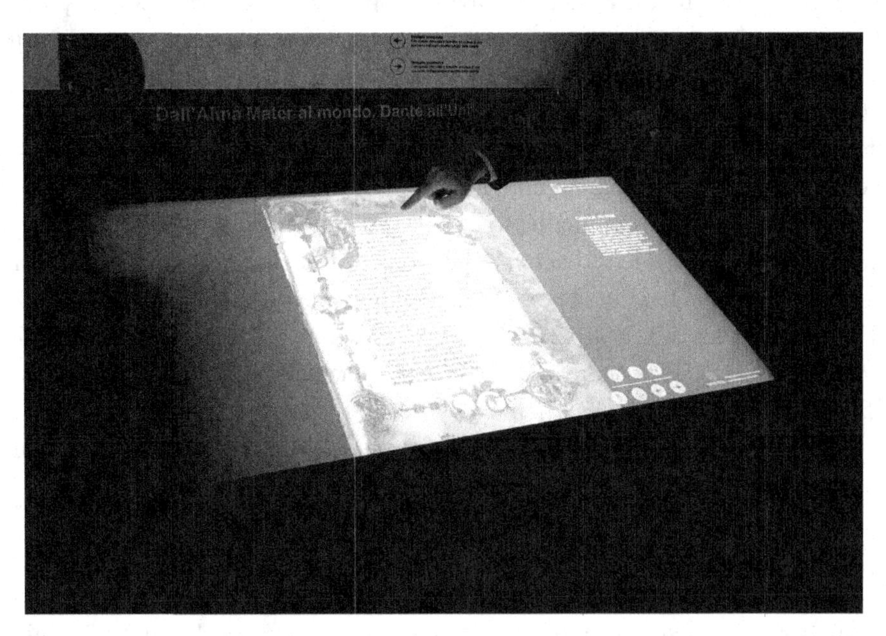

Figure 7. The digital replica of the Manuscript no. 589, page 137, with the custom shader in ISLe's GUI, representing all the features of the parchment, inks and gold leaf

The manuscript measures mm 273x187 and it is made of parchment, written with iron gallic ink and it includes many miniatures. Three pages were digitized (pages 1, 69 and 137), as they represent the initial verses of Hell, Purgatory and Heaven. Decorations and initial letters aside of the digitized pages were featuring gold leaf layers. This material acquires metal behaviors, as the reflected energy is different from the parchment's one. The shader included a further map to replicate this feature with high albedo, which means the surface reflects most of the radiation that hits it, in a metallic effect. To display with significant realism this kind of behavior proper of gilding, at first the specular properties and the material portions of the manuscript where the gold leaf was layered were extracted exploiting again the photometric stereo technique.

To successfully replicate this gold leaf portions, the reflection map generated in *nLights* was filtered to isolate all the pixels resulting in a high albedo value. The parchment, the ink's albedo and the specular gold components could be separated, generating a mask map to be later used in the production of the visualization shader in Unity 3D. Finally, to render the iron gallic ink and the gold leaf, maps with dedicated alpha channels were added to activate portions of the page influenced by the Fresnel effect and the filtered specular maps. During the exhibit in Bologna the digitized pages run on three different, 55 inches wide touch screens, at a 3840x2160 px max resolution, connected to workstations equipped with Core i7 9700 at 3 GHz CPUs and 8 GB RAM, implemented with a graphic card ASUS Phoenix Radeon RX 550 and GDDR5 4GB video memory.

Also, more complex objects were studied and represented through the presented approach (Figure 8), including a heterogeneous set of cases such as:

1. A *Porcupinefish* (Diodon Antennatus) undergone to complete taxidermy treatment: bounding box 35 × 19 × 25 cm wide, highly specular skin and tiny details.
2. A *Globe by astronomer Horn d'Arturo*: bounding box 31 × 31 × 46 cm wide, highly reflective regular surface with a considerable Fresnel effect.
3. A *bust of the scientist, military, geologist Luigi Ferdinando Marsili* (1658–1730): bounding box 41 × 67 × 99 cm wide, made in Carrara marble a highly translucent, non-homogeneous material at the scale of the measurement process.
4. A *statue representing Hercules* (bounding box 100 × 90 × 275 cm wide, made in sandstone).

Figure 8. The four specimens belonging to the Sistema Museale di Ateneo (SMA, Bologna, Italy) at the University of Bologna: from left, a Porcupinefish, a Globe by astronomer Horn d'Arturo, the bust of Luigi Ferdinando Marsili and, to the right, statue representing Hercules

The digitized 3D models of these specimens, which were still not exhibited, were an important test field for the replication of complex geometries. The originals were chosen from the Sistema Museale di Ateneo (SMA, Bologna, Italy) of the University of Bologna hosted inside Palazzo Poggi. The museum hosts various collections dedicated to geography and nautical sciences, military architecture collected by Luigi Ferdinando Marsili, physics, natural history, chemistry, human anatomy, along with the collection of fossils and dissected animals gathered by the naturalist Ulisse Aldrovandi. For these case studies and how they were digitized using smartphones after a precise color calibration to get faithful texture maps, please refer to Apollonio et al. (2021).

FUTURE RESEARCH DIRECTIONS

After a careful analysis of the results coming from some assessment surveys that were carried out through questionnaires during the mentioned exhibitions, the general appreciation of the *ISLe* visualization system revealed an estimation much higher than expected, both from museum visitors and scholars and restorers who tried it out. These gratifying outcomes prove how is more and more important the accurate dissemination of cultural assets to involve people in engaging activities to promote human-centered society, learning from the past. The whole *ISLe* framework pushes these principles also opening new perspectives. For example, the design of a cloud-based platform supporting the easy creation of a visualization context for any cultural object and able to offer the same functionalities presented in this chapter or, in other words, a framework offered to the users in an unattended manner, without prepared museum staff strongly involved in the customization of the RTR system, would be a further step in the development of a shared approach in the digitization of CH. This could also be part of a wider infrastructure supporting Heritage Sciences and could be reasonably interconnected to other existing services.

Figure 9. The Porcupinefish 3D model navigated on a Microsoft Surface Go tablet PC, which runs the ISLe visualization runtime

A future direction involves also the technical outlook, since the porting of the RTR pipeline on a more standard, open-source, and web-based environment would be more effective in creating dedicated systems for museums. This open approach would surely foster the data management for large collections, with a general perspective of huge number of artifacts easily digitized and visualized through a common platform in an automatic or semi-automatic manner.

CONCLUSION

In this chapter, some perspectives on how to spread knowledge and culture by reproducing and analytically visualize artifacts belonging to CH collections were introduced and substantiate through a workflow culminating in an application called *ISLe, InSight Leonardo*. 3D digital artefacts on the *ISLe* platform can be explored with high versatility on many devices, spanning from museums kiosks to mobile devices (Figure 9), revealing hidden details and telling the object's story, following a path of remarks and notes overlaid to the replicas of original drawings or manuscripts. In terms of economic efforts, the *ISLe* pipeline proved to be reasonably affordable for museums and cultural institutions: costs depend on devices adopted to digitally acquire the CH objects and on computers used to display in kiosk mode the replicas, with implications on visual quality that is scalable according to curators needs.

The digital artifacts, especially when visualized under different light sources by simply touching some buttons layered on the GUI, express a level of visual quality suitable both for a general audience (e.g., museum visitors that will experience the model through the mentioned RTR systems, that generally require low-poly 3D models), and for field experts (e.g., scholars that will definitely consult a high-poly version, which can be enabled by software according to the final device's hardware specifications).

Eventually, coming back to the *Society 5.0* paradigm with its three pillars of data, information and knowledge, the introduced platform is a novel way to spread and disseminate knowledge, translating data into information for all, considering information itself as data that got meanings by selecting, processing and visualizing knowledge for the particular purpose of Cultural Heritage dissemination.

However, the achieved results highlighted that further progresses are still possible and desirable on some unsolved issues. The extension of case studies to a wider set of samples, for instance, could lead to identify and address additional problems occurring with the consolidated devices and workflow; another possible research frontier is the complete automation of the proposed pipeline, through procedural algorithms and Deep Learning applications for derivative models construction, leading to a research that will hopefully broaden the horizon of the research in the field.

REFERENCES

Agus, M., Marton, F., Bettio, F., Hadwiger, M., & Gobbetti, E. (2018). Data-Driven Analysis of Virtual 3D Exploration of a Large Sculpture Collection in Real-World Museum Exhibitions. *J. Comput. Cult. Herit., 11*, 1–20.

Akenine-Möller, T., Haines, E., & Hoffman, N. (2018). Real-Time Rendering (4th ed.). Taylor & Francis. doi:10.1201/b22086

Apollonio, F. I., Fantini, F., Garagnani, S., & Gaiani, M. (2021). A Photogrammetry-Based Workflow for the Accurate 3D Construction and Visualization of Museums Assets. *Remote Sensing*, *13*(486), 1–40.

Apollonio, F. I., Foschi, R., Gaiani, M., & Garagnani, S. (2021). How to Analyze, Preserve, and Communicate Leonardo's Drawing? A Solution to Visualize in RTR Fine Art Graphics Established from "the Best Sense". *J. Comput. Cult. Herit.*, *14*(3), 1–30.

Apollonio, F. I., & Gaiani, M. (2020). Digitalizzare l'Uomo vitruviano. In P. Salvi, A. Mariani, & V. Rosa (Eds.), *Leonardo da Vinci e l'Accademia di Brera* (pp. 55–62). Silvana Editoriale.

Bartell, F. O., Dereniak, E. L., & Wolfe, W. L. (1980). The theory and measurement of bidirectional reflectance distribution function (BRDF) and bidirectional transmittance distribution function (BTDF). *Proceedings of the Society for Photo-Instrumentation Engineers*, *257*, 154–160. doi:10.1117/12.959611

Berns, R. S., & Tongbo, C. (2012). Updated Practical Total Appearance Imaging of Paintings. *Proceedings of the IS&T Archiving Conference*, 162-167.

Berthelot, M., Nony, N., Gugi, L., Bishop, A., & De Luca, L. (2015). The Avignon Bridge: A 3D reconstruction project integrating archaeological, historical and geomorphological issues. *The International Archives of the Photogrammetry, Remote Sensing and Spatial Information Sciences*, *40*(W4), 223–227. doi:10.5194/isprsarchives-XL-5-W4-223-2015

Burke, V., Jørgensen, D., Jørgensen, F. A. (2020). Museums at Home: Digital Initiatives in Response to COVID-19. *Norsk museumstidsskrift*, *6*(2), 117-123.

Burley, B. (2012). Physically-based shading at Disney, course notes. In SIGGRAPH '12 Courses. ACM.

Burley, B. (2015). Extending Disney's Physically Based BRDF with Integrated Subsurface Scattering. In SIGGRAPH'15 Courses. ACM.

Cox, B. D., & Berns, R. S. (2015). Imaging artwork in a studio environment for computer graphics rendering. Proceeding of SPIE, 939803.

CultLab3D. (n.d.). https://www.cultlab3d.de/

Deguchi, A., Hirai, C., Matsuoka, H., Nakano, T., Oshima, K., Tai, M., & Tani, S. (2020). What Is Society 5.0? In Hitachi-UTokyo Laboratory (Ed.), Society 5.0. Springer.

DIME4HERITAGE. (n.d.). https://cordis.europa.eu/project/id/302799

Donner, C., & Wann Jensen, H. (2005). Light diffusion in multi- layered translucent materials. *ACM Transactions on Graphics*, *24*(3), 1032–1039. doi:10.1145/1073204.1073308

DOORS. (n.d.). https://cordis.europa.eu/project/id/101036071

Eberly, D. H. (2015). *3D Game Engine Design: A Practical Approach to Real-Time Computer Graphics*. Taylor & Francis, CRC Press.

Elias, N., Mas, N., & Cotte, P. (2011). Review of several optical non-destructive analyses of an easel painting. Complementarity and crosschecking of the results. *Journal of Cultural Heritage*, *12*(4), 335–345. doi:10.1016/j.culher.2011.05.006

Europeana, D. S. I. (n.d.). *2 - Access to Digital Resources of European Heritage - Deliverable D4.4.* Report on ENUMERATE Core Survey 4. Available online: https://pro.europeana.eu/files/Europeana_Professional/Projects/Project_list/ENUMERATE/deliverables/DSI-2_Deliverable%20D4.4_Europeana_Report%20on%20ENUMERATE%20Core%20 Survey%204.pdf

Europeana. (n.d.). https://www.europeana.eu/en/about-us

Fassi, F., Mandelli, A., Teruggi, S., Rechichi, F., Fiorillo, F., & Achille, C. (2016). Lecture Notes in Computer Science: Vol. 9769. *VR for Cultural Heritage. Augmented Reality, Virtual Reality, and Computer Graphics. AVR 2016*. Springer.

Ferwerda, J. A., & Darling, B. A. (2013). Tangible Images: Bridging the Real and Virtual Worlds. In CCIW 2013 proceedings. Lecture Notes in Computer Science, 7786. Springer.

Fukuyama, M. (2018, July). Society 5.0: Aiming for a New Human-Centered Society. *Japan Spotlight*, 47-50.

Gaiani, M., Apollonio, F. I., Bacci, G., Ballabeni, A., Bozzola, M., Foschi, R., Garagnani, S., & Palermo, R. (2020). Seeing inside drawings: a system for analysing, conserving, understanding and communicating Leonardo's drawings. In *Leonardo in Vinci. At the origins of the Genius* (pp. 207–239). Giunti.

Gaiani, M., Apollonio, F. I., & Fantini, F. (2019). Evaluating smartphones color fidelity and metric accuracy for the 3D documentation of small artefacts. *The International Archives of the Photogrammetry, Remote Sensing and Spatial Information Sciences*, *42*(W11), 539–547. doi:10.5194/isprs-archives-XLII-2-W11-539-2019

Gaiani, M., Apollonio, F. I. & Fantini, F. (2020). Una metodología inteligente para la digitalización de colecciones museísticas. *EGA Expresión Gráfica Arquitectónica, 25*(38), 170–181.

Gaiani, M., & Ballabeni, A. (2018). *SHAFT (SAT & HUE Adaptive Fine Tuning), a new automated solution for target-based color correction. In Colour and Colorimetry Multidisciplinary Contributions. XIVB, Maggioli Editore*. ITA.

Gardner, A., Tchou, C., Hawkins, T., & Debevec, P. (2003). Linear Light Source Reflectometry. *ACM Transactions on Graphics*, *22*(3), 749–758. doi:10.1145/882262.882342

Google Arts and Culture. (n.d.). https://artsandculture.google.com/

Guarnera, D., Guarnera, G. C., Ghosh, A., Denk, C., & Glencross, M. (2016). BRDF Representation and Acquisition. *Computer Graphics Forum*, *35*(2), 625–650. doi:10.1111/cgf.12867

Hanrahan, P., & Krueger, W. (1993). Reflection from layered surfaces due to subsurface scattering. In *Proceedings of the 20th Annual Conference on Computer Graphics and Interactive Techniques (SIGGRAPH 93)*. ACM. 10.1145/166117.166139

Hasegawa, T., Tsumura, N., Nakaguchi, T. & Iino, K. (2011). Photometric Approach to Surface Reconstruction of Artist Paintings. *Journal of Electronic Imaging, 20*(1), 11.

Heitz, E. (2014). Understanding the masking-shadowing function in microfacet-based BRDFs. *Journal of Computer Graphics Techniques*, *3*(2), 32–91.

Ishii, M., Moryiama, T., Toda, M., Kohmoto, K., & Saito, M. (2008). Color degradation of textiles with natural dyes and blue scale standards exposed to white LED lamps: evaluation for effectiveness as museum lighting. *Journal of Light and Visual Environment*, *32*(4), 370-378.

Japan Business Federation (Keidanren). (2016), Toward Realization of the New Economy and Society. In Reform of the Economy and Society by the Deepening of "Society 5.0". Author.

Kist, C. (2020). Museums, Challenging Heritage and Social Media During COVID-19. *Museum & Society*, *18*(3), 345–348. doi:10.29311/mas.v18i3.3539

Malzbender, T., Gelb, D., & Wolters, H. (2001). Polynomial texture maps. In Proceedings of Siggraph 01. Computer Graphics (SIGGRAPH 01: 28th International Conference on Computer Graphics and Interactive Techniques. ACM.

McCamy, C. S., Marcus, H., & Davidson, J. (1976). A color-rendition chart. *Journal of Applied Photographic Engineering*, *2*, 95–99.

Menna, F., Nocerino, E., Morabito, D., Farella, E. M., Perini, M., & Remondino, F. (2017). An open source low-cost automatic system for image-based 3D digitization. *The International Archives of the Photogrammetry, Remote Sensing and Spatial Information Sciences*, *42*(W8), 155–162. doi:10.5194/isprs-archives-XLII-2-W8-155-2017

Mudge, M., Malzbender, T., Chalmers, A., Scopigno, R., Davis, J., Wang, O., Gunawardane, P., Ashley, M., Doerr, M., Proenca, A., & Barbosa, J. (2008). Image-based empirical information acquisition, scientific reliability, and long-term digital preservation for the natural sciences and cultural heritage. Eurographics (Tutorials), 2, 4.

Negri, M., & Marini, G. (2020). *Le 100 Parole dei Musei* [The 100 Words of Museums]. Marsilio.

Nicodemus, F. E. (1965). Directional reflectance and emissivity of an opaque surface. *Applied Optics*, *4*(7), 767–775. doi:10.1364/AO.4.000767

Papas, M., de Mesa, K., & Wann Jensen, H. (2014). A Physically-Based BSDF for Modeling the Appearance of Paper. *Computer Graphics Forum*, *33*(4), 133–142. doi:10.1111/cgf.12420

Pescarin, S., D'Annibale, E., Fanini, B., & Ferdani, D. (2018). Prototyping on site Virtual Museums: The case study of the co-design approach to the Palatine hill in Rome (Barberini Vineyard) exhibition. *Proceedings of the 3rd Digital Heritage International Congress (DigitalHERITAGE)*, 1–8. 10.1109/DigitalHeritage.2018.8810135

Petrelli, D. (2019). Making virtual reconstructions part of the visit: An exploratory study. *Digital Applications in Archaeology and Cultural Heritage*, *15*, e00123. doi:10.1016/j.daach.2019.e00123

Petrelli, D., Not, E., Damala, A., van Dijk, D., & Lechner, M. (2014). meSch – Material Encounters with Digital Cultural Heritage. Digital Heritage. Progress in Cultural Heritage: Documentation, Preservation, and Protection. EuroMed 2014. Lecture Notes in Computer Science, 8740.

Piccablotto, G., Aghemo, C., Pellegrino, A., Iacomussi, P., & Radis, M. (2015). Study on conservation aspects using LED technology for museum lighting. *Energy Procedia*, *78*, 1347–1352. doi:10.1016/j.egypro.2015.11.152

Pierdicca, R., Frontoni, E., Zingaretti, P., Sturari, M., Clini, P., & Quattrini, R. (2015). Lecture Notes in Computer Science: Vol. 9254. *Advanced Interaction with Paintings by Augmented Reality and High-Resolution Visualization: A Real Case Exhibition. Augmented and Virtual Reality. AVR 2015*. Springer.

Pitzalis, D., & Pillay, R. (2009). Il sistema IIPimage: Un nuovo concetto di esplorazione di immagini ad alta risoluzione. [The IIPimage system: A New Concept of Exploration of High Resolution Images]. *Archeologia e Calcolatori*, (Supp. 2.), 239–244.

Ponchio, F., Corsini, M., & Scopigno, R. (2018). A compact representation of relightable images for the web. In *Proceedings of the 23rd International ACM Conference on 3D Web Technology. ACM*. 10.1145/3208806.3208820

Relio[2]. (n.d.). www.relio.it

Remondino, F., Rizzi, A., Barazzetti, L., Scaioni, M., Fassi, F., Brumana, R., & Pelagotti, A. (2011). Review of Geometric and Radiometric Analyses of Paintings. *The Photogrammetric Record*, *26*(136), 439–461. doi:10.1111/j.1477-9730.2011.00664.x

Salama, R., & ElSayed, M. (2018). Basic elements and characteristics of game engine. *Global Journal of Computer Sciences: Theory and Research*, *8*(3), 126–131. doi:10.18844/gjcs.v8i3.4023

Schlick, C. (1994). An Inexpensive BRDF Model for Physically Based Rendering. *Computer Graphics Forum*, *13*(3), 233–246. doi:10.1111/1467-8659.1330233

Tenenbaum, J. M. (2006). AI Meets Web 2.0: Building the Web of Tomorrow, Today. *AI Magazine*, *27*(4), 47.

Tominaga, S., & Tanaka, N. (2008). Spectral Image Acquisition, Analysis, and Rendering for Art Paintings. *Journal of Electronic Imaging, 17*(4), 13.

Torrance, K. E., & Sparrow, E. M. (1967). Theory for off-specular reflection from roughened surfaces. *Journal of the Optical Society of America*, *57*(9), 1105–1114. doi:10.1364/JOSA.57.001105

Unity3D. (n.d.). http://www.unity3d.com

Vandermeulen, B., Hameeuw, H., Watteeuw, L., Van Gool, L., & Proesmans, M. (2018). Bridging Multi-light & Multi-Spectral images to study, preserve and disseminate archival documents. *Archiving Conference*, *2018*(1), 64–69. doi:10.2352/issn.2168-3204.2018.1.0.15

Walter, B., Marschner, S. R., Li, H., & Torrance, K. E. (2007). Microfacet models for refraction through rough surfaces. In *Proceedings of the 18th Eurographics conference on Rendering Techniques*. Eurographics Association.

Wandell, B. A., & Farrell, J. E. (1993). Water into wine: Converting scanner RGB to tristimulus XYZ. Device-Indep. *Color Imaging Imaging Syst. Integr*, *1909*, 92–100.

Wang, J., Zhao, S., Tong, X., Snyder, J., & Guo, B. (2008). Modeling Anisotropic Surface Reflectance with Example-Based Microfacet Synthesis. *ACM Transactions on Graphics*, *27*(3), 3. doi:10.1145/1360612.1360640

Watteeuw, L., Hameeuw, H., Vandermeulen, B., Van der Perre, A., Boschloos, V., Delvaux, L., Proesmans, M., Van Bos, M., & Van Gool, L. (2016). Light, shadows and surface characteristics: The multispectral Portable Light Dome. *Applied Physics. A, Materials Science & Processing*, *122*(11), 976. doi:10.100700339-016-0499-4

Westin, S. H., Arvo, J., & Torrance, K. E. (1992). Predicting reflectance functions from complex surfaces. *Proceedings of the SIGGRAPH 92*, 255–264. 10.1145/133994.134075

Witikon. (n.d.). http://witikon.eu/

Woodham. R.J. (1980). Photometric Method for Determining Surface Orientation from Multiple Images. *Opt. Eng.*, *19*(1), 191139.

Zarrad, A. (2018). *Game engine solutions. Simulation and gaming*. IntechOpen. doi:10.5772/intechopen.71429

KEY TERMS AND DEFINITIONS

Cultural Heritage: The cultural heritage (CH) is the legacy of assets of a society that is inherited from past generations. It includes tangible culture (such as monuments, books, works of art, artifacts, etc.), intangible culture (such as traditions, language, and knowledge), and natural heritage (including culturally significant landscapes, and biodiversity).

Digital Replica: A digital replica is a faithful copy of an original artifact in the digital domain, including its appearance, its morphology and how it is meant to interact with possible users.

LED: A light-emitting diode (LED) is a semiconductor light source that emits light when current flows through it. In order to replicate in the digital domain light behaviors, LED lights are often used to digitize CH collections in museums, as they are less dangerous than other illuminants.

Photometric Stereo: In computer vision, photometric stereo is a technique used to get an accurate estimation of a surface's normals by observing the object under different lighting conditions, since light reflected by a surface is dependent on the orientation of it in relation to the light source.

Real-Time Rendering: In computer graphics, Real-Time Rendering (RTR) is a specific visualization pipeline that consists of three conceptual stages: the application stage, the geometry stage, and the rasterizing stage. The outcome of this pipeline is an interactive visual representation of a scene, which

Shader: A shader is a computer application that analytically simulates the appropriate levels of light, darkness, and color during the rendering of a 3D scene. It is used to faithfully replicate materials and surfaces in digital visualization.

Total Appearance: A 3D solution feature in which the digital reproduction of the visual properties of a surface is a faithful transposition of the physical properties of the replicated object through a process that covers the mathematical validation of the properties of the surface, an accurate simulation of light reflections and a perceptual final image corrected for the human eye.

Chapter 21
Digital Explorations in Archive Drawings:
A Project for Cannaregio Ovest in Venice by Luciano Semerani, 1978

Starlight Vattano
https://orcid.org/0000-0002-4510-874X
Università Iuav di Venezia, Italy

ABSTRACT

The chapter shows some of the outcomes of a research project begun in 2021 in collaboration with the Archivio Progetti Iuav of Venice, with the aim of disseminating the drawings, documents, and projects preserved. On the basis of the documentary collection including pieces, projects, models, together with a conspicuous repository of photographs and reproductions, the research deepens a little-explored aspect of an unbuilt Venice, circumscribing the investigation scope to the 20th century masters of architecture who contributed in rethinking the urban form of the lagoon city, such as Luciano Semerani's project for the sestiere of Cannaregio Ovest in 1978. The discussion on the Venetian structural system, the urban trace, and the architectural configuration is re-established in a dialogue between its history and its contemporaneity. This is achieved starting from the digital models and virtual tours with in-depth texts that integrate the information actions with respect to the qualities of the architectures and urban spaces activated and consulted with the exploration of the model.

INTRODUCTION

The digitisation process of the different graphic and documentary information, preserved at the Archivio Progetti Iuav, addresses the communication of heritage, elaborating scenarios consistent with the need to promote forms of knowledge easily shared. This information can be made available on interactive online platforms, through the use of immersive techniques and technologies for data processing and restitution.

DOI: 10.4018/978-1-6684-4854-0.ch021

They can combine theoretical and design aspects of the investigated drawings while providing new narrative strategies for immersive fruition.

Contextually to the fervent debate on the use of digital technologies for an increasingly open access to the cultural heritage, some outcomes of the broader research started in 2021 are shown, which saw the collaboration between the Project Support Lab of Iuav and the Archivio Progetti Iuav developed around the common objective of enhancing the documentary, graphic and photographic heritage contained therein by means of accessible, sharable and popular digital solutions. The subject of the research was the International Competition organised by Iuav in 1978[1] during which the city of Venice became a field of testing and intellectual speculation on themes such as the regeneration of historic city centres, the relationship between contemporary design and the existing fabric, the rethinking of infrastructural connections (by land and sea) between the lagoon city and the mainland, and the question of living in relation to building typology, public spaces and services.

The area of intervention concerned the sestiere of Cannaregio Ovest, which the ten protagonists of the Competition considered as a model of possibility capable of assuming a new social, economic and urban development (Figure 1). A place from which industrial and technological changes would provide interference and stimuli for a new architectural vision. The ten proposals show approaches and strategies that differ in terms of intervention scale, typological choices on architectural organisms, infrastructural network, connection (physical or visual) with the historical city, design references and fields of investigation developed.

BACKGROUND

The current reflection posed on Digital Heritage determines a constant rethinking on the ways of visualisation, communication and dissemination of cultural heritage across different research fields and disciplines "from museography to computer graphics, from archaeology to design, from art history to engineering, from archives to statistics, etc. It is therefore an overarching term, encompassing many ICT topics and heritage themes" (Pescarin, 2016, p. 1), increasingly leading to an overlap and interconnection between different knowledge. The era of large 'digital meta-collections' is taking shape in an increasingly massive way, bringing into its domain not only groupings of digital information pertaining to specific institutions, but also and above all their interoperability and ability to define a dissemination of knowledge increasingly relying on shared and networked exploration (Windhager et al., 2016).

Along the lines of the definitions given by UNESCO to the concepts of tangible and intangible cultural heritage (UNESCO, 2003), looking at the use and development of digital technologies for the preservation and enhancement of heritage makes it possible to develop ways of accessing even those cultural assets that are difficult to reach or even just to consult, based on the assumption that preserving heritage means documenting, protecting, reconstructing and disseminating (Skublewska-Paszkowska et al., 2022).

There has long been talk of increasingly democratic ways of access aimed at an open science, making people increasingly aware of their heritage and thus of the need to safeguard it through participatory processes: "the application of technology in a democratic manner could refer to access to the technology for all, as well as use of the technology to actually serve democratic purposes. This prompts us [...] to consider whether the user driven approaches to the collection of digital heritage data in some ways represents the Heritage community taken control of the digital agenda" (Laing, 2020, p. 2).

The treatment of open and shared data contributes to the definition of a democratic dimension of the research dissemination in manifold fields. In this direction, 3D management and modeling tools take on a creative meaning, insofar as the communication and sharing of heritage, as is also the case in archaeology, establishes transparent processes aimed at the dissemination of knowledge, by means of increasingly accessible visualization and manipulation technologies (Opgenhaffen et al., 2021).

It is precisely in the possibility of open access that online management of cultural property plays a prominent and growing role in people's activities. Dedicated online platforms are participating in the process of reformulating the cultural good and the social interest in interacting with it. The increasingly cogent goal on the part of museums has now become to reach a considerable number of users, taking into account born-digital activities (Karp, 2004).

Rahaman (2018) introduces the topic of "Heritage interpretation" considered as a tool for effective learning, communication and information management. The focus is therefore on the attitude and empathy that the individual user develops toward the asset, considering "'digital heritage interpretation' as a process rather than as a tool to present or communicate with end-users" (Rahaman, 2018, p. 208).

At the present time, the numerous collections of the Archivio Progetti Iuav are catalogued on the web page of the specific section, according to a structure of links through which it is possible to access general information on individual architects, architecture competitions, projects and installations of the Biennale Architettura. Consultation of the individual items listed online leads to brief descriptions summarising the topics, biographies and types of documents indexed, accessible from the web or on request.

The dedicated catalogue section allows the document structure to be viewed either by entering keywords or by entering the names of the collections. In both modes, the archival units show the document type, the reference to the specific collection and in some cases, the attached images.

The online graphics gallery contains original drawings and often photographs of models not always available for consultation, unless requested directly from the Archivio Progetti. In the light of the current methods of accessing and consulting the material, of the possibilities of visualisation and interaction with the graphic information actually present in the archive for external users, the research described in this paper proposes an in-depth interpretation, digitisation and graphic reconstruction activity through the elaboration of digital models, which can also be explored in immersive mode, providing multiple and unprecedented readings of an unbuilt and imagined Venice. The digital reconstructions, by making known design experiences and lines of research adopted, shed light on the design work of architects, including those coming from different backgrounds and strongly linked to the issues and methods of design within the city's historical passages, becoming an integral material of the documentary apparatus already present, but with an even more open, consultable and sharable mode of access.

The act of watching and visualizing now takes on an increasingly holistic and virtual sense of the reality perception: "an ideal VR system provides humans experiences that are indistinguishable from an experience that could be real" (Wang et al., 2020, p. 258). It seems to be evident, now more than ever, to wonder about the visual qualities of digital objects, their relationship to reality and different modes of visualization.

In this regard, consider the definition given by Sherman and Craig (2003) on virtual reality thought of in four domains or "key elements": the virtual world, the immersion, the sensory feedback, and the interactivity. Virtual space is defined by the two scholars as an imaginary space composed through a specific device or instrumentation capable of returning multiple viewpoints experienceable through immersive perception, involving both physical and mind dimensions.

A fundamental element in immersive experience is that of "sensory feedback", based on the direct response of the users who simulate the physical reality in the virtual one and, through the recording of movements or reactions, provide more integrated and heterogeneous outputs from time to time. Interactivity is the fourth key element, commented on by Sherman and Craig (2003). This aspect is described as an additional form or ability to mutate the point of view and thus the degree of explorability and knowledge of the observed object.

GRAPHIC AND RECONSTRUCTING POSSIBILITIES

Issues, Controversies, Problems

The research project aims to return a digital graphic corpus that recounts the wealth of information contained in the archive by means of additional unpublished images, video tours and explorations in immersive mode, taking into account the transversal possibilities of innovative techniques and technologies in the field of outcome dissemination. A further intention is to foster the development of immersive and integrated technologies, extending the thematic sphere to museum dimensions, in response to the issues of the preservation and enhancement of the Italian historical heritage and the cultural, architectural and historical value of Italian-made products, looking at Venice as a catalyst for hi-tech strategies to highlight the Italian way of designing and building the city.

In relation with the premises on which the international seminar that took place in the lagoon city in 1978 was configured, the currently available archive documents suffer from a condition of indeterminateness, at times of incompleteness, typical of those project drawings conceived to be further modified, manipulated and altered again. Therefore, the documentary apparatus of the projects interpreted and returned digitally in the research, as well as in the specific case of Luciano Semerani's work, shows numerous incomplete or missing information, the character of which establishes different issues from time to time, on an architectural, urban and landscape scale; issues intimately linked to the cultural sedimentation and design experience of the ten protagonists.

In this sense, in the phase of graphic interpretation and digital elaboration, it was necessary to make integrations between the original drawings and the information deduced from other sources (realised works, portfolios, private archives).

From graphic, composition and design comparisons it was possible to trace coherent solutions dropped into the digital model, resolving the information missing in the initial drawings. This was a critical reading, a necessary integration of the graphic corpus provided by the Archivio Progetti Iuav.

Only by the presence of these graphic and design interferences was it possible to define the becoming of an unbuilt place, such as the possible images for the sestiere of Cannaregio Ovest (Figure 2).

The digital narratives reveal a 'possible Venice', a field of intellectual speculation on several fronts: the project of the dreamed city; the relationship between Venetian memory and contemporary space; the images of an architecture telling its utopias through drawing. The following images, therefore, are to be seen as digital re-elaborations containing, on the one hand, the information captured during the project phase in the competition boards; on the other, as graphic places in which suggestions and quotations have been accumulated, results of the comparison and critical reading of the drawing, and of the project.

Figure 1. Scheme with the identification of the sestiere of Cannaregio Ovest in Venice, the station and the Lista di Spagna
Source: S. Vattano 2022.

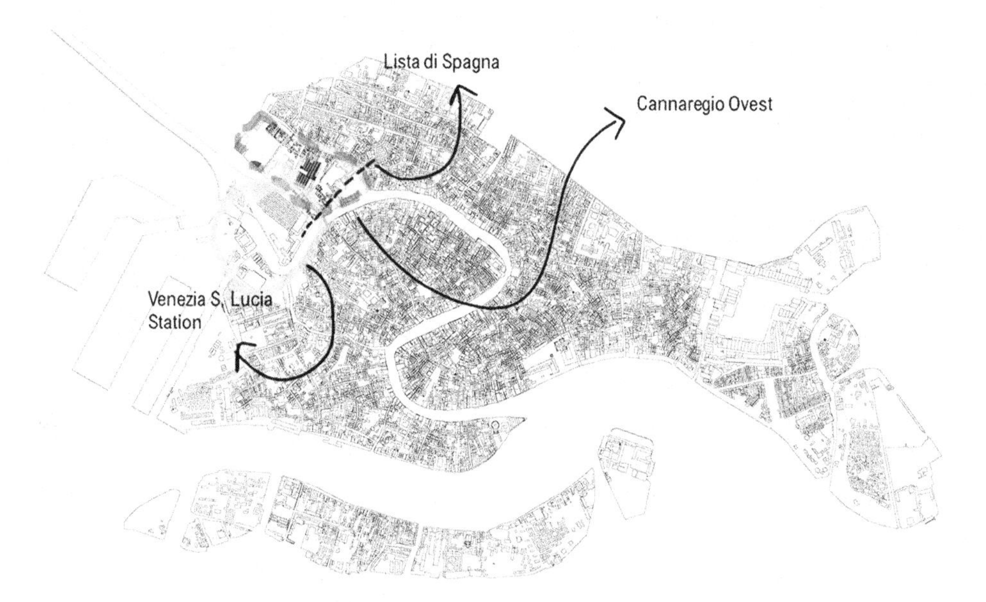

Methodology and Digital Outcomes

The research on the ten projects presented at the International Seminar was structured with respect to three work phases. The first part of the research concerned the retrieval of the graphic information and bibliographic sources necessary for digital processing, consultation and subsequent bibliographic organisation. The definition of the case studies was carried out by means of collaboration and ongoing dialogue with the management of the Archivio Progetti Iuav.

Following the collection of the iconographic sources, the interpretation and restitution of the selected projects involved an initial moment of in-depth graphic analysis.

In this phase, the drawings of plans, vertical sections, elevations, perspective and axonometric views were elaborated, useful for the construction of the digital model, including compositive, constructive and spatial information.

The work process for the development of digital models was performed from the interpretation, reconstruction and management of the 2D model using Rhinoceros software.

From the two-dimensional structure, the 3D model was obtained, placed both in the Cannaregio sestiere and in the city of Venice digital reconstruction. The next step involved the management of materials, lighting, and rendering by Twinmotion software, within which clips were made of the project's paths and spatialities.

Other exports of the model portions for visualization in the immersive environment were obtained from the same software and experienced with HTC Vive viewers.

The 360-degree panoramas obtained by Twinmotion were subsequently edited into panoramic virtual tours with the Pano2VR software, which allowed the inclusion of additional graphic, textual and interactive content (links to online pages, videos or animated images).

Figure 2. The ten-project proposal presented at the international convention for the rethinking of Cannaregio Ovest, 1978
Source: S. Vattano, 2022.

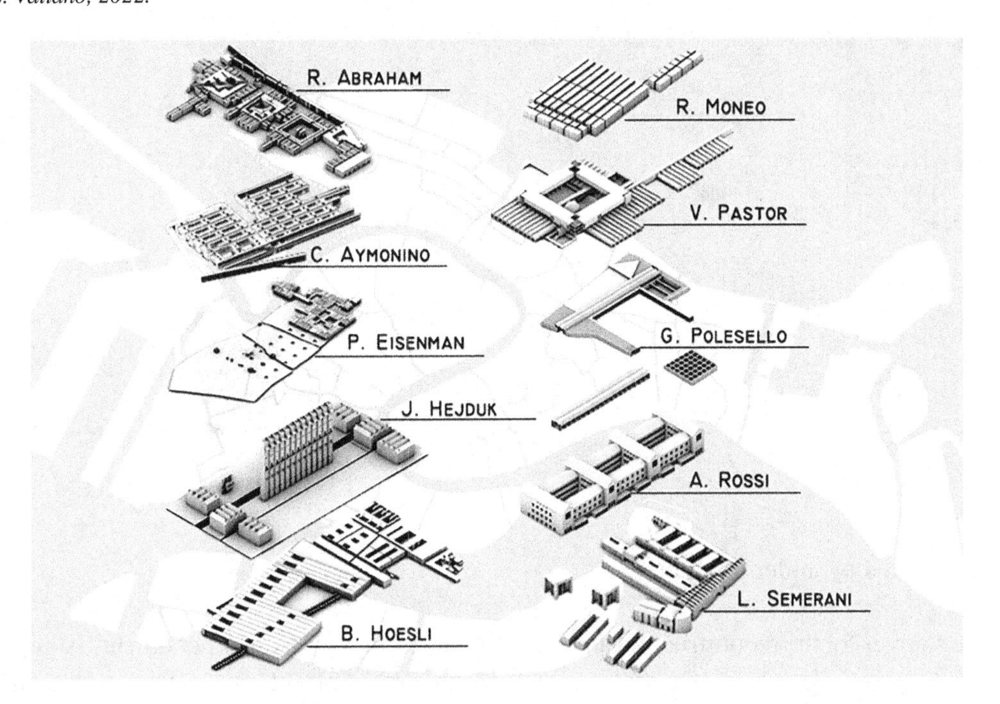

The explorable models were published on the Sketchfab platform, in which summary information about the architect, the building, and the historical context was included. The digital products produced were reprocessed to create explanatory videos, processed with DaVinciResolve software and storytelling systems designed to be enjoyed by a wide and heterogeneous range of users.

The digital models were processed to be managed and explored in different forms (Figure 3).

For this reason, in the last phase, different forms of 'digital objects' were produced that were able to provide a level of information that could be consulted from fixed or mobile devices and on open source platforms in different formats and viewing modes, depending on the type of use: images, virtual tours that could be explored in standard and immersive modes, and videos.

The parameters against which the ten projects were read, interpreted and compared, and therefore specifically that of Luciano Semerani proposed here, concern:

a) the suggestions that characterized the design principle;
b) the design references (literary, cinematographic, aesthetic-philosophical ones);
c) the graphic-narrative structure adopted by the designer for its representation, especially by means of the method of representation adopted;
d) the structural elements (urban, architectural, landscape) that defined the design matrices;
e) the formal and material relationship between the proposed intervention and its immediate surroundings;
f) the level of depth (scale and detail) of the project.

Figure 3. Example of shareable digital model in an open source platform (Sketchfab)
Source: S. Vattano, 2021

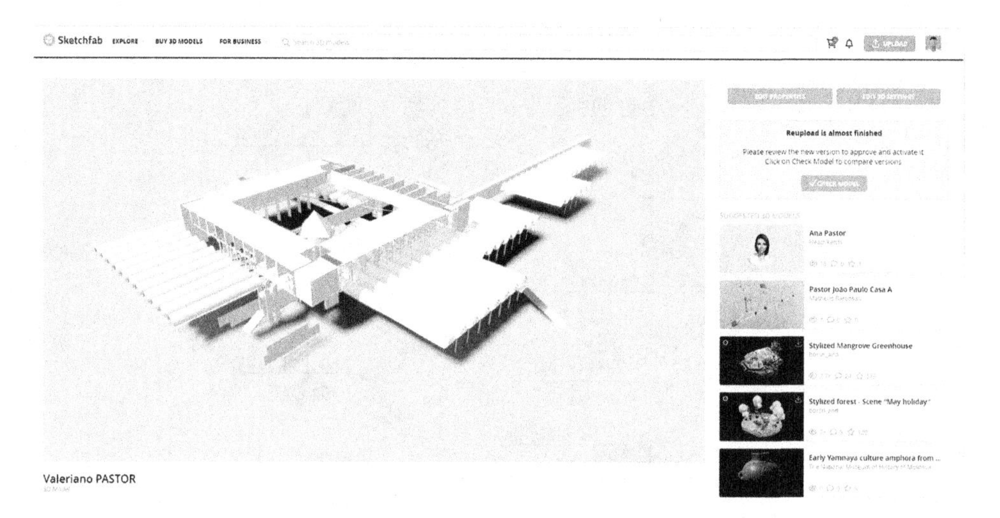

This is followed by in-depth studies:

g) on the sources for the reconstruction and possible integration with other documents (sources found in other archives, designers' biographies, work analysis, possible interviews for a better understanding of the project);
h) on the restitution method (choice of viewpoint in rendered views, type of representation chosen, highlighting of the most significant elements);
i) on the management of the model (surface treatment, color management);
j) on the critical aspects encountered in the restitution phase (lack of graphic information, incomplete designs, discrepancies between original elaborations).

At the end of the brief description, for each individual project reconstructed, the textual apparatus is accompanied by some *ex-post* reflections on the relationship between the archive document, the digital processing and the new images to support the knowledge integrated by the new graphic-narrative information, a trace of which can be consulted on the Archivio Progetti Iuav website[2].

Representation Methods

The digitisation, the construction of digital models, the synthesis of compositional principles through the graphic schematisation of case studies defines the basic material for the realisation of video tours, also in immersive visualisation (Figures 4, 5).

Some of the elements identified and organised in the descriptive phase of the project concern: the definition of the area of intervention, the functional organisation of the spaces, the internal/external relationship of the architecture, the relationship with the context, the interaction between the original drawings and the digital model. The project narrative is presented through different types of digital products:

k) digital models;
l) interactive and immersive virtual tours;
m) 360° images;
n) views obtained from the model (perspectives, axonometric exploded and perspective sections);
o) descriptive videos;
p) sharable and open digital models published on open-source platforms.

Figure 4. Example of online and interactive virtual tour visualization (Pano2Vr)
Source: S. Vattano, 2021.

Figure 5. Example of 360° panoramic views from the digital model of the project of L. Semerani
Source: S. Vattano, 2021.

Moreover, as already mentioned, since we are dealing with imagined places and evocative suggestions of concepts oscillating between the abstract and the utopian, the graphic re-elaborations show monochrome digital models, a methodological approach adopted to emphasise the project's condition of incompleteness, understood as a place of theoretical speculation and convergence of spatial visualities.

The digital model, in the absence of chromatic and material definition, maintains the value of the possibility of visualisation that the eye can continue to experience in the attempt to reach the project's completion.

As in the case of the other competition proposals, reading the drawings of Luciano Semerani's project reveals some of the criticalities and/or discrepancies in both the graphics and design of the object of study.

These moments of knowledge and therefore in-depth examination of the peculiarities, gaps and inevitable conjectural condition of the archive drawings, constitute the nucleus of the reflections involved in the digitisation and construction of the digital model.

The synthesis of compositional features, both on an urban and architectural scale, was carried out through the graphic schematisation of the elements that characterise Semerani's proposal. Therefore, in order to delineate and organise the descriptive character of the images produced and their explorative-narrative attribute, some of the aspects inherent to the in-depth study and observation of the project were identified: the definition of the intervention area, the functional organisation of the spaces, the internal/external relationship of the architecture, the relationship with the context, the interaction between the original drawings and the digital model.

Semerani dwells on the urban context as the object of study and takes the city as the system of delimitation and orientation of architecture, opening up the field to a broader vision on the question of building the city.

The complexity provided by the compositional classification shows the dynamic correlation between phenomena and causes, thus introducing the role of the time and technical variable. In this sense, the architect highlights certain aspects, both urban and building, that participate in the definition of his proposal: the connection between the historic and contemporary city; the empty space recalling historical urban episodes; the re-proposition of building typologies recalling the Venetian memory; the suggestion of traditional building techniques and the enhancement of pre-existence.

Figure 6. 3D model visualization with the insertion of the archive drawing of the Semerani's project plan
Source: S. Vattano, 2022.

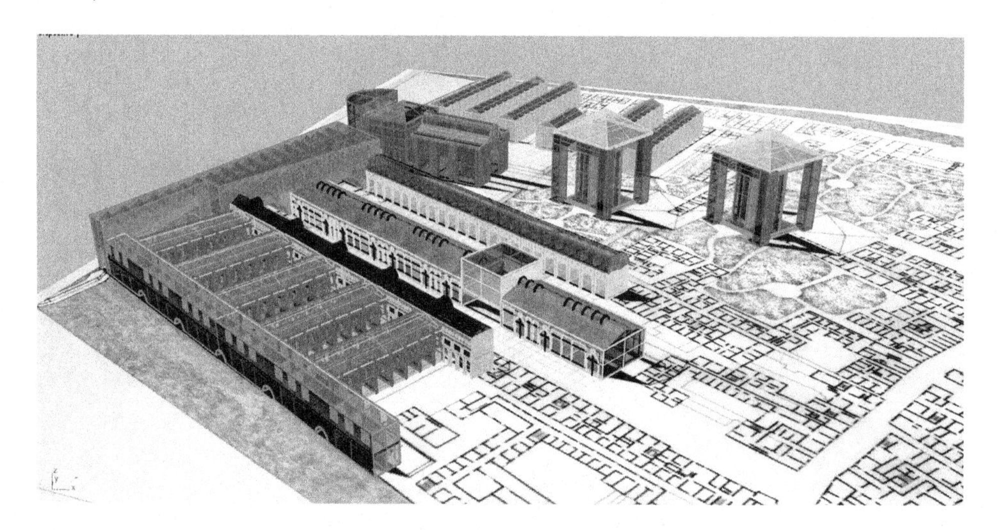

The choices regarding the methods of representation adopted for the interpretation of the drawings and their graphic restitution were conveyed by the same methods chosen by Semerani, the key elements characterising the imagined places, the viewpoints and spaces represented in the archive drawings, the textual descriptions not always accompanied by the drawings, and again sketches with annotations containing the rapidity and configurative immediacy of the project forms.

The graphic reconfiguration of archive documents has therefore taken shape with the elaboration of views from the model and digital hybridisations (Figure 6), meaning by 'hybridisations', combinations of images, integrations between the digital model and the archive drawing capable of amplifying the perception of the imagined spatialities and providing an increasingly integrated degree of knowledge.

Explorability and Visualisation of the Digital Model

Digital models were processed to be managed and explored in different forms. Their level of information was deepened on the basis of the type of digital object and the type of communicative use: images, virtual tours that can be explored in standard and immersive mode, and videos (Figure 7).

Figure 7. The ghosted view mode of the 3D digital model of the Semerani's hypothesis
Source: S. Vattano, 2022.

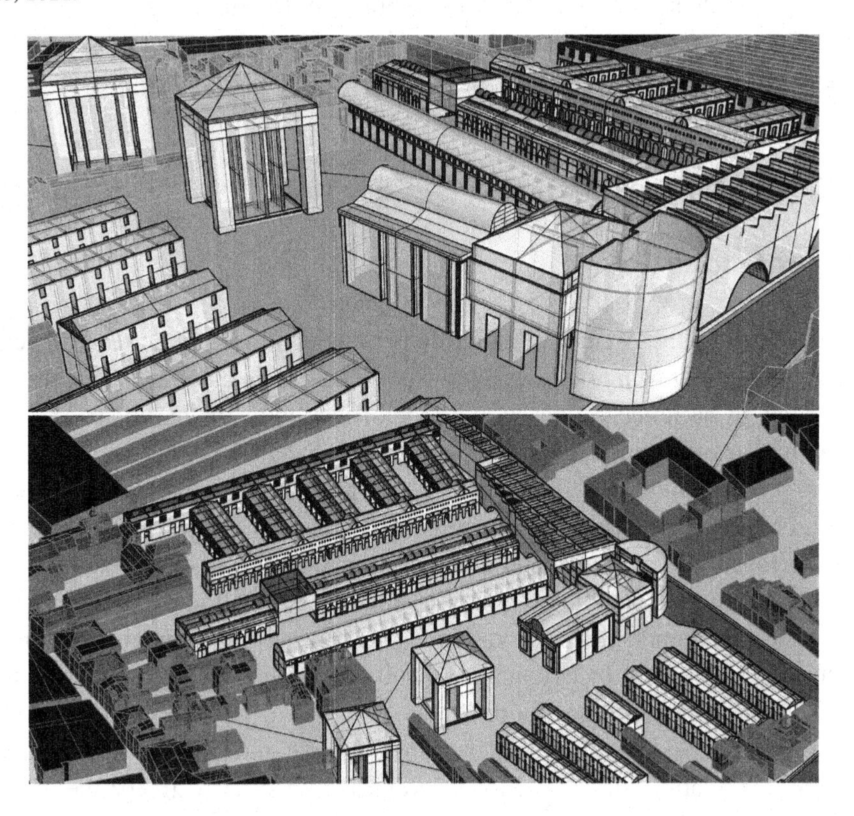

An interactive mapping that collects and synthesises the ten proposals of the competition in Cannaregio is dynamically developed in a video tour, supplemented by 360° panoramic images, explanatory clips extracted from the digital models, historical-constructive and design information on the individual case studies. As in the other cases, also for the study conducted on Semerani the design elements highlighted in the virtual tours and in the 360° images dealt with specific moments of the urban and architectural reflection: the definition of the intervention area, the functional organisation of the spaces, the internal/external relationship of the architecture, the relationship with the context, the interaction between the original drawings and the digital model.

The video montage, intended as a dynamic narrative tool, focused on the operations of tracing, isolating and describing the constituent elements of the project in a dynamic manner, for the construction of a hybrid narrative between the digital model and the archive document. The question is linked to another consideration, that of the video as a process of approach, of overcoming absence. Its *eidos*, in the original Platonic sense (= from the Greek "form", "aspect" from which also derives *εἴδωλον* (*éidolon*) then, in Italian, "idol", understood according to the meaning of "simulacrum", "figure"), corresponds to the act in its unfolding, in its simultaneity, in the brevity of the captured image, of the memory, of the evidence.

The video narration therefore contains a brief description of the project interpreted graphically and returned digitally, of which the following are summarised: the suggestions that characterised the design principle; the design references; the graphic-narrative structure adopted by Semerani for its restitution, above all by means of the chosen method of representation; the structural elements (urban, architectural, landscape) that defined the project matrixes; the formal and material relationship between the proposed intervention and its immediate surroundings; the level of detail (scale and details) of the project.

LUCIANO SEMERANI, CANNAREGIO OVEST, 1978

Graphic Interpretations

The documents provided by the Archivio Progetti Iuav on which the graphic reconstruction was based are: 1 axonometry with the drawing of the ground floor plan, at the base and the elevation; 1 coloured sketch of the complex of buildings proposed, with identification of the paved spaces, internal and external; 1 board with plan and axonometric sketches (Figures 8-10).

Among the ten projects presented, Semerani's develops a reasoning strongly related to the role of permanence and architectural tradition. In this sense, this case study offers the possibility to look, on the one hand, at the heterogeneity of the techniques adopted by the architect in the restitution of the idea; on the other hand, at the choices made on the representation of the relationship between the new and the existing, to the extent that Semerani works on the isolation of certain elements often quoting previous projects or tradition. Semerani's project returns, as well, a degree of depth on housing types and uses, among the most developed in the competition.

The points developed in Semerani's project deal with the role of residences and popular structures in urban development; the relationship between waterways and pedestrian streets; the ways of functional structuring of the city; the settlement techniques of construction and the building typology.

Figure 8. Original drawings of the Semerani's project proposal for Cannaregio Ovest, 1978
Source: © Università Iuav di Venezia, Archivio Progetti, fondo La Biennale di Venezia, 2022.

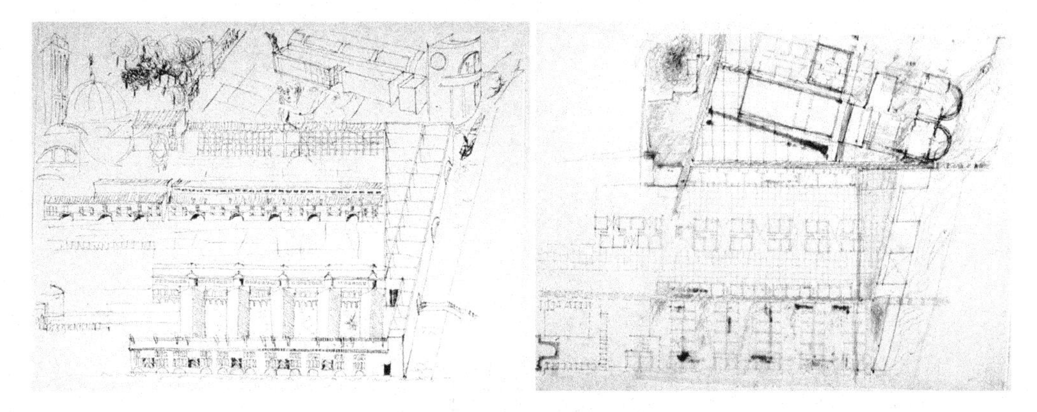

The architect focuses on the complexity guaranteed by the impossibility of a typological classification that requires a dynamic correlation between phenomena and causes, thus introducing the role of the new time variables. Semerani reinterprets and relocates some architectural citations, as is the case with the reinterpretation of the Basilica in Vicenza, or the citation of Diocletian's Palace in Spalato and the iron and glass architecture of Paxton's Crystal Palace. These are episodes that question the "artifact island," introduced by Semerani himself, in order to trace the specificities of the new fabric.

In fact, the theme of historicity, of the monument as a citation, of relational systems or unfinished prototypes are explored in their attitude to "[...] conform according to the particular functional and formal needs of the context" (Rosa 1983, p. 30). In this sense, the project conceived by Semerani raises three questions: that of the relationship between urban theory and architectural theory; that of the scaling of design themes; that of the meaning of the urban context. Settlement types are organized into mixed-function buildings and serial systems of parallel blocks against party wall, terraced or core blocks around a specific architectural object. Ultimately, the theme of cultural references intervenes, the object to which architecture refers and the function it performs in the sense of delimiting and orienting the reference (Dal Co, 1980). Looking at Saverio Muratori's urban theory, Semerani recognizes some specific facts for the city of Venice, including the role of permanence in the body of cultural references. In the case of Cannaregio, he addresses the system of the urban structure of Venice and therefore, on the one hand, the project for the new area of the Tronchetto Novissime, of 1964 and on the other hand, the "heterology" of the city. It is an interpretation that takes into account of the complex interrelation between urban and architectural events, from which theories and techniques of design are connected and verified (Figures 10, 11).

In the place of this modality of observation and perception of the design question on an urban scale, the architect recognizes some architectural "facts", which are concretized in the citations of the Diocletian's Palace in Spalato and in the hospital of Le Corbusier for Venice of 1963, from which he derives the compactness of the plan; as well as in the revival of the Crystal Palace by Paxton, to which they look for the use of construction techniques and function. Semerani's design reflections take shape in the historical multiplicity, orographic layering, and water furrowed paths, giving room for uncertainty, rethinking, and crisis only in the place and extent of excavation, in the degrees of necessity peculiar to a specific place and no other (Figures 12, 13).

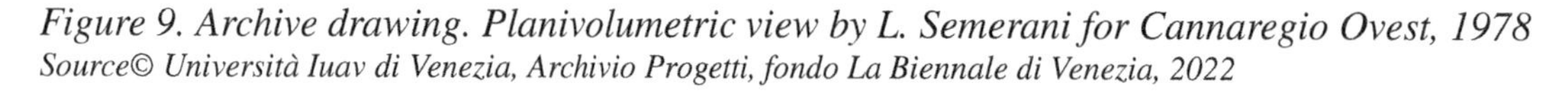

Figure 9. Archive drawing. Planivolumetric view by L. Semerani for Cannaregio Ovest, 1978
Source© Università Iuav di Venezia, Archivio Progetti, fondo La Biennale di Venezia, 2022

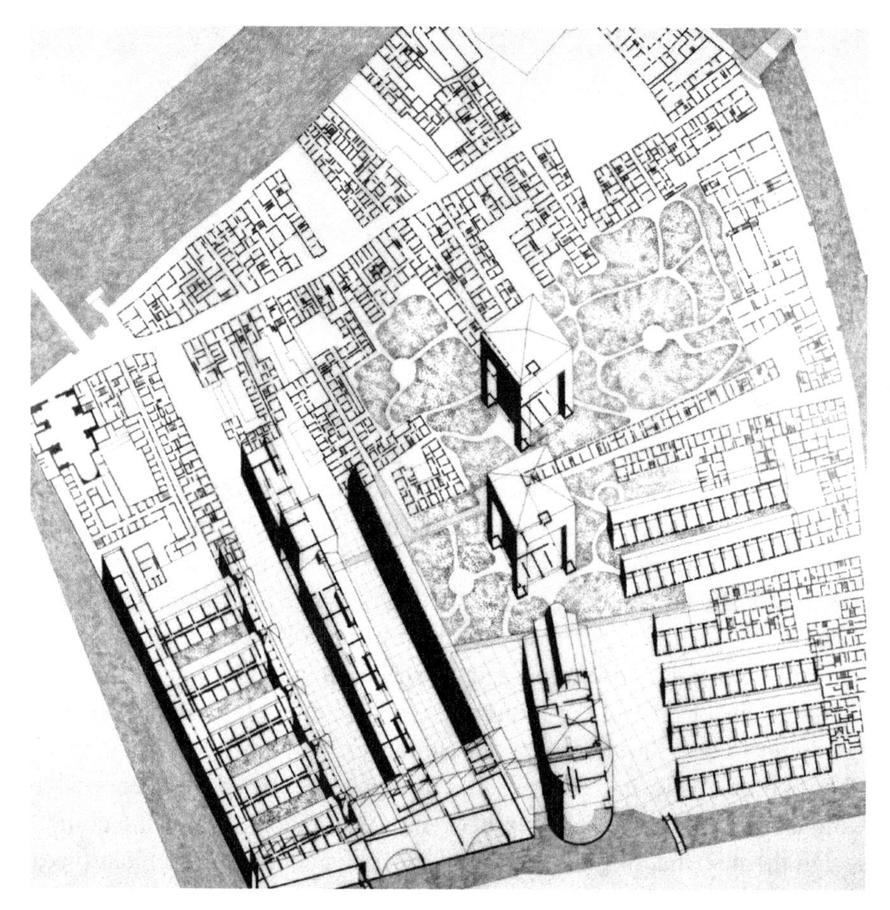

Figure 10. Plan of Venice and identification of the sestiere of Cannaregio Ovest (highlighted in white). Area of intervention
Source: S. Vattano, 2022.

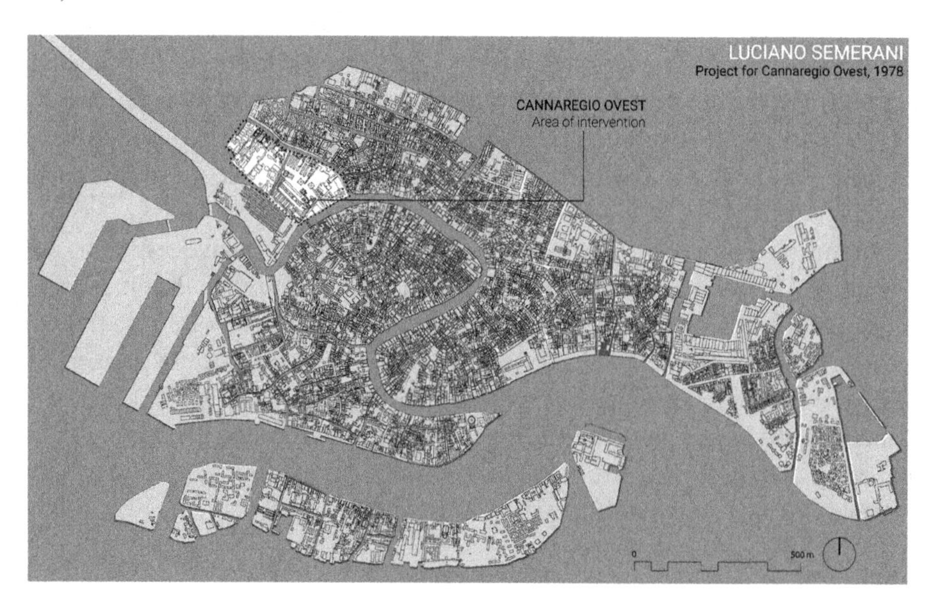

Figure 11. On the left, the plan of the Semerani's project proposal within the sestiere of Cannaregio Ovest and the identification of monumental buildings and the building typologies. On the right, the 3D digital model of the Semerani's project (highlighted in orange)
Source: S. Vattano, 2022.

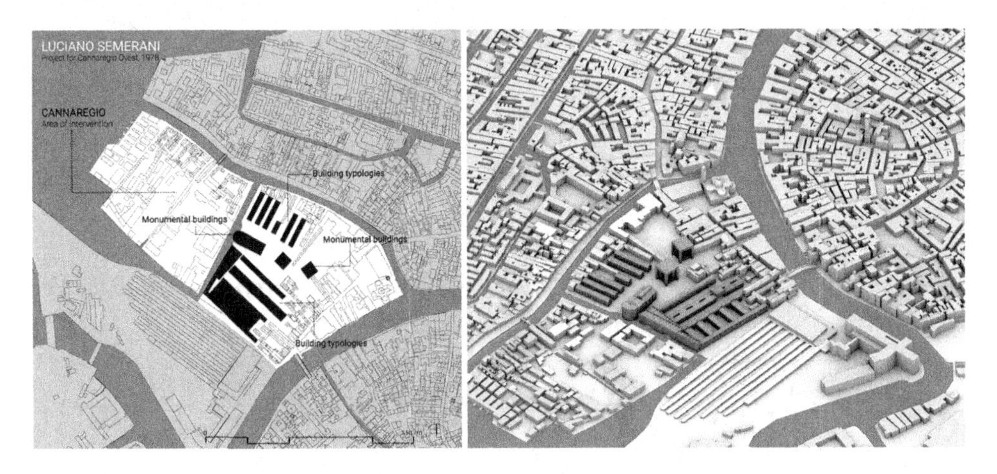

Figure 12. 3D model visualization in rendered mode with the insertion of the archive drawing of the Semerani's project planivolumetric view
Source: S. Vattano, 2022.

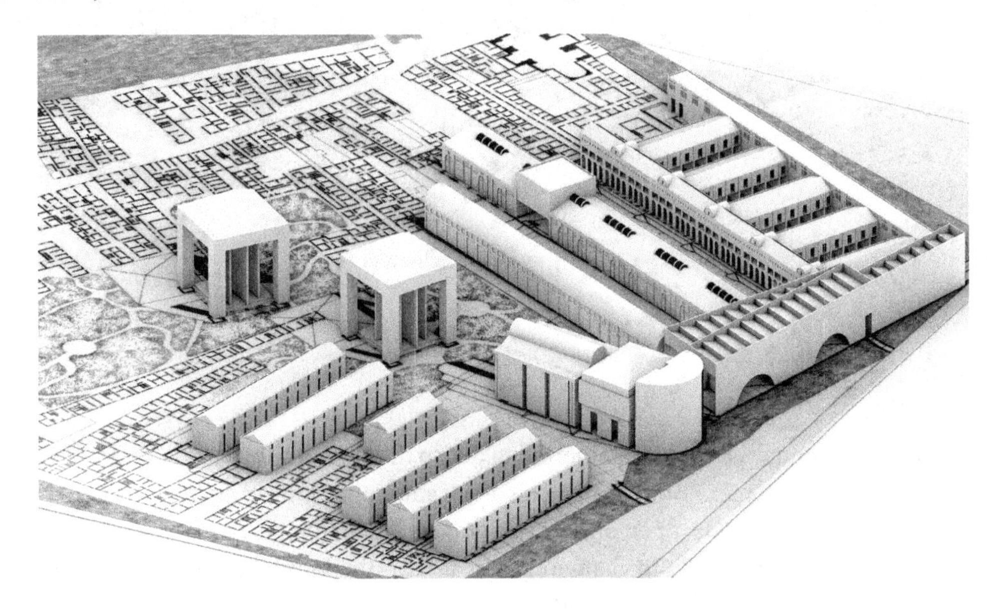

As is the case in his proposal for Cannaregio Ovest, in later projects his solutions will shape the past. As occurs for the port of Trieste, in 1990, in which the large building, pierced and curved, refers back to Adolf Loos' Babylon; or again, the insertion of the basilica type, included in an earlier project, in 1982, for the Catena fair, in Mantua, of which Semerani will repropose the planimetric layout: the scanning of two spaces with a rectangular base, with a stepped apse (Figure 14).

Figure 13. Top view and axonometric cross sections from the 3D digital model of the Semerani's project
Source: S. Vattano, 2022.

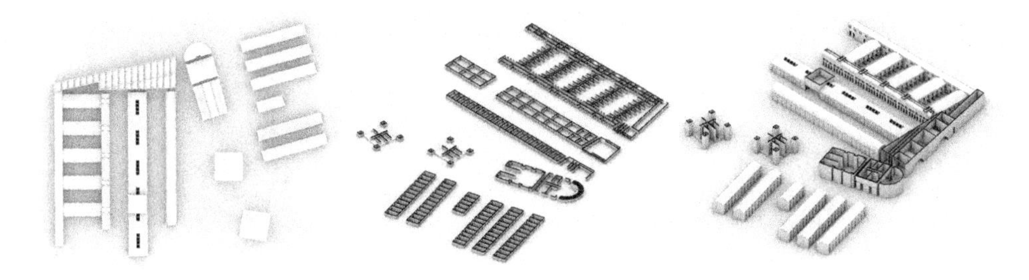

Figure 14. Axonometric view from the 3D digital model of the Semerani's proposal within the Venice model
Source: S. Vattano, 2022.

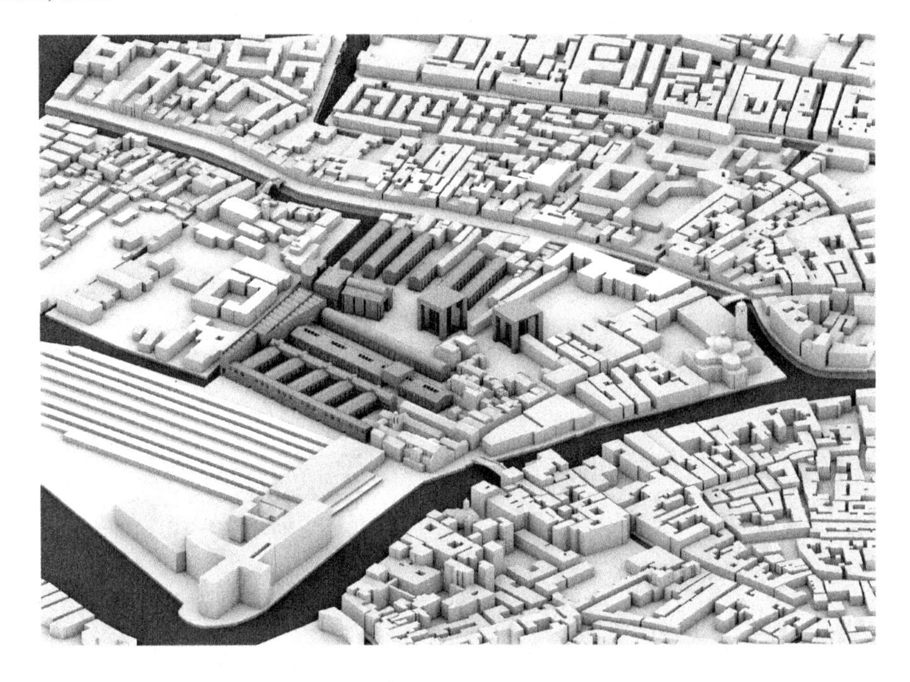

From the 1965-83 Cattinara Hospital, Semerani takes up the shed roofing of the block arranged with the long front on Rio della Crea; in the hypothesis for Cannaregio, however, he will not recall to the portico system, but to a solid façade, open only at two arches. Again, in the Town Hall of Osoppo, which followed the international convention in Cannaregio by a year, the building will be found covered by a large, ribbed vault that will recall the basilica-theater overlooking Rio della Crea.

The elements identified are some monumental buildings that recall previous projects and building types attested on Rio di Cannaregio and Rio de la Crea, the two pavilions with a quadrilateral plan, arranged on the same axis that contains the basilica-theater. The latter, arranged with an apse terminal on the canal, seems to configure the solution of the large exedra designed for the Parco Urbano di Bologna of 1984 or probably the rectangular-based pavilion of the Osoppo Town Hall of 1979, covered by a ribbed vault, as a citation to the Basilica of Vicenza (Semerani, 1990, p. 58). In an off-scale, a trapezoidal-plan building closes the courtyard, in-line, and terraced building types. The water entrances take up the model of a council house that is also an archetype (the Marinaressa) (Semerani, 1990). The roof is the shed roof of the Cattinara hospital, which returns, above, the porticoed scansion of the courtyard types. The Venetian canal system responds to the building types and their diversification. Thus, the specificities of the Venetian fabric and quotations from the past contribute to the restitution of a new order to the islands: "thus on the canal that separates the island of San Giobbe from the railway station are located courtyard houses, on the inner salizzada an in-line house with Palladian internal stairs; on the campo that is reinvented terraced houses. This is a general response to be explored, to the problem, in the Venetian context, of the monuments-residence relationship" (Semerani, 1990, p. 64).

The code built for the urban analogy (the Pesaro Cemetery), in 1979, establishes a cross-reference to the operation proposed by Semerani in Cannaregio, in the citation of some cultural references, as in the case of the «wise disposition of bodies on the landfills, typical of Hellenistic cities» (Semerani 2000, p. 54). A year after the proposal for Cannaregio, in the Pesaro Cemetery, the isolated monument with a quadrilateral base and the basilica-theater are found again.

Once again, as in the Osoppo Town Hall, Semerani will make present the architectural patterns proper to the public buildings of the Friulian-Venetian city: the Palazzo della Ragione in Padua, the Basilica in Vicenza, and the Municipal Palace in Udine. To the juxtaposition of autonomous figures, Semerani follows up a real montage by parts that define, in the operations of translation and rotation with respect to the boundary paths of the urban mesh, the unity of the whole (Figure 15).

Figure 15. Axonometric view from the 3D digital model of the Semerani's proposal
Source: S. Vattano, 2022.

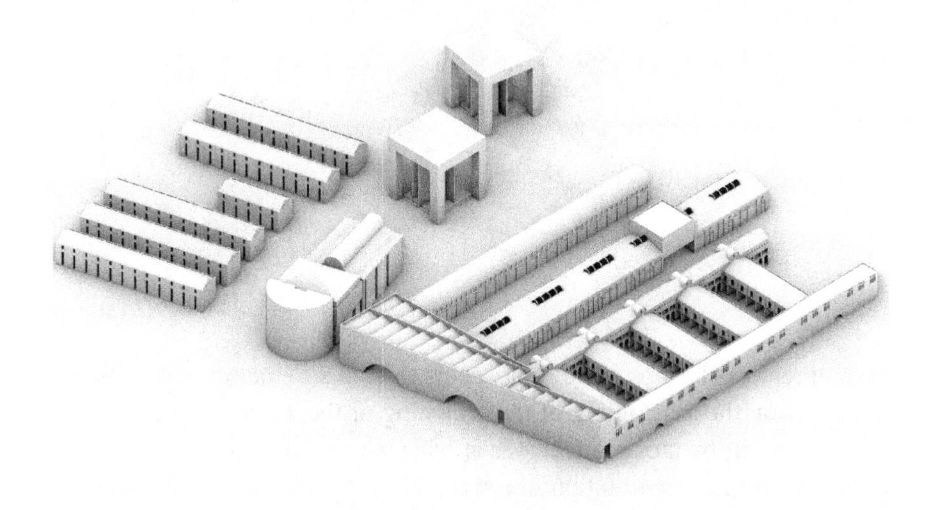

SOLUTIONS AND RECOMMENDATIONS

The chromatic expression that characterizes some of the project drawings, both in double projection and axonometry, intervenes in the accentuation of a "determined will"[3], to emphasize the definition of architecture which, according to Semerani, is never the development of an image, but sedimentation of successive images. Thus, in the alternation between the different techniques and methods of representation, the imaginative process embodies the combined role of dematerializing solidity, clarity in the constructive layout and the definition of a language.

Semerani notes that "the significance of a project lies at once in the object to which it referred and in the function that the project performs in the sense of delimiting and orienting the reference. Therefore, if the techniques of design, the problems of composing are in themselves neutral, agnostic, the significance of a project lies entirely in distinguishing a spoon from a city, a city from a city" (Dal Co, 1980, p. 154).

To describe his proposal, Semerani makes a model characterized by three perspex planes, each one showing: the models of the new intervention; the axonometry of the pre-existing buildings; and finally, the plan of Diocletian's palace in Spalato and the Crystal Palace, of 1851.

Figure 16. Axonometric view from the 3D digital model of the Semerani's proposal
Source: S. Vattano, 2022.

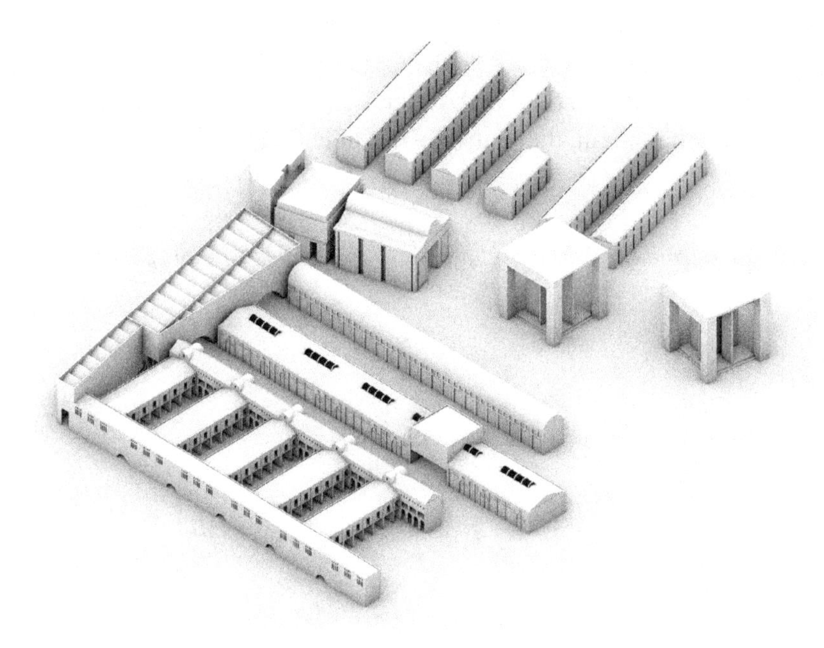

In this montage of represented architecture, it seems to read a Semerani who two decades later, recalling one of his walks on the scaffolding of the crowning of the Torre Velasca, wrote: "[...] I felt so clearly that the *machine* built by the architect was at once the appearance of a character, the unveiling of a symbol, the montage of a representation, the realization of an image. Reflection, mirroring, echoing, not copying, not reproduction, perhaps imitation, but, precisely, like the son on the father, like the marble on the fossil shell, concretion, metamorphosis, continuation in short or long times of the process" (Semerani, 1990, p. 10).

By the possibilities offered within the field of hypothesis, such as that which could constitute an international design competition in Venice, Semerani thought it important to test both the significance and the "scope of typological knowledge" inevitably linked to the sphere of abstraction on which the competition drew, proposing to develop three issues "[...] that of the relationship between urban theory and theory of architecture, that of the scalarity of design themes, and that of the significance, for architectural design, of the assumption of the urban context" (Dal Co, 1980, p. 154).

The graphic elaborations proposed by Semerani illustrate these reflections by grafting, to the design theme, cultural references (Figure 16). The first quotation given with a diagram on the urban structure of Venice is that of the Novissime project, at the conclusion of the Canal Grande, in the area of Piazzale Roma and Tronchetto. It adds that of the hospital designed by Le Corbusier, in 1963, which becomes an integral part of the proposed layout for Cannaregio.

Figure 17. Perspective rendered view from the 3D digital model of the Semerani's proposal
Source: S. Vattano, 2022.

A number of annotations organized by areas are made on the notes on the design choices: 1) the canal system; 2) the pedestrian system; and 3) settlement types. The reflection on the first aspect concerns the identification of a new perimeter crossing of the lagoon viability, starting from the connection between the station and the San Geremia area; the characters of the settlement types depend on the different hierarchical organization of the canals. This reflection is accompanied by a diagram in which Le Corbusier's hospital is related to Rio della Crea, Rio di Cannaregio and Rio di San Giobbe. Relative to the second issue, the pedestrian system, Semerani shift the focus from the canal system by virtue of proceeding inward into the sestiere, from the "island-manufatto-attributes" of the urban scale to the structures arranged on Lista di Spagna, the urban commercial axis. The identification of the three project cornerstones (the areas of S. Giobbe and S. Geremia and the ponte degli Scalzi) constitutes the first stage in understanding the reordering needs of S. Giobbe, in which the main places of relationship at the scale of Cannaregio are arranged (Figure 17).

Among the elements graphically isolated in the descriptive scheme of the third question posed by the two architects, that on settlement types, once again Le Corbusier's hospital polarizes, along with the railway station, the portion of land affected by the project. Under the heading "the island-manufacture" some reflections on the vocation of artifacts with different functions are given. One category: multifunctionality. The areas of the station and that of the former slaughterhouse leave space for references:

"the iron-and-glass architecture of the great world expositions, Le Corbusier's hospital compacted like Diocletian's palace in Spalato" (Dal Co, 1980, p. 158).

Concluding the reasoning, as we read from the annotations made on the sketches, the equipment system is the device that defines the convergence of architecture and island, also admitting that the limits of the site and those of the architectures become a single element. Other schemes, in the speed of their fulfilment, take up, on the one hand, the question of the structure of the routes on the Canal Grande and Lista di Spagna, with mixed-function buildings; on the other one, the serial systems of parallel blocks against party wall and terraced, long courtyard, and buildings with central courtyard.

Figure 18. Perspective rendered view from the 3D digital model of the Semerani's proposal
Source: S. Vattano, 2022.

The first solution develops two blocks tilted with respect to the axis of the project layout that define two urban park areas arranged outward. In the second solution, long and open courtyard systems fit between the station and S. Geremia area, arranged parallel to the parallel blocks against party walls systems along Lista di Spagna. Two orthogonal axes determine the layout of the system with respect to the parallel to the parallel blocks against party walls building and according to the development of the Rio di Cannaregio. In this solution, the architectures are organized "around a rectangular field (or a basilica)", as stated in the project annotations (Figures 18, 19).

The graphic material produced by Semerani for the settlement modules, as well as for the other architectural facts of the project "can only be generic because only a dialectical reading of the context can bring those deformations that shape the settlement module" [Rosa 1983, p. 30].

All these reflections, together with the theoretical assumptions promote a physical structuring of figuration of the relations between places, in which the dimension of collective cultural spaces dialogues with that of the space-theater and the space to be inhabited (Figures 20 - 22).

Figure 19. Perspective view from the 3D digital model of the Semerani's proposal
Source: S. Vattano, 2021.

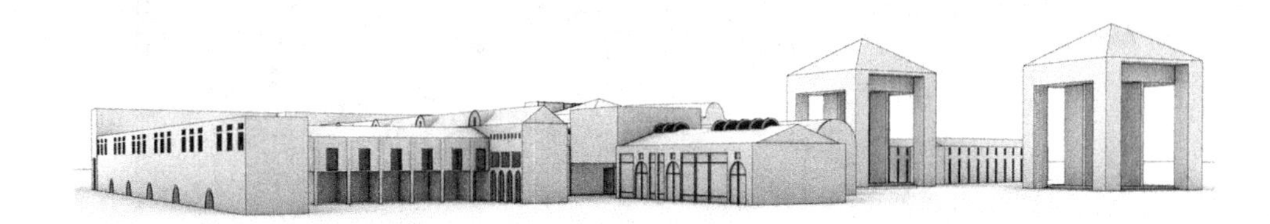

FUTURE RESEARCH DIRECTIONS

Considering the research experience carried out, the reading of the drawings kept in the Archivio Progetti Iuav, the digitization and subsequent processing of the digital models give access to unpublished images that describe, at different scales, multiple issues related to urban texture, form, functions, metaphorical meanings and the relationship of the city with the modern. Luciano Semerani's case study offers the possibility of developing a research method that is both transversal to different modes of representation and simultaneously open to content implementation. The digital model and the collection of information return an image of the city of Venice, the unbuilt one, which concerns not only the projects of the international competition that is the subject of this research, but also the vast unbuilt architectural heritage of the 20th century. In fact, the research is continuing with the collaboration of the Archivio Progetti Iuav, with the in-depth study and elaboration of interactive and searchable digital models obtained from the graphic interpretation conducted on the projects submitted for the Biennale Architettura of 1985 for the Ponte dell'Accademia, Ca' Venier dei Leoni and the 1988 Italian Pavilion. This is the creation of a shareable and integrated graphic corpus that returns images of an unbuilt Venice on the Canal Grande.

The digital models, by making known design experiences and adopted lines of research, shed light on the design making of architects from different backgrounds and strongly connected to the issues and methods of design within the historical pieces of the city, establishing a rich relationship of reciprocity between the archival drawings and the digital model and opening the field to a broader research on the Italian question of building the city.

On the track of the research initiated in collaboration with the direction of the Archivio Progetti, the research is developing by still looking at the vast documentary heritage held in the Archives on the unbuilt architecture in Venice. The theme can be located in the context of the 1985 Biennale and in some of the most significant competitions on the track of the Canal Grande, dwelling in particular on project proposals defining a mapping of otherwise unexplored architectural events of Venetian memory. In fact, the research is continuing with the collaboration of the Archivio Progetti Iuav, with the in-depth study and elaboration of interactive and searchable digital models obtained from the graphic interpretation conducted on the projects submitted for the Biennale Architettura of 1985 for the Ponte dell'Accademia, Ca' Venier dei Leoni and the 1988 Italian Pavilion.

The digital reconstructions created as part of this research are formalized into a body of informative and explorable elements that return a heterogeneous narrative about the unbuilt in Venice. The possibilities offered by the open consultation of the data make the digital products an additional resource for both users, broadly speaking, and the scholarly community.

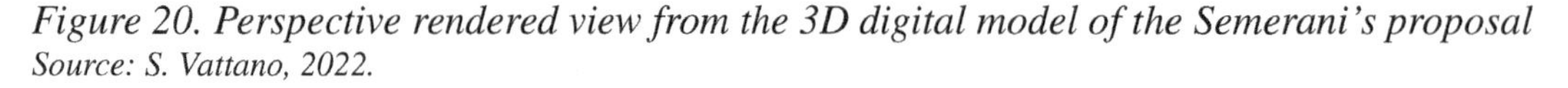

Figure 20. Perspective rendered view from the 3D digital model of the Semerani's proposal
Source: S. Vattano, 2022.

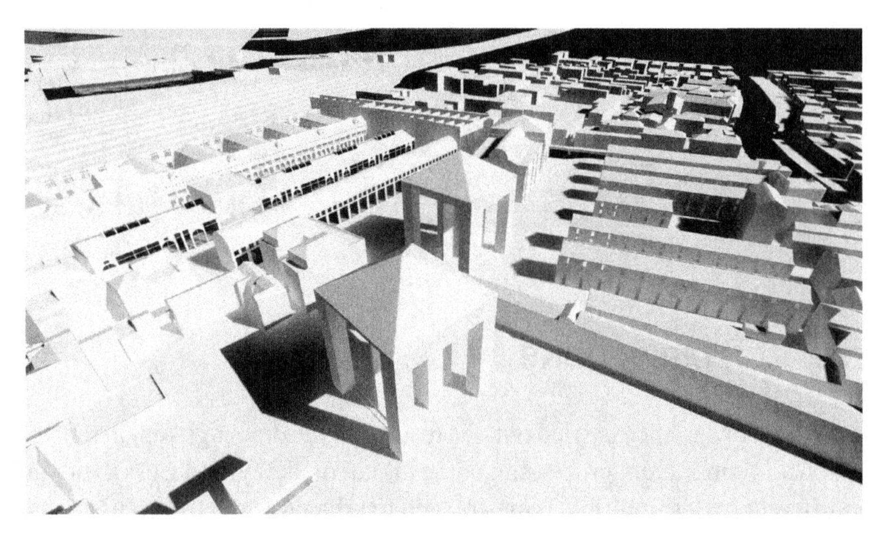

The consultation activities of digital reinterpretations imply a direct involvement of the observer in the acquisition of knowledge about the archival material by means of different forms of representation to which the research presented here relies. Part of the research published on the online site of the Archivio Progetti Iuav has made it possible to activate additional forms of sharing the graphic heritage and digital interpretations that take place in the context of workshops, seminar activities, and dissemination of knowledge by means of special visualization tool. In this sense, these mixed modes of visualization and exploration have been considered as a starting point in international multi-university collaborative competitions, providing new materials for consultation and knowledge of the heritage, in addition to the one already in the archive.

With the aim of implementing the current modes of access and consultation of the material, of visualization and interaction with the graphic information in the archive for external users, the research project is conducting in-depth interpretation, digitization, and graphic reconstruction through the elaboration of digital models, which can also be explored in immersive mode, that provide multiple and unseen readings of an imagined and unbuilt Venice.

CONCLUSION

Reading once again through new images the unbuilt architectural project produces visual solutions that, while investigating the reality of the urban context, also call into question the cultural and functional alterations occurred over time. The possibility of observing the drawing/model from viewpoints that can be explored by means of the technological and visual opportunities offered by digital technology is offered to third-party observers by attributing further meanings and figurations to the documentary and graphic material preserved at the Archivio Progetti Iuav.

In this sense, the in-depth case studies, thanks to the digital tools employed and the ways in which the reconstructed models can be visualized, including in immersive mode, have been the subject of various forms of exploration. The creation of virtual tours, obtained with 360° images and generated by rendering software, provided direct and interactive access to digital environments in which it is possible to consult in-depth texts, short videos of the models or links to websites to also consolidate the integration between the archival document and the digital product.

The possibility of using open-access platforms, for the publication of reconstructed models, also responds to the need for a solid connection between new generations and cultural heritage, a topic also put in place by the EU, with the launch of the European Year of Cultural Heritage already in 2018.

Figure 21. Perspective view from the 3D digital model of the Semerani's proposal
Source: S. Vattano, 2022.

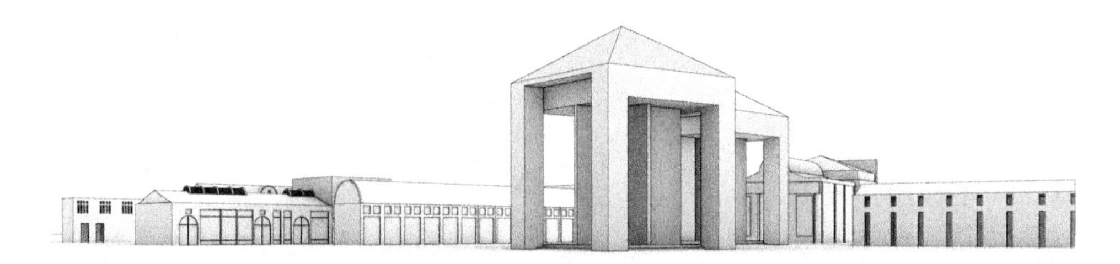

New digital products redefine how cultural heritage, whether urban, architectural or graphic-documentary, can be accessed, returning it in an open form available for digital media use. Some of the results of the research developed here, including videos of digital environments, renderings of models, in-depth report boards, have been published on the web pages of the Iuav Project Archive, as part of the collections and research contained therein[4].

Semerani's investigated project, redrawn and elaborated in digital forms that are easily usable and accessible, is identified within a mapping of the lagoon city that constitutes the starting point and at the same time the identification of the themes addressed, urban dynamics and lived spaces. Interactive virtual tours make it possible to trace paths of knowledge by the user who comes into contact with the places of the Venetian project, through immersive experiences of an unexplored reality.

The development of this kind of research activity opens up new questions about the richness of the nourished themes present in the archival collection, in parallel addressed to further transversal disciplinary articulations: landscape, architectural design, communication.

Reconstructing memory through the valorization of archival drawings represents an analytical way of bringing to light the cultural, theoretical and technical quality of building the Italian city, starting from Venice, understood as a case study and starting point for the rethinking of new imaginaries, new ways of telling and building the spaces of history within the contemporary city; in this sense, a new reflection on the images of Venice, provides multiple keys to the interpretation of an emblem of made in Italy.

From the perspective of an increasingly shared and accessible enjoyment of cultural heritage, graphic interpretations of an unbuilt and little-known Venice inevitably confront the question of the time-and-space location of the project, "a given representation of an architectural environment, executed under particular time-space conditions and the result of a specific attitude of thought [...] is subject to being studied over time by different observers each of whom ascribes to each sign meanings that are a function of his/her knowledge, his/her aspirations, and his/her own judgment of the representational techniques used" (De Rubertis, 2010, p. 121).

Therefore, it is a matter of constructing a narrative through images about some moments of Venetian architecture, of the possible memory that in the contemporary digital transition establishes a hybrid form of heritage reading: on the one hand, the archival document, in its meaning as a source of knowledge; on the other hand, the explorable and immersive digital dimension that amplifies the perception of space by offering itself in its evidence quality to users.

Figure 22. Axonometric view from the 3D digital model of the Semerani's proposal within Venice model
Source: S. Vattano, 2022.

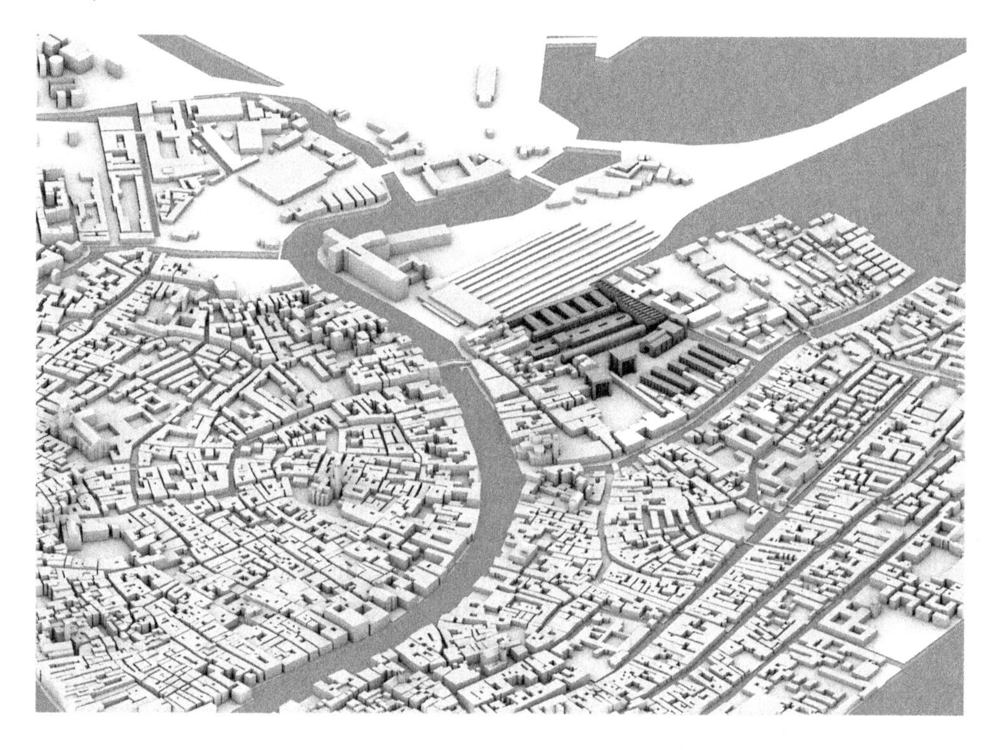

Research Results

The reconstruction of the ten projects presented as part of the international competition on the rethinking of the Cannaregio Ovest sestiere brings to light a little-explored face of Venice, the unbuilt city, of the project left on paper containing multiple issues. Some of the outcomes of the research, as reported in the preceding paragraphs, involve the development of explanatory videos on the ten projects, with the inclusion of texts, schemes of analysis of the intervention site, combinations of archival drawings and digital model, providing heterogeneous narratives on the ten reconstructed proposals. The overall video has been posted on the Archivio Progetti Iuav - Video page (a, b). An online exhibit (c, d), a virtual tour with a summary of the main elements that define the ten projects was posted on the Archivio Progetti Iuav - Petit Tour page (a). It consists of 24 images that briefly describe (two images per project) the main features of each project, accompanied by short text descriptions. The integration between the new reconstructions and the archival images published in this section allow open access to the consultation of high-resolution images and to complement some of those existing in the online section of the archive.

a) Video of the models available at the Archivio Progetti Iuav web page (number #8): http://www5.iuav.it/homepage/webgraphics/IUAV-PAGINE.INTERNE/IUAV-MOSTREONLINE/video.htm, Retrieved April 13, 2022.
b) Vattano S. (cur.) (2022), Video – 10 immagini per Venezia, available at: https://www.youtube.com/embed/FED4r4rFb_A, Retrieved April 13, 2022.
c) Petit-Tour of the models available at the Archivio Progetti Iuav web page (number #50): http://www5.iuav.it/homepage/webgraphics/IUAV-PAGINE.INTERNE/IUAV-MOSTREONLINE/petit-tour.htm, Retrieved April 13, 2022.
d) Vattano S. (cur.) (2022), Petit-Tour – 10 immagini per Venezia. Narrare il non costruito nelle esplorazioni digitali dei 10 progetti per Cannaregio Ovest, 1978, available at: http://www5.iuav.it/homepage/webgraphics/IUAV-PAGINE.INTERNE/IUAV-MOSTREONLINE/10IMMAGINI/10immagini.htm, Retrieved April 9, 2022.

ACKNOWLEDGMENT

We thank the Archivio Progetti Iuav (http://www.iuav.it/ARCHIVIO-P/) for the kind permission of the documentary sources provided during the research and included in the digital models of the following images: 6, 8, 9, 12.

REFERENCES

Ajroldi, C. (Ed.). (1987). *Dieci progetti come occasione di studio: architetture di Carlo Aymonino, Guido Canella, Paul Chemetov, Costantino Dardi, Vittorio De Feo, Vittorio Gregotti, Gianugo Polesello, Alberto Samonà, Luciano Semerani, Alvaro Siza Vieira*. Officina.

Amistadi, L. (Ed.). (2005). *Saper credere in architettura: quaranta domande a Luciano Semerani*. Clean.

Ch'ng, E., Li, Y., Cai, S., & Leow, F. T. (2020). The effects of VR environments on the acceptance, experience, and expectations of cultural heritage learning. *Journal on Comput Cult Herit*, *13*(1), 1–20. doi:10.1145/3352933

Cipresso, P., Giglioli, I. A. C., Raya, A., & Riva, G. (2018). The past, present, and future of virtual and augmented reality research: A network and cluster analysis of the literature. *Frontiers in Psychology*, *9*, 1–20. doi:10.3389/fpsyg.2018.02086 PMID:30459681

Dal Co. F. (1980). 10 immagini per Venezia. Venezia, Italy: OfficinaEdizioni.

Exhibition Catalogue. (2013). *Luciano Semerani: l'architetto nelle città*. Torino, Italy: Accademia University Press.

Grilli, E., & Remondino, F. (2019). Classification of 3D Digital Heritage. *Remote Sensing*, *11*(7), 1–23. doi:10.3390/rs11070847

Karp, C. (2004). Digital Heritage in Digital Museums. *Museum International*, *56*(1-2), 45–51. doi:10.1111/j.1350-0775.2004.00457.x

Laing, R. (2020). Built heritage modelling and visualisation: The potential to engage with issues of heritage value and wider participation. *Developments in the Built Environment.*, *4*, 1–7. Retrieved March 13, 2022, from. doi:10.1016/j.dibe.2020.100017

Li, Y., & Ch'ng, E. (2022). A Framework for Sharing Cultural Heritage Objects in Hybrid Virtual and Augmented Reality Environments. In Visual Heritage: Digital Approaches in Heritage Science. Springer. doi:10.1007/978-3-030-77028-0_23

Marinković, B., Šegan Radonjić, M., Novaković, M., & Ognjanović, Z. (2022). Digital Documentation Management of Cultural Heritage. In S. D'Amico & V. Venuti (Eds.), *Handbook of Cultural Heritage Analysis*. Springer. doi:10.1007/978-3-030-60016-7_74

Opgenhaffen, L., Lami, M., & Mickleburgh, H. (2021). Art, Creativity and Automation. From Charters to Shared 3D Visualization Practices. *Open Archaeology*, *7*(1), 1648–1659. doi:10.1515/opar-2020-0162

Pavlidis, G., & Koutsoudis, A. (2022). 3D Digitization of Tangible Heritage. In S. D'Amico & V. Venuti (Eds.), *Handbook of Cultural Heritage Analysis*. Springer. doi:10.1007/978-3-030-60016-7_47

Pescarin, S. (2016). Digital heritage into practice. *SCIRES, 6*(1), 1-4. http://www.sciresit.it/article/view/12003

Rahaman, H. (2018). Digital heritage interpretation: A conceptual framework. *Digital Creativity*, *29*(2-3), 208–234. doi:10.1080/14626268.2018.1511602

Rosa, G. (Ed.). (1983). *Semerani+Tamaro. La città e i progetti*. Kappa.

Rushton, H., & Schnabel, M. A. (2022). Immersive Architectural Legacies: The Construction of Meaning in Virtual Realities. In Visual Heritage: Digital Approaches in Heritage Science. Springer. doi:10.1007/978-3-030-77028-0_13

Rushton, H., Silcock, D., Schnabel, M. A., Moleta, T., & Aydin, S. (2018). Moving images in digital heritage: architectural heritage in virtual reality. *AMPS series, 14*, 29-39.

Semerani, L. (1983). *Progetti per una città*. FrancoAngeli.

Semerani, L. (1987). *Lezioni di composizione architettonica*. Arsenale.

Semerani, L. (1991). *Passaggio a nord-est: itinerari attorno ai progetti di Luciano Semerani e Gigetta Tamaro*. Electa.

Semerani, L. (2000). *Luciano Semerani e Gigetta Tamaro: Architetture e progetti*. Skira.

Semerani, L. (2000). *L'altro moderno*. Allemandi.

Semerani, L. (2007). *L'esperienza del simbolo: lezioni di teoria e tecnica della progettazione architettonica*. Clean.

Semerani, L. (2013). *Incontri e lezioni: attrazione e contrasto tra le forme*. Clean.

Semerani, L. (2020). *Il ragazzo dell'IUAV*. LetteraVentidue.

Sherman, W. R., & Craig, A. B. (2003). *Understanding Virtual Reality: Interface, Application, and Design*. Morgan Kaufmann Publishers, Inc.

Silcock, D., Rushton, H., Rogers, J., & Schnabel, M. A. (2018). Tangible and intangible digital heritage: creating virtual environments to engage public interpretation. In *Computing for a better tomorrow, 36th annual conference on education and research in computer aided architectural design in Europe, 2*. Lodz University of Technology. 10.52842/conf.ecaade.2018.2.225

Skublewska-Paszkowska, M., Milosz, M., Powroznik, P., & Lukasik, E. (2022). 3D technologies for intangible cultural heritage preservation-literature review for selected databases. *Heritage Science, 10*(3), 1–24. doi:10.118640494-021-00633-x PMID:35003750

Testi, G. (Ed.). (1980). *Progetto Realizzato*. Marsilio Editori.

UNESCO. (2003). *Text of the Convention for the Safeguarding of the Intangible Cultural Heritage*. Retrieved April 8, 2022 from https://ich.unesco.org/en/convention

Wang, X. (2020). Digital Heritage. In H. Guo, M. F. Goodchild, & A. Annoni (Eds.), *Manual of Digital Earth*. Springer. doi:10.1007/978-981-32-9915-3_17

Windhager, F., Federico, P., Mayr, E., Schreder, G., & Smuc, M. (2016). A Review of Information Visualization Approaches and Interfaces to Digital Cultural Heritage Collections. In *Proceedings of the 9th Forum Media Technology*. St. Pölten, Austria: CEUR-WS. Retrieved March 7, 2022 from https://www.researchgate.net/publication/313442197_A_Review_of_Information_Visualization_Approaches_and_Interfaces_to_Digital_Cultural_Heritage_Collections

KEY TERMS AND DEFINITIONS

Archive Drawing Digitisation: Reading and integration of the missing information by means of models elaborated with an information level suited to their exploratory possibilities and to be published on online and open source platforms enhancing the accessibility of the range of users.

Digital Exploration: The valorisation and online dissemination of the graphic heritage preserved in the archives aims at the construction of memory by means of digital interpretations and elaborations telling new places and imaginaries.

Digital Hybridisations: Combinations of images, integrations between the digital model and the archive drawing capable of amplifying the perception of the imagined spatialities and providing an increasingly integrated degree of knowledge.

Digital Knowledge: Increasingly massive way of collecting data able of bringing into its domain not only groupings of digital information not only belonging to specific institutions or entities, but also their interoperability to create a dissemination of knowledge based on shared and networked exploration.

Digital Objects: Items produced to provide a level of information that can be consulted from fixed or mobile devices and on open source platforms in different formats and viewing modes, depending on the type of use: images, virtual tours, explorable in standard and immersive modes, and videos.

Typological Knowledge: Linked to the sphere of abstraction, this kind of knowledge develops three issues: the relationship between urban theory and theory of architecture; the scalarity of design themes; the significance of the assumption of the urban context, for architectural project.

Unbuilt Project Views: A given representation of an architectural environment, executed under particular time-space conditions, a result of a specific attitude of thought that is subject to being studied over time by different observers who can ascribes to each sign meanings as a function of own knowledge, aspirations, and judgment on the representational techniques used.

ENDNOTES

1 Ten groups of architects participated in the convention, which were headed by: Raimund Abraham, Carlo Aymonino, Peter Eisenman, John Hejduk, Bernhard Hoesli, Rafael Moneo, Valeriano Pastor, Gianugo Polesello, Aldo Rossi, and Luciano Semerani.

2 Vattano S. (cur.), *10 immagini per Venezia. Narrare il non costruito nelle esplorazioni digitali dei 10 progetti per Cannaregio Ovest, 1978*, in Petit Tour, © Archivio Progetti Iuav. Dispoibile al link: http://www5.iuav.it/homepage/webgraphics/IUAV-PAGINE.INTERNE/IUAV-MOSTREONLINE/10IMMAGINI/10immagini.htm, Retrieved April 4, 2022.

3 In an interview Giovanni Fraziano did with Gigetta Tamaro about the design method adopted in her work with Luciano Semerani, the architect responds to the question, "Many times, the realization of the diversity between the two characters in front of me has been cleared by Semerani with a precise but sibylline phrase: Gigetta comes from sculpture... I, on the other hand, from painting. What does this coming explicitly from sculpture and painting mean more explicitly?", Gigetta Tamaro replied, "Luciano always says that it is also necessary to use color... because color in his opinion helps to accentuate a certain will" (Semerani, 2000, p. 68). For the full interview, please refer to the text: Semerani, L. (2000). *Tamaro G., Luciano Semerani e Gigetta Tamaro: Architetture e progetti*. Milano, Italy: Skira, 67-71.

4 For more information, see the Archivio Progetti Iuav web pages available at the following links: http://www5.iuav.it/homepage/webgraphics/IUAV-PAGINE.INTERNE/IUAV-MOSTREONLINE/video.htm, Retrieved April 1, 2022; "10 immagini per Venezia": https://www.youtube.com/embed/FED4r4rFb_A, Retrieved April 4, 2022; "10 immagini per Venezia. Narrare il non costruito nelle esplorazioni digitali dei 10 progetti per Cannaregio Ovest, 1978": http://www5.iuav.it/homepage/webgraphics/IUAV-PAGINE.INTERNE/IUAV-MOSTREONLINE/10IMMAGINI/10immagini.htm, Retrieved April 4, 2022.

Section 4

Healthcare and Fragile People

Chapter 22
Introducing Mixed Reality for Clinical Uses

Giuseppe Emmanuele Umana
https://orcid.org/0000-0002-1573-431X
Gamma Knife Center, Ospedale Cannizzaro, Italy

Paolo Palmisciano
Gamma Knife Center, Ospedale Cannizzaro, Italy

Nicola Montemurro
https://orcid.org/0000-0002-3686-8907
Azienda Ospedaliera Universitaria Pisana, Italy

Gianluca Scalia
https://orcid.org/0000-0002-9465-2506
Garibaldi Hospital, Italy

Dragan Radovanovic
University of Belgrade, Serbia

Kevin Cassar
University of Malta, Malta

Stefano Maria Priola
https://orcid.org/0000-0002-5153-6230
Northern Ontario School of Medicine University, Canada

Igor Koncar
University of Belgrade, Serbia

Predrag Stevanovic
University of Belgrade, Serbia

Mario Travali
Cannizzaro Hospital, Italy

ABSTRACT

The advent of mixed reality (MR) has revolutionized human activities on a daily basis, striving for augmenting professional and social interactions at all levels. In medicine, MR tools have been developed and tested at an increasing rate over the years, playing a promising role in assisting physicians while improving patient care. In this chapter, the authors present their initial experience in introducing different MR algorithms in routine clinical practice from their implementation in several neurosurgical procedures to their use during the COVID-19 pandemic. A general summary of the current literature on MR in medicine has also been reported.

DOI: 10.4018/978-1-6684-4854-0.ch022

INTRODUCTION

Mixed reality (MR) allows to visualize the real world and holographic 3D objects at simultaneously. In medicine, the real-time interaction of these two elements may allow improved understanding of the human anatomy. Detailed knowledge of each patient's represents the mainstay for optimal surgical planning. This is especially true in neurosurgery, where the target diseases are frequently located into deep sited and critical regions of the brain, needed to be treated with concurrent preservation of adjacent functionally-intact brain structures to ensure patient's optimal quality-of-life and survival. While the currently available technology in daily neurosurgical practice allows physicians only to look at a distant monitors during real-time operations, mixed reality (MR) allows to visualize directly "through" the patient's anatomy. In this way, the patient's anatomy is perceived with a greater detail, and the surgical planning may be devised within the surgeon's mind with more confidence. The hybrid visualization of virtual objects and real-world's anatomy represents the most recent advancement, which further offers the ability to customize the surgical planning and to share it with the surgical team, the patients, and their family. In this chapter we present the MR applications performed at the department of neurosurgery of the Cannizzaro Hospital, Catania, Italy, and in the two departments of C19-ICU unit and clinic for Vascular and Endovascular Surgery of the University Clinical-Hospital Centre "Dr Dragiša Mišović-Dedinje", Belgrade, Serbia.

BACKGROUND

The hardware that we use is HoloLens 2 (HL2) (Microsoft TM). It is a head-mounted display unit connected with a remote cloud for images reconstruction and audio-video storage (Figure 1).

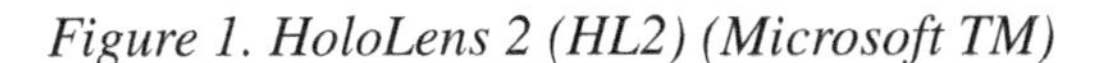
Figure 1. HoloLens 2 (HL2) (Microsoft TM)

The headset can be adjusted on the user head and tilted up, down, forward, or backward through a posterior crown (Davies, 2015). The anterior part of the device contains several sensors and their related hardware, including the processors, cameras and projection lenses. The visor is dark-coloured and includes two transparent combiner lenses, in whose inferior portion are shown the projected images (Kipman & Juarez, 2015). The device's settings can be personalized based on the user's vision characteristics (the interpupillary distance (IPD), or accustomed vision of the user) (Hachman, 2015; Hollister, 2015). Two 3D audio speakers are site close to the user's ears, allowing to simultaneously listen sounds from the real world and virtual reality. Using head-related transfer functions (HRTF), the HL2 produces binaural audio, offering a real-life virtual experience (Holmdahl, 2015; Microsoft., 2020).

The frontal portion of the device contains two buttons, one at the right for volume control and one at the left for brightness control. The posterior portion has a concave power button and 5 lightning spots, showing battery levels (Munzer et al., 2019). Connectivity include Wi-Fi 802.11 ac, Bluetooth 5.0 e USB C and overall technical characteristics can be summarized as follows: transparent holographic lenses, resolution 2k 3: 2 LED light engine, radians> 2.5k, eye-based rendering optimization of the display for 3D eye position, tracking for head and eyes, depth by Azure Kinect sensor, accelerometer, gyroscope, magnetometer, camera: 8MP still images, 1080p30 video (Hempel, 2015; Holmdahl, 2015; Microsoft, 2015; Rubino, 2018).

MIXED REALITY TESTS IN NEUROSURGERY

Informed Consent

MR offers the opportunity to record video and audio during the holographic representation. This is of special interest during informed consent as it can enhance and augment patient's understanding of surgical risks and goals. The possibility to record the discussion between physicians, patients and family members may be also useful to minimize the risk of possible legal issue related to patient's misunderstanding of the therapeutic plan.

Head Tumor

At our institution, we use MR for preoperative and intraoperative navigation during brain tumor surgery. In the preoperative phase, the holographic rendering offers an immersive experience, increasing the understanding of the patient's brain anatomy and contributing to highlight the safer surgical corridor and the adjacent critical brain areas. During surgery, MR can be combined with the standard optic navigation, used as comparison to evaluate the accuracy of MR holographic rendering. In a period of 1 year, we tested MR in 53 brain tumor patients and compared its performance to the standard optic navigation in terms of real-time "geolocation" within the brain anatomy. The holographic navigation proved to be accurate, reliable, and with equal accuracy compared to the standard optic navigation. The only limitation we encountered was characterized by the overlapping of the hologram, which is manual, based on the visual overlapping of the 3D object over the patient's anatomy, and, thus, operator dependent (Figure 2).

Interestingly, the device allows to scroll the dataset not only with the MR reconstruction, but in any spatial orientation (Figure 3).

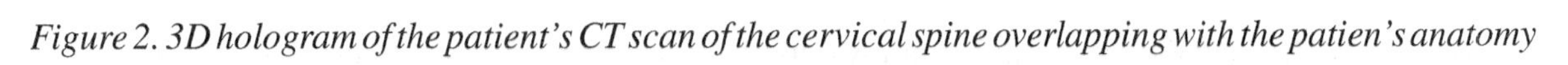

Figure 2. 3D hologram of the patient's CT scan of the cervical spine overlapping with the patien's anatomy

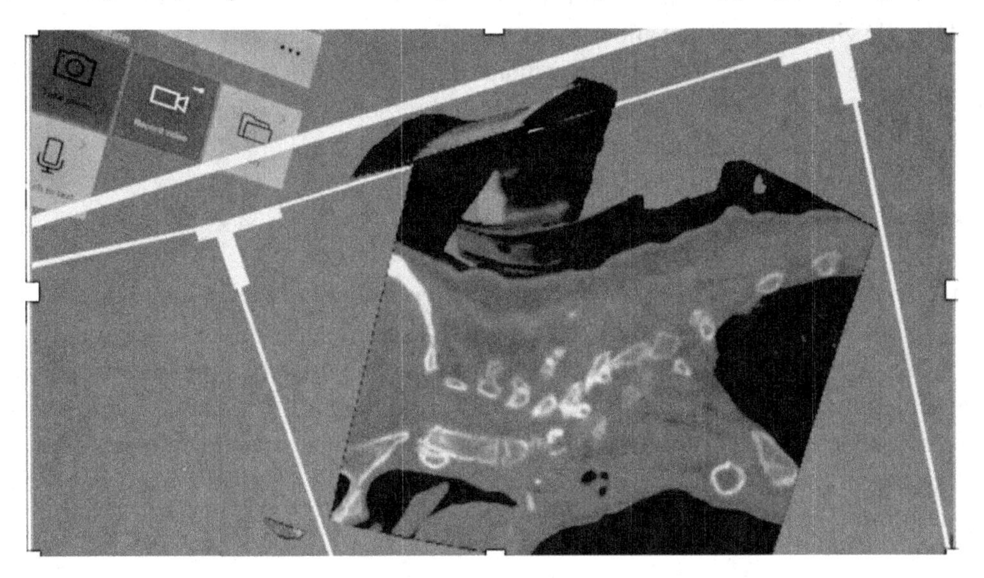

Figure 3. Real-time scrolling of the MR 3D hologram overlapped with the patient's anatomy

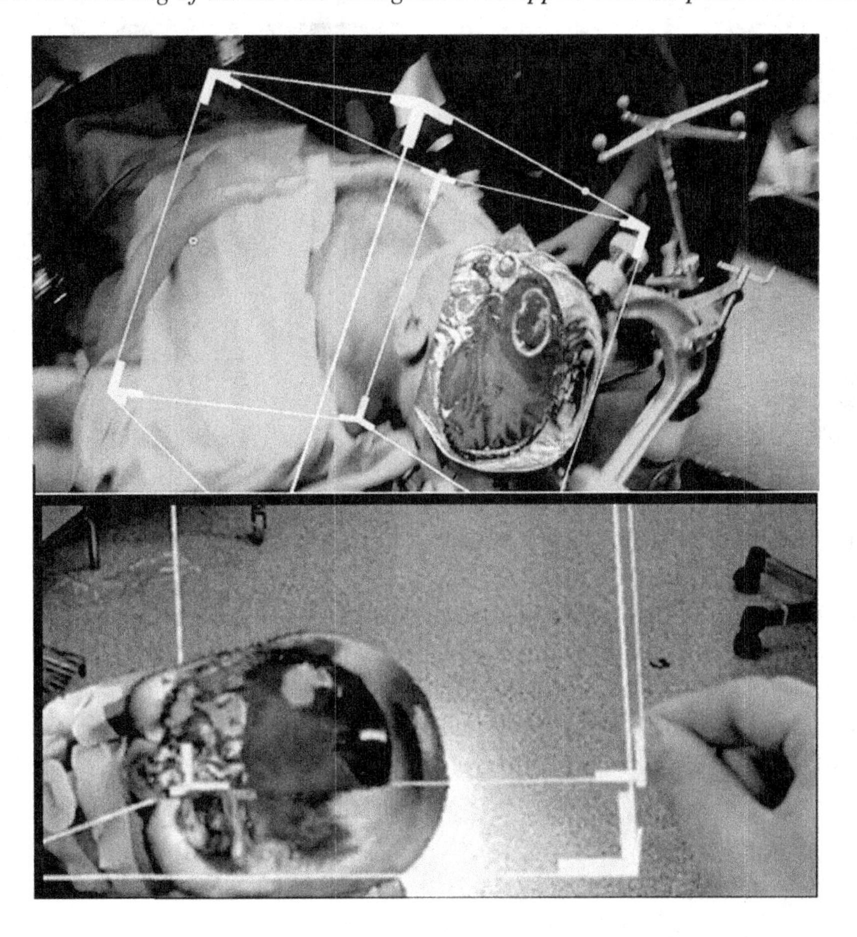

To improve the visibility of the hologram, we used greenlight in our OR. Greenlight offers a good visibility of the 3D virtual object and allow to safely perform surgery in real-time with minor chromatic limitations (e.g., the blood is perceived as dark-blue colored). On the other hand, the use of the direct illumination with white light, prevents to visualize the hologram, thus forcing the surgeon to move away from the white light field while using the MR navigation.

Head Trauma

Head trauma is one of the most important causes of major disability in the young population (Montemurro et al., 2020). In selected cases, decompressive craniectomy (i.e., removal of part of the skull) is required as a life-saving procedure to reduce the intracranial compression of the brain by post-traumatic intracranial hematomas. In survivor patients, cranioplasty (e.g., surgical placement of custom-made skull prostheses) is performed to reconstruct the skull as a cosmetic and therapeutic measure (Montemurro et al., 2021). Prostheses are custom-made, aimed to perfectly fit with the patient's anatomy to be correctly positioned at the level of the bone defect (Figure 4).

Figure 4. CT scanning of the skull prosthesis to plan th surgical implant

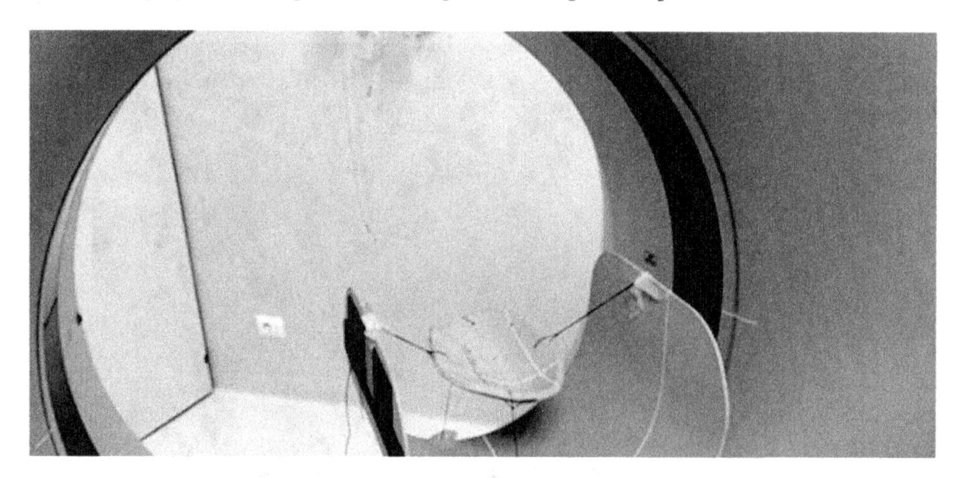

MR rendering can be used alongside templates to evaluate the accuracy of the prosthesis before surgery and during the operation (Figure 5).

Giving that cosmetic results is one of the main goals in these patients, the ability to share with the patients the expected outcome before the surgery is of outmost importance, and helps to increase patients' confidence in undergoing this type of surgery. MR is also a useful tool to improve the visualization of fracture classification of the base of the skull and safely plan best treatment strategies (Umana et al., 2022) (**Figure 6**).

Figure 5. A) modified "n" shaped skin flap scar and the phantom prosthesis over it. The peculiarity of this skin flap is represented by the vascular sparing of the afferent vessels. Note the posterior end of the incision, that finish at the level of the parietal prominence, securing blood supply from the posterior auricular artery and from the superficial temporal artery as well. This nuance, prevent necrosis of the posterior portion of the scar and its related complications; B) final intraoperative positioning of the custom-bone cranioplasty. C1-D1-E1) 3D MR rendering of the bone defect; C2-E2) intraoperative overlapping of the 3D MR object of the skull and the planned cranioplasty simulation; C3-D3-E3) skin flap scar before the cranioplasty procedure; C4-D4-E4) soft tissue MR rendering of the bone defect; D2) holographic intraoperative navigation of the cranioplasty prosthesis offers a similar visial feedback of the usual phantom; E) post-cranioplasty MR skin and temporalis muscle reconstruction rendering

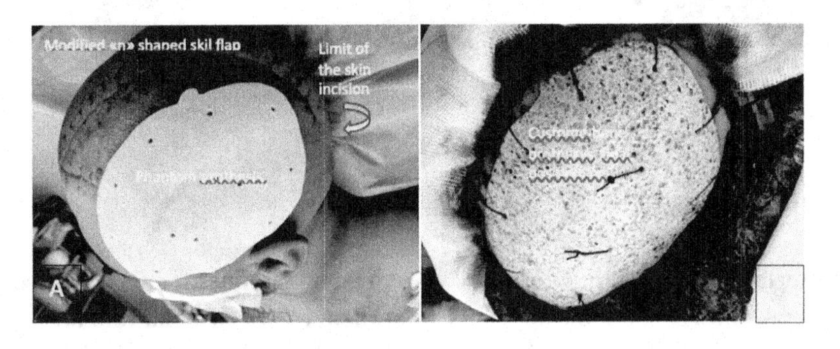

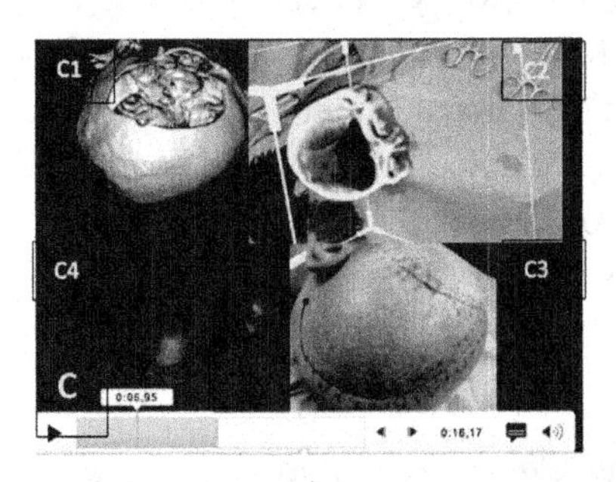

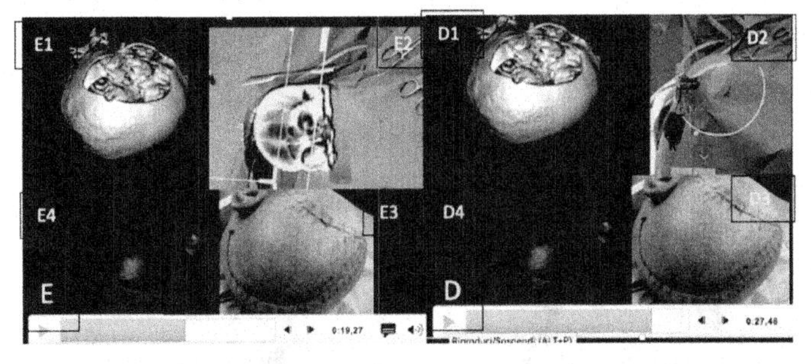

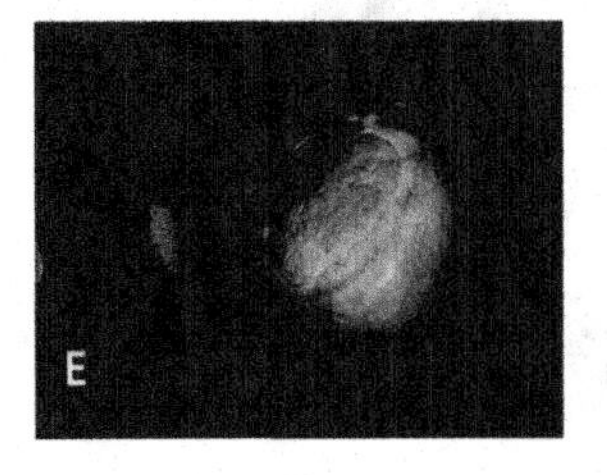

Figure 6. 3D holographic classification of anterior skull base fractures

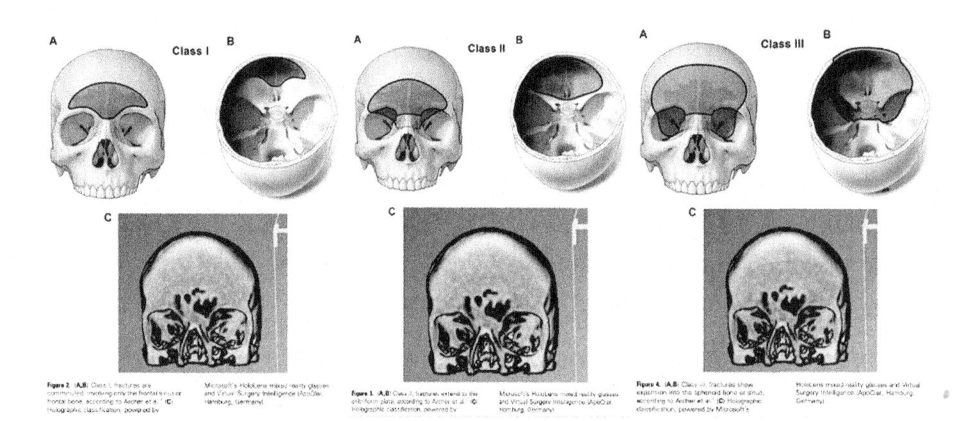

Cerebrovascular Surgery

Aneurysms and arteriovenous malformations (AVM) represent a unique challenge for neurosurgeons. The surgical plan must take into account the spatial development of the vascular dilatation or malformation to customize the surgical approach accordingly. MR allows to see through the patient's head both before surgery and during patient's positioning, so to allow to identify the exact location of the target disease. This is currently performed using 3D reconstruction visualized on a computer after prior elaboration by a neuroradiologist. The possibility to visualize the 3D objects, rendering the aneurysm or the AVM directly "in" the patient, further improves the quality of the information available for the surgeon in the preoperative settings, likely to improve performances and outcomes (Figure 7).

Figure 7. MR rendering is helpful to improve the three-dimensional perception of the vascular malformation in the planning phase, both for brain vascular malformation and stroke. The figure shows (*) the occlusion of the internal carotid artery in its extracranial tract, in a case of tandem occlusion of a stroke patient

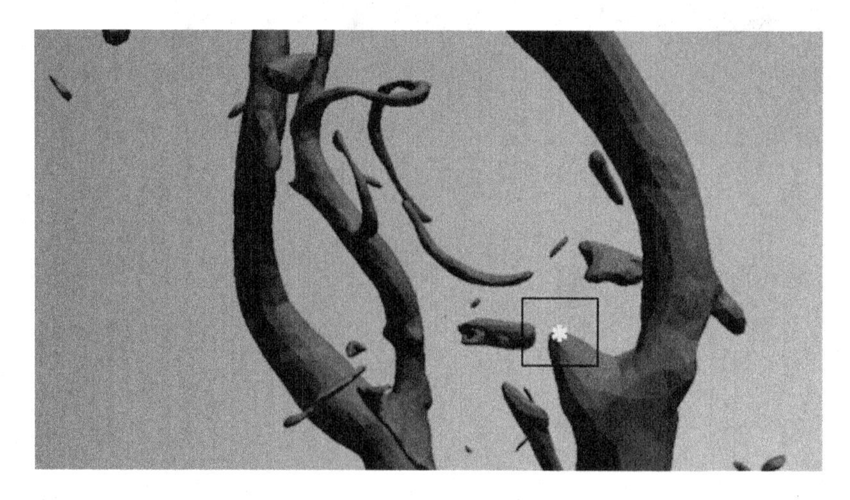

Spine

Pilot studies in spine surgery have been conducted for cervical spine reconstruction in complex degenerative and traumatic cases, and for preoperative spinal level identification to plan surgical strategies. The possibility to visualize radiologic datasets through the patient during surgery offers the ability to modify the surgical technique and reduce the surgical invasiveness. However, at the present stage, the overlapping of the hologram over the patient requires the use skin markers and integration with intraoperative imaging (**Figure 8**).

Figure 8. MR used for intraoperative level definition. Figure 8A shows a 3D MR rendering of the spine overlapped over the patient after positioning. Figure 8B shows the overlapping of the CT scan dataset, that offers the possibility to scroll the CT of the spine directly through the patient, after her final surgical positioning. Figure 8C shows the intraoperative X-ray control that documents the correspondence of the cervical level with the three modalities, as shown by the inserted needle (*)

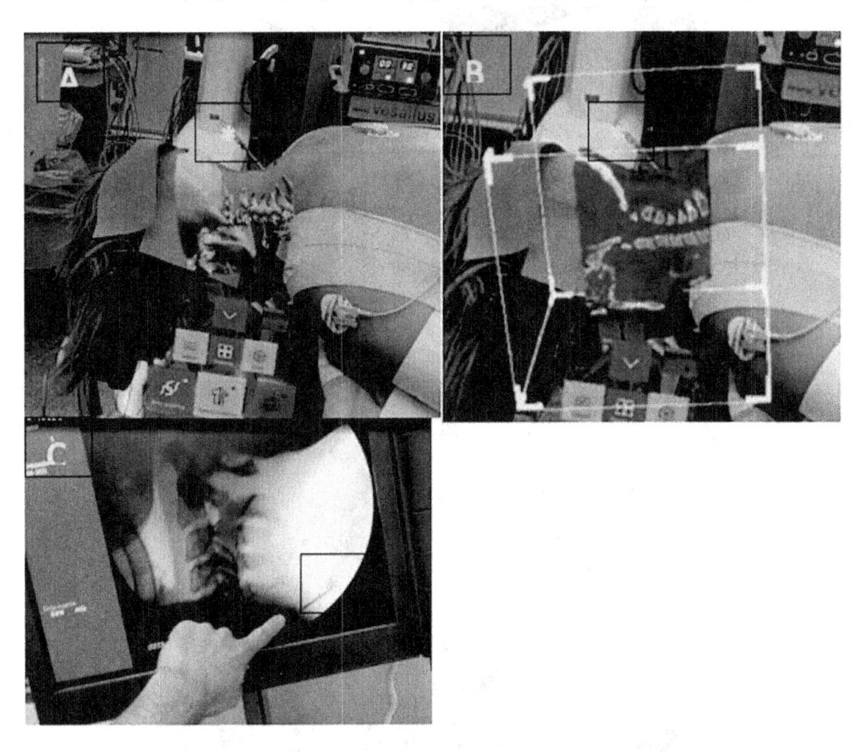

Remote Proctoring

With the advances of both surgical techniques and devices, new approaches and new instrumentations are a common and positive component of the modern clinical practice. To speed up the learning curve, experienced surgeons with peculiar expertise in selected techniques move to hosting hospitals, also in other countries, to support the surgical equip in the early stages of new types of surgeries. New surgical routes and new devices are usually associated to a less familiar anatomy, and thus proctors play a crucial role during their early implementations. Experienced surgeons with minor knowledge of that particular approach may promptly improve their new skills if supported in real-time. MR can play an important role in proctoring activities, reducing the need of moving physically into different hospital. This could be of special interest not only in the actual pandemic situation but above all to support surgeons in remote areas.

Figure 9. (A) 3D computer-based simulation, showing the entry point definition based on anatomical landmarks, to both the right and left, Evans' evaluation, simulation of the catheter placement (B-F), bilaterally obtaining grade A positioning (G) (Umana et al., 2021)

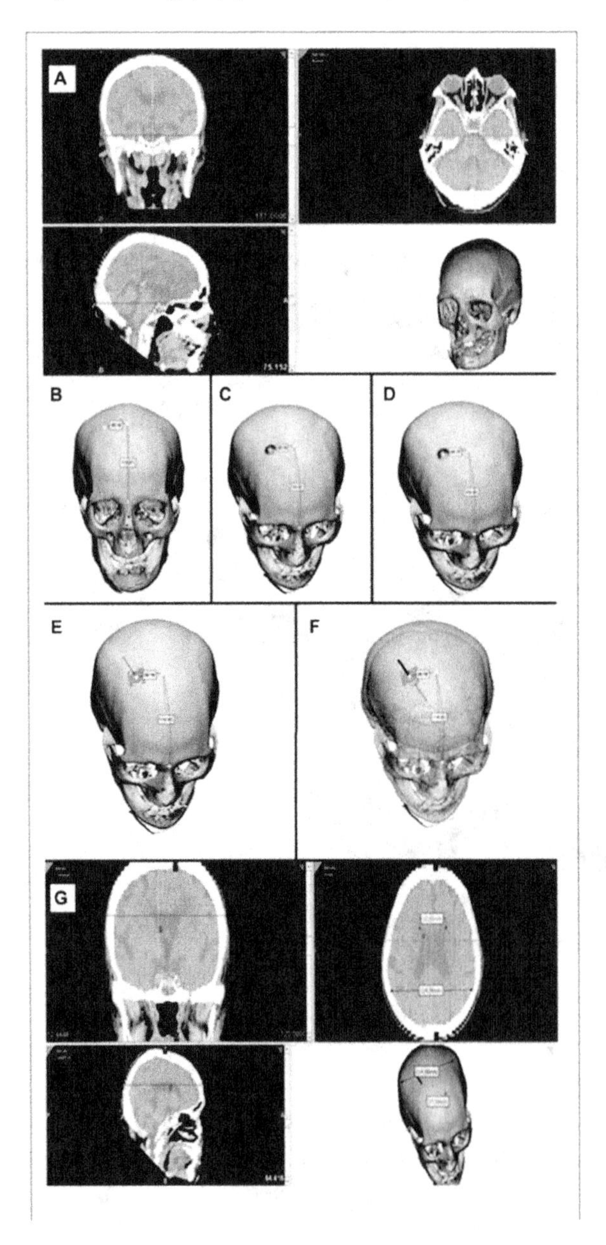

Figure 10. Finceramica plaster model navigation after merging the patient's CT scan (already treated for post-traumatic craniotomy, but with no midline shift). (A,B). Superficial tracing for navigation registration (BrainLab) of the plaster model. (C) Entry point definition with a holographic representation of the device. (D). Positioning of the real device on the defined entry point. (E) Overlap of the holographic device with the real one, to match the entry point. (F) Introduction of the navigated stylet inside the device. (G). Its representation with the standard navigation. (Umana et al., 2021)

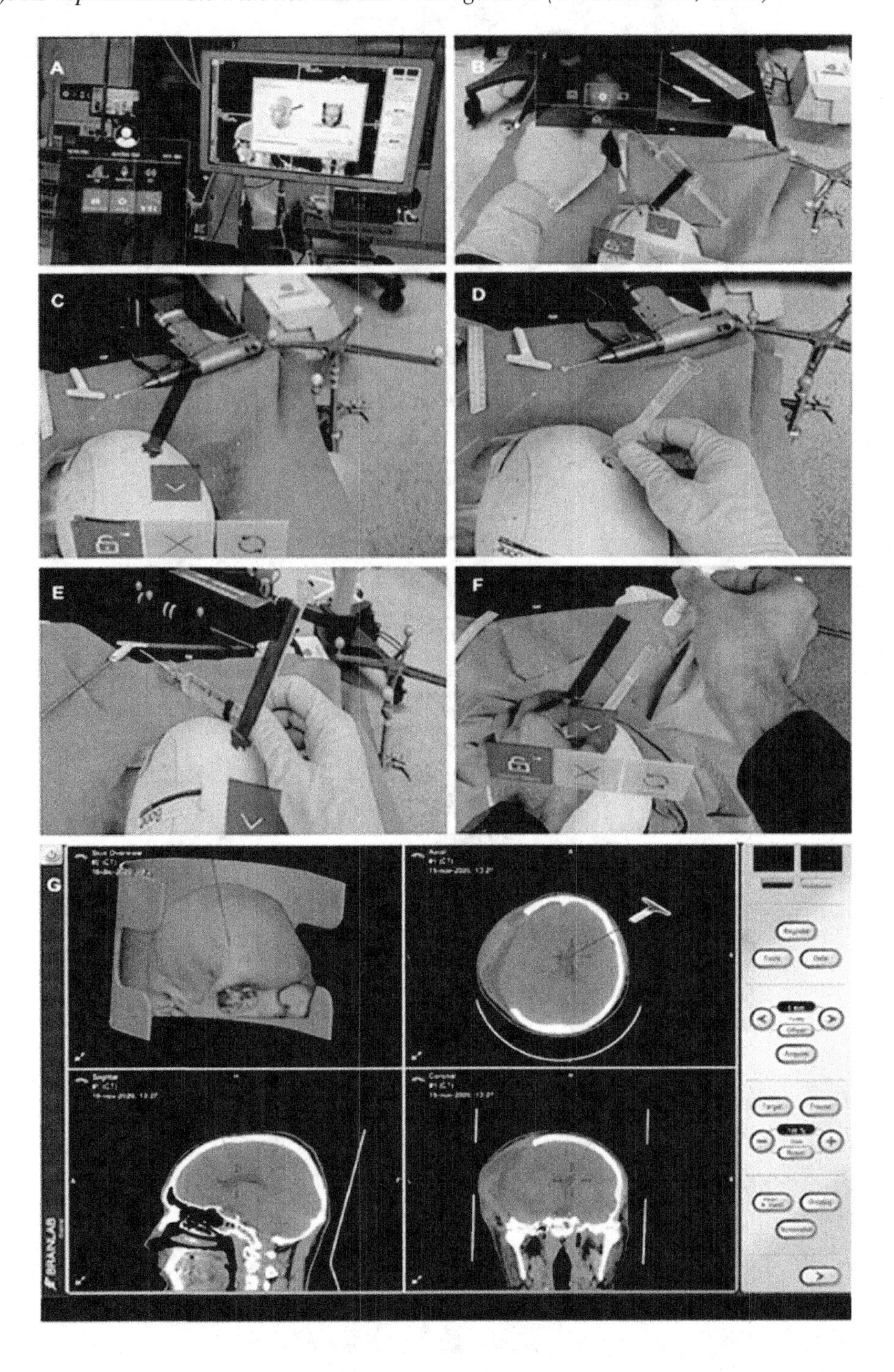

Figure 11. Augmented reality-based simulation (ApoQlar) to locate the entry point. A real ruler was overlapped on the hologram based on both MRI and CT scans; the actual device was overlapped over its holographic reconstruction, and the navigated stylet was introduced to improve familiarity with the procedure and technology (Umana et al., 2021)

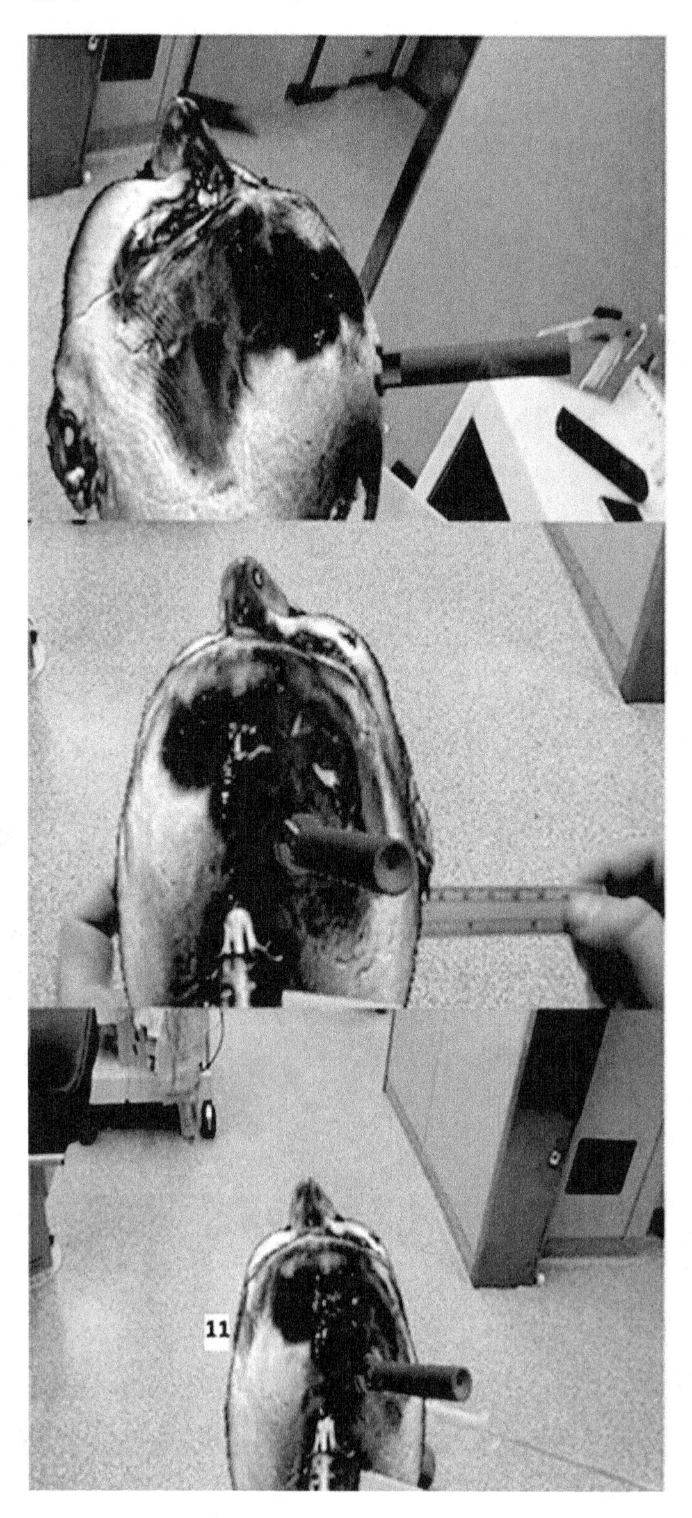

Laboratory Simulation

MR can be proficiently used for surgical simulation, and this is of benefit for new devices, needed to be preliminarily tested. We applied MR technology for a novel device for external ventricular drain placement. Compared to other technologies like standard optical navigation, plaster model, computer-based simulation for geometrical considerations, MR offered similar results (Umana et al., 2021) (Figures 9-11).

Telemedicine

The complex organization of the healthcare network, contemplate the institution of hub and spoke centers. This is associated with an increased need for remote advice for high specialty disciplines. There are several platforms that allow to share dataset and to record the consultation, which can later be printed and added to the patient's files. MR allows to visualize datasets in a more familiar way, offering to the colleagues located at spoke centers the possibility to chat with specialist in a conversation similar to an in-person one. The possibility to share over distances the elaboration of the images and to focus the critical situation with other colleagues in other hospitals also improves the understanding of both the spoke center's doctors and the hub center's specialist. The efficacy of the remote consultation can only benefit from the perception of being in the same place of the patient, side-by-side with other colleagues requiring immediate support, as augmented with MR tools (Montemurro, 2022).

Education

The main advantage of MR is to offer an immersive experience, which greatly assists the learning process in medical education. Learning in an immersive environment helps simplify the spatial relation between the anatomical structures, and this strongly supports neuroanatomy study. In particular, the main difficulty on neuroanatomy is represented by the presence of a large number of complex structures contained in such a small intracranial space. The possibility to highlight with colors and to enlarge the radiological images is an important tool to boost the learning curve and to maintain a deep-seated learning of these concepts.

MIXED REALITY TESTS IN SARS-COV-2 INTENSIVE CARE UNIT

The COVID-19 pandemic has overstrained even the most developed healthcare systems (Moghadas et al., 2020; WHO, 2022). Within a short period of time, traditional hospital treatment has faced numerous challenges. The protection of health and safety of healthcare professionals is a key priority for maintaining the quality of care provided to patients, as well as for preserving the health systems' capability of providing care on a large scale (The PanSurg Collaborative Group, 2020). The uncommonly large number of COVID-19 patients in a short period of time has led to an excessive engagement of healthcare professionals, which has in turn resulted in the problem being compounded by chronic stress and exhaustion at work, with an even more frequent incidence of the burnout syndrome (Lacy & Chan, 2018; Munzer et al., 2019; Raudenská et al., 2020).

The existing shifts doctors need to carry-out, the infection transmission risk, and the long duration of the pandemic impose the need for reduction of the staff's exposure in SARS-CoV-2 highly contagious environment. Young and less experienced clinicians have accessed the red zones in large numbers. Consequently, the most demanding patients in the COVID-19 intensive care unit (C19-ICU) often aren't treated by the most experienced physicians. The risk of communication errors is higher than in the pre-pandemic period. Problems have also been detected in the continuity of the treatment, the sharing of data and information on the patients' condition is usually reduced to the basic information, which is not always sufficient. The communication between physicians of red zone with the outside staff is difficult, while the need for such communication is important. Difficult decisions are to be made, which directly affects the patient's survival and success of treatment. Another important logistic problem is the large use of protective equipment (PPE). The risk of shortages is constantly present (Fadela, 2020). There is an urgent need for new strategies to optimize the use of the PPE and protect the healthcare professionals and patients from the spread of COVID-19 (The Lancet, 2020; Tomasi et al., 2020).

In order to improve the quality and the efficiency of the services delivered to patients, everyday medical practice has in the past few years been enhanced with the introduction of innovative technological tools, which allow a less stressful engagement of medical staff, an improvement of the professional comfort and of the job productivity (Uohara et al., 2020). In our department Microsoft's Hololens 2 (HL2) was tested for potential clinical uses. It allows the use of artificial intelligence (AI) and mixed/augmented reality (MR/AR) in everyday practice. MR/AR represents a combination of the physical and virtual worlds, in which users can interact with virtual objects while keeping their presence in the physical (real) world. HL2 produces and provides virtual visual, auditory and tactile experiences along with the possibility of interaction with the holographic enhanced environment. As opposed to the virtual reality which enables the user to interact with entirely artificial environment, the augmented (mixed) reality generates 3D computer objects on real-world surfaces, thus providing mixed stereoscopic visualization. While observing the real world, the user can manipulate digital contents in the form of holograms generated by the device (Proniewska et al., 2021). This creates a stereoscopic image resulting from combining a three-dimensional virtual model made up of images taken from diagnostic devices (CT, MRI, PET etc.), with real-world surfaces (Proniewska et al., 2021).

There are many studies on the use of augmented reality for surgeries in the operating theatre, in emergency wards and in medical education (Mishra et al., 2022; Munzer et al., 2019; Ogdon, 2019; Schoeb et al., 2020; Sirilak & Muneesawang, 2018; Umana et al., 2021, 2022).

We have found studies in literature which present the experiences with the use of HL2 for reducing healthcare professionals' exposure to aerosol formation procedures in the London Imperial College nephrology ward during the COVID-19 pandemic (Levy et al., 2021; Martin et al., 2020; Tomasi et al., 2020). So far, we found no studies on the usability of HL2 in treating C19-ICU patients.

THE GOAL OF THE HL2-ICU RESEARCH

A pilot study was carried out in the C19-ICU unit of the University Clinical-Hospital Centre "Dr Dragiša Mišović-Dedinje", Belgrade, Serbia, in May 2021. The primary goal was to investigate whether the use of HL2 device in a C19-ICU environment may reduce time to exposure to C19-ICU red zone doctors. A secondary goal was to assess the usability and acceptability of this new technology. The measure of the anesthesiologists' exposure to the SARS-CoV-2 was taken to be the average time per-doctor per-shift spent in the intensive care unit. The device usability and acceptability were assessed by sharing a survey with the involved medical teams.

Before the introduction of the HL2 device, the clinical practice consisted of the entry of the whole team of doctors into a 20-bed C19-ICU. The team consisted of 3 or 4 doctors depending on the shift. During an 8-hour shift they took breaks according to their needs and the circumstances. All the work related to patients was carried out in C19-ICU. Prior to the start of the pilot study, physicians were trained to work with HL2 during April 2021. During the pilot study they were not aware that time spent in C19-ICU is measured for both, regular and HL2 groups. The measurement was blind, performed by persons who were not aware of its purpose. The measure of the doctor exposure to the COVID-19 infectious agent zone was taken to be the total time spent in the C19-ICU during the morning shift, as well as the number of C19-ICU entries. These variables were recorded for each doctor respectively for a 30-day period, i.e. 15 days without, and 15 days with the use of the HL2 device. After completion of the study, average time per-doctor per-measurement was calculated. The anesthesiologists' impressions on the usability and acceptability of this new technology, and their satisfaction, were assessed through a survey (enclosure 1). During the use of the HL2 device, one doctor from the team entered the C19-ICU red zone with the device on (Figure 12).

In this setting, the morning ward rounds started with one doctor of the morning shift inside the C19-ICU zone, while the rest of the team would take part from the green zone via the Microsoft Teams platform. Both, doctor in C19-ICU and the team members outside the red zone had a direct view of the patients, monitoring devices, the parameters on the ventilators or other life support devices. In addition to that, device's MR technology enables all the participants to see laboratory results, radiological diagnostic findings, captured notes with personal observations of the previous shift physicians – all in form of holograms, placed around the patient (placed in patient proximity) (Figure 13).

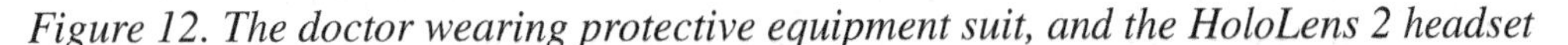

Figure 12. The doctor wearing protective equipment suit, and the HoloLens 2 headset

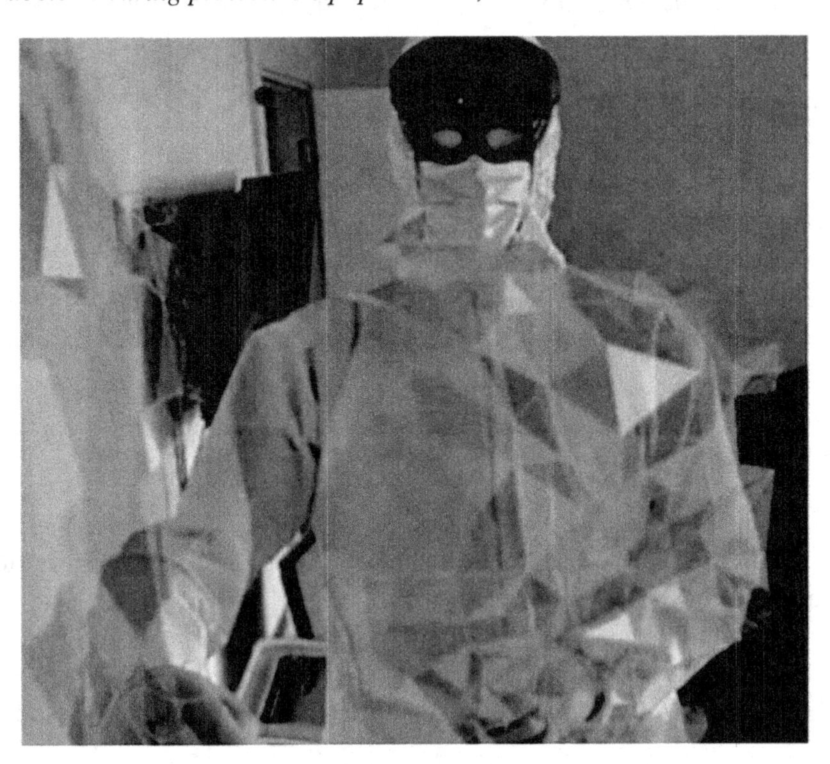

Figure 13. A holographic presentation of radiological scans with a direct view of the patient

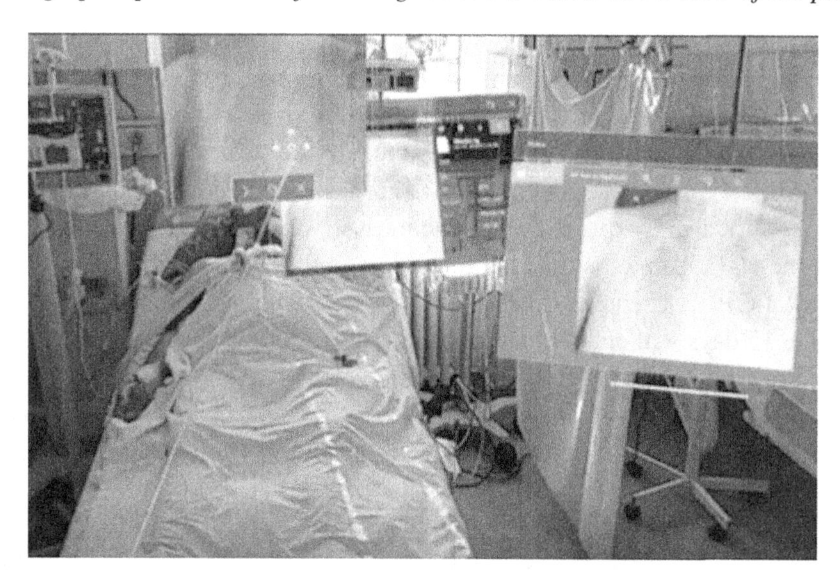

In this way all the team members were able to take an active part in decision-making. The green-zone team members would then write the therapy and the daily information updates for the patients and share pointers for further analyses required. We measured the time each physician spent in the red zone with the use of the Hololens2 device. After the conclusion of the 15-day of test with the HL2 device, the employees were asked to complete a survey, which purpose was to investigate their previous familiarity with the mixed reality characteristics, the users' satisfaction with the device and their view of the benefits of its application and the possibility of its daily use, both during the pandemic and in other areas of medicine as well.

Statistical Analysis

Results are presented as count (%), means ± standard deviation with 95% confidence interval. Groups are compared using linear mixed model and paired samples t test. All p values less than 0.05 were considered significant. All data were analyzed using SPSS 20.0 (IBM Corp. Released 2011. IBM SPSS Statistics for Windows, Version 20.0. Armonk, NY: IBM Corp.) and R 3.4.2. (R Core Team (2017). R: A language and environment for statistical computing. R Foundation for Statistical Computing, Vienna, Austria. URL https://www.R-project.org/.).

C19-ICU Results

The study included 21 doctors (7 males, and 14 females, who were evaluated in regular mode (without the HL2) and with HL2. In Regular mode, 155 measurements were performed, while in HoloLens mode, 104 measurements regarding the ICU were performed.

The average time spent in regular mode for the 155 measurement was 364±71 (95% CI 353-375) minutes, minimum 215-maximum 530; while in HL2 mode, in 104 measurements the average time spent in ICU was 288±59 (95% 276-299) minutes, minimum 134, maximum 420. Using linear mix model, with time as dependent, scale identity matrix for repeated covariance type, mode (regular or Hololens) as factor variable and subject ID as random effect, significant difference of 76 (95% CI 59-93) minutes in average (21% average decrease) was obtained between the modes (p<0.001). Figure 14 reveals the difference between the modalities used in the time spent in C19-ICU.

Figure 14. Difference in the time spent in C19-ICU between the modalities used

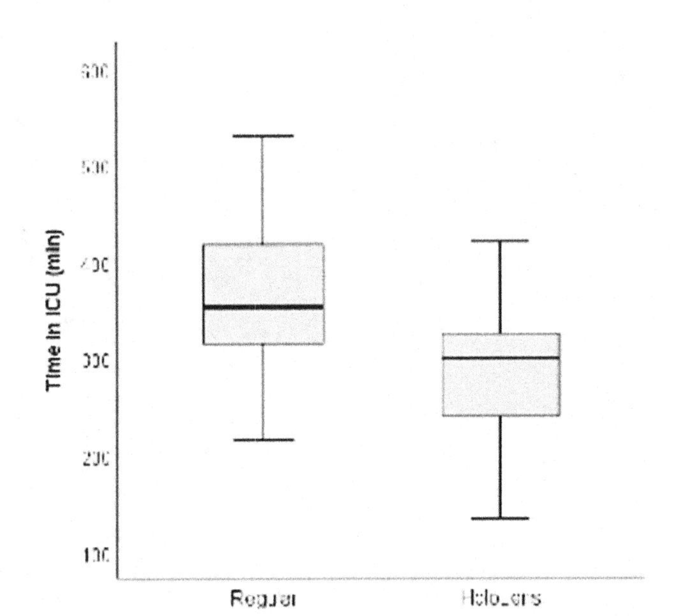

Each doctor time is averaged in Regular and Hololens mode and the average change for each participant is showed in Figure 15. The average time per doctor significantly decreased (p<0.001) of 74±52 (95% CI 50-97) minutes (20,5% average decrease), from 361±45 (95% CI 341-381) minutes in regular mode to 287±33 (95% CI 272-302) in Hololens mode.

Before the use of HL 2, only one doctor had any experience with the mixed reality technology. At the end of the study fourteen doctors completed the questionnaire, and the results are shown in Table 1.

Large majority thought that the use of the HL2 device is not complicated to use, does not interfere with or impact their routine work, but rather helps them in making better informed decisions and easier resolving of any critical incidents in C19-ICU. Eleven of 14 anesthesiologists think that this work method contributes to better communication among colleagues. Overall satisfaction with the use of the HL2 device in treating critically ill COVID- 19 patients was reported by 85.7% of the surveyed.

Figure 15. Time spent by each doctor using without (Regular) and with (HoloLens) the use of the HoloLens

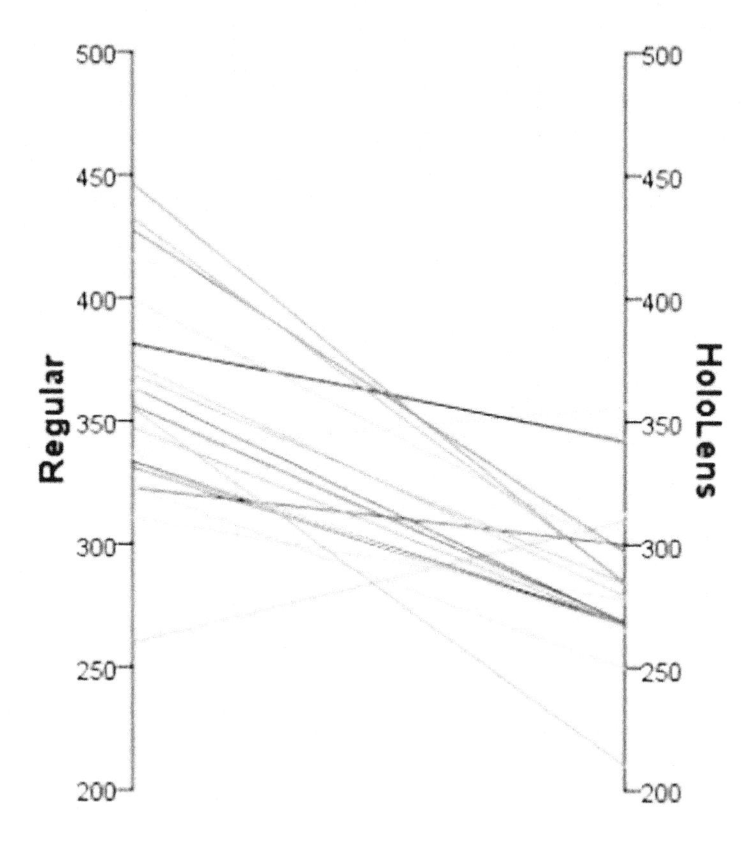

Table 1. The results of the survey conducted among the doctors who used Hololens 2 device.

Question	Yes	No	Don't know
Was the use of HoloLens 2 device complicated for you?	5 (35.7%)	9 (64.3%)	0 (0%)
Do you think that the use of HoloLens 2 device helped you in making decisions in your work?	12 (85.7%)	0 (0%)	2 (14.3%)
Do you think that the use of HoloLens 2 device contributes to better treatment of patients?	12 (85.7%)	1 (7.1%)	1 (7.1%)
Do you think that the use of HoloLens 2 device could also be used in other fields of medicine?	13 (92.9%)	0 (0%)	1 (7.1%)
Would you like to use the HoloLens 2 device in your daily work in the future?	11 (78.6%)	0 (0%)	3 (21.4%)
Do you think that the use of HoloLens 2 device contributes to better communication among colleagues?	11 (78.6%)	1 (7.1%)	2 (14.3%)
Had you been familiar withe the mixed reality technology before the use of the HoloLens 2 device?	1 (7.1%)	13 (92.9%)	0 (0%)
Are you satisfied with the use of the HoloLens 2 device during treatment of patients infected with COVID-19?	12 (85.7%)	0 (0%)	2 (14.3%)

VASCULAR SURGERY

Decision making and performing procedure of abdominal aortic aneurysm repair in a complex patient

Modern vascular surgery has evolved into independent surgical specialization that is dynamic, supported by technology and show different options. As life expectancy is extended, incidence of vascular diseases is increasing, together with patient's complexity. Fragile patients with several comorbidities in advanced age are very difficult to deal with, both for surgery and anesthesiology. One of the most frequent vascular diseases is abdominal aortic aneurysm that usually develop in the subrenal segment of abdominal aorta. Rupture of such an aneurysm is a devastating, life threatening complication and traditional medicine has only one solution to prevent rupture – surgical repair. Abdominal aneurysm surgery is a complex procedure, and its difficulty increases progressively with the proximal extension of the aneurysm aortic section, where visceral or renal branches take origin, or with the more distal site with the involvement of the iliac and femoral arteries.

As a response to these challenges different vascular and endovascular techniques arose providing less invasive options to treat such patients in acceptable manner. Abdominal aortic aneurysm repair can be performed either by opening abdominal cavity and replacing the aneurysm with synthetic (Dacron-poliester) lumen or by implanting a stent graft (combination of stent and Dacron conduit) into the aneurysm. Endovascular techniques are providing treatment usually from the groin or axillar access, or both, without opening abdominal or thoracic cavities. Procedures performed from remote access and control, like that one where stent graft is implanted in the abdominal aorta controlled from the groin, need to be plan thoroughly. On the other hand, planning but also the execution of such procedures depends on technology, especially imaging technology. Nowadays, very frequently image fusion technique is used when preoperative images are fused with intraoperative ones, to improve the accuracy of procedure, to reduce the radiation exposure and the quantity of contrast medium used during the operation. Accurate image processing and interpretation are crucial for preoperative planning, but also during multidisciplinary team meeting assessment. Presentation of vascular problems and decision making depends on image transfer and image analysis. It is not uncommon that doubled or even tripled exams are performed to patients just because image transfer is not possible. If adequate image transfer is possible than making multidisciplinary teams on event international levels are possible.

A challenging case is presented, discussed in a multidisciplinary meeting between equips in different countries and aided by MR for the data sharing (Figure 16). By using HL2 technology, the case was presented to an expert colleague from Malta, prof Cassar, who offered his contribution by sharing his experience in treating such complex patients.

The reported case is a patient with abdominal aortic aneurysm that extends in the juxtarenal section, classified as juxtarenal abdominal aortic aneurysm. In addition, the patient was affected by cardio and respiratory comorbidities and hostile abdomen (multiple previous abdominal operations), that makes new abdominal procedures risky and technically challenging. By using MR, all aspects of the problem were presented in detail, with very clear and detailed image presentation that made discussion and decision very professional and realistic. It was possible to go through images repeatedly, to augment them and communication was efficient. The surgical procedure was then performed and during procedure it was possible to continue to share information thanks to MR to show the intraoperative findings and discuss the final result. After discussion among vascular surgeon, interventional radiologists, anesthesiologist present in the operating room and professor, expert, from remote location, it was decided to treat these patients with endovascular means by using "chimney" parallel graft for one renal artery and to implant stent graft in the infrarenal aorta. For this procedure we used bilateral groin and right brachial incision. Communication and presentation of the patient was performed by operator, who was already in the operating room by the patient in anesthesia (patient signed informed consent that this technology will be used for such procedure). By wearing HoloLens glasses operator was able also to assist images, see them repeatedly and have interactive and immediate consultation making his work easier during the operation by comparing previous images with image on the screen, replacing image fusion technology. In addition, wearing such glasses had no interruption on real work besides digital reality that was happing in his peripheral vision.

Figure 16. Prof Cassar with HoloLens based in Malta and observing images presented in the operating room in Belgrade, where anonymized angiography of abdominal aortic aneurysm are presented. Moving, changing and controlling of the images was done by operator standing by the patient in general anesthesia. Patient signed informed consent that this technology will be used for such procedure

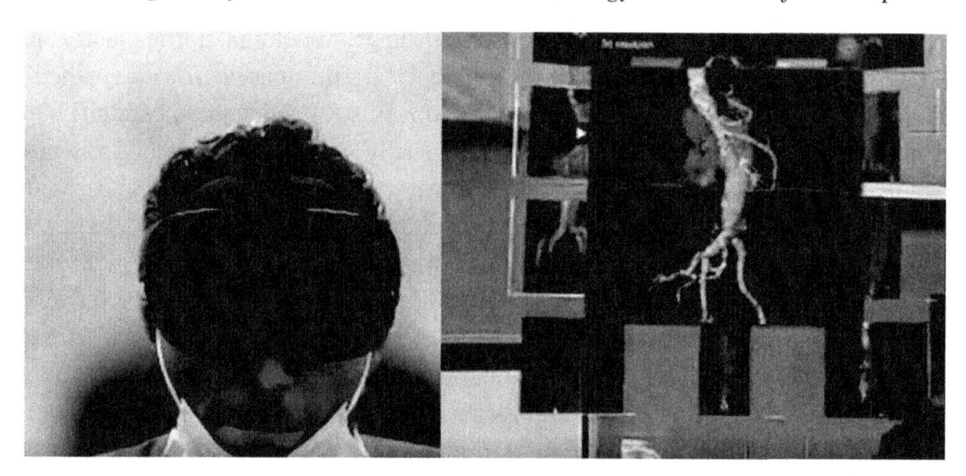

Figure 17. Multiplanar vision of MDCT angiography guided by linear ray made by finger of the operator oriented towards the image

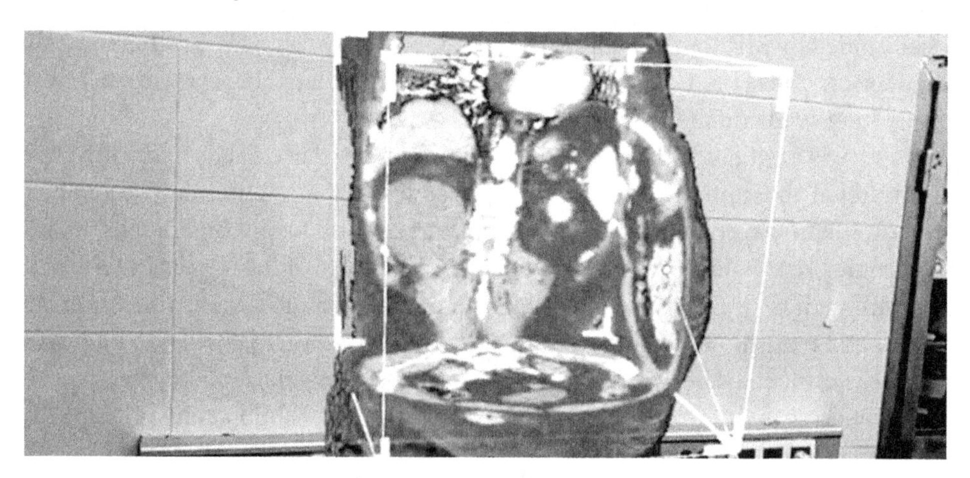

DISCUSSION

The time spent by doctor in C19-ICU was certainly reduced, in several ways. Fewer doctors had to stay in the infectious area at the same time, the number of entries were reduced, while the average time spent there was reduced by 21%, i.e. by 76 minutes. Less exposure to a highly virulent aerosol reduces the likelihood of transmitting the infection to a doctor, while long-term benefits can be expected in terms of reducing the physical and mental strain to which physicians at C19-ICU are exposed.

The possibility to carry out all the administrative work from a distance, such as daily patient information updates, issuing requests in the healthcare information system, and any other analysis, which take a lot of time for physicians in the red zone, enables them to devote more time to the patients. Joint decision-making with senior doctor taking part from a distance should improve the quality of treatment and provide better solutions to numerous problems, but also lessen the pressure felt by anesthesiologists in the red zone in case of making the hardest decisions and clear the doubts relating to treatment methods. In this way it is also possible to organize expert consultative examinations from a distance, with specialists not working in the same institution, which is all too often required in the COVID-19 pandemic (Montemurro et al., 2021).

The HL2 device enables audio and video communication to the team outside of the red zone. At the same time, the patients, their radiological scans and results can also be seen by physicians from a distance, which raises the quality of assessment of the patient's condition and contributes to making better treatment plans. All this enables the physicians to prescribe the essential therapy outside the infectious zone and then bring the lists into the zone, thus saving time to physicians in the red zone.

Using HL2 in red zone enables medical staff to achieve even more than pulling-in the team of doctors through the video call for consultation, brainstorming and decision-making. The device itself allows the user to operate with ease while wearing it.

As the medical staff needs to be equipped with the viro-protection gear all the time (sometimes even wearing 2 pairs of gloves) it is almost impossible to use any kind of modern electronic device in covid19 red-area, through which the person could access the latest medical- related data (blood analysis, X-Ray/CT-scan/MRI imagery), as phones and tablets rely on touchscreen technology, and devices with keyboard. "Handsfree" moment of HL2 device enables the user to overcome these impediments, as the device itself offers such UI (User Interface) which is consumed with ease (outside of the main view field important for primary work), while the in-build AI (Arteficial Intelligence) recognizes the hand gestures the user makes while controlling device.

In addition to hand-gestures, the device AI also observes the eye-movement and detects when user is focusing on particular option which is offered in the UI, with extra-option of selecting the option by eye-blinking.

Furthermore, the HL2 is also equipped with Speech-Recognition technology as another set of AI-powered tools which can be used for UI selection and environment control, but more importantly – for dictating the notes for each patient which can be stored, shared, or even preserved in same spatial environment for all other users which will use the device in the same workspace.

As mentioned in the previous bullet, the device itself is empowered with such equipment and Machine Learning algorithms that it can map the surrounding space, recognize its characteristics, and even "remember" the entire room as the spatial environment in which it operates multiple times, if the room is re-entered or event if the device is powered-off. This opened-up such use cases which increased the work efficiency tremendously while scaling-down the exposure to virus the medical staff had.

The mixed reality generated by HL2 provides a unique possibility to position by the patient's bed the current and past scans, laboratory results and other analyses which remain there, and may be immediately revisited with each new use of the device. As the doctors from the nightshift were doing the final patient visits, they captured the medical context of each patient by placing the holograms above or beside the patient: holographic radiographic and laboratory findings, hologram of captured notes with personal observations, and optional holographic indicators if some patient needs to be prioritized or handled with special care and preparing the work environment entirely for the next shift of doctors (Rubino, 2018).

Finally, when a doctor in the morning shift enters the red-zone, all the necessary pieces of medical context for each patient are in place, and all the handoff between the shifts had already been done with Augmented Reality features this device brought-in. This saves the time otherwise lost in searching for the scans of each individual patient in the information system, which previously reduced the time spent with the patient.

The survey carried out with the engaged physicians has demonstrated that most anesthesiologists are satisfied with their experience. Most of the staff feel more secure and satisfied when using this technology, which we suppose results from the possibility to share responsibility with other physicians from a distance. This does not jeopardize the care and treatment of patients, but rather improves the continuity of treatment through an easier exchange of information with the colleagues who are about to enter the red zone. Many instances of consultative decision-making related to further activities in the red zone certainly improve the quality of treatment and reduce the responsibility of red-zone physicians, and the mental pressure they feel as a result.

This way of work organization may help in a pandemic, when there are not enough physicians for working in intensive care units, with the most critically ill patients. There is not a country in the world with enough health professionals trained to work with the most serious, critically ill patients in conditions such as these imposed by the COVID-19 pandemic.

Given that a great majority of doctor (92.9%) find themselves for the first time before a device emitting mixed reality, it was necessary to carry out training before the first use. According to the results of the survey, 64.7% users found that the use of HoloLens 2 was not complicated. Finally, most of the surveyed, 78.6% of them, would like to use the HL2 device in their everyday work, and as many as 92.9% see the application of this technology in medical practice even beyond the COVID-19 pandemic.

The use of HL2 in clinical practice is subject to certain limitations. They include battery duration limits, the internet connection stability and rate issues (which can be overcome with mobile internet routers), while some users were dissatisfied with this technology for not being in direct contact with the patient, which can negatively affect the physician-patient relationship (Microsoft., 2020; Rubino, 2018).

The COVID-19 pandemic has been present on a global scale since December 2019, responsible for 271 963 258 cumulative cases and over 5 331 019 cumulative deaths, and that it has imposed new conditions on medical work organization as well (Afonso et al., 2021). These conditions have been present for a while now and have been leading to an ever higher incidence of the burnout syndrome in the staff working with the most critically ill patients in intensive care units. Besides physical exhaustion, we can by no means disregard the mental issues caused by the high fatality rates in the most seriously ill COVID-19 patients. The long- term consequences for the staff working in intensive care units are yet to be seen, but this is not an optimistic issue. For that reason, the most recent technological devices are being brought into play to reduce both the physical and mental engagement of the medical professionals. Although this pilot study was carried out in a short period of time, it still points towards steps to be taken to resolve certain problems facing doctors working in C19- ICU and suggest future work organization methods. The results of our study are optimistic, but they require an additional, more comprehensive, and stricter evaluation.

LIMITATIONS AND FUTURE DIRECTIONS

The use of MR is in its infancy and, of course, Hololens 2 shows several limitations. The superimposition of the hologram on the patient's anatomy, at the moment, is manual or semi-automatic, making this process dependent on the operator, negatively affecting its accuracy. Furthermore, the interaction with real objects, which should be recognized by the hologram, is still unresolved. These two main limitations represent the biggest obstacle to introducing Hololens 2 into daily practice. Improvement of the device, which should include a version dedicated to clinical use, should help overcome these and other minor problems.

CONCLUSION

Although this pilot study was carried out in a short period of time, it shows key points that should be addressed to solve the actual issues to help doctors facing working in C19- ICU and to improve the efficiency of the work chain. The results of our study are optimistic, but they require an additional, more comprehensive, and stricter evaluation. In a pandemic, any way to reduce health workers' exposure to the viral environment is significant. HL2 is just one of the logistic solutions, although its use and role in medicine has yet to be shown beyond the conditions imposed by the pandemic. MR promises to disrupt of clinical practice in several contexts. From education to intraoperative navigation, MR improves the familiarity with the anatomy of the patients, favoring safer treatments. The only limitations to its introduction for clinical use are technical, but the great interest on this technology is expected to quickly achieve the required improvements in the next future, changing our practice, offering better treatments to our patients.

REFERENCES

Afonso, A. M., Cadwell, J. B., Staffa, S. J., Zurakowski, D., & Vinson, A. E. (2021). Burnout Rate and Risk Factors among Anesthesiologists in the United States. *Anesthesiology, 134*(5), 683–696. doi:10.1097/ALN.0000000000003722 PMID:33667293

Davies, C. (2015). *HoloLens hands-on: Building for Windows Holographic.* Academic Press.

Fadela, C. (2020). *Shortage of personal protective equipment endangering health workers worldwide.* World Health Organization. https://www.who.int/news/item/03-03-2020-shortage-of-personal-protective-equipment-endangering-health-workers-worldwide

Hachman, M. (2015). *Developing with HoloLens: Decent hardware chases Microsoft's lofty augmented reality ideal.* Academic Press.

Hempel, J. (2015). *Project HoloLens: Our Exclusive Hands-On With Microsoft's Holographic Goggles*. Academic Press.

Hollister, S. (2015). *Microsoft HoloLens Hands-On: Incredible*. Amazing, Prototype-y as Hell.

Holmdahl, T. (2015). *BUILD 2015: A closer look at the Microsoft HoloLens hardware*. Academic Press.

Kipman, A., & Juarez, S. (2015). *Developing for HoloLens*. Microsoft.

Lacy, B. E., & Chan, J. L. (2018). Physician Burnout: The Hidden Health Care Crisis. *Clinical Gastroenterology and Hepatology*, *16*(3), 311–317. doi:10.1016/j.cgh.2017.06.043 PMID:28669661

Levy, J. B., Kong, E., Johnson, N., Khetarpal, A., Tomlinson, J., Martin, G. F., & Tanna, A. (2021). The mixed reality medical ward round with the MS HoloLens 2: Innovation in reducing COVID-19 transmission and PPE usage. *Future Healthcare Journal*, *8*(1), e127–e130. doi:10.7861/fhj.2020-0146 PMID:33791491

Martin, G., Koizia, L., Kooner, A., Cafferkey, J., Ross, C., Purkayastha, S., Sivananthan, A., Tanna, A., Pratt, P., & Kinross, J. (2020). Use of the HoloLens2 Mixed Reality Headset for Protecting Health Care Workers During the COVID-19 Pandemic: Prospective, Observational Evaluation. *Journal of Medical Internet Research*, *22*(8), e21486. doi:10.2196/21486 PMID:32730222

Microsoft. (2015). *Introducing the Microsoft HoloLens Development Edition*. https://www.microsoft.com/it-it/hololens

Microsoft. (2020). *Microsoft HoloLens - The Science Within - Spatial Sound with Holograms*. https://gr.pinterest.com/pin/microsoft-hololens-the-science-within-spatial-sound-with-holograms--280982464230744292/

Mishra, R., Narayanan, M. D. K., Umana, G. E., Montemurro, N., Chaurasia, B., & Deora, H. (2022). Virtual Reality in Neurosurgery: Beyond Neurosurgical Planning. *International Journal of Environmental Research and Public Health*, *19*(3), 1719. doi:10.3390/ijerph19031719 PMID:35162742

Moghadas, S. M., Shoukat, A., Fitzpatrick, M. C., Wells, C. R., Sah, P., Pandey, A., Sachs, J. D., Wang, Z., Meyers, L. A., Singer, B. H., & Galvani, A. P. (2020). Projecting hospital utilization during the COVID-19 outbreaks in the United States. *Proceedings of the National Academy of Sciences of the United States of America*, *117*(16), 9122–9126. doi:10.1073/pnas.2004064117 PMID:32245814

Montemurro, N. (2022). Telemedicine: Could it represent a new problem for spine surgeons to solve? *Global Spine Journal*, (6), 1306–1307. Advance online publication. doi:10.1177/21925682221090891 PMID:35363083

Montemurro, N., Condino, S., Cattari, N., D'Amato, R., Ferrari, V., & Cutolo, F. (2021). Augmented Reality-Assisted Craniotomy for Parasagittal and Convexity En Plaque Meningiomas and Custom-Made Cranio-Plasty: A Preliminary Laboratory Report. *International Journal of Environmental Research and Public Health*, *18*(19), 9955. doi:10.3390/ijerph18199955 PMID:34639256

Montemurro, N., Santoro, G., Marani, W., & Petrella, G. (2020). Posttraumatic synchronous double acute epidural hematomas: Two craniotomies, single skin incision. *Surgical Neurology International*, *11*, 435. doi:10.25259/SNI_697_2020 PMID:33365197

Munzer, B. W., Khan, M. M., Shipman, B., & Mahajan, P. (2019). Augmented Reality in Emergency Medicine: A Scoping Review. *Journal of Medical Internet Research*, *21*(4), e12368. doi:10.2196/12368 PMID:30994463

Ogdon, D. C. (2019). HoloLens and VIVE Pro: Virtual Reality Headsets. *Journal of the Medical Library Association: JMLA*, *107*(1). Advance online publication. doi:10.5195/jmla.2019.602

Proniewska, K., Pręgowska, A., Dołęga-Dołęgowski, D., & Dudek, D. (2021). Immersive technologies as a solution for general data protection regulation in Europe and impact on the COVID-19 pandemic. *Cardiology Journal*, *28*(1), 23–33. doi:10.5603/CJ.a2020.0102 PMID:32789838

Raudenská, J., Steinerová, V., Javůrková, A., Urits, I., Kaye, A. D., Viswanath, O., & Varrassi, G. (2020). Occupational burnout syndrome and post-traumatic stress among healthcare professionals during the novel coronavirus disease 2019 (COVID-19) pandemic. *Best Practice & Research. Clinical Anaesthesiology*, *34*(3), 553–560. doi:10.1016/j.bpa.2020.07.008 PMID:33004166

Rubino, D. (2018). *Microsoft HoloLens - Here are the full processor, storage and RAM specs*. https://www.windowscentral.com/microsoft-hololens-processor-storage-and-ram

Schoeb, D. S., Schwarz, J., Hein, S., Schlager, D., Pohlmann, P. F., Frankenschmidt, A., Gratzke, C., & Miernik, A. (2020). Mixed reality for teaching catheter placement to medical students: A randomized single-blinded, prospective trial. *BMC Medical Education*, *20*(1), 510. doi:10.118612909-020-02450-5 PMID:33327963

Sirilak, S., & Muneesawang, P. (2018). A New Procedure for Advancing Telemedicine Using the HoloLens. *IEEE Access: Practical Innovations, Open Solutions*, *6*, 60224–60233. doi:10.1109/ACCESS.2018.2875558

The Lancet. (2020). COVID-19: protecting health-care workers. *The Lancet, 395*(10228), 922. doi:10.1016/S0140-6736(20)30644-9

The PanSurg Collaborative Group. (2020). *The three vital lessons Italian hospitals have learned in fighting covid-19*. Heal Serv J.

Tomasi, S. O., Umana, G. E., Scalia, G., & Winkler, P. A. (2020). In Reply: Rongeurs, Neurosurgeons, and COVID-19: How Do We Protect Health Care Personnel During Neurosurgical Operations in the Midst of Aerosol-Generation From High-Speed Drills? *Neurosurgery*, *87*(2), E166–E166. doi:10.1093/neuros/nyaa213 PMID:32385489

Umana, G. E., Pucci, R., Palmisciano, P., Cassoni, A., Ricciardi, L., Tomasi, S. O., Strigari, L., Scalia, G., & Valentini, V. (2022). Cerebrospinal Fluid Leaks After Anterior Skull Base Trauma: A Systematic Review of the Literature. *World Neurosurgery*, *157*, 193–206.e2. doi:10.1016/j.wneu.2021.10.065 PMID:34637942

Umana, G. E., Scalia, G., Yagmurlu, K., Mineo, R., Di Bella, S., Giunta, M., Spitaleri, A., Maugeri, R., Graziano, F., Fricia, M., Nicoletti, G. F., Tomasi, S. O., Raudino, G., Chaurasia, B., Bellocchi, G., Salvati, M., Iacopino, D. G., Cicero, S., Visocchi, M., & Strigari, L. (2021). Multimodal Simulation of a Novel Device for a Safe and Effective External Ventricular Drain Placement. *Frontiers in Neuroscience*, *15*, 690705. Advance online publication. doi:10.3389/fnins.2021.690705 PMID:34194297

Uohara, M. Y., Weinstein, J. N., & Rhew, D. C. (2020). The Essential Role of Technology in the Public Health Battle Against COVID-19. *Population Health Management*, *23*(5), 361–367. doi:10.1089/pop.2020.0187 PMID:32857014

WHO. (2022). *WHO Coronavirus (COVID-19) Dashboard (2021)*. WHO.

KEY TERMS AND DEFINITIONS

Burnout: Physical, emotional, or mental exhaustion, accompanied by decreased motivation, lowered performance and negative attitudes towards oneself and others.

COVID-19: Viral infection spread in the year 2020 responsible for a worldwide pandemic.

Intensive Care Unit: Special department of a hospital catering to patients with severe or life-threatening illnesses and injuries, which require constant care, close supervision from life support equipment and medication in order to ensure normal bodily functions.

Mixed Reality: Science within the realm of virtual reality, allowing direct interactions between the physical and digital worlds through intuitive 3D human, computer, and environmental actions.

Neurosurgery: Medical subspecialty focusing on the treatment of patients with pathologies involving the brain and/or the spine.

Proctoring: In medicine, objective evaluation of a physician's clinical competence by someone serving as a proctor who represents and is responsible to the medical staff.

Telemedicine: Distribution of health-related services and information via electronic information and telecommunication technologies.

Chapter 23
From Virtual Reality to 360° Videos:
Upgrade or Downgrade? The Multidimensional Healthcare VR Technology

Francesca Borghesi
Applied Technology for Neuro-Psychology Lab, Istituto Auxologico Italiano (IRCCS), Milan, Italy

Valentina Mancuso
Faculty of Psychology, eCampus University, Novedrate, Italy

Elisa Pedroli
Applied Technology for Neuro-Psychology Lab, Istituto Auxologico Italiano (IRCCS), Milan, Italy

Pietro Cipresso
Applied Technology for Neuro-Psychology Lab, Istituto Auxologico Italiano (IRCCS), Milan, Italy & Department of Psychology, University of Turin, Italy

ABSTRACT

This chapter aims to describe the multidimensional virtual reality tools applied to healthcare: in particular the comparison between virtual reality traditional tolls and the 360° videos. The VR traditional devices could differ in terms of specific graphics (2D/3D), display devices (head mounted display), and tracking/sensing tools. Although they are ecological tools, they have several problems such as cybersickness, high-cost software, and psychometric issues. Instead, the 360° videos can be described as an extension of virtual reality technology: they are immersive videos or spherical videos that give the opportunity to immerse the subject in authentic natural environments, being viewed via an ordinary web browser in that a user can pan around by clicking and dragging. The comparison between those two technologies stems from the question if 360° videos could solve and overcome the problems related to virtual reality and be an effective and more ecological alternative.

DOI: 10.4018/978-1-6684-4854-0.ch023

INTRODUCTION

Among all the Virtual Reality (VR) technologies, a concurrent emerging trend are 360° videos, which are also known as immersive videos or spherical videos. This chapter will go into the advantages and disadvantages of the VR spectrum of technologies: it will highlight when VR potential ends and where those of 360° videos start. 360° videos could be a more innovative cheaper, ecological and realistic tool than VR devices: they take the opportunity to immerse yourself in authentic natural environments, being viewed via an ordinary web browser or mobile device where a user can pan around by clicking and dragging the environment. A 360-degree video is made up of a series of 360-degree images with a predetermined time interval between them. Each 360-degree image is a panorama taken with an omnidirectional camera or a combination of cameras to cover the entire horizontal field of view (i.e., 360-degree FOV). They could be an innovative smart and ecological alternative to VR tools because they could solve and overcome cybersickness problems, high cost of software and the use of VR 3D helmets.

BACKGROUND

In the last 10 years, VR technology has innovated the world of healthcare: it has been rendering the experimental and clinical fields, as well as assessment and rehabilitation settings, more environmentally friendly, appealing, and personalized (Cipresso et al., 2018). Although VR has great possibilities, it also presents some technical and psychometric limitations: first, when used, most VR systems must be connected to a computer and require an external tracking device. As a result, they are difficult to utilize for the cords that restrict the user's movements. Tethering also necessitates synchronization of the various peripherals (HMD, joystick, gloves, etc.) which necessitates a sophisticated technical setup that increases the system's latency. Moreover, it is challenging to create effective VR experiences since it necessitates the creation of 3D models and interfaces, the specification of user-environment interaction models, and the integration of external devices such as sensors and gloves. Furthermore, most of the times the development of VR environments needs information and engineering intervention because software needs to be customized each time, per each experiment, and this requires huge efforts and high cost in terms of development. Finally, psychometric problems regard the lack of literature about the test-rest and usability of VR tools: only few studies have looked at usability, demonstrating that those tools are actually usable, easy to learn, challenging and engaging, and devoid of major adverse effects (Borgnis et al., 2022; Freeman et al., 2017). These example barriers can be removed with the use of 360° videos. Although, how they are made, they do not allow proper interaction, by linking them together it may be possible to create this illusion: for example, with software like *Insta VR*, it is possible to allow navigation in any environment: by putting a *link* or a *hotspot* on a door it is possible to change the environment and give the appearance to change rooms. This feature has innumerable applications. In this way, a clinician could also create ecological testing or improve existing ones. For example, personality is traditionally assessed through self-reports questionnaire: the most famous is the Big Five questionnaire which assesses personality among five dimensions (extraversion, agreeableness, conscientiousness, emotional stability, and openness to experience) (Caprara et al., 1993). Cipresso and Riva (2015) have already proposed to assess personality using 360° videos, creating different situations where each component is posed either in multiple situations with a single choice or in only one situation with multiple choices. One example consists in immersing the user in a path in the forest with a crossroads, where he/she, through a narrative in the virtual environment, must choose a visible street (closeness to experiences) from the one side or a dark street (openness to experiences) on the other side. This method could also be implemented to build different assessment settings, like for example testing emotional regulation, and achieve not only many clinical purposes but also marketing choices.

MAIN FOCUS OF THE CHAPTER

Virtual Reality Traditional Devices

Virtual reality (VR) is a computer-generated technology that allows users to be completely immersed in a virtual environment, often achieved by taking over the entirety of a participant's peripheral field-of-view, using special devices such as head-mounted displays (HMD), headphones and gloves (Liu et. al., 2019; Sutherland et al., 2019). VR could offer a "convincing illusion and a sensation of being inside an artificial world that exists only in the computer" (Tieri et al., 2018): from being an observer of an action, the subject transforms into the protagonist, performing actions while overcoming physical, economic, and safety barriers. To create such illusions, VR requires a position tracker that detects the user's movements and attributes in real time, allowing the virtual contents to be updated in real time through a 3D visualization system (e.g. a computer, a smartphone) (Riva et al., 2020). VR relies on devices like a HMD that can fill the field of view with an image and track every small movement you make while looking around. Audio is almost always included with video, and headphones are used to provide the sound.

VR has three distinct characteristics: interaction, immersion, and imagination. *Interaction* refers to the possibility for the user to interact with the virtual scene. It could be "embodied": for example, new devices (such as Leap Motion) allow you to use your hands to grasp objects in a virtual scene, making interaction more natural. *Immersion* refers to the objective level of sensory fidelity that a VR system provides (Slater & Wilbur, 1997) and can range from non-immersive to fully immersive, depending on which technologies are involved. At last, *imagination* refers to the use of multi-dimensional perception information provided by VR scenes, which guarantee the same feelings as the real world (Riva et al., 2020).

VR employs a set of input tools (keyboard, mouse, trackball, and joystick) that the computer integrates and updates in real time to obtain information on the subject's actions. This enables the creation of a dynamic three-dimensional world, which is then returned to the subject via sophisticated data (output tools). Three types of VR can be distinguished based on these output tools (Milgram and Kishino, 1994; Ijsselsteijn and Riva, 2003):

Immersive: it concerns the sound, visualization, movement and tactile devices (3D helmet, gloves and sensory trackers) that isolate the perceptive channels of the subject by immersing him on a sensory level. Immersive scenarios are usually provided by means of the use of HMDs or cave automatic virtual environments (CAVE) which entirely cover the user' field of view.

Semi-Immersive: it is determined by rooms equipped with devices and surround rear-projection screens that reproduce the stereoscopic images of the computer and project them on the walls, with different shapes and degrees of convexity. Usually, users can interact with the screens by means of a joypad or other peripherals.

Non-Immersive: it is determined by a monitor that acts as a "window" through which the user sees the world in 3D; interaction with the virtual world can be done through the mouse, the joystick or other peripherals such as gloves.

Immersive and semi-immersive devices, in particular, can provide the sensation of being immersed in a virtual environment, whereas non-immersive devices can simply display content based on their position and use (Ventura et al., 2019). According to the "predicting coding" paradigm, VR simulates a real-life scenario, providing real-time adaptations and multisensorial feedbacks as a result of the user's actions (Riva et al., 2019), a mechanism that resembles how the brain acts to create an internal model (or matrix) of the body and surrounding space.

Since its ecological and smart use, VR has been used in different fields, as gaming (Zyda, 2005; Meldrum et al., 2012), military training (Alexander et al., 2017), architectural design (Song et al.,2017), education (Englund et al., 2017), learning and social skills training (Schmidt et al., 2017), simulations of surgical procedures (Gallagher et al., 2005), assistance to the elderly or psychological treatments are other fields in which VR is bursting strongly (Freeman et al., 2017; Neri et al., 2017). About the lastly named clinical application, a recent work of Cipresso and collaborators (2018) showed that VR in the last 10 years has developed rapidly and immensely in the field of health care: it is noted its inclusion in the top 10 list of rehabilitation and clinical neurology categories (about 10% of the total production in the last 5 years). It's also worth noting that, when taken together, neuroscience and neurology have grown from about 12% to about 18.6% in the last five years. With the turn of the millennium, VR research shifted firmly into the clinical-VR period, with a focus on rehabilitation, neurosurgery, and a new phase of therapeutic and laparoscopic skills (Cipresso et al., 2018). In the healthcare field, the use of VR has positive and negative outcomes as the Figure 1 shows:

Figure 1. VR positive and negative healthcare use

VR POSITIVE HEALTHCARE USE

- A Realistic environments
- Co-registration
- Neurofisiological devices
- Ecological environments
- Costumized and personalized therapy
- Home health-care

VR NEGATIVE HEALTHCARE USE

- Cyber-sickness Refresh rate (frame rate) of on-screen images
- Tethering problems
- Psychometric problems
- High cost of VR head-set and Software development

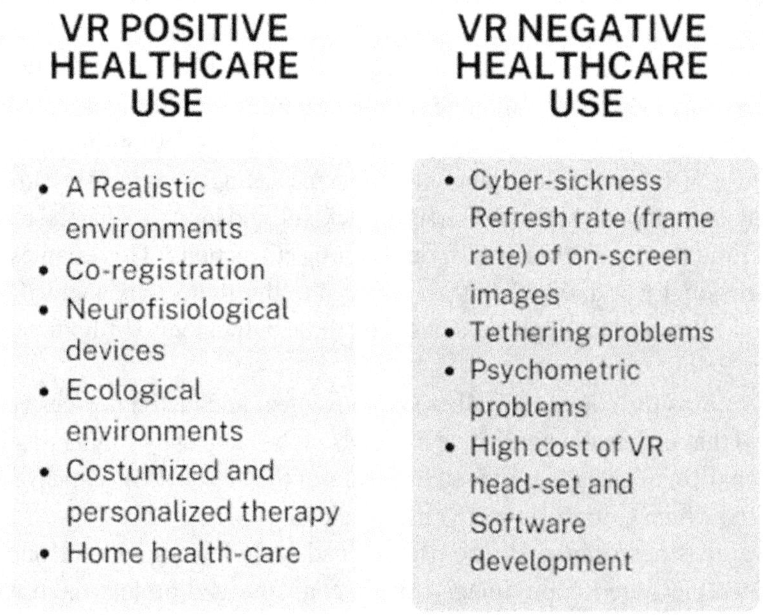

VR Positive Healthcare Use

VR is a type of human-computer interaction system that immerses patients and healthy subjects in an immersive and interactive simulated environment that they believe is similar to real-world objects and events (Lombard & Ditton, 1997), giving them a natural, immediate, direct, and palpable experience (Josman et al., 2008). As a result, by simulating everyday contexts, VR allows for the design and creation of realistic spatial and temporal scenarios, circumstances, or objects. Virtual Environment (VE) appears to be environmentally and ecologically friendly: it is made up of a number of computer-simulated scenarios (in which subjects could interact interactively with 3D items in real-time) (Bohil et al., 2011; Campbell et al., 2009; Parson, 2015; Parsons et al., 2013).

Individuals can see the movements of a virtual avatar portrayed coherently with their real body in a first-person perspective, which is one of the fundamental aspects of VR: individuals can see the movements of a virtual avatar portrayed coherently with their real body in a first-person perspective, which is one of the fundamental aspects of VR (Vogeley et al., 2004). This connection between the virtual and real bodies creates a sense of "presence" in the virtual body, which causes real-world muscular, physiological, and cognitive responses (Banakou et al., 2018; Burin et al., 2019a; Osimo et al., 2015). Subjects' actual behaviors change as a result of the virtual body's visible movements, which are fueled by a sense of "presence" (Burin et al., 2019b). Furthermore, even when no actual motions are present, watching an avatar perform aerobic exercise has a positive impact on cognitive components and their neural substrates (Burin et al., 2019a). These examples suggest that VR can create a strong sense of immersion and presence by utilizing a brain mechanism known as embodied simulation, which encourages users' active participation and involvement by allowing physical and emotional interaction with the environment via the digital medium.

Another advantage of VR is its multidimensional signal recording: it is possible to co-register multiple neurophysiological signals with a detection device, such as Magnetic Resonance Imaging (MRI), Electroencephalogram (EEG), Magnetoencephalography (MEG), Electrocardiogram (ECG), Kinematics gait analysis, Eye Tracker (ET) and stimulation devices, i.e Non-Invasive Brain Stimulation (NIBS) and Neurofeedback (Mishra et al., 2021).

The study of the alteration of the psychobiological state of mind during VR induced dynamic perceptions is known to use neurophysiological recordings in conjunction with VR devices to map neuronal activity. Researchers are experimenting with various neuroimaging modalities to study cognition using the VR experience. For example, EEG was used by researchers to study the responses evoked by the presentation of specific three-dimensional (3D) virtual tunnels with navigational images (Leroy et al., 2017), and to characterize the neural generators of the brain oscillations in a virtual tennis court, related to motor imagery (Cebolla et al., 2017). Furthermore, Roberts and collaborators (2019) combined an OPM-based MEG system with a VR system for the first time, recording alpha oscillations and visual evoked fields in the presence of VR simulations. VR systems can also be used to conduct various experiments in neuroprosthetics, motor control, and coordination activity. Bach and colleagues (2012), for example, created and tested an MRI-compatible VR-based setup that can produce illusory sensations through visual and tactile stimulations. The experiment used illusory ownership experiences for a virtual limb to demonstrate MRI response acquisition. The right-hemispheric activation of the cerebellum, as well as bilateral activity of the ventral premotor cortex, secondary somatosensory cortex, extrastriate visual cortex, and right-hemispheric activation of the cerebellum, were observed in a highly susceptible subject. In addition, functional imaging techniques, such as functional MRI (fMRI), can also be incorporated in VR stimulations. Together with VR devices, these brain scanning modalities can be used to get comparative data of brain response (such as power spectrum density values together with corresponding topo maps, event-related potential, and responses) in pre-activity, activity, and post-activity sessions of VR-based cognitive tasks. Functional neuroimaging using fMRI together with VR can be used to reveal the underlying neural correlates of motor functions in neurodegenerative diseases, such as freezing behavior in Parkinson's disease (PD). Shine and colleagues (Shine et al., 2011) demonstrated a real-time VR-based walking task that was observed with fMRI both with and without dopaminergic medication. Instead of measuring cardio and pupil size, Zen and colleagues (2021) used VR to stimulate participants' fear of heights (FoH) by creating four different VR scenarios, including a virtual scene from the VR game "Richie's plank experience" and a realistic stimulus of basketball hitting. The synchronized eye movement (EMO), pupil, and ECG of 17 healthy subjects with an even mix of men and women have been recorded for FoH analysis (Zen et al., 2021). Olonsky and collaborators used an eye tracker integrated into the VR Display to collect a variety of eye movement metrics that could indicate a user's knowledge or memory of a specific word. They discovered that eye movement and pupil radius have a strong relationship with user memory, and that a variety of other metrics can be used to classify the state of word understanding (Orlonsky et al., 2019). For NIBS, it has been shown that combining it with VR can be beneficial in rehabilitation settings (Mancuso et al., 2020; Stramba-Badiale et al., 2020), but it has also been used to induce embodiment for an artificial hand (Bassolino et al., 2018), to treat spider phobia (Notzon et al., 2015), and in interventions in a variety of populations, including children with cerebral palsy, post-stroke patients (Massetti et al., 2017). Another excellent example of neurofeedback is DEEP, a VR video game created on the fly that uses respiratory-based biofeedback to help young people manage subclinical stress and anxiety symptoms. Almost all of the participants had significantly lower arousal levels after playing DEEP, demonstrating the effectiveness of this biofeedback video game in controlling emotions (Colombo et al., 2021; Weerdmeester et al., 2017).

As can be seen from the majority of the studies discussed above, VR is well developed in clinical settings for assessment and rehabilitation (Borgnis et al., 2022, Cipresso et al, 2018, Freeman et al., 2017, Riva et al., 2020; Tuena et al., 2021). Clinicians can observe their patients in a realistic setting: patients can in fact interact with computer-simulated items and surroundings in three dimensions and in real time, just as they would in real life (Climent et al., 2010). As a result, VR may make assessing functional behaviors in people with functional deficiencies easier (Tarnanas et al., 2013), allowing clinicians to see their patients in real-life situations (Climent et al., 2010). Traditional neuropsychological assessment procedures could be integrated into VR to improve their reliability and psychometric validity (Riva, 1997; Rizzo et al., 2001). VR assessments appear to be more effective than traditional paper and pencil tests in some cases: for example, Borgnis and collaborators (2022) recently demonstrated that this efficacy is specific to the disease studied and the active executive function. In Schizophrenia, for example, VR-based evaluation tools revealed a deficiency in executing virtual grocery shopping activities owing to deficits in planning, mental flexibility, rule compliance, split attention, and problem-solving (Aubin et al., 2018). VR-based assessment (i.e. V-MET) for OCD appears to be more sensitive to effects and changes in the executive sphere of patients and controls than traditional tests: patients made more mistakes, acted inefficiently (e.g., not using the market map), and scored higher in split attention and perseverance (persevering in errors is a clear sign of reduced flexibility) (La Paglia, et al., 2014). VR assessment tools can predict superior scores on neuropsychological tests in people with mood problems, implying that it might employ a finer index of EFs than a traditional paper and pencil exam (Horlyck et al., 2021).

Also in rehabilitation settings, VR-based tools allow for treating patients in ecologically relevant environments (Morganti, 2004; Morganti et al., 2007). These VEs enable patients to interact with life-like contexts and reproduce complex emotional and cognitive experiences (such as planning and organizing practical actions, attention shift) resembling everyday life situations (Castelnuovo et al., 2003) in ecologically valid and controlled environments. Secondly, VR-based tools provide personalized therapy depending on the abilities and requirements of patients (Lo Priore et al., 2003; Rand et al., 2009). The ability to customize situations in real-time, concentrating on the patient's features and requests (Parsons, 2015; Parsons et al., 2011) ensures the potential of tailoring scenarios in real-time, with important ramifications, especially in terms of rehabilitation (Castelnuovo et al., 2003; Rand et al., 2009). Individuals received instant feedback on their dual-task performance and had the opportunity to improve it: clinicians may track subjects' development in real time and tailor instruction to the needs and progress of the patients (Lo Priore et al., 2003).

In addition, VR allows people to perform activities at a distance, in the comfort and safety of their own homes (Dores et al., 2012). This finding is significant because it addresses two critical clinical issues: long waiting lists for health-care services and difficulties transferring patients between their homes and health-care facilities. VR provides also fun and engaging experiences with different tasks (e.g., exploration, challenges) that motivate patients to complete them. Conversely, patients perceive conventional rehabilitation methods, consisting of repeated behaviors without immediate feedback, as repetitive and boring (Castelnuovo et al., 2003; Rand et al., 2009, Mancuso et al., 2020). In terms of effects rehabilitation, several researchers found promising outcomes in treating many aspects of this construct in adult stroke, Traumatic Brain Injury (TBI), and Mild Cognitive Impairment patients. These studies found that VR-based therapy improved patients' performance more than traditional therapy. Patients, for example, increased their performance in planning, organizing, problem-solving, and multitasking tasks (Jacoby et al., 2013). Importantly, the study found that individuals were able to transfer rehabilitative outcomes from the VR therapy to real-world function, both in comparable activities (such as grocery shopping) and in new Instrumental Activities of Daily Living (IADL) tasks (such as cooking) (Jacoby et al., 2013; Carelli et al., 2008; Klinger et al., 2004; Rizzo & Kim, 2005). It's conceivable because the VR simulation tasks resembled daily activities more closely than those employed in traditional treatment (Liao et al., 2019, 2020).

VR Negative Healthcare Use

Although the use of VR has been innovatively useful within healthcare, it has operative and psychometric problems. They can be divided into three major issues when it comes to the operational ones: cyber-sickness, refresh rate (frame rate) of on-screen images and tethering problems (Riva et al., 2020). The most common cause of nausea and discomfort associated with virtual reality locomotion is *cyber-sickness*, which occurs when the virtual scene does not directly translate to real-life motion. Dizziness, headaches, sweating, nausea, and vomiting have all been linked to the use of VEs in some studies (Armstrong et al., 2013). This finding has to be taken into account in specific clinical settings, such as TBI patients, where headaches are a prevalent complaint. As a result, VR might worsen these symptoms, lowering the test findings' validity significantly. Another important factor is consumers' technology experience, particularly among the elderly, because poor exam performance could be attributed to a lack of understanding of how VR works (Parsons & Phillips, 2016). Anyway, to overcome this problem, the clinician could propose the subjects training with the tool before the real test (a familiarization phase), to maximize familiarity with the technological platform (Parsons and Phillips, 2016). Instead, the *frame rate of on-screen images* regards hardware (display technology) or software (low-end PC) limitations: a threshold of at least 60 frames per second is reported as a minimum to avoid discomfort and nausea over time (Sutherland et al., 2019). At last, the *tethering problems* regards the fact that most VR systems must be connected to a computer and require an external tracking device. VR can be used with input devices that allow the user to communicate with the VE and output devices that allow the user to see, hear, smell, or touch everything that occurs in the VE, according to the literature. Input devices, more specifically the keyboard, mouse, trackball, and joystick are simple desktop input devices that allow the user to issue continuous and discrete commands or movements to the environment. However, the input devices are the most problematic for the tethering use of VR: they are difficult to utilize for cables that restrict the user's motions. Furthermore, tethering also necessitates synchronization of the various peripherals (HMD, joystick, gloves, etc.) which necessitates a sophisticated technical setup that increases the system's latency (the delay cumulated from data acquisition on the patients to multimodal outputs) (Cipresso et al, 2018; Riva et al., 2020).

In addition to practical issues, there are also problems with statistical analysis. Although the results of using VR for assessment and rehabilitation are promising, there are a number of psychometric issues to be addressed: construct validity, discriminant validity, usability, and test re-test reliability that are not examined in all studies (Borgnis et al, 2022). As a result of these measures, the evaluation and rehabilitation techniques used were more reliable, effective, and generalizable. Overall, the majority of these VR-based assessment instruments have high construct validity, with significant correlations between the principal outcome measures and the scores of current standardized paper-and-pencil tests (Borgnis et al., 2022; Freeman et al., 2017; Plotnik et al., 2017). Despite those promising results, the literature highlighted the lack of studies on two other critical components for an instrument exploitable in a clinical setting: usability and test-retest reliability. Only a few studies have looked at usability, with results indicating that tools were usable, easy to learn, challenging and engaging, and devoid of major adverse effects (Aubin et al., 2018; Borgnis et al., 2021; Voinescu et al., 2021; Rand et al., 2005; Pedroli et al., 2013; Kizony et al., 2003; Shen et al., 2020). Several studies have highlighted the importance of evaluating usability and user experience when designing VR-based products over the years, especially for issues related to dizziness (Tuena et al., 2020; Pedroli et al., 2013; Pedroli et al., 2019; Sauer et al., 2020). Previous research has shown that cybersickness can cause users to have unpleasant experiences, impairing their performance and reducing the validity of test results (Armstrong et al., 2013). For example, headaches are a prevalent symptom in the TBI group; thus, a VR tool that could increase these symptoms would have an impact on performance, lowering the validity of the test results. In terms of the final critical components the idea of test-retest reliability plays a significant role especially in the therapeutic setting, where clinicians must longitudinally monitor patients, such as along a rehabilitative course. VR test-rest reliability will require more research to evaluate how to assess the consistency of results, to repeat the same test on the same sample at a different point in time, and to compare a test with its parallel forms. Borgnis and collaborators (2022) recently showed that out of 100 Executive Functions VR assessment and rehabilitation studies only one (Plotnik et al., 2020) had a test-retest measure. Other statistical concerns include the lack of a control group (Liao et al., 2019) and the small sample sizes (between 20 and 50 participants), despite the fact that the experimental groups were well matched for the main sociodemographic characteristics. Therefore, the following studies will have to expand the sample and introduce another treatment to confirm these promising results.

Lastly, the development of VR-based tools with complex VEs and tasks requires numerous specialized technological skills and high costs (Parsons, 2015). Those problems stem from the fact that some innovative VR head-mounted sets are too expensive and the use and the creation of the VEs needs information and engineering intervention. Expensive VR headset (i.e 6.000$), like Varjo XR-3 and Varjo VR-3, has integrated eye & hand tracking as the fastest and most accurate eye tracking integrated into a headset with a 200 Hz frame rate and built in Ultraleap hand tracking; a full frame Bionic Display with the highest resolution (over 70 ppd) and the widest field of view (115°) for unmatched realism; perfect comfort with a 3-point precision fit headband, active cooling, and ultra-wide optical design ensure comfortable usage even during longer VR sessions. These pricey VR headsets are still useful because they provide a more immersive reality than real life and, thanks to their ergonomics, also reduce dizziness. They are, however, almost never used in research due to their high cost. Furthermore, in the past, VR development was primarily focused on hardware solutions, whereas today's virtual solutions are primarily focused on software. Hardware has evolved into a commodity that is frequently available at a low cost but with a low image resolution. On the other hand, software needs to be customized each time, per each experiment, and this requires huge efforts in terms of development. Researchers in Augmented Reality (AR) and VR today need to be able to adapt software in their labs (Cipresso et al., 2018). It is true that the VEs can be created ad hoc using commercial software or free software available. An example of three major free software more used are Unity, ALTspace VR and Cospace Edu and in all, it is possible to use and create new VR environments. Although most of the time the intervention of computer technicians is required because programming skills are required and, even if virtual environments are already present, they are incompatible with clinical rehabilitation settings and thus cannot be modified further (Cipresso et al., 2018).

THE 360-DEGREE REVOLUTION

Among all the VR technologies, a concurrent emerging trend are 360° videos, which is also known as immersive videos or spherical videos. A 360-degree video is made up of a series of 360-degree images with a predetermined time interval between them. Each 360-degree image is a panorama taken with an omnidirectional camera or a combination of cameras to cover the entire horizontal field of view (i.e., 360-degree FOV). A 360-degree image is flattened in an equirectangular format in which longitude lines are projected to vertical straight lines of constant spacing, as shown in Figure 2. Latitude lines are mapped to horizontal straight lines with constant spacing in the same way. In other words, 360° video/image seamlessly surrounds and occupies the viewer's entire vision, as opposed to traditional 2-dimensional (2D) video/image, which only covers a limited plane.

They are designed to be seen through VR headsets, just as VR games and other interactive experiences, but they can also be watched on flat-screen devices, such as a phone or a computer, by dragging the viewpoint with a mouse or a finger.

For how they are realized, you can't directly engage with the story, although you can pick where to gaze. This gives users the ability to shape their own interpretation of the story, which is referred to as *agency*. As a director, it is possible to design an experience for the audience with the goal of directing their attention in certain directions, but ultimately, it is up to them to decide where they will look and how they will spend their time. One of the most significant differences between 360 and 2D video is thus agency, which aids in the generation of presence, which is what gives you the impression of being in the environment. And consistently with VR, 360° videos are capable of creating a sense of presence: i.e., the illusion of "being there" created by a virtual environment and is capable of eliciting strong emotional responses as it gives the user the feeling of being physically present, as well as the illusion of interacting and reacting as if they were in the real world (Chirico et al., 2017; Slater, 2003; Ventura et al., 2019).

Figure 2. 360° video/image

Users have a point of view in any 360° video, but it is possible to choose between first-person and third-person perspectives. The term "first person point of view" refers to the audience seeing through the eyes of a character as if they were the character themselves. When utilized properly, this perspective may be a powerful tool for conveying emotions. The term "third-person point of view" instead refers to the viewer's perception of the story from a distance rather than taking part in it. When opposed to the first-person perspective, this can be a good method because there are fewer chances of completely breaking presence. On the other hand, too much third person can occasionally result in a ghost effect. This is the sensation of not having a direct relationship with your surroundings, despite the fact that you are physically there in it.

VR and 360° videos can also be distinguished based on how their environments are build: VR is based on computer-generated environments created using special computer programs such as, for example, Unity, allowing the creation of all conceivable scenarios, from real to completely fictional, and the possibility to interact with them in numerous ways. On the other hand, these environments require a high level of technical knowledge, including programming skills. 360° video, on the other hand, is a real recording made by a special camera that can be viewed through a head-mounted display (such as a computer-generated video), but with fewer opportunities for interaction. Thus, unlike VR video games, which feature computer-generated characters and environments, 360 videos are created by using 360 cameras to capture live action in the actual world. They are different from 2D video in that they show the full world rather than just one area of it.

Beyond the end-user technologies that enable 360 content creation and consumption, video sharing websites (e.g., YouTube) and social media platforms (e.g., Facebook) enable users to publish and disseminate such content. Streaming 360-degree videos has become a popular VR application as it has become easier to create, distribute, and consume personalized 360-degree videos than VR ones.

Although, for how they are made, they do not allow proper interaction, by linking them together it may be possible to create this illusion: for example, with software like *Insta VR or 3D-vista* it is possible to allow navigation in any environment: by putting a *link* or a *hotspot* on a door it is possible to change the environment and give the appearance to change rooms. This feature has innumerable applications. In this way, 360° videos, besides being easy to use and cost-efficient, can provide a higher sense of presence with the impression that touching people, objects and animals is possible and that the spectator may be touched in return.

In this way, a clinician could also create ecological testing or improve existing ones. In fact, unfortunately, self-reported measures are susceptible to a variety of biases arising from the subjects' self-judgment and self-consciousness in front of the investigator. Many psychometric methods have been used to identify and delete possible instances of lying and deceitfulness in order to eliminate any deceiving biases in order to deal with these issues. Building tests with a higher rate of ecological validity could be one solution to these issues. In a perfect world, you'd be able to evaluate behaviors in specific tasks in everyday situations. Unfortunately, a number of issues prevent researchers from doing so. For starters, conducting an invasive experiment in participants' real-life situations would raise numerous ethical concerns, so it is preferable to create prototypical situations under the supervision of a researcher.

For example, personality is traditionally assessed through self-reports questionnaire: the most famous is the Big Five questionnaire which assess personality among five dimensions (extraversion, agreeableness, conscientiousness, emotional stability, and openness to experience) (Caprara et al., 1993). Cipresso and Riva (2015) have already proposed to assess personality using 360° videos, creating different situations where each component is posed either in multiple situations with a single choice or in only one situation with multiple choice. One example consists in immersing the user in a path in the forest with a crossroads, where he/she, through a narrative in the virtual environment, must choose a visible street (closeness to experiences) from the one side or a dark street (openness to experiences) on the other side. This method could also be implemented to build different assessment settings, like for example testing emotional regulation, and achieve not only many clinical purposes but also marketing choices.

Therefore, 360° videos offer the opportunity to immerse yourself in authentic natural environments that are known to evoke positive emotions (Chirico et al., 2017; Li et al, 2021; Liszio et al., 2018; Yeo et al., 2020).

Some studies, for instance, compared 360-degree videos versus other devices in their effectiveness in eliciting emotions in terms of both physiological and psychological responses. In this regard, Higuera-Trujillo and Lopez-Tarruell (2017) compared pictures, a 360° panorama, VR scenario and a real scenario. By conducting a comparative validation, the study's main goal is to determine which of the three formats provides the closest approximation of experience in physical environments.

The results showed that, in terms of psychological and physiological responses, the 360° panoramic and VR formats were more strongly related to the physical environment than photographs. While it is the most commonly used format in environmental-behavioral studies, the pictures currently appear to be the least appropriate display option available, particularly when comparing different VR technologies. Going deeper into the differences between immersive technologies, 360° panorama produced the most accurate results in terms of psychological responses, while VR format achieved the closest approach to physical life situations in terms of physiological reactions, which could be attributed to the effect of interactivity on the sense of presence (Higuera-Trujillo & Lopez-Tarruell, 2017). Moreover, Chirico and Gaggioli (2019) randomly assigned 50 people (25 women and 25 men) to one of two scenarios: a real-life contemplative scene (a panoramic view of a lake) or immersive 360-degree footage of the same landscape. The type and valence of emotions, as well as participants' sense of presence, were then compared across conditions. The emotions elicited by virtual and natural environments were not significantly different, according to the findings. The only exceptions were anger and amusement, which were significantly higher in the natural condition and significantly lower in the virtual condition, respectively. These findings could imply that there are not only a few degrees of separation between VR and reality, but also that they are not on the same experience continuum. However, in the context of comparison between immersive technologies, Brivio et al., 2021 did not find any differences in terms of positive and negative affect, heart rate and feeling of presence between 360° and VR conditions.

In recent years, some authors have begun to present neuropsychological stimuli using 360° environments (immersive photographs or videos) delivered via smartphones (Serino et al., 2017). Participants can be immersed in everyday scenarios from a first-person perspective using 360 ° technologies. Serino and colleagues investigated active visual perception in patients with frontal lobe damage using a 360° version of the Picture Interpretation Test (PIT) (Rosci et al., 2005). Nineteen PD patients and 19 healthy controls underwent a traditional neuropsychological evaluation to obtain a global cognitive profile (Montreal Cognitive Assessment - MoCa) and level of executive functioning via TMT and FAS (Nasreddine et al., 2005). In addition, participants were evaluated using PIT 360°, which consisted of two phases: "familiarization" and "experimentation."

Participants were immersed in a 360° environment (meeting room) during the familiarization phase, where they had to find objects and answer questions. Subjects were immersed in a 360° environment that replicated Favretto's painting during the experiment. Participants had to figure out what was going on in the scene as quickly as possible. Both traditional neuropsychological assessment and PIT 360° revealed different performances in PD patients compared to HC, according to the findings. PD patients took longer to provide a correct interpretation of the scene proposed, gave significantly more detailed descriptions of the scene, and appeared to be more prone to distractor interference, despite the fact that the percentage of participants who failed to correctly interpret the scene was similar between the two groups.As a result, PD patients had more trouble focusing on the most important elements for a correct interpretation of the scene. These findings support Luria's theory, implying that this test can detect active visual perception deficits. Furthermore, results revealed a significant correlation between PIT 360° indexes and traditional executive function tests: TMT and FAS (convergent validity).

Realdon and colleagues (2019) found similar results in detecting executive impairments in Sclerosis Multiple Sclerosis in the following study (Realdon et al., 2019). A total of 39 MS patients and healthy controls were evaluated using a standard neuropsychological battery and PIT 360°. Patients with MS performed significantly worse on the PIT 360 ° than the control group in terms of interpretation time and number of elements named, which was consistent with previous research on executive impairments in MS. Even though the global cognitive level and standard neuropsychological tests of executive functioning (TMT and FAS) were still in the non-pathological range, PIT 360° was able to successfully differentiate these two groups. As a result, these findings suggested that PIT 360° was a highly sensitive ecological tool for detecting EF deficits since the early clinical stages. As a result of these findings, it was concluded that PIT 360° is an ecological tool that is highly sensitive for detecting EF deficits even in the early stages of MS. Another example is the app built and validated in Pedroli et al. (2021) for motor rehabilitation at home. This iPad app takes advantage of the potential of 360° videos to help frail patients improve their balance through a series of exercises that increase in difficulty. The app contains ten sessions that should be played three times a week for three weeks. The usability study's findings are very encouraging, and patients are eager to try out this app at home as a motor rehabilitation guide.

Overall, 360-degree videos are a cheaper and easy to use technology. 360-degree cameras, such as insta360, Theta, GoPro OmniAll are in fact available with less than $500 USD and only need their application to be downloaded for its use.

Despite these advantages, there are also much more expensive cameras such as Insta 360 pro which, beside its cost, also requires stitching 360-degree panoramic pictures. Stitching means combining all images with overlapping sections to create a single panoramic or high-resolution image. Image stitching requires very precise overlaps and identical exposure settings for the best results. Algorithms are required for compositing surface creation, pixel alignment, image alignment, and distinctive feature recognition to serve as alignment reference points for software. Moreover, literature using 360 degrees video has addressed the VR psychometrics problems (i.e usability and construct validity): most of the studies had tested the usability (Pedroli et al., 2020; Pedroli et al., 2022) and also the construct validity, comparing 360 degrees outcomes with traditional testing (Realdon et al., 2019; Pieri et al., 2021).

FUTURE DIRECTION

It's likely that in the future, people in the healthcare field will be able to experience VR in ways that they can't yet imagine. However, there are plenty of impending technological advancements to look forward to in the near future, including VR technology that may be faster, lighter, and less expensive. As a result, 360° videos may be the best candidate for widespread adoption of standalone and mobile-based VR platforms without the need for a computer.

In addition, advances in smartphone technology (such as better cameras and processors) will allow us to enjoy smoother Augmented Reality (AR) and Virtual Reality (VR) experiences on our devices. Thanks to 5G wireless networks, people will be able to enjoy these technologies from anywhere on the planet. There are a few significant VR technological advancements on the horizon, including LiDAR (Light Detection and Ranging), which is a technique capable of creating a 3D map of the environment (Kim et al., 2021).

It will be possible to create more realistic AR experiences on our phones thanks to this technology. LiDAR technology is now available on the iPhone 12 and iPad Pro, and other devices are likely to follow suit in the near future.

Rather than appearing as a flat image, it can give AR projects a sense of depth. It also supports occlusion, which means that any real-world object in front of the AR item should obstruct the view of it by definition. Future research could look into the use of this technology in healthcare settings, such as home rehabilitation via tablet or mobile phone.

CONCLUSIONS

VR provides a fantastic opportunity for researchers to create prototypical situations that can aid in the understanding of personality traits. Unfortunately, VR is generally difficult to manage, and many researchers are put off by technological barriers such as the requirement for programming skills, which are often lacking among psychological researchers. Even though the cost of technology has steadily decreased over the last decade, the cost of creating virtual environments remains high because it necessitates the use of programmers and 3D graphics experts. A complete immersive virtual reality system was estimated to cost around $10,000 per unit in the past (including head-mounted display of medium quality, graphics workstations, sensors and devices for interaction). In addition, the cost of developing (or purchasing) therapeutic virtual environments had to be factored into these costs. Some of these barriers, however, have been partially overcome in recent years. First and foremost, the development of clinical protocols for a variety of diseases has resulted in the conduct of controlled trials demonstrating the efficacy of cyber therapy (Holden, 2005; Riva, 2003; Riva et al, 2020). Hardware costs, on the other hand, have dropped dramatically. The cost of an immersive VR system consisting of a laptop with a good graphics card and a wearable head-mounted display with motion-tracking sensors is approximately 1,500 Euros, thanks to the incredible expansion of the 3D gaming industry and LCD displays. The cost of software, the need of a computer and an external tracking device, on the other hand, continues to be a barrier to market expansion.

These limitations could be overcome by the use of 360-degree videos, known also as spherical videos which are recorded using a camera with multiple camera lenses or a rig of multiple cameras.

The use of multiple lenses allows for simultaneous recording of all directions, effectively giving the camera a full view of what is going on around it. Unlike traditional VR content, which requires a specific platform and programming skills, 360-degree videos can be easily recorded by a clinical team using specific cameras (e.g., GoPro Max, Insta360 One, or Ricoh Theta SC) that cost less than $500 USD, allowing for the quick development of ecologically oriented content that is directly connected to the patient's real-world environment. Moreover, YouTube and Facebook directly support these kinds of videos, making it simple to share developed content with patients and their families using common smartphones.

Although they are not designed to allow proper interaction, by linking them together, it may be possible to create this illusion: for example, with software such as Insta VR or 3D-vista, navigation in any environment can be enabled: by placing a link or a hotspot on a door, the environment can be changed, and the appearance of changing rooms can be created. This feature can be used in a variety of ways. 360° videos can thus provide a greater sense of presence by giving the impression that touching people, objects, and animals is possible, and that the spectator may be touched in return, in addition to being simple to use and cost-effective.

These features make 360° videos a potential technology in the healthcare sector undoubtedly with possibility of expansion on the market.

As a result, we believe that 360° videos will be the best candidate for the widespread adoption of standalone and mobile-based VR platforms that do not require the use of a computer.

ACKNOWLEDGMENT

This research received no specific grant from any funding agency in the public, commercial, or not-for-profit sectors.

REFERENCES

Alexander, T., Westhoven, M., & Conradi, J. (2017). Virtual Environments for Competency-Oriented. *Education + Training*, *498*, 23–29. doi:10.1007/978-3-319-42070-7_3

Apps, M. A. J., & Tsakiris, M. (2014). The free-energy self: A predictive coding account of self-recognition. *Neuroscience and Biobehavioral Reviews*, *41*, 85–97. doi:10.1016/j.neubiorev.2013.01.029 PMID:23416066

Armstrong, C. M., Reger, G. M., Edwards, J., Rizzo, A. A., Courtney, C. G., & Parsons, T. D. (2013). Validity of the Virtual Reality Stroop Task (VRST) in active duty military. *Journal of Clinical and Experimental Neuropsychology*, *35*(2), 113–123. doi:10.1080/13803395.2012.740002 PMID:23157431

Aubin, G., Béliveau, M.-F., & Klinger, E. (2018). An exploration of the ecological validity of the Virtual Action Planning–Supermarket (VAP-S) with people with schizophrenia. *Neuropsychological Rehabilitation*, *28*(5), 689–708. doi:10.1080/09602011.2015.1074083 PMID:26317526

Bach, F., Çakmak, H.K., Maaß, H., Bekrater-Bodmann, R., Foell, J., Diers, M., Trojan, J., Fuchs, X., & Flor, H. (2012). *Illusory Hand Ownership Induced by an MRI Compatible Immersive Virtual Reality Device*. Academic Press.

Banakou, D., Kishore, S., & Slater, M. (2018). Virtually being Einstein results in an improvement in cognitive task performance and a decrease in age bias. *Frontiers in Psychology*, *9*, 917. doi:10.3389/fpsyg.2018.00917 PMID:29942270

Bassolino, M., Franza, M., Bello Ruiz, J., Pinardi, M., Schmidlin, T., Stephan, M. A., Solcà, M., Serino, A., & Blanke, O. (2018). Non-invasive brain stimulation of motor cortex induces embodiment when integrated with virtual reality feedback. *The European Journal of Neuroscience*, *47*(7), 790–799. doi:10.1111/ejn.13871 PMID:29460981

Bohil, C. J., Alicea, B., & Biocca, F. A. (2011). Virtual reality in neuroscience research and therapy. *Nature Reviews. Neuroscience*, *12*(12), 752–762. doi:10.1038/nrn3122 PMID:22048061

Borgnis, F., Baglio, F., Pedroli, E., Rossetto, F., Isernia, S., Uccellatore, L., Riva, G., & Cipresso, P. (2021). EXecutive-Functions Innovative Tool (EXIT 360°): A Usability and User Experience Study of an Original 360°-Based Assessment Instrument. *Sensors (Basel)*, *21*(17), 5867. doi:10.339021175867 PMID:34502758

Borgnis, F., Baglio, F., Pedroli, E., Rossetto, F., Uccellatore, L., Oliveira, J., Riva, G., & Cipresso, P. (2022). Available Virtual Reality-Based Tools for Executive Functions: A Systematic Review. *Frontiers in Psychology*, 13. PMID:35478738

Brivio, E., Serino, S., Negro Cousa, E., Zini, A., Riva, G., & De Leo, G. (2021). Virtual reality and 360° panorama technology: A media comparison to study changes in sense of presence, anxiety, and positive emotions. *Virtual Reality (Waltham Cross)*, *25*(2), 303–311. doi:10.100710055-020-00453-7

Burin, D., Kilteni, K., Rabuffetti, M., Slater, M., & Pia, L. (2019b). Body ownership increases the interference between observed and executed movements. *PLoS One*, *14*(1), e0209899. doi:10.1371/journal.pone.0209899 PMID:30605454

Burin, D., Yamaya, N., Ogitsu, R., & Kawashima, R. (2019a). Virtual training leads to real acute physical, cognitive, and neural benefits on healthy adults: Study protocol for a randomized controlled trial. *Trials*, *20*(1), 20. doi:10.118613063-019-3591-1 PMID:31511036

Calvert, G. A., & Thesen, T. (2004, January). Multisensory integration: Methodological approaches and emerging principles in the human brain (2014). *Journal of Physiology, Paris*, *98*(1-3), 191–205. Advance online publication. doi:10.1016/j.jphysparis.2004.03.018 PMID:15477032

Campbell, Z., Zakzanis, K. K., Jovanovski, D., Joordens, S., Mraz, R., & Graham, S. J. (2009). Utilizing Virtual Reality to Improve the Ecological Validity of Clinical Neuropsychology: An fMRI Case Study Elucidating the Neural Basis of Planning by Comparing the Tower of London with a Three-Dimensional Navigation Task. *Applied Neuropsychology*, *16*(4), 295–306. doi:10.1080/09084280903297891 PMID:20183185

Caprara, G. V., Barbaranelli, C., Borgogni, L., & Perugini, M. (1993). The "Big Five Questionnaire": A new questionnaire to assess the five factor model. *Personality and Individual Differences*, *15*(3), 281–288. doi:10.1016/0191-8869(93)90218-R

Carelli, L., Morganti, F., Weiss, P. L., Kizony, R., & Riva, G. (2008). A virtual reality paradigm for the assessment and rehabilitation of executive function deficits post stroke: Feasibility study. *2008 Virtual Rehabilitation*, 99–104. doi:10.1109/ICVR.2008.4625144

Castelnuovo, G., Lo Priore, C., Liccione, D., & Cioffi, G. (2003). Virtual Reality based tools for the rehabilitation of cognitive and executive functions: The V-STORE. *PsychNology Journal*, *1*, 310–325.

Cebolla, A. M., Palmero-Soler, E., Leroy, A., & Cheron, G. (2017). EEG Spectral Generators Involved in Motor Imagery: A swLORETA Study. *Frontiers in Psychology*, *8*, 2133. doi:10.3389/fpsyg.2017.02133 PMID:29312028

Chirico, A., Cipresso, P., Yaden, D. B., Biassoni, F., Riva, G., & Gaggioli, A. (2017). Effectiveness of immersive videos in inducing awe: An experimental study. *Scientific Reports*, *7*(1), 1–11. doi:10.103841598-017-01242-0 PMID:28450730

Chirico, A., & Gaggioli, A. (2019). When virtual feels real: Comparing emotional responses and presence in virtual and natural environments. *Cyberpsychology, Behavior, and Social Networking*, *22*(3), 220–226. doi:10.1089/cyber.2018.0393 PMID:30730222

Cipresso, P., Giglioli, I. A. C., Raya, M. A., & Riva, G. (2018). The past, present, and future of virtual and augmented reality research: A network and cluster analysis of the literature. *Frontiers in Psychology*, *9*, 2086. doi:10.3389/fpsyg.2018.02086 PMID:30459681

Cipresso, P., & Riva, G. (2015). Ecological Settings by Means of Virtual Reality. The Wiley handbook of personality assessment, 240.

Climent, G., Banterla, F., & Iriarte, Y. (2010). Virtual reality, technologies and behavioural assessment. *AULA Ecol. Evaluation Atten. Processes*, *2010*, 19–28.

Colombo, D., Díaz-García, A., Fernandez-Álvarez, J., & Botella, C. (2021). Virtual reality for the enhancement of emotion regulation. *Clinical Psychology & Psychotherapy*, *28*(3), 519–537. doi:10.1002/cpp.2618 PMID:34048621

Dores, A. R., Carvalho, I. P., Barbosa, F., Almeida, I., Guerreiro, S., Oliveira, B., de Sousa, L., & Caldas, A. C. (2012). Computer-Assisted Rehabilitation Program – Virtual Reality (CARP-VR): A Program for Cognitive Rehabilitation of Executive Dysfunction. doi:10.1007/978-3-642-31800-9_10

Englund, C., Olofsson, A. D., & Price, L. (2017). Teaching with technology in higher education: Understanding conceptual change and development in practice. *Higher Education Research & Development, 36*(1), 73–87. doi:10.1080/07294360.2016.1171300

Freeman, D., Reeve, S., Robinson, A., Ehlers, A., Clark, D., Spanlang, B., & Slater, M. (2017). Virtual reality in the assessment, understanding, and treatment of mental health disorders. *Psychological Medicine, 47*(14), 2393–2400. doi:10.1017/S003329171700040X PMID:28325167

Gallagher, A. G., Ritter, E. M., Champion, H., Higgins, G., Fried, M. P., Moses, G., Smith, C. D., & Satava, R. M. (2005). Virtual reality simulation for the operating room: Proficiency-based training as a paradigm shift in surgical skills training. *Annals of Surgery, 241*(2), 364–372. doi:10.1097/01.sla.0000151982.85062.80 PMID:15650649

Higuera-Trujillo, J. L., Maldonado, J. L. T., & Millán, C. L. (2017). Psychological and physiological human responses to simulated and real environments: A comparison between Photographs, 360 Panoramas, and Virtual Reality. *Applied Ergonomics, 65*, 398–409. doi:10.1016/j.apergo.2017.05.006 PMID:28601190

Holden, M. K. (2005). Virtual environments for motor rehabilitation [review]. *Cyberpsychology & Behavior, 8*(3), 187–211. doi:10.1089/cpb.2005.8.187 PMID:15971970

Hørlyck, L. D., Obenhausen, K., Jansari, A., Ullum, H., & Miskowiak, K. W. (2021). Virtual reality assessment of daily life executive functions in mood disorders: Associations with neuropsychological and functional measures. *Journal of Affective Disorders, 280*, 478–487. doi:10.1016/j.jad.2020.11.084 PMID:33248416

Ijsselsteijn, W., & Riva, G. (2003). Being there: the experience of presence in mediated environments. In G. Riva, F. Davide, & W. A. IJsselsteijn (Eds.), *Studies in new technologies and practices in communication. Being there: concepts, effects and measurement of user presence in synthetic environments* (pp. 3–16). IOS Press.

Jacoby, M., Averbuch, S., Sacher, Y., Katz, N., Weiss, P. L., & Kizony, R. (2013). Effectiveness of Executive Functions Training Within a Virtual Supermarket for Adults With Traumatic Brain Injury: A Pilot Study. *IEEE Transactions on Neural Systems and Rehabilitation Engineering, 21*(2), 182–190. doi:10.1109/TNSRE.2012.2235184 PMID:23292820

Josman, N., Hof, E., Klinger, E., Marié, R. M., Goldenberg, K., Weiss, P. L., & Kizony, R. (2006). Performance within a virtual supermarket and its relationship to executive functions in post-stroke patients. *2006 International Workshop on Virtual Rehabilitation*, 106-109. 10.1109/IWVR.2006.1707536

Kim, I., Martins, R. J., Jang, J., Badloe, T., Khadir, S., Jung, H. Y., Kim, H., Kim, J., Genevet, P., & Rho, J. (2021). Nanophotonics for light detection and ranging technology. *Nature Nanotechnology, 16*(5), 508–524. doi:10.103841565-021-00895-3 PMID:33958762

Kizony, R., Katz, N., & Weiss, P. L. (2003). Adapting an immersive virtual reality system for rehabilitation. *Computer Animation and Virtual Worlds, 14*, 261–268.

Klinger, E., Chemin, I., Lebreton, S., & Marie, R. M. (2004). A virtual supermarket to assess cognitive planning. *Annual Review of Cybertherapy and Telemedicine*, *2*, 49–57.

La Paglia, F., la Cascia, C., Cipresso, P., Rizzo, R., Francomano, A., Riva, G., & la Barbera, D. (2014). Psychometric Assessment Using Classic Neuropsychological and Virtual Reality Based Test: A Study in Obsessive-Compulsive Disorder (OCD) and Schizophrenic Patients. doi:10.1007/978-3-319-11564-1_3

Leroy, A., Cevallos, C., Cebolla, A. M., Caharel, S., Dan, B., & Cheron, G. (2017). Short-term EEG dynamics and neural generators evoked by navigational images. *PLoS One*, *12*(6), 12. doi:10.1371/journal.pone.0178817 PMID:28632774

Li, H., Yang, Z., Zhang, X., Wang, H., Liu, H., Cao, Y., & Zhang, G. (2021). Access to Nature via Virtual Reality: A Mini Review. *Frontiers in Psychology*, *12*, 4324. doi:10.3389/fpsyg.2021.725288 PMID:34675840

Liao, Y.-Y., Chen, I.-H., Lin, Y.-J., Chen, Y., & Hsu, W.-C. (2019). Effects of Virtual Reality-Based Physical and Cognitive Training on Executive Function and Dual-Task Gait Performance in Older Adults With Mild Cognitive Impairment: A Randomized Control Trial. *Frontiers in Aging Neuroscience*, *11*, 162. Advance online publication. doi:10.3389/fnagi.2019.00162 PMID:31379553

Liao, Y.-Y., Tseng, H.-Y., Lin, Y.-J., Wang, C.-J., & Hsu, W.-C. (2020). Using virtual reality-based training to improve cognitive function, instrumental activities of daily living and neural efficiency in older adults with mild cognitive impairment. *European Journal of Physical and Rehabilitation Medicine*, *56*(1). Advance online publication. doi:10.23736/S1973-9087.19.05899-4 PMID:31615196

Liszio, S., & Masuch, M. (2019). Interactive immersive virtual environments cause relaxation and enhance resistance to acute stress. *Annual Review of Cybertherapy and Telemedicine*, *17*, 65–71.

Liu, Y., Tan, W., Chen, C., Liu, C., Yang, J., & Zhang, Y. (2019). A Review of the Application of Virtual Reality Technology in the Diagnosis and Treatment of Cognitive Impairment. *Frontiers in Aging Neuroscience*, *11*, 280. doi:10.3389/fnagi.2019.00280 PMID:31680934

Lo Priore, C., Castelnuovo, G., Liccione, D., & Liccione, D. (2003). Experience with V-STORE: Considerations on Presence in Virtual Environments for Effective Neuropsychological Rehabilitation of Executive Functions. *Cyberpsychology & Behavior*, *6*(3), 281–287. doi:10.1089/109493103322011579 PMID:12855084

Lombard, M., & Ditton, T. (2006). At the Heart of It All: The Concept of Presence. *Journal of Computer-Mediated Communication, 3*(2). doi:10.1111/j.1083-6101.1997.tb00072.x

Mancuso, V., Stramba-Badiale, C., Cavedoni, S., & Cipresso, P. (2022). Biosensors and Biofeedback in Clinical Psychology. *Comprehensive Clinical Psychology*, 28–50. doi:10.1016/B978-0-12-818697-8.00002-9

Mancuso, V., Stramba-Badiale, C., Cavedoni, S., Pedroli, E., Cipresso, P., & Riva, G. (2020). Virtual Reality Meets Non-invasive Brain Stimulation: Integrating Two Methods for Cognitive Rehabilitation of Mild Cognitive Impairment. *Frontiers in Neurology*, *11*, 566731. Advance online publication. doi:10.3389/fneur.2020.566731 PMID:33117261

Massetti, T., Crocetta, T. B., Da Silva, T. D., Trevizan, I. L., Arab, C., Caromano, F. A., & Monteiro, C. B. de M. (2017). Application and outcomes of therapy combining transcranial direct current stimulation and virtual reality: A systematic review. *Disability and Rehabilitation. Assistive Technology, 12*(6), 551–559. doi:10.1080/17483107.2016.1230152 PMID:27677678

Meldrum, D., Glennon, A., Herdman, S., Murray, D., & McConn-Walsh, R. (2012). Virtual reality rehabilitation of balance: Assessment of the usability of the Nintendo Wii ® Fit Plus. *Disability and Rehabilitation. Assistive Technology, 7*(3), 205–210. doi:10.3109/17483107.2011.616922 PMID:22117107

Milgram, P., & Kishino, F. (1994). A Taxonomy of Mixed Reality Visual Displays. *IEICE Transactions on Information and Systems, 77*, 1321–1329.

Mishra, S., Kumar, A., Padmanabhan, P., & Gulyás, B. (2021). Neurophysiological Correlates of Cognition as Revealed by Virtual Reality: Delving the Brain with a Synergistic Approach. *Brain Sciences, 11*(1), 51. doi:10.3390/brainsci11010051 PMID:33466371

Morganti, F. (2004). Virtual interaction in cognitive neuropsychology. *Studies in Health Technology and Informatics, 99*, 55–70. PMID:15295146

Morganti, F., Gaggioli, A., Strambi, L., Rusconi, M. L., & Riva, G. (2007). A virtual reality extended neuropsychological assessment for topographical disorientation: A feasibility study. *Journal of Neuroengineering and Rehabilitation, 4*(1), 26–26. doi:10.1186/1743-0003-4-26 PMID:17625011

Nasreddine, Z. S. (2005). The Montreal cognitive assessment: A brief screening tool for mild cognitive impairment. *Journal of the American Geriatrics Society, 53*, 695–699. doi:10.1111/j.1532-5415.2005.53221.x PMID:15817019

Neri, S. G., Cardoso, J. R., Cruz, L., Lima, R. M., de Oliveira, R. J., Iversen, M. D., & Carregaro, R. L. (2017). Do virtual reality games improve mobility skills and balancemeasurements in community-dwelling older adults? Systematic review andmeta-analysis. *Clinical Rehabilitation, 31*(10), 1292–1304. doi:10.1177/0269215517694677 PMID:28933612

Notzon, S., Deppermann, S., Fallgatter, A. J., Diemer, J., Kroczek, A. M., Domschke, K., Zwanzger, P., & Ehlis, A. (2015). Psychophysiological effects of an iTBS modulated virtual reality challenge including participants with spider phobia. *Biological Psychology, 112*, 66–76. doi:10.1016/j.biopsycho.2015.10.003 PMID:26476332

Orlosky, J., Huynh, B., & Hollerer, T. (2019). Using eye tracked virtual reality to classify understanding of vocabulary in recall tasks. *Proceedings - 2019 IEEE International Conference on Artificial Intelligence and Virtual Reality, AIVR 2019*, 66–73. 10.1109/AIVR46125.2019.00019

Osimo, S. A., Pizarro, R., Spanlang, B., & Slater, M. (2015). Conversations between self and self as Sigmund Freud—A virtual body ownership paradigm for self counselling. *Scientific Reports, 5*(1), 1–14. doi:10.1038rep13899 PMID:26354311

Parsons, T. D. (2015). Virtual Reality for Enhanced Ecological Validity and Experimental Control in the Clinical, Affective and Social Neurosciences. *Frontiers in Human Neuroscience, 9*. Advance online publication. doi:10.3389/fnhum.2015.00660 PMID:26696869

Parsons, T. D., Courtney, C. G., Arizmendi, B. J., & Dawson, M. E. (2011). Virtual Reality Stroop Task for Neurocognitive Assessment. *Studies in Health Technology and Informatics*, *163*, 433–439. PMID:21335835

Parsons, T. D., Courtney, C. G., & Dawson, M. E. (2013). Virtual reality Stroop task for assessment of supervisory attentional processing. *Journal of Clinical and Experimental Neuropsychology*, *35*(8), 812–826. doi:10.1080/13803395.2013.824556 PMID:23961959

Parsons, T. D., & Phillips, A. S. (2016). Virtual reality for psychological assessment in clinical practice. *Practice Innovations (Washington, D.C.)*, *1*(3), 197–217. doi:10.1037/pri0000028

Pedroli, E., Cipresso, P., Greci, L., Arlati, S., Mahroo, A., Mancuso, V., Boilini, L., Rossi, M., Stefanelli, L., Goulene, K., Sacco, M., Stramba-Badiale, M., Riva, G., & Gaggioli, A. (2020). A new application for the motor rehabilitation at home: Structure and usability of Bal-App. *IEEE Transactions on Emerging Topics in Computing*, *9*(3), 1290–1300. doi:10.1109/TETC.2020.3037962

Pedroli, E., Cipresso, P., Serino, S., Albani, G., & Riva, G. (2013). A Virtual Reality Test for the Assessment of Cognitive Deficits: Usability and Perspectives. *Proceedings of the ICTs for Improving Patients Rehabilitation Research Techniques*. 10.4108/icst.pervasivehealth.2013.252359

Pedroli, E., La Paglia, F., Cipresso, P., la Cascia, C., Riva, G., & la Barbera, D. (2019). A Computational Approach for the Assessment of Executive Functions in Patients with Obsessive–Compulsive Disorder. *Journal of Clinical Medicine*, *8*(11), 1975. doi:10.3390/jcm8111975 PMID:31739514

Pedroli, E., Mancuso, V., Stramba-Badiale, C., Cipresso, P., Tuena, C., Greci, L., Goulene, K., Stramba-Badiale, M., Riva, G., & Gaggioli, A. (2022). Brain M-App's Structure and Usability: A New Application for Cognitive Rehabilitation at Home. *Frontiers in Human Neuroscience*, *16*, 898633. doi:10.3389/fnhum.2022.898633 PMID:35782042

Petkova, V. I., & Ehrsson, H. H. (2008). If I were you: Perceptual illusion of body swapping. *PLoS One*, *3*(12), e3832. doi:10.1371/journal.pone.0003832 PMID:19050755

Pieri, L., Serino, S., Cipresso, P., Mancuso, V., Riva, G., & Pedroli, E. (2022). The ObReco-360: A new ecological tool to memory assessment using 360 immersive technology. *Virtual Reality (Waltham Cross)*, *26*(2), 639–648. doi:10.100710055-021-00526-1

Plotnik, M., Ben-Gal, O., Doniger, G. M., Gottlieb, A., Bahat, Y., Cohen, M., Kimel-Naor, S., Zeilig, G., & Beeri, M. S. (2020). Multimodal immersive trail making – virtual reality paradigm to study cognitive-motor interactions. *bioRxiv*. Advance online publication. doi:10.1101/2020.05.27.118760

Plotnik, M., Doniger, G. M., Bahat, Y., Gottleib, A., ben Gal, O., Arad, E., Kribus-Shmiel, L., Kimel-Naor, S., Zeilig, G., Schnaider-Beeri, M., Yanovich, R., Ketko, I., & Heled, Y. (2017). Immersive trail making: Construct validity of an ecological neuropsychological test. *2017 International Conference on Virtual Rehabilitation (ICVR)*, 1–6. 10.1109/ICVR.2017.8007501

Rand, D., Katz, N., Shahar, M., Kizony, R., & Weiss, P. L. (2005). The virtual mall: A functional virtual environment for stroke rehabilitation. *Annual Review of Cybertherapy and Telemedicine: A Decade of VR*, *3*, 193–198.

Rand, D., & Rukan, S. B.-A., Weiss, P. L., & Katz, N. (2009). Validation of the Virtual MET as an assessment tool for executive functions. *Neuropsychological Rehabilitation*, *19*(4), 583–602. doi:10.1080/09602010802469074 PMID:19058093

Realdon, O., Serino, S., Savazzi, F., Rossetto, F., Cipresso, P., Parsons, T. D., Cappellini, G., Mantovani, F., Mendozzi, L., Nemni, R., Riva, G., & Baglio, F. (2019). An ecological measure to screen executive functioning in MS: The Picture Interpretation Test (PIT) 360°. *Scientific Reports*, *9*(1), 1–8. doi:10.103841598-019-42201-1 PMID:30952936

Riva, G. (1997). *Virtual Reality in Neuro-Psycho-Physiology: Cognitive, Clinical and Methodological Issues in Assessment and Treatment* (Vol. 44). IOS Press.

Riva, G. (2003). Applications of virtual environments in medicine. *Methods of Information in Medicine*, *42*(5), 524–534. PubMed

Riva, G., Mancuso, V., Cavedoni, S., & Stramba-Badiale, C. (2020). Virtual reality in neurorehabilitation: A review of its effects on multiple cognitive domains. *Expert Review of Medical Devices*, *17*(10), 1035–1061. doi:10.1080/17434440.2020.1825939 PMID:32962433

Riva, G., Wiederhold, B. K., & Mantovani, F. (2019). Neuroscience of Virtual Reality: From Virtual Exposure to Embodied Medicine. *Cyberpsychology, Behavior, and Social Networking*, *22*(1), 82–96. doi:10.1089/cyber.2017.29099.gri PMID:30183347

Rizzo, A. A., Buckwalter, J. G., Bowerly, T., Zaag, C. V., Humphrey, L., Neumann, U., Chua, C., Kyriakakis, C., Rooyen, A. V., & Sisemore, D. (2000). The Virtual Classroom: A Virtual Reality Environment for the Assessment and Rehabilitation of Attention Deficits. *Cyberpsychology, Behavior, and Social Networking*, *3*(3), 483–499. doi:10.1089/10949310050078940

Rizzo, A. A., & Kim, G. J. (2005). A SWOT Analysis of the Field of Virtual Reality Rehabilitation and Therapy. *Presence (Cambridge, Mass.)*, *14*(2), 119–146. doi:10.1162/1054746053967094

Roberts, G., Holmes, N., Alexander, N., Boto, E., Leggett, J., Hill, R. M., Shah, V., Rea, M., Vaughan, R., Maguire, E. A., Kessler, K., Beebe, S., Fromhold, T. M., Barnes, G. R., Bowtell, R., & Brookes, M. J. (2019). Towards OPM-MEG in a virtual reality environment. *NeuroImage*, *199*, 408–417. doi:10.1016/j.neuroimage.2019.06.010 PMID:31173906

Rosci, C. E., Sacco, D., Laiacona, M., & Capitani, E. (2004). Interpretation of a complex picture and its sensitivity to frontal damage: A reappraisal. *Neurological Sciences*, *25*(6), 322–330. doi:10.100710072-004-0365-6 PMID:15729495

Sauer, J., Sonderegger, A., & Schmutz, S. (2020). Usability, user experience and accessibility: Towards an integrative model. *Ergonomics*, *63*(10), 1207–1220. doi:10.1080/00140139.2020.1774080 PMID:32450782

Schmidt, M., Beck, D., Glaser, N., & Schmidt, C. (2017). A Prototype Immersive, Multi-user 3D Virtual Learning Environment for Individuals with Autism to Learn Social and Life Skills: A Virtuoso DBR Update. doi:10.1007/978-3-319-60633-0_15

Serino, S., Pedroli, E., Tuena, C., De Leo, G., Stramba-Badiale, M., Goulene, K., Mariotti, N. G., & Riva, G. (2017). A Novel Virtual Reality-Based Training Protocol for the Enhancement of the "Mental Frame Syncing" in Individuals with Alzheimer's Disease: A Development-of-Concept Trial. *Frontiers in Aging Neuroscience*, *9*, 240. doi:10.3389/fnagi.2017.00240 PMID:28798682

Shen, J., Xiang, H., Luna, J., Grishchenko, A., Patterson, J., Strouse, R. V., Roland, M., Lundine, J. P., Koterba, C. H., Lever, K., Groner, J. I., Huang, Y., & Lin, E. D. (2020). Virtual Reality–Based Executive Function Rehabilitation System for Children With Traumatic Brain Injury: Design and Usability Study. *JMIR Serious Games*, *8*(3), 8. doi:10.2196/16947 PMID:32447275

Shine, J. M., Ward, P. B., Naismith, S. L., Pearson, M., & Lewis, S. J. (2011). Utilising functional MRI (fMRI) to explore the freezing phenomenon in Parkinson's disease. *Journal of Clinical Neuroscience*, *18*(6), 807–810. doi:10.1016/j.jocn.2011.02.003 PMID:21398129

Slater, M. (2003). A note on presence terminology. *Presence Connect, 3*(3), 1-5.

Slater, M., & Wilbur, S. (1997). A Framework for Immersive Virtual Environments (FIVE): Speculations on the Role of Presence in Virtual Environments. *Presence (Cambridge, Mass.)*, *6*(6), 603–616. doi:10.1162/pres.1997.6.6.603

Song, H., Chen, F., Peng, Q., Zhang, J., & Gu, P. (2017). Improvement of userexperience using virtual reality in open-architecture product design. *Proceedings of the Institution of Mechanical Engineers. Part B, Journal of Engineering Manufacture*, 232.

Stramba-Badiale, C., Mancuso, V., Cavedoni, S., Pedroli, E., Cipresso, P., & Riva, G. (2020). Transcranial Magnetic Stimulation Meets Virtual Reality: The Potential of Integrating Brain Stimulation With a Simulative Technology for Food Addiction. *Frontiers in Neuroscience*, *14*, 720. Advance online publication. doi:10.3389/fnins.2020.00720 PMID:32760243

Sutherland, J., Belec, J., Sheikh, A., Chepelev, L., Althobaity, W., & Chow, B. J. W. (2019). Applying Modern Virtual and Augmented Reality Technologies to Medical Images and Models. Journal of Digital Imaging, 1(32), 38–53. doi:10.100710278-018-0122-7

Sutter, C., Drewing, K., & Müsseler, J. (2014). Multisensory integration in action control. *Frontiers in Psychology*, *5*, 544. doi:10.3389/fpsyg.2014.00544 PMID:24959154

Tarnanas, I., Schlee, W., Tsolaki, M., Müri, R., Mosimann, U., & Nef, T. (2013). Ecological Validity of Virtual Reality Daily Living Activities Screening for Early Dementia: Longitudinal Study. *JMIR Serious Games*, *1*(1), e1. doi:10.2196/games.2778 PMID:25658491

Tieri, G., Morone, G., Paolucci, S., & Iosa, M. (2018). Virtual reality in cognitive and motor rehabilitation: Facts, fiction and fallacies. *Expert Review of Medical Devices*, *15*(2), 107–117. doi:10.1080/17434440.2018.1425613 PMID:29313388

Tuena, C., Mancuso, V., Stramba-Badiale, C., Pedroli, E., Stramba-Badiale, M., Riva, G., & Repetto, C. (2021). Egocentric and allocentric spatial memory in mild cognitive impairment with real-world and virtual navigation tasks: A systematic review. *Journal of Alzheimer's Disease*, *79*(1), 95–116. doi:10.3233/JAD-201017 PMID:33216034

Tuena, C., Pedroli, E., Trimarchi, P. D., Gallucci, A., Chiappini, M., Goulene, K., Gaggioli, A., Riva, G., Lattanzio, F., Giunco, F., & Stramba-Badiale, M. (2020). Usability Issues of Clinical and Research Applications of Virtual Reality in Older People: A Systematic Review. *Frontiers in Human Neuroscience, 14*, 93. doi:10.3389/fnhum.2020.00093 PMID:32322194

Ventura, S., Brivio, E., Riva, G., & Baños, R. M. (2019). Immersive Versus Non-immersive Experience: Exploring the Feasibility of Memory Assessment Through 360° Technology. *Frontiers in Psychology, 10*, 2509. Advance online publication. doi:10.3389/fpsyg.2019.02509 PMID:31798492

Vogeley, K., May, M., Ritzl, A., Falkai, P., Zilles, K., & Fink, G. R. (2004). Neural correlates of first-person perspective as one constituent of human self-consciousness. *Journal of Cognitive Neuroscience, 16*(5), 817–827. doi:10.1162/089892904970799 PMID:15200709

Voinescu, A., Fodor, L.-A., Fraser, D. S., Mejias, M., & David, D. (2019). Exploring the Usability of Nesplora Aquarium, a Virtual Reality System for Neuropsychological Assessment of Attention and Executive Functioning. *2019 IEEE Conference on Virtual Reality and 3D User Interfaces (VR)*, 1207–1208. 10.1109/VR.2019.8798191

Weerdmeester, J., van Rooij, M., Harris, O., Smit, N., Engels, R. C. M. E., & Granic, I. (2017). Exploring the Role of Self-efficacy in Biofeedback Video Games. *Extended Abstracts Publication of the Annual Symposium on Computer-Human Interaction in Play*, 453–461. 10.1145/3130859.3131299

Yeo, N. L., White, M. P., Alcock, I., Garside, R., Dean, S. G., Smalley, A. J., & Gatersleben, B. (2020). What is the best way of delivering virtual nature for improving mood? An experimental comparison of high definition TV, 360 video, and computer generated virtual reality. *Journal of Environmental Psychology, 72*, 101500. doi:10.1016/j.jenvp.2020.101500 PMID:33390641

Zyda, M. (2005). From visual simulation to virtual reality to games. *Computer, 38*(9), 25–32. doi:10.1109/MC.2005.297

ADDITIONAL READING

Blair, C., Walsh, C., & Best, P. (2021). Immersive 360° videos in health and social care education: A scoping review. *BMC Medical Education, 21*(1), 590. doi:10.118612909-021-03013-y PMID:34819063

Cipresso, P., & Serino, S. (2014). *Virtual Reality: Technologies, Medical Applications and Challenges*. Nova Science Publishers, Inc.

Hu, H. Z., Feng, X. B., Shao, Z. W., Xie, M., Xu, S., Wu, X. H., & Ye, Z. W. (2019). Application and Prospect of Mixed Reality Technology in Medical Field. *Current Medical Science, 39*(1), 1–6. doi:10.100711596-019-1992-8 PMID:30868484

Kim, O., Pang, Y., & Kim, J. H. (2019). The effectiveness of virtual reality for people with mild cognitive impairment or dementia: A meta-analysis. *BMC Psychiatry, 19*(1), 219. doi:10.118612888-019-2180-x PMID:31299921

Rosenberg, R. S., Baughman, S. L., & Bailenson, J. N. (2013). Virtual superheroes: Using superpowers in virtual reality to encourage prosocial behavior. *PLoS One*, *8*(1), e55003. doi:10.1371/journal.pone.0055003 PMID:23383029

Slater, M. (2009). Place illusion and plausibility can lead to realistic behaviour in immersive virtual environments. *Philosophical Transactions of the Royal Society of London. Series B, Biological Sciences*, *364*(1535), 3549–3557. doi:10.1098/rstb.2009.0138 PMID:19884149

Zheng, R., Wang, T., Cao, J., Vidal, P. P., & Wang, D. (2021). Multi-modal physiological signals based fear of heights analysis in virtual reality scenes. *Biomedical Signal Processing and Control*, *70*, 102988. Advance online publication. doi:10.1016/j.bspc.2021.102988

KEY TERMS AND DEFINITIONS

3D Headset Mounted: Through the virtual reality headsets (Valve, Oculus Quest, Reverb G2, Samsung Odissey, Vive Cosmos, Oculus rift) subjects can have the option to enter any virtual world, domain, or experience you can imagine.

Cyber Sickness: The most prevalent cause of nausea and discomfort associated with virtual reality locomotion, which happens when the virtual scene does not immediately transfer to real-life motion.

Healthcare: The prevention, diagnosis, treatment, amelioration, or cure of disease, illness, injury, and other physical and mental disabilities in individuals is referred to as health care or healthcare.

Immersive Virtual Reality: It refers to the use of sound, imagery, movement, and tactile instruments to isolate the subject's perceptual channels by immersing him/her in a sensory experience.

Non-Immersive Virtual Reality: It is determined by a monitor that serves as a "window" through which the user may observe the environment in three dimensions, through the mouse, the joystick or other peripherals such as gloves.

Semi-Immersive Virtual Reality: Rooms with equipment and surround rear-projection screens that recreate the computer's stereoscopic visuals and project them on the walls.

Spherical Videos: Spherical videos, or 360° videos, may generate powerful emotional responses by giving users the sensation of being physically there, as well as the illusion of engaging and reacting as if they were in the actual world.

Usability: Usability indicates if the tools are usable, easy to learn, challenging and engaging, and devoid of major adverse effects.

Chapter 24
Virtual Representations for Cybertherapy:
A Relaxation Experience for Dementia Patients

Francesca Maria Ugliotti
https://orcid.org/0000-0001-5370-339X
Politecnico di Torino, Italy

ABSTRACT

The development of serious games has enabled new challenges for the healthcare sector in psychological, cognitive, and motor rehabilitation. Thanks to virtual reality, stimulating and interactive experiences can be reproduced in a safe and controlled environment. This chapter illustrates the experimentation conducted in the hospital setting for the non-pharmacological treatment of cognitive disorders associated with dementia. The therapy aims to relax patients of the agitation cluster through a gaming approach through the immersion in multisensory and natural settings in which sound and visual stimuli are provided. The study is supported by a technological architecture including the virtual wall system for stereoscopic wall projection and rigid body tracking.

INTRODUCTION

In the third millennium, the discipline of drawing, in its various forms of representation, is facing a significant challenge in improving the quality of human life. The introduction and the consolidated diffusion of new means of expression and communication have enabled new perspectives of use that embrace very different fields and domains. The concept of Digital Twin contributes enormously to the vision of a Smart Healthcare (Tian et al., 2019) and truly Smart Hospital, where infrastructure, services and people are the crucial aspects to be managed as schematised in Figure 1. The main idea is to have a building with a digital brain and systems that can connect with users' needs and meet them in every respect using advanced approaches and digital technologies. This vision makes the hospital an ideal

DOI: 10.4018/978-1-6684-4854-0.ch024

place for experimentation as it must master a multitude of challenges simultaneously. Hospitals must reduce labor and operating costs, enable their staff to work more efficiently, optimize space efficiency, and comply with changing regulations without compromising a positive patient experience (Siemens Switzerland Ltd, 2022). In order to deal with these tasks, they are increasingly leveraging digitalization and technological innovations to build resiliency, enhance productivity and meet their strategic objectives. In this framework, high-tech solutions involving cyberspace, advanced robotics, 5G networks, Internet of Things (IoT) and Artificial Intelligence (AI) are therefore brought into play in a synergetic way to evolve and drive future growth and innovation for the healthcare experience. The tools are exploited to provide a virtual-based field to ensure the broadest participation in even more processes revolutionizing hospitals on the human, financial and operational levels. The focus is to seek the machine's efficiency and improve the quality of healthcare treatments and services delivery for the patient's well-being by combining advanced medical concepts and state-of-the-art devices. Within this context, telehealth and Remote Patient Monitoring (RPM) and Virtual Reality (VR) (Riva, 2002) find their place and acceptance by the patient, especially during and after the COVID-19 pandemic. A Smart Hospital is thus a hospital that uses technology to improve the quality of life of its users across the board, overcoming physical and spatial barriers. In this framework, the relation between three-dimensional digital representation, also derived from parametric models, and Virtual Reality technologies, represents an innovative frontier by enabling multi-dimensional scenarios and levels of interdisciplinary collaboration. This combination can be put to the service of the health sector in several areas: the technical knowledge and design, the usability of spaces and information (Ugliotti et al., 2019), training for nurses and caregivers, patient awareness and entertainment, diagnostics and prevention, and physical and psychophysical rehabilitation (Lányi, 2006). Therefore, they can embrace very different points of view of the hospital occupants, from managers and technicians to medical staff and patients and their families.

Figure 1. Smart Hospital framework

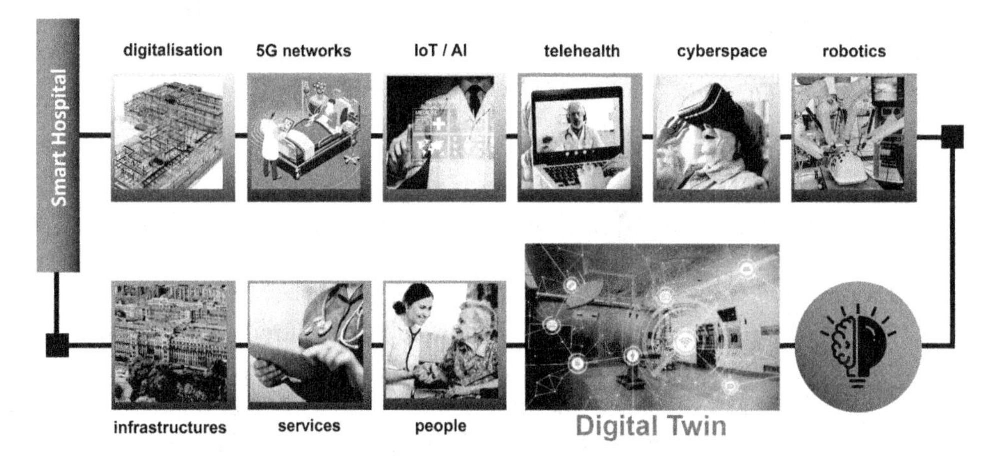

In the context of the scenario outlined, the chapter illustrates the application of Virtual Reality for therapeutic purposes. Compared to telehealth, which uses new technologies to provide health services remotely, cybertherapy employs technology to change the attitudes and behavior of its users, with long-term cognitive and bio-physiological effects (Emmelkamp, 2011). Cybertherapy was born in the United

States in the late 1980s, thanks to the interest and funding of the US Department of Defence. It refers to the different forms of clinical assessment and therapy that are based on the prevalent use of new technologies (Wiederhold & Wiederhold, 2004). In recent years, preliminary investigations have been conducted in this field, focusing on preserving residual skills and executive functions by reproducing of everyday contexts and actions. Through virtual therapeutics (Spiegel, 2020), i.e., Virtual or Augmented Reality experiences used alone or in combination with other tools or with traditional medicines, specific health needs of patients are being addressed. The application fields mainly concern clinical psychology and cognitive and motor rehabilitation (Borgnis et al., 2022). Among psychological impairments, solutions to control anxiety (Opris et al., 2012) and stress, phobias (Morina et al. 2015) and addiction, obesity and eating disorders are studied. The effectiveness of *COVID Feel Good* (COVID Feel Good, 2022), a simple virtual therapy experience able to reduce the anxiety and depression and overcomes the psychological burden generated by the Coronavirus pandemic (Riva et al., 2021), is currently in the news. The clinical trial was conducted by the Italian company *Become* in collaboration with the Università Cattolica and the Istituto Auxologico Italiano. The innovative approach of Immersive Virtual Telepresence (IVT) tools (Riva et al., 2004) makes it possible to improve the performance of rehabilitation and reintegration programs (Rizzo & Kim, 2005) for certain impaired functions. It allows these deficits to be managed, overcome or reduced and enables the patient to benefit from therapeutic support in environments where disorders commonly develop. Cognitive training (Pedroli et al., 2013) based on digital tasks and exercises can be an excellent solution for involving participants in structured mental activities and improving their cognitive functions, especially if the motivational and playful aspect is emphasized. For example, stroke patients (Huygelier et al., 2021) reported less fatigue when using a robotic device to navigate a virtual plane displayed on a regular computer monitor than without this visual feedback (Mirelman et al., 2009). A walk in the park (Moyle et al., 2016), searching for products on the shelves of a supermarket (Zygouris et al., 2015), executing a recipe (Valve Corporation, 2022; Foloppe et al., 2018; X-Tech Blog, 2022), are some activities that require the implementation of choices and strategies. RelieVRx (AppliedVR Inc., 2021) is the first and only Food and Drug Administration (FDA)-authorized at-home immersive Virtual Reality non-pharmacological pain treatment. It is a prescription-use immersive system intended to provide adjunctive treatment based on cognitive behavioral therapy skills (Garcia et al., 2021) and other evidence-based behavioral methods for patients with a diagnosis of Chronic Lower Back-Pain (CLBP). This significant utilization variability is supported by the faithful three-dimensional rendering of human body parts or an avatar that can interact with situations and objects to achieve different goals depending on the subjects involved. The possibility of customizing the scenes and activities, making them more engaging, and adapting the therapy according to patients' feedback on performance are advantages of experimenting with the new visualization method. Volumetric and 360-degree videos (O'Sullivan et al., 2018) are moreover used as low-cost newer technologies in terms of time and resources for development as they are spherical recordings captured by sophisticated cameras with omnidirectional lenses that can collect images from all over the scene. They offer the opportunity to immerse oneself in authentic natural environments that are known to evoke positive emotions (Yeo et al., 2020; Yu et al. 2018; Chirico & Gaggioli, 2019; Lee et al., 2021). For these reasons, they are especially suitable for home-based rehabilitation activities (Pedroli et al, 2022; Pedroli et al, 2021; Boilini et al., 2020) and make sharing content easy with patients and their families on smartphones and tablets beside head-mounted displays. The literature review shows that these innovative digital applications may represent a revolution in cognitive screening methodology within the clinical setting of Dementia (Robert et al., 2014). Preliminary experiences of serious games to assess executive functions (Castelnuovo et.al, 2003), attention, memory and visuospatial orientation are detected (Zucchella et al., 2014; Serrani, 2014; Manera et al., 2015; Dudzinski et al., 2016; Valladares-Rodriguez et al., 2018). The *Cognitive Virtual Stimulation* application presented in this research is created to offer a relaxation/rehabilitation therapy besides an entertainment experience with multisensory stimuli. The serious game developed is used to set up a clinical study.

Research Background

The background of this study is the CANP – la CAsa Nel Parco project. It proposes e-health solutions for management processes, telemedicine and telehealth. The aim is to support the accessibility and interoperability of information to achieve better services, decentralise care, rationalise resources, and improve care pathways. The European Commission indirectly funds it through the POR FESR 2014/2020 regional program within the preparatory actions for realizing the future Turin Health, Research, and Innovation Park. In this context, the activities of the Politecnico di Torino have been focused on the design and development of a service ecosystem to support patients, family members/caregivers and professionals to improve patients' quality of life, as shown in Figure 2. On the one hand, the *Re-Hub* platform has been developed for mapping the healthcare facilities present on the territory. It integrates a recommendation system that allows the user to compare them in a personalized way concerning objective aspects linked to pathology and medical needs and subjective aspects, i.e. territorial services, functional characteristics of preference and user perception. This system, which adopts the logic of booking engines, allows operators and patients to make more effective use of rehabilitation facilities thanks to exploring indicators of interest. On the other hand, a set of Virtual Reality applications for the health sector are promoted within the *Virtual-Hub* platform. These solutions are characterized by a high degree of innovation and creativity, as evidenced by the limited availability of studies in the literature. The approach outline three synergistic lines of research: (i) Design – the creation of multidimensional scenarios for the realization of hospital environments, healthcare facilities and private homes that are increasingly functional and comfortable concerning the needs of users; (ii) Awareness raising: engaging users through interactive experiences to promote consciousness and disseminate information (Ugliotti, 2020); (iii) Cybertherapy – using digital models to test innovative relaxation/rehabilitation therapies for patients (De Luca & Ugliotti, 2020; Fardello, 2020). This last thread is detailed in the following.

Figure 2. Application to raise awareness among patients and their families regarding home hospitalisation

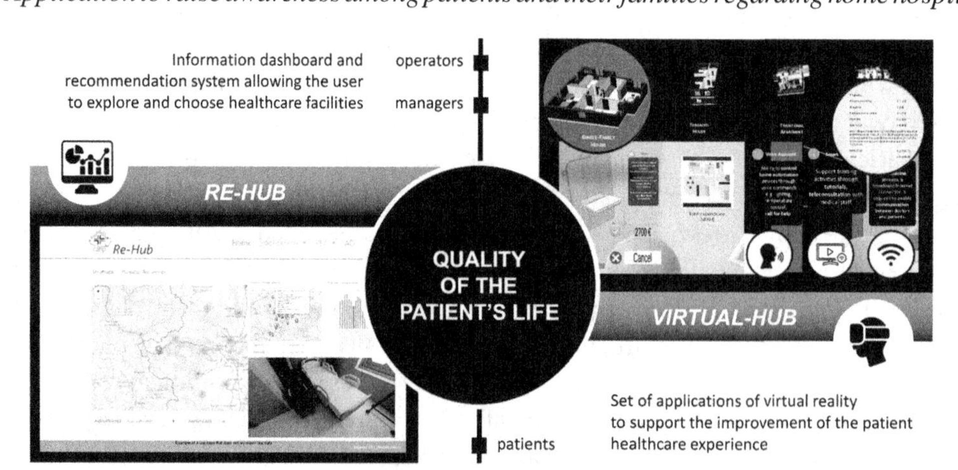

METHODOLOGY

This specific contribution is part of broader research investigating the conditions that can make cybertherapy experimentation effective. As shown in Figure 3, different aspects are considered: such as pathology, patient target (e.g., age, gender, social background), level of self-sufficiency (e.g. autonomous experience, caregiver-assisted or physician-guided), degree of immersiveness (fully immersive, semi-immersive) and place where to carry out the experience (e.g. patient's home, dedicated rooms in hospital or nursing home). This chapter aims to illustrate a virtual experience evaluated for pathologies that lead to cognitive impairment from the generated matrix of combinations.

Dementia is a comprehensive definition that includes various neurocognitive disorders, often in a progressive manner accompanied by a worsening of the quality of life of both patient and caregiver. It impairs various brain functions, such as memory, language, reasoning, orientation and the ability to perform complex problems. These cognitive dysfunctions are associated with personality and behavioral changes, including irritability, anxiety, depression, insomnia and apathy. The pharmacological search for a molecule that can modify the natural history of the disease has met with multiple failures (ADI, 2018). The only possible therapy is the symptomatic one. The focus has therefore shifted toward non-pharmacological interventions (Abraha, 2017). These are indicated as the first line of treatment, especially in psychological and behavioral disorders associated with Dementia. Non-pharmacological interventions can be classified as (i) sensory stimulation interventions (acupuncture, aromatherapy, chromotherapy, touch/massotherapy, light therapy, ortho therapy, music and dance therapy, transcutaneous electrical stimulation, and multisensory stimulation/snoezelen); (ii) cognitive-affective interventions such as reminiscence therapy; and (iii) other behavioral management interventions such as pet therapy. Among sensory stimulation techniques, only music therapy has been shown in the literature to be effective, especially in reducing agitation, aggression and anxiety. There is still a long way to go in this field, overcoming the current limitations of the studies conducted so far (i.e. low sample size, unclear definition of study designs, heterogeneity of intervention protocols, short duration of treatment and follow-up period).

Figure 3. Methodological investigation matrix

	Pathology	Target	Self-sufficiency	Immersiveness	Place
Case 1	Dementia	Senior	Doctor-guided	Semi-immersive	Hospital
Case 2	Dementia	Senior	Staff-guided	Semi-immersive	Nursing home
Case 3	Dementia	Senior	Caregiver-assisted	Full-immersive	Patient home
Case N	...	...	...	...	...

The methodological approach adopted involves an investigation path by successive stages of development to evaluate the involvement of Virtual Reality as an element of support for the no-pharmacological management of Dementia-associated behavior disorders. The primary objective is the relaxation of patients through immersion in peaceful environments. It is then possible to introduce interaction elements until an actual cognitive rehabilitation activity can be set up. As a first step, the reaction of patients to the use of Virtual Reality in the hospital context is evaluated through a guided experience by specialized operators and assisted by medical staff. In a further stage, the application in nursing homes and then in the patient's living environment can be considered. Depending on the specific scenario, several technologies can be tested.

In the context of the CANP project, the study promotes a semi-immersive guided experience with Virtual Wall technology in a hospital setting. A Virtual Reality application has been implemented, and a clinical trial has been designed in collaboration with the Geriatrics Department of the Molinette complex of Turin, as described in the following.

Cognitive Virtual Stimulation Application

At the base of this study is the hypothesis that, as happens in reality for physical spaces, virtual environments can also provide positive experiences for Dementia and Alzheimer's patients. Through cybertherapy, it is being tested how quiet and comfortable places can help reduce agitation and anxiety, making users feel better and improving their mood. The *Cognitive Virtual Stimulation* application is implemented using three-dimensional representations and the *Unity* game engine platform (Unity Technologies, 2022) to create interactive content with real-time animations. It is designed to run with the Virtual Wall system. The semi-immersive navigation shown in Figure 4 is allowed through 3D goggles and the Xbox joypad. As mentioned above, the experience is guided considering the target user's age, around 80 years old, and ability to use new technologies. An operator leads the patient on this virtual walk and initiates the different activities.

Figure 4. Semi-immersive guided cybertherapy experience for Dementia patients

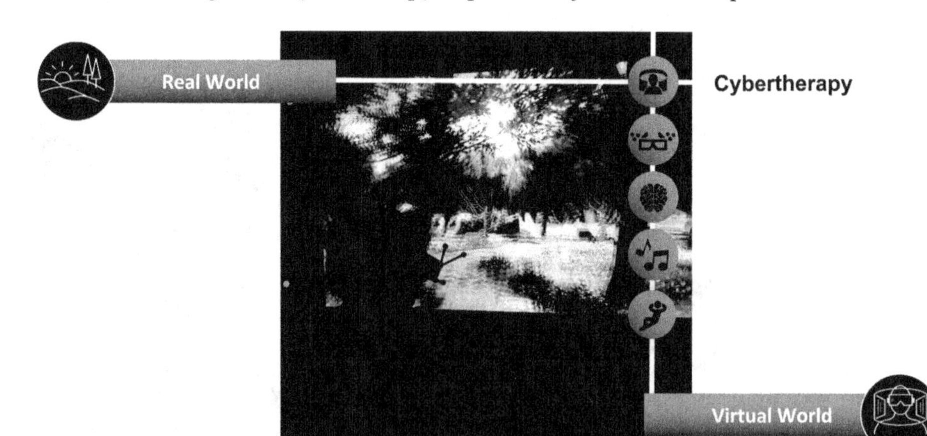

The virtual settings are designed to promote a gradual transition from the real to the virtual context and stimulate different brain areas to slow down the degradation process triggered by the disease. The experience unfolds based on three different scenarios, as shown in Figure 5: the geriatrics ward of the Molinette hospital, an indoor and an outdoor environment, a multisensory room and a natural setting, respectively. The first two scenes come from BIM models, while the third is only graphic modelling. In fact, in the framework of the CANP project, solutions are also analyzed for the digitalization of the built environment, to make information about the building and the assets contained therein usable, and for the participatory design of increasingly high-performance environments available to patients. The indoor setting was originally derived from an actual study and design activity (Carvajal Talero, 2019). This sequence of virtual spaces is conceived to activate different emotions in the patient, of discovery in the initial phase, relaxation and subsequent interaction in the following ones.

The session is started by walking along the corridor of the hospital ward. It recalls the route taken by the patient to reach the place of experimentation. The user's recognition of a familiar environment establishes a phase of acclimatization by establishing a relationship of confidence with the patient, who does not feel disoriented and, therefore, in trouble. The scene is rendered realistically, even if beautified by additional furnishing and natural elements. The path is accompanied by background music to further isolate the participant from the surrounding context. At the end of the corridor, on the right, there is the multisensory room that activates the second set and initiates the appropriate experimental activities.

The indoor scene represents the virtual transposition of a multisensory room designed according to the *Snoezelen* method - a Dutch term deriving from the contraction of the verbs "sniffen" (to explore) and "doezelen" (to relax). The design principles provide a welcoming and relaxing environment. The approach is also known in the literature under the term Multi-Sensory Environment (MSE) (Collier et al., 2017). It is used among the innovative non-pharmacological therapies for patients suffering not only from Dementia but also from other cognitive disabilities and psychiatric pathologies (van Weert et al., 2005). The initial phase of environmental exploration is passive. The objective is to achieve the relaxation of the patient acting in particular through the features of the space such as furniture elements, colors, and soothing sounds to transmit positive energy. The operator accompanies the patient to discover these characteristics. Lighting effects are one of the distinctive aspects achieved by employing bubble lamps, fiber optics, and spotlights. Soft items are placed on the floor, such as pillows or beanbags, simulating comfortable seating in correspondence with a large monitor that reproduces the dynamic vision of an aquarium with colorful fish moving slowly and drawing repetitive trajectories among the corals. Water is also present in the room by bubble tubes that move vertically at a slow and constant speed. They are used as a complementary focal instrument to encourage visual tracking, color recognition, physical movement and hand-eye coordination. Two windows are placed from which well-known Italian points of interest, such as views of the Mole Antonelliana and the Basilica di Superga, can be appreciated. This strategy is used to maintain a connection with reality and open up the environment, not forcing the patient into an utterly locked place that could have triggered feelings of fear and discomfort. Cause and effect items are implemented to allow the individual to control elements within the environment, although not directly as discussed later and return visual- acoustic effects. For example, it is possible to experience different color schemes of the setting by adjusting the room's lighting from cold blue-violet tones to warm yellow-orange tones. Then the patient is subjected to visual and audio cognitive stimuli in the following phase. The focus is on distracting the patient from interrupting possible agitated behavior. The *Memory Train*'s first activity involves scrolling on the screen previously used to show the aquarium a sequence of images evocative of the user's years of youth and maturity, with famous people, familiar places, and advertising slogans. For example, black and white images of the carousel, wooden school desks and some of Totò's films are shown. These inputs aim to evoke lost memories and emotions in the patient, who may express interest in sharing and recounting them, finding peace and serenity. A second activity encourages the cognitive sphere through music therapy, thanks by listening to a choice of very well-known songs and playing different rhythmical musical instruments (saxophone, electric guitar, drums and violin). From the indoor setting, there is a link to the outdoor one.

The outdoor scene consists of a natural landscape characterized by trees, paths, leaves, streams, a waterfall, and small shrubs. Some animals, from cats to lions, cows to giraffes, arouse amazement thanks to their unusual location. Small squirrels are moving under the trees, and little birds are chirping. The presence of the wind is simulated through the gentle fluttering of tree branches. Here again, the presence of water is used as sound therapy for healing. There is a pool of water in which the shadows of moving fish can be clearly recognized. As the user approaches, the natural running sound of the water becomes louder and louder to emphasize the realistic perception of the scene. Exploration is more free form than before, and the patient's involvement within a pleasant environment is assessed.

Figure 5. Cognitive Virtual Stimulation application structure

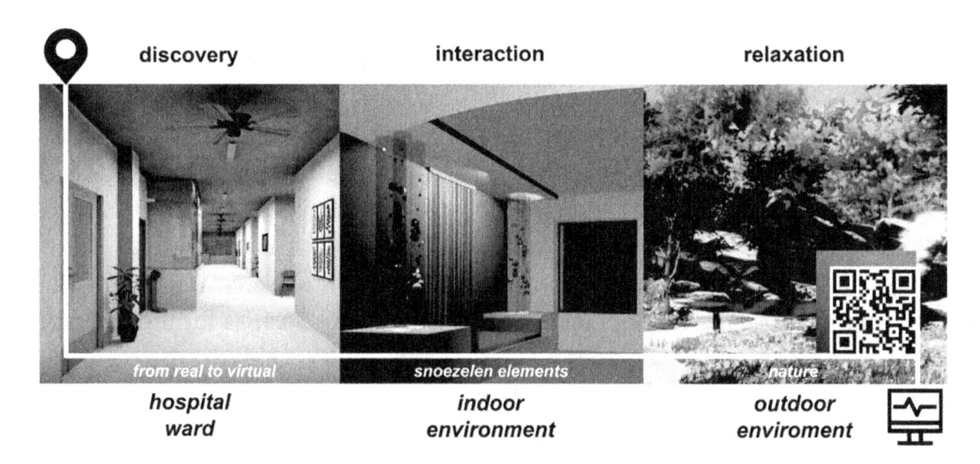

Virtual Wall Virtual Reality Technology

The *Cognitive Virtual Stimulation* application is designed to be executed with the Virtual Wall technology, which integrates stereoscopic wall projection and rigid body tracking tools. It is part of the Cave Automatic Virtual Environment (CAVE) visualization system invented at the University of Illinois at Chicago in 1992 (Cruz-Neira et al., 1992). It consists of a cube-shaped VR room in which the walls, floors and ceilings are projection screens. This VR theatre gives back the feeling of being immersed in the virtual scenario projected on the screens by a correct reading of the spaces, volumes and distances simulated with a scale of 1:1 (Cruz-Neira et al., 1993). The Virtual Wall solution instead involves only one screen for a semi-immersive experience. The graphical environment is realized by transmitting the three-dimensional model to the screen using a high precision stereoscopic holographic projector. The user can perceive its positioning in the three-dimensional space and interact with the screen through devices equipped with reflective markers. These sensors are detected by the cameras connected to the workstation transmitting a tracking signal processed by the software program. The higher the signal accuracy, the greater the user tracking in the surrounding space. In our specific study, 3D glasses, an Xbox joypad and four infrared cameras are used, as shown in Figure 6. For the Virtual Wall system to be installed, it is necessary to have a room of adequate size, with a free wall of at least 2.70x3.10 m. It is essential to be able to darken the room completely. The equipment must be carefully installed and calibrated in the specific place where it will be used to be effective. Some tracking problems may occur due to reflective surfaces, such as tiles, transparent surfaces and cabinet panels, and non-removable light sources. The virtual environment will be more engaging and realistic if the physical room is confined, free of clutter, with neutral tones.

Figure 6. Virtual Wall Virtual Reality technology

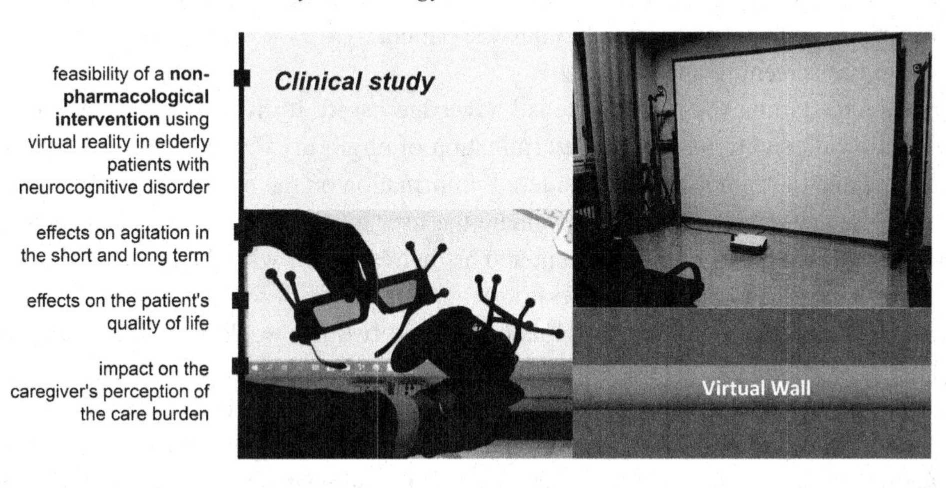

Clinical Trial

At the same time as designing the *Cognitive Virtual Stimulation* application, the opportunity arose to carry out a clinical trial as part of the CANP project. The prospective observational study concerns the use of Virtual Reality and the Virtual Wall system to manage agitation in Dementia patients. Precisely, it is planned to subject the older adults with Dementia to a cycle of twice-weekly sessions of semi-immersive virtual therapy for two months. The overview of the technological infrastructure deployed is outlined in the following section. The main primary and secondary objectives of the study are as follows.

- To assess the feasibility of a non-pharmacological intervention using Virtual Reality in the elderly patient with moderate-to-severe major neurocognitive disorder and agitation.
- To rate the enjoyment of different types of virtual environments.
- To evaluate the effects on agitation in the short and long term.
- To consider the effects on a patient's quality of life.
- To value the impact on the caregiver's perception of the care load.

It is expected to enroll fifty people referred to the outpatient of the S.C.U. Geriatrics department, who are 75 years of age or older and have been diagnosed with moderate-severe major neuro-cognitive disorder according to the diagnostic criteria of the DSM-V (American Psychiatric Association, 2013). The level of cognitive impairment is assessed through the Mini-Mental State Examination (MMSE) test (Folstein et al., 1975) and Clinical Dementia Rating (CDR) parameter (Morris, 1997). Irritability, aggressiveness, aberrant motor activity and sleep disturbances are some of the behavioral impairments considered. They are primarily concerned with the agitation cluster. The place of experimentation is the hospital. Depending on the requirements for using the Virtual Wall mentioned before, a suitable room was selected at the hospital ward of *Servizio di Ospedalizzazione A Domicilio della S.C. Geriatria e Malattie Metaboliche dell'Osso – PO Molinette* of the *Azienda Ospedaliero Universitaria Città della Salute e della Scienza di Torino* and the set-up and configuration was carried out. The trail was then carefully engineered to make the individual guided virtual experience replicable and comparable in its modalities for all participating subjects. The study protocol comprises five main phases in which different professionals are involved depending on the actions to be carried out. At least two operators and one doctor participate in the study in addition to the patient.

Phase One: Pre-screening and enrolment

The first step to starting the programme is knowledge-based. It involves the assessment of the patient's initial clinical conditions and the determination of eligibility for inclusion in the study. Mainly the drug therapy and the care load are considered. Information on the patient's biography and the level of digital competence is also essential to evaluate the user feedback during the therapy session. This activity consists of an interview of the patient and his/her caregiver with the doctor.

Phase Two: Semi-immersive therapy sessions

Each patient undergoes a program of sixteen sessions, two per week for eight consecutive weeks. Each session lasts 15 minutes, and two experimenters and a doctor are present in addition to the participant. The caregiver can attend the session but should not speak or have eye contact with the patient. It is believed that communicating with a familiar person can distract the patient and affect the session, the enjoyment and the possible outcome of the experience. The subject is welcomed into the experimental environment and made to feel comfortable, allowing him/her to adapt to the situation. He/she is seated in the centre of the room in front of the Virtual Wall screen and is asked to wear an accelerometric sensor placed on the patient's wrist to monitor vital parameters and 3D glasses with markers. To make the session as experiential as possible, the operator positions himself immediately behind the patient and leads him through the virtual walk by moving with the controls of the Xbox joypad. The doctor sits next to the patient and observes his reaction. A second operator also observes the session from the back of the room. The environment is darkened, and the feeling for the patient is that of being alone. The same application is provided to the user for all 16 sessions. Although this may sound repetitive, it is essential for the validation of the study. Moreover, the state of confusion caused by the disease makes the patient forget the contents of the treatment from one session to the next, thus making the experience as attractive and stimulating as the first time. As an experimenter leads the session, it was possible to define precisely all aspects characterising the virtual walk-in terms of actions to be performed and timing and overall duration. The start of the route with the linear exploration of the hospital ward corridor (entrance and exit) is used as an acclimatisation time. It, therefore, has a short duration of 1 minute. The multisensory room is assigned a total of 11 minutes. The gradual discovery of the different elements is 4 minutes long and ends with the user observing the aquarium. The visual stimuli proposed in the *Memory Train* activity take up to 3 minutes. In comparison, the sound stimuli with listening to the instruments and a famous song of the user's choice are set aside for 4 minutes. In the natural landscape, 3 minutes are spent. The experience ends with returning to the hospital corridor, where everything started.

Phase Three: Significant session data recording

At the same time as the virtual therapy session takes place, the patient's reactions are recorded, and observation sheets are filled in by the doctor and the second operator together. At the beginning and immediately at the end of the session, the patient will be measured for blood pressure, heart rate, and respiratory rate. Accelerometric measurements, which are significant for the quantity of movement the subject makes during the session, and heart rate are recorded to check the state of relaxation of the subject. These parameters are measured via the smartwatch-type accelerometer sensor worn on the patient's wrist. The collected data are downloaded using dedicated applications as Comma-Separated Values (CSV) files and made available for analysis. Then these values are correlated with the sequence of the different virtual environments and the visual and sound inputs proposed in the virtual walk. In addition to the trend, the standard deviation, median, mode, maximum, minimum and mean values are also calculated for each dataset. The experimenter also notes aspects of non-verbal communication and body movements: open/closed eyes, the posture assumed in the chair, e.g., whether the back is against the backrest, whether the legs are crossed. These elements help to understand the patient's state of relaxation, whether at ease or in a state of agitation and restlessness. The patient can also be filmed during the session to observe better these elements such as facial expressions, gestures, and behaviour. The patient's reaction to the visual and auditory stimuli is scored, in particular, if the attitude is passive and no interaction is established or if the patient asks questions or comments, telling about their personal experience and verbalising their state of mind. Another significant element is the active listening during the music part, where it is observed whether the patient keeps the rhythm, whistles or sings. At the end of the intervention, the doctor conducts an informal interview with the patient to assess satisfaction. In particular, the presence of emotions such as confusion, fear, sadness, and the appearance of any side effects like discomfort, vertigo, dizziness, and headache due to the use of the technology are checked. The level of gradability of the non-immersive virtual reality tool is assessed by constructing a Likert scale (Joshi et al., 2015). A PDF summary monitoring report is generated using the data storage system discussed in Phase Five based on information collected during each session. The computer language used is Javascript. The experimenter manually loads the punctual information into the system, while the heart rate and accelerometer data are processed from the external datasets recorded by the smartwatch. An example is given in Figure 7. The first part of the report presents a graphical representation of the patient's heart rate and accelerometer fluctuations. Both graphs show on the x-axis the time values, i.e. the time instants in which the recordings are made, while the y-axis shows the values of the monitored data expressed in their unit of measurement. In the graphic restitution of the diagram, vertical reference lines are implemented at specific time instants. These identify the transition between one virtual setting and another, considering the different inputs proposed. They, therefore, correspond to the timing described in Phase Two, i.e. minutes 0, 1, 5, 8, 12, and 15. The second part of the report contains a radar chart indicating: enjoyment, relaxation, level of personal involvement and outcome of interaction with sound and visual input. The two datasets refer to the multisensory and natural rooms. This type of graph is used to quickly visualise the effect and compare the results for the two different environments. It is not possible to establish a priori which of the two virtual environments can return more satisfying sensations. The report's last section is dedicated to collecting a short descriptive text representative of the individual experience. It includes any aspects not considered and evaluated utilising appropriate scales that may prove helpful in understanding the phenomena.

Figure 7. Summary session monitoring report

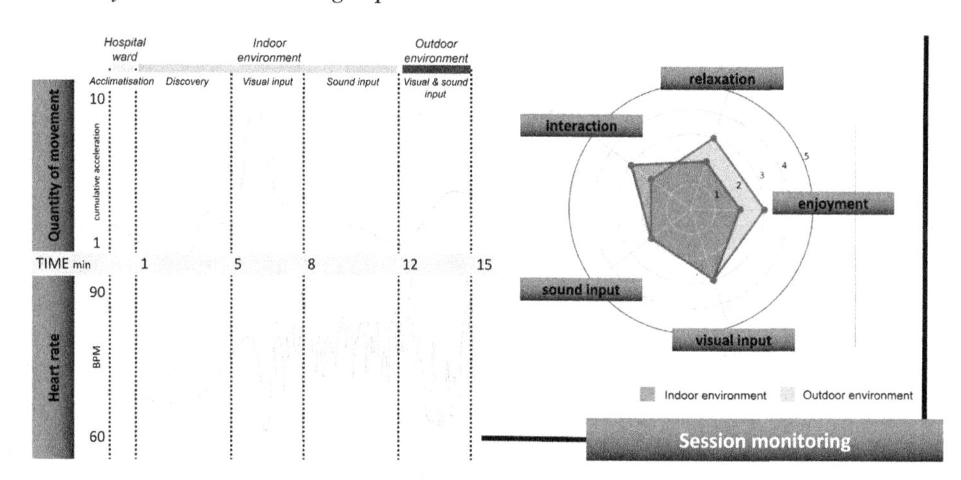

Phase Four: Baseline, conclusion, post-conclusion assessment

During the program, three moments of clinical evaluation are identified. They are conducted by the doctor through a structured interview individually with both the patient and the caregiver. Patients are evaluated at baseline (t0), at the end of 8 weeks of non-pharmacological treatment (t1), and one month after the conclusion of treatment (t2) in order to detect the feasibility of the intervention with Virtual Reality, the immediate and long-term effects of an innovative non-pharmacological approach on agitation. The data collection is done through structured interviews and the completion of validated scales. The evaluation includes cognitive and functional assessment of the patient, measurement of agitation, also considering indirect aspects such as the impact on quality of life and sleep characteristics of the patient, pharmacotherapy and caregiver burden. With regard to cognitive status, Mini-Mental State Examination (MMSE) test (Folstein et al., 1975) and Clinical Dementia Rating (CDR) parameter (Morris, 1997) are considered. The Cohen-Mansfield Agitation Inventory (CMAI) index (Cohen Mansfield, 1986) and the Pittsburgh Sleep Quality Index (PSQI) self-report questionnaire (Buysse et al., 1989) are assessed for the state of agitation, and the EQ-5D-5L for the patient's quality of life (Herdman et al., 2011). Finally, the caregiver's care burden is assessed through the Caregiver Burden Inventory (CBI) scale (Novak & Guest, 1989).

Phase Five: Reporting and data analytics

As anticipated in the previous steps, many parameters must be monitored to validate the effectiveness of the therapy. For this reason, a structured data collection system was deemed appropriate. The systematization of the dataset makes it possible to analyze the information of test subjects' patients, compare results and perform statistical processing through a generalized linear model for repeated measurements. The logical architecture of the computer network is of the client-server type. The Google Chrome web browser is used with desktop visualisation via local address for the user interface. The entry, consultation and updating of the characteristic patient and related session data are managed through dedicated sections created through HyperText Markup Language (HTML) documents. A local server interprets client input requests and communicates with the database. The XAMPP control panel is used to activate the Apache and MySQL modules. Communication between the local server and the database takes place via a specific Personal Home Page Language (PHP). The user can display the output via client-processed HTML pages. The process is schematised in Figure 8. When a new patient is registered, a unique identification code is assigned, which will represent the participant throughout the entire care process. A specific section is implemented in the client to visualise the database without the possibility of modification. It allows the doctor to use the system easily while preserving the integrity of the data.

Figure 8. Logical client-server-database integration process

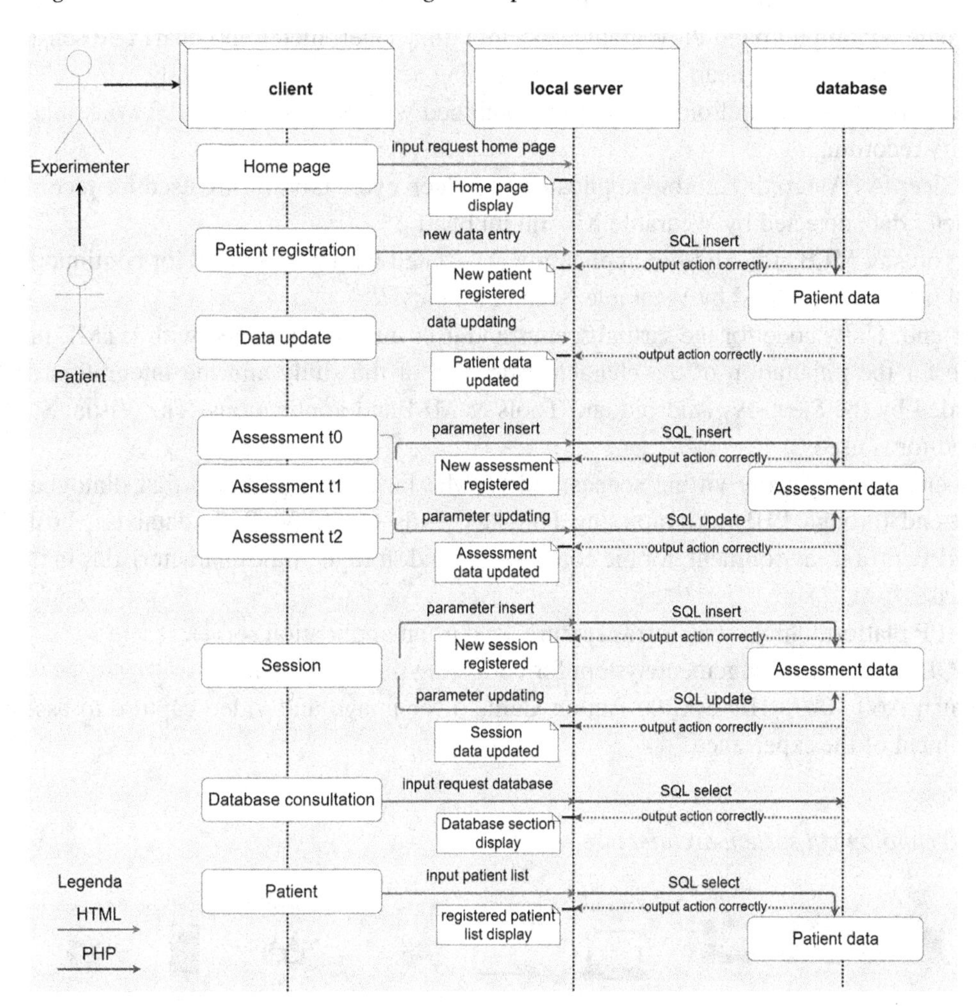

Technological System Architecture

The pilot study is supported by a system of technologies and platforms for testing non-pharmacological interventions using Virtual Reality to manage of cognitive disorders associated with Dementia. The architecture includes several layers of technology, communication and data storage, as shown in Figure 9.

- Virtual Reality Application: designed with Unity software in C#. The application is launched via Unity software. The configuration and the operation require the use of the Virtual Wall system.
- Unity: software for creating and launching virtual scenes.
- Virtual Wall: a system for interactive Virtual Reality based on stereoscopic wall projection equipped with fixed components, transportable if necessary and complete with tools for tracking rigid bodies.
- Optitrack Motive: rigid body tracking software.
- MiddleVR: software for receiving and processing tracking signals by superimposing them on the virtual scene's three-dimensional model.
- Wearable Xiaomi mi band 2: wearable system with accelerometer and heart rate sensors used to monitor movement and heart rate while performing activities non-invasively.
- App Android Mi Fit: Android application combined with Xiaomi mi band 2 Wearable device for activity recording.
- App Sleep As Android: Android application for sleep cycle monitoring used for recording accelerometer data detected by Wearable Xiaomi mi band 2.
- App Tools & Mi Band: Android application associated with Mi Fit used for continuous monitoring of heart rate detected by Wearable Xiaomi mi band 2.
- Front-end: Unity code for the virtual scenes and user interface realized with HTML markup language for the imputation of the characteristic data of the study and the integration of the data recorded by the Sleep As Android and Tools & Mi Band applications. The Visual Studio Code text editor is used.
- Back-end: Unity for the virtual scenes and HTML JavaScript platform that dialogues with the Front-end through PHP programming language calls to the MySQL database, hosted by the XAMPP virtual environment, for the consultation and storage of the characteristics in the MySQL database.
- XAMPP platform: high-level infrastructure used as an application server.
- MySQL: database management system for collecting the study data.
- Apeman A60 1080p HD Sports: Action Camera for image and video capture to assess patient enjoyment of the experience.

Figure 9. Technological system architecture

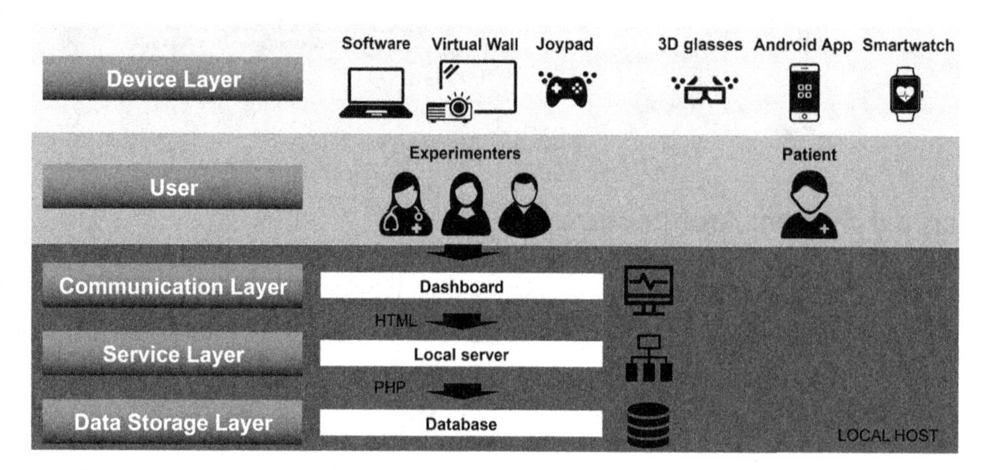

Pre-Clinical Trial Session

The *Cognitive Virtual Stimulation* application and the prototype configuration of the system have been fully realized and are available for a clinical study. However, the trial has been temporarily suspended due to the COVID-19 epidemiological emergency. In this situation, it was impossible to access the geriatrics department's areas and recruit patients from the target group in question, as it was considered unsafe to have them come to the hospital twice a week. However, a limited test sample is available thanks to some preliminary sessions conducted in concomitance with the installation of the Virtual Wall technology at the Molinette Hospital in Turin. Fifteen patients underwent the first session of the trial on a voluntary basis. No critical issues or adverse reactions emerged at this stage that hindered the session's running. Despite the age target considered, around 80 years of age, the subjects involved were willing to carry out the virtual session. Some did not quite understand why they had to undergo this experience; on the other hand, no one shied away from doing it. They readily agreed to wear the 3D goggles for body tracking and the smartwatch to monitor heart rate and movement. Nobody decided to interrupt the session before the 15-minute time limit. The three-dimensional settings enriched with sound and interactive elements were generally assessed as pleasant. All participants recognized the correspondence of the virtual corridor with the real one of the hospitals they had just walked through to reach the test site, even though it had been decorated with greenery and paintings that were not there. Viewing the aquarium definitely aroused strong interest and entertainment in the multisensory room. Observing the movement of the water and the fishes caught their attention. At the same time, the sequence of pictures proposed in the Memory Train stimulated the patient's openness and willingness to recall moments of lived life. In some cases, the participant reported personal events more or less related to the presented photographs. It was noted that the content of the input was not necessarily related to the patient's stories. Almost no one recognised the Mole and Superga outside the windows present in the room. However, the images were associated with monuments of collective interest familiar to the participant's life. Almost all participants generally achieved a beginning feeling of relaxation in correspondence with the delivery of the sound stimuli. As anticipated, music therapy is widely affirmed in this field. So, it proved to be even when the delivery took place in a virtual setting. It can be noted that all participants showed a more laid-back attitude while listening to the music. The position in the chair is less rigid and more comfortable; the legs are crossed, and the hands and feet keep the rhythm. The patient started to sing in limited cases, more frequently moving his lips or whistling. It was not possible to clearly identify the patients' preferences concerning virtual settings. Thanks to the natural elements, the outdoor setting is indeed rated positively by many users. However, it has been found that being in an environment with a vast horizon can make it difficult for patients, who may even feel disoriented. In contrast, the indoor environment provides a protected environment as it is closed and well-defined, so the user does not feel discomfort. Also, patients with hallucination disorders experienced the presence of shadows often associated with people in the natural environment. For this reason, demented patients with these alterations were excluded from the study. On the other hand, however, the user detected a complete feeling of immersion in the natural environment. This aspect may be due to the gradual patient acclimatisation to the virtual context. For example, some subjects shifted their heads, intending to dodge tree branches during their walk. The observation of the non-verbal language expressed by the patient is fundamental to verifying the performance of the experience both in terms of the quality of the graphic representation and possible calming effects. Only in the case of a drug-sedated patient the session was not very useful as the subject's attention span was minimal, and he tended to fall asleep. Notably, one patient, who was not very communicative during the session, manifested a condition of total relaxation by whistling a song while virtually walking in the natural landscape setting guided by the operator, as shown in the video shown in Figure 10. This episode in which the most significant involvement was found encourages further feasibility studies and experiments related to the use of virtual technologies.

Figure 10. Feeling of total relaxation achieved in the natural environment

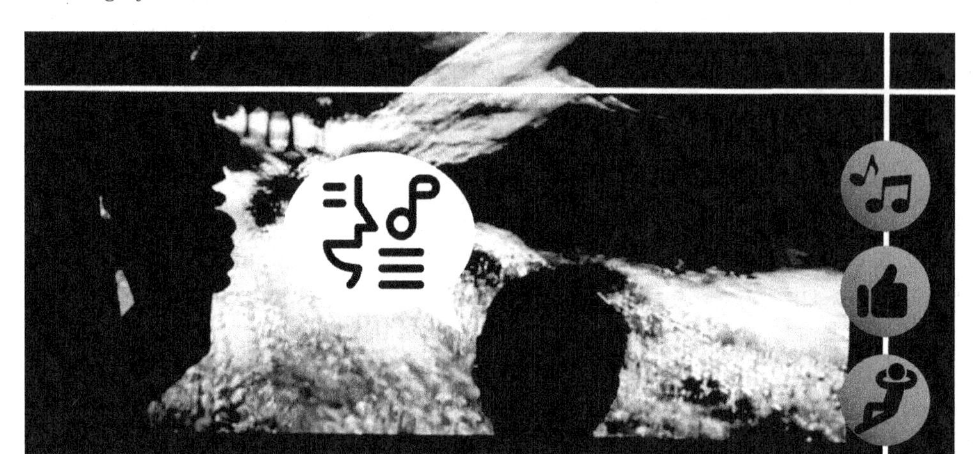

FUTURE RESEARCH DIRECTIONS

The experimentation mentioned above constitutes the first methodological step in a broader research project introducing the use of Virtual Reality for the treatment of cognitive-behavioural disorders. In the event of positive, or at any rate not negative, results from the clinical trial carried out in the hospital setting, new paths of investigation would open up. On the one hand, it is possible to evaluate the customisation of the application to provide an experience characterised by the most significant elements of each participant's life and respond to specific needs. Starting from the *Cognitive Virtual Stimulation* application, for instance, it is possible to replace the images proposed in the *Memory Train* activity with photographs from the patient's private sphere so that they can recognise themselves in different situations at different times in their lifetime. On the other hand, it is planned to set up an actual cognitive rehabilitation programme with the gradual enrichment of the proposed sensory activities and the introduction of a section dedicated to the performance of some basic and more advanced memory exercises (De Luca & Ugliotti, 2020; Carvajal Talero, 2019). For example, the caregiver can ask the patient to remember a meaningful date for him/her or the current date and compose it virtually using a slider. Alternatively, patients can exercise their minds by recomposing the order of a sequence of colours or a jigsaw puzzle. The use of representation, shapes and colours becomes essential to make the experience enjoyable. The patient can also become autonomous in exploring space and performing cognitive exercises. After an initial period of using cybertherapy in a hospital setting, it is expected that the application can be extended to nursing homes and home settings. It is essential to remodel the experience using less expensive and less complex technologies in this latter case. Dementia appears to be just one of the possible pathologies for these studies and activities.

CONCLUSION

This chapter presents the most exciting results of cyberspace exploitation obtained within the CANP research project framework. The therapy is evaluated for the non-pharmacological treatment of Dementia-related cognitive disorders. The *Cognitive Virtual Stimulation* application proposes a multi-sensory indoor scenario and a natural outdoor setting to evaluate the effectiveness of Virtual Reality for the relaxation of the elderly patient with moderate-to-severe major neurocognitive disorder and agitation. Solutions involving serious games are currently evaluated to improve patients' quality of life through innovative, captivating experiences. The implementation as a diagnostic and therapeutic tool offers innovative possibilities for the understanding, evaluation and rehabilitation of numerous cognitive, psychiatric and motor disorders. The study examines Dementia and considers a target group of senior users. However, this experiential approach is considered extraordinarily versatile and can embrace a broad spectrum of case histories and pathologies. From the elderly to children, from care to entertainment. Creating a specialized virtual technology center would undoubtedly enrich the hospital's range of services and be of interest to multiple hospital departments. It is also essential to assess the impact of research on the ability to design actions for the future, considering the development and maturation time of studies and prototypes and the characteristics of possible users. In particular, the target audience is not the patient and older adult of today but the one who will be able to take up interactivity and deal more naturally with technological devices in the medium to long term. The multidisciplinary approach applied to the healthcare context can contribute to the patient's well-being from the perspective of the Society 5.0 and foster new market and research fronts. In this scenario, the role of representation is put at the service of the Smart Hospital to extend and enrich the processes of participatory design and user experience activities.

ACKNOWLEDGMENT

The European Commission indirectly supported this research titled "la CAsa Nel Parco" (CANP) through the POR FESR 2014/2020 Piedmont regional program [ID CANP 9455-1572-71680]. The author is pleased to thank the S.C. Geriatria e Malattie Metaboliche dell'Osso – Ospedalizzazione A Domicilio of the Molinette Hospital under the supervision of Dr Renata Marinello for the collaboration to the realisation of the clinical trial, and the biomedical master's student Mario Fardello and the research fellows Riccardo Levante, Francesco Alotto and Daniela De Luca for permission to exhibit part of their work.

REFERENCES

Abraha, I., Rimland, J.M., Trotta F.M., Dell'Aquila, G., Cruz-Jentoft, A., Petrovic, M., Gudmundsson, A., Soiza, R., O'Mahony, D., Guaita, A., & Cherubini, A. (2017). Systematic review of of non-pharmacological interventions to threat behavioural disturbances in older patients with dementia. The SENATOR-On Top series. *BMJ Open*. . doi:10.1136/bmjopen-2016-012759

Alzheimer's Disease International (ADI). (2018). World Alzheimer Report 2018. *The state of the art of dementia research: New frontiers*. https://www.alz.co.uk/research/WorldAlzheim erReport2018.pdf

American Psychiatric Association. (2013). *DSM-5 Task Force. Diagnostic and statistical manual of mental disorders: DSM-5* (5th ed.). American Psychiatric Publishing, Inc. doi:10.1176/appi.books.9780890425596

Applied V. R. Inc. (2021). *RelieVRx*. https://www.relievrx.com/

Boilini, L., Rossi, M., Stefanelli, L., Goulene, K., Sacco, M., Stramba-Badiale, M., Riva, G., & Gaggioli, A. (2020). A new application for the motor rehabilitation at home: structure and usability of Bal-App. *IEEE Transactions on Emerging Topics in Computing, 9*(3), 1290-1300. doi:10.1109/TETC.2020.3037962

Borgnis, F., Baglio, F., Pedroli, E., Rossetto, F., Uccellatore, L., Oliveira, J., Riva, G., & Cipresso, P. (2022). Available Virtual Reality-Based Tools for Executive Functions: A Systematic Review. *Frontiers in Psychology, 13*. doi:10.3389/fpsyg.2022.833136

Buysse, D.J., Reynolds III, C.F., Monk, T.H., Berman, S.R., & Kupfer, D.J. (1989). The Pittsburgh Sleep Quality Index: A new instrument for psychiatric practice and research. *Journal of Psychiatric Research, 28*(2), 193-213. doi:10.1016/0165-1781(89)90047-4

CANP. (2018). *CANP - la casa nel parco* [CANP – The House in the Park]. http://casanelparco-project.it/

Carvajal Talero, J. R. (2019). *Virtual Reality in Treatments for Alzheimer's patients* [Master's thesis]. Politecnico di Torino.

Castelnuovo, G., Lo Priore, C., Liccione, D., & Cioffi, G. (2003). Virtual Reality based tools for the rehabilitation of cognitive and executive functions: The V-STORE. *PsychNology Journal, 1*(3), 310–325.

Chirico, A., & Gaggioli, A. (2019). When virtual feels real: Comparing emotional responses and presence in virtual and natural environments. *Cyberpsychology, Behavior, and Social Networking, 22*(3), 220–226. doi:10.1089/cyber.2018.0393 PMID:30730222

Cohen Mansfield, J. (1986). Agitated behaviors in the elderly. II. Preliminary results in the cognitively deteriorated. *Journal of the American Geriatrics Society, 34*(10), 722-7) doi:10.1111/j.1532-5415.1986.tb04303.x

Collier, L., & Jakob, A. (2017) The Multisensory Environment (MSE) in Dementia Care: Examining Its Role and Quality From a User Perspective. *HERD, 10*(5), 9-51. doi:10.1177/1937586716683508

COVID Feel Good. (2022). https://www.covidfeelgood.com/

Cruz-Neira, C., Sandin, D. J., & DeFanti, T. A. (1993). Surround-Screen Projection-based Virtual Reality: The Design and Implementation of the CAVE. In *SIGGRAPH'93: Proceedings of the 20th Annual Conference on Computer Graphics and Interactive Techniques* (pp.135–142). Association for Computing Machinery. 10.1145/166117.166134

Cruz-Neira, C., Sandin, D.J., DeFanti, T.A., Kenyon, R.V., & Hart, J.C. (1992). The CAVE: Audio Visual Experience Automatic Virtual Environment. *Communications of the ACM, 35*(6), 64-72. doi:10.1145/129888.129892

De Luca, D., & Ugliotti, F. M. (2020). Virtual Reality to Stimulate Cognitive Behavior of Alzheimer's and Dementia Patients. In Augmented Reality, Virtual Reality, and Computer Graphics. AVR 2020 (pp.101-113). Springer. doi:10.1007/978-3-030-58468-9_8

Dudzinski, E. (2016). Using the Pool Activity Level instrument to support meaningful activity for a person with dementia: a case study. *British Journal of Occupational Therapy*, *79*(2), 65–68. doi:10.1177/0308022615600182

Emmelkamp, P. M. (2011). Effectiveness of cybertherapy in mental health: A critical appraisal. *Annual Review of Cybertherapy and Telemedicine*, *2011*, 3–8. PMID:21685633

Fardello, M. (2020). *Sperimentazione di terapie innovative che utilizzano la realtà virtuale per la gestione non farmacologica dei disturbi del comportamento associati a demenza* [Experimentation of innovative therapies that use virtual reality for the non-pharmacological management of behavioral disorders associated with dementia] [Master's thesis]. Politecnico di Torino.

Foloppe, D. A., Richard, P., Yamaguchi, T., Etcharry-Bouyx, F., & Allain, F. (2018). The potential of virtual reality-based training to enhance the functional autonomy of Alzheimer's disease patients in cooking activities: a single case study. Neuropsychological Rehabilitation, 28(5), 709–733. doi:10.1080/09602011.2015.1094394

Folstein, M. F., Folstein, S. E., & McHugh, P. R. (1975). Mini-mental state. A practical method for grading the cognitive state of patients for the clinician. *Journal of Psychiatric Research*, *12*(3), 189–198. doi:10.1016/0022-3956(75)90026-6 PMID:1202204

Garcia, L.M., Birckhead, B.J., Krishnamurthy, P., Sackman, J., Mackey, I.G, Robert, G.L., Salmasi, V., Maddox, T., & Darnall, B.D (2021). An 8-week self-administered at-home behavioral skills-based virtual reality program for chronic low back pain: double-blind, randomized, placebo-controlled trial conducted during COVID-19. *Journal of Medical Internet Research*, *23*(2). doi:10.2196/26292

Herdman, M., Gudex, C., Lloyd, A., Janssen, M.F., Kind, P., Parkin, D., Bonsel, G., & Badia, X. (2011). Development and preliminary testing of the new five-level version of EQ-5D (EQ-5D-5L). *Qual Life Res.*, *20*(10). doi:10.1007/s11136-011-9903-x

Huygelier, H., Mattheus, E., Abeele, V. V., van Ee, R., & Gillebert, C. R. (2021). The Use of the Term Virtual Reality in Post-Stroke Rehabilitation: A Scoping Review and Commentary. *Psychologica Belgica*, *61*(1), 145–162. doi:10.5334/pb.1033

Joshi, A., Saket, K., Chandel, S., & Dinesh Kumar, P. (2015). Likert Scale: Explored and Explained. *Current Journal of Applied Science and Technology*, *7*(4), 396-403. doi:10.9734/BJAST/2015/14975

Lányi, C. S. (2006). Virtual reality in healthcare. In N. Ichalkaranje, A. Ichalkaranje, & L. Jain (Eds.), *Intelligent Paradigms for Assistive and Preventive Healthcare* (Vol. 19, pp. 88–116). Springer. doi:10.1007/11418337_3

Lee, N., Choi, W. & Lee, S. (2021) Development of an 360-degree virtual reality video-based immersive cycle training system for physical enhancement in older adults: a feasibility study. *BMC Geriatrics*, *21*, 325. doi:10.1186/s12877-021-02263-1

Manera, V., Petit, P. D., Derreumaux, A., Orvieto, I., Romagnoli, M., Lyttle, G., David, R., & Robert, P. H. (2015). "Kitchen and cooking", a serious game for mild cognitive impairment and Alzheimer's disease: a pilot study. *Frontiers in Aging Neuroscience*, *7*, 24. doi:10.3389/fnagi.2015.00024

Mirelman, A., Bonato, P., & Deutsch, J. E. (2009). Effects of Training With a Robot-Virtual Reality System Compared With a Robot Alone on the Gait of Individuals After Stroke. *Stroke*, *40*(1), 169–174. doi:10.1161/STROKEAHA.108.516328

Morina, N., Ijntema, H., Meyerbröker, K., & Emmelkamp, P. M. (2015). Can virtual reality exposure therapy gains be generalized to real-life? A meta-analysis of studies applying behavioral assessments. *Behaviour Research and Therapy*, *74*, 18–24. doi:10.1016/j.brat.2015.08.010 PMID:26355646

Morris, J. C. (1997). Clinical Dementia Rating: A reliable and valid diagnostic and staging measure for dementia of the Alzheimer type. *International Psychogeriatrics*, *9*(S1), 173176. doi:10.1017/S1041610297004870 PMID:9447441

Moyle, W., Jones, C., Sung, B., & Dwan, T. (2016). *Alzheimer's Australia Victoria The Virtual Forest project: Impact on engagement, happiness, behaviours & mood states of people with dementia*. Griffith University.

Novak, M., Guest, C. (1989). Application of a multidimentional caregiver burden inventory. *The Gerontologist*, *29*(6), 798-803. . doi:10.1093/geront/29.6.798

O'Sullivan, B., Alam, F., & Matava, C. (2018) Creating Low-Cost 360-Degree Virtual Reality Videos for Hospitals: A Technical Paper on the Dos and Don'ts. *J Med Internet Res*, *20*(7). doi:10.2196/jmir.9596

Opris, D., Pintea, S., Garcia-Palacios, A., Botella, C., Szamoskozi, S., & David, D. (2012). Virtual reality exposure therapy in anxiety disorders: A quantitative meta-analysis. *Depression and Anxiety*, *29*(2), 85–93. doi:10.1002/da.20910 PMID:22065564

Pedroli, E., Cipresso, P., Greci, L., Arlati, S., Mahroo, A., Mancuso, V., Boilini, L., Rossi, M., Stefanelli, L., Goulene, K., Sacco, M., Stramba-Badiale, M., Riva, G., & Gaggioli, A. (2021). A New Application for the Motor Rehabilitation at Home: Structure and Usability of Bal-App. IEEE Transactions on Emerging Topics in Computing, 9(3), 1290-1300. doi:10.1109/TETC.2020.3037962

Pedroli, E., Cipresso, P., Serino, S., Albani, G., & Riva, G. (2013). A Virtual Reality Test for the Assessment of Cognitive Deficits: Usability and Perspectives. *Proceedings of the ICTs for Improving Patients Rehabilitation Research Techniques*. 10.4108/icst.pervasivehealth.2013.252359

Pedroli, E., Mancuso, V., Stramba-Badiale, C., Cipresso, P., Tuena, C., Greci, L., Goulene, K., Stramba-Badiale, M., Riva, G., & Gaggioli, A. (2022). Brain M-App's Structure and Usability: A New Application for Cognitive Rehabilitation at Home. *Front. Hum. Neurosci*, *16*. doi:10.3389/fnhum.2022.898633

Riva, G. (2002). Virtual reality for health care: the status of research. CyberPsychology & Behavior, 219-225. doi:10.1089/109493102760147213

Riva, G., Bernardelli, L., Castelnuovo, G., Di Lernia, D., Tuena, C., Clementi, A., Pedroli, E., Malighetti, C., Sforza, F., Wiederhold, B. K., & Serino, S. (2021). A Virtual Reality-Based Self-Help Intervention for Dealing with the Psychological Distress Associated with the COVID-19 Lockdown: An Effectiveness Study with a Two-Week Follow-Up. *International Journal of Environmental Research and Public Health*, *18*(15). . doi:10.3390/ijerph18158188

Riva, G., Morganti, F., & Villamira, M. (2004). *Immersive Virtual Telepresence: virtual reality meets eHealth* (Vol. 99). Studies in Health Technology and Informatics. doi:10.3233/978-1-60750-943-1-255

Riva, G., Wiederhold, B. K., & Mantovani, F. (2019). Neuroscience of Virtual Reality: From Virtual Exposure to Embodied Medicine. *Cyberpsychol Behav Soc Netw*, *22*(1), 82-96. doi:10.1089/cyber.2017.29099.gri

Rizzo, A. A., & Kim, G. J. (2005). *A SWOT analysis of the field of VR rehabilitation and therapy presence Teleoper* (Vol. 14). Virtual Environ. doi:10.1162/1054746053967094

Robert, P.H., König, A., Amieva, H., Andrieu, S., Bremond, F., Bullock, R., Ceccaldi, M., Dubois, B., Gauthier, S., Kenigsberg P.A., Nave, S., Orgogozo, J.M., Piano, J., Benoit, M., Touchon, J., Vellas, B., Yesavage, J., & Manera V. (2014). Recommendations for the use of Serious Games in people with Alzheimer's Disease, related disorders and frailty. *Frontiers in Aging Neuroscience*, *6*, 54. . doi:10.3389/fnagi.2014.00054

Serrani, D. (2014). Virtual reality training improves spatial navigation disorientation in dementia patients. *Dementia (London)*, 189–195.

Siemens Switzerland Ltd. (2022). *Whitepaper. Hospitals harness digitalization to reach new levels of operational performance.* https://new.siemens.com/global/en/markets/healthcare/smart-hospitals/documents-resources/smart-hospitals-whitepaper.html

Spiegel, B. (2020). *VRx: how virtual therapeutics will revolutionize medicine*. Basic Books New York.

Tian, S., Yang, W., Grange, J. M. L., Wang, P., Huang, W., & Ye, Z. (2019). Smart healthcare: making medical care more intelligent. Global Health Journal, 3(3), 62-65. doi:10.1016/j.glohj.2019.07.001

Ugliotti, F. M. (2020). Increase the awareness of ALS patients through a Virtual Reality Application. In *EDULEARN20 Proceedings, 12th International Conference on Education and New Learning Technologies* (pp. 4403-4408). IATED Academy. 10.21125/edulearn.2020.1166

Ugliotti, F. M., Osello, A., Levante, R., & Urbina, E. N. B. (2019). Digital evolution of representation implemented at the Galliera Hospital in Genova [Evoluzione digitale della rappresentazione applicata all'Ospedale Galliera di Genova]. Riflessioni. L'arte del disegno/il disegno dell'arte [Reflections. The art of drawing/the drawing of art] (pp. 1775-1780). Gangemi Editore spa.

Unity Technologies. (2022). https://unity.com/

Valladares-Rodriguez, S., Fernández-iglesias, M., Anido-Rifòn, L., Facal, D., & Pèrez-Rodriguez, R. (2018). Episodix: A serious game to detect cognitive impairment in senior adults. A psychometric study. *PeerJ*, *6*. . doi:10.7717/peerj.5478

Valve Corporation. (2022). *The Cooking Game VR.* https://store.steampowered.com/app/857180/The_Cooking_Game_VR/

van Weert, J.C., van Dulmen, A.M., Spreeuwenberg, P.M., Ribbe, M.W., & Bensing, J.M. (2005) Behavioral and mood effects of snoezelen integrated into 24-hour dementia care. *Journal of the American Geriatrics Society*, *53*(1), 24-33. . doi:10.1111/j.1532-5415.2005.53006.x

Wiederhold, B. K., & Wiederhold, M. D. (2004). *The future of cybertherapy: improved options with advanced technologies Cybertherapy*. IOS Press.

X-Tech Blog. (2022). *Let's Start Cooking With Virtual Reality!* http://x-tech.am/lets-start-to-cook-virtual-reality-cooking-lessons/

Yeo, N. L., White, M. P., Alcock, I., Garside, R., Dean, S. G., Smalley, A. J., & Gatersleben, B. (2020). What is the best way of delivering virtual nature for improving mood? An experimental comparison of high definition TV, 360 video, and computer generated virtual reality. *Journal of Environmental Psychology*, *72*, 101500. doi:10.1016/j.jenvp.2020.101500 PMID:33390641

Yu, C.-P., Lee, H.-Y., & Luo, X.-Y. (2018). The effect of virtual reality forest and urban environments on physiological and psychological responses. *Urban Forestry and Urban Greening*, *35*, 106-114. doi:10.1016/j.ufug.2018.08.013

Zucchella, C., Sinforiani, E., Tassorelli, C., Cavallini, E., Tost-Pardell, D., Grau, S., Pazzi, S., Puricelli, S., Bernini, S., Bottiroli, S., Vecchi, T., Sandrini, G., & Nappi, G. (2014). Serious games for screening pre-dementia conditions: From virtuality to reality? A pilot project. *Functional Neurology*, *29*(3), 153–158. PMID:25473734

Zygouris, S., Giakoumis, D., Votis, K., Doumpoulakis, S., Ntovas, K., Segkouli, S., Karagiannidis, C., Tzovaras, D., & Tsolaki, M. (2015). Can a virtual reality cognitive training application fulfill a dual role? Using the virtual supermarket cognitive training application as a screening tool for mild cognitive impairment. J Alzheimers Dis, 44(4), 1333-47. doi:10.3233/JAD-141260

KEY TERMS AND DEFINITIONS

Cave Automatic Virtual Environment (CAVE): Surround-screen projection-based virtual reality. Theater immersive visualization system was invented at the University of Illinois in 1992, consisting of a cube-shaped VR room in which the walls, floors and ceilings are projection screens.

Cognitive Stimulation Therapy: Treatment which aims to improve cognitive skills and quality of life of people through activities such as categorisation, word association, discussion of current affairs and executive functions.

Cyberspace: Digital space navigable in virtual mode by people from different realities communicating with each other within a computerised world of digital networks.

Cybertherapy: Different forms of clinical assessment and treatment that use new experiential technologies as their primary intervention tool.

Dementia: Global deterioration of the cognitive state, often in a progressive manner that impairs various brain functions, such as memory, language, reasoning, orientation, and the ability to perform complex problems. These cognitive dysfunctions are associated with personality and behavioural changes, including irritability, anxiety, depression, insomnia, and apathy.

Digital Twin: Virtual representation of an object or system that spans its lifecycle, is updated from real-time data, and uses simulation, machine learning and reasoning to help decision-making.

Multi-Sensory Environment: Dedicated space or room where sensory stimulation can be controlled (intensified or reduced), presented in isolation or combination, packaged for active or passive interaction, and matched to fit the perceived motivation, interests, leisure, relaxation, therapeutic and educational needs of the user.

Serious Game: Games designed for educational purposes.

Smart Hospital: Hospital building with a digital brain and systems that can connect with users' needs.

Virtual Wall: Screen projection-based virtual reality part of the CAVE visualisation system. It involves only one screen for a semi-immersive experience and the use of peripherals equipped with markers such as 3D glasses and a joypad.

Chapter 25
Human Fragilities Supported by the Digital Social World

Nicola Rimella
https://orcid.org/0000-0001-7990-1423
Politecnico di Torino, Italy

Edoardo Patti
Politecnico di Torino, Italy

Francesco Alotto
Olivetti SpA, Italy

ABSTRACT

Technological progress must aim at creating Society 5.0 by developing tools to support people. This contribution aims to show how modern technologies and their integration into society can support people with fragility. In particular, the authors present the prototype of a technology that the Turin Polytechnic has developed to provide an IoT device control tool for people with motor neuron degeneration. This, through the use of eye-trackers and building information models (BIM), allows the navigation of models in virtual reality and interaction with different devices and services. Furthermore, the use of micro-services and the use of standard exchange formats allow easy integration with different services. The authors want to show how it is possible to build applications that, by bridging the real and the visual, can restore autonomy and quality of life to the frailest people.

INTRODUCTION

The condition of human fragility can be identified in all those individuals who fall into a state of physiological vulnerability like older people or people suffering from chronic diseases that debilitate the body. Modern technological development and the mindset characterizes by the 5.0 society, have contributed to the emergence of tools to support humans, not only to perform actions automatically or with less effort, but also to communicate and have new experiences. To have a better understanding of the direction in

DOI: 10.4018/978-1-6684-4854-0.ch025

which research is currently heading and the potential developments in the coming years in connection with autonomy and the quality of life of the frailest people, a literature review is conducted on various topics. The analysis of the literature makes it clearer how topics that may appear to be separate and not integrable are indispensable for the construction of a digital social world capable of supporting fragile human beings.

In this chapter, the authors focus in particular on the analysis and development of systems to support sufferers of motor neuron degeneration. Among the diseases that affect motor neurons is Amyotrophic Lateral Sclerosis (ALS) (Vichi, 2016). This disease affects one in three of every 100,000 people in the world between the ages of 50-65 but can occur even earlier. One of the first symptoms that occur with the onset of the disease is depression, due to the inability to perform simple actions and communicate efficiently with the people around them (S. Zarei, 2015). To support people affected by this frailty, the SIRIO technology was born, developed by the departments of structural engineering, construction and geotechnics (DISEG) and automatic and computer science (DAUIN) of the Polytechnic of Turin (F. Alotto, 2019). This technology was developed to allow, through the use of eye-trackers, virtual models and IoT devices, simple actions and virtual experiences to be performed just by moving the eyes. Interaction with the virtual environment is done through the use of the Building Information Model (BIM), It is defined as "a digital representation of physical and functional characteristics of a facility. [...] It serves as a shared knowledge resource for information about a facility forming a reliable basis for decisions during its lifecycle from inception onward" (BuildingSMARTalliance, 2007). This is used as a starting point for the implementation of Digital Twins (DT) (Grieves, 2014) which combine static building data with dynamic data derived from sensors and actuators.

The format that is used to allow the exchange of information and the integration with the different services that make up the system is the .ifc (Industry Foundation Classes) format (BuildingSMART International, 2022). This format enables the import of building geometries into different applications to improve visualization, and it also permits the use of BIM model data for various purposes, such as the management of IoT devices.

Thanks to the great development of Industry 4.0, Internet of Things (IoT) devices today can connect people with people, people with objects, or objects with objects. In recent years is also born out the concept of the Social Internet of Things (SIoT), i.e. the ability of connected objects to autonomously understand the context in which they operate and support human decisions (Rho Seungmin, 2019). The IoT devices implemented in the SIRIO system can be commanded by users to turn on lights or change the temperature of the room, and data is sent to a telemedicine system that collects the user's vital parameters and allows the doctor to make examinations without having to physically visit the patient. The Healthcare sector is one of the sectors where IoT technology has grown strongly (S. R. Islam, 2015), this is also caused by the wide range of parameters that need to be constantly monitored to enable efficient telemedicine services such as glucose, saturation, blood pressure or ECG. Constant monitoring of these parameters can be of great help in diagnosis and emergency interventions.

Currently, people with motor neuron degeneration use devices to increase their communication skills. The most common devices are Speech Generating Devices (SGDs) or devices for Eye-Tracking and Eye-Gaze detection. Generally, eye-detection can be used by any type of patient with a movement inability, and they are mainly based on infrared systems that hit the eye, illuminating it. In this way, it is possible to trace corneal movements using the Corneal Centre Pupil Reflection (PCCR).

The use of digital models and game engine software enables the possibility to use eye-tracking devices to navigate environments in Virtual Reality (VR), thereby allowing different experiences and navigating in different places that otherwise could not be visited by those with this type of fragility.

The authors will also analyse the concept of Metaverse as a new type of sociality that provides an immersive experience based on augmented reality technology, creates a mirror image of the real world based on digital twin technology (Huansheng Ning H. W., 2021). Today, the concept of the Metaverse is mainly used in connection with blockchain, decentralisation and tokeconomics, but the applications of these virtual worlds are manifold.

This contribution aims to show how the application of BIM models, Digital Twins, IoT, Metaverse, and OpenSource services can contribute strongly to the support of fragility.

LITERATURE REVIEW

For the literature review, the authors questioned how the research was moving towards developing solutions that connect the virtual world, consisting of digital objects and social groups (e.g. social networks), with the physical world, consisting of people and artefacts.

During this section, it is explained how the use of applications that create a bridge between the virtual and physical worlds can significantly improve the quality of life of frail people, who are often unable to interact in the physical world.

This connection between the virtual and real world is also used by the SIRIO system, which will be better presented later in this contribution. The system uses IoT devices and a VR application to allow the fragile user to connect with the Metaverse and the digital twin in his room, in order to access both digital content, useful for his inclusiveness and sociality, and tools that can increase their autonomy in performing everyday gestures.

Figure 1. Virtual and Real Worlds connection

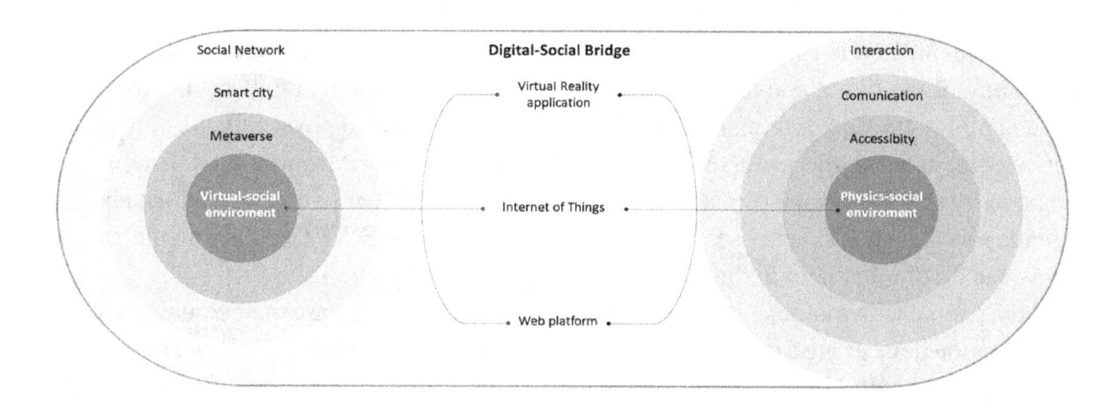

In Figure 1, the two circles represent the physical and virtual worlds respectively, while in the middle there are the technology what the author have defined as digital social bridges, i.e. tools that enable communication between the real and the virtual world. The physical world is the place where normally people interact with objects, have access to products and services and communicate with each other; in the virtual world, it is possible to do the same things using web services which connect us and allow interaction, communication and access to services as in the real world.

The purpose of the literature review is to show how research in recent years has moved in the same direction as the proposed SIRIO system, using innovative technologies to connect the real and virtual worlds.

The literature review that is proposed selects articles that deal with different ways of applying the digital social bridge. To this end, tools and services for creating virtual worlds and digital relationships, grouped as Virtual-Social tools, are related to different types of digital social bridges.

Virtual-Social Tools:

- BIM, the Building Information Model is the most reliable way to recreate buildings and built environments in virtual form, a BIM model will be used as the basis for creating the DT used for the SIRIO application;
- Metaverse, a Metaverse is a virtual world in which users can interact with objects and people through virtual reality;
- DigitalTwin, by combining dynamic and static data, the digital twin becomes the virtual pair of a real object, allowing us to monitor building data, simulate different scenarios or help us understand real objects;
- Smart City, including applications that use city services to provide benefits to our community;
- Social-network, these tools are now part of our lives, however the concept of social networking is still evolving by increasing the possible interactions between people.

Digital Social Bridge:

- IoT, the Internet of thinks enables real-world data to be sent to end-users and, via actuators, it is also possible to enable interaction between humans and objects;
- Web application, which includes applications that use the web to exchange data and make services accessible to end users;
- Smart application, including applications with community benefits;
- Healthcare application, applications relevant to the field covered in this contribution;
- Virtual Reality application, end-user applications using virtual reality to improve communication.

The topics described above will be discussed more in detail later in this section.

After looking for articles discussing the topics described above, the authors analyse and read them in order to better understand the topics covered and the purpose of the research. The presence of topics inside the articles was valued by entering a T in the case of 'True' and an F in the case of 'False'. In addition, by entering the date of publication of the various articles, it was possible to estimate the research trend in the area over the last five years.

To be able to examine the connection between the two themes described previously, contributions that had at least three of the overall topics and at least one topic in each were selected.

Table 1. Contribute selected for the review

		VIRTUAL-SOCIAL TOOLS					DIGITAL SOCIAL BRIDGE				
ARTICLE	DATE	DT	BIM	SMART CITY	METAVERSE	Social network	HEALTHCARE application	WEBPLATFORM	IOT	Smart application	VR
(Yue Pan, 2021)	2021	T	T	F	F	F	F	F	T	F	F
(Angel Ruiz-Zafra, 2022)	2022	F	T	F	F	F	F	T	T	T	F
(Yuxiang Zhang, 2020)	2020	F	T	F	F	F	F	F	F	T	T
(Hee-soo Choi, 2017)	2017	F	F	F	T	F	F	T	F	T	T
(Huansheng Ning H. W., 2021)	2021	F	F	F	T	T	F	T	F	F	T
(Lik-Hang Lee, 2021)	2021	T	T	F	T	T	F	T	F	F	T
(R. Craig Lefebvre, 2013)	2013	F	F	F	F	T	T	T	F	F	F
(Tali Hatukaa, 2020)	2020	F	F	T	T	T	F	F	F	T	F
(Vittorio Miori, 2017)	2017	F	F	F	F	T	T	T	T	T	F
(Kyoungro Yoon, 2021)	2021	T	F	F	T	F	F	F	T	F	T
(Haihan Duan, 2021)	2021	T	F	F	T	F	F	F	F	T	T
(Rho Seungmin, 2019)	2019	F	F	F	F	T	F	T	T	F	F
(Awais Ahmad, 2019)	2019	F	F	F	F	T	F	T	T	F	F
(Ben Falchuk, 2018)	2018	F	F	F	T	T	F	F	T	F	F
(Vaia Moustaka, 2019)	2019	F	F	T	F	T	F	F	T	F	F
(Qichao Xu, 2021)	2021	F	F	F	F	T	T	F	T	F	F
(Aamir Hussain, 2015)	2015	F	F	T	F	T	T	T	T	T	F

continues on following page

Table 1. Continued

		VIRTUAL-SOCIAL TOOLS					DIGITAL SOCIAL BRIDGE				
ARTICLE	DATE	DT	BIM	SMART CITY	METAVERSE	Social network	HEALTHCARE application	WEBPLATFORM	IOT	Smart application	VR
(Arrigo Palumbo, 2021)	2021	F	F	F	F	T	T	T	T	T	F
(Juan Manuel Davila Delgado, 2021)	2021	T	T	F	F	F	F	F	T	F	F
(Calin Boje, 2020)	2020	T	T	F	F	F	F	F	T	F	F
(David Jones, 2020)	2020	T	T	F	F	F	T	F	F	F	F
(Min Deng, 2021)	2021	T	T	T	F	F	F	F	T	F	F
(Weiwen Cui, 2021)	2021	T	T	F	F	F	F	F	T	F	F
(Jaime Ibarra Jimenez, 2019)	2019	T	F	F	F	T	T	F	T	F	F
(Nathan Moore, 2019)	2019	F	F	F	F	T	T	F	F	T	T
(Maninder Jeet Kaur, 2019)	2020	T	F	T	F	F	T	F	T	F	F
(SIAVASH H. KHAJAVI, 2019)	2019	T	T	F	F	F	F	F	T	F	F
(Yue Han, 2021)	2021	T	F	T	T	F	F	F	T	F	T
(Dessislava Petrova-Antonova, 2021)	2021	T	F	T	F	F	F	F	T	F	F
(Abubaker Basheer Abdalwhab Altohami, 2021)	2021	F	T	F	F	F	F	T	T	F	F
(M. Mazhar Rathore, 2018)	2018	F	F	T	F	F	F	T	T	F	F
(Rubén Alonso, 2019)	2019	T	T	F	F	F	F	F	T	F	F

Table 1. shows the references, years of publication and topics discussed in the 32 papers selected for the literature review.

From Chart 1.1 in Figure 2, it is possible to observe that in 2021 there was a considerable increase in the number of articles dealing with the selected topics, meaning that in recent years research has been spending a

great deal of time on the search in that topic, especially in the areas of Digital Twin, IoT, Virtual Reality and Metaverse.

Chart 1.2 shows the number of topics globally present in the review. This chart highlights that IoT is the topic most present, in twenty-three of the articles in the sample.

What the authors trying to explore through this review is the connection that exists between these topics, and show how research in recent years is focusing more on investigating concrete applications that combine these different areas.

Figure 2. Chart 1.1 - Research trends over the past five years. Chart 1.2 - Number of topics in the review.

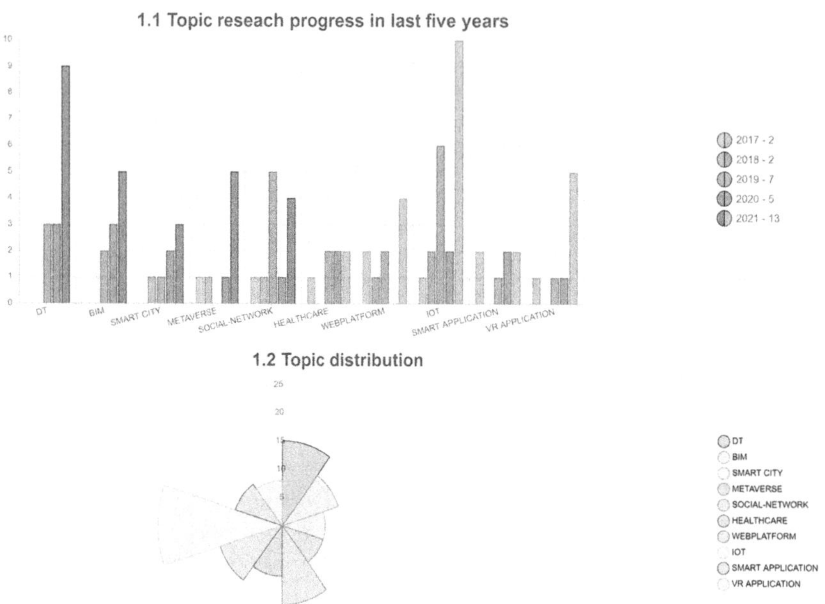

Virtual-Social World

The Virtual-Social Tools category includes tools and services that are used to virtualize objects, buildings, entire cities, or social relationships.

The Building Information Model (BIM) (BuildingSMARTalliance, 2007) is the static database of the building and combines geometric and geolocation data with alphanumeric data which are entered like parameters and linked to the different assets that make up the building. This information can be related to the different rooms in the building to facilitate facility management, or to the materials in the envelope to enable energy simulations. A BIM model is generated using BIM authoring software offered by major software houses such as Autodesk, Graphisoft, or Nemetschek. This software uses proprietary formats for the realization of the model that does not allow it to be used in third-party applications. For this reason, the Industry Foundation Classes (IFC) format was introduced.

The .ifc file format is a standard for describing the digital assets of a building environment and is certified by ISO 16739-1:2018. Several versions of the IFC formats currently exist and are being continuously updated to allow for better export and management of the BIM model. The latest certified stable version is IFC4, which is used for the creation of the digital twin in the SIRIO application.

For the creation of a digital twin is essential to have a high certainty of data relating to the assets that make up the building, a digital twin in fact must satisfy three essential characteristics: "physical products in Real Space, virtual products in Virtual Space, and the connections of data and information that ties the virtual and real products together" (Grieves, 2014). Considering this necessary high reliability, it is obvious that the digital twin implementation of a building should start from a BIM model as it closely reflects the way the building was constructed. To create an efficient DT it is necessary to link the dynamic data, coming from the real world, with the static data in the BIM model. Depending on the different usage of the DT, the level of reliability of the data associated with the digital model must have changed. As explained by (Calin Boje, 2020), the best way to develop a DT is to start from the BIM model and integrate the parameter required to read the dynamic data into it.

This is possible by defining standards in the IFC model as shown in (Angel Ruiz-Zafra, 2022) where the BIM model is related to data collected by an IoT system.

Standardization in modeling and processes is a topic also applicable to Smart Cities. The definition of a Smart City has been subject to discussion for many years, as shown by (Vito Albino, 2015) who in his contribution analyses various interpretations of the smart city concept concluding that the definition of a smart city must include the qualities of people and communities as well as ICTs.

The integration of Geographical Information System (GIS) tools with ICTs systems in smart cities enables the localization of city data on a map system. As far as city models are involved, there are also exchange formats that can be used to connect building models, in IFC format, with GIS models and their connections to smart city services. This format is called CityGML, a standard exchange format for entire city models. In (Ruben de Laat, 2010) is shown an interesting integration between BIM and GIS through the CityGML format and it is explained how the use of online BIM-based servers, BIMserver in this case, can allow the integration of third-party services with three-dimensional models.

In (Tali Hatukaa, 2020), the authors make a more actual analysis of smart cities and smart urbanism issues and challenges. The article explains how smart city concepts are evolving in major Israeli cities and how private companies and public administrations are increasing services for citizens, as shown in "Israel's Smart City Map 2019" shown in Figure 3.

This map was created by Creators, a Tel Aviv-based organization dedicated to supporting emerging start-ups (Creators, 2022), and CityZoom, a non-profit organization founded by the Ministry of the Economy, the Ministry of the Interior, Digital Israel, and in partnership with Tel Aviv University and Atidim Hi-tech Park, Peres center of peace and innovation.

The idea of a virtual place where different users can access tools and services by creating a community is completely similar to the Metaverse concept.

Figure 3. City Map Israel's Smart City Map 2019 that show different services implemented by private and public stakeholders

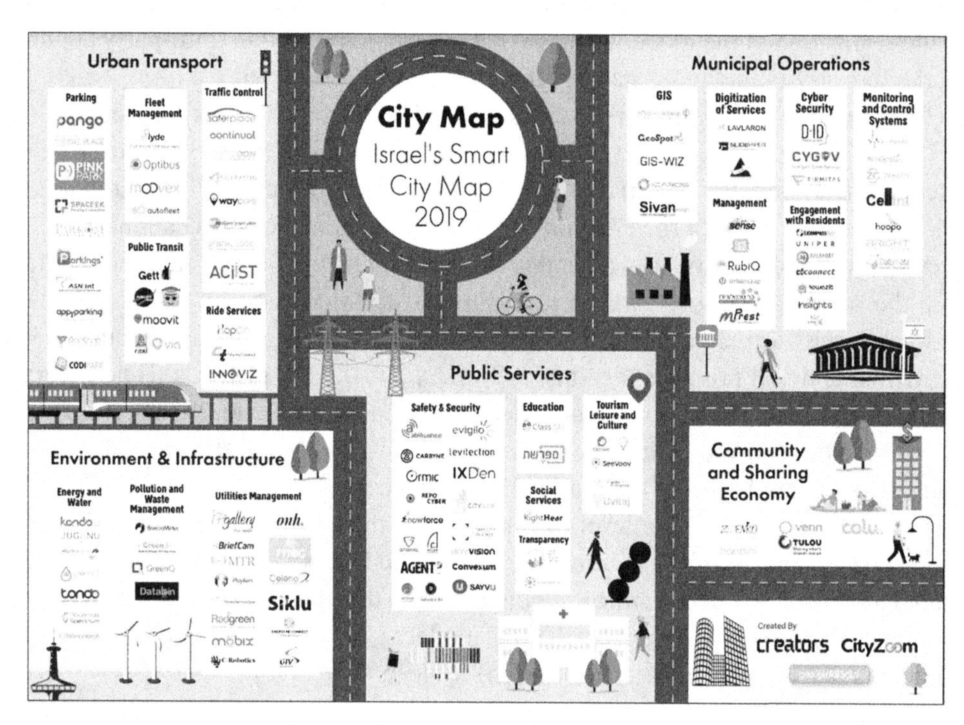

In recent years, the Metaverse has been greatly developed, especially in connection with blockchain and decentralized finance, but the concept of the Metaverse was theorized by Neal Stephenson in his book "Snow Crash" back in 1992 (Stephenson, 1992). The Metaverse is defined by (Huansheng Ning, 2021) as "a new type of Internet application and social form that integrates a variety of new technologies".

Currently, the most popular Metaverse applications use virtual reality to provide more immersive and social experiences for their users and allow them to interact with services and other users located in the Metaverse. Some of the most popular Metaverse are for example Decentraland, Sandbox, Roblox, or horizon. The last one was created by Mark Zuckerberg when he changed the name of the company from 'Facebook' to 'Meta', precisely to recall the Metaverse concept.

A fundamental part of the Metaverse is sociality, a new form of social network that will allow us to interact with people, things and services in virtual environments, becoming owners of our digital content thanks to the use of BlockChain technology.

The last topic that makes up the Virtual-Social tools described above is the use of social networks to facilitate the exchange of information. This type of sociability is especially crucial for frail people, for which real-world sociability is often difficult, if not impossible in cases of advanced illness.

As shown in his contribution (R. Craig Lefebvre, 2013) the use of social networks can bring benefits to disease prevention, detection, and treatment. The possibility for a frail person to be able to communicate with groups of people with similar problems helps and prevents a progression of illness, often accelerated by loneliness and depression. The use of social networks in the field of healthcare is still in development but a great step forward has already been made with telemedicine that can connect patients with doctors and other hospital staff, and consequently allow remote monitoring, as shown in (Arrigo Palumbo, 2021).

Several of these virtualisation technologies are used in the SIRIO application to allow the user to navigate the DT of his or her home or to escape from everyday life and explore other places. There is also an initial implementation in the direction of the Metaverse, combining the concepts of gamification with the concepts of smartcity and virtualisation.

Real World Bridge

By real world bridge, the author is referring to all those applications, tools, or services in which virtuality is united with reality. This section analyses some of the applications that combine the topics discussed in the Virtual-social world section with reality through the digital social world.

In (Awais Ahmad, 2019) the authors make an overview of the potential of the Socio-cyber network to define human behavior's using big data. In recent years, the cost of smart devices has decreased significantly, and devices have become accessible to everyone. Smart devices collect a large amount of data and can communicate with us and other objects. In their contribution, the authors show how the analysis of collectible data from social networks and smart devices can help us predict our behavior, subsequently driving our everyday choices. These devices are part of the Internet of Things (IoT) world that includes a wide range of devices capable of connecting consumer and/or industrial objects on the Internet, bringing together information and managing the devices via software to increase efficiency, create new services or bring health benefits and safety (S. M. Riazul Islam, 2015). The potential interactions and the activities of devices increase when they are connected to the Internet as cyber-physical systems (CPSs) (Awais Ahmad, 2019). CPSs, therefore, play a fundamental role on the possible interactions between humans and objects, they create a bridge between heterogeneous objects allowing them to perform actions or send data via service-oriented architectures, in this way they guarantee data heterogeneity.

As also explained (Rho Seungmin, 2019) n more recent years IoT devices can connect by forming relational structures as in social networks. This type of interaction is called the Social Internet of Things and is defined as an IoT system where things are capable of establishing social relationships with other objects, autonomously concerning humans. This type of new application, made possible by the social communication of objects, can support humans not only with automatic actions but also in the decision-making process.

Another tool that could be used as a bridge to allow people to interact within the virtual world is the use of web platforms and web services. The applications of web platforms are many and the authors will focus on the analysis of service-oriented platforms that allow the integration of different tools to support frailty. An example is provided by (Aamir Hussain, 2015) where the authors show the process for realizing a web application to support elderly people in Smart Cities. The authors propose a people-centric sensing framework where the frail person can access services in the social network and healthcare context and constantly communicate their location and biometric data collected by smart devices with the hospital.

For frail people, the possibility of accessing to a social network of other people to ask for support or for communicating biometric values instantly are high-value services that can save their lives and increase their quality of life.

Another interesting smart application based on the use of IoT to support Amyotrophic Lateral Sclerosis (ALS) patients is proposed by (Arrigo Palumbo, 2021) where the system 'Smart solutions for health Monitoring and independent mobiLity for Elderly and disabled people' (SIMpLE) is described.

This is a Cloud-Based system to enable the monitoring of the health conditions of people with ALS. In their contribution, the authors describe the aim of SIMpLE: "The SIMpLE system aims to ensure a better quality of life and promote greater social inclusion for all people with severe disabilities" (Arrigo Palumbo, 2021), the SIMpLE system was developed during the past 2 years of the Covid-SARS 2019 pandemic, a situation in which frail people have been in great difficulty. Through SIMpLE, the monitoring of vital parameters is improved and, in case of any anomaly behavior, the system can call for help.

In the SIRIO application, described in this contribution, immersiveness due to virtual reality is added to the various IoT services used in the applications discussed above. Classic virtual reality applications use head-mounted displays (HMDs) to immerse the user in virtual reality, one example is shown in (Nathan Moore, 2019). This contribution shows how immersion in a virtual environment that emulates reality can be used to train healthcare workers in hospital first aid. The authors show how through immersiveness due to virtual reality it is possible to significantly improve the learning of healthcare personnel.

However, virtual reality is only one part of the Reality-virtuality continuum (Paul Milgram, 1994) specifically in this contribution an application of Augmented Virtuality will be presented as the user navigates within the virtual space but can augment virtuality with IoT devices in the room.

Figure 4. Chart that show the dependences between the different topics

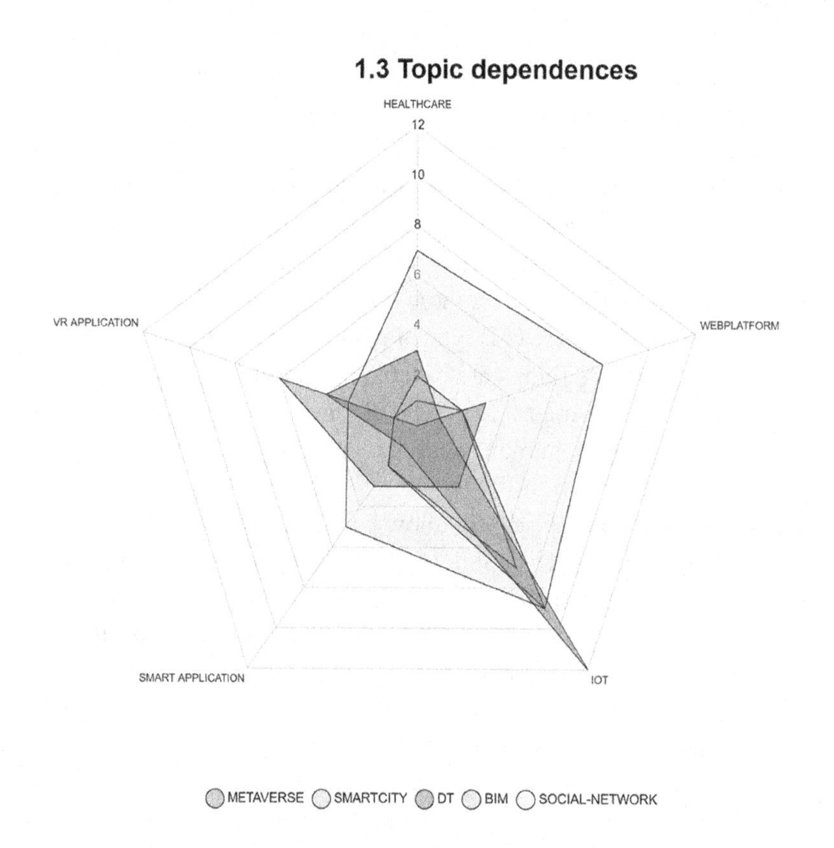

Chart 1.3 in Figure 4 shows the relationships that exist between the elements of the virtual-social world, datasets in the chart, and the different bridges that have been analyzed, represented on the tips of the pentagon, the values represent the number of articles analyzing the two topics. The chart immediately shows the strong relationship that exists between Smart City, DT and BIM with the IoT world. The Metaverse topic, on the other hand, is closely related to virtual reality applications, and social networks appear to be the most transversal topic and therefore the one that most easily bridges with reality.

METHODOLOGY

In this chapter, the technology and architecture of the SIRIO system are presented; the designed software can control IoT devices in a room through the use of virtual reality. The users interact with the software through the use of an eye-tracer to control the mouse cursor in a way that can also be used by more advanced ALS patients.

The architecture of the SIRIO system is shown in figure 5, where the different layers are shown. The system is composed of 3 layers: i) Digital-Twin Layer, ii) Services Layer, and iii) Application Layer.

The Digital Twin Layer is composed of two main subsections which include Virtual Assets and IoT devices. the virtual asset subsection includes all geometry and data files linked to 3D models. The application uses different types of files: .ifc files containing both the 3D geometries of the BIM model and the data of the assets that make it up, .gltf files containing only geometries, and text files containing information on non-BIM based models. The IoT Devices subsection, on the other hand, includes all devices connected to the system that send or receive information from the application users. These devices can be divided into Sensors for biomedical monitoring, such as heart rate monitors, Indoor sensors that allow room conditions such as temperature and humidity to be observed, and Smart home devices, which include all smart devices that can be controlled by the application.

The Services Layer includes all the services that enable the system to have interaction between the components of the Digital Twin Layer and the application layer. Inside it, therefore, there are the services that enable the reading of models and the creation of the user interface, together with the services required to control IoT devices and the user's management.

The last layer is the application layer where there are the final applications that can be used by the end users. The Hospital Information System (HIS) is the web platform capable to connect with SIRIO to download patient data and deliver telemedicine services, and the use of the real-time connection with the patient allows monitoring of the disease progression and its evaluation. The Home-Control Patient App is an application that patients can use directly to control their smart devices and access the various services designed by the authors. Finally, the application layer includes the third-party applications, i.e. all such applications that aim to fit into the SIRIO system to increase the services dedicated to frail persons and their caregivers. In the remainder of the chapter, the different layers are analyzed in detail.

Figure 5. System architecture layers

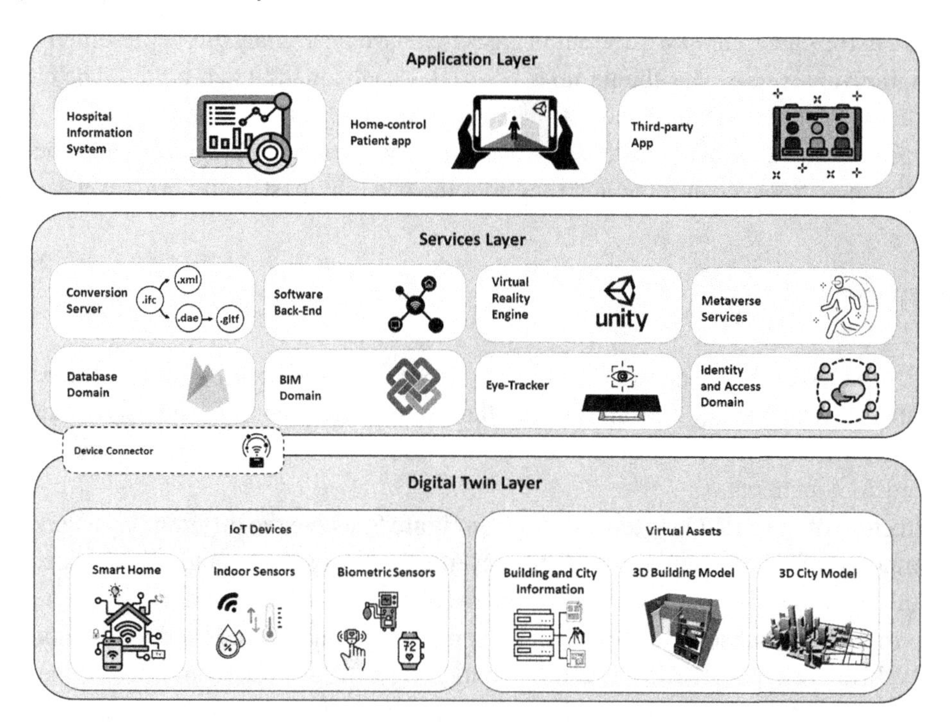

Digital Twin Layer

The DT layer is the lowest layer of the SIRIO system and includes, as explained above, the digital models and IoT devices that can be used by end-users. The digital models that are used to control IoT devices inside the SIRIO system are derived from a .ifc file. An IFC contains all the BIM information useful for managing the device information, but the correct implementation of the BIM model is a time-consuming and labor-intensive task to extract the correct information.

The creation of the model begins with an analysis of the available input data to improve the knowledge of the building. This data is often derived from files created with Computer-Aided Design (CAD) or, if no design files are available, the creation of the model must begin with an on-site survey to ensure that it replicates closely the real environment. After reliably collecting the data, the authors moved on to the definition of the assets that make up the environment, the objects that can be queried when using the application and the relative levels of information needed (17412-1:2021, 2021) are defined, which allow the definition of the data that will be associated with the assets of the BIM model, the individual objects that make up the building. The process of assembling the information model, whether of a building or an entire city, is carried out using BIM authoring software, i.e. software specialized in the 3D modeling and management of building information.

The parameters that can be incorporated within the BIM model also include the data required to: interact with IoT devices, download data and update the status of actuators in the environment. To better control the information stored inside the .ifc files that are exported by the BIM authoring software, appropriate Property Sets are created, which are parameter clusters that contain the information to integrate the services connected to sensors and actuators.

The IoT devices that are used by the system can be divided into three types: biomedical sensors, which are not present inside the BIM model, environmental sensors and smart devices, which are also modeled and managed in the BIM. The heterogeneity of the types of sensors and actuators used is controlled through the use of a device connector, a software that can translate the requests that are to be sent to the different devices and consequently allow the connection of different technologies.

Services Layer

The Services Layer is the central part of the schema in Figure 4 and encompasses all the services that are required to be able to connect models, IoT devices, databases, conversion processes, accesses and accounts, and external services that can be integrated via the designed microservices architecture.

The use of eye-tracking devices basically allows to control the cursor on the screen. For the realisation of the prototype, various alternatives were considered, from the most common toobi eye tracker 5, a device that can be connected to a windows operating system and is mainly used for gaming, up to all-in-one medical devices such as the Sagittarius M.U. 15 from the company Mesa Ideas, used for the final prototype.

The application that is used by the fragile user is developed using the game engine software Unity3D (Technologies, 2022). The software allows applications to be developed on different types of support: mobile devices, stand-alone PC, WebApplication and others. Using Unity3D, the user interface (UI), which is controlled by using an eye-tracking device, is created, and the Back-End part of the application is programmed, which allows the different services to be updated and linked to the desired final output. The Back-End also sends the requests to the device connector, which, by translating them, interrogates the relevant sensors and actuators.

BIM models in .ifc format are uploaded to a cloud server called BIM domain, using the service offered by BIMServer. BIMServer (BIMserver, 2022) is an OpenSource service that uploads .ifc models to a 'BIM based' server that is able, through the use of the functions provided by the BIMserver community to perform actions typical of collaborative BIM strategies: versioning control, feature surveys, etc.

The models in .ifc format are not directly readable by the application created in Unity3D, and therefore some conversions must be carried out via the Conversion server before they can be used.

The first conversion takes place through the use of IfcConverter (IfcOpenShell, 2022), an OpenSource converter that allows IFC files to be converted into various formats, both geometric and textual. To preserve the data contained in the BIM model, the first conversion is made from .ifc to .xml a textual format capable of preserving the properties associated with the unique guid code of the individual elements that make up the IFC. The guid is a unique code that is created when the IFC is exported, the same code is used to rename the elements when the conversion from .ifc to .dae is performed using again IfcConverter. To facilitate the reading and loading of geometries, the .dae file is converted further into .gltf format using the COLLADA2GLTF (Group, 2022) OpenSource library.

The data about the different sensors, the actions that are performed inside the application and the medical information related to the fragile user that is using the application are instead managed by the Database Domain, which exposes REST web services to access and update data. The Database Domain consists of 2 main parts: IoT4SLA database, Statistical database. IoT4SLA is the main database where data related to the different users of the system are stored.

Disease-related data, such as medical records, treatments and controls, are associated in the database with a unique user code, which is also associated with information about the user's location and the devices that are installed and controllable from the user's room.

Statistical database, on the other hand, is the database that is used to collect historical data on the use of the application, data recorded by environmental sensors and the progress of the user's illness. These data are analyzed using data analytics tools to understand if the user had an increase in autonomy due to the actions performed by the application. The data analyzed in this way is then stored in the IoT4SLA database so the results can be visualized.

The Identity and Access Domain is the service that allows different users to connect to the SIRIO system. The identity and access domain manages logins, last accesses and eventually the possibility to use multiplayer services so that friends and other relatives can make virtual visits to their dear ones.

Metaverse services make possible the connection with other virtual worlds and enable the connection of several people in the same virtual environment so the avatars can interact with each other. Inside the application Home-control patient app there is a 'Show-room' in which users can choose other virtual models to navigate. The various digital models in the Show-room are linked to different functionalities to allow: guided tours inside virtual museums, streaming services to watch films with friends or relatives, and games of various kinds.

Application Layer

The last layer that makes up the SIRIO system is the application layer, within this layer there are the different applications that can connect to the SIRIO system.

The HIS is the WebApp that connects SIRiO to the local hospital system. It allows doctors and caregivers to interact with patients to make remote visits, monitor the patient's condition in real-time, and use data from statistical analyses to evaluate the progress of the disease and the quality of life of the frail user. The HIS is also used to register new patients, manage hospital rooms, and modify therapies associated with individual patients.

The Home-Control Patient App is the application that is used directly by the frail user. This has been developed using Unity3D and tracks the movement of the cursor on the monitor to perform the different actions and interactions. Figure 6 shows the home of the application that is displayed after logging in, and from which three different scenes can be accessed: i) House, ii) Explore, iii) Dashboard.

Figure 6. Home of the application

To be able to use the application, it is necessary to connect an eye-tracking device to the PC on which the application is run; this makes it possible for the computer to understand where the user is looking and consequently move the cursor on the monitor. Using this mechanism, it was possible to set up the execution of the functions connected to the different buttons, with which the user can interact in the application, only after the button concerned has been looked at for more than 2 seconds. On the other hand, the buttons that allow movement and navigation in the model do not have this time gap for selection and therefore user only needs to look at them to move the view or move the avatar in the model.

Clicking on the "Home" button shows in figure 6 the user loads the IFC model of the house that is stored on BIMserver and linked it to the user ID. From this scene, it is possible to interact with various sensors and actuators that have been modelled inside the BIM model, and navigate the model in first-person mode.

When an IoT device is selected automatically, the software, reading the element's guid and searches within the xml file extracted from the BIM model for information related to that asset. Figure 7 shows an example of the implementation of a smart light bulb

Figure 7. Action execution example

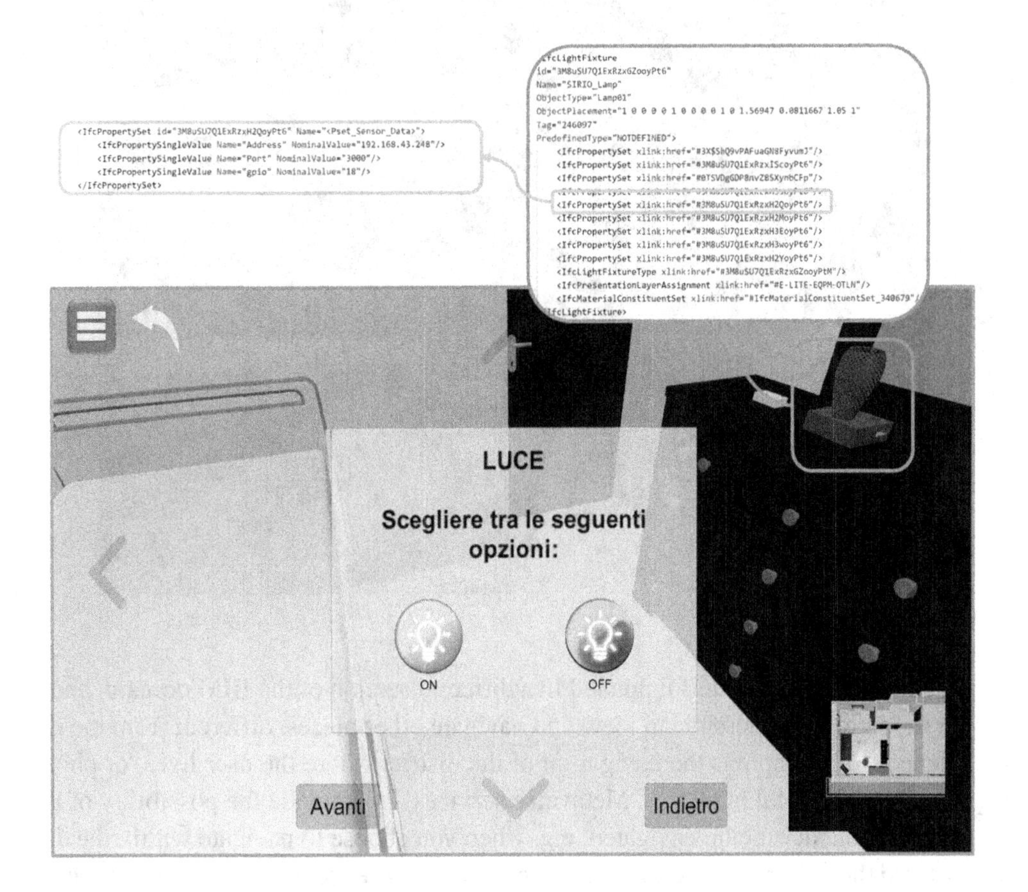

Looking at the 3D object located above the room drawer for 3 seconds shows a panel with 2 buttons, interaction with these buttons allows the intelligent light bulb to be switched on or off. This is possible thanks to a programming of the Back-End which by searching for the selected guid can link back to the correct Property Set in the IFC file, in the example "Pset_Sensor_Data".

Using the parameters inserted in the BIM model and the ones defined inside the application by the user, the software builds the web request necessary to perform the actions on the smart objects.

The second button in figure 6, "Explore", allows the user access to the model showroom, discussed in the previous section. In this part of the application, different kinds of virtual models can be loaded, as shown in figure 8.

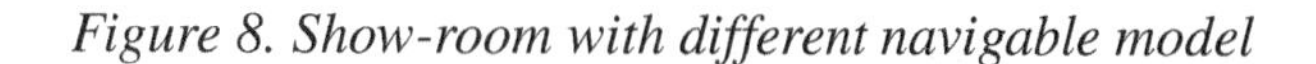

Figure 8. Show-room with different navigable model

The Show-room reads the models uploaded in a different section of the BIM domain, and by selecting the model of interest it is possible to view and navigate other places, different from the dwelling or place of residence. An example is the navigation of the district where the user lives, or photo-realistic models representing historical buildings. Metaverse services also provide the possibility of linking different services to the models being navigated, e.g. when you choose to navigate within the district, you will also be offered the possibility of driving a car to better explore the model. This approach is designed to provide greater stimulation to users, offering them the possibility of escaping, at least virtually, from their fragile situation.

The last button in figure 6 is the Dashboard. This button allows users to interact directly with different IoT devices without the virtual reality option, monitor environmental and biomedical data, and connect with medical staff for regular visits.

RESULTS

The result of this contribution is the development of a system that bridges the real and virtual worlds with the aim of restoring autonomy and quality of life to the most fragile people. The application acts as a Digital-Social Bridge by converting the interactions and sociability of the virtual world into real-world actions and communication.

Prototyping of the SIRiO system began in late 2021, continuing the work of (F. Alotto, 2019) and then testing in collaboration with the Molinette Hospital in Turin. The authors decided to use a Raspberry pi model 4 to manage sensors and actuators and an all-in-one device as an optical communicator and hardware to execute the home-control patient app software. Figure 8 shows the prototype created in the Drawing for the future laboratory of the Politecnico di Torino. Image A at number 1 of figure 9 shows the case of the IoT system made using 3D printing, the case is made of PLA, an insulating plastic material. Image B shows the components of the IoT system placed inside the case, number 2 is the Raspberry pi model 4, an electronic board able to manage the different devices of the system, number 3 is a temperature and humidity sensor model DHT 11, while numbers 4 and 5 are a relay model srd 05vdc and a 220V bulb. Image C in figure 9 at number 6 shows the display of the optical communicator with the software running and at number 7 the eye-tracker is connected to the all-in-one device.

The all-in-one communicator in the image corresponds to the model Sagittarius M.U. 15 from the company Mesa Ideas.

The test was performed by a girl suffering from ALS for more than 10 years in the most advanced stage of the disease. During the tests, feedback regarding the use of the interface, the commands that can be used and the activities for which the use of the digital-social bridge is necessary were collected from the patient.

FUTURE RESEARCH DIRECTIONS

The micro-service system developed for SIRiO and exposed in this contribution enables the connection of several heterogeneous smart devices. One of the first upgrades that will be carried out to the system will be the addition of sensors and actuators of different types so that users can perform more and more actions and improve their monitoring by doctors and caregivers.

Moreover, the system has great possibilities for development when considering the current growth of the Metaverse. By increasing the concept of digital identity and by integrating public administration (PA) services into a Metaverse based on smart cites, fragile people can be made more independent. In fact, by connecting PA services to a virtual city, fragile users would be able to: attend schools, visit museums, socialize and even work virtually.

Figure 9. System prototype and patient testing

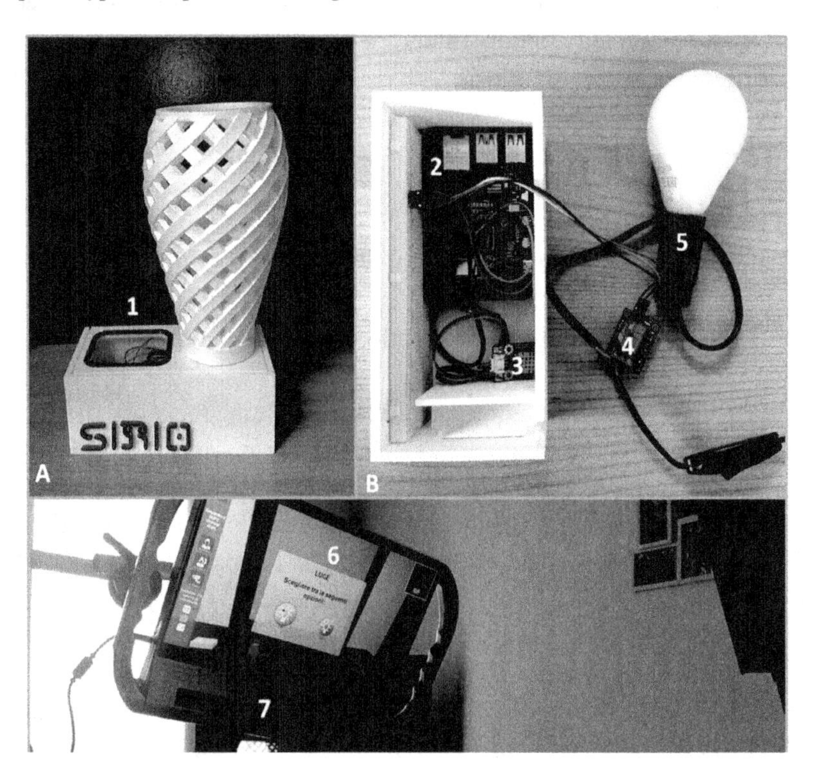

CONCLUSION

This article discussed how new technologies and their possible applications can act as bridges, enabling the most fragile users to communicate and perform simple actions in the digital social world. In particular, a solution developed by the Politecnico di Torino was presented, in which BIM-VR-IoT are integrated to create a smart system that supports ALS patients even in the most advanced stage of the disease. The proposed solution, in addition to increasing the level of autonomy and quality of life of the most fragile people, can track users' biometric data, such as heart rate, blood pressure and saturation, communicating this data to doctors and caregivers.

The use of digital models and virtual reality, combined with the opportunities offered by modern IoT systems, provide a more immersive and engaging experience than classical home automation systems.

The aim of society 5.0 is to use modern technology to support human beings, the project presented in this contribution has the same objective and it is our task to drive technological progress towards supporting the most fragile persons.

The project "An Algorithm for Progressive Neurodegenerative Motor Neuron Disease to Automate Hospital Rooms/Home Environments by Combining BIM and IoT through VR", was carried out in the framework of the Proof of Concept (PoC) Instrument initiative implemented by LINKS foundation, with the support of LIFTT, with funds from Compagnia di San Paolo.

The authors also thank Mesa Ideas srl for their support and the provision of technologies necessary for the research.

ACKNOWLEDGMENT

This contribution presents the work that has been done for the project "An Algorithm for Progressive Neurodegenerative Motor Neuron Disease to Automate Hospital Rooms/Home Environments by Combining BIM and IoT through VR". The authors would like to thank the LINKS Foundation, with the support of LIFTT, with funds from Compagnia di San Paolo, which permitted the design of this proof of concept (PoC). Special thanks go to the company Mesa Ideas, which believed in the project and in the spirit of the research, providing its knowledge and technology that allowed the prototype to be realised.

REFERENCES

17412-1:2021, U. E. (2021). *Building Information Modelling - Livello di fabbisogno informativo - Parte 1: Concetti e principi*.

Aamir Hussain, R. W. (2015). Health and emergency-care platform for the elderly and disabled people in the Smart City. *Journal of Systems and Software*110.

Abubaker Basheer Abdalwhab Altohami, N. A. (2021). Investigating Approaches of Integrating BIM, IoT, and Facility Management for Renovating Existing Buildings: A Review. MDPI – Sustainability, 13.

Alotto, A. A. (2019). *A software solution for ALS patients to automate hospital rooms by combining BIM, Virtual Reality and IoT*. Dept. of Biomedical Eng., Politecnico di Torino.

Angel Ruiz-Zafra, K. B. (2022). IFC+: Towards the integration of IoT into early stages of building design. *Automation in Construction*, 136.

Arrigo Palumbo, N. I. (2021). SIMpLE A Mobile Cloud-Based System for Health Monitoring of People with ALS. *MDPI Sensors 2021*.

Awais Ahmad, M. B.-A. (2019). Socio-cyber network: The potential of cyber-physical system to define human behaviors using big data analytics. *Future Generation Computer Systems*, 92.

Ben Falchuk, S. L. (2018). The Social Metaverse - Battle for privacy. *IEEE Technology and Society Magazine*.

BIMserver. (2022). Retrieved from bimserver.org: The open source BIM

BuildingSMART International. (2022). *About - What is openBIM?* Retrieved from BuildingSmart: https://www.buildingsmart.org/about/openbim/openbim-definition/

BuildingSMARTalliance. (2007). National building information modeling standard (vol. 1). National Institute of Building Sciences.

Calin Boje, A. G. (2020). Towards a semantic Construction Digital Twin: Directions for future research. *Automation in Construction*, 114.

Craig Lefebvre, R. A. S. (2013). Digital Social Networks and Health. Social Media as a Tool in Medicine. American Heart Association, Inc.

Creators. (2022). Retrieved from https://www.creatorspad.com/about

Cui, W. D. H. (2021). Digital twin-driven building projects: a new paradigm towards real-time information convergence throughout the building lifecycle. In *UIA 2021 RIO: 27th World Congress of Architects*. ACSA.

David Jones, C. S. (2020). Characterising the Digital Twin: A systematic literature review. *CIRP Journal of Manufacturing Science and Technology*, 29.

Min Deng, C. C. (2021). From bim to digital twins: a systematic review of the evolution of intelligent building representations in the aec-fm industry. *ITcon Vol. 26 - Journal of Information Technology in Construction*.

Dessislava Petrova-Antonova, S. I. (2021). *Digital Twin Modeling of Smart Cities. Human Interaction, Emerging Technologies and Future Applications III*. Springer.

Duan, H. J. L. (2021). Metaverse for Social Good: A University Campus Prototype. Multimedia. Cornell University.

Grieves, M. (2014). *Digital twin: Manufacturing excellence through virtual factory replication*. White paper.

Group, T. K. (2022). *COLLADA2GLTF*. Retrieved from pagina github: https://github.com/KhronosGroup/COLLADA2GLTF

Han, Y. D. N. (2021). A Dynamic Resource Allocation Framework for Synchronizing Metaverse with IoT Service and Data. Computer Science and Game Theory. Cornell University.

Hee-soo Choi, S.-h. K. (2017). A content service deployment plan for metaverse museum exhibitions-Centering on the combination of beacons and HMDs. *International Journal of Information Management*.

IfcOpenShell. (2022). *IfcConvert: An application for converting ifc geometry into several file formats*. Retrieved from IfcOpenShell: http://ifcopenshell.org/ifcconvert

Islam, S. R. D. K.-S. (2015). The internet of things for health care: a comprehensive survey. IEEE Access, 3, 678–708.

Jaime Ibarra Jimenez, H. J. (2019). *Health Care in the Cyberspace: Medical Cyber-Physical System and Digital Twin Challenges*. Digital Twin Technologies and Smart Cities.

Juan Manuel Davila Delgado, L. O. (2021). Digital Twins for the built environment: Learning from conceptual and process models in manufacturing. *Advanced Engineering Informatics*, 49.

Kyoungro Yoon, S.-K. K.-H. (2021). Interfacing Cyber and Physical Worlds: Introduction to IEEE 2888 Standards. In *International Conference on Intelligent Reality (ICIR)*. IEEE.

Lik-Hang Lee, T. B. (2021). *All One Needs to Know about Metaverse: A Complete Survey on Technological Singularity, Virtual Ecosystem, and Research Agenda. Computer and society*. Cornell University.

Maninder Jeet Kaur, V. P. (2019). *The Convergence of Digital Twin, IoT, and Machine Learning: Transforming Data into Action*. Digital Twin Technologies and Smart Cities.

Mazhar, M., & Rathore, A. P.-H. (2018). Exploiting IoT and big data analytics: Defining Smart Digital City using realtime urban data. *Sustainable Cities and Society*, 40.

Nathan Moore, S. Y. (2019). ALS-SimVR: Advanced Life Support Virtual Reality Training Application. *VRST 2019 - 25th ACM Symposium on Virtual Reality Software and Technology.*

Ning, H. H. W. (2021). A Survey on Metaverse: The State-of-the-art, Technologies, Applications, and Challenges. Cornell University.

Paul Milgram, H. T. (1994). *Augmented Reality: A class of displays on the reality-virtuality continuum. SPIE* (Vol. 2351). Telemanipulator and Telepresence Technologies.

Rho Seungmin, Y. C. (2019). Social Internet of Things: Applications, architectures and protocols. *Future Generation Computer Systems.*

Riazul, S. M., & Islam, D. K. (2015). The Internet of Things for Health Care: A Comprehensive Survey. Computer science. *IEEE Access: Practical Innovations, Open Solutions.*

Rubén Alonso, M. B. (2019). *SPHERE: BIM Digital Twin Platform* (Vol. 20). MDPI - Proceeding.

Ruben de Laat, L. v. (2010). Integration of BIM and GIS: The Development of the CityGML GeoBIM Extension. *Advances in 3D Geo-Information Sciences.*

Seungmin Rho, Y. C. (2019). Social Internet of Things: Applications, architectures and protocols. *Future Generation Computer Systems*, 92.

Siavash H. Khajavi, N. H. (2019). Digital Twin: Vision, Benefits, Boundaries, and Creation for Buildings. *IEEE Access, 7.*

Stephenson, N. (1992). *Snow Crash.* Academic Press.

Tali Hatukaa, H. Z. (2020). From smart cities to smart social urbanism: A framework for shaping the socio-technological ecosystems in cities. *Telematics and Informatics*, 55.

TechnologiesU. (2022). *Unity*. Retrieved from https://unity.com/

Vaia Moustaka, Z. T. (2019). Enhancing social networking in smart cities: Privacy and security bonderlines. *Technological Forecasting and Social Change*, 142.

Vichi, M. (2016). *Amyotrophic Lateral Sclerosis in Italy: characteristics and geographical distribution of first hospitalizations in the year 2005.* Academic Press.

Vito Albino, U. B. (2015). Smart Cities: Definitions, Dimensions, Performance, and Initiatives. *Journal of Urban Technology*, 22.

Vittorio Miori, D. R. (2017). *Improving life quality for the elderly through the Social Internet of Things (SIoT). Global Internet of Things Summit (GIoTS).* IEEE.

Xu, Q. Z. S. (2021). Fast Containment of Infectious Diseases With E-Healthcare Mobile Social Internet of Things. IEEE Internet of Things Journal, 8(22).

Yue Pan, L. Z. (2021). A BIM-data mining integrated digital twin framework for advanced project management. *Automation in Construction*, 124.

Zarei, K. C. (2015). A comprehensive review of amyotrophic lateral sclerosis. *Surgical Neurology International.*

Zhang, Y. R. C. (2020). Developing a visually impaired older people Virtual Reality (VR) simulator to apply VR in the aged living design workflow. *International Conference Information Visualisation (IV).*

ADDITIONAL READING

Ben Falchuk, S. L. (2018). The Social Metaverse Battle for privacy. *IEEE Technology and Society Magazine.*

David Jones, C. S. (2020). Characterising the Digital Twin: A systematic literature review. *CIRP Journal of Manufacturing Science and Technology*, 29.

Kshetri, N. (2022). *Web 3.0 and the Metaverse Shaping Organizations' Brand and Product Strategies.* IEEE Computer Society. doi:10.1109/MITP.2022.3157206

Liupengfei Wu, W. L. (2022). Linking permissioned blockchain to Internet of Things (IoT)-BIM platform for off-site production management in modular construction. *Computers in Industry.*

KEY TERMS AND DEFINITIONS

Building Information Model: It is a data-rich, object oriented, intelligent, and parametric digital representation of the facility, from which views and appropriate data for various users' needs can be extracted and analyzed to generate information that can be used to make decisions and improve the process of delivering the facility.

Building Information Modelling: It is a method based on a building model containing any information about the construction. In addition to 3D object-based models, it contains information about specifications, building elements specifications, economy, and programs.

Human Fragility: The condition of human fragility can be identified in all those individuals who fall into a state of physiological vulnerability. In particular, we can think on older people or people suffering from chronic diseases that debilitate the body.

Interoperability: Is defined as the ability of two or more systems or components to exchange information and to use the information that has been exchanged.

Metaverse: The metaverse is defined as the potential evolution of the Internet. A metaverse is a virtual world in which users can interact and communicate with each other and the surrounding virtual environment. The term is a portmanteau of 'Meta', in which from the Greek μετά, meta, meaning 'after' or 'beyond' and universe.

Smart City: A city can be called a smart city when its investments ensure sustainable economic development and a different quality of life.

Virtual Reality: Virtual reality is a technology that aims to immerse users in computer-generated environments using head-mounted displays (HMDs), wired gloves, cybert suits or other devices.

Chapter 26
Digital Twin for Amyotrophic Lateral Sclerosis:
A System for Patient Engagement

Matteo Del Giudice
Politecnico di Torino, Italy

Roberta Surian
Politecnico di Torino, Italy

Anna Osello
Politecnico di Torino, Italy

ABSTRACT

This chapter focuses on the context in which patients such as those with Amyotrophic Lateral Sclerosis (ALS) are placed and what possibilities information and communication technologies (ICTs) offer to keep them in touch with the world to reach Society 5.0. In particular, the authors intend to show how the healthcare sector can use digital twin (DT) through elements of augmented virtuality (AR) and building information modelling (BIM) to create interactive interfaces that can solve, in part, problems involving frail patients but at the same time allowing their monitoring. Interconnection is possible through a gamification approach. In addition, a solution that considers the user (patient) involvement and that aims at its increase through interaction with alternative places to their home so as to stimulate them to keep an active mind and the degree of fun in a limiting condition is proposed.

INTRODUCTION

In recent years, the adoption of innovative technologies for the development of Smart Cities and Smart Society has been studied by many researchers to refine the management of the built environment related to healthcare facilities. This goal achievement involves the use of interdisciplinary information considering both medical personnel for patient management and technical one for facility management. In this

DOI: 10.4018/978-1-6684-4854-0.ch026

context, the concept of Digital Health can be defined as "*the cultural transformation of how disruptive technologies that provide digital and objective data accessible to both caregivers and patients leads to an equal level doctor-patient relationship with shared decision-making and the democratization of care*" (Meskó et al., 2017). This transformation involves the adoption of many technologies capable of handling large amounts of data that must then be interpreted by qualified personnel. The inclusion of such innovative methods and tools can often be counterproductive if not accompanied by an awareness of the goals to be achieved and the challenges to be met, such as the overcoming of digital divide.

The elaboration of virtual environments, able to provide scenarios of interaction between the user and the surrounding environment, poses several challenges related to technological innovation and human behavior. Gamification approach is one of the main strategies adopted. It can be defined as "*the use of game design elements in non-game contexts*" (Deterding et al., 2011). It is based on the use of technologies of the gaming world in contexts whose interest is to increase user involvement. The idea behind gamification is to exploit the motivational possibilities of entertainment games in other spheres (i.e., health sector, building sector) by leveraging the features of a certain application to make user engagement more motivating. Using a Gamification approach, it is possible to harness a predisposition to motivation in performing an action or a task. This is since games can put users in favorable conditions, stimulating them and increasing their involvement.

In the age of connection, the information society (society 4.0) needs to use technological advancement and ICTs to solve social problems and overcome the fragilities of people with disabilities through an integration of cyberspace and physical space. Through this paradigm shift, it is possible to speak of a Society 5.0 in which people are connected with cyberspace and, through artificial intelligence (AI) algorithms, can overcome the limitations of physical space to improve the quality of life for all citizens. Through an approach to gamification and Serious Games (SGs), it is possible to trigger this change by relating people with ICTs. (Narvaez Rojas, 2021).

Many researchers are investigating these arguments and a broad range of consumer applications for monitoring and managing one's own health and well-being are available on the market. One important sector is SGs for health (Wattanasoontorn et al., 2013), games used to drive health-related outcomes. The majority of these are "health behavior changes games" (Baranowski et al., 2008) or "health games" (Kharrazi et al., 2012) affecting the health behaviors of health care receivers (and not e.g., training health care providers) (Wattanasoontorn et al., 2013). The activities on which developed applications focus concern physical activity, nutrition, and stroke rehabilitation, with an about equal share of (a) "exergames" or "active video games" directly requiring physical activity as input, (b) behavioral games focusing specific behaviors, (c) rehabilitation games guiding rehabilitative movements, and (d) educational games targeting belief and attitude change as a precondition to behavior change (Kharrazi et al., 2012).

BACKGROUND

Gamification in Building Sector

With the advancement of computer and information technologies in recent years, innovative methods such as Building Information Modelling (BIM) and visualization tools such as Virtual Reality (VR), Augmented Reality (AR), and Mixed Reality (MR) have been applied widely across the Architecture, Engineering, Construction, and Operation (AECO) industry (Springer, 2022). In the context of the AECO sector, SGs have been applied to construction health and safety education, professional skill training, evacuation training, behavioral analysis, emergency studies, and decision-making processes

Game features and mechanisms have been integrated with these visualization applications, expanding their impact. Many analogies can be found between the building process and the role-playing game ecosystem in which there are various missions that can be performed by different actors who may use different tools. Each tool is needed to solve a certain task/activity in a certain time. Gamification associated with the world of construction therefore requires the development of a shared digital platform among the various actors involved to achieve a given project objective (Bottiglieri, M. A., 2019). While the development of a 3D information model is essential, it is also important that the information can reach the end user to enable him/her to perform a certain activity, thanks in part to interoperability.

Moving to Industry 4.0 and intelligence construction, Digital Twins (Grieves, 2014) is another evidence for improving the interactive experience between information models and end users. Therefore, the immersive visualization environment (VR and MR) can be made from the gathered data of existing buildings, thus, to develop a dynamic interactive gamificative BIM environment to enhance owners and end users experience by using various headsets.

The proposed innovation in this contribution takes part into the Smart Advanced Modelling for Care (SAM4Care) project (Osello, 2018), which aims to unite two souls: i) the management of heterogeneous data domains related to the building and urban scale; ii) the patients' care who, due to neurodegenerative diseases, are no longer able to please their personal needs through the body movement. For these reasons, the Proof of Concept (PoC) project was born related to the elaboration of an integrated and open-source system, based on a special algorithm for the connection of BIM models with Internet of Things (IoT) databases through VAR. The proposed innovation is based on the creation of a Digital Twin for Amyotrophic Lateral Sclerosis (DTALS) to identify its role and impact as a game changer in the disease management. The DTALS involves the development of a virtual replica related to physical reality. The development of this system is based on a user-centered approaches, increasing emphasis on engagement utilizing processes, increasing collaboration in program development, testing, and data sharing to achieve higher quality, more sustainable outcomes oriented to people welfare (Fleming et al, 2016).

Figure 1. Suggestive image of the investigated health monitoring system

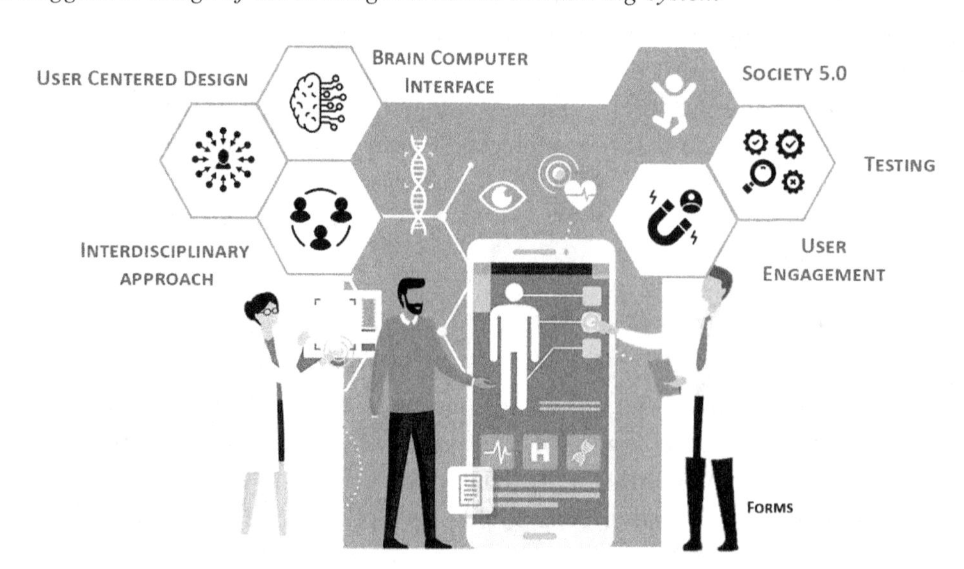

For this contribution, the selected case study is a patient suffering from a neurodegenerative disease, placed in a residential environment converted to an inpatient room. In the latter, a network of sensors necessary for data acquisition is set up.

Human-Computer Interaction and Reality-Virtuality Continuum for Digital Health

E-health was born in 1990, then in 2010 with the digitisation of healthcare, following the pandemic emergency of 2019, the necessity of customising monitoring processes to better follow patients has emerged even more. Furthermore, the cost of treating patients with serious conditions has increased, along with the need to modernise hospitals and medical facilities (Meskó, Drobni, Bényei, Gergely, & Győrffy, 2017). As a result of this global event, the processing of medical data and an efficient system is simultaneously the subject of studies that have led to the coining of the term Digital Health (E-health). It can thus be defined as "*the proper use of technology for improving the health and wellbeing of people at individual and population levels, as well as enhancing the care of patients through intelligent processing of clinical and genetic data*" (Fatehi, Samadbeik, & Azar, 2020). Thus, the concept of the Digital Twin in relation to the patient has been increasingly approached in order to improve treatment, monitoring and diagnostic methods. In this way, the patient, between the medical sector and digital technologies, becomes the focus of Digital Healthcare. Essential technologies in this development are Artificial Intelligence (AI), Machine Learning (ML) and the use of sensors, which combine ICT technologies to solve or alleviate patients' problems. The interest of Digital Healthcare is to provide a better administrative and clinical service, but also to help patients to better manage their health, especially in the most sensitive cases, by making them active participants in making decisions about their own health (Edirippulige & Senanayake, 2020) (Meskó, Drobni, Bényei, Gergely, & Győrffy, 2017). Obviously, patients need to be prepared to participate in the management of their medical conditions and it is only through collaboration that the success of these initiatives can be achieved.

Several solutions have focused their attention on Human-Computer Interaction (HCI) to generate a final product to approach the growing adoption of Digital Health. Typically, HCI consists of a study phase, followed by a design phase, an implementation phase, and a hardware and software analysis phase. In order to bring a viable product to life, it is of primary importance that one first studies the intended audience. This is important because following a demand and/or need, a product can be designed to meets the needs of the end user. HCI has been shown to stimulate the human cognitive faculties, defined as the whole range of information that we are able to remember and process in order to use a system (Alonso-Valerdi, 2017). The simpler the stimuli are, the more they meet the end user's need, the easier the platform will be to use and the easier it will be to use. In the course of studies, different technologies that can stimulate the cognitive load of the end-user have been evaluated: sound synthesis, kinesthetic and tactile feedback and eye-tracking that captures the movement of the cornea. In particular, the latter method caters for a whole audience of people with illnesses or disabilities. Eye-tracking technologies capture images as a result of the optical reflection of the cornea: a pattern of infrared light is projected and it hits the eyes and locates the cornea, so its movements can be followed. This makes it easy to understand how eye movement tracking can control a computer as well as to understand, at least in terms of statistics, the degree of interest and attention of the user. Every eye-tracking technology requires an initial calibration process. Eye-tracking as a substitute for mouse pointing is the most widespread way of exploiting this technology (Björnsson et al, 2020).

The eye-gaze usage is often combined with the use of Virtual Reality (VR). VR is commonly understood as a technology that totally immerses the user in a 'synthetic' environment. It has been shown that patients who took an active part in research involving the use of Virtual Reality (VR) were able to actively experience the research, also providing positive feedback: the experience instilled in them a reduction in pain and anxiety (Meskó et al, 2017). The VR system can be distinguished into:

- *Non-immersive system*: uses a screen or computer on which the virtual environment is displayed.
- *Fully immersive system*: gives users a more realistic simulation experience because the person is totally immersed in the virtual environment and has to use a technological system consisting of sensors, visors and controllers.

VR is thus a technology that plays an active and effective part in the therapy of patients. VR - like Augmented Reality (AR) and Augmented Virtuality (AV), which together define the spectrum of Mixed Reality (MR) with the reality-virtuality continuum) - is reality in which the real and virtual worlds are represented together on a single screen (Cáceres, Carrasco, & Ríos, 2018).

Figure 2. Representation of Reality-Virtuality Continuum (Heliyon, 2018)

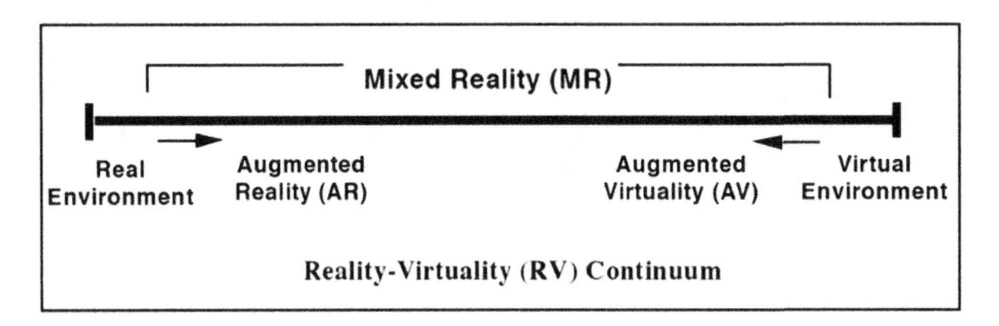

AR can be defined as "*augmenting natural feedback to the operator with simulated cues*" and is closer to the real world, whereas AV is closer to the virtual world (Cáceres, Carrasco, & Ríos, 2018). In AR, interactions in the virtual world are reflected in the physical world (Edirippulige & Senanayake, 2020). In the medical field, AR has been used for medical imaging, superimposing data via imaging techniques. Virtual Environments (VE) are very versatile especially for those with severe paralysis: they attempt to reconstruct circumstances and situations through which patients can gradually adapt to their condition, but at the same time can maintain and train the abilities they still own (Beaudouin-Lafon, 1993). While some studies have started to conceive the idea of creating a Digital Twin of the patient himself, with this work the authors want to conceive the Digital Twin of what surrounds the patient, the place where he/she lives to provide a virtual representation of it: so, he/she could communicate with physical reality (Meskó et al, 2017).

The Reality-Virtuality continuum offers the opportunity to establish an ideal connection between the physical world and the virtual world, but as virtual information overlaps with reality and vice versa, it becomes increasingly complex to establish the boundary between AR and AV. In fact, each environment becomes contaminated with information from different domains that increase the quality of the developed Digital Twin. The study initially considered AV as the main area of investigation of the gaming environment, exploiting the virtual model as a graphical interface connecting the two worlds. Promising as gamification for health and well-being may be, the essential question remains whether gamified interventions are effective in driving behavior change, health, and well-being, and more specifically, whether they manage to do so via intrinsic motivation (Johnson et al, 2016).

Having defined a precise target group to be addressed, the demand arises from the patients' need to acquire more autonomy, but at the same time there is a demand to shorten the distance between medium and patient. For patients with motor neuron deficiency, the use of eye-gaze or eye-detection technology is vital, and it is the best solution to make them communicate (Björnsson et al, 2020). Popular technologies include *Tobii*, *EagleEyes* and so on; making full use of eye-detection technology to control a computer. The greatest benefit of this kind of technology lies in the possibility of interacting with Graphic Computer Interfaces (GUIs), being able, for example, to take advantage of keyboards to communicate. The wink is the most widely used way to replace the mouse click. An important factor not to be forgotten and to be evaluated with the daily use of such a technology, such as the one proposed here, is eye fatigue, followed by cognitive fatigue. With this in mind, one must try to optimise the quality of navigation as much as possible and make sure that they are used in such a way that the experience is not frustrating. One can assess this through Fittz's Law, which measures the time it takes to perform an action on the screen in a given area over a given period of time (Björnsson et al, 2020).

In this sense, HCI has been used as a means of expanding the senses for those who are deprived of them due to pathologies, by stimulating each person's cognitive perception. In the construction of the platform, medical and engineering disciplines were united, so as to create strong communication between professionals who could mutually bring points of discussion to light in order to build a valid platform for the doctors themselves, but above all for patients. In this case, a technology was studied that would be able to meet a specific target population: people with total body paralysis, i.e., ALS patients. The main symptoms are asymmetry and spasticity of movements, atrophy of muscles, speech disorders that later lead to dysphagia and failure to breathe, the main cause of death for these patients. Consequent to the onset of primary symptoms is the development of depressive states (in 4 – 56% of cases) that arise because of the inability to perform most of the daily life actions. The combination of these issues contributes to the progressive decrease in patients' quality of life. For these reasons, research focused on the need to define a specific technology for neuro-degenerative diseases, on the optimization of life quality and on maintaining patients' autonomy as much as possible (Zarei et al, 2015). The crucial point is the strong social impact they have, degrading the freedom of action (Alotto et al, 2019) (Eid et al, 2016). This condition makes the patient conscious from one side, but on the other side he/she cannot make movements or communicate verbally due to a complete paralysis of almost all voluntary muscles in the body, except for the eye muscles. Patients suffering from neurodegenerative diseases or unable to move can try to keep their minds and cognitive abilities active. The technology, however, can be aimed at a broader audience including all those who cannot move their limbs or part of them, so as to provide them with an alternative that allows them to perform certain actions of daily living independently. With this in mind, we wanted to focus on an adaptive interface - personal health as intuitive and simple as possible. This was achieved by translating operations with objects into a 'language' that could interface with VR and physical reality in order to bring the actions requested as input to the interface back to reality. More appropriately, then, it will speak of AV.

The conceptual and structural model thus conceived was submitted to the attention of healthcare professionals in order to better understand what might interest them in the care of patients on the one hand, but also to interface between professionals working together to provide an innovative technology useful to patients. At the same time, 20 ALS patients were interviewed by *ad hoc* forms to better understand their needs and wishes. In this way, the two end-users were involved in the design of the platform.

METHODOLOGY

In the present work, the authors focused their attention on patients suffering from Amyotrophic Lateral Sclerosis (ALS) that mainly affects subjects between 50 and 65 years old (Zarei et al., 2015). This type of Disease limits the actions of daily life since it affects Motor Neurons (MND), preventing more and more that the patient movements, breathe and speak independently. Different technologies have been created over the years to overcome the deficits of this category and they have turned their gaze on possible solutions that would allow patients to communicate. Among these, Eye-Tracking is the most widespread technology: it allows to track the Pupil Corneal Center (PCC) movement providing a method to interact with electronic devices.

This paper defines the essential features of an interface for the development of a communication medium (based on Human-Computer Interaction) for this type of patients, to help them overcome the barriers of their disease. Visualization, interaction, and analysis of the person's behavior in the developed digital environment are the requirements for improving the healthcare system through the development of digital assistants, reducing the distance between doctor and patient. Leveraging the Digital Health phenomenon, the aim is to develop an interface that allows the virtual navigation within patient's home (and beyond) to perform simple actions that are reflected in physical reality. There is precisely where a subject suffering from ALS in the more advanced stages of the disease has no control. Through this innovation, patients regain a new autonomy by becoming an active part of the proposed technological system. The invention implements a novel algorithm that, starting from the information contained in a BIM model, integrates IoT technology to control indoor automation. The 3D model exploration allows the navigation of virtual environments to patients suffering from motor neuron degradation (e.g., ALS) and enables interaction with common objects such as electrical systems, doors and windows.

Figure 3. Adopted workflow for the development of a DTALS platform

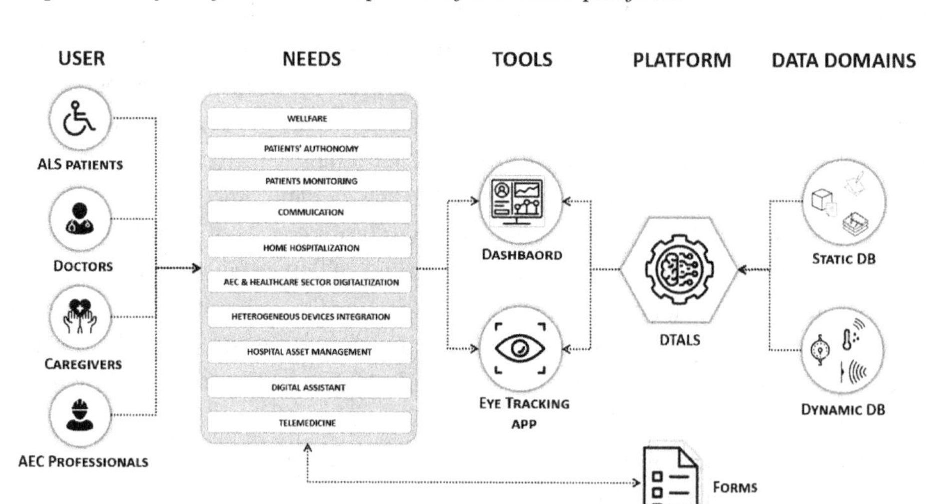

Figure 3 shows the relationships between users' needs and the tools hypothesized to fulfill them by interacting with the DTALS platform.

In the healthcare sector, in addition to ALS patients, many actors (e.g., doctors, caregivers, AEC professionals, owners) are involved in their management aimed at either improving the quality of life of patients in healthcare or home environments. For this reason, the development of an interactive digital platform can be used for multiple purposes, encompassing both healthcare and also real estate management aspects. Clearly, each user explicits different needs that therefore involve the development of different tools that can leverage the heterogeneous information collected in the data domains.

The elaboration of the DT prototype of a home or of a health facility, in which the patient resides, stems from information requirements identified from a census based on the drafting of specific questionnaires. The final users (e.g., patients, doctors) contribute to outline the characteristics of the proposed innovative system, defining their desires and needs that the PoC intends to fulfill. In this way end users become an integral part of the proposed system achieving the concept of Society 5.0. The DT of the selected home or healthcare facility provides an AV experience and establishes a link between virtual and real. In this way, the concept of DT expands from the building model to the patient, who through his avatar regains some skills lost due to his/her condition.

The concept model of the DTALS, characterized by three main parts "*physical products in Real Space, virtual products in Virtual Space, and the connections of data and information that ties the virtual and real products together*" (Grieves, 2014). Therefore, the development of the DTALS platform has to:

- Enable the collection of static data (BIM domain) and dynamic data (IoT database);
- Allow the data interchange and the connection between the BIM and IoT devices using gamification engine (e.g., Unity 3D);
- Provide the necessary services to patients for improving their quality of life;
- Facilitate their monitoring and the useful applications for medical équipe.

In particular two applications were selected and investigated for the DTALS: the Patiet App and the Medical App.

Patient App

The application aims to increase autonomy to patients by allowing them, through non-immersive virtual reality, to perform many tasks independently in the home and/or hospital environment. This app allows users to:

- Navigate within a virtual space to patients suffering from motor neuron impairment (they are incapacitated in limb movement, while having intact cognition and control of eye movements);
- Interact with common objects such as electrical systems, doors, windows by users;

The BIM model, 3D graphic building database, is enriched with specific information for the subsequent application in Unity aimed for the interaction of the patient with the surrounding environment.

Thanks to the parametric approach of the BIM methodology, 3D parametric models have been developed by using the BIM authoring platform *Autodesk Revit®*. As the DTALS platform was conceived as a scalable web open-source platform, the BIM model were initially exported in the *.ifc* OpenBIM format and the uploaded in the BIM domain using a BIM based server that is essential to manage models in the correct way and it allows to modify, replace and implement models, check files versions to understand what changes have been made and to approve or not such changes. The model navigation in a VR environment was developed in Unity3D engine. As it unfortunately does not support *.ifc* data-format, BIM models have been converted in *.obj* format and *.glb* format. Through this interoperability process, a GUI was developed for enabling the model uploading process in the BIM Domain and for produce a reliable, updatable BIM model.

In designing the platform for patients, the idea was to exploit the power of the icon model, launched on the market by Mackintosh Finder, so as to make familiarisation with it more immediate (Alonso-Valerdi et al., 2017). But, again, attention was paid to trying to make the gap between medium and patient smaller, which is often important. In this way, the interaction between the two actors is shorter, as the patient can directly call the doctor or nurse, while the doctor himself can independently access the real-time patient data

Figure 4. Sample of the proposed DTALS infrastructure for patient use in the indoor environment

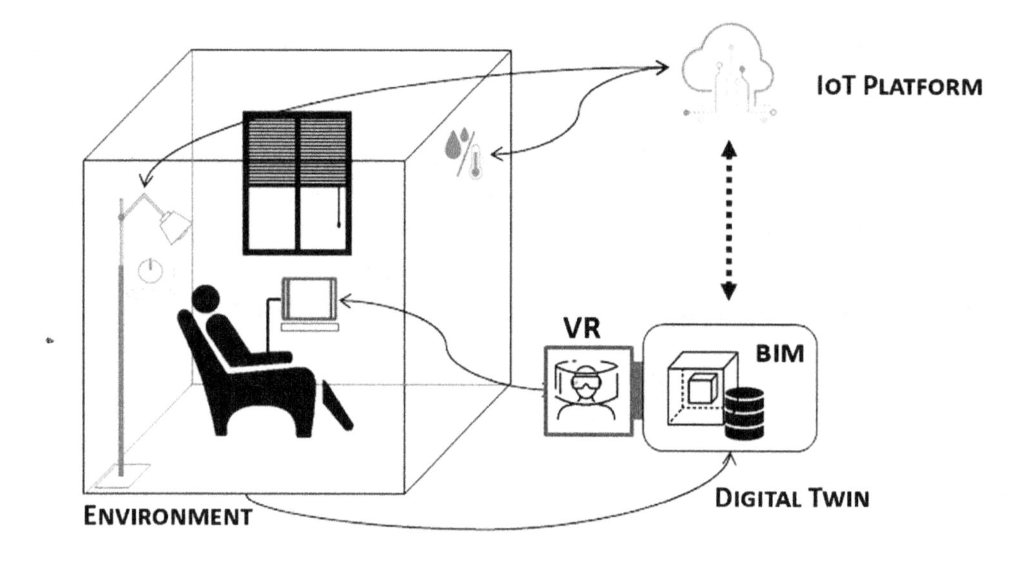

Medical App

The application aims to visualize patients' health parameters, collecting and transmitting environmental/medical data in compliance with current regulations through IoT with plug and play integration of new services. Figure 4 show a sample of the GUIs proposed for the DTALS.

A Web platform is being developed for physicians and caregivers where the patient's history, diagnosis and treatment can be placed. In addition, a track is kept of the interactions between the patient and the environment to allow to understand, for example, his activity, his progress in the pathology and his rehabilitation process, in a broad sense. This is not merely about consulting users on drafts but also about a deep understanding of user needs and preferences, and actively involving users in design processes from the outset. By contrast, a user-centered design would begin with users, to understand issues such as how and when they would be willing to use the internet for mental wellbeing, and to explore their current behavior, needs, and preferences. Compared to the Patient app, this application is still in the early stages of study and needs more investigation.

After completing the DTALS prototype, the testing phase begins to assess the positive aspects and critical issues for improvement by directly involving patients. In this regard, two form were developed to involve both patients in direct use and medical practitioners to define the content of the telemedicine application. The latter were also asked questions related to the *Patient App* to improve the usability of the application. Through both the surveys and the testing phase of the *Patient App*, the key aspect was to involve users by making them become active participants in the project.

The use of serious gaming and gamification, enhanced telepresence, and increased use of persuasive technology are promising in this regard. Moreover, the routine assessment of engagement may help to further develop the field and monitor progress toward this goal.

Figure 5. Examples of Medical App dashboard

Survey of User Needs – Questionnaire

A specific questionnaire was drawn up for ALS patients in order to get to know the user and his/her needs better so that the *Patient App* could be calibrated according to the feedback obtained.

The user was asked the following questions: (i) Age; (ii) Gender; (iii) City; (iv) Do you know what Virtual Reality (VR) is? (Yes, no); (v) Have you ever used Virtual Reality (VR)? (Yes, Yes – immersive VR, Yes – non-immersive VR, No); (vi) Do you think that VR could be an added-value? If Yes, explain how (Yes, No, I don't know); (vii) Do you know what an eye-tracker is? (Yes, No); (viii) Have you ever used an eye-tracker? (Yes, No); (ix) Would you like to be able to perform certain actions of daily living by yourself (e.g., turn on/off the light or the TV)? (Yes, No); (x) If you were alone in your room, which of these daily actions would you like to be able to perform? (Turn on/off the light, Turn on/off the television, Check if there is anyone in the room around you, Call the nurse, Open/close the shutter, Manage room temperature); (xi) Here you can list other actions you would like to be able to do: (free short answer).

The first three more general questions were asked to identify the type of user. Questions (iv) to (viii) were introduced to investigate his/her knowledge and experience (if any) with Virtual Reality and Eye-tracker, so as to begin to understand how well trained the average ALS patient is and consequently whether training is necessary. Questions (ix) to (xi) were introduced to investigate the user's needs and wishes.

Further questions were asked to those users who already had a communicator: (i) Your communicator is composed by: (All-in-one, Computer and eye-tracker, Tablet and eye-tracker); (ii) If you have an eye-tracker, indicate its brand (free short answer); (iii) How long do you use the communicator per day? (5 min, 15 min, 30 min, 45 min, 1 hour, more than 1 hour); (iv) How do you get on with the communicator you have? (1, 2, 3, 4, 5); (v) If you have a communication system, does it let you to perform everyday actions independently? (Yes, no); (vi) Do you feel that the interface (i.e., the way you can navigate) of your communicator slows down the execution of certain actions? (Yes, No, I don't know).

Furthermore, DTALS prototype was tested with a patient of 36 years old, affected by ALS. This candidate was selected with the help of a remarkable Italian hospital. She was allowed to fully use the *Patient App* and to navigate the virtual world. At the end of the experience, she was given a post-experience questionnaire to evaluate her degree of engagement. She was asked: (i) How much did you enjoy navigating virtual environments using your eyes? (Rate from 1-very little to 5-very much); (ii) How useful did you find using the application to perform actions such as switching on the light? (Rate from 1-very little to 5-very much); (iii) How important to you is the ability to play a video game with your eyes? (Rate from 1-very little to 5-very much); (iv) How intuitive did you find using the application? (Rate from 1-very little to 5-very much); (v) How do you rate the application in general as an aid in performing simple actions? (e.g., switching on lights)? (Rate from 1-very little to 5-very much); (vi) How do you rate the application in general as a game and pastime? (Rate from 1-very little to 5-very much); (vii) Which aspects of the application do you think could be improved? (Movement commands, Main menu, Real world interactions, Virtual world interactions, Navigable models, Graphic, Other...); (viii) Did the application make you tired? (Yes - more than programs I use, Yes - as other programs I use, Not so much, No); (ix) How realistic did you find navigation in the virtual environment? (Rate Apartment, Colosseo, Neighborhood and Cloister from 1-very little to 5-very much).

Specifically, some questions were asked in order to understand the strengths and weaknesses of DTALS to work on: (i) Do you find the possibility of being able to choose the speed of movement useful in using the application? (Yes - I change velocity due to the model, Yes but I would like to go faster, No - I used 0.5, No – I used 1, No – I used 1.5); (ii) How do you rate the speed of rotation (rate from 1-very little to 5-very much); (iii) Are you satisfied with the method of switching on the lights? (Yes, I would like to turn them on with the virtual switch, I would like to turn them on trough a dedicated screen, no); (iv) How do you rate the navigation of the neighborhood with the car? (Difficult, Easy, Funny, I would like to do more things); (v) Do you have recommendations on how to improve the application interface? (Free short answer).

Survey of Physician's Needs – Questionnaire

As mentioned before, a key role for patients is played by the healthcare personnel who treat and monitor them, especially since many of these patients are followed thanks to Telemedicine. These questions were submitted to the medical staff to define the contents of the *Medical App*.

Again, the role played by the specialists who participated in the survey was put into context. Here there are the general questions that were asked: (i) What is your job? (Doctor, Nurse, Psychologist, Physiotherapist, Speech Therapist, Other...); (ii) In what type of structure do you work? (Hospital, RSA, Clinic, Other...); (iii) Do you work: (Alone, In a multidisciplinary team); (iv) In which city do you work? (Short open answer).

Regarding VR, the authors would like to understand how trained the medical field is, so they ware asked: (i) Do you know what Virtual Reality (VR) is? (Yes, no); (ii) Have you ever used VR? (Yes, Yes - immersive, Yes - non-immersive, No); (iii) Do you think that the integration of Virtual Reality in the medical field could be an added value? (Yes, if Yes how would you use it?, No).

In a second module the technology was presented, after that they were asked: (i) Do you think our technology could be useful to you? (Yes, No); (ii) Could your facility accommodate our technology (BIM-VR-IoT algorithm)? (Yes, No, don't know); (iii) Does your facility have a domotic control system? (Yes, No, don't know); (iv) Does your facility already adopt digital building models? (Yes, No, I don't know); (v) Is it possible to install additional devices (e.g., IoT devices) in your facility? (Yes, No, I don't know).

Then the focus was on ALS patients: (i) How many years have you been working with ALS patients?; (ii) How many ALS patients do you follow?; (iii) How many patients (with ALS or similar diseases) are at home?; (iv) How many patients (with ALS or similar diseases) are in health care facilities?; (v) How many of your patients use an optical communicator?; (vi) How much time do you (on average) dedicate to an ALS patient? (5 min, 15 min, 30 min, 45 min, 1 hr, More than 1 hr); (vii) How much time do you spend on an ALS patient does NOT relate to medication or medical treatment? (5 min, 15 min, 30 min, 45 min, 1 hour, More than 1 hour); (viii) Do you think it would be useful to collect data in one platform? (Yes, No, if Yes which data do you think could be prioritised?); (ix) Which data do you think could be collected automatically and could be useful to you for ALS patient care? (Short open answer); (x) Which parameters do you wish could be controlled in the ALS patient's environment? (Short open answer); (xi) Do you think it would be useful for you if the patient could perform daily actions with much more autonomy? (Yes, No); (xii) Based on your experience, in which (simple) actions could the ALS patient be more autonomous? (Short open answer); (xiii) Would it be pleasant for the ALS patient to regain more autonomy in actions of daily living? (Yes, No).

RESULTS

Patient App

To meet patients' needs, a dedicated Graphic User Interface (GUI) was therefore built: the *Patient App*. The author solution allows the creation of a scalable system that controls indoor automation through IoT devices by storing information in a BIM model. The physical reality surrounding patients suffering from motor neuron degradation is represented in the BIM 3D model and it can be explored by using an Eye-Tracker. The interaction with the environment is permitted through IoT devices that convert actions into the real-world. In this way, the DTALS is born. By providing the ability to more autonomously control the space around them, patients can regain more control over even simple actions that they were unfortunately no longer able to perform. In addition, they break routines to which they had been adapted to.

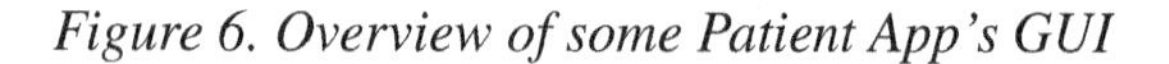
Figure 6. Overview of some Patient App's GUI

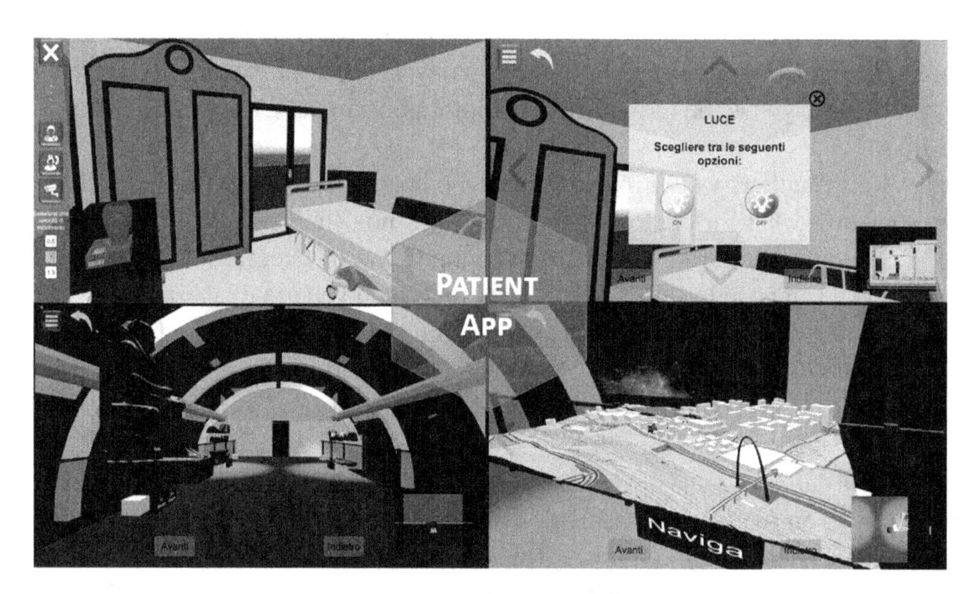

Data were collected from a sample of 20 ALS patients. Questionnaires showed that 35% of patients were male, the 15% of them were aged between 30 – 40 years old and 40 – 50 years old, 20% were older than 70 years old and the most affected age groups were between 50 – 60 years and 60 – 70 years old with 25% respectively. The most frequent region of origin was Piedmont.

The 80% of patients had never used Virtual Reality, only 20% had interfaced with it: one person used immersive VR, and another didn't know the type of VR used, while two people indicated non-immersive VR. Despite this low number, 60% of interviewed said they knew what an eye-tracker was but 70% had never used one.

Only 25% surveyed own an eye-tracker, 80% of which were all-in-one devices. The most popular brands of eye-trackers.

Unanimously, ALS patient reported using the communicator for more than 1 hour per day, expressing the fact that a communication system in 80% of cases enables them to carry out daily life actions and it is the only way they can have to communicate with others. Despite talking about more than most, 100% of patients still express a desire for more independence in daily actions. With this in mind, examples of activities were given, and candidates expressed their interest as shown in Figure 7.

Figure 8 reports which kind of actions patients would like to do, excluding the one previously suggested by the authors.

Since people with ALS can't interact with others, excluding doctors and caregivers, with the technology here presented, the goal was to investigate the emotional state of their experience. Whether interaction with the outside world and with other people could be a source of relief for them. The 100% believe that having the opportunity to interact with the world outside the room could provide them with a sense of well-being and the 89.5% is interested in visiting places with VR. A special interest emerged for natural environments (sea, mountains) but also places of cultural interaction such as cinemas, museums, cities and monumental landmarks.

Figure 7. Daily life actions suggested in the questionnaire

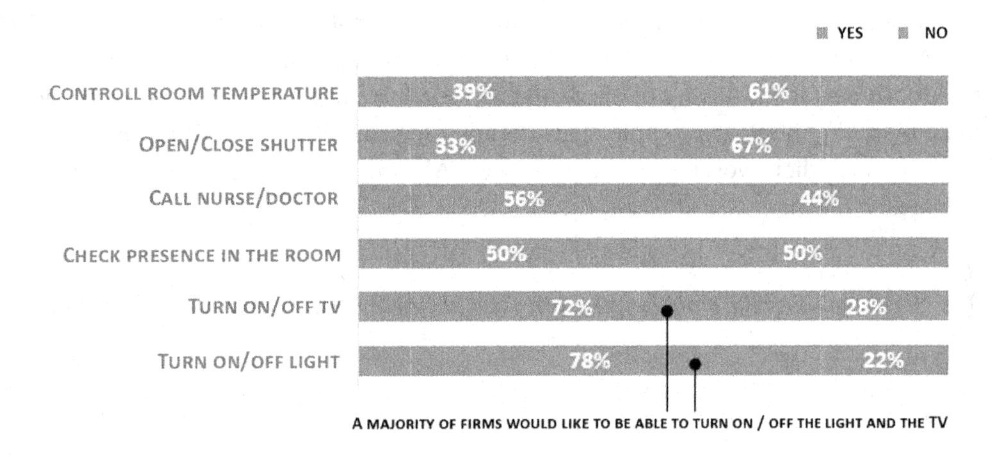

Figure 8. Summary of actions that ALS patients would like to perform

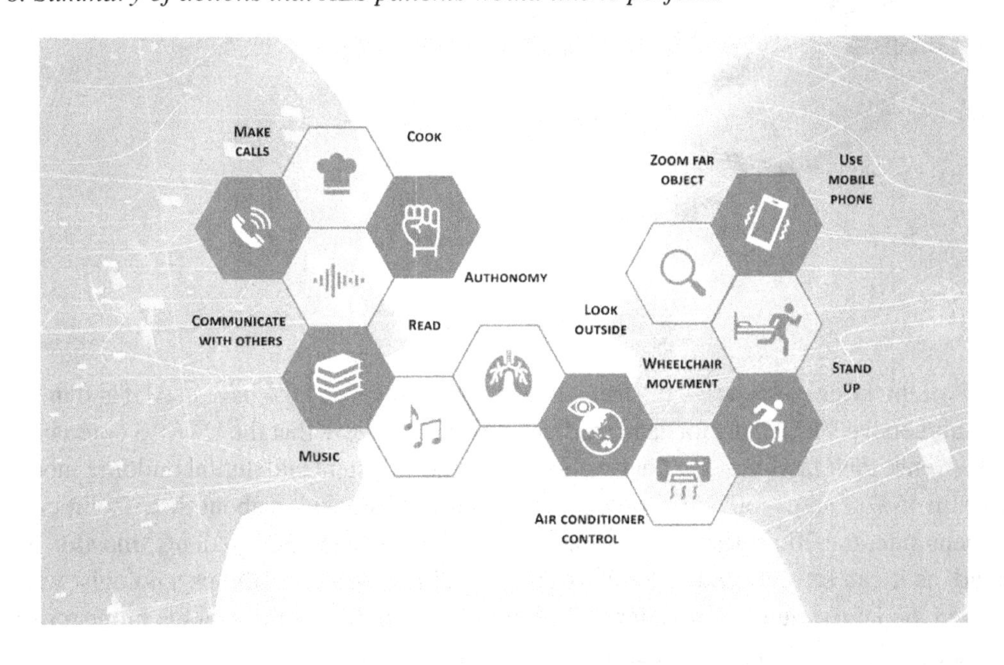

Medical App

The Medical App has been investigated in a form of mock-up to envision the Hospital Information System (HIS). The author solution allows to improve patients and environments monitoring, becoming more independent, to control home automation devices and improving patients' quality of life. Doctors can monitor patients by checking their health parameter over the time, administrating pharmaceutical therapies, managing diagnoses and updating follow-ups.

The healthcare personnel involved in the survey carry out their duties in hospital, only 5% work in a rehabilitation centre and on average they are part of a multidisciplinary team. Given that many people interact with ALS patients, it is clear that research to improve communication and autonomy of patients implies an improvement in the quality of life of all people involved, which also benefits the facilities that accommodate them. Figure 9 shows a pie chart defining the role of healthcare professionals, together with the map in which their working-area can be seen. Most of them are concentrated in Northern of Italy and are mostly doctors.

Figure 9. Pie chart depicting the main health professionals involved in the survey and their areas of work

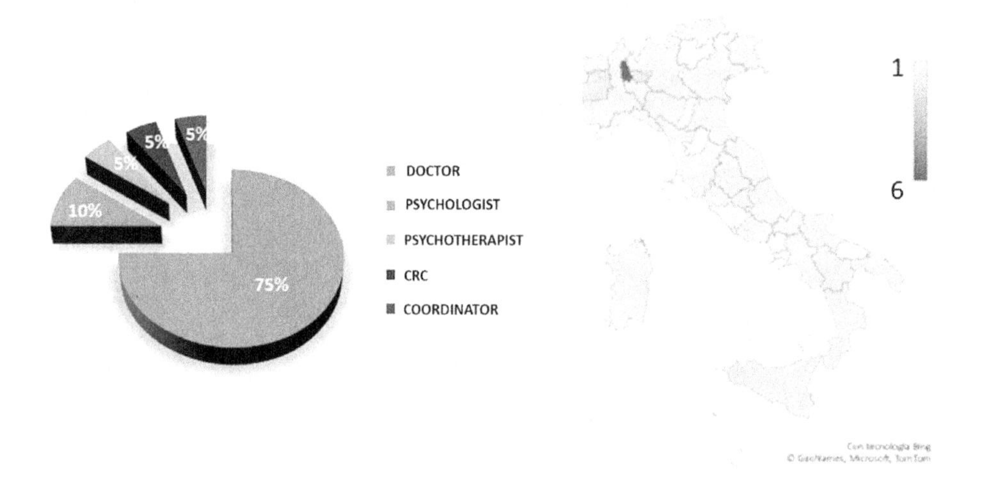

The 90% of interviewed knew what Virtual Reality was, however, 60% have used it and in 25% cases it was non-immersive VR. All doctors and medical personnel believe that the DTALS could be useful to their work, even if their facilities don't have a domotic control system and digital building models aren't adopted yet in 45% of case – 30% of people didn't know the information about their facilities.

On patient side, the 20 professionals involved follow more than 2500 patients, most of whom are homebound, as it can be seen from Figure 10. Among all professionals, the psychologist was the one who followed the largest number of patients. The orange line indicates the average number of patients, hospitalised and non-hospitalised, that each professional manages.

It was found that doctors spend about 45 minutes per ALS patient in 40% of cases, only 20% require care that may last for more than 1 hour. Not all the time that is devoted to each patient is for the drugs administration or medical treatments, in fact 45% of professionals dedicate 30 minutes to other treatments and patient care. The 65% of them were interested in centralizing the information of each patient, with a view to simplifying their treatment and optimizing time. Asking what type of data, they would like to control and/or collect automatically on a single platform, the main themes were identified and summarized in Figure 11.

Figure 10. Combined graph of domiciled and hospitalized patients that each healthcare figure cares for

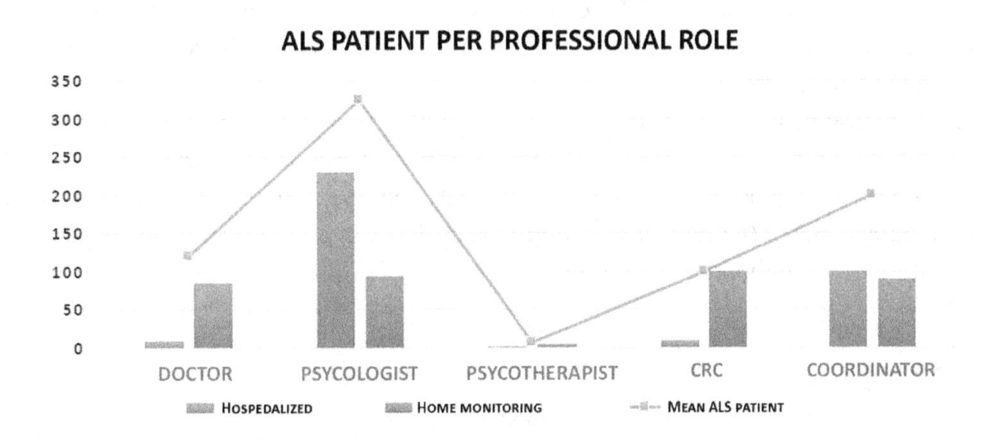

Figure 11. Summary of data health professionals consider useful for ALS patient care

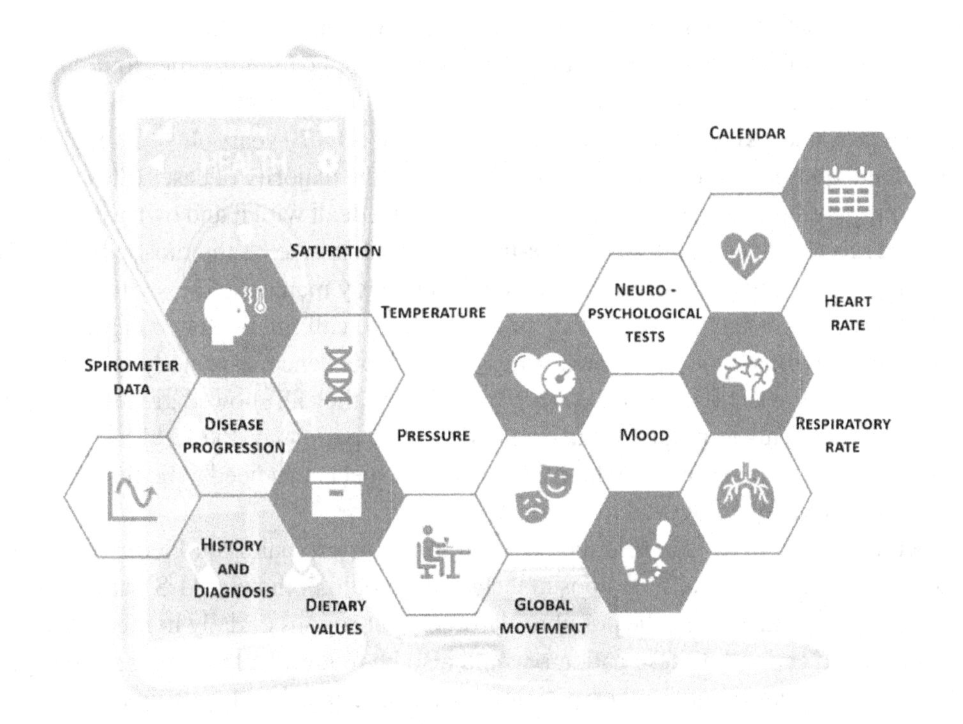

Patient Post-Experience – Questionnaire

The DTALS technology test candidate enjoyed navigating through virtual environments using her eyes, in fact she gave 5 points to the intuitiveness of the *Patient App*. In addition, she liked being able to do certain actions (such as switching on/off the light), affecting the physical reality around her. It was not relevant for her to use a visual communicator to play a video game, but overall, she rated the use of the App with 5 points.

One of the most important aspects of this experience was to ensure that user's eyes wouldn't be excessively tired. She didn't experience a sense of fatigue after using it, but she recommended to introduce the stand-by mode to rest eyes.

The virtual environments she visited were all rated with 4 pts, while the representation of her flat reached 5 pts outlining excellent feedback on the 3D representation realism. As far as the use of the controls was concerned, the adjustable speed was the best solution in her opinion to navigate different models; the rotation speed of the controls was more than sufficient, and she found the machine introduction for navigating the neighbourhood pleasant.

CONCLUSIONS

The development of the DTALS prototype was an opportunity to define the interactions between the construction sector, healthcare, and defining new frontiers for companies producing medical support tools for patients with disabilities. Thanks to several technologies on the market, the innovation proposed by the ongoing research activity highlighted how transversal and multidisciplinary skills can be an added value for technological innovation that must have a practical application in order to reduce the deficits that this type of illness causes to patients.

Above all respondents, ALS affects mainly males of about 50 – 70 years old, coming from the North of Italy. The patient questionnaire therefore concluded that in the majority of cases (80%) Virtual Reality is a known technology, however, only a minority have actually dealt with it and own a communicator. Of all the brands suggested, Tobii and Eyetech eye-tracker were found to be the most popular. The preference of those interviewed was for actions such as independently managing the switching on and off of objects such as the light or television, as well as being able to call the doctor or caregiver. In general, all feedback addressed the possibility of acquiring more independence, especially motor independence - as one might expect. Presenting the *Patient App* to the users, they all showed great interest in using it, providing an added impetus to progress with the research. In particular, the field test carried out with the patient scored 4.25/5 overall and also highlighted some aspects that need to be further explored with a view to perfecting the technology.

Of the doctors interviewed, 75% held the occupation of doctor, mainly in Piedmont and Lombardy regions. From their point of view, the proposed technology could both help ALS patients as they agree with the authors that by introducing the possibility to carry out actions of daily living independently, the time caregivers spend caring for these patients would also be reduced. Furthermore, the psychological aspect should not be underestimated either: patients often find themselves lonely and committing time to manage themselves more consciously and independently would help them regain a sense of well-being.

Technological innovation is reducing the distance between patients and doctors, improving the quality of life of the former and improving the efficiency of the latter. With the introduction of methods and tools related to Digital Health, new roles are being created involving the younger generations in the medical field. Interdisciplinary knowledge must be increasingly sought after to offer solutions in which the synergy between different skills is the real added value for the achievement of a Smart Society. Heterogeneous, static, and dynamic data domains may contribute to define a new concept of Digital Reality in which built environment and person constitute are two complementary systems that are continuously improved through new frontiers for data management such as Artificial Intelligence (AI) and Machine Learning (ML). Certainly, the use of DTs in medical care is still in its infancy. It is desirable that the development of these first DTALS applications will improve patients' lifestyles, defining new paradigms linked to the concept of remote living (e.g. metaverse) to improve not only the healthcare aspect but also the social and personal life aspect. Improving people's quality of life through the adoption of ICTs is one of the principles of Society 5.0. Through multidisciplinary, in the near future it will be possible to achieve results that will increasingly place man in the foreground, who will acquire greater autonomy, without being bound by his physical being, having free access to cyberspace.

FUTURE RESEARCH DIRECTIONS

In the future, a digital assistant, GIO, will be introduced to assist the user in interacting with the interface. Secondly, further tests will be carried out by involving other patients in the testing of the interface, integrating post-experience feedback to improve the user interface (doctor and patient). With a view to the future, the possibility of creating a community, within which all owners of the Patient App will be able to interact with each other, will also be set up.

ACKNOWLEDGMENT

The project "An Algorithm for Progressive Neurodegenerative Motor Neuron Disease to Automate Hospital Rooms/Home Environments by Combining BIM and IoT through VR" was carried out in the framework of the Proof of Concept (PoC) Instrument initiative implemented by LINKS foundation, with the support of LIFTT, with funds from Compagnia di San Paolo.

All the authors are pleased to thank the involved people who actively joined the project, showing true interest in the proposed technology.

Special thanks are extended to the company Mesa Ideas, which provided the hardware technology to support the research.

REFERENCES

Alonso-Valerdi, L. M., & Mercado-García, V. R. (2017). Enrichment of Human-Computer Interaction in Brain-Computer Interfaces via Virtual Environments. *Computational Intelligence and Neuroscience*, 1-12.

Alotto, F. (2019). A software solution for ALS patients to automate hospital rooms by combining BIM, Virtual Reality and IoT. Dept. of Biomedical Eng., Politecnico di Torino.

Baranowski, T., Buday, R., Thompson, D. I., & Baranowski, J. (2008). Playing for real: Video games and stories for health-related behavior change. *American Journal of Preventive Medicine*, *34*(1), 74–82. doi:10.1016/j.amepre.2007.09.027 PMID:18083454

Beaudouin-Lafon, M. (1993). An overview of human-computer interaction. Biochimie, Société française de biochimie et biologie moléculaire, 75(5), 321-329.

Björnsson, B., Borrebaeck, C., & Elander, N. e. (2020). Digital twins to personalize medicine. *Genome Medicine*, *12*(4), 1–4. PMID:31892363

Bottiglieri, M. A. (2019). *Realtime Interactive Model: AEC gamification from Unity to Forge and reverse flow*. Autodesk University.

Cáceres, E., Carrasco, M., & Ríos, S. (2018). Evaluation of an eye-pointer interaction device for human-computer interaction. *Heliyon*, *4*(3), 1–28. doi:10.1016/j.heliyon.2018.e00574 PMID:29862340

Deterding, S. D. (2011). From game design elements to gamefulness: defining" gamification". In *Proceedings of the 15th international academic MindTrek conference: Envisioning future media environments* (p. 9-15). Tampere, Finland: Association for Computing Machinery. 10.1145/2181037.2181040

Edirippulige, S., & Senanayake, B. (2020). Opportunities and Challenges in Digital Healthcare Innovation. In Professional Practices for Digital Healthcare (p. 97-112). IGI Global.

Eid, M. A., Giakoumidis, N., & El Saddik, A. (2016). A novel eye-gaze-controlled wheelchair system for navigating unknown environments: Case study with a person with ALS. *IEEE Access: Practical Innovations, Open Solutions*, *4*, 558–573. doi:10.1109/ACCESS.2016.2520093

Fatehi, F., Samadbeik, M., & Azar, K. (2020). What is digital health? Review of definitions. In A. Värri (Ed.), *Integrated Citizen Centered Digital Health and Social Care: Citizens as Data Producers and Service co-Creators* (pp. 67–71). IOS Press BV. doi:10.3233/SHTI200696

Fleming, T. M., de Beurs, D., Khazaal, Y., Gaggioli, A., Riva, G., Botella, C., Baños, R. M., Aschieri, F., Bavin, L. M., Kleiboer, A., Merry, S., Lau, H. M., & Riper, H. (2016). Maximizing the Impact of e-Therapy and Serious Gaming: Time for a Paradigm Shift. *Frontiers in Psychiatry*, *7*, 1–7. doi:10.3389/fpsyt.2016.00065 PMID:27148094

Johnson, D., Deterding, S., Kuhn, K.-A., Staneva, A., Stoyanov, S., & Hides, L. (2016). Gamification for health and wellbeing: A systematic review of the literature. *Internet Interventions: the Application of Information Technology in Mental and Behavioural Health*, *6*, 89–106. doi:10.1016/j.invent.2016.10.002 PMID:30135818

Kharrazi, H., Faiola, A., & Defazio, J. (2009). *Health care game design: behavioral modeling of serious gaming design for children with chronic diseases* (Vol. 5613). Human-Computer Interaction. Interacting in Various Application Domains Lecture Notes in Computer Science.

Meskó, B., Drobni, Z., Bényei, É., Gergely, B., & Győrffy, Z. (2017). Digital health is a cultural transformation of traditional healthcare. *mHealth, 3*(38), 1-8.

Milgram, P., Takemura, H., & Utsumi, A. &. (1995). Augmented reality: a class of displays on the reality-virtuality continuum. *Proc. SPIE 2351, Telemanipulator and Telepresence Technologies*, 282-292.

Narvaez Rojas, C., Alomia Peñafiel, G., Loaiza Buitrago, D. F., & Tavera Romero, C. A. (2021). Society 5.0: A Japanese Concept for a Superintelligent Society. *Sustainability*, *13*(12), 6567. doi:10.3390u13126567

Osello, A. (2018). *BIM Virtual and Augmented Reality in the health field between technical and terapeutic use. In RAPPRESENTAZIONE/MATERIALE/IMMATERIALE. 40° Convegno Internazionale dei Docenti della Rappresentazione*. Gangemi Editore.

Wattanasoontorn, V., Boada, I., García, R., & Sbert, M. (2013). Serious games for health. *Entertainment Computing*, *4*(4), 231–247. doi:10.1016/j.entcom.2013.09.002

Wu, T.-Y., Chang, Y.-C., Chen, S.-T., & Chiang, I.-T. (2012). A Preliminary Study on Using Augmented Virtuality to Improve Training for Intercollegiate Archers. In *2012 IEEE Fourth International Conference On Digital Game And Intelligent Toy Enhanced Learning* (p. 212-216). Takamatsu, Japan: IEEE Xplore. 10.1109/DIGITEL.2012.58

Zarei, S., Carr, K., Reiley, L., Diaz, K., Guerra, O., Altamirano, P., Pagani, W., Lodin, D., Orozco, G., & Chinea, A. (2015). A comprehensive review of amyotrophic lateral sclerosis. *Surgical Neurology International*, *6*(1), 1–23. doi:10.4103/2152-7806.169561 PMID:26629397

KEY TERMS AND DEFINITIONS

Building Information Model: It is a data-rich, object oriented, intelligent, and parametric digital representation of the facility, from which views and appropriate data for various users' needs can be extracted and analyzed to generate information that can be used to make decisions and improve the process of delivering the facility.

Building Information Modelling: It is a method based on a building model containing any information about the construction. In addition to 3D object-based models, it contains information about specifications, building elements specifications, economy, and programs.

Digital Twin: It is a virtual replica of reality and contains three main parts: physical products in real space, virtual products in virtual space, and the connections of data and information that ties the virtual and real products together.

E-Health: Is an emerging field in the intersection of medical informatics, public health and business, referring to health services and information delivered or enhanced through the Internet and related technologies. In a broader sense, the term characterizes not only a technical development, but also a state-of-mind, a way of thinking, an attitude, and a commitment for networked, global thinking, to improve health care locally, regionally, and worldwide by using information and communication technology.

Interoperability: Is defined as the ability of two or more systems or components to exchange information and to use the information that has been exchanged.

User Engagement: The quality of user experience characterised by the depth of an actor 's investment when interacting with a digital system.

Compilation of References

3D content in Europeana task force. (2020). https://pro.europeana.eu/project/3d-content-in-europeana

Aamir Hussain, R. W. (2015). Health and emergency-care platform for the elderly and disabled people in the Smart City. *Journal of Systems and Software*110.

Abbott, J. M., Klein, B., & Ciechomski, L. (2008). Best practices in online therapy. *Journal of Technology in Human Services*, *26*(2–4), 360–375. doi:10.1080/15228830802097257

Abdellatif, A. J., McCollum, B., & McMullan, P. (2018, March). Serious games: Quality characteristics evaluation framework and case study. In *2018 IEEE Integrated STEM Education Conference (ISEC)* (pp. 112-119). IEEE. 10.1109/ISECon.2018.8340460

Aboagye, E., Yawson, J., & Appiah, K. (2020). COVID-19 and E-learning: The challenges of students in tertiary institutions. *Social Education Research*, 109-115.

Abraha, I., Rimland, J.M., Trotta F.M., Dell'Aquila, G., Cruz-Jentoft, A., Petrovic, M., Gudmundsson, A., Soiza, R., O'Mahony, D., Guaita, A., & Cherubini, A. (2017). Systematic review of of non-pharmacological interventions to threat behavioural disturbances in older patients with dementia. The SENATOR-On Top series. *BMJ Open*. . doi:10.1136/bmjopen-2016-012759

Abubaker Basheer Abdalwhab Altohami, N. A. (2021). Investigating Approaches of Integrating BIM, IoT, and Facility Management for Renovating Existing Buildings: A Review. MDPI – Sustainability, 13.

Afonso, A. M., Cadwell, J. B., Staffa, S. J., Zurakowski, D., & Vinson, A. E. (2021). Burnout Rate and Risk Factors among Anesthesiologists in the United States. *Anesthesiology*, *134*(5), 683–696. doi:10.1097/ALN.0000000000003722 PMID:33667293

Aghimien, D., Aigbavboa, C., Oke, A., & Koloko, N. (2018). *Digitalisation in construction industry: Construction professionals perspective*. Academic Press.

Agostino, D., Arnaboldi, M., & Lampis, A. (2020). Italian state museums during the COVID-19 crisis: From onsite closure to online openness. *Museum Management and Curatorship*, *35*(4), 362–372. doi:10.1080/09647775.2020.1790029

Aguiar Costa, A., & Grilo, A. (2015). BIM-Based E-Procurement: An Innovative Approach to Construction E-Procurement. *TheScientificWorldJournal*, *2015*, 1–15. doi:10.1155/2015/905390 PMID:26090518

Aguirre Montero, A., & López-Sánchez, J. (2021). Intersection of Data Science and Smart Destinations: A Systematic Review. *Frontiers in Psychology*, *12*, 12. doi:10.3389/fpsyg.2021.712610 PMID:34393952

Agus, M., Marton, F., Bettio, F., Hadwiger, M., & Gobbetti, E. (2018). Data-Driven Analysis of Virtual 3D Exploration of a Large Sculpture Collection in Real-World Museum Exhibitions. *J. Comput. Cult. Herit.*, *11*, 1–20.

Aheleroff, S., Xu, X., Zhong, R., & Lu, Y. (2021). Digital Twin as a Service (DTaas) in Industry 4.0: An Architecture Reference Model. *Advanced Engineering Informatics*.

Ahmad Yusri, M. H., Johan, M. A., Khusaini, N. S., & Ramli, M. H. M. (2022). Preservation of cultural heritage: A comparison study of 3D modelling between laser scanning, depth image, and photogrammetry methods. *Jixie Gongcheng Xuebao, 19*(2), 125–146.

Ahmad, M. (2020). Categorizing Game Design Elements into Educational Game Design Fundamentals. In I. Deliyannis (Ed.), *Game Design and Intelligent Interaction*. IntechOpen. doi:10.5772/intechopen.89971

Ahmed, S. (2019). A review on using opportunities of augmented reality and virtual reality in construction. *Organization Technology and Management in Construction an International Journal, 11*(1), 1839–1852.

Ajroldi, C. (Ed.). (1987). *Dieci progetti come occasione di studio: architetture di Carlo Aymonino, Guido Canella, Paul Chemetov, Costantino Dardi, Vittorio De Feo, Vittorio Gregotti, Gianugo Polesello, Alberto Samonà, Luciano Semerani, Alvaro Siza Vieira*. Officina.

Akçayır, M., & Akçayır, G. (2017). Advantages and challenges associated with augmented reality for education: A systematic review of the literature. *Educational Research Review, 20*, 1–11.

Akenine-Möller, T., Haines, E., & Hoffman, N. (2018). Real-Time Rendering (4th ed.). Taylor & Francis. doi:10.1201/b22086

Akour, I., Al-Maroof, R., Alfaisal, R., & Salloum, S. (2022). A conceptual framework for determining metaverse adoption in higher institutions of gulf area: An empirical study using hybrid SEM-ANN approach. *Computers and Education: Artificial Intelligence, 3*.

Al Fatta, H., Maksom, Z., & Zakaria, M. H. (2018). *Game-based learning and gamification: Searching for definitions. International Journal of Simulation: Systems, Science and Technology*. doi:10.5013/IJSSST.a.19.06.41

Alchalabi, A. E., Shirmohammadi, S., Eddin, A. N., & Elsharnouby, M. (2018). FOCUS: Detecting ADHD Patients by an EEG-Based Serious Game. *IEEE Transactions on Instrumentation and Measurement, 67*(7), 1512–1520. doi:10.1109/TIM.2018.2838158

Alexander, C. (1967). *Notes on the synthesis of form*. Harvard University Press.

Alexander, T., Westhoven, M., & Conradi, J. (2017). Virtual Environments for Competency-Oriented. *Education + Training, 498*, 23–29. doi:10.1007/978-3-319-42070-7_3

Alonso-Valerdi, L. M., & Mercado-García, V. R. (2017). Enrichment of Human-Computer Interaction in Brain-Computer Interfaces via Virtual Environments. *Computational Intelligence and Neuroscience,* 1-12.

Alotto, A. A. (2019). *A software solution for ALS patients to automate hospital rooms by combining BIM, Virtual Reality and IoT*. Dept. of Biomedical Eng., Politecnico di Torino.

Alotto, F. (2019). A software solution for ALS patients to automate hospital rooms by combining BIM, Virtual Reality and IoT. Dept. of Biomedical Eng., Politecnico di Torino.

Alsawaier, R. S. (2018). The effect of gamification on motivation and engagement. *The International Journal of Information and Learning Technology, 35*(1), 56–79. doi:10.1108/IJILT-02-2017-0009

Alsehri, A. (2021). The Effectiveness of a Micro-Learning Strategy in Developing the Skills of Using Augmented Reality Applications among Science Teachers in Jeddah. *International Journal of Educational Research Review, 6*(2), 176–183. doi:10.24331/ijere.869642

Aluede, O., Imhonde, H., & Eguavoen, A. (2006). Academic, career and personal needs of Nigerian university students. *Journal of Instructional Psychology*, *33*(1), 50–57.

Alves, A., Schmidt, A., Carthcat, K., & Hostins, R. (2015). Exploring technological innovation towards inclusive education: Building digital games - an interdisciplinary challenge. *Procedia: Social and Behavioral Sciences*, *174*, 3081–3086. doi:10.1016/j.sbspro.2015.01.1043

Alzheimer's Disease International (ADI). (2018). World Alzheimer Report 2018. *The state of the art of dementia research: New frontiers*. https://www.alz.co.uk/research/WorldAlzheim erReport2018.pdf

Amann, S., & Heinsius, J. (2021). Research for CULT Committee – Cultural and creative sectors in post-Covid-19 Europe: crisis effects and policy recommendations. European Parliament, Policy Department for Structural and Cohesion Policies.

American Psychiatric Association. (2000). Diagnostic and statistical manual of mental disorders, DSM-5 (4th ed.). doi:10.1176/appi.books.9780890423349

American Psychiatric Association. (2013). Diagnostic and statistical manual of mental disorders, DSM-5 (5th ed.). doi:10.1176/appi.books.9780890425596

Amin, D., & Govilkar, S. (2015). Comparative Study of Augmented Reality SDK's. *International Journal on Computational Science & Applications*, *5*(1), 11–26.

Amistadi, L. (Ed.). (2005). *Saper credere in architettura: quaranta domande a Luciano Semerani*. Clean.

Ammar, M., Russello, G., & Crispo, B. (2018). Internet of Things: A survey on the security of IoT frameworks. *Journal of Information Security and Applications*, *38*, 8–27. doi:10.1016/j.jisa.2017.11.002

Anastasovitis, E., Ververidis, D., Nikolopoulos, S., & Kompatsiaris, I. (2017). Digiart: Building new 3D cultural heritage worlds. In *2017 3DTV Conference: The True Vision-Capture, Transmission and Display of 3D Video (3DTV-CON)* (pp. 1-4). IEEE.

Anderson, E. F., McLoughlin, L., Liarokapis, F., Peters, C., Petridis, C., & de Freitas, S. (2009). Serious games in cultural heritage. *The 10th International Symposium on Virtual Reality, Archaeology and Cultural Heritage VAST - State of the Art Reports*.

Andrianaivo, L. N., D'Autilia, R. & Palma, V. (2019). Architecture recognition by means of convolutional neural networks. *ISPRS Arch. XLII-2-W15*, 77-84

Angel Ruiz-Zafra, K. B. (2022). IFC+: Towards the integration of IoT into early stages of building design. *Automation in Construction*, 136.

Anilakumari, M. C. (2012). *Developing a multimedia remedial tracking package for Dysgraphia among primary school students with Specific learning disabilities*. School of Pedagogical Sciences, Mahatma Gandhi University. http://hdl.handle.net/10603/25931

Anitec-Assinform. (2021). *Il Digitale in Italia 2021, Mercati, Dinamiche* (Vol. 1). Policy.

Antonelli, C., Cassioli, D., Franchi, F., Graziosi, F., Marotta, A., Pratesi, M., . . . Santucci, F. (2018, July). The city of l'aquila as a living lab: the incipict project and the 5g trial. In *2018 IEEE 5G World Forum (5GWF)* (pp. 410-415). IEEE.

Apolito, P. (2014). *Ritmi di festa. Corpo, danza, socialità*. Il Mulino.

Apollonio, F. I., Fantini, F., Garagnani, S., & Gaiani, M. (2021). A Photogrammetry-Based Workflow for the Accurate 3D Construction and Visualization of Museums Assets. *Remote Sensing*, *13*(486), 1–40.

Apollonio, F. I., Foschi, R., Gaiani, M., & Garagnani, S. (2021). How to Analyze, Preserve, and Communicate Leonardo's Drawing? A Solution to Visualize in RTR Fine Art Graphics Established from "the Best Sense". *J. Comput. Cult. Herit.*, *14*(3), 1–30.

Apollonio, F. I., & Gaiani, M. (2020). Digitalizzare l'Uomo vitruviano. In P. Salvi, A. Mariani, & V. Rosa (Eds.), *Leonardo da Vinci e l'Accademia di Brera* (pp. 55–62). Silvana Editoriale.

Apollonio, F., Gaiani, M., & Bertacchi, S. (2019). Managing Cultural Heritage with Integrated Services Platform. *ISPRS-International Archives of the Photogrammetry, Remote Sensing and Spatial Information Sciences*, *4211*, 91–98.

Appio, F., Frattini, F., Petruzzelli, A., & Neirotti, P. (2021). Digital Transformation and Innovation Management: A Synthesis of Existing Research and an Agenda for Future Studies. *Journal of Product Innovation Management*, *38*(1), 4–20. doi:10.1111/jpim.12562

Applied V. R. Inc. (2021). *RelieVRx*. https://www.relievrx.com/

Apps, M. A. J., & Tsakiris, M. (2014). The free-energy self: A predictive coding account of self-recognition. *Neuroscience and Biobehavioral Reviews*, *41*, 85–97. doi:10.1016/j.neubiorev.2013.01.029 PMID:23416066

Arayici, Y., Fernando, T., Munoz, V., & Bassanino, M. (2018). Interoperability specification development for integrated BIM use in performance based design. *Automation in Construction*, *85*, 167–181. doi:10.1016/j.autcon.2017.10.018

ArCO Ontology. Available online: http://wit.istc.cnr.it/arco

Ardila, A. (2014). *Aphasia Handbook*. Department of Communication Sciences and Disorders, Florida International University.

Arenghi, A., Garofolo, I., & Sormoen, O. (Eds.). (2016). Accessibility As A Key Enabling Knowledge for Enhancement of Cultural Heritage. Franco Angeli.

Ariffin, M. M., Halim, F. A. A., & Sugathan, S. K. (2017). Mobile application for Dyscalculia children in Malaysia. *Proceedings of the 6th International Conference on Computing and Informatics-ICOCI*, 467-472. https://repo.uum.edu.my/id/eprint/22891

Ariffin, M. M., Halim, F. A. A., Arshad, N. I., Mehat, M., & Hashim, A. S. (2019). Calculic kids© mobile app: the impact on educational effectiveness of dyscalculia children. *International Journal of Innovative Technology and Exploring Engineering, 8*(8S), 701-705. https://H11200688S1919©BEIESP

Armstrong, C. M., Reger, G. M., Edwards, J., Rizzo, A. A., Courtney, C. G., & Parsons, T. D. (2013). Validity of the Virtual Reality Stroop Task (VRST) in active duty military. *Journal of Clinical and Experimental Neuropsychology*, *35*(2), 113–123. doi:10.1080/13803395.2012.740002 PMID:23157431

Armstrong, D., Djemame, K., & Kavanagh, R. (2017). Towards energy aware cloud computing application construction. *Journal of Cloud Computing*, *6*(1), 14. doi:10.118613677-017-0083-2

Arrigo Palumbo, N. I. (2021). SIMpLE A Mobile Cloud-Based System for Health Monitoring of People with ALS. *MDPI Sensors 2021*.

Asensio, J. A., Criado, J., Padilla, N., & Iribarne, L. (2019). Emulating home automation installations through component-based web technology. *Future Generation Computer Systems*, *93*, 777–791. doi:10.1016/j.future.2017.09.062

Ashworth, J. (2020). *Virtual Reality for Family Education, Exercise and Entertainment*. https://www.parentmap.com/article/virtual-reality-family-education-exercise-and-entertainment

Asnar-Díaz, I., Rodríguez-García, A. M., & Romero-Rodríguez, J. M. (2018). La tecnología móvil de Realidad Virtual en educación: unarevisión del estado de la literatura científica en España [Virtual Reality mobile technology in education: a review of the state of the scientific literature in Spain]. *Revista de Educación Mediática y TIC, 7*(1), 256–274. doi:10.21071/edmetic.v7i1.10139

Atik, G., & Yalçın, I. (2010). Counseling needs of educational sciences students at the Ankara University. *Procedia: Social and Behavioral Sciences, 2*(2), 1520–1526. doi:10.1016/j.sbspro.2010.03.228

Aubin, G., Béliveau, M.-F., & Klinger, E. (2018). An exploration of the ecological validity of the Virtual Action Planning–Supermarket (VAP-S) with people with schizophrenia. *Neuropsychological Rehabilitation, 28*(5), 689–708. doi:10.1080/09602011.2015.1074083 PMID:26317526

Avgerinakis, K. (2018). V4design for enhancing architecture and video game creation. In *2018 IEEE International Symposium on Mixed and Augmented Reality Adjunct* (pp. 305-309). IEEE.

Awais Ahmad, M. B.-A. (2019). Socio-cyber network: The potential of cyber-physical system to define human behaviors using big data analytics. *Future Generation Computer Systems*, 92.

Ayodele, E. O. (2011). Abandonment of construction projects in Nigeria: Causes and effects. *Journal of Emerging Trends in Economics and Management Science, 2*(2), 142–145. doi:10.10520/EJC133887

Bach, F., Çakmak, H.K., Maaß, H., Bekrater-Bodmann, R., Foell, J., Diers, M., Trojan, J., Fuchs, X., & Flor, H. (2012). *Illusory Hand Ownership Induced by an MRI Compatible Immersive Virtual Reality Device*. Academic Press.

Badri, A., Boudreau-Trudel, B., & Souissi, A. S. (2018). Occupational health and safety in the industry 4.0 era: A cause for major concern? *Safety Science, 109*, 403–411. doi:10.1016/j.ssci.2018.06.012

Bagheri, B., Yang, S., Kao, H.-A., & Lee, J. (2015). Cyber-physical Systems Architecture for Self-Aware Machines in Industry 4.0 Environment. *IFAC-PapersOnLine, 48*(3), 1622–1627. doi:10.1016/j.ifacol.2015.06.318

Bagnara, S., & Pozzi, S. (2014). Interaction design e riflessione. In Le ragioni del design. FrancoAngeli.

Baiardi, L., & Ferreira, E. A. A. (2020). The integrated project for the redevelopment of a historic building: an example of BIM and IoT integration to manage the comfort of the building. In C. M. Bolognesi & C. Santagati (Eds.), *Impact of Industry 4.0 on Architecture and Cultural Heritage* (pp. 261–282). IGI Global. doi:10.4018/978-1-7998-1234-0.ch011

Baker, K. D., & Ray, M. (2011). Online counseling: The good, the bad, and the possibilities. *Counselling Psychology Quarterly, 24*(4), 341–346. doi:10.1080/09515070.2011.632875

Balado, J., Frías, E., González-Collazo, S. M., & Díaz-Vilariño, L. (2022). New Trends in Laser Scanning for Cultural Heritage. In D. Bienvenido-Huertas & J. Moyano-Campos (Eds.), *New Technologies in Building and Construction* (pp. 167–186). Springer. doi:10.1007/978-981-19-1894-0_10

Ballarin, M., Balletti, C., & Vernier, P. (2018). Replicas in cultural heritage: 3D printing and the museum experience. *The International Archives of the Photogrammetry, Remote Sensing and Spatial Information Sciences, 2018*(42), 55–62. doi:10.5194/isprs-archives-XLII-2-55-2018

Balletti, C., & Ballarin, M. (2019). An Application of Integrated 3D Technologies for Replicas in Cultural Heritage. *ISPRS International Journal of Geo-Information, 2019*(8), 285. doi:10.3390/ijgi8060285

Balletti, C., Bertellini, B., Gottardi, C., & Guerra, F. (2019). Geomatics techniques for the enhancement and preservation of cultural heritage. *The International Archives of the Photogrammetry, Remote Sensing and Spatial Information Sciences, XLII-2*(W11), 133–140. doi:10.5194/isprs-archives-XLII-2-W11-133-2019

Balletti, C., Bertellini, B., Gottardi, C., & Guerra, F. (2021). Geomatics techniques for the enhancement and preservation of Cultural Heritage. *The International Archives of the Photogrammetry, Remote Sensing and Spatial Information Sciences*, *42*, 133–140.

Balzani, M., & Maietti, F. (2020). Data acquisition protocols and semantic modelling of the historical-architectural heritage: the INCEPTION project. Digital Strategies for Cultural Heritage, 2(1), 83-95.

Banakou, D., Kishore, S., & Slater, M. (2018). Virtually being Einstein results in an improvement in cognitive task performance and a decrease in age bias. *Frontiers in Psychology*, *9*, 917. doi:10.3389/fpsyg.2018.00917 PMID:29942270

Banfi, F., Bolognesi, C., Bonini, J., & Mandelli, A. (2021). The virtual historical reconstruction of the cerchia dei navigli of Milan: From historical archives, 3D survey and HBIM to the virtual visual storytelling. *The International Archives of the Photogrammetry, Remote Sensing and Spatial Information Sciences*, *XLVI-M-1-2021*, 39–46. doi:10.5194/isprs-archives-XLVI-M-1-2021-39-2021

Banfi, F., & Previtali, M. (2021). Human–Computer Interaction Based on Scan-to-BIM Models,Digital Photogrammetry, Visual Programming Language andeXtended Reality (XR). *Applied Sciences (Basel, Switzerland)*, *11*(13), 6109. doi:10.3390/app11136109

Bangor, A., Kortum, P., & Miller, J. (2009). Determining what individual SUS scores mean: Adding an adjective rating scale. *Journal of Usability Studies*, *4*, 114–123.

Barak, A., Hen, L., Boniel-Nissim, M., & Shapira, N. (2008). A comprehensive review and a meta-analysis of the effectiveness of Internet-based psychotherapeutic interventions. *Journal of Technology in Human Services*, *26*(2–4), 109–160. doi:10.1080/15228830802094429

Barak, A., Klein, B., & Proudfoot, J. G. (2009). Defining Internet-supported therapeutic interventions. *Annals of Behavioral Medicine*, *38*(1), 4–17. doi:10.100712160-009-9130-7 PMID:19787305

Baranda, J., Mangues-Bafalluy, J., Zeydan, E., Vettori, L., Martínez, R., Li, X., Garcia-Saavedra, A., Chiasserini, C. F., Casetti, C., Tomakh, K., Kolodiazhnyi, O., & Bernardos, C. J. (2020). On the Integration of AI/ML-based scaling operations in the 5Growth platform. *In Proceedings IEEE Conference on Network Function Virtualization and Software Defined Networks (NFV-SDN),* 105-109.

Baran, E., & AlZoubi, D. (2020). Human-Centered Design as a Frame for Transition to Remote Teaching during the COVID-19 Pandemic. *Journal of Technology and Teacher Education*, *28*(2), 365–372.

Baranowski, T., Buday, R., Thompson, D. I., & Baranowski, J. (2008). Playing for real: Video games and stories for health-related behavior change. *American Journal of Preventive Medicine*, *34*(1), 74–82. doi:10.1016/j.amepre.2007.09.027 PMID:18083454

Barazzetti, L., Brumana, R., della Torre, S., Gusmeroli, G., & Schiantarelli, G. (2018). Point clouds turned into finite elements: The umbrella vault of Castel Masegra. *IOP Conference Series. Materials Science and Engineering*, *364*, 364. doi:10.1088/1757-899X/364/1/012087

Barba, S., Barbarella, M., Di Benedetto, A., Fiani, M., Gujski, L., & Limongiello, M. (2019). Accuracy Assessment of 3D Photogrammetric Models from an Unmanned Aerial Vehicle. *Drones (Basel)*, *2019*(3), 79. doi:10.3390/drones3040079

Barba, S., Di Filippo, A., Cotella, V., Ferreyra, C., & Amalfitano, S. (2021). A SLAM Integrated Approach for Digital Heritage Documentation. *Culture and Computing. Interactive Cultural Heritage and Arts. HCII*, *2021*, 27–39.

Barbero, A., Vergari, R., Ugliotti, F. M., Del Giudice, M., Osello, A., & Manzone, F. (2021). *Automated semantic and syntactic BIM data validation using visual programming language.* doi:10.30682/tema0702m

Bardelli, P. G., Garda, E., Mangosio, M., Mele, C., & Ostorero, C. (2010). *Il teatro Regio di Torino da Carlo Mollino ad oggi*. Dario Flaccovio Editore.

Barker, G. G., & Barker, E. E. (2021). Online therapy: Lessons learned from the COVID-19 health crisis. *British Journal of Guidance & Counselling*, *50*(1), 66–81. doi:10.1080/03069885.2021.1889462

Barnett, J. E. (2005). Online counseling: New entity, new challenges. *The Counseling Psychologist*, *33*(6), 872–880. doi:10.1177/0011000005279961

Baron, R. A., Branscombe, N. R., & Byrne, D. (2009). *Social psychology* (12th ed.). Pearson.

Bartell, F. O., Dereniak, E. L., & Wolfe, W. L. (1980). The theory and measurement of bidirectional reflectance distribution function (BRDF) and bidirectional transmittance distribution function (BTDF). *Proceedings of the Society for Photo-Instrumentation Engineers*, *257*, 154–160. doi:10.1117/12.959611

Bartle, R. A. (2009). From MUDs to MMORPGs: The history of virtual worlds. In J. Hunsinger, L. Klastrup, & M. Allen (Eds.), *International Handbook of Internet Research* (pp. 23–39). Springer Netherlands. doi:10.1007/978-1-4020-9789-8_2

Bassolino, M., Franza, M., Bello Ruiz, J., Pinardi, M., Schmidlin, T., Stephan, M. A., Solcà, M., Serino, A., & Blanke, O. (2018). Non-invasive brain stimulation of motor cortex induces embodiment when integrated with virtual reality feedback. *The European Journal of Neuroscience*, *47*(7), 790–799. doi:10.1111/ejn.13871 PMID:29460981

Bathje, G. J., Kim, E., Rau, E., Bassiouny, M. A., & Kim, T. (2014). Attitudes toward face-to-face and online counseling: Roles of self-concealment, openness to experience, loss of face, stigma, and disclosure expectations among Korean college students. *International Journal for the Advancement of Counseling*, *36*(4), 408–422. doi:10.100710447-014-9215-2

Batty, M. (2013). *The New Science of Cities*. The MIT Press. doi:10.7551/mitpress/9399.001.0001

Batubara, M. H., Mesran, Sihite, A. H., & Saputra, I. (2017). Aplikasi Pembelajaran Teknik Mesin Otomotif Kendaraan Ringan Dengan Metode Computer Assisted Instruction. *Informasi Dan Teknologi Ilmiah*, *12*(2), 266–270.

Baumann, M. A., MacLean, K. E., Hazelton, T. W., & McKay, A. (2010). Emulating human attention-getting practices with wearable haptics. *2010 IEEE Haptics Symposium*, 149–156. 10.1109/HAPTIC.2010.5444662

Beaudouin-Lafon, M. (1993). An overview of human-computer interaction. Biochimie, Société française de biochimie et biologie moléculaire, 75(5), 321-329.

Beech, J., & Chadwick, S. (2013). *The Business of Sport Management*. Pearson.

Belardi, P., & Bori, S. (2018) Sistina Experience. *Paesaggio urbano*, *4*, 41-49.

Ben Falchuk, S. L. (2018). The Social Metaverse - Battle for privacy. *IEEE Technology and Society Magazine*.

Benjamin, W., & Tiedemann, R. (2015). Das Passagen-Werk: Vol. Band 5, Ed. 7, 934, Ed. 7 (7. Auflage). Suhrkamp.

Bentkowska-Kafel, A., Denard, H., & Baker, D. (Eds.). (2012). *Paradata and transparency in virtual heritage*. Ashgate Publishing, Ltd.

Berkeley, G. (1710). A Treatise Concerning the Principles of Human Knowledge. Academic Press.

Bernardello, R., Borin, P., Panarotto, F., Giordano, A., & Valluzzi, M. R. (2020). BIM representation and classification of masonry pathologies using semi-automatic procedure. In J. Kubica, A.Kwiecień & L. Bednarz (Eds.). *Brick and Block Masonry - From Historical to Sustainable Masonry: Proceedings of the 17th International Brick/Block Masonry Conference (17thIB2MaC 2020)*, July 5-8, 2020, Kraków, Poland (1st ed.). CRC Press

Berns, R. S., & Tongbo, C. (2012). Updated Practical Total Appearance Imaging of Paintings. *Proceedings of the IS&T Archiving Conference*, 162-167.

Berthelot, M., Nony, N., Gugi, L., Bishop, A., & De Luca, L. (2015). The Avignon Bridge: A 3D reconstruction project integrating archaeological, historical and geomorphological issues. *The International Archives of the Photogrammetry, Remote Sensing and Spatial Information Sciences*, *40*(W4), 223–227. doi:10.5194/isprsarchives-XL-5-W4-223-2015

Bertin, J. (1970). Le graphique. *Communications (Englewood)*, 15.

Bertin, J. (2010). *Semiology of Graphics: Diagrams, Networks, Maps* (1st ed.). Esri Press.

Bevan, R. (2016). *The Destruction of Memory: Architecture at War*. Reaktion Books Ltd.

Bevilacqua, M. G., Russo, M., Giordano, A., & Spallone, R. (2022, March). 3D Reconstruction, Digital Twinning, and Virtual Reality: Architectural Heritage Applications. In *2022 IEEE Conference on Virtual Reality and 3D User Interfaces Abstracts and Workshops* (pp. 92-96). IEEE.

Bhatt, J., & Verma, H. K. (2015). Design and development of wired building automation systems. *Energy and Building*, *103*, 396–413. doi:10.1016/j.enbuild.2015.02.054

Biagini, C. (2007). BIM strategies in architectural project management. In *GRAPHICA 2007. Desafio da era digital: Ensino e Tecnologia*. UFPR.

Bianchini, C., Inglese C. & Ippolito, A. (2016). Il contributo della Rappresentazione nel Building Information Modeling (BIM) per la gestione del costruito [The contribution of Representation in Building Information Modeling (BIM) for building management]. *DisegnareCon*, *16*(9), 10.1-10.9.

Bianchini, C., Inglese, C., Ippolito, A., Maiorino, D., & Senatore, L. J. (2019). Building Information Moldeling (BIM): Great Misunderstanding or Potential opportunities for the Design disciplines? In Architecture and Design: Breakthroughs in Research and Practice (pp. 365-386). IGI Global.

Bianchini, C. (2014). Survey, modeling, interpretation as multidisciplinary components of a Knowledge System. *SCIRES-ITSClentific RESearch and Information Technology*, *4*(1), 15–24.

Bianchini, C., Inglese, C., & Ippolito, A. (2016). The role of BIM (Building Information Modeling) for representation and managing of built and historic artifacts. *Disegnarecon*, *9*(16), 10–11.

Bihanic, D. (2015). New Challenges for Data Design. Springer.

Bimberg, P., Weissker, T., & Kulik, A. (2020). On the Usage of the Simulator Sickness Questionnaire for Virtual Reality Research. *2020 IEEE Conference on Virtual Reality and 3D User Interfaces Abstracts and Workshops (VRW)*, 464-467. 10.1109/VRW50115.2020.00098

BIMerr Ontology. (n.d.). Available online: https://bimerr.iot.linkeddata.es/

BIMserver. (2022). Retrieved from bimserver.org: The open source BIM

BIMserver. (n.d.). *BIMserver*. Retrieved from BIMserver: https://github.com/opensourceBIM/BIMserver

BIMServer.Center. (n.d.). *BIMServer.Center*. Retrieved from BIMServer.Center: https://bimserver.center/en

Bitelli, G., Balletti, C., Brumana, R., Barazzetti, L., D'Urso, M. G., Rinaudo, F., & Tucci, G. (2017). Metric documentation of Cultural Heritage: Research directions from the Italian GAMHER project. *The International Archives of the Photogrammetry, Remote Sensing and Spatial Information Sciences*, *XLII-2*(W5), 83–90. doi:10.5194/isprs-archives-XLII-2-W5-83-2017

Björnsson, B., Borrebaeck, C., & Elander, N. e. (2020). Digital twins to personalize medicine. *Genome Medicine, 12*(4), 1–4. PMID:31892363

Blaster, B., Ladner, R., & Burgstahler, S. (2016). *Lesson Learned: Engaging Students With Disabilities on a National Scale.* University of Washington.

Blessinger, P., & Wankel, C. (2012). Innovative Approaches in Higher Education: An Introduction to Using Immersive Interfaces. In C. Wankel & P. Blessinger (Eds.), *Increasing Student Engagement and Retention Using Immersive Interfaces: Virtual Worlds, Gaming, and Simulation* (Vol. 6, Part C, pp. 3–14). Emerald Group Publishing Limited. doi:10.1108/S2044-9968(2012)000006C003

Bloom, B. S., Engelhart, M. D., Furst, E. J., Hill, W. H., & Krathwohl, D. R. (1956). *Taxonomy of educational objectives: the classification of educational goals. In Handbook I. Cognitive Domain.* Longmans, Green.

Bocconcino, M. M., & Manzone, F. (2019), Sistemi informativi e strumenti grafici per la manutenzione di manufatti complessi [Information systems and graphic tools for the maintenance of complex buildings]. In Ingegno e costruzione nell'epoca della complessità. Politecnico di Torino.

Bocconcino, M. M., Lo Turco, M., Vozzola, M., & Rabbia, A. (2021). Intelligent Information Systems for the representation and management of the city. Urban survey and design for resilience. PROJECT, 5, 90-107.

Bocconcino, M. M. (2018). La tecnologia BIM per il controllo strutturale - Metodi e strumenti grafici per il monitoraggio strutturale e per la manutenzione di manufatti complessi. In *Controlli strutturali: metodologia e applicazione* (pp. 123–142). Maggioli.

Bocconcino, M. M. (2019). Graphic information and visual communication: tools for simplifying knowledge. In *Riflessioni - L'arte del disegno/Il disegno dell'arte* (pp. 1427–1434). Gangemi.

Bocconcino, M. M. (2022). *Mappe "nd" che mostrano ciò che non si vede, per un'immaginazione del concreto - Sistemi informativi e prospettive future che già sono il presente ["nD" maps that show what cannot be seen, for an imagination of the concrete - Information systems and future perspectives that are already the present]* (Vol. 3). Edifir.

Bohil, C. J., Alicea, B., & Biocca, F. A. (2011). Virtual reality in neuroscience research and therapy. *Nature Reviews. Neuroscience, 12*(12), 752–762. doi:10.1038/nrn3122 PMID:22048061

Boje, C., Guerriero, A., Kubicki, S., & Rezgui, Y. (2020). Towards a semantic Construction Digital Twin: Directions for future research. *Automation in Construction, 114*, 103179. doi:10.1016/j.autcon.2020.103179

Bolognesi, C. M., & Signorini, M. (2021). Digital Twins: combined surveying praxis for modelling. In *ARQUEOLÓGICA 2.0–9th International Congress on Archaeology, Computer Graphics, Cultural Heritage and Innovation. GEORES–3rd GEOmatics and pREServation.* (pp. 275-280). Editorial Universitat Politècnica de València.

Bolognesi, C., & Aiello, D. A. A. (2020). From Digital Survey to a Virtual tale. Virtual reconstruction of the convent on Santa Maria delle Grazie in Milan. In C. M. Bolognesi & C. Santagati (Eds.), *Impact of Industry 4.0 on Architecture and Cultural Heritage* (pp. 49–75). IGI Global. doi:10.4018/978-1-7998-1234-0.ch003

Bonsma, P., Bonsma, I., Ziri, A. E., Iadanza, E., Maietti, F., Medici, M., Ferrari, F., Sebastian, R., Bruinenberg, S., & Lerones, P. M. (2018), Handling huge and complex 3D geometries with Semantic Web technology, Florence Heri-Tech – The Future of Heritage Science and Technologies. *IOP Conf. Series: Materials Science and Engineering, 364.*

Borda, A., & Bowen, J. (2017). Smart Cities and Cultural Heritage: A Review of Developments and Future Opportunities. In *Proceedings of EVA London* (pp.9-18). London, UK: BCS London. 10.14236/ewic/EVA2017.2

Borgnis, F., Baglio, F., Pedroli, E., Rossetto, F., Uccellatore, L., Oliveira, J., Riva, G., & Cipresso, P. (2022). Available Virtual Reality-Based Tools for Executive Functions: A Systematic Review. *Frontiers in Psychology, 13*. doi:10.3389/fpsyg.2022.833136

Borgnis, F., Baglio, F., Pedroli, E., Rossetto, F., Isernia, S., Uccellatore, L., Riva, G., & Cipresso, P. (2021). EXecutive-Functions Innovative Tool (EXIT 360°): A Usability and User Experience Study of an Original 360°-Based Assessment Instrument. *Sensors (Basel), 21*(17), 5867. doi:10.339021175867 PMID:34502758

Borgnis, F., Baglio, F., Pedroli, E., Rossetto, F., Uccellatore, L., Oliveira, J., Riva, G., & Cipresso, P. (2022). Available Virtual Reality-Based Tools for Executive Functions: A Systematic Review. *Frontiers in Psychology*, 13. PMID:35478738

Borin, P., Giordano, A., & Campagnolo, D. (2021). Scan-Vs-Bim Analysis for Historical Buildings. In R. P. Suárez & N. M. Dorta (Eds.), *Redibujando el futuro de la Expresión Gráfica aplicada a la edificación / Redrawing the future of Graphic Expression applied* (pp. 1257–1272). Tirant Humanidades.

Börner, K., Sanyal, S., & Vespignani, A. (2007), Network Science. Annual Review of Information Science & Technology, 41, 537-607. doi:10.1126cience.1167742

Bottiglieri, M. A. (2019). *Realtime Interactive Model: AEC gamification from Unity to Forge and reverse flow*. Autodesk University.

Bottino, R. M., Ferlino, L., Ott, M., & Tavella, M. (2007). Developing strategic and reasoning abilities with computer games at primary school level. *Computers & Education, 49*(4), 1272–1286. doi:10.1016/j.compedu.2006.02.003

Bowen, J. P., & Giannini, T. (2021). Digitality: A reality check. *Proceedings of EVA London, 2021*, 12–19.

Bozzelli, G., Raia, A., Ricciardi, S., De Nino, M., Barile, N., Perrella, M., & Palombini, A. (2019). An integrated VR/AR framework for user-centric interactive experience of cultural heritage: The ArkaeVision project. *Digital Applications in Archaeology and Cultural Heritage, 15*, e00124. doi:10.1016/j.daach.2019.e00124

Brettel, M., Friederichsen, N., Keller, M., & Rosenberg, M. (2014). How Virtualization, Decentralization and Network Building Change the Manufacturing Landscape: An Industry 4.0 Perspective. *International Journal of Information and Communication Engineering, 8*(1), 37–44.

Brivio, E., Serino, S., Negro Cousa, E., Zini, A., Riva, G., & De Leo, G. (2021). Virtual reality and 360° panorama technology: A media comparison to study changes in sense of presence, anxiety, and positive emotions. *Virtual Reality (Waltham Cross), 25*(2), 303–311. doi:10.100710055-020-00453-7

Brooke, J. (1996). SUS - A quick and dirty usability scale. *Usability Eval. Ind., 189*, 4–7.

Brown, M., McCormack, M., Reeves, J., Brooks, C., & Grajek, S. (2020). *2020 EDUCAUSE Horizon report, teaching and learning edition*. EDUCAUSE. https://www.educause.edu/horizon-report-2020

Brumana, R., Oreni, D., Barazzetti, L., Cuca, B., Previtali, M., & Banfi, F. (2020). Survey and Scan to BIM Model for the Knowledge of Built Heritage and the Management of Conservation Activities. In B. Daniotti, M. Gianinetto, & S. Della Torre (Eds.), *Digital Transformation of the Design, Construction and Management Processes of the Built Environment. Research for Development* (pp. 391–400). Springer. doi:10.1007/978-3-030-33570-0_35

Brumana, R., Stanga, C., & Banfi, F. (2022). Models and scales for quality control: Toward the definition of specifications (GOA-LOG) for the generation and re-use of HBIM object libraries in a Common Data Environment. *Applied Geomatics, 14*(S1), 151–179. doi:10.100712518-020-00351-2

Bruno, N., Bianchi, G., Roncella, R., & Zerbi, A. (2015). An open-HGIS project for the city of Parma: database structure and map registration. In M. A. Brovelli, M. Minghini, & M. Negretti (Eds.), Geomatics Workbooks (Vol. 12, pp. 189–203). Academic Press.

Bruno, N., Mikolajewska, S., Roncella, R., & Zerbi, A. (2022). Integrated processing of photogrammetric and laser scanning data for frescoes restoration. *International Archives of Photogrammetry, Remote Sensing and Spatial Information Sciences, XLVI-2/W1-2022*, 105-112.

Bruno, N., Rechichi, F., Achille, C., Zerbi, A., Roncella, R., & Fassi, F. (2020). Integration of historical GIS data in a HBIM system. *Int. Arch. Photogramm. Remote Sens. Spatial Inf. Sci.*, 427–434.

Brusaporci, S. (2017b). *The importance of being honest. 3D Printing: Breakthroughs in Research and Practice*. Hershey, PA: IGI Global.

Brusaporci, S., Graziosi, F., Franchi, F., & Maiezza, P. (2018). Remediating the historical city. Ubiquitous augmented reality for cultural heritage enhancement. In *International and Interdisciplinary Conference on Digital Environments for Education, Arts and Heritage* (pp. 305-313). Springer.

Brusaporci, S. (2020). Toward smart heritage: Cultural challenges in digital built heritage. In *Applying Innovative Technologies in Heritage Science* (pp. 271–296). IGI Global. doi:10.4018/978-1-7998-2871-6.ch013

Brusaporci, S. (Ed.). (2017a). *Digital Innovations in Architectural Heritage Conservation: Emerging Research and Opportunities: Emerging Research and Opportunities*. IGI Global. doi:10.4018/978-1-5225-2434-2

Brusaporci, S., Graziosi, F., Franchi, F., Maiezza, P., & Tata, A. (2021). Mixed reality experiences for the historical storytelling of cultural heritage. In *From Building Information Modelling to Mixed Reality* (pp. 33–46). Springer. doi:10.1007/978-3-030-49278-6_3

Buhalis, D., & Karatay, N. (2022). Mixed Reality (MR) for Generation Z in Cultural Heritage Tourism Towards Metaverse. In J. L. Stienmetz, B. Ferrer-Rosell, & D. Massimo (Eds.), *Information and Communication Technologies in Tourism 2022. ENTER 2022* (pp. 16–27). Springer. doi:10.1007/978-3-030-94751-4_2

Buiatti, E. (2014). *Forma Mentis. Neuroergonomia sensoriale applicata alla progettazione*. FrancoAngeli.

BuildingSMART International. (2022). *About - What is openBIM?* Retrieved from BuildingSmart: https://www.buildingsmart.org/about/openbim/openbim-definition/

BuildingSMARTalliance. (2007). National building information modeling standard (vol. 1). National Institute of Building Sciences.

Buono, M., & Giugliano, G. (2021). Systems and models of intelligent connection and interaction for society 5.0. *Revista Iberoamericana Académico-Científica Universitaria de Humanidades. Arte y Cultura, 9*, 195–208.

Burin, D., Kilteni, K., Rabuffetti, M., Slater, M., & Pia, L. (2019b). Body ownership increases the interference between observed and executed movements. *PLoS One, 14*(1), e0209899. doi:10.1371/journal.pone.0209899 PMID:30605454

Burin, D., Yamaya, N., Ogitsu, R., & Kawashima, R. (2019a). Virtual training leads to real acute physical, cognitive, and neural benefits on healthy adults: Study protocol for a randomized controlled trial. *Trials, 20*(1), 20. doi:10.118613063-019-3591-1 PMID:31511036

Burke, B. (2020). *Gartner Top Strategic Technology Trends for 2022*. Academic Press.

Burke, V., Jørgensen, D., Jørgensen, F. A. (2020). Museums at Home: Digital Initiatives in Response to COVID-19. *Norsk museumstidsskrift, 6*(2), 117-123.

Burkey, B. (2022). From Bricks to Clicks: How Digital Heritage Initiatives Create a New Ecosystem for Cultural Heritage and Collective Remembering. *The Journal of Communication Inquiry*, *46*(2), 185–205. doi:10.1177/01968599211041112

Burley, B. (2012). Physically-based shading at Disney, course notes. In SIGGRAPH '12 Courses. ACM.

Burley, B. (2015). Extending Disney's Physically Based BRDF with Integrated Subsurface Scattering. In SIGGRAPH' 15 Courses. ACM.

Burry, M. (2020). Seeking an Urban Philosophy: Carlo Ratti and the Senseable City. *Architectural Design*, *90*(3), 32–37. doi:10.1002/ad.2565

Butterworth, B., & Kovas, Y. (2013). Understanding neurocognitive developmental disorders can improve education for all. *Science*, *19*(340), 300–305. doi:10.1126cience.1231022 PMID:23599478

Buttigli, M. (1629). *Descrittione dell'apparato fatto, per honorare la prima, e solenne entrata in Parma della serenissima prencipessa, Margherita di Toscana, duchessa di Parma*. Piacenza, & c., Appresso Seth, & Erasmo Viotti.

Buysse, D.J., Reynolds III, C.F., Monk, T.H., Berman, S.R., & Kupfer, D.J. (1989). The Pittsburgh Sleep Quality Index: A new instrument for psychiatric practice and research. *Journal of Psychiatric Research*, *28*(2), 193-213. doi:10.1016/0165-1781(89)90047-4

Byrne, D. (2008). Heritage as social action. In G. Fairclough, R. Harrison, J. H. Jameson, & J. Schofield (Eds.), *The heritage reader* (pp. 149–173). Routledge.

Bystrom, K. E., Barfield, W., & Hendrix, C. (1999). A conceptual model of the sense of presence in virtual environments. *Presence (Cambridge, Mass.)*, *8*(2), 241–244. doi:10.1162/105474699566107

Cáceres, E., Carrasco, M., & Ríos, S. (2018). Evaluation of an eye-pointer interaction device for human-computer interaction. *Heliyon*, *4*(3), 1–28. doi:10.1016/j.heliyon.2018.e00574 PMID:29862340

Calvert, G. A., & Thesen, T. (2004, January). Multisensory integration: Methodological approaches and emerging principles in the human brain (2014). *Journal of Physiology, Paris*, *98*(1-3), 191–205. Advance online publication. doi:10.1016/j.jphysparis.2004.03.018 PMID:15477032

Campbell, Z., Zakzanis, K. K., Jovanovski, D., Joordens, S., Mraz, R., & Graham, S. J. (2009). Utilizing Virtual Reality to Improve the Ecological Validity of Clinical Neuropsychology: An fMRI Case Study Elucidating the Neural Basis of Planning by Comparing the Tower of London with a Three-Dimensional Navigation Task. *Applied Neuropsychology*, *16*(4), 295–306. doi:10.1080/09084280903297891 PMID:20183185

Can, S., & Zeren, Ş. G. (2019). The role of Internet addiction and basic psychological needs in explaining the academic procrastination behavior of adolescents. *Çukurova Üniversitesi Eğitim Fakültesi Dergisi*, *48*(2), 1012–1040. 10.14812/cufej.544325

CANP. (2018). *CANP - la casa nel parco* [CANP – The House in the Park]. http://casanelparco-project.it/

Cantrell, B., & Mekies, A. (2018). *Codify: Parametric and Computational Design in Landscape Architecture*. Routledge. doi:10.4324/9781315647791

Canullo, G. (2016). I Bifolchi e l'eucarestia. La cappella maggiore della chiesa di Santa Maria delle Vergini a Macerata. *Journal of the Section of Cultural Heritage*.

Canzoneri, E., Ferrè, E. R., & Haggard, P. (2014). Combining proprioception and touch to compute spatial information. *Experimental Brain Research*, *232*(4), 1259–1266. doi:10.100700221-014-3842-z PMID:24468725

Caprara, G. V., Barbaranelli, C., Borgogni, L., & Perugini, M. (1993). The "Big Five Questionnaire": A new questionnaire to assess the five factor model. *Personality and Individual Differences*, *15*(3), 281–288. doi:10.1016/0191-8869(93)90218-R

Cardone, V. (2008). Modelli grafici dell'architettura e del territorio [Graphic Models of architecure and the territory]. CUES.

Cardos¸, R. A. I., David, O. A., & David, D. O. (2017). Virtual reality exposure therapy in flight anxiety: A quantitative meta-analysis. *Computers in Human Behavior*, *72*, 371–380. doi:10.1016/j.chb.2017.03.007

Cardozo, T. M., & Papadopoulos, C. (2021). Heritage artefacts in the COVID-19 Era: The aura and authenticity of 3D models. *Open Archaeology*, *7*(1), 519–539. doi:10.1515/opar-2020-0147

Carelli, L., Morganti, F., Weiss, P. L., Kizony, R., & Riva, G. (2008). A virtual reality paradigm for the assessment and rehabilitation of executive function deficits post stroke: Feasibility study. *2008 Virtual Rehabilitation*, 99–104. doi:10.1109/ICVR.2008.4625144

Carter, T., Seah, S. A., Long, B., Drinkwater, B., & Subramanian, S. (2013). UltraHaptics: Multi-point mid-air haptic feedback for touch surfaces. *UIST 2013 - Proceedings of the 26th Annual ACM Symposium on User Interface Software and Technology*, 505–514.

Carvajal Talero, J. R. (2019). *Virtual Reality in Treatments for Alzheimer's patients* [Master's thesis]. Politecnico di Torino.

Casey, J. A. (1995). Developmental issues for school counselors using technology. *Elementary School Guidance & Counseling, 30*(1), 26-34. https://www.jstor.org/stable/42871189

Castellini, C., Mattioli, S., Signorini, C., Cotozzolo, E., Noto, D., Moretti, E., Brecchia, G., Dal Bosco, A., Belmonte, G., Durand, T., De Felice, C., & Collodel, G. (2019). Effect of Dietary n-3 Source on Rabbit Male Reproduction. *Oxidative Medicine and Cellular Longevity*, *16*(70), 531-549.

Castells, M. (1996). *The Rise of the Network Society*. Blackwell.

Castelnuovo, G., Lo Priore, C., Liccione, D., & Cioffi, G. (2003). Virtual Reality based tools for the rehabilitation of cognitive and executive functions: The V-STORE. *PsychNology Journal*, *1*, 310–325.

Catalogo Generale dei Beni Culturali. (n.d.). https://catalogo.beniculturali.it

Cavallari Murat, A. (1968), Forma urbana e architettura nella Torino Barocca. Dalle premesse classiche alle conclusioni neoclassiche, Unione tipografico-editrice torinese.

Çebi, E. (2009). *University students' attitudes toward seeking psychological help: Effects of perceived social support, psychological distress, prior help-seeking experience and gender* [Doctoral Dissertation]. Middle East Technical University. https://open.metu.edu.tr/handle/11511/18743

Cebolla, A. M., Palmero-Soler, E., Leroy, A., & Cheron, G. (2017). EEG Spectral Generators Involved in Motor Imagery: A swLORETA Study. *Frontiers in Psychology*, *8*, 2133. doi:10.3389/fpsyg.2017.02133 PMID:29312028

Cecchini, C., Cundari, M. R., Palma, V., & Panarotto, F. (2019). Data, Models and Visualization: Connected Tools to Enhance the Fruition of the Architectural Heritage in the City of Padova. In C. L. Marcos (Ed.) Graphic Imprints. Springer, 633-646.

Celaschi, F. (2017). Advanced design-driven approaches for an Industry 4.0 framework: The human-centred dimension of the digital industrial revolution. *Strategic Design Research Journal*, *10*(2), 97–104.

Censis. (2021). *Il valore economico e sociale dei videogiochi in Italia* [The Economic and Social Value of video games in Italy]. Primo Rapporto IIDEA-CENSIS.

Centofanti, M. (2010). Della natura del modello architettonico. In S. Brusaporci (Ed.), *Sistemi informativi integrati per la tutela, la conservazione e la valorizzazione del patrimonio architettonico e urbano* [Integrated information systems for the protection, conservation and enhancement of the architectural and urban heritage] (pp. 43–54). Gangemi.

Ch'ng, E., Li, Y., Cai, S., & Leow, F. T. (2020). The effects of VR environments on the acceptance, experience, and expectations of cultural heritage learning. *Journal on Comput Cult Herit, 13*(1), 1–20. doi:10.1145/3352933

Champion, E., & Rahaman, H. (2019). 3D digital heritage models as sustainable scholarly resources. *Sustainability, 11*(8), 2425. doi:10.3390u11082425

Champion, E., & Rahaman, H. (2020). Survey of 3D digital heritage repositories and platforms. *Virtual Archaeology Review, 11*(23), 1–15. doi:10.4995/var.2020.13226

Checa, D., & Bustillo, A. (2019). A review of immersive virtual reality serious games to enhance learning and training. *Multimedia Tools and Applications, 79*, 1–27.

Chen, Q., García de Soto, B., & Adey, B. T. (2018). Construction automation: Research areas, industry concerns and suggestions for advancement. *Automation in Construction, 94*, 22–38. doi:10.1016/j.autcon.2018.05.028

Chester, A., & Glass, C. A. (2006). Online counselling: A descriptive analysis of therapy services on the Internet. *British Journal of Guidance & Counselling, 34*(2), 145–160. doi:10.1080/03069880600583170

Chiabrando, F., Sammartano, G., Spanò, A., & Spreafico, A. (2019). Hybrid 3D Models: When Geomatics Innovations Meet Extensive Built Heritage Complexes. *ISPRS International Journal of Geo-Information, 8*(3), 124. doi:10.3390/ijgi8030124

Chiapparini, A. (2012). *Communication and Cultural Heritage: Communication as Effective Tool for Heritage Conservation and Enhancement* (Unpublished doctoral dissertation). Politecnico di Milano University.

Childs, J., & Taylor, Z. (2022). The Internet and K-12 Education: Capturing Digital Metrics During the COVID-19 Era. Technology, Knowledge and Learning.

Chirico, A., Cipresso, P., Yaden, D. B., Biassoni, F., Riva, G., & Gaggioli, A. (2017). Effectiveness of immersive videos in inducing awe: An experimental study. *Scientific Reports, 7*(1), 1–11. doi:10.103841598-017-01242-0 PMID:28450730

Chirico, A., & Gaggioli, A. (2019). When virtual feels real: Comparing emotional responses and presence in virtual and natural environments. *Cyberpsychology, Behavior, and Social Networking, 22*(3), 220–226. doi:10.1089/cyber.2018.0393 PMID:30730222

Chong, H. T., Lim, C. K., Ahmed, M. F., Tan, K. L., & Mokhtar, M. B. (2021). Virtual reality usability and accessibility for cultural heritage practices: Challenges mapping and recommendations. *Electronics (Basel), 10*(12), 1430. doi:10.3390/electronics10121430

Choromański, K., Łobodecki, J., Puchała, K., & Ostrowski, W. (2019). Development of virtual reality application for cultural heritage visualization from multi-source 3D data. *The International Archives of the Photogrammetry, Remote Sensing and Spatial Information Sciences, XLII–2*(W9), 261–267. doi:10.5194/isprs-archives-XLII-2-W9-261-2019

Christou, S. (2010). Virtual Reality in Education. In *Affective, Interactive and Cognitive Methods for E-Learning Design: Creating an Optimal Education Experience*. IGI Global. doi:10.4018/978-1-60566-940-3.ch012

Chui, K., Lytras, M., & Visvizi, A. (2018). Energy Sustainability in Smart Cities: Artificial Intelligence, Smart Monitoring, andOptimization of Energy Consumption. *Energies, 11*(11), 2869. doi:10.3390/en11112869

Chung, N., Han, H., & Joun, Y. (2015). Tourists' intention to visit a destination: The role of augmented reality (AR) application for a heritage site. *Computers in Human Behavior, 50*, 588–599. doi:10.1016/j.chb.2015.02.068

CIDOC-CRM Ontology. (n.d.). Available online: http://www.cidoc-crm.org/

Cipresso, P., & Riva, G. (2015). Ecological Settings by Means of Virtual Reality. The Wiley handbook of personality assessment, 240.

Cipresso, P., Giglioli, I. A. C., Raya, A., & Riva, G. (2018). The past, present, and future of virtual and augmented reality research: A network and cluster analysis of the literature. *Frontiers in Psychology, 9*, 1–20. doi:10.3389/fpsyg.2018.02086 PMID:30459681

Ciribini, A.L. (2013). *L' information modeling e il settore delle costruzioni* [Information modeling and the construction sector]. IIM and BIM. Santarcangelo di Romagna: Maggioli Publisher.

Ciurea, C., Pocatilu, L., & Gheorghe Filip, F. (2020). Using Modern Information and Communication Technologies to Support the Access to Cultural Values. *Journal of System and Management Sciences, 10*(2), 1–20.

Clark, D. (2006). *Games and E-Learning*. Caspian Learning Ltd.

Climent, G., Banterla, F., & Iriarte, Y. (2010). Virtual reality, technologies and behavioural assessment. *AULA Ecol. Evaluation Atten. Processes, 2010*, 19–28.

Cobb, S. V. G. (2007). Virtual environments supporting learning and communication in special needs education. *Topics in Language Disorders, 27*(3), 211–225. doi:10.1097/01.TLD.0000285356.95426.3b

Cohen Mansfield, J. (1986). Agitated behaviors in the elderly. II. Preliminary results in the cognitively deteriorated. *Journal of the American Geriatrics Society, 34*(10), 722-7) doi:10.1111/j.1532-5415.1986.tb04303.x

Colchester, K., Hagras, H., Alghazzawi, D., & Aldabbagh, G. (2017). A survey of artificial intelligence techniques employed for adaptive educational systems within E-learning platforms. *Journal of Artificial Intelligence and Soft Computing Research, 7*(1), 47–64. doi:10.1515/jaiscr-2017-0004

Collier, L., & Jakob, A. (2017) The Multisensory Environment (MSE) in Dementia Care: Examining Its Role and Quality From a User Perspective. *HERD, 10*(5), 9-51. doi:10.1177/1937586716683508

Collodel, G., Masini, M., Signorini, C., Moretti, E., Castellini, C., Noto, D., & Innocenti, A. (2019). Antioxidants, Dietary Fatty Acids, and Sperm: A Virtual Reality Applied Game for Scientific Dissemination. Oxidative Medicine and Cellular Longevity.

Colombo, D., Díaz-García, A., Fernandez-Álvarez, J., & Botella, C. (2021). Virtual reality for the enhancement of emotion regulation. *Clinical Psychology & Psychotherapy, 28*(3), 519–537. doi:10.1002/cpp.2618 PMID:34048621

Colon, Y., & Stren, S. (2011). Counseling groups online: Theory and framework. In R. Kraus, G. Stricker, & C. Speyer (Eds.), *Online counseling: A handbook for mental health professionals* (2nd ed., pp. 183–202). Academic Press. doi:10.1016/B978-0-12-378596-1.00010-1

Coluccelli, G., Loffredo, V., Monti, L., Spada, M. R., Franchi, F., & Graziosi, F. (2018). 5G Italian MISE Trial: Synergies Among Different Actors to Create a "5G Road". In *2018 IEEE 4th International Forum on Research and Technology for Society and Industry (RTSI)* (pp. 1-4). IEEE.

Colucci, E., De Ruvo, V., Lingua, A., Matrone, F., & Rizzo, G. (2020). HBIM-GIS Integration: From IFC to CityGML Standard for Damaged Cultural Heritage in a Multiscale 3D GIS. *Applied Sciences (Basel, Switzerland), 2020*(10), 1356. doi:10.3390/app10041356

Colucci, E., Xing, X., Kokla, M., Mostafavi, M. A., Noardo, F., & Spanò, A. (2021). Ontology-Based Semantic Conceptualisation of Historical Built Heritage to Generate Parametric Structured Models from Point Clouds. *AppliedSciences*, *11*(6), 2813.

Commission recommendation on a common European data space for cultural heritage. (2021). https://digital-strategy.ec.europa.eu/en/news/commission-proposes-common-european-data-space-cultural-heritage

Cooperation on advancing digitisation of cultural heritage. (2019). *Digital Day 2019*. Retrieved from https://ec.europa.eu/digital-single-market/en/news/eu-member-states-sign-cooperate-digitising-cultural-heritage

CORDIS. (2021). *How digital technologies can play a vital role for the preservation of Europe's cultural heritage.* https://cordis.europa.eu/article/id/413473-how-digital-technologies-can-play-a-vital-role-for-the-preservation-of-cultural-heritage

Cordts, M., Omran, M., Ramos, S., Rehfeld, T., Enzweiler, M., Benenson, R., Franke, U., Roth, S., & Schiele, S. (2016). The Cityscapes Dataset for Semantic Urban Scene Understanding. In *IEEE Conference on Computer Vision and Pattern Recognition* (CVPR), 3213-3223.

Cortiella, C., & Horowitz, S. H. (2014). *The State of Learning Disabilities: Facts, Trends and Emerging Issues* (3rd ed.). National Center for Learning Disabilities. https://www.ncld.org/wp-content/uploads/2014/11/2014-State-of-LD

Cosovic, M., & Ramic-Brkic, B. (2020). Application of Game-Based Learning in Cultural Heritage. *Proceedings of the 2nd International Workshop on Visual Pattern Extraction and Recognition for Cultural Heritage Understanding.*

Council of Europe. (2000). *European Landscape Convention.* Retrieved on April 27, 2022 from https://rm.coe.int/1680080621

Council of Europe. (2005). *Framework Convention on the Value of Cultural Heritage for Society.* Retrieved on April 27, 2022 from https://rm.coe.int/1680083746

Coupry, C., Noblecourt, S., Richard, P., Baudry, D., & Bigaud, D. (2021). BIM-based Digital Twin and XR Devices to Improve Maintenance Procedures in Smart Buildings: A Literature Review. *Applied Sciences (Basel, Switzerland)*, *11*(15), 6810. Advance online publication. doi:10.3390/app11156810

COVID Feel Good. (2022). https://www.covidfeelgood.com/

Cox, B. D., & Berns, R. S. (2015). Imaging artwork in a studio environment for computer graphics rendering. Proceeding of SPIE, 939803.

Craig Lefebvre, R. A. S. (2013). Digital Social Networks and Health. Social Media as a Tool in Medicine. American Heart Association, Inc.

Creators. (2022). Retrieved from https://www.creatorspad.com/about

Creswell, J. W. (2007). *Qualitative inquiry and research design: Choosing among five approaches* (2nd ed.). Sage.

Creswell, J. W. (2009). *Research design: Qualitative, quantitative, and mixed methods approaches* (3rd ed.). Sage.

Crusco, A. H., & Wetzel, C. G. (1984). The Midas Touch: The Effects of Interpersonal Touch on Restaurant Tipping. *Personality and Social Psychology Bulletin*, *10*(4), 512–517. doi:10.1177/0146167284104003

Cruz-Neira, C., Sandin, D.J., DeFanti, T.A., Kenyon, R.V., & Hart, J.C. (1992). The CAVE: Audio Visual Experience Automatic Virtual Environment. *Communications of the ACM*, *35*(6), 64-72. doi:10.1145/129888.129892

Cruz-Neira, C., Sandin, D. J., & DeFanti, T. A. (1993). Surround-Screen Projection-based Virtual Reality: The Design and Implementation of the CAVE. In *SIGGRAPH'93: Proceedings of the 20th Annual Conference on Computer Graphics and Interactive Techniques* (pp.135–142). Association for Computing Machinery. 10.1145/166117.166134

Cuesta, A. (2016). Evacuation Modeling Trends. Springer.

Cuesta, A., Abreu, O., & Alvear, D. (2016). *Evacuation Modeling Trends*. Springer.

Cui, W. D. H. (2021). Digital twin-driven building projects: a new paradigm towards real-time information convergence throughout the building lifecycle. In *UIA 2021 RIO: 27th World Congress of Architects*. ACSA.

CultLab3D. (n.d.). https://www.cultlab3d.de/

D'Urso, C. (2011). Information Integration for Facility Management. *IT Professional*, *13*(6), 48–53. doi:10.1109/MITP.2011.100

Dal Co. F. (1980). 10 immagini per Venezia. Venezia, Italy: OfficinaEdizioni.

Dall'Osso, G. (2021). Haptic Rhythmics for Mediation Design Between Body and Space. *DIID*, *74*(74). Advance online publication. doi:10.30682/diid7421d

Dallasega, P., Rauch, E., & Linder, C. (2018). Industry 4.0 as an enabler of proximity for construction supply chains: A systematic literature review. *Computers in Industry*, *99*, 205–225. doi:10.1016/j.compind.2018.03.039

David Jones, C. S. (2020). Characterising the Digital Twin: A systematic literature review. *CIRP Journal of Manufacturing Science and Technology*, 29.

Davies, C. (2015). *HoloLens hands-on: Building for Windows Holographic.* Academic Press.

Davis, S., Nesbitt, K., & Nalivaiko, E. (2014). A Systematic Review of Cybersickness. *Proceedings of the 2014 Conference on Interactive Entertainment*, 1–9.

De Felice, F., Travaglioni, M., & Petrillo, A. (2021). Innovation Trajectories for a Society 5.0. *Data*, *6*(11), 115. doi:10.3390/data6110115

De Fino, M., Bruno, S., & Fatiguso, F. (2022). Dissemination, assessment and managementof historic buildings by thematic virtual tours and 3D models. *Virtual Archaeology Review*, *13*(26), 88–102. doi:10.4995/var.2022.15426

De Luca, D., & Ugliotti, F. M. (2020). Virtual Reality to Stimulate Cognitive Behavior of Alzheimer's and Dementia Patients. In Augmented Reality, Virtual Reality, and Computer Graphics. AVR 2020 (pp.101-113). Springer. doi:10.1007/978-3-030-58468-9_8

De Luca, D., Ugliotti, F.M., (2021). *Digital meets Opera: an interactive service model towards accessibility and sustainability*. DN.

De Luca, L. (2020, October). Towards the Semantic-aware 3D Digitisation of Architectural Heritage: The" Notre-Dame de Paris" Digital Twin Project. In *Proceedings of the 2nd Workshop on Structuring and Understanding of Multimedia heritAge Contents* (pp. 3-4). Academic Press.

De Luca, V. (2016, luglio). Oltre l'interfaccia: Emozioni e design dell'interazione per il benessere. *MD Journal,* 106–119.

De Marco, R., & Dell'Amico, A. (2020) Connecting the Territory between Heritage and Information: Databases and Models for the Cultural Heritage Routes. In *Proceedings UID 2020 CONNECTING | drawing for weaving relationships* (pp. 258-277). Franco Angeli.

De Marco, R., & Pettineo, A. (2021). The recognition of Heritage qualities from feature-based digital procedures in the analysis of historical urban contexts. *Int. Arch. Photogramm. Remote Sens. Spatial Inf. Sci., XLVI-2/W1-2022*, 175–182.

Deguchi, A., Hirai, C., Matsuoka, H., Nakano, T., Oshima, K., Tai, M., & Tani, S. (2020). What Is Society 5.0? In Hitachi-UTokyo Laboratory (Ed.), Society 5.0. Springer.

Deguchi, A., Hirai, C., Matsuoka, H., Nakano, T., Oshima, K., Tai, M., & Tani, S. (2020). What is society 5.0. *Society, 5*, 1–23.

De-Marcos, L., Garcia-Lopez, E., & Garcia-Cabot, A. (2016). On the effectiveness of game-like and social approaches in learning: Comparing educational gaming, gamification & social networking. *Computers & Education, 95*, 99–113. doi:10.1016/j.compedu.2015.12.008

Demirdöğen, G., Işık, Z., & Arayici, Y. (2020). Lean Management Framework for Healthcare Facilities Integrating BIM, BEPS and Big Data Analytics. *Sustainability, 12*(17), 7061. doi:10.3390u12177061

Dessislava Petrova-Antonova, S. I. (2021). *Digital Twin Modeling of Smart Cities. Human Interaction, Emerging Technologies and Future Applications III*. Springer.

Deterding, S. D. (2011). From game design elements to gamefulness: defining" gamification". In *Proceedings of the 15th international academic MindTrek conference: Envisioning future media environments* (p. 9-15). Tampere, Finland: Association for Computing Machinery. 10.1145/2181037.2181040

Di Giulio, R., Maietti, F., Piaia, E., Medici, M., Ferrari, F., & Turillazzi, B. (2017). Integrated data capturing requirements for 3D semantic modelling of Cultural Heritage: The INCEPTION protocol. *The International Archives of the Photogrammetry, Remote Sensing and Spatial Information Sciences, XLII-2*(W3), 251–257. doi:10.5194/isprs-archives-XLII-2-W3-251-2017

Di Giuseppantonio Di Franco, P., Galeazzi, F., & Vassallo, V. (2018). *Authenticity and cultural heritage in the age of 3D digital reproductions*. McDonald Institute for Archaeological Research.

di Milano, P. (2004). Reconciling construction innovation and standardisation on major projects. *Engineering, Construction, and Architectural Management, 11*(5), 366–372. doi:10.1108/09699980410558566

Di Natale, A. F., Repetto, C., Riva, G., & Villani, D. (2020). Immersive virtual reality in K-12 and higher education: A 10-year systematic review of empirical research. *British Journal of Educational Technology, 51*(6), 2006–2033. doi:10.1111/bjet.13030

Diara, F., & Rinaudo, F. (2019). From reality to parametric models of Cultural Heritage assets for HBIM. *The International Archives of the Photogrammetry, Remote Sensing and Spatial Information Sciences, XLII-2*(W15), 413–419. doi:10.5194/isprs-archives-XLII-2-W15-413-2019

Diara, F., & Rinaudo, F. (2021). ARK-BIM: Open-Source Cloud-Based HBIM Platform. *Applied Sciences (Basel, Switzerland), 11*(18), 8770. doi:10.3390/app11188770

Dickey, M. D. (2005). Brave new (interactive) worlds: A review of the design affordances and constraints of two 3D virtual worlds as interactive learning environments. *Interactive Learning Environments, 13*(1–2), 121–137. doi:10.1080/10494820500173714

Dilberoglu, U., Gharehpapagh, B., Yaman, U., & Dolen, M. (2017). The Role of Additive Manufacturing in the Era of Industry 4.0. *Procedia Manufacturing, 11*, 545–554. doi:10.1016/j.promfg.2017.07.148

Diller, S., Rhea, D., & Shedroff, N. (2005). *Making Meaning: How Successful Businesses Deliver Meaningful Customer Experiences*. New Riders. https://sfx.ethz.ch/sfx_nebis?url_ver=Z39.88-2004&ctx_ver=Z39.88-2004&ctx_enc=info:ofi/enc:UTF-8&rfr_id=info:sid/sfxit.com:opac_856&url_ctx_fmt=info:ofi/fmt:kev:mtx:ctx&sfx.ignore_date_threshold=1&rft.object_id=1000000000297576&svc_val_fmt=info:ofi/fmt:kev:mtx:sch_svc&

DIME4HERITAGE. (n.d.). https://cordis.europa.eu/project/id/302799

Doerr, M. (2003). The CIDOC CRM- an ontological approach to semantic interoperability of metadata. *AI Magazine*, *24*(3), 75.

Donato, V., Lo Turco, M., & Bocconcino, M. M. (2017). BIM-QA/QC in the architectural design process. *Architectural Engineering and Design Management*, 1-16. doi:10.1080/17452007.2017.1370995

Donner, C., & Wann Jensen, H. (2005). Light diffusion in multi- layered translucent materials. *ACM Transactions on Graphics*, *24*(3), 1032–1039. doi:10.1145/1073204.1073308

DOORS. (n.d.). https://cordis.europa.eu/project/id/101036071

Dores, A. R., Carvalho, I. P., Barbosa, F., Almeida, I., Guerreiro, S., Oliveira, B., de Sousa, L., & Caldas, A. C. (2012). Computer-Assisted Rehabilitation Program – Virtual Reality (CARP-VR): A Program for Cognitive Rehabilitation of Executive Dysfunction. doi:10.1007/978-3-642-31800-9_10

Douglas, W. Rae. (2003). *City: Urbanism and Its End*. Yale University Press. https://sfx.ethz.ch/sfx_lib4ri?url_ver=Z39.88-2004&ctx_ver=Z39.88-2004&ctx_enc=info:ofi/enc:UTF-8&rfr_id=info:sid/sfxit.com:opac_856&url_ctx_fmt=info:ofi/fmt:kev:mtx:ctx&sfx.ignore_date_threshold=1&rft.object_id=1000000000473628&svc_val_fmt=info:ofi/fmt:kev:mtx:sch_svc&

Dourish, P. (2004). *Where the action is: the foundations of embodied interaction*. MIT Press.

Drachen, A., & Canossa, A. (2009, September). Towards gameplay analysis via gameplay metrics. *Proceedings of the 13th international MindTrek conference: Everyday life in the ubiquitous era*, 202-209. 10.1145/1621841.1621878

Duan, H. J. L. (2021). Metaverse for Social Good: A University Campus Prototype. Multimedia. Cornell University.

Dudzinski, E. (2016). Using the Pool Activity Level instrument to support meaningful activity for a person with dementia: a case study. *British Journal of Occupational Therapy*, *79*(2), 65–68. doi:10.1177/0308022615600182

Duran, G., & Uygur, S. (2022). A Comprehensive Systematic Literature Review About Smartness in Tourism. In Handbook of Research on Digital Communications, Internet of Things, and the Future of Cultural Tourism. IGI Global. doi:10.4018/978-1-7998-8528-3.ch011

Eberly, D. H. (2015). *3D Game Engine Design: A Practical Approach to Real-Time Computer Graphics*. Taylor & Francis, CRC Press.

Edirippulige, S., & Senanayake, B. (2020). Opportunities and Challenges in Digital Healthcare Innovation. In Professional Practices for Digital Healthcare (p. 97-112). IGI Global.

Eid, M. A., Giakoumidis, N., & El Saddik, A. (2016). A novel eye-gaze-controlled wheelchair system for navigating unknown environments: Case study with a person with ALS. *IEEE Access: Practical Innovations, Open Solutions*, *4*, 558–573. doi:10.1109/ACCESS.2016.2520093

Elias, N., Mas, N., & Cotte, P. (2011). Review of several optical non-destructive analyses of an easel painting. Complementarity and crosschecking of the results. *Journal of Cultural Heritage*, *12*(4), 335–345. doi:10.1016/j.culher.2011.05.006

El-Said, O., & Aziz, H. (2022). Virtual Tours a Means to an End: An Analysis of Virtual Tours' Role in Tourism Recovery Post COVID-19. *Journal of Travel Research, 61*(3), 528–548. doi:10.1177/0047287521997567

Emmelkamp, P. M. (2011). Effectiveness of cybertherapy in mental health: A critical appraisal. *Annual Review of Cybertherapy and Telemedicine, 2011*, 3–8. PMID:21685633

Empler, T. (2018). Musei tradizionali, musei virtuali. La funzione divulgativa delle ICT [Traditional museums, virtual museums. The divulgation role of ICT]. *DisegnareCon, 11*(21), 13.1-13.9.

Empler, T., Caldarone, A., & Rossi, M. L. (2021). BIM Survey. Critical Reflections on the Built Heritage's Survey. In C. Bolognesi & D. Villa (Eds.), *From Building Information Modelling to Mixed Reality* (pp. 109–122). Springer.

Emuze, F. A. (2018). Productivity of digital fabrication in construction: Cost and time analysis of a robotically built wall. *Automation in Construction, 92*, 297–311. doi:10.1016/j.autcon.2018.04.004

Englund, C., Olofsson, A. D., & Price, L. (2017). Teaching with technology in higher education: Understanding conceptual change and development in practice. *Higher Education Research & Development, 36*(1), 73–87. doi:10.1080/07294360.2016.1171300

Erdost, T. (2004). *Trust and self-disclosure in the context of computer mediated communication* [Doctoral Dissertation]. Middle East Technical University. https://open.metu.edu.tr/handle/11511/14126

Ericsson, K. A. (2006). Protocol Analysis and Expert Thought: Concurrent Verbalizations of Thinking during Experts' Performance on Representative Tasks. In K. A. Ericsson, N. Charness, P. J. Feltovich, & R. R. Hoffman (Eds.), *The Cambridge handbook of expertise and expert performance* (pp. 223–241). Cambridge University Press. doi:10.1017/CBO9780511816796.013

Ericsson, K. A., & Simon, H. A. (1993). *Protocol Analysis: Verbal Reports as Data*. MIT Press. doi:10.7551/mitpress/5657.001.0001

Erkan, S., Çankaya, Z. C., Terzi, S., & Özbay, Y. (2011). Üniversite psikolojik danışma ve rehberlik merkezlerinin incelenmesi. *Mehmet Akif Ersoy Üniversitesi Eğitim Fakültesi Dergisi, 11*(22), 174–198. https://dergipark.org.tr/en/pub/maeuefd/issue/19395/206011

Erkan, S., Yaşar, Ö., Cihangir-Çankaya, Z., & Terzi, Ş. (2012). University students' problem areas and psychological help-seeking willingness. *Education in Science, 37*(164), 94–107. http://egitimvebilim.ted.org.tr/index.php/EB/article/view/402

ESPON. (2021). *HERIWELL – Cultural Heritage as a Source of Societal Well-being in European Regions Draft findings on the linkages between cultural heritage and societal well-being*. https://www.espon.eu/HERIWELL

Etzi, R., Ferrise, F., Bordegoni, M., Zampini, M., & Gallace, A. (2018). The Effect of Visual and Auditory Information on the Perception of Pleasantness and Roughness of Virtual Surfaces. *Multisensory Research, 31*(6), 501–522. doi:10.1163/22134808-00002603 PMID:31264615

Etzi, R., & Gallace, A. (2016). The arousing power of everyday materials: An analysis of the physiological and behavioral responses to visually and tactually presented textures. *Experimental Brain Research, 2016*(234), 1659–1666. doi:10.100700221-016-4574-z PMID:26842855

Europa Nostra. (2020). *Covid-19 & Beyond. Challenges and Opportunities for Cultural Heritage*. https://www.europanostra.org/wp-content/uploads/2020/10/20201014_COVID19_Consultation-Paper_EN.pdf

European Commission, Directorate-General for Research and Innovation. (2021). *Horizon Europe: strategic plan 2021-2024*. https://data.europa.eu/doi/10.2777/083753

European Commission. (2018). *Digital Economy and Society Index Report 2018. Integration of Digital Technologies.* European Commission.

European Commission. (2020). *Basic principles and tips for 3D digitisation of cultural heritage.* https://digital-strategy.ec.europa.eu/en/library/basic-principles-and-tips-3d-digitisation-cultural-heritage

European Commission. (2020). *Basic principles and tips for 3D digitisation of tangible cultural heritage for cultural heritage professionals and institutions and other custodians of cultural heritage.* https://digital-strategy.ec.europa.eu/en/library/basic-principles-and-tips-3d-digitisation-cultural-heritage

European Commission. (2020). *Digital Economy and Society Index (DESI) 2020. Integration of digital technology.* European Commission.

European Commission. (2021a). *Recovery plan for Europe.* https://ec.europa.eu/info/strategy/recovery-plan-europe_en

European Commission. (2021b). *Europe's Digital Decade: digital targets for 2030.* https://ec.europa.eu/info/strategy/priorities-2019-2024/europe-fit-digital-age/europes-digital-decade-digital-targets-2030_en

European Commission. (2022). *Study on quality in 3D digitisation of tangible cultural heritage.* https://digital-strategy.ec.europa.eu/en/library/study-quality-3d-digitisation-tangible-cultural-heritage

Europeana Data Model Ontology. (n.d.). Available online: https://pro.europeana.eu/page/edm-documentation

Europeana, D. S. I. (n.d.). *2 - Access to Digital Resources of European Heritage - Deliverable D4.4.* Report on ENUMERATE Core Survey 4. Available online: https://pro.europeana.eu/files/Europeana_Professional/Projects/Project_list/ENUMERATE/deliverables/DSI-2_Deliverable%20D4.4_Europeana_Report%20on%20ENUMERATE%20Core%20Survey%204.pdf

Europeana. (n.d.). https://www.europeana.eu/en/about-us

Eurostat. (2022). *Impact of Covid-19 crisis on construction.* Obtenido de https://ec.europa.eu/eurostat/statistics-explained/index.php?title=Impact_of_Covid-19_crisis_on_construction

Evans, J. (2009). Online counselling and guidance skills: A practical resource for trainees and practitioners. *Sage (Atlanta, Ga.).* Advance online publication. doi:10.4135/9781446216705

Exhibition Catalogue. (2013). *Luciano Semerani: l'architetto nelle città.* Torino, Italy: Accademia University Press.

Expert Group on Digital Cultural Heritage and Europeana. (n.d.). https://digital-strategy.ec.europa.eu/en/policies/europeana-digital-heritage-expert-group

Fadela, C. (2020). *Shortage of personal protective equipment endangering health workers worldwide.* World Health Organization. https://www.who.int/news/item/03-03-2020-shortage-of-personal-protective-equipment-endangering-health-workers-worldwide

Fanea-Ivanovici, M., & Pan, M. C. (2020). From Culture to Smart Culture. How Digital Transformations Enhance Citizens' Well-Being Through Better Cultural Accessibility and Inclusion. *IEEE Access: Practical Innovations, Open Solutions, 8*, 37988–38000. doi:10.1109/ACCESS.2020.2975542

Fan, L., Xinmin, L., Bingcheng, W., & Li, W. (2017). Interactivity, Engagement, and Technology Dependence: Understanding Users' Technology Utilisation Behaviour. *Behaviour & Information Technology, 36*(2), 113–124. doi:10.1080/0144929X.2016.1199051

Faraj, G., & Micsik, A. (2021). Representing and Validating Cultural Heritage Knowledge Graphs in CIDOC-CRM Ontology. *Future Internet, 13*(11), 277. doi:10.3390/fi13110277

Fardello, M. (2020). *Sperimentazione di terapie innovative che utilizzano la realtà virtuale per la gestione non farmacologica dei disturbi del comportamento associati a demenza* [Experimentation of innovative therapies that use virtual reality for the non-pharmacological management of behavioral disorders associated with dementia] [Master's thesis]. Politecnico di Torino.

Farhah Saidin, N., Abd Halim, N. D., & Yahayal, N. (2015). A Review of Research on Augmented Reality in Education. *International Education Studies*, *8*(13), 1–8.

Fass, D. (2015). Affordances and Safe Design of Assistance Wearable Virtual Environment of Gesture. *Procedia Manufacturing*, *3*, 866–873. doi:10.1016/j.promfg.2015.07.343

Fassi, F., Achille, C., Mandelli, A., Rechichi, F., & Parri, S. (2015). A New idea of BIM system for visualization, web sharing and using huge complex 3d models for facility management. *The International Archives of the Photogrammetry, Remote Sensing and Spatial Information Sciences*, *XL-5*(W4), 359–366. doi:10.5194/isprsarchives-XL-5-W4-359-2015

Fassi, F., Fregonese, L., Adami, A., & Rechichi, F. (2017). BIM systems for the conservation and preservation of the mosaics of San Marco in Venice. *The International Archives of the Photogrammetry, Remote Sensing and Spatial Information Sciences*, *XLII-2*(W5), 229–236. doi:10.5194/isprs-archives-XLII-2-W5-229-2017

Fassi, F., Mandelli, A., Teruggi, S., Rechichi, F., Fiorillo, F., & Achille, C. (2016). Lecture Notes in Computer Science: Vol. 9769. *VR for Cultural Heritage. Augmented Reality, Virtual Reality, and Computer Graphics. AVR 2016.* Springer.

Fatehi, F., Samadbeik, M., & Azar, K. (2020). What is digital health? Review of definitions. In A. Värri (Ed.), *Integrated Citizen Centered Digital Health and Social Care: Citizens as Data Producers and Service co-Creators* (pp. 67–71). IOS Press BV. doi:10.3233/SHTI200696

Fathy, D., Mohamed, H. E., & Ehab, A. (2022). Fans Behave as Buyers? Assimilate Fan-based and Team-based Drivers of Fan Engagement. *Journal of Research in Interactive Marketing*, *16*(3), 329–345. doi:10.1108/JRIM-04-2021-0107

Fayad, M. E., Hamza, H., & Sanchez, H. (2003). A pattern for an effective class responsibility collaborator (CRC) cards. *Proceedings Fifth IEEE Workshop on Mobile Computing Systems and Applications*, 584–587. 10.1109/IRI.2003.1251469

Fenichel, M. A. (2011). Online behavior, communication, and experience. In R. Kraus, G. Stricker, & C. Speyer (Eds.), *Online counseling: A handbook for mental health professionals* (2nd ed., pp. 3–18). Academic Press. doi:10.1016/B978-0-12-378596-1.00001-0

Fenichel, M. A., Suler, J., Barak, A., Zelvin, E., Jones, G., Munro, K., Meunier, V., & Walker-Schmucker, W. (2002). Myths and realities of online clinical work. *Cyberpsychology & Behavior*, *5*(5), 481–497. doi:10.1089/109493102761022904 PMID:12448785

Ferguson, B. (2012). The emergence of games for health. *Games for Health Journal*, *1*(1), 1–2. doi:10.1089/g4h.2012.1010 PMID:26196423

Ferrari, F., Maietti, F., & Balzani, M. (2019), INCEPTION – Patrimonio Culturale Inclusivo in Europa mediante la modellazione semantica 3D [INCEPTION - Inclusive Cultural Heritage in Europe through 3D semantic modelling]. In *Il Simposio UID di internazionalizzazione della ricerca. Patrimoni culturali, Architettura, Paesaggio e Design tra ricerca e sperimentazione didattica* [The UID Symposium of Internationalization of Research. Cultural Heritage, Architecture, Landscape and Design between research and didactic experimentation]. Firenze: DIDApress.

Ferriss, T. (2013, April 3). *The First-Ever Quantified Self Notes (Plus: LSD as Cognitive Enhancer?)*. The Blog of Author Tim Ferriss. https://tim.blog/2013/04/03/the-first-ever-quantified-self-notes-plus-lsd-as-cognitive-enhancer

Ferwerda, J. A., & Darling, B. A. (2013). Tangible Images: Bridging the Real and Virtual Worlds. In CCIW 2013 proceedings. Lecture Notes in Computer Science, 7786. Springer.

Field, A. (2009). *Discovering statistics using SPSS* (3rd ed.). Sage.

Field, T. (2001). *Touch.* MIT Press. doi:10.7551/mitpress/6845.001.0001

Fleming, T. M., de Beurs, D., Khazaal, Y., Gaggioli, A., Riva, G., Botella, C., Baños, R. M., Aschieri, F., Bavin, L. M., Kleiboer, A., Merry, S., Lau, H. M., & Riper, H. (2016). Maximizing the Impact of e-Therapy and Serious Gaming: Time for a Paradigm Shift. *Frontiers in Psychiatry, 7*, 1–7. doi:10.3389/fpsyt.2016.00065 PMID:27148094

Florio, R., Catuogno, R., & Della Corte, T. (2019). Integrated methodologies for the knowledge and regeneration of the Paestum site. The role of the nature between the temples and the sea. *Sustainable Mediterranean Construction*, 93-101.

Florio, R., Catuogno, R., & Della Corte, T. (2019). The interaction of knowledge as though field experimentation of the integrated survey. The case of Sacristy of Francesco Solimena in the church of San Paolo Maggiore in Naples. *SCIRES. SCIentific RESearch and Information Technology., 9*(2), 69–84.

Foloppe, D. A., Richard, P., Yamaguchi, T., Etcharry-Bouyx, F., & Allain, F. (2018). The potential of virtual reality-based training to enhance the functional autonomy of Alzheimer's disease patients in cooking activities: a single case study. Neuropsychological Rehabilitation, 28(5), 709–733. doi:10.1080/09602011.2015.1094394

Folstein, M. F., Folstein, S. E., & McHugh, P. R. (1975). Mini-mental state. A practical method for grading the cognitive state of patients for the clinician. *Journal of Psychiatric Research, 12*(3), 189–198. doi:10.1016/0022-3956(75)90026-6 PMID:1202204

Font Sentias, J. (2020). El mapping de Sant Climent de Taüll [The mapping of the Sant Climent de Taüll]. *Mnenòsine*, 10.

Fox, J., & Bailenson, J. N. (2009). Virtual self–modeling: The effects of vicarious reinforcement and identification on exercise behaviors. *Media Psychology, 12*(1), 1–25. doi:10.1080/15213260802669474

Franchi, F., Marotta, A., Rinaldi, C., Graziosi, F., Fratocchi, L., & Parisse, M. (2022). What Can 5G Do for Public Safety? Structural Health Monitoring and Earthquake Early Warning Scenarios. *Sensors (Basel), 22*(8), 3020. doi:10.339022083020 PMID:35459005

Frascari, M. (2017). An Age of Paper. In A. Goodhouse (Ed.), *When is the digital in architecture?* (pp. 24–31). Canadian Center for Architecture.

Frazer, J. H. (1993). The architectural relevance of cybernetics. *Systems Research, 10*(3), 43–48. doi:10.1002res.3850100307

Freeman, A., Adams Becker, S., Cummins, M., Davis, A., & Hall Giesinger, C. (2017). *NMC/CoSN Horizon Report: 2017 K-12 Edition.* New Media Consortium; Consortium for School Networking. https://files.eric.ed.gov/fulltext/ED588803.pdf

Freeman, D., Reeve, S., Robinson, A., Ehlers, A., Clark, D., Spanlang, B., & Slater, M. (2017). Virtual reality in the assessment, understanding, and treatment of mental health disorders. *Psychological Medicine, 47*(14), 2393–2400. doi:10.1017/S003329171700040X PMID:28325167

Fregonese, L., Taffurelli, L., Adami, A., Chiarini, S., Cremonesi, S., Helder, J., & Spezzoni, A. (2017). Survey and modelling for the BIM of Basilica of San Marco in Venice. *The International Archives of the Photogrammetry, Remote Sensing and Spatial Information Sciences, XLII-2*(W3), 303–310. doi:10.5194/isprs-archives-XLII-2-W3-303-2017

Frey, M. (2007). CabBoots: Shoes with integrated guidance system. *Proceedings of the 1st International Conference on Tangible and Embedded Interaction - TEI '07*, 245. 10.1145/1226969.1227019

Fricker, P. (2021). *The Relevance of Computational Design Thinking in Landscape Architecture. A Pedagogy of Data-Informed Design Processes across Scales* [PhD Thesis]. ETH Zurich. doi:10.3929/ethz-b-000495639

Fricker, P., & Munkel, G. (2013). How to Teach New Tools in Landscape Architecture in the Digital Overload. In *Computation and Performance – Proceedings of the 31st ECAADe Conference – Volume 2*. Faculty of Architecture, Delft University of Technology. http://cumincad.scix.net/cgi-bin/works/Show?ecaade2013_028

Fricker, P., Hovestadt, L., Fritz, O., Dillenburger, B., & Braach, M. (2007). Organised Complexity. *Predicting the Future: 25th ECAADe Conference Proceedings,* 695-701. http://papers.cumincad.org/cgi-bin/works/paper/ecaade2007_118

Fricker, P., Kotnik, T., & Borg, K. (2020). Computational Design Pedagogy for the Cognitive Age. *Anthropologic: Architecture and Fabrication in the Cognitive Age*, 695–692.

Fricker, P. (2018). *The Real Virtual or the Real Real: Entering Mixed Reality*. Wichmann Verlag., doi:10.14627/537642044

Fricker, P., Kotnik, T., & Piskorec, L. (2019). *Structuralism: Patterns of Interaction Computational Design Thinking across Scales*. Wichmann Verlag. doi:10.14627/537663026

Fricker, P., & Munkel, G. (2015). Data Mapping – Explorative Big Data Visualization in Landscape Architecture. In E. Buhmann (Ed.), *Peer Reviewed Proceedings Digital Landscape Architecture 2015* (pp. 141–150). Wichmann.

Fridolf, K., & Nilsson, D. (2011). People's Subjective Estimation of Fire Growth: An Experimental Study of Young Adults. *Fire Safety Science*, *10*, 161–172.

Frima, F. K. (2020). Penerapan Praktikum Jarak Jauh Pada Topik Pertumbuhan Mikroba Dalam Masa Darurat Covid-19 Di Institut Teknologi Sumatera. *Jurnal Pendidikan Sains (Jps), 8*(2), 102. doi:10.26714/jps.8.2.2020.102-109

Fukuyama, M. (2018, July). Society 5.0: Aiming for a New Human-Centered Society. *Japan Spotlight,* 47-50.

Fuller, R. B. (1969). *Operating manual for spaceship Earth*. Southern Illinois University Press.

Gaiani, M., Apollonio, F. I. & Fantini, F. (2020). Una metodología inteligente para la digitalización de colecciones museísticas. *EGA Expresión Gráfica Arquitectónica, 25*(38), 170–181.

Gaiani, M., Apollonio, F. I., Bacci, G., Ballabeni, A., Bozzola, M., Foschi, R., Garagnani, S., & Palermo, R. (2020). Seeing inside drawings: a system for analysing, conserving, understanding and communicating Leonardo's drawings. In *Leonardo in Vinci. At the origins of the Genius* (pp. 207–239). Giunti.

Gaiani, M., Apollonio, F. I., & Fantini, F. (2019). Evaluating smartphones color fidelity and metric accuracy for the 3D documentation of small artefacts. *The International Archives of the Photogrammetry, Remote Sensing and Spatial Information Sciences*, *42*(W11), 539–547. doi:10.5194/isprs-archives-XLII-2-W11-539-2019

Gaiani, M., & Ballabeni, A. (2018). *SHAFT (SAT & HUE Adaptive Fine Tuning), a new automated solution for target-based color correction. In Colour and Colorimetry Multidisciplinary Contributions. XIVB, Maggioli Editore*. ITA.

Gallace, A. (2022). Cervelli reali in mondi virtuali: psicologia e neuroscienze del metaverso [Real brains in virtual worlds: psychology and neuroscience of metaverses]. In Metaverso. Mondadori Editore.

Gallace, A. (2012). Living with touch: Understanding tactile interactions. *The Psychologist, 25*, 3–5.

Gallace, A., & Girondini, M. (2022). Social touch in virtual reality. *Current Opinion in Behavioral Sciences*, *43*, 249–254. doi:10.1016/j.cobeha.2021.11.006

Gallace, A., Ngo, M. K., Sulaitis, J., & Spence, C. (2012). Multisensory presence in virtual reality: Possibilities & limitations. In G. Ghinea, F. Andres, & S. Gulliver (Eds.), *Multiple sensorial media advances and applications: New developments in MulSeMedia* (pp. 1–40). IGI Global. doi:10.4018/978-1-60960-821-7.ch001

Gallace, A., & Spence, C. (2008). The cognitive and neural correlates of "tactile consciousness": A multisensory perspective. *Consciousness and Cognition, 17*(1), 370–407. doi:10.1016/j.concog.2007.01.005 PMID:17398116

Gallace, A., & Spence, C. (2010). The science of interpersonal touch: An overview. *Neuroscience and Biobehavioral Reviews, 34*(2), 246–259. doi:10.1016/j.neubiorev.2008.10.004 PMID:18992276

Gallace, A., & Spence, C. (2014). *Touch with the Future: The Sense of Touch from Cognitive Neuroscience to Virtual Reality*. Oxford University Press. doi:10.1093/acprof:oso/9780199644469.001.0001

Gallace, A., & Spence, C. (2016). Social touch. In *Affective Touch and the Neurophysiology of CT Afferents* (pp. 227–238). Springer New York. doi:10.1007/978-1-4939-6418-5_14

Gallace, A., Tan, H. Z., & Spence, C. (2007). The body surface as a communication system: The state of the art after 50 years. *Presence (Cambridge, Mass.), 16*(6), 655–676. doi:10.1162/pres.16.6.655

Gallagher, A. G., Ritter, E. M., Champion, H., Higgins, G., Fried, M. P., Moses, G., Smith, C. D., & Satava, R. M. (2005). Virtual reality simulation for the operating room: Proficiency-based training as a paradigm shift in surgical skills training. *Annals of Surgery, 241*(2), 364–372. doi:10.1097/01.sla.0000151982.85062.80 PMID:15650649

Gandolfi, V. (1980). *Il teatro Farnese di Parma* [The Farnese Theatre in Parma]. Luigi Battei.

Garcia Carrizosa, H., Diaz, J., Krall, R., & Sisinni Ganly, F. (2019). Cultural differences in ARCHES: A European participatory research project—working with mixed access preferences in different cultural heritage sites. *The International Journal of the Inclusive Museum, 12*(3), 33–50.

Garcia, L.M., Birckhead, B.J., Krishnamurthy, P., Sackman, J., Mackey, I.G, Robert, G.L., Salmasi, V., Maddox, T., & Darnall, B.D (2021). An 8-week self-administered at-home behavioral skills-based virtual reality program for chronic low back pain: double-blind, randomized, placebo-controlled trial conducted during COVID-19. *Journal of Medical Internet Research, 23*(2). doi:10.2196/26292

Gardner, A., Tchou, C., Hawkins, T., & Debevec, P. (2003). Linear Light Source Reflectometry. *ACM Transactions on Graphics, 22*(3), 749–758. doi:10.1145/882262.882342

Gartner. (2022). *Top Strategic Technology Trends for 2022*. Gartner.

Gazzola, V., Spezio, M. L., Etzel, J. A., Castelli, F., Adolphs, R., & Keysers, C. (2012). Primary somatosensory cortex discriminates affective significance in social touch. *Proceedings of the National Academy of Sciences of the United States of America, 2012*(109), 9688. doi:10.1073/pnas.1113211109 PMID:22665808

Georgopoulos, A. (2017). Data Acquisition for geometric Documentation of Cultural Heritage. In M. Ioannides & N. Magnenat-Thalmann (Eds.), *Mixed Reality and Gamification for Cultural Heritage* (pp. 29–73). Springer. doi:10.1007/978-3-319-49607-8_2

Georgopoulos, A., & Stathopoulou, E. K. (2017). Data acquisition for 3D geometric recording: state of the art and recent innovations. In M. Vincent, V. López-Menchero Bendicho, M. Ioannides, & T. Levy (Eds.), *Heritage and Archaeology in the Digital Age. Quantitative Methods in the Humanities and Social Sciences* (pp. 1–26). Springer. doi:10.1007/978-3-319-65370-9_1

Getty Architecture & Art Thesaurus. (n.d.). Available online: https://www.getty.edu/research/tools/vocabularies/aat/

Ghani, I., Rafi, A., & Woods, P. (2020). The effect of immersion towards place presence in virtual heritage environments. *Personal and Ubiquitous Computing*, *24*(6), 861–872. doi:10.100700779-019-01352-8

Giannakoulopoulos, A., Pergantis, M., Poulimenou, S. M., & Deliyannis, I. (2021). Good Practices for Web-Based Cultural Heritage Information Management for Europeana. *Information (Basel)*, *12*(5), 179. doi:10.3390/info12050179

Giannetti, S., Lodovisi, A., Incerti, M., Grassivaro, A., & Sardo, A. (2019). Esperienze di projection mapping per la valorizzazione delle facciate dipinte nei territori estensi. In P. Belardi (Ed.), *Riflessioni. L'arte del disegno e disegno dell'arte, Atti del 41° Convegno Internazionale dei Docenti della Rappresentazione* [Reflections. The art of drawing and art design, Proceedings of the 41st International Conference of Representation Teachers] (pp. 1621-1628). Gangemi Editore.

Gibson, J. J. (1979). The ecological approach to visual perception [Un approccio ecologico alla percezione visiva]. Houghton Mifflin.

Gibson, J. J. (1974). *The perception of the visual world*. Greenwood Press.

Gibson, J. J. (1977). The Theory of Affordances. In *R. E. Shaw & J. Bransford (A c. Di), Perceiving, Acting, and Knowing*. Lawrence Erlbaum Associates.

Gibson, J. J. (1986). *The ecological approach to visual perception*. L. Erlbaum.

Gibson, W. (2002). *Cyberspace*. Heyne.

Ginzarly, M., & Jordan Srour, F. (2022). Cultural heritage through the lens of COVID-19. *Poetics*, *92*(A), 101622.

Giordano, A. (2017). Mapping Venice. From visualizing Venice to visualizing cities. In B. A. Piga & R. Salerno (Eds.), *Urban design and representation. A multidisciplinary and multisensory approach* (pp. 143–151). Springer.

Giordano, A. (2019). New Interoperable Tools to Communicate Knowledge of Historic Cities and Their Preservation and Innovation. In A. Luigini (Ed.), *Proceedings of the 1st International and Interdisciplinary Conference on Digital Environments for Education, Arts and Heritage. EARTH 2018. Advances in Intelligent Systems and Computing*, vol 919. Springer, Cham, 34-43.

Giordano, A., Borin, P., Cundari, M. R., & Panarotto, F. (2014). La chiesa degli Eremitani a Padova: rilievo, documentazione, storia [The church of the Eremitani in Padua: Survey, documentation, history]. In P. Giandebiaggi & C. Vernizzi (Eds.), *Italian Survey & International Experience, Atti del 36° Convegno Internazionale dei Docenti della Rappresentazione* (pp. 869–876). Gangemi Editore.

Giordano, A., Friso, I., Borin, P., Monteleone, C., & Panarotto, F. (2018). Time and Space in the History of Cities. In S. Münster, K. Friedrichs, F. Niebling, & A. Seidel-Grzesińska (Eds.), *Digital Research and Education in Architectural Heritage. UHDL DECH 2017 2017. Communications in Computer and Information Science* (Vol. 817, pp. 47–62). Springer.

Girardin, F., Vaccari, A., Gerber, A., Birderman, A., & Ratti, C. (2009). Quantifying Urban Attractiveness from the Distribution and Density of Digital Footprints. *International Journal of Spatial Data Infrastructures Research*, 4.

Gireesh Kumar, T. K. (2021). Designing a Comprehensive Information System for Safeguarding the Cultural Heritage: Need for Adopting Architectural Models and Quality Standards. *Library Philosophy and Practice*, *2021*(5392), 1-19.

Girot, C. (2016). *The course of landscape architecture: A history of our designs on the natural world, from prehistory to the present*. Thames & Hudson.

Glaessgen, E., & Stargel, D. (2012). The Digital Twin Paradigm for Future NASA and U.S. Air Force Vehicles. *Structure and Dynamics*.

Google Arts and Culture. (n.d.). https://artsandculture.google.com/

Gorini, A., Gaggioli, A., Vigna, C., & Riva, G. (2008). A second life for eHealth: Prospects for the use of 3-D virtual worlds in clinical psychology. *Journal of Medical Internet Research, 10*(3), e21. doi:10.2196/jmir.1029 PMID:18678557

Governo Italiano. (2021). *Piano Nazionale di Ripresa e Resilienza.* National Recovery and Resilience Plan.

Gözüm, A., Metin, S., Uzun, H., & Karaca, N. (2022). Developing the Teacher Self-Efficacy Scale in the Use of ICT at Home for Pre-school Distance Education During Covid-19. In D. Ifenthaler (Ed.), *Technology, Knowledge and Learning*. Springer.

Grabarczyk, P., & Pokropski, M. (2016). Perception of Affordances and Experience of Presence in Virtual Reality. *Avant (Torun), 7*(2), 25–44. doi:10.26913/70202016.0112.0002

Graesser, A. C. (2017). Reflections on Serious Games BT - Instructional Techniques to Facilitate Learning and Motivation of Serious Games. Springer International Publishing. doi:10.1007/978-3-319-39298-1_11

Graham, K., Chow, L., & Fai, S. (2018). Level of detail, information and accuracy in Building Information Modelling of existing and heritage buildings. *Journal of Cultural Heritage Management and Sustainable Development, 8*(4), 495–507. doi:10.1108/JCHMSD-09-2018-0067

Grassini, S., & Laumann, K. (2020). Are modern head-mounted displays sexist? A systematic review on gender differences in HMD-mediated virtual reality. *Frontiers in Psychology, 7*, 11. doi:10.3389/fpsyg.2020.01604 PMID:32903791

Greengard, S. (2019). *Virtual reality*. Mit Press.

Greenleaf, W. J., & Tovar, M. A. (1994). Augmenting reality in rehabilitation medicine. *Artificial Intelligence in Medicine, 6*, 289-299.

Gregor, S. (2013). Internet of Things (IoT): A vision, architectural elements, and future directions. *Future Generation Computer Systems, 29*(7), 1645–1660. doi:10.1016/j.future.2013.01.010

Grieves, M. (2014). *Digital twin: Manufacturing excellence through virtual factory replication*. White paper.

Grieves, M., & Vickers, J. (2017). Digital Twin: Mitigating Unpredictable, Undesirable Emergent Behavior in Complex Systems. In *Transdisciplinary Perspectives on Complex Systems* (pp. 85–113). Springer International Publishing. doi:10.1007/978-3-319-38756-7_4

Griffiths, F., Lindenmeyer, A., Powell, J., Lowe, P., & Thorogood, M. (2006). Why are health care interventions delivered over the Internet? A systematic review of the published literature. *Journal of Medical Internet Research, 8*(2), e10. doi:10.2196/jmir.8.2.e10 PMID:16867965

Grilli, E., & Remondino, F. (2020). Machine Learning Generalisation across Different 3D Architectural Heritage. ISPRS International Journal of Geo–Information, 9(6).

Grilli, E., & Remondino, F. (2019). Classification of 3D Digital Heritage. *Remote Sensing, 11*(7), 1–23. doi:10.3390/rs11070847

Grilli, E., & Remondino, F. (2020). Machine learning generalisation across different 3D architectural heritage. *ISPRS International Journal of Geo-Information, 9*(6), 1–19.

Group, T. K. (2022). *COLLADA2GLTF*. Retrieved from pagina github: https://github.com/KhronosGroup/COLLADA2GLTF

Grynszpan, O., Martin, J.-C., & Nadel, J. (2008). Multimedia interfaces for users with high functioning autism: An empirical investigation. *International Journal of Human-Computer Studies, 66*(8), 628–639. doi:10.1016/j.ijhcs.2008.04.001

Gualandi, M. L., Scopigno, R., Wolf, L., Richards, J., Heinzelmann, M., Hervas, M. A., Vila, L., & Zallocco, M. (2016). ArchAIDE-archaeological automatic interpretation and documentation of cEramics. In C. E. Catalano, & L. De Luca (Eds.), *EUROGRAPHICS Workshop on Graphics and Cultural Heritage* (pp. 1-4). The Eurographics Association.

Guarnera, D., Guarnera, G. C., Ghosh, A., Denk, C., & Glencross, M. (2016). BRDF Representation and Acquisition. *Computer Graphics Forum, 35*(2), 625–650. doi:10.1111/cgf.12867

Gunay, H. B., Shen, W., & Newsham, G. (2019). Data analytics to improve building performance: A critical review. *Automation in Construction, 97*, 96–109. doi:10.1016/j.autcon.2018.10.020

Güneri, O. Y. (2006). Counseling services in Turkish universities. *International Journal of Mental Health, 35*(1), 26–38. doi:10.2753/IMH0020-7411350102

Güneri, O. Y., Aydin, G., & Skovholt, T. (2003). Counseling needs of students and evaluation of counseling services at a large urban university in Turkey. *International Journal for the Advancement of Counseling, 25*(1), 53–63. doi:10.1023/A:1024928212103

Gupta, M. (2020). *What is Digitization, Digitalization, and Digital Transformation?* ARC Advisory Group. https://www.arcweb.com/blog/what-digitization-digitalization-digital-transformation

Gupta, S., Kaushik, P., & Suma, D. (2015). Game Play Evaluation Metrics. *International Journal of Computers and Applications, 117*(3), 32–35. doi:10.5120/20538-2902

Gurjanov, A., Zakoldaev, A., Shukalov, A., & Zharino, I. (2020). The smart city technology in the super-intellectual Society 5.0. *Journal of Physics: Conference Series, 1679*(3), 032029. doi:10.1088/1742-6596/1679/3/032029

Gurwin, G. (2019). *AR Technology is Letting Children With Disabilities Play Sports.* VR Fitness Insider. https://www.vrfitnessinsider.com/ar-technology-is-letting-children-with-disabilities-play-sports/

Haans, A., & IJsselsteijn, W. A. (2009). The Virtual Midas Touch: Helping Behavior After a Mediated Social Touch. *IEEE Transactions on Haptics, 2*(3), 136–140. doi:10.1109/TOH.2009.20 PMID:27788077

Hachman, M. (2015). *Developing with HoloLens: Decent hardware chases Microsoft's lofty augmented reality ideal.* Academic Press.

Haleema, A., Javaid, M., Qadri, M., & Suman, R. (2022). Understanding the role of digital technologies in education: A review. *Sustainable Operations and Computers*, 275-285.

Hall, E. T. (1966). *The Hidden Dimension.* Anchor Books.

Halpern, O. (2017). Architecture as Machine: The smart city deconstructed. In When is the digital in architecture (pp. 123–175). Sternberg Press.

Hamilton, D., McKechnie, J., Edgerton, E., & Wilson, C. (2021). Immersive virtual reality as a pedagogical tool in education: A systematic literature review of quantitative learning outcomes and experimental design. *Journal of Computers in Education, 8*(1), 1–32. doi:10.100740692-020-00169-2

Han, Y. D. N. (2021). A Dynamic Resource Allocation Framework for Synchronizing Metaverse with IoT Service and Data. Computer Science and Game Theory. Cornell University.

Han, D. I., Claudia tom Dieck, M., & Jung, T. (2018). User experience model for augmented reality applications in urban heritage tourism. *Journal of Heritage Tourism, 13*(1), 46–61. doi:10.1080/1743873X.2016.1251931

Handani, S. W., Suyanto, M., & Sofyan, A. F. (2016). Penerapan Konsep Gamifikasi pada E-Learning Untuk Pembelajaran Animasi 3 Dimensi. *Jurnal Telematika, 9*(1), 42–53. doi:10.2214/ajr.181.6.1811716b

Hanrahan, P., & Krueger, W. (1993). Reflection from layered surfaces due to subsurface scattering. In *Proceedings of the 20th Annual Conference on Computer Graphics and Interactive Techniques (SIGGRAPH 93)*. ACM. 10.1145/166117.166139

Hasegawa, T., Tsumura, N., Nakaguchi, T. & Iino, K. (2011). Photometric Approach to Surface Reconstruction of Artist Paintings. *Journal of Electronic Imaging, 20*(1), 11.

Hassani, F. (2015). Documentation of Cultural Heritage. Techniques, potentials and constraints. *The International Archives of the Photogrammetry, Remote Sensing and Spatial Information Sciences, 40*(5), 207–214. doi:10.5194/isprsarchives-XL-5-W7-207-2015

Hassan, R. (2004). *Media, Politics and the Network Society*. McGraw-Hill Education.

Hassenzahl, M. (2010). Experience design: Technology for all the right reasons. In J. M. Carroll (Ed.), *Synthesis lectures on human-centered informatics* (pp. 1–95). Morgan & Claypool. doi:10.1007/978-3-031-02191-6

Hawas, S., & Marzouk, M. (2017). In Y. Arayaci, J. Counsell, L. Mahdjoubi, G. Nagy, & K. Dewidar (Eds.), *Integrating Value Map with Building Information Modelling Approach for Documenting Historic Buildings in Egypt* (pp. 62–72). Heritage Building Information Modelling.

Hedman, P., Srinivasan, P. P., Mildenhall, B., Barron, J. T., & Debevec, P. (2021). Baking Neural Radiance Fields for Real-Time View Synthesis. In *Proceedings of the IEEE/CVF International Conference on Computer Vision*, 5875-5884.

Hee-soo Choi, S.-h. K. (2017). A content service deployment plan for metaverse museum exhibitions-Centering on the combination of beacons and HMDs. *International Journal of Information Management.*

Hein, A., Böhm, M., & Krcmar, H. (2018). *Platform Configurations within Information Systems Research: A Literature Review on the Example of IoT Platforms*. Multikonferenz Wirtschaftsinformatik.

Hein, R. M., Wienrich, C., & Latoschik, M. E. (2021). A systematic review of foreign language learning with immersive technologies. *AIMS Electronics and Electrical Engineering, 5*(2), 117–145. doi:10.3934/electreng.2021007

Heitz, E. (2014). Understanding the masking-shadowing function in microfacet-based BRDFs. *Journal of Computer Graphics Techniques, 3*(2), 32–91.

Hempel, J. (2015). *Project HoloLens: Our Exclusive Hands-On With Microsoft's Holographic Goggles*. Academic Press.

Hennig-Thurau, T., Ravid, S. A., & Sorenson, O. (2021). The Economics of Filmed Entertainment in the Digital Era. *Journal of Cultural Economics, 45*(2), 1–14. doi:10.100710824-021-09407-6

Hensel, M. U. (2015). *Grounds and envelopes reshaping architecture and the built environment*. Routledge, Taylor & Francis Group.

Herdman, M., Gudex, C., Lloyd, A., Janssen, M.F., Kind, P., Parkin, D., Bonsel, G., & Badia, X. (2011). Development and preliminary testing of the new five-level version of EQ-5D (EQ-5D-5L). *Qual Life Res., 20*(10). doi:10.1007/s11136-011-9903-x

Hermann, M., Pentek, T., & Otto, B. (2016). Design Principles for Industrie 4.0 Scenarios. *2016 49th Hawaii International Conference on System Sciences (HICSS)*, 3928–3937. 10.1109/HICSS.2016.488

Herrera, G., Alcantua, F., Jordan, R., Blanquer, A., Labajo, G., & De Pablo, C. (2005). Development of Symbolic play through the use of Virtual Reality tools in children with Autistic Spectrum Disorders: Two case studies. *Autism*. Advance online publication. doi:10.1177/1362361307086657 PMID:18308764

Herrera, G., Jordan, R., & Vera, L. (2006). Abstract concept and imagination teaching through virtual reality in people with autism spectrum disorders. *Technology and Disability, 18*(4), 173–180. doi:10.3233/TAD-2006-18403

Hertenstein, M. J., Keltner, D., App, B., Bulleit, B. A., & Jaskolka, A. R. (2006). Touch communicates distinct emotions. *Emotion (Washington, D.C.)*, *6*(3), 528–533. doi:10.1037/1528-3542.6.3.528 PMID:16938094

Higuera-Trujillo, J. L., Maldonado, J. L. T., & Millán, C. L. (2017). Psychological and physiological human responses to simulated and real environments: A comparison between Photographs, 360 Panoramas, and Virtual Reality. *Applied Ergonomics*, *65*, 398–409. doi:10.1016/j.apergo.2017.05.006 PMID:28601190

Holden, M. K. (2005). Virtual environments for motor rehabilitation [review]. *Cyberpsychology & Behavior*, *8*(3), 187–211. doi:10.1089/cpb.2005.8.187 PMID:15971970

Hollister, S. (2015). *Microsoft HoloLens Hands-On: Incredible*. Amazing, Prototype-y as Hell.

Holmdahl, T. (2015). *BUILD 2015: A closer look at the Microsoft HoloLens hardware*. Academic Press.

Holmes, C., & Foster, V. (2012). A preliminary comparison study of online and face-to-face counseling: Client perceptions of three factors. *Journal of Technology in Human Services*, *30*(1), 14–31. doi:10.1080/15228835.2012.662848

Horizon Europe strategic plan 2021-2024. (n.d.). *European Commission*. https://ec.europa.eu/commission/presscorner/detail/en/ip_21_1122

Hørlyck, L. D., Obenhausen, K., Jansari, A., Ullum, H., & Miskowiak, K. W. (2021). Virtual reality assessment of daily life executive functions in mood disorders: Associations with neuropsychological and functional measures. *Journal of Affective Disorders*, *280*, 478–487. doi:10.1016/j.jad.2020.11.084 PMID:33248416

Hossain, A., & Nadeem, A. (n.d.). *Towards digitizing the construction industry: State of the art of Construction 4.0*. Academic Press.

Howard, P. (2003). *Heritage: management, interpretation, identity*. Continuum. doi:10.5040/9781350933941

Huang, K. T., Ball, C., Francis, J., Ratan, R., Boumis, J., & Fordham, J. (2019). Augmented versus virtual reality in education: An exploratory study examining science knowledge retention when using augmented reality/virtual reality Mobile applications. *Cyberpsychology, Behavior, and Social Networking*, *22*(2), 105–110. doi:10.1089/cyber.2018.0150 PMID:30657334

Huang, R., Yang, J., & Zheng, L. (2013). The Components and Functions of Smart Learning Environments for Easy, Engaged and Effective Learning The Demands on Rebuilding Learning Environments in Information Society. *International Journal for Educational Media and Technology*, *7*(1), 4–14.

Huffman, K. L., & Giordano, A. (2019). Visualizing Venice to Visualizing Cities - Advanced Technologies for Historical Cities Visualization. In F. Niebling, S. Münster & H. Messemer (Eds.), *Research and Education in Urban History in the Age of Digital Libraries: Second International Workshop, UHDL 2019*, Dresden, Germany, October 10–11, 2019. Springer, 171-187.

Huffman, K. L., Giordano, A., & Bruzelius, C. (Eds.). (2017). *Visualizing Venice: Mapping and Modeling Time and Change in a City*. Taylor & Francis.

Hull, J., & Ewart, I. (2020). Conservation data parameters for BIM-enabled heritage asset management. *Automation in Construction, 119*.

Hurkxkens, I. G. (2017). Robotic Landscapes: Developing Computational Design Tools Towards Autonomous Terrain Modeling. *ACADIA 2017: Disciplines & Disruption: Proceedings of the 37th Annual Conference of the Association for Computer Aided Design in Architecture (ACADIA)*, 292-297. http://papers.cumincad.org/cgi-bin/works/paper/acadia17_292

Huygelier, H., Mattheus, E., Abeele, V. V., van Ee, R., & Gillebert, C. R. (2021). The Use of the Term Virtual Reality in Post-Stroke Rehabilitation: A Scoping Review and Commentary. *Psychologica Belgica*, *61*(1), 145–162. doi:10.5334/pb.1033

Iadanza, E., Maietti, F., Medici, M., Ferrari, F., Turillazzi, B., & Di Giulio, R. (2020). Bridging the Gap between 3D Navigation and Semantic Search. The INCEPTION platform. *IOP Conference Series. Materials Science and Engineering*, *949*(1), 012079.

Iadanza, E., Maietti, F., Ziri, A. E., Di Giulio, R., Medici, M., Ferrari, F., Bonsma, P., & Turillazzi, B. (2019). Semantic Web Technologies Meet BIM for Accessing and Understanding Cultural Heritage. *The International Archives of the Photogrammetry, Remote Sensing and Spatial Information Sciences*, *42*(W9), 381–388.

Ibanes-Etxeberria, A., G'omez-Carrasco, C. J., Fontal, O., & García-Ceballos, S. (2020). Virtual environments and augmented reality applied to heritage education. An evaluative study. *Applied Sciences (Basel, Switzerland)*, *10*(7), 2352. doi:10.3390/app10072352

Ibáñez, M., Uriarte Portillo, A., Zatarain Cabada, R., & Barron Estrada, M. (2020). Impact of augmented reality technology on academic achievement and motivation. *Computers & Education*, *145*, 1–9.

Ibrahim, N., & Ali, N. M. (2018). A conceptual framework for designing virtual heritage environment for cultural learning. *Journal on Computing and Cultural Heritage*, *11*(2), 1–27. doi:10.1145/3117801

ICOMOS Australia. (2013). *The Burra Charter. Practice Note_Interpretation*. Retrieved on April 27, 2022 from https://australia.icomos.org/wp-content/uploads/Practice-Note_Interpretation.pdf

ICOMOS. (1964). *International charter for the conservation and restoration of monuments and sites*. Retrieved on April 27, 2022 from https://www.icomos.org/charters/venice_e.pdf

ICOMOS. (2008). *Charter for the Interpretation and Presentation of Cultural Heritage Sites*. Retrieved on April 27, 2022 from http://icip.icomos.org/downloads/ICOMOS_Interpretation_Charter_ENG_04_10_08.pdf

ICOMOS. (2013). *The Burra Charter*. Retrieved on April 27, 2022 from http://portal.iphan.gov.br/uploads/ckfinder/arquivos/The-Burra-Charter-2013-Adopted-31_10_2013.pdf

IDEO. (2014). *Design Thinking for Educators toolkit*. Retrieved March 10, 2020 from http://www.ideo.com/work/toolkit-for-educators

IfcOpenShell. (2022). *IfcConvert: An application for converting ifc geometry into several file formats*. Retrieved from IfcOpenShell: http://ifcopenshell.org/ifcconvert

Iggo, A., & Andres, K. H. (1982). Morphology of cutaneous receptors. *Annual Review of Neuroscience*, *5*(1), 1–31. doi:10.1146/annurev.ne.05.030182.000245 PMID:6280572

Ijsselsteijn, W., & Riva, G. (2003). Being there: the experience of presence in mediated environments. In G. Riva, F. Davide, & W. A. IJsselsteijn (Eds.), *Studies in new technologies and practices in communication. Being there: concepts, effects and measurement of user presence in synthetic environments* (pp. 3–16). IOS Press.

ImmersionVR. (2022). *VR for Education - The Future of Education*. https://immersionvr.co.uk/about-360vr/vr-for-education/

Incerti, M., D'Amico, S., Giannetti, S., Lavoratti, G., & Velo, U. (2018). Le digital humanities per lo studio e la comunicazione di beni culturali architettonici: Il caso dei mausolei di Teodorico e Galla Placidia in Ravenna [Digital humanities for the study and communication of architectural cultural heritage: the case of the mausoleums of Theodoric and Galla Placidia in Ravenna]. *Archeologia e Calcolatori*, *29*, 297–316.

Indriani, R., Sugiarto, B., & Purwanto, A. (2016). Pembuatan Augmented Reality Tentang Pengenalan Hewan Untuk Anak Usia Dini Berbasis Android Menggunakan Metode Image Tracking Vuforia. *Seminar Nasional Teknologi Informasi Dan Multimedia*, 73–78.

Inman, D. P., Loge, K., & Leavens, J. (1997). VR education and rehabilitation. *Communications of the ACM*, *40*(8), 53–58. doi:10.1145/257874.257886

International Data Corporation. (2022). *Wearables Deliver Double-Digit Growth for Both Q4 and the Full Year 2021, According to IDC*. IDC. https://www.idc.com/getdoc.jsp?containerId=prUS48935722

Ippoliti, E., & Casale, A. (2018). Rappresentare, comunicare, narrare. In A. Luigini & C. Panciroli (Eds.), Ambienti digitali per l'educazione all'arte e al patrimonio. Franco Angeli, 128-150.

Ishii, M., Moryiama, T., Toda, M., Kohmoto, K., & Saito, M. (2008). Color degradation of textiles with natural dyes and blue scale standards exposed to white LED lamps: evaluation for effectiveness as museum lighting. *Journal of Light and Visual Environment*, *32*(4), 370-378.

Islam, S. R. D. K.-S. (2015). The internet of things for health care: a comprehensive survey. IEEE Access, 3, 678–708.

ISO. (2010) Human-centered design for interactive systems. ISO 9241-210:2010 (E).

ISO. (2010). *ISO 9241-210:2010. Ergonomics of human-system interaction — Part 210: Human-centred design for interactive systems*.

Israilidis, J., Odusanya, K., & Mazhar, M. (2019). Exploring knowledge management perspectives in smart city research: A review and future research agenda. *International Journal of Information Management*.

Itani, O., & Hollebeek, L. (2021). Light at the end of the tunnel: Visitors' virtual reality (versus in-person) attraction site tour-related behavioral intentions during and post-COVID-19. *Tourism Management*, 84.

Jacoby, M., Averbuch, S., Sacher, Y., Katz, N., Weiss, P. L., & Kizony, R. (2013). Effectiveness of Executive Functions Training Within a Virtual Supermarket for Adults With Traumatic Brain Injury: A Pilot Study. *IEEE Transactions on Neural Systems and Rehabilitation Engineering*, *21*(2), 182–190. doi:10.1109/TNSRE.2012.2235184 PMID:23292820

Jaime Ibarra Jimenez, H. J. (2019). *Health Care in the Cyberspace: Medical Cyber-Physical System and Digital Twin Challenges*. Digital Twin Technologies and Smart Cities.

Jakobsen, M. R., Sahlemariam Haile, Y., Knudsen, S., & Hornbæk, K. (2013). Information Visualization and Proxemics: Design Opportunities and Empirical Findings. *IEEE Transactions on Visualization and Computer Graphics*, *19*(12), 2386–2395. doi:10.1109/TVCG.2013.166 PMID:24051805

Jan, J. F. (2018). Application of open-source software in community heritage resources management. *International Journal of Geo-Information*, *7*, 426.

Japan Business Federation (Keidanren). (2016), Toward Realization of the New Economy and Society. In Reform of the Economy and Society by the Deepening of "Society 5.0". Author.

Jean-Caurant, A., & Doucet, A. (2020). Accessing and investigating large collections of historical newspapers with the NewsEye platform. In *Proceedings of the ACM/IEEE Joint Conference on Digital Libraries in 2020* (pp. 531-532). ACM.

Jeffs, T. L. (2009). Virtual reality and special needs. *Themes in Science and Technology Education*, *2*(1-2), 253-268. https://eric.ed.gov/?id=EJ1131319

Jerald, J. (2015). *The VR book: Human-centered design for virtual reality*. Morgan & Claypool Publishers. doi:10.1145/2792790

Jetter, H. C., Reiterer, H., & Geyer, F. (2014). Blended Interaction: Understanding natural human–computer interaction in post-WIMP interactive spaces. *Personal and Ubiquitous Computing*, *18*(5), 1139–1158. doi:10.100700779-013-0725-4

Jin. (1978). *Visibility Through fire smoke*. Academic Press.

Jin, X., Wah, B. W., Cheng, X., & Wang, Y. (2015). Significance and Challenges of Big Data Research. *Big Data Research*, *2*(2), 59–64. doi:10.1016/j.bdr.2015.01.006

Johnson, D., Deterding, S., Kuhn, K.-A., Staneva, A., Stoyanov, S., & Hides, L. (2016). Gamification for health and wellbeing: A systematic review of the literature. *Internet Interventions: the Application of Information Technology in Mental and Behavioural Health*, *6*, 89–106. doi:10.1016/j.invent.2016.10.002 PMID:30135818

Johnson-Glenberg, M. C. (2018). Immersive VR and Education: Embodied Design Principles That Include Gesture and Hand Controls. *Frontiers in Robotics and AI*, *5*, 81. doi:10.3389/frobt.2018.00081 PMID:33500960

Jones, D., Snider, C., Nassehi, A., Yon, J., & Hicks, B. (2020). Characterising the Digital Twin: A systematic literature review". *CIRP Journal of Manufacturing Science and Technology*, *29*, 36–52. doi:10.1016/j.cirpj.2020.02.002

Joseph, C. (2020). Augmented reality and virtual reality to aid students with learning disability: A review. *International Journal of Scientific & Technology Research*, *9*(2), 6475–6478. http://www.ijstr.org/paper-references.php?ref=IJSTR-1219-26850

Joshi, A., Saket, K., Chandel, S., & Dinesh Kumar, P. (2015). Likert Scale: Explored and Explained. *Current Journal of Applied Science and Technology*, *7*(4), 396-403. doi:10.9734/BJAST/2015/14975

Josman, N., Hof, E., Klinger, E., Marié, R. M., Goldenberg, K., Weiss, P. L., & Kizony, R. (2006). Performance within a virtual supermarket and its relationship to executive functions in post-stroke patients. *2006 International Workshop on Virtual Rehabilitation*, 106-109. 10.1109/IWVR.2006.1707536

Juan Manuel Davila Delgado, L. O. (2021). Digital Twins for the built environment: Learning from conceptual and process models in manufacturing. *Advanced Engineering Informatics*, 49.

Jusuf, H. (2016). Penggunaan Gamifikasi dalam Proses Pembelajaran [The Use of Gamification in the Learning Process]. *Jurnal TICOM*, *5*(1), 1–6.

Kaldeli, E., García-Martínez, M., Isaac, A., Scalia, P. S., Stabenau, A., Almor, I. L., & Herranz, M. (2022). Europeana Translate: Providing multilingual access to digital cultural heritage. In *Proceedings of the 23rd Annual Conference of the European Association for Machine Translation* (pp. 297-298). European Association for Machine Translation.

Kang, S. (2020). Going Beyond Just Watching: The Fan Adoption Process of Virtual Reality Spectatorship. *Journal of Broadcasting & Electronic Media*, *64*(3), 499–518. doi:10.1080/08838151.2020.1798159

Kang, S. H., & Gratch, J. (2010). Virtual humans elicit socially anxious interactants' verbal self-disclosure. *Computer Animation and Virtual Worlds*, *21*(May), 473–482. doi:10.1002/cav.345

Karp, C. (2004). Digital Heritage in Digital Museums. *Museum International*, *56*(1-2), 45–51. doi:10.1111/j.1350-0775.2004.00457.x

Karuei, I., MacLean, K. E., Foley-Fisher, Z., MacKenzie, R., Koch, S., & El-Zohairy, M. (2011). Detecting vibrations across the body in mobile contexts. *Proceedings of the SIGCHI Conference on Human Factors in Computing Systems*, 3267–3276. 10.1145/1978942.1979426

Katifori, A., Roussou, M., Perry, S., Palma, G., Drettakis, G., Vizcay, S., & Philip, J. (2018). The EMOTIVE Project - Emotive Virtual Cultural Experiences through Personalized Storytelling. CIRA@EuroMed.

Keating, S. J., Leland, J. C., Cai, L., & Oxman, N. (2017). Toward site-specific and self-sufficient robotic fabrication on architectural scales. *Science Robotics*, *2*(5), eaam8986. Advance online publication. doi:10.1126cirobotics.aam8986 PMID:33157892

Kemp, G., Smith, M., & Segal, J. (2020). *Learning Disabilities and Disorders*. HelpGuide, Trusted nonprofit guide to mental health & wellness. https://www.helpguide.org/articles/autism-learning-disabilities/learning-disabilities-and-disorders.htm

Kennedy, R. S., Drexler, J. M., Compton, D. E., Stanney, K. M., Lanham, D. S., & Harm, D. L. (2003). Configural scoring of simulator sickness, cybersickness, and space adaptation syndrome: Similarities and differences. In L. J. Hettinger & M. W. Haas (Eds.), *Virtual and adaptive environments: Applications, implications, and human performance issues* (pp. 247–278). Lawrence Erlbaum Associates Publishers.

Kennedy, R. S., Lane, N. E., Berbaum, K. S., & Lilienthal, M. G. (1993). Simulator sickness questionnaire: An enhanced method for quantifying simulator sickness. *The International Journal of Aviation Psychology*, *3*(3), 203–220. doi:10.120715327108ijap0303_3

Kepczynska-Walczak, A. &Walczak, B.M. (2015). Built heritage perception through representation of its atmosphere. *Experiential Simulation, 1*.

Kerski, J. J. (2022). Online, Engaged Instruction in Geography and GIS Using IoT Feeds, Web Mapping Services, and Field Tools within a Spatial Thinking Framework. *Geography Teacher (Erie, Pa.)*, *19*(3), 93–101. doi:10.1080/19338341.2022.2070520

Kersten, T. P., Tschirschwitz, F., Deggim, S., & Lindstaedt, M. (2018). Virtual reality for cultural heritage monuments–from 3D data recording to immersive visualisation. In *Euro-Mediterranean Conference* (pp. 74-83). Springer. 10.1007/978-3-030-01765-1_9

Kettner, R., Bader, P., Kosch, T., Schneegass, S., & Schmidt, A. (2017). Towards pressure-based feedback for non-stressful tactile notifications. *Proceedings of the 19th International Conference on Human-Computer Interaction with Mobile Devices and Services - MobileHCI '17*, 1–8. 10.1145/3098279.3122132

Khan, I., Melro, A., Carla, A., & Oliveira, L. (2020). Systematic review on gamification and cultural heritage dissemination. *Journal of Digital Media & Interaction*, *3*(8), 19–41.

Khan, M. F., Hussain, M. A., Ahsan, K., Saeed, M., Naddem, A., & Ali, S. A. (2017). Augmented reality based spelling assistance to dysgraphia students. *Journal of Basic and Applied Sciences*, *13*, 500–507. doi:10.6000/1927-5129.2017.13.82

Kharrazi, H., Faiola, A., & Defazio, J. (2009). *Health care game design: behavioral modeling of serious gaming design for children with chronic diseases* (Vol. 5613). Human-Computer Interaction. Interacting in Various Application Domains Lecture Notes in Computer Science.

Kibuku, R., Ochieng, D., & Wausi, A. (2020). E-learning challenges faced by universities in Kenya: A literature review. *Electronic Journal of E-Learning*, 150-161.

Kim, H. K., Park, J., Choi, Y., & Choe, M. (2018). Virtual reality sickness questionnaire (VRSQ): Motion sickness measurement index in a virtual reality environment. *Applied Ergonomics*, *69*, 66–73. doi:10.1016/j.apergo.2017.12.016 PMID:29477332

Kim, I., Martins, R. J., Jang, J., Badloe, T., Khadir, S., Jung, H. Y., Kim, H., Kim, J., Genevet, P., & Rho, J. (2021). Nanophotonics for light detection and ranging technology. *Nature Nanotechnology*, *16*(5), 508–524. doi:10.103841565-021-00895-3 PMID:33958762

Kim, J. J.-Y. (2004). The Organizational Complex: Architecture, Media, and Corporate Space [review]. *Technology and Culture*, *45*(2), 468–470. doi:10.1353/tech.2004.0072

Kincade, E. A., & Kalodner, C. R. (2004). The use of groups in college and university counseling centers. In J. L. DeLucia-Waack, D. A. Gerrity, C. R. Kalodner, & M. T. Riva (Eds.), *Handbook of group counseling and psychotherapy* (pp. 366–377). Sage., doi:10.4135/9781452229683.n26

King, E., Smith, M. P., Wilson, P. F., & Williams, M. A. (2021). Digital Responses of UK Museum Exhibitions to the COVID-19 Crisis. *Curator*, *64*(3), 487–504.

King, L., Stark, J. F., & Cooke, P. (2016). Experiencing the Digital World: The Cultural Value of Digital Engagement with Heritage. *Heritage & Society*, *9*(1), 76–101. doi:10.1080/2159032X.2016.1246156

Kipman, A., & Juarez, S. (2015). *Developing for HoloLens*. Microsoft.

Kist, C. (2020). Museums, Challenging Heritage and Social Media During COVID-19. *Museum & Society*, *18*(3), 345–348. doi:10.29311/mas.v18i3.3539

Kitchin, R. (2014). Big Data, new epistemologies and paradigm shifts. *Big Data & Society*, *1*(1), 12. doi:10.1177/2053951714528481

Kizony, R., Katz, N., & Weiss, P. L. (2003). Adapting an immersive virtual reality system for rehabilitation. *Computer Animation and Virtual Worlds*, *14*, 261–268.

Klanten, R. (2010). *Data flow 2: Visualizing information in graphic design*. Gestalten.

Klee, P. (1959). *Teoria della forma e della figurazione* (Vol. 1). Feltrinelli.

Klinger, E., Chemin, I., Lebreton, S., & Marie, R. M. (2004). A virtual supermarket to assess cognitive planning. *Annual Review of Cybertherapy and Telemedicine*, *2*, 49–57.

Kodratoff, Y. (1989). *Introduction to Machine Learning*. Elsevier.

KOMINFO. (2017). *Survei Penggunaan TIK Serta Implikasinya terhadap Aspek Sosial Budaya Masyarakat 2017*. BPPSDM KOMINFO.

Kontogianni, G., Koutsaftis, C., Skamantzari, M., Chrysanthopoulou, C., & Georgopoulos, A. (2017). Utilising 3D realistic models in serious games for cultural heritage. *International Journal of Computational Methods in Heritage Science*, *1*(2), 21–46. doi:10.4018/IJCMHS.2017070102

Korro Bañuelos, J., Rodríguez Miranda, Á., Valle-Melón, J. M., Zornoza-Indart, A., Castellano-Román, M., Angulo-Fornos, R., Pinto-Puerto, F., Acosta Ibáñez, P., & Ferreira-Lopes, P. (2021). The Role of Information Management for the Sustainable Conservation of Cultural Heritage. *Sustainability*, *13*(8), 4325. doi:10.3390u13084325

Kowalewski, B., & Girot, C. (2021). *The Site Visit: Towards a Digital in Situ Design Tool*. Wichmann Verlag., doi:10.14627/537705022

Koydemir, S., Erel, Ö., Yumurtacı, D., & Şahin, G. N. (2010). Psychological help-seeking attitudes and barriers to help-seeking in young people in Turkey. *International Journal for the Advancement of Counseling*, *32*(4), 274–289. doi:10.100710447-010-9106-0

Krath, J., Schürmann, L., & Von Korflesch, H. F. (2021). Revealing the theoretical basis of gamification: A systematic review and analysis of theory in research on gamification, serious games and game-based learning. *Computers in Human Behavior, Vol.*, *125*(106963), 2–33. doi:10.1016/j.chb.2021.106963

Kritzinger, W., Karner, M., Traar, G., Henjes, J., & Sihn, W. (2018). Digital Twin in Manufacturing: A Categorical Literature Review and Classification. *IFAC-Pap*, 1016-1022.

Kritzinger, W., Karner, M., Traar, G., Henjes, J., & Sihn, W. (2018). Digital Twin in manufacturing. *IFAC, 51*(11), 1016–1022.

Kuligowski, E., Peacock, R., & Hoskins, B. (2010). *A Review of Building Evacuation Models* (2nd ed.). NIST Pubs.

Kulkarni, S. (2019). *Virtual And Augmented Realities Landscape In Indian Higher Education*. Adaption of social media in academia. https://www.asmaindia.in/blog/virtual-and-augmented-reality-landscape-in-indian-higher-education

Kuriyan, N. M., & James, J. (2018). Prevalence of learning disability in India: A need for mental health awareness programme. *Conference: First National Conference on Mental Health Education*. doi: 10.4103/0253-

Kuznetsov, S., & Paulos, E. (2010), Rise of the Expert Amateur: DIY Projects, Communities, and Cultures. In *Proceedings of the 6th Nordic Conference on Human-Computer Interaction: Extending Boundaries*. Association for Computing Machinery.

Kyaw, B. M., Saxena, N., Posadzki, P., Vseteckova, J., Nikolaou, C. K., George, P. P., Divakar, U., Masiello, I., Kononowicz, A. A., Zary, N., & Tudor Car, L. (2019). Virtual reality for health professions education: Systematic review and meta-analysis by the digital health education collaboration. *Journal of Medical Internet Research, 21*(1), 12959. doi:10.2196/12959 PMID:30668519

Kye, B., Han, N., Kim, E., Park, Y., & Jo, S. (2021). Educational applications of metaverse: Possibilities and limitations. *Journal of Educational Evaluation for Health Professions, 18*, 32. doi:10.3352/jeehp.2021.18.32 PMID:34897242

Kyoungro Yoon, S.-K. K.-H. (2021). Interfacing Cyber and Physical Worlds: Introduction to IEEE 2888 Standards. In *International Conference on Intelligent Reality (ICIR)*. IEEE.

La Paglia, F., la Cascia, C., Cipresso, P., Rizzo, R., Francomano, A., Riva, G., & la Barbera, D. (2014). Psychometric Assessment Using Classic Neuropsychological and Virtual Reality Based Test: A Study in Obsessive-Compulsive Disorder (OCD) and Schizophrenic Patients. doi:10.1007/978-3-319-11564-1_3

Lacy, B. E., & Chan, J. L. (2018). Physician Burnout: The Hidden Health Care Crisis. *Clinical Gastroenterology and Hepatology, 16*(3), 311–317. doi:10.1016/j.cgh.2017.06.043 PMID:28669661

Laing, R. (2020). Built heritage modelling and visualisation: The potential to engage with issues of heritage value and wider participation. *Developments in the Built Environment., 4*, 1–7. Retrieved March 13, 2022, from. doi:10.1016/j.dibe.2020.100017

Lanier, J. (2017). *Dawn of the New Everything: A Journey Through Virtual Reality*. Henry Holt and Co., Inc.

Lányi, C. S. (2006). Virtual reality in healthcare. In N. Ichalkaranje, A. Ichalkaranje, & L. Jain (Eds.), *Intelligent Paradigms for Assistive and Preventive Healthcare* (Vol. 19, pp. 88–116). Springer. doi:10.1007/11418337_3

Lasi, H., Fettke, P., Kemper, H.-G., Feld, T., & Hoffmann, M. (2014). Industry 4.0. *Business & Information Systems Engineering, 6*(4), 239–242. doi:10.100712599-014-0334-4

Lazer, D., Pentland, A., Adamic, L., Aral, S., Barabási, A., Brewer, D., Christakis, N., Contractor, N., Fowler, J., Gutmann, M., Jebara, T., King, G., Macy, M., Roy, D., & Van Alstyne, M. (2009). Computational Social Science. *Science, 323*(5915), 721-723.

Leach, N. (2018). Informational Cities. In P. F. Yuan & N. Leach (Eds.), *Informational Cities* (pp. 104–113). Tongji University Press Co., Ltd.

Learning Disabilities Association of America. (2018). *Types of Learning Disabilities*. LDA: Core Principles for the Identification and Support of Individuals with Learning Disabilities. https://ldaamerica.org/types-of-learning-disabilities/

Lederman, S. J., & Klatzky, R. L. (2009). Haptic perception: A tutorial. *Attention, Perception & Psychophysics, 71*(7), 1439–1459. doi:10.3758/APP.71.7.1439 PMID:19801605

Lee, J., Kim, J., Ahn, J., & Woo, W. (2019). Context-aware risk management for Architectural Heritage using Historic Building Information Modeling and Virtual Reality. *Journal of Cultural Heritage,* 242–252.

Lee, L., Braud, T., Zhou, P., Wang, L., Xu, D., Lin, Z., . . . Hui, P. (2021). *From Internet and Extended Reality to Metaverse: Technology Survey, Ecosystem and Future Directions.* arXiv preprint arXiv:2110.05352.

Lee, N., Choi, W. & Lee, S. (2021) Development of an 360-degree virtual reality video-based immersive cycle training system for physical enhancement in older adults: a feasibility study. *BMC Geriatrics, 21*, 325. doi:10.1186/s12877-021-02263-1

Lefebvre, H. (2019). *Elementi di Ritmanalisi. Introduzione alla conoscenza dei ritmi.* LetteraVentidue.

Leroy, A., Cevallos, C., Cebolla, A. M., Caharel, S., Dan, B., & Cheron, G. (2017). Short-term EEG dynamics and neural generators evoked by navigational images. *PLoS One, 12*(6), 12. doi:10.1371/journal.pone.0178817 PMID:28632774

Lestiyo, I. (2021). *Gaming Report.* Unity Technologies.

Levy, J. B., Kong, E., Johnson, N., Khetarpal, A., Tomlinson, J., Martin, G. F., & Tanna, A. (2021). The mixed reality medical ward round with the MS HoloLens 2: Innovation in reducing COVID-19 transmission and PPE usage. *Future Healthcare Journal, 8*(1), e127–e130. doi:10.7861/fhj.2020-0146 PMID:33791491

Li, J., & Yang, H. (2017). A Research on Development of Construction Industrialization Based on BIM Technology under the Background of Industry 4.0. *MATEC Web of Conferences, 100*, 02046. 10.1051/matecconf/201710002046

Li, Y., & Ch'ng, E. (2022). A Framework for Sharing Cultural Heritage Objects in Hybrid Virtual and Augmented Reality Environments. In Visual Heritage: Digital Approaches in Heritage Science. Springer. doi:10.1007/978-3-030-77028-0_23

Liao, Y.-Y., Chen, I.-H., Lin, Y.-J., Chen, Y., & Hsu, W.-C. (2019). Effects of Virtual Reality-Based Physical and Cognitive Training on Executive Function and Dual-Task Gait Performance in Older Adults With Mild Cognitive Impairment: A Randomized Control Trial. *Frontiers in Aging Neuroscience, 11*, 162. Advance online publication. doi:10.3389/fnagi.2019.00162 PMID:31379553

Liao, Y.-Y., Tseng, H.-Y., Lin, Y.-J., Wang, C.-J., & Hsu, W.-C. (2020). Using virtual reality-based training to improve cognitive function, instrumental activities of daily living and neural efficiency in older adults with mild cognitive impairment. *European Journal of Physical and Rehabilitation Medicine, 56*(1). Advance online publication. doi:10.23736/S1973-9087.19.05899-4 PMID:31615196

Licaj, A. (2018). *Information Visualization. Disciplina liquida intersoggettiva* [Doctoral dissertation]. University of Genoa. doi:10.15167/licaj-ami_phd2018-05-09

Li, F. (2018). The digital transformation of business models in the creative industries: A holistic framework and emerging trends. *Technovation.*

Li, H., Yang, Z., Zhang, X., Wang, H., Liu, H., Cao, Y., & Zhang, G. (2021). Access to Nature via Virtual Reality: A Mini Review. *Frontiers in Psychology, 12*, 4324. doi:10.3389/fpsyg.2021.725288 PMID:34675840

Lik-Hang Lee, T. B. (2021). *All One Needs to Know about Metaverse: A Complete Survey on Technological Singularity, Virtual Ecosystem, and Research Agenda. Computer and society*. Cornell University.

Lim, M. Y., & Aylett, R. (2007). Narrative Construction in a Mobile Tour Guide. In M. Cavazza & S. Donikian (Eds.), *Virtual Storytelling. Using Virtual Reality Technologies for Storytelling. ICVS 2007* (Vol. 4871, pp. 51–62). Springer.

Lim, V., Frangakis, N., Tanco, L. M., & Picinali, L. (2018). PLUGGY: A pluggable social platform for cultural heritage awareness and participation. In *Advances in Digital Cultural Heritage* (pp. 117–129). Springer.

Liszio, S., & Masuch, M. (2019). Interactive immersive virtual environments cause relaxation and enhance resistance to acute stress. *Annual Review of Cybertherapy and Telemedicine*, *17*, 65–71.

Liu, Y., Tan, W., Chen, C., Liu, C., Yang, J., & Zhang, Y. (2019). A Review of the Application of Virtual Reality Technology in the Diagnosis and Treatment of Cognitive Impairment. *Frontiers in Aging Neuroscience*, *11*, 280. doi:10.3389/fnagi.2019.00280 PMID:31680934

Li, X., Yi, W., Chi, H., Wang, W., & Chan, A. P. C. (2018). A critical review of virtual and augmented reality (VR/AR) applications in construction safety. *Automation in Construction*, *86*, 150–162.

Lo Priore, C., Castelnuovo, G., Liccione, D., & Liccione, D. (2003). Experience with V-STORE: Considerations on Presence in Virtual Environments for Effective Neuropsychological Rehabilitation of Executive Functions. *Cyberpsychology & Behavior*, *6*(3), 281–287. doi:10.1089/109493103322011579 PMID:12855084

Lo Turco, M., & Bocconcino, M. M. (2017). Esattezza, molteplicità e integrazione nell'Information Modeling & Management [Exactitude, multiplicity and integration in Information Modelling & Management]. TECHNE, 13, 267-277.

Lo Turco, M., Piumatti, P., Calvano, M., Giovannini, E. C., Mafrici, N., Tomalini, A., & Fanini, B. (2019). Interactive Digital Environments for Cultural Heritage and Museums. Building a digital ecosystem to display hidden collections. *Disegnarecon*, *12*(23), 7.1-7.11.

Lo Turco, M., & Bocconcino, M. M. (2015). La rappresentazione operativa - Operative representation. In *BIM* (pp. 131–160). Massimiliano Lo Turco.

Lo Turco, M., Bocconcino, M. M., Cangialosi, G., & Serini, M. (2015). Dal disegno di progetto al modello di cantiere: le radici del FM. In *AR BIM GIS, a cura di Anna Osello* (pp. 126–139). Gangemi.

Lo Turco, M., Bocconcino, M. M., Vozzola, M., Giovannini, E., & Tomalini, A. (2021). *Il BIM per il Construction Management. Il caso della Domus Eleganza a Milano [BIM for Construction Management. The case of the Domus Eleganza in Milan]. 3D Modeling & BIM 2021 - Digital Twin.*

Löffler, D., Tscharn, R., Schaper, P., Hollenbach, M., & Mocke, V. (2019). Tight Times: Semantics and Distractibility of Pneumatic Compression Feedback for Wearable Devices. *Proceedings of Mensch Und Computer 2019 on - MuC'19*, 411–419. doi:10.1145/3340764.3340796

Logothetis, S., Karachaliou, E., Valari, E., & Styli, E. (2018). Open source cloud-based technologies for BIM. *The International Archives of the Photogrammetry, Remote Sensing and Spatial Information Sciences*, *42*, 607–614. doi:10.5194/isprs-archives-XLII-2-607-2018

Löken, L. S., Wessberg, J., Morrison, I., McGlone, F., & Olausson, H. (2009). Coding of pleasant touch by unmyelinated afferents in humans. *Nature Neuroscience*, *12*(5), 547–548. doi:10.1038/nn.2312 PMID:19363489

Lombard, M., & Ditton, T. (2006). At the Heart of It All: The Concept of Presence. *Journal of Computer-Mediated Communication*, *3*(2). doi:10.1111/j.1083-6101.1997.tb00072.x

Longstaff, P. H., Armstrong, N. J., Perrin, K., Parker, W. M., & Hidek, M. A. (2010). Building resilient communities: A preliminary framework for assessment. *Homeland Security Affairs*, *6*(3).

Loos, A., & Weigel, C. (2018). I-MEDIA-CITIES: Automatic Metadata Enrichment of Historic Media Content. In Metrology for Archaeology and Cultural Heritage (MetroArchaeo) (pp. 351-356). IEEE.

López, F. J., Lerones, P. M., Llamas, J., Gómez-García-Bermejo, J., & Zalama, E. (2018). Linking HBIM graphical and semantic information through the Getty AAT: Practical application to the Castle of Torrelobatón. *IOP Conf. Ser.: Mater. Sci. Eng., 364*.

Lovreglio, R., Duan, X., & Rahouti, A. (2021). Comparing the effectiveness of fire extinguisher virtual reality and video training. *Virtual Reality, 25*, 133–145.

Lusa, S., Rahmanto, Y., & Priyopradono, B. (2020). The Development Of Web 3d Application For Virtual Museum Of Lampung Culture. *Psychology and Education Journal, 57*(9), 188–193.

M'Closkey, K., & VanDerSys, K. (2017). *Dynamic Patterns: Visualizing Landscapes in a Digital Age* (1st ed.). Routledge., doi:10.4324/9781315681856

Macarulla, M., Casals, M., Forcada, N., & Gangolells, M. (2017). Implementation of predictive control in a commercial building energy management system using neural networks. *Energy and Building, 151*, 511–519. doi:10.1016/j.enbuild.2017.06.027

Macaruso, P., Hook, P. E., & McCabe, R. (2006). The efficacy of computer-based supplementary phonics programs for advancing reading skills in at-risk elementary students. *Journal of Research in Reading, 29*(2), 162–172. doi:10.1111/j.1467-9817.2006.00282.x

Macrì, E., & Cristofaro, C. L. (2021). The Digitalisation of Cultural Heritage for Sustainable Development: The Impact of Europeana. In P. Demartini, L. Marchegiani, M. Marchiori, & G. Schiuma (Eds.), *Cultural Initiatives for Sustainable Development. Contributions to Management Science* (pp. 373–400). Springer.

Madriaga, A., & Rubio, I. (2012). *Videojuegos y discapacidad. El reto de la inclusion* [Video games and Disability. The challenge of inclusion]. Ministerio de Sanidad, Servicios Sociales e Igualdad. Secretarìa de Estado de Servicios Sociales e Igualdad. Instituto de Mayores y Servicios Sociales.

Mahroum, S., Ferchachi, N., & Gomes, A. (2018). Inside the black box. Journey mapping digital innovation in government. INSEAD. The Business School for the World.

Maier, P., Klinker, G., & Tonis, M. (2009). *Augmented reality for teaching spatial relations*. Conference Ofthe International Journal Of Arts & Sciences.

Maietti, F., & Tasselli, N. (2020). Connessioni digitali. Integrazione dati in ambiente BIM per l'intervento sul costruito esistente [Digital connections. Data integration in BIM environment for the intervention on Existing Buildings]. In *Proceedings of 42° Convegno Internazionale dei Docenti delle discipline della Rappresentazione - Congresso della Unione Italiana Disegno* [Proceedings of 42 ° International Conference of Teachers of the disciplines of Representation - Congress of the Italian Drawing Union] (pp. 585-598). FrancoAngeli.

Maietti, F., Di Giulio, R., Medici, M., Ferrari, F., Ziri, A. E., Turillazzi, B., & Bonsma, P. (2020). Documentation, Processing, and Representation of Architectural Heritage Through 3D Semantic Modelling: The INCEPTION Project. In C. Bolognesi & C. Santagati (Eds.), *Impact of Industry 4.0 on Architecture and Cultural Heritage* (pp. 202–238). IGI Global.

Maietti, F., & Ferrari, F. (2021). Un Competence Centre europeo per la conservazione, il restauro e la valorizzazione del patrimonio culturale. Il progetto 4CH. In R. A. Genovese (Ed.), *Il patrimonio culturale tra la transizione digitale, la sostenibilità ambientale e lo sviluppo umano* [Cultural heritage between the digital transition, environmental sustainability and human development] (pp. 203–216). Giannini Editore.

Makransky, G., Terkildsen, T. S., & Mayer, R. E. (2019). Adding immersive virtual reality to a science lab simulation causes more presence but less learning. *Learning and Instruction*, *60*, 225–236. doi:10.1016/j.learninstruc.2017.12.007

Maldonado, T. (1997). *Critica alla ragione informatica*. Feltrinelli.

Mallen, M. J., & Vogel, D. L. (2005). Introduction to the major contribution: Counseling psychology and online counseling. *The Counseling Psychologist*, *33*(6), 761–775. doi:10.1177/0011000005278623

Mallen, M. J., Vogel, D. L., & Rochlen, A. B. (2005). The practical aspects of online counseling. *The Counseling Psychologist*, *33*(6), 776–818. doi:10.1177/0011000005278625

Mallen, M. J., Vogel, D. L., Rochlen, A. B., & Day, S. X. (2005). Online counseling: Reviewing the literature from a counseling psychology framework. *The Counseling Psychologist*, *33*(6), 819–871. doi:10.1177/0011000005278624

Mallgrave, H. F. (2015). *L'empatia degli spazi*. Raffaello Cortina Editore.

Malzbender, T., Gelb, D., & Wolters, H. (2001). Polynomial texture maps. In Proceedings of Siggraph 01. Computer Graphics (SIGGRAPH 01: 28th International Conference on Computer Graphics and Interactive Techniques. ACM.

Mancuso, V., Stramba-Badiale, C., Cavedoni, S., & Cipresso, P. (2022). Biosensors and Biofeedback in Clinical Psychology. *Comprehensive Clinical Psychology*, 28–50. doi:10.1016/B978-0-12-818697-8.00002-9

Mancuso, V., Stramba-Badiale, C., Cavedoni, S., Pedroli, E., Cipresso, P., & Riva, G. (2020). Virtual Reality Meets Non-invasive Brain Stimulation: Integrating Two Methods for Cognitive Rehabilitation of Mild Cognitive Impairment. *Frontiers in Neurology*, *11*, 566731. Advance online publication. doi:10.3389/fneur.2020.566731 PMID:33117261

Manera, V., Petit, P. D., Derreumaux, A., Orvieto, I., Romagnoli, M., Lyttle, G., David, R., & Robert, P. H. (2015). "Kitchen and cooking", a serious game for mild cognitive impairment and Alzheimer's disease: a pilot study. *Frontiers in Aging Neuroscience*, *7*, 24. doi:10.3389/fnagi.2015.00024

Maniello, D. (2014). *Realtà aumentata in spazi pubblici. Tecniche base di video mapping* [Augmented reality in public spacers. Basic techniques for video mapping]. Le Penseur.

Maniello, D. (2020). Digital anastylosis for digital augmented spaces: spatial Augmented reality applied to Cultural Heritage. In M. Lo Turco, E. C. Giovannini, & N. Mafrici (Eds.), *Digital & Documentation. Digital strategies for Cultural Heritage* (Vol. 2, pp. 141–151). Pavia University Press.

Maninder Jeet Kaur, V. P. (2019). *The Convergence of Digital Twin, IoT, and Machine Learning: Transforming Data into Action*. Digital Twin Technologies and Smart Cities.

Manzotti, A., Cerritelli, F., Esteves, J. E., Lista, G., Lombardi, E., La Rocca, S., Gallace, A., McGlone, F. P., & Walker, S. C. (2019, October). Dynamic touch reduces physiological arousal in preterm infants: A role for c-tactile afferents? *Developmental Cognitive Neuroscience*, *39*, 100703. doi:10.1016/j.dcn.2019.100703 PMID:31487608

Maral, P., & Pande, N. (2020). Progressive development of posttraumatic stress disorder and its holistic evolution of natural treatments. In R. Nicholson (Ed.), *Natural Healing as Conflict Resolution* (1st ed., pp. 73–99). IGI Global.

Marinković, B., Šegan Radonjić, M., Novaković, M., & Ognjanović, Z. (2022). Digital Documentation Management of Cultural Heritage. In S. D'Amico & V. Venuti (Eds.), *Handbook of Cultural Heritage Analysis*. Springer. doi:10.1007/978-3-030-60016-7_74

Mariotti S. (2021). The Use of Serious Games as an Educational and Dissemination Tool for Archaeological Heritage Potential and Challenges for the Future. *Consolidation, 1.* . doi:10.30687/mag/2724-3923/2021/03/005

Marra, A. (2017). Il complesso monumentale di Santa Chiara a Napoli: un modello innovativo per la conoscenza e la. *Conoscere, conservare, valorizzare. Il patrimonio religioso culturale, 3*, 141-146.

Martin-Brualla, R., Radwan, N., Sajjadi, M. S. M., Barron, J. T., Dosovitskiy, A., & Duckworth, V. (2021). NeRF in the Wild: Neural Radiance Fields for Unconstrained Photo Collections. *Proceedings of the IEEE Computer Society Conference on Computer Vision and Pattern Recognition*, 7206–7215.

Martinez, S., Jardon, A., Navarro, J. M., & Gonzalez, P. (2008). Building industrialization: Robotized assembly of modular products. *Assembly Automation, 28*(2), 134–142. doi:10.1108/01445150810863716

Martin, G., Koizia, L., Kooner, A., Cafferkey, J., Ross, C., Purkayastha, S., Sivananthan, A., Tanna, A., Pratt, P., & Kinross, J. (2020). Use of the HoloLens2 Mixed Reality Headset for Protecting Health Care Workers During the COVID-19 Pandemic: Prospective, Observational Evaluation. *Journal of Medical Internet Research, 22*(8), e21486. doi:10.2196/21486 PMID:32730222

Martín-Lerones, P., Olmedo, D., López-Vidal, A., Gómez-García-Bermejo, J., & Zalama, E. (2021). BIM Supported Surveying and Imaging Combination for Heritage Conservation. *Remote Sensing, 13*(84), 1584.

Martin, R. (2003). *The organizational complex: Architecture, media, and corporate space*. MIT.

Massetti, T., Crocetta, T. B., Da Silva, T. D., Trevizan, I. L., Arab, C., Caromano, F. A., & Monteiro, C. B. de M. (2017). Application and outcomes of therapy combining transcranial direct current stimulation and virtual reality: A systematic review. *Disability and Rehabilitation. Assistive Technology, 12*(6), 551–559. doi:10.1080/17483107.2016.1230152 PMID:27677678

Massie, T. H., & Salisbury, K. J. (1994). PHANToM haptic interface: a device for probing virtual objects. American Society of Mechanical Engineers, Dynamic Systems and Control Division.

Matrone, F., Grilli, E., Martini, M., Paolanti, M., Pierdicca, R., & Remondino, F. (2020). Comparing Machine and Deep Learning Methods for Large 3D Heritage Semantic Segmentation. ISPRS International Journal of Geo–Information, 9(9), 1-22.

Mattar, J., Ramos, D. K., & Lucas, M. R. (2022). DigComp-Based Digital competence Assessment Tools: Literature Review and Instrument Analysis. *Education and Information Technologies*, 1–25.

Matthews, T., Dey, A. K., Mankoff, J., Carter, S., & Rattenbury, T. (2004). A toolkit for managing user attention in peripheral displays. *Proceedings of the 17th Annual ACM Symposium on User Interface Software and Technology - UIST '04*, 247. 10.1145/1029632.1029676

Matthies, D. J. C., Müller, F., Anthes, C., & Kranzlmüller, D. (2013). ShoeSoleSense: Proof of concept for a wearable foot interface for virtual and real environments. *Proceedings of the 19th ACM Symposium on Virtual Reality Software and Technology - VRST '13*, 93. 10.1145/2503713.2503740

Ma, X., Xiong, F., Olawumi, T. O., Dong, N., & Chan, A. P. C. (2018). Conceptual Framework and Roadmap Approach for Integrating BIM into Lifecycle Project Management. *Journal of Management Engineering, 34*(6), 05018011. doi:10.1061/(ASCE)ME.1943-5479.0000647

Mayer, R. E., Mautone, P., & Prothero, W. (2002). Pictorial aids for learning by doing in a multimedia geology simulation game. *Journal of Educational Psychology, 94*(1), 171–185. doi:10.1037/0022-0663.94.1.171

Mazhar, M., & Rathore, A. P.-H. (2018). Exploiting IoT and big data analytics: Defining Smart Digital City using realtime urban data. *Sustainable Cities and Society*, 40.

McArthur, J. J., & Bortoluzzi, B. (2018). Lean-Agile FM-BIM: A demonstrated approach. *Facilities, 36*(13/14), 676–695. doi:10.1108/F-04-2017-0045

McCamy, C. S., Marcus, H., & Davidson, J. (1976). A color-rendition chart. *Journal of Applied Photographic Engineering, 2*, 95–99.

McCauley, M. E., & Sharkey, T. J. (1992). Cybersickness: Perception of Self-Motion in Virtual Environments. *Presence (Cambridge, Mass.), 1*(3), 311–318. doi:10.1162/pres.1992.1.3.311

McGonigal, J. (2011). *Reality is Broken : Why games Make Us Better and How They Can Change the World.* The Pinguin Press.

McKinsey & Company. (2017). *Reinventing construction: A route to higher productivity.* McKinsey & Company.

McKinsey & Company. (2020). *Rise of the platform era: The next chapter in construction technology.* https://www.mckinsey.com/industries/private-equity-and-principal-investors/our-insights/rise-of-the-platform-era-the-next-chapter-in-construction-technology

McLuhan, M., & Baltes, M. (2001). *Das Medium ist die Botschaft = The medium is the message* (Vol. 154). Verlag der Kunst.

Meadows, D. H., Meadows, D. L., Randers, J., & Behrens, W. W. (1972). *Limits to Growth.* New American Library.

Meegan, E., Murphy, M., Keenaghan, G., Corns, A., Shaw, R., Fai, S., Scandura, S., & Chenaux, A. (2021). Virtual Heritage Learning Environments. In Proceedings Digital Heritage. Progress in Cultural Heritage: Documentation, Preservation, and Protection. EuroMed 2020 (pp. 427–437). Academic Press.

Mehta, V. (2013). Evaluating Public Space. *Journal of Urban Design, 19*(1), 53–88. doi:10.1080/13574809.2013.854698

Meldrum, D., Glennon, A., Herdman, S., Murray, D., & McConn-Walsh, R. (2012). Virtual reality rehabilitation of balance: Assessment of the usability of the Nintendo Wii ® Fit Plus. *Disability and Rehabilitation. Assistive Technology, 7*(3), 205–210. doi:10.3109/17483107.2011.616922 PMID:22117107

Mellon, J. G. (2008). Urbanism, Nationalism and the Politics of Place: Commemoration and Collective Memory. *Canadian Journal of Urban Research, 17*(1), 58–77.

Mendes, R. N., & Grando, R. C. (2008). The computer game SimCity 4 and its pedagogical potential in math classes. *Revista Zetetiké, 2*(16), 118–176.

Menges, A. (2011). *Computational design thinking.* Wiley.

Menna, F., Nocerino, E., Morabito, D., Farella, E. M., Perini, M., & Remondino, F. (2017). An open source low-cost automatic system for image-based 3D digitization. *The International Archives of the Photogrammetry, Remote Sensing and Spatial Information Sciences, 42*(W8), 155–162. doi:10.5194/isprs-archives-XLII-2-W8-155-2017

Merchant, Z., Goetz, E. T., Cifuentes, L., Keeney-Kennicutt, W., & Davis, T. J. (2014). Effectiveness of virtual reality-based instruction on students' learning outcomes in K-12 and higher education: A meta-analysis. *Computers & Education, 70*, 29–40. doi:10.1016/j.compedu.2013.07.033

Merriam, S. B. (2009). *Qualitative research: A guide to design and implementation.* Jossey-Bass.

Meskó, B., Drobni, Z., Bényei, É., Gergely, B., & Győrffy, Z. (2017). Digital health is a cultural transformation of traditional healthcare. *mHealth, 3*(38), 1-8.

Meyer, E., Grussenmeyer, P., Perrin, J. P., Durand, A., & Drap, P. (2007). A web information system for the management and the dissemination of Cultural Heritage data. *Journal of Cultural Heritage*, *8*(4), 396–411.

Microsoft. (2015). *Introducing the Microsoft HoloLens Development Edition.* https://www.microsoft.com/it-it/hololens

Microsoft. (2020). *Microsoft HoloLens - The Science Within - Spatial Sound with Holograms.* https://gr.pinterest.com/pin/microsoft-hololens-the-science-within-spatial-sound-with-holograms--280982464230744292/

Microsoft. (2021). *What is cloud computing?* Retrieved from Azure Microsoft: https://azure.microsoft.com/en-gb/overview/what-is-cloud-computing/#uses

Mihelj, M., Novak, D., & Beguš, S. (2014). *Virtual Reality Technology and Applications* (Vol. 68). Springer Netherlands., doi:10.1007/978-94-007-6910-6

Mikolajewska. (2021). *Tecnologie digitali integrate per la conoscenza, la conservazione e la valorizzazione del patrimonio culturale storico. Il teatro Farnese di Parma* [Integrated digital technologies for the knowledge, conservation and enhancement of historical cultural heritage. The Farnese Theatre of Parma] [Doctoral dissertation]. University of Parma, Parma, Italy.

Mikolajewska, S., & Zerbi, A. (2019). Uno specchio dell'arte: il proscenio e l'affresco sulla parete di fondo del teatro Farnese di Parma [Mirror of the art: the proscenium and the fresco on the back wall of the Farnese Theatre in Parma]. In P. Belardi (Ed.), *Riflessioni: l'arte del disegno, il disegno dell'arte, Atti del 41° Convegno Internazionale dei Docenti della Rappresentazione* (pp. 1027–1034). Gangemi Editore.

Milgram, P., & Kishino, F. (1994). A Taxonomy of Mixed Reality Visual Displays. *IEICE Transactions on Information and Systems*, *77*, 1321–1329.

Milgram, P., Takemura, H., & Utsumi, A. &. (1995). Augmented reality: a class of displays on the reality-virtuality continuum. *Proc. SPIE 2351, Telemanipulator and Telepresence Technologies*, 282-292.

Milgram, S. (1974). *Obedience to Authority; An Experimental View*. Harpercollins.

Miller, J. K., & Gergen, K. J. (1998). Life on the line: The therapeutic potentials of computer-mediated conversation. *Journal of Marital and Family Therapy*, *24*(2), 189–202. doi:10.1111/j.1752-0606.1998.tb01075.x PMID:9583058

Miłosz, M., Montusiewicz, J., Kęsik, J., Żyła, K., Miłosz, E., Kayumov, R., & Anvarov, N. (2022). Virtual scientific expedition for 3D scanning of museum artifacts in the COVID-19 period – The methodology and case study. *Digital Applications in Archaeology and Cultural Heritage*, *26*, e00230.

Min Deng, C. C. (2021). From bim to digital twins: a systematic review of the evolution of intelligent building representations in the aec-fm industry. *ITcon Vol. 26 - Journal of Information Technology in Construction.*

Mirelman, A., Bonato, P., & Deutsch, J. E. (2009). Effects of Training With a Robot-Virtual Reality System Compared With a Robot Alone on the Gait of Individuals After Stroke. *Stroke*, *40*(1), 169–174. doi:10.1161/STROKEAHA.108.516328

Mishra, R., Narayanan, M. D. K., Umana, G. E., Montemurro, N., Chaurasia, B., & Deora, H. (2022). Virtual Reality in Neurosurgery: Beyond Neurosurgical Planning. *International Journal of Environmental Research and Public Health*, *19*(3), 1719. doi:10.3390/ijerph19031719 PMID:35162742

Mishra, S., Kumar, A., Padmanabhan, P., & Gulyás, B. (2021). Neurophysiological Correlates of Cognition as Revealed by Virtual Reality: Delving the Brain with a Synergistic Approach. *Brain Sciences*, *11*(1), 51. doi:10.3390/brainsci11010051 PMID:33466371

Mitchell, T. (1997). *Machine Learning*. McGraw Hill.

Moens, B., van Noorden, L., & Leman, M. (2010). D-Jogger: Syncing Music with Walking. *Proceedings of SMC Conference 2010,* 451–456. http://hdl.handle.net/1854/LU-1070528

Moghadas, S. M., Shoukat, A., Fitzpatrick, M. C., Wells, C. R., Sah, P., Pandey, A., Sachs, J. D., Wang, Z., Meyers, L. A., Singer, B. H., & Galvani, A. P. (2020). Projecting hospital utilization during the COVID-19 outbreaks in the United States. *Proceedings of the National Academy of Sciences of the United States of America*, *117*(16), 9122–9126. doi:10.1073/pnas.2004064117 PMID:32245814

Mohammadi, M., Rashidi, M., Mousavi, V., Karami, A., Yu, Y., & Samali, B. (2021). Case study on accuracy comparison of digital twins developed for a heritage bridge via UAV photogrammetry and terrestrial laser scanning. *Proceedings of the 10th International Conference on Structural Health Monitoring of Intelligent Infrastructure*, *10*, 1-8.

Mohammed, G. S., Wakil, K., & Nawroly, S. S. (2018). The Effectiveness of Microlearning to Improve Students' Learning Ability. *International Journal of Educational Research Review*, *3*(3), 32–38. doi:10.24331/ijere.415824

Moloney, J., Spehar, B., Globa, A., & Wang, R. (2018). The affordance of virtual reality to enable the sensory representation of multi-dimensional data for immersive analytics: From experience to insight. *Journal of Big Data*, *5*(1), 53. doi:10.118640537-018-0158-z

Montemurro, N. (2022). Telemedicine: Could it represent a new problem for spine surgeons to solve? *Global Spine Journal*, (6), 1306–1307. Advance online publication. doi:10.1177/21925682221090891 PMID:35363083

Montemurro, N., Condino, S., Cattari, N., D'Amato, R., Ferrari, V., & Cutolo, F. (2021). Augmented Reality-Assisted Craniotomy for Parasagittal and Convexity En Plaque Meningiomas and Custom-Made Cranio-Plasty: A Preliminary Laboratory Report. *International Journal of Environmental Research and Public Health*, *18*(19), 9955. doi:10.3390/ijerph18199955 PMID:34639256

Montemurro, N., Santoro, G., Marani, W., & Petrella, G. (2020). Posttraumatic synchronous double acute epidural hematomas: Two craniotomies, single skin incision. *Surgical Neurology International*, *11*, 435. doi:10.25259/SNI_697_2020 PMID:33365197

Moretti, M. (2016). *Senso e paesaggio. Analisi percettive e cartografie tematiche in ambiente GIS* [Sense and landscape. Perceptual Analysis and Thematic cartography in a GIS environment]. Franco Angeli.

Morganti, F. (2004). Virtual interaction in cognitive neuropsychology. *Studies in Health Technology and Informatics*, *99*, 55–70. PMID:15295146

Morganti, F., Gaggioli, A., Strambi, L., Rusconi, M. L., & Riva, G. (2007). A virtual reality extended neuropsychological assessment for topographical disorientation: A feasibility study. *Journal of Neuroengineering and Rehabilitation*, *4*(1), 26–26. doi:10.1186/1743-0003-4-26 PMID:17625011

Morhenn, V., Beavin, L. E., & Zak, P. J. (2012). Massage increases oxytocin and reduces adrenocorticotropin hormone in humans. *Alternative Therapies in Health and Medicine*, *18*(6), 11–18. PMID:23251939

Morina, N., Ijntema, H., Meyerbröker, K., & Emmelkamp, P. M. (2015). Can virtual reality exposure therapy gains be generalized to real-life? A meta-analysis of studies applying behavioral assessments. *Behaviour Research and Therapy*, *74*, 18–24. doi:10.1016/j.brat.2015.08.010 PMID:26355646

Morris, J. C. (1997). Clinical Dementia Rating: A reliable and valid diagnostic and staging measure for dementia of the Alzheimer type. *International Psychogeriatrics*, *9*(S1), 173176. doi:10.1017/S1041610297004870 PMID:9447441

Mortara, M., Catalano, C. E., Bellotti, F., Fiucci, G., Houry-Panchetti, M., & Petridis, P. (2014). Learning cultural heritage by serious games. *Journal of Cultural Heritage*, *15*(3), 318–325. doi:10.1016/j.culher.2013.04.004

Moseley, G. L., Gallace, A., & Spence, C. (2012). Bodily illusion in health and disease: Physiological and clinical perspectives and the concept of a cortical body matrix. *Neuroscience and Biobehavioral Reviews, 36*(1), 34–46. doi:10.1016/j.neubiorev.2011.03.013 PMID:21477616

Moyano, J., Nieto-Julián, J. E., Bienvenido-Huertas, D., & Marín-García, D. (2020). Validation of close-range photogrammetry for architectural and archaeological heritage: Analysis of point density and 3D mesh geometry. *Remote Sensing, 12*(21), 3571.

Moyle, W., Jones, C., Sung, B., & Dwan, T. (2016). *Alzheimer's Australia Victoria The Virtual Forest project: Impact on engagement, happiness, behaviours & mood states of people with dementia*. Griffith University.

Mudge, M., Malzbender, T., Chalmers, A., Scopigno, R., Davis, J., Wang, O., Gunawardane, P., Ashley, M., Doerr, M., Proenca, A., & Barbosa, J. (2008). Image-based empirical information acquisition, scientific reliability, and long-term digital preservation for the natural sciences and cultural heritage. Eurographics (Tutorials), 2, 4.

Muktamath, V. U., Priya, R. H., & Chand, S. (2021). Types of specific learning disability. *IntechOpen,* 1-20. doi:10.5772/intechopen.100809

Mulholland, P., Wolff, A., Kilfeather, E., Maguire, M., & O'Donovan, D. (2016). Modelling Museum Narratives to Support Visitor Interpretation. In Bordoni, L., Mele, F. & Sorgente, A. (Eds.), Artificial Intelligence for Cultural Heritage. Cambridge Scholars Publishing, 3–22.

Müller, T., Evans, A., Schied, C., & Keller, A. (2022). Instant Neural Graphics Primitives with a Multiresolution Hash Encoding. *ACM Transactions on Graphics. Association for Computing Machinery, 41*(4), 1–15.

Münster, S., Apollonio, F. I., Bell, P., Kuroczynski, P., Di Lenardo, I., Rinaudo, F., & Tamborrino, R. (2019). Digital cultural heritage meets digital humanities. *The International Archives of the Photogrammetry, Remote Sensing and Spatial Information Sciences, 42*(W15), 813–820. https://doi.org/10.5194/isprs-archives-XLII-2-W15-813-2019

Münster, S., Utescher, R., & Ulutas Aydogan, S. (2021). Digital topics on cultural heritage investigated: How can data-driven and data-guided methods support to identify current topics and trends in digital heritage? *Built Heritage, 5*(1), 1–13.

Munzer, B. W., Khan, M. M., Shipman, B., & Mahajan, P. (2019). Augmented Reality in Emergency Medicine: A Scoping Review. *Journal of Medical Internet Research, 21*(4), e12368. doi:10.2196/12368 PMID:30994463

Murphy, M., McGovern, E., & Pavia, S. (2011). Historic Building Information Modelling - Adding Intelligence to laser and image-based surveys. *International Archives of the Photogrammetry, Remote Sensing and Spatial Information Sciences, 38*(5).

Mustaqim, I., & Kurniawan, N. (2017). Pengembangan Media Pembelajaran Berbasis Augmented Reality [Augmented Reality-Based Learning Media Development]. *Edukasi Elektro, 1*(1), 26–48.

Narvaez Rojas, C., Alomia Peñafiel, G., Loaiza Buitrago, D. F., & Tavera Romero, C. A. (2021). Society 5.0: A Japanese Concept for a Superintelligent Society. *Sustainability, 13*(12), 6567. doi:10.3390u13126567

Nasreddine, Z. S. (2005). The Montreal cognitive assessment: A brief screening tool for mild cognitive impairment. *Journal of the American Geriatrics Society, 53*, 695–699. doi:10.1111/j.1532-5415.2005.53221.x PMID:15817019

Nathan Moore, S. Y. (2019). ALS-SimVR: Advanced Life Support Virtual Reality Training Application. *VRST 2019 - 25th ACM Symposium on Virtual Reality Software and Technology.*

National Association of Special Education Teachers. (2022). *Introduction to Learning Disabilities.* https://www.naset.org/fileadmin/user_upload/LD_Report/LD_Report_1_Intro_to_LD.doc.pdf

National Center for Education Statistic. (2020). *Students with Disabilities.* The Condition of Education. https://nces.ed.gov/programs/coe/indicator_cgg.asp

National Institute of Mental Health. (2018). *Autism Spectrum Disorder.* U.S. Department of Health and Human Services. https://www.nimh.nih.gov/sites/default/files/documents/health/publications/autism-spectrum-disorder/19-mh-8084-autismspectrumdisorder.pdf

National Survey of Children's Health. (2007). *Data query from the child and adolescent health measurement initiative.* http://childhealthdata.org/browse/survey/results?q=1219

Nazir, S., Gallace, A., Manca, D., & Overgard, K. I. (2016). Immersive virtual environment or conventional training? Assessing the effectiveness of different training methods on the performance of industrial operators in an accident scenario. In P. M. Ferreira Martins Arezes & P. V. Rodrigues de Carvalho (Eds.), *Ergonomics and Human Factors in Safety Management.* CRC Press.

Neelamkavil, J. (2009). *Automation in the Prefab and Modular Construction Industry.* Academic Press.

Negri, M., & Marini, G. (2020). *Le 100 Parole dei Musei* [The 100 Words of Museums]. Marsilio.

Negroponte, N. (1998). Being digital. Knopf.

Negroponte, N. (1970). *The architecture machine: Toward a more human environment.* The Massachusetts Institute of Technology MIT.

Negroponte, N. (1975). *Soft architecture machines.* MIT Press.

NEMO - Network of European Museum Organizations. (2021). *Follow-up survey on the impact of the COVID-19 pandemic on museums in Europe.* https://www.ne-mo.org/fileadmin/Dateien/public/NEMO_documents/NEMO_COVID19_FollowUpReport_11.1.2021.pdf

Neri, S. G., Cardoso, J. R., Cruz, L., Lima, R. M., de Oliveira, R. J., Iversen, M. D., & Carregaro, R. L. (2017). Do virtual reality games improve mobility skills and balancemeasurements in community-dwelling older adults? Systematic review andmeta-analysis. *Clinical Rehabilitation, 31*(10), 1292–1304. doi:10.1177/0269215517694677 PMID:28933612

NetConsulting Cube. (2020). *Mercato Digital Workspace In Italia, 2017-2022E.* Author.

Ng, W. (2012). Can we teach digital natives digital literacy? *Computers & Education, 59*(3), 1065–1078. doi:10.1016/j.compedu.2012.04.016

Niccolucci, F., & Felicetti, A. (2018). A CIDOC CRM-based Model for the Documentation of Heritage Sciences. *2018 3rd Digital Heritage International Congress (DigitalHERITAGE) held jointly with 2018 24th International Conference on Virtual Systems & Multimedia (VSMM 2018),* 1-6.

Niccolucci, F. (2017). Documenting archaeological science with CIDOC CRM. *International Journal on Digital Libraries, 18*, 223–231.

Niccolucci, F., Felicetti, A., & Hermon, S. (2022). Populating the Data Space for Cultural Heritage with Heritage Digital Twins. *Data, 7*(8), 105.

Nicodemus, F. E. (1965). Directional reflectance and emissivity of an opaque surface. *Applied Optics, 4*(7), 767–775. doi:10.1364/AO.4.000767

Nicoletti, R., & Borghi, A. M. (2007). *Il controllo motorio.* Il Mulino.

Nicolson, R. I., Fawcett, A. J., & Nicolson, M. K. (2000). Evaluation of a computer-based reading intervention in infant and junior schools. *Journal of Research in Reading*, *23*(2), 194–209. doi:10.1111/1467-9817.00114

Nielsen, N., & Budiu, R. (2001). *Success Rate: The Simplest Usability Metric*. N/G Nielsen Norman Group.

Niewohner, N., Asmar, L., Wortmann, F., Roltgen, D., Kuhn, A., & Dumitrescu, R. (2019). Design fields of agile innovation management in small and medium sized enterprises. *Procedia CIRP*, *84*, 826–831. doi:10.1016/j.procir.2019.04.295

Ninaus, M., Moeller, K., McMullen, J., & Kiili, K. (2017). Acceptance of Game-Based Learning and Intrinsic Motivation as Predictors for Learning Success and Flow Experience. *International Journal of Serious Games*, *4*(3), 15–30. doi:10.17083/ijsg.v4i3.176

Ning, H. H. W. (2021). A Survey on Metaverse: The State-of-the-art, Technologies, Applications, and Challenges. Cornell University.

Nofal, E. (2019). *Phygital Heritage. Communicating Built Heritage Information through the Integration of Digital Technology into Physical Reality* (Unpublished doctoral dissertation). KU Leuven, Arenberg Doctoral School.

Nofal, E. (2019). *Phygital Heritage: Communicating Built Heritage Information through the Integration of Digital Technology into Physical Reality* [PhD Thesis]. KU Leuven.

Nofal, E., Reffat, R., & Vande Moere, A. (2017) Phygital heritage: An approach for heritage communication Immersive Learning. In *Proceedings of Research Network Conference* (pp. 220-229). Verlag der Technischen Universität Graz.

Nofal, E., Reffat, R., Boschloos, V., Hameeuw, H., & Vande Moere, A. (2018). Evaluating the role of tangible interaction to communicate tacit knowledge of built heritage. *Heritage*, *1*(2), 414–436.

Noriska, N. J., Widyaningrum, R., & Nursetyo, K. I. (2021). Pengembangan Microlearning pada Mata Kuliah Difusi Inovasi Pendidikan di Prodi Teknologi Pendidikan [Microlearning Development in Diffusion of Educational Innovation Courses in Educational Technology Study Program]. *Jurnal Pembelajaran Inovatif*, *4*(1), 100–107. doi:10.21009/JPI.041.13

Norman, D. (1988). The psychology of everyday things [Psicopatologia degli oggetti quotidiani]. Basic Books.

Norman, D. (2013). The psychology of everyday things [Il design degli oggetti quotidiani]. Basic Books.

Notzon, S., Deppermann, S., Fallgatter, A. J., Diemer, J., Kroczek, A. M., Domschke, K., Zwanzger, P., & Ehlis, A. (2015). Psychophysiological effects of an iTBS modulated virtual reality challenge including participants with spider phobia. *Biological Psychology*, *112*, 66–76. doi:10.1016/j.biopsycho.2015.10.003 PMID:26476332

Novak, M., Guest, C. (1989). Application of a multidimentional caregiver burden inventory. *The Gerontologist*, *29*(6), 798-803. . doi:10.1093/geront/29.6.798

Nugraha, A. C., Khairudin, M., & Hertanto, D. B. (2017). Rancang Bangun Game Edukasi Sebagai Media Pembelajaran Mata Kuliah Praktik Teknik Digital [Design of Educational Games as Learning Media for Digital Engineering Practice Courses]. *Jurnal Edukasi Elektro*, *1*(1), 92–98. doi:10.21831/jee.v1i1.15121

Nugraha, H., Rusmana, A., Khadijah, U., & Gemiharto, I. (2021). Microlearning Sebagai Upaya dalam Menghadapi Dampak Pandemi pada Proses Pembelajaran [Microlearning as an Effort to Deal with The Pandemic Impact on The Learning Process]. *Kajian Dan Riset Dalam Teknologi Pembelajaran*, *8*(3), 225–236. doi:10.17977/um031v8i32021p225

Nuñez, M., Quirós, R., Nuñez, I., Carda, J. B., & Camahort, E. (2008). Collaborative Augmented Reality for Inorganic Chemistry Education. *WSEAS International Conference. Proceedings. Mathematics and Computers in Science and Engineering*, *5*(January), 271–277.

O'Sullivan, B., Alam, F., & Matava, C. (2018) Creating Low-Cost 360-Degree Virtual Reality Videos for Hospitals: A Technical Paper on the Dos and Don'ts. *J Med Internet Res*, *20*(7). doi:10.2196/jmir.9596

Oesterreich, T. D., & Teuteberg, F. (2016). Understanding the implications of digitisation and automation in the context of Industry 4.0: A triangulation approach and elements of a research agenda for the construction industry. *Computers in Industry*, *83*, 121–139. https://doi.org/10.1016/j.compind.2016.09.006

Ogdon, D. C. (2019). HoloLens and VIVE Pro: Virtual Reality Headsets. *Journal of the Medical Library Association: JMLA*, *107*(1). Advance online publication. doi:10.5195/jmla.2019.602

Ogunsemi, D. R., & Jagboro, G. O. (2006). Time-cost model for building projects in Nigeria. *Construction Management and Economics*, *24*(3), 253–258. https://doi.org/10.1080/01446190500521041

Oiha, A., Jebelli, H., & Sharifironizi, M. (2023). Understanding Students' Engagement in Learning Emerging Technologies of Construction Sector: Feasibility of Wearable Physiological Sensing Systems-Based Monitoring. In *Proceedings of the Canadian Society of Civil Engineering Annual Conference 2021*. Springer Nature Singapore.

Ok, M. W., Kim, M. K., Kang, E. Y., & Bryant, B. R. (2016). How to find good apps: An evaluation rubric for instructional apps for teaching students with learning disabilities. *Intervention in School and Clinic*, *51*(4), 244–252. doi:10.1177/1053451215589179

Olmos-Raya, E., Ferreira-Cavalcanti, J., Contero, M., Castellanos, M. C., Giglioli, I. A. C., & Alcañiz, M. (2018). Mobile virtual reality as an educational platform: A pilot study on the impact of immersion and positive emotion induction in the learning process. *Eurasia Journal of Mathematics, Science and Technology Education*, *14*(6), 2045–2057.

Opgenhaffen, L., Lami, M., & Mickleburgh, H. (2021). Art, Creativity and Automation. From Charters to Shared 3D Visualization Practices. *Open Archaeology*, *7*(1), 1648–1659. doi:10.1515/opar-2020-0162

Oppio, A., & Dell'Ovo, M. (2021). Cultural Heritage Preservation and Territorial Attractiveness: A Spatial Multidimensional Evaluation Approach. In P. Pileri & R. Moscarelli (Eds.), *Cycling & Walking for Regional Development. Research for Development* (pp. 105–125). Springer.

Opris, D., Pintea, S., Garcia-Palacios, A., Botella, C., Szamoskozi, S., & David, D. (2012). Virtual reality exposure therapy in anxiety disorders: A quantitative meta-analysis. *Depression and Anxiety*, *29*(2), 85–93. doi:10.1002/da.20910 PMID:22065564

Organisation for Economic Co-operation and Development. (2016). *Innovating Education and Educating for Innovation: The Power of Digital Technologies and Skills*. OECD Publishing. doi:10.1787/9789264265097-en

Orlosky, J., Huynh, B., & Hollerer, T. (2019). Using eye tracked virtual reality to classify understanding of vocabulary in recall tasks. *Proceedings - 2019 IEEE International Conference on Artificial Intelligence and Virtual Reality, AIVR 2019*, 66–73. 10.1109/AIVR46125.2019.00019

Osello, A. (2012). *Il futuro del disegno con il BIM per ingegneri e architetti* [The future of drawing with BIM for engineers and architects]. Dario Flaccovio editore s.r.l.

Osello, A. (2015). BIM and Interoperability for Cultural Heritage through ICT. In *Handbook of Research on Emerging Digital Tools for Architectural Surveying, Modeling, and Representation* (pp. 281-298). IGI Global.

Osello, A. (2018). *BIM Virtual and Augmented Reality in the health field between technical and terapeutic use. In RAPPRESENTAZIONE/MATERIALE/IMMATERIALE. 40° Convegno Internazionale dei Docenti della Rappresentazione*. Gangemi Editore.

Osello, A. (2018). BIM, Virtual and Augmented Reality in the health field between technical and therapeutic use. In R. Salerno (Ed.), *Rappresentazione/Materiale/Immateriale, Atti del 40° Convegno Internazionale dei Docenti della Rappresentazione* (pp. 1535–1538). Gangemi Editore.

Osimo, S. A., Pizarro, R., Spanlang, B., & Slater, M. (2015). Conversations between self and self as Sigmund Freud—A virtual body ownership paradigm for self counselling. *Scientific Reports*, *5*(1), 1–14. doi:10.1038rep13899 PMID:26354311

Ott, M., & Pozzi, F. (2010). Towards a new era for Cultural Heritage Education: Discussing the role of ICT. *Computers in Human Behavior.*

Ottoni, F. (2008). From geometrical and crack survey to static analysis method: the case study of Santa Maria del Quartiere dome in Parma (Italy). In D. D'Ayala & E. Fodde (Eds.), *Structural Analysis of Historical Construction* (Vol. I, pp. 697–704). Taylor & Francis Group. doi:10.1201/9781439828229.ch79

Owolabi, M. S., Omowonuola, A. A., Lawal, O. A., Dosoky, N. S., Collins, J. T., Ogungbe, I. V., & Setzer, W. N. (2017). Phytochemical and bioactivity screening of six Nigerian medicinal plants. *Journal of Pharmacognosy and Phytochemistry*, *6*(6), 1430–1437.

Oxman, R. (2008). Digital architecture as a challenge for design pedagogy: Theory, knowledge, models and medium. *Design Studies*, *29*(2), 99–120. doi:10.1016/j.destud.2007.12.003

Pagano, A., Palombini, A., Bozzelli, G., De Nino, M., Cerato, I., & Ricciardi, S. (2020). ArkaeVision VR Game: User Experience Research between Real and Virtual Paestum. *Applied Sciences (Basel, Switzerland)*, *2020*(10), 3182. doi:10.3390/app10093182

Page, B. J. (2004). Online group counseling. In J. L. DeLucia-Waack, D. A. Gerrity, C. R. Kalodner, & M. T. Riva (Eds.), *Handbook of group counseling and psychotherapy* (pp. 609–620). Sage. doi:10.4135/9781452229683.n43

Paladini, A., Dhanda, A., Reina Ortiz, M., Weigert, A., Nofal, E., Min, A., Gyi, M., Su, S., Van Balen, K., & Santana Quintero, M. (2019). Impact Of Virtual Reality Experience On Accessibility Of Cultural Heritage. *The International Archives of the Photogrammetry, Remote Sensing and Spatial Information Sciences*, *XLII-2*(W11), 929–936.

Palestini, C. & Basso, A. (2017). The photogrammetric survey methodologies applied to low cost 3D virtual exploration in multidisciplinary field. *International Archives of the Photogrammetry, Remote Sensing and Spatial Information Sciences - ISPRS Archives*, 42(2W8), 195–202.

Palma, V. (2019). Towards deep learning for architecture: a monument recognition mobile app. *International Archives of the Photogrammetry, Remote Sensing and Spatial Information Sciences - ISPRS Archives* XLII-2/W9, 551–556.

Palmerini, L., Rocchi, L., Mellone, S., Valzania, F., & Chiari, L. (2011). Feature Selection for Accelerometer-Based Posture Analysis in Parkinson's Disease. *IEEE Transactions on Information Technology in Biomedicine*, *15*(3), 481–490. doi:10.1109/TITB.2011.2107916 PMID:21349795

Panchuk, D., Klusemann, M. J., & Hadlow, S. M. (2018). Exploring the effectiveness of immersive video for training decision-making capability in elite, youth basketball players. *Frontiers in Psychology*, *9*(27), 2315. doi:10.3389/fpsyg.2018.02315 PMID:30538652

Pandey, A. (2018). *15 Types Of Microlearning For Formal And Informal Learning In The Workplace.* https://elearningindustry.com/types-of-microlearning-formal-informal-learning-workplace-15

Panero, J., & Zelnik, M. (1979). *Human Dimension and Interior Space. A Source Book of Design Reference Standards.* Whitney Library of Design.

Panou, C., Ragia, L., Dimelli, D., & Mania, K. (2018). An Architecture for Mobile Outdoors Augmented Reality for Cultural Heritage. *ISPRS International Journal of Geo-Information, 7*(463), 2–24.

Panwala, S., Shaikh, A. S., Ghare, A., Kazi, S., Khan, M., & Rangwala, M. (2017). *Augmented Realilty for Educational Enhancement.* University of Mumbai. http://ir.aiktclibrary.org:8080/xmlui/handle/123456789/2052

Papas, M., de Mesa, K., & Wann Jensen, H. (2014). A Physically-Based BSDF for Modeling the Appearance of Paper. *Computer Graphics Forum, 33*(4), 133–142. doi:10.1111/cgf.12420

Papastergiou, M. (2009). Digital game-based learning in high school Computer Science education: Impact on educational effectiveness and student motivation. *Computers & Education, 1*(52), 1-12. doi:10.1016/j.compedu.2008.06.004

Papathanasiou-Zuhrt, D., Weiss-Ibanez, D. F., & Di Russo, A. (2017). The gamification of heritage in the unesco enlisted medieval town of Rhodes. GamiFIN, 60-70.

Parisi, A. R. (2021). *Valutazione della maturità digitale delle PMI di costruzioni a supporto della transizione verso industria 4.0.* Politecnico di Milano.

Parisi, P., Turco, M., & Giovanni, E. (2019). The value of knowledge through H-BIM models: Historic documentation with a semantic approach. *The International Archives of the Photogrammetry, Remote Sensing and Spatial Information Sciences, 42*(W9), 581–588. doi:10.5194/isprs-archives-XLII-2-W9-581-2019

Park, K., Sinha, U., Barron, J., Bouaziz, S., Goldman, D. B., Seitz, S., & Martin-Brualla, R. (2022). Nerfies: *Deformable Neural Radiance Fields. IEEE/CVF International Conference on Computer Vision (ICCV-2021)*, 5845–5854.

Parmaxi, A. (2020). Virtual reality in language learning: A systematic review and implications for research and practice. *Interactive Learning Environments*, 1–13, 1049–4820. doi:10.1080/10494820.2020.1765392

Parnow, J. (2015). *Micro Visualization.* Retrieved from https://microvis.info/

Parrinello, S., & Cioli, F. (2018). Un progetto di recupero per il complesso monumentale di Usolye nella regione della Kama Superiore. *Restauro Archeologico, 26*(1), 92-111.

Parrinello, S., & Dell'Amico, A. (2019). Experience of Documentation for the Accessibility of Widespread Cultural Heritage. *Heritage, 2*(1), 1032–1044.

Parrinello, S., Picchio, F., De Marco, R., & Dell'Amico, A. (2019). Documenting The Cultural Heritage Routes. The Creation Of Informative Models Of Historical Russian Churches On Upper Kama Region. *The International Archives of the Photogrammetry, Remote Sensing and Spatial Information Sciences, XLII-2*(W15), 887–894.

Parsons, S., Mitchell, P., & Leonard, A. (2005). Do adolescents with autistic spectrum disorders adhere to social conventions in virtual environments? *Autism: an International Journal of Research and Practise., 9*(1), 95–117. doi:10.1177/1362361305049032 PMID:15618265

Parsons, T. D. (2015). Virtual Reality for Enhanced Ecological Validity and Experimental Control in the Clinical, Affective and Social Neurosciences. *Frontiers in Human Neuroscience, 9.* Advance online publication. doi:10.3389/fnhum.2015.00660 PMID:26696869

Parsons, T. D., Courtney, C. G., Arizmendi, B. J., & Dawson, M. E. (2011). Virtual Reality Stroop Task for Neurocognitive Assessment. *Studies in Health Technology and Informatics, 163*, 433–439. PMID:21335835

Parsons, T. D., Courtney, C. G., & Dawson, M. E. (2013). Virtual reality Stroop task for assessment of supervisory attentional processing. *Journal of Clinical and Experimental Neuropsychology, 35*(8), 812–826. doi:10.1080/13803395.2013.824556 PMID:23961959

Parsons, T. D., & Phillips, A. S. (2016). Virtual reality for psychological assessment in clinical practice. *Practice Innovations (Washington, D.C.), 1*(3), 197–217. doi:10.1037/pri0000028

Patacas, J., Dawood, N., & Kassem, M. (2020). BIM for facilities management: A framework and a common data environment using open standards. *Automation in Construction, 120.*

Patton, M. Q. (2002). *Qualitative research and evaluation methods* (3rd ed.). Sage.

Paul Milgram, H. T. (1994). *Augmented Reality: A class of displays on the reality-virtuality continuum. SPIE* (Vol. 2351). Telemanipulator and Telepresence Technologies.

Pavan, A., Mancini, M., Lo Turco, M., Pola, A., Mirarchi, C., Rigamonti, G., & Bocconcino, M. M. (2017). *Applicazione dell'approccio INNOVance per le imprese di costruzione* [Application of the INNOVance appraoch for construction companies]. Edilstampa.

Pavan, A., Mirarchi, C., & Giani, M. (2017). *BIM: metodi e strumenti. Progettare, costruire e gestire nell'era digitale.* Tecniche Nuove.

Pavlidis, G., & Koutsoudis, A. (2022). 3D Digitization of Tangible Heritage. In S. D'Amico & V. Venuti (Eds.), *Handbook of Cultural Heritage Analysis.* Springer. doi:10.1007/978-3-030-60016-7_47

Pedroli, E., Cipresso, P., Greci, L., Arlati, S., Mahroo, A., Mancuso, V., Boilini, L., Rossi, M., Stefanelli, L., Goulene, K., Sacco, M., Stramba-Badiale, M., Riva, G., & Gaggioli, A. (2020). A new application for the motor rehabilitation at home: Structure and usability of Bal-App. *IEEE Transactions on Emerging Topics in Computing, 9*(3), 1290–1300. doi:10.1109/TETC.2020.3037962

Pedroli, E., Cipresso, P., Serino, S., Albani, G., & Riva, G. (2013). A Virtual Reality Test for the Assessment of Cognitive Deficits: Usability and Perspectives. *Proceedings of the ICTs for Improving Patients Rehabilitation Research Techniques.* 10.4108/icst.pervasivehealth.2013.252359

Pedroli, E., La Paglia, F., Cipresso, P., la Cascia, C., Riva, G., & la Barbera, D. (2019). A Computational Approach for the Assessment of Executive Functions in Patients with Obsessive–Compulsive Disorder. *Journal of Clinical Medicine, 8*(11), 1975. doi:10.3390/jcm8111975 PMID:31739514

Pedroli, E., Mancuso, V., Stramba-Badiale, C., Cipresso, P., Tuena, C., Greci, L., Goulene, K., Stramba-Badiale, M., Riva, G., & Gaggioli, A. (2022). Brain M-App's Structure and Usability: A New Application for Cognitive Rehabilitation at Home. *Frontiers in Human Neuroscience, 16*, 898633. doi:10.3389/fnhum.2022.898633 PMID:35782042

Peinado-Santana, S., Hernández-Lamas, P., Bernabéu-Larena, J., Cabau-Anchuelo, B., & Martín-Caro, J. A. (2021). Public works heritage 3D model digitisation, optimisation and dissemination with free and open-source software and platforms and low-cost tools. *Sustainability, 13*(23), 13020.

Pellas, N., Mystakidis, S., & Kazanidis, I. (2021). Immersive virtual reality in K-12 and higher education: A systematic review of the last decade scientific literature. *Virtual Reality (Waltham Cross), 25*(3), 835–861. doi:10.100710055-020-00489-9

Pelliccio, A., Saccucci, M., & Grande, E. (2017). HT_BIM: Parametric modelling for the assessment of risk in historic centers. *DISEGNARECON, 10*(18), 1–12.

Peng, H., Ma, S., & Spector, J. M. (2019). Personalized Adaptive Learning: An Emerging Pedagogical Approach Enabled by a Smart Learning Environment. In Lecture Notes in Educational Technology (pp. 171–176). Springer International Publishing. doi:10.1007/978-981-13-6908-7_24

Perticarini, M., Marzocchella, V., & Mataloni, G. (2021). A Cycle Path for the safeguard of Cultural Heritage: Augmented reality and New LiDAR Technologies. In A. Arena, M. Arena, D. Mediati, & P. Raffa (Eds.), *CONNETTERE/ CONNECTING: un disegno per annodare e tessere / drawing for weaving relationships* (pp. 2571–2579).

Pescarin, S. (2016). Digital heritage into practice. *SCIRES, 6*(1), 1-4. http://www.sciresit.it/article/view/12003

Pescarin, S. (2016). Digital heritage into practice. *SCIRES-IT-SCIentific RESearch and Information Technology*, *6*(1), 1–4.

Pescarin, S., D'Annibale, E., Fanini, B., & Ferdani, D. (2018). Prototyping on site Virtual Museums: The case study of the co-design approach to the Palatine hill in Rome (Barberini Vineyard) exhibition. *Proceedings of the 3rd Digital Heritage International Congress (DigitalHERITAGE)*, 1–8. 10.1109/DigitalHeritage.2018.8810135

Peters, B., & Peters, T. (2018). Introduction—Computing the Environment. In *Computing the Environment—Digital Design Toosl for Simulation and Visualisation of Sustainable Architecture* (pp. 1–13). John Wiley & Sons.

Peters, E. A. (2018). Coloring the Temple of Dendur. *Metropolitan Museum Journal*, *53*, 8–23. doi:10.1086/701737

Petkova, V. I., & Ehrsson, H. H. (2008). If I were you: Perceptual illusion of body swapping. *PLoS One*, *3*(12), e3832. doi:10.1371/journal.pone.0003832 PMID:19050755

Petrelli, D., Not, E., Damala, A., van Dijk, D., & Lechner, M. (2014). meSch – Material Encounters with Digital Cultural Heritage. Digital Heritage. Progress in Cultural Heritage: Documentation, Preservation, and Protection. EuroMed 2014. Lecture Notes in Computer Science, 8740.

Petrelli, D. (2019). Making virtual reconstructions part of the visit: An exploratory study. *Digital Applications in Archaeology and Cultural Heritage*, *15*, e00123. doi:10.1016/j.daach.2019.e00123

Petti, L., Trillo, C., & Makore, C. (2019). Towards a Shared Understanding of the Concept of Heritage in the European Context. *Heritage*, 2531-2544.

Piano Nazionale di Ripresa e Resilienza [National Recovery and Resilience Plan]. (2021). Available online: https://www.mise.gov.it/index.php/it/68-incentivi/2042324-piano-nazionale-di-ripresa-e-resilienza-i-progetti-del-mise

Piccablotto, G., Aghemo, C., Pellegrino, A., Iacomussi, P., & Radis, M. (2015). Study on conservation aspects using LED technology for museum lighting. *Energy Procedia*, *78*, 1347–1352. doi:10.1016/j.egypro.2015.11.152

Picchio, F., De Marco, R., Dell'Amico, A., Doria, E., Galasso, F., La Placa, S., Miceli, A. & Parrinello, S. (2020). Urban analysis and modelling procedures for the management of historic centres. Bethlehem, Solikamsk, Kotor and Santo Domingo. *Paesaggio urbano, 2020*(2), 103-115.

Picchio, F., Bercigli, M., & De Marco, R. (2018). Digital Scenarios and Virtual Environments for the Representation of Middle Eastern Architecture. In C. L. Marcos (Ed.), *Graphic Imprints The Influence of Representation and Ideation Tools in Architecture* (pp. 541–556). Springer.

Piccone, G. (1978). *The Castle of Baia - History, Legend and Poetry*. Società Editrice Napoletana.

Picon, A. (2017). Histories of the digital: Information, computer and communication. In When is the Digital in Architecture? (pp. 80–99). Sternberg Press.

Pienimäki, M., Kinnula, M., & Livari, N. (2021). Finding fun in non-formal technology education. *International Journal of Child-Computer Interaction*.

Pierdicca, R., Paolanti, M., Matrone, F., Martini, M., Morbidoni, C., Malinverni, E. S., Frontoni, E., & Lingua, A. M. (2020). Point Cloud Semantic Segmentation Using a Deep Learning Framework for Cultural Heritage. Remote Sensing, 12(6), 1-23.

Pierdicca, R., Frontoni, E., Zingaretti, P., Sturari, M., Clini, P., & Quattrini, R. (2015). Lecture Notes in Computer Science: Vol. 9254. *Advanced Interaction with Paintings by Augmented Reality and High-Resolution Visualization: A Real Case Exhibition. Augmented and Virtual Reality. AVR 2015*. Springer.

Pieri, L., Serino, S., Cipresso, P., Mancuso, V., Riva, G., & Pedroli, E. (2022). The ObReco-360: A new ecological tool to memory assessment using 360 immersive technology. *Virtual Reality (Waltham Cross), 26*(2), 639–648. doi:10.100710055-021-00526-1

Pitoyo, M. D., Sumardi, S., & Asib, A. (2020). Gamification-based assessment: The washback effect of quizizz on students' learning in higher education. *International Journal of Language Education, 4*(1), 1–10. doi:10.26858/ijole.v4i2.8188

Pitzalis, D., & Pillay, R. (2009). Il sistema IIPimage: Un nuovo concetto di esplorazione di immagini ad alta risoluzione. [The IIPimage system: A New Concept of Exploration of High Resolution Images]. *Archeologia e Calcolatori*, (Supp. 2.), 239–244.

Pixelplex. (2020). *VR/AR in Education and Training*. https://pixelplex.io/blog/ar-and-vr-in-education-and-training/

Ploch, C. J., Bae, J. H., Ju, W., & Cutkosky, M. (2016). Haptic skin stretch on a steering wheel for displaying preview information in autonomous cars. *2016 IEEE/RSJ International Conference on Intelligent Robots and Systems (IROS)*, 60–65. 10.1109/IROS.2016.7759035

Plotnik, M., Doniger, G. M., Bahat, Y., Gottleib, A., ben Gal, O., Arad, E., Kribus-Shmiel, L., Kimel-Naor, S., Zeilig, G., Schnaider-Beeri, M., Yanovich, R., Ketko, I., & Heled, Y. (2017). Immersive trail making: Construct validity of an ecological neuropsychological test. *2017 International Conference on Virtual Rehabilitation (ICVR)*, 1–6. 10.1109/ICVR.2017.8007501

Plotnik, M., Ben-Gal, O., Doniger, G. M., Gottlieb, A., Bahat, Y., Cohen, M., Kimel-Naor, S., Zeilig, G., & Beeri, M. S. (2020). Multimodal immersive trail making – virtual reality paradigm to study cognitive-motor interactions. *bioRxiv*. Advance online publication. doi:10.1101/2020.05.27.118760

Pocobelli, D., Boehm, J., Bryan, P., Still, J., & Grau-Bové, J. (2018). BIM for Heritage Science: A review. *Heriage Science, 6*(1).

Pohl, H., Brandes, P., Ngo Quang, H., & Rohs, M. (2017). Squeezeback: Pneumatic Compression for Notifications. *Proceedings of the 2017 CHI Conference on Human Factors in Computing Systems - CHI '17*, 5318–5330. 10.1145/3025453.3025526

Ponchio, F., Corsini, M., & Scopigno, R. (2018). A compact representation of relightable images for the web. In *Proceedings of the 23rd International ACM Conference on 3D Web Technology. ACM*. 10.1145/3208806.3208820

Poobrasert, O., & Gestubtim, W. (2013). Development of assistive technology for students with dyscalculia. *2nd International Conference in E-Learning and E-Technologies Education*, 60-63. 10.1109/ICeLeTE.2013.6644348

Pordelan, N., Sadeghi, A., Abedi, M. R., & Kaedi, M. (2018). How online career counseling changes career development: A life design paradigm. *Education and Information Technologies, 23*(6), 2655–2672. doi:10.100710639-018-9735-1

Porras, D., Carrasco, J., Carrasco, P., González-Aguilera, D., & Lopez Guijarro, R. (2021). Drone Magnetometry in Mining Research. An Application in the Study of Triassic Cu–Co–Ni Mineralizations in the Estancias Mountain Range, Almería (Spain). *Drones (Basel), 5*(4), 151. doi:10.3390/drones5040151

Portalés, C., Sebastián, J., Alba, E., Sevilla, J., Gaitán, M., Ruiz, P., & Fernández, M. (2018). Interactive tools for the preservation, dissemination and study of silk heritage—An introduction to the silknow project. *Multimodal Technologies and Interaction, 2*(2), 28.

Pradibta, H., Harijanto, B., & Wibowo, D. W. (2016). Penerapan Augmented Reality Sebagai Alternatif Media Pembelajaran. *Smartics, 2*(2), 43–48. https://ejournal.unikama.ac.id/index.php/jst/article/view/1693

Prensky. (2001). Digital natives, digital immigrants. *On the Horizon*, 1-10.

Prensky, M. (2001). *Digital Game-based Learning*. McGraw-Hill.

Previati, A. (2020). *The Importance of Fan Engagement and Fan Management in Sports*. Academic Press.

Prieto, J., Lacasa, P., & Martínez-Borda, R. (2022). *Approaching metaverses: Mixed reality interfaces in youth media platforms*. New Techno-Humanities.

Principles of Seville. (2012). Retrieved on April 27, 2022 from http://sevilleprinciples.com/

Prof. Dr Graafland, A. (Ed.). (2012). Architecture, technology & design. Digital Studio for Research in Design, Visualization and Communication.

Promoter. (2021). *Digitalmeetsculture*. Obtenido de Digitalmeetsculture: https://www.digitalmeetsculture.net/article/eu-commission-recommendation-to-accelerate-the-digitisation-of-cultural-heritage-assets/

Proniewska, K., Pręgowska, A., Dołęga-Dołęgowski, D., & Dudek, D. (2021). Immersive technologies as a solution for general data protection regulation in Europe and impact on the COVID-19 pandemic. *Cardiology Journal*, *28*(1), 23–33. doi:10.5603/CJ.a2020.0102 PMID:32789838

Pronzati, A. (2021). *Micro-viz: modelli per la progettazione grafica di visualizzazioni dati in spazi ridotti.* https://www.politesi.polimi.it/handle/10589/175020?mode=complete

PROV-O. (n.d.). *The PROV Ontology*. Available online: https://www.w3.org/TR/prov-o/

Prusty, A., Yeh, C. J., Sengupta, R., & Smith, A. (2021). Dyscalculia: Difficulties in Making Arithmetical Calculation. In Handbook of Research on Critical Issues in Special Education for School Rehabilitation Practices. IGI Global. doi:10.4018/978-1-7998-7630-4.ch023

Pulcrano, M. (2020). Modelli digitali interconnessi per ampliare la conoscenza e migliorare la fruizione del patrimonio costruito [Digital models interconnected to expand knowledge and improve the use of cultural heritage]. In *Proceedings of 42° Convegno Internazionale dei Docenti delle discipline della Rappresentazione - Congresso della Unione Italiana Disegno* [Proceedings of 42 ° International Conference of Teachers of the disciplines of Representation - Congress of the Italian Drawing Union] (pp. 2604-2621). FrancoAngeli.

Pusparisa, Y. (2020). *Pengguna Smartphone diperkirakan Mencapai 89% Populasi pada 2025 | Databoks*. https://databoks.katadata.co.id/datapublish/2020/09/15/pengguna-smartphone-diperkirakan-mencapai-89-populasi-pada-2025

Puyuelo, M., Higón, J. L., Merino, L., & Contero, M. (2013). Experiencing Augmented Reality as an Accessibility Resource in the UNESCO Heritage Site Called "La Lonja". *Procedia Computer Science*, *25*, 171–178.

Radianti, J., Majchrzak, T. A., Fromm, J., & Wohlgenannt, I. (2020). A systematic review of immersive virtual reality applications for higher education: Design elements, lessons learned, and research agenda. *Computers & Education*, *147*, 103778. doi:10.1016/j.compedu.2019.103778

Rahaman, H. (2018). Digital heritage interpretation: A conceptual framework. *Digital Creativity*, *29*(2-3), 208–234. doi:10.1080/14626268.2018.1511602

Ramm, R., Heinze, M., Kühmstedt, P., Christoph, A., Heist, S., & Notni, G. (2022). Portable solution for high-resolution 3D and color texture on-site digitization of cultural heritage objects. *Journal of Cultural Heritage*, *53*, 165–175.

Ramsgard Thomsen, M., Tamke, M., Nicholas, P., & Ayres, P. (2020). *CITA complex modelling*. Academic Press.

Rand, D., Katz, N., Shahar, M., Kizony, R., & Weiss, P. L. (2005). The virtual mall: A functional virtual environment for stroke rehabilitation. *Annual Review of Cybertherapy and Telemedicine: A Decade of VR*, *3*, 193–198.

Rand, D., & Rukan, S. B.-A., Weiss, P. L., & Katz, N. (2009). Validation of the Virtual MET as an assessment tool for executive functions. *Neuropsychological Rehabilitation*, *19*(4), 583–602. doi:10.1080/09602010802469074 PMID:19058093

Rantakari, J., Inget, V., Colley, A., & Häkkilä, J. (2016). Charting Design Preferences on Wellness Wearables. *Proceedings of the 7th Augmented Human International Conference 2016 on - AH '16,* 1–4. 10.1145/2875194.2875231

Rapetti, N. (2019). Strumenti e metodi per la progettazione InfraBIM. In A. Osello, A. Fonsati, N. Rapetti, & F. Semeraro (Eds.), InfraBIM. Il BIM per le infrastrutture (pp. 45-63). Roma: Gangemi Editore.

Rapp, A., & Cena, F. (2016). Personal informatics for everyday life: How users without prior self-tracking experience engage with personal data. *International Journal of Human-Computer Studies*, *94*, 1–17. doi:10.1016/j.ijhcs.2016.05.006

Ratajczak, J., Riedl, M., & Matt, D. T. (2019). BIM-based and AR Application Combined with Location-Based Management System for the Improvement of the Construction Performance. *Building (London)*, *9*(5), 118. doi:10.3390/buildings9050118

Raudenská, J., Steinerová, V., Javůrková, A., Urits, I., Kaye, A. D., Viswanath, O., & Varrassi, G. (2020). Occupational burnout syndrome and post-traumatic stress among healthcare professionals during the novel coronavirus disease 2019 (COVID-19) pandemic. *Best Practice & Research. Clinical Anaesthesiology*, *34*(3), 553–560. doi:10.1016/j.bpa.2020.07.008 PMID:33004166

Realdon, O., Serino, S., Savazzi, F., Rossetto, F., Cipresso, P., Parsons, T. D., Cappellini, G., Mantovani, F., Mendozzi, L., Nemni, R., Riva, G., & Baglio, F. (2019). An ecological measure to screen executive functioning in MS: The Picture Interpretation Test (PIT) 360°. *Scientific Reports*, *9*(1), 1–8. doi:10.103841598-019-42201-1 PMID:30952936

Rebenitsch, L., & Owen, C. (2014). Individual variation in susceptibility to cybersickness. In *Proceedings of the 27th annual ACM symposium on User interface software and technology* (pp. 309-317). ACM. 10.1145/2642918.2647394

Regia-Corte, T., Marchal, M., Cirio, G., & Lécuyer, A. (2013). Perceiving affordances in virtual reality: Influence of person and environmental properties in perception of standing on virtual grounds. *Virtual Reality (Waltham Cross)*, *17*(1), 17–28. doi:10.100710055-012-0216-3

Relio². (n.d.). www.relio.it

Remondino, F., Barazzetti, L., Nex, F., Scaioni, M., & Sarazzi, D. (2011). UAV photogrammetry for mapping and 3D modelling– current status and future perspectives. *The International Archives of the Photogrammetry, Remote Sensing and Spatial Information Sciences*, *38*(1).

Remondino, F., Rizzi, A., Barazzetti, L., Scaioni, M., Fassi, F., Brumana, R., & Pelagotti, A. (2011). Review of Geometric and Radiometric Analyses of Paintings. *The Photogrammetric Record*, *26*(136), 439–461. doi:10.1111/j.1477-9730.2011.00664.x

Repetto Málaga, L., & Brown, K. (2019). Museums as Tools for Sustainable Community Development: Four Archaeological Museums in Northern Peru. *Museum International*, *71*(3-4), 60–75.

Rho Seungmin, Y. C. (2019). Social Internet of Things: Applications, architectures and protocols. *Future Generation Computer Systems*.

Riavis, V. (2019). Discovering Architectural Artistic Heritage Through the Experience of Tactile Representation: State of the Art and New Development. *DisegnareCon, 12*(23), 10.1-10.9.

Riazul, S. M., & Islam, D. K. (2015). The Internet of Things for Health Care: A Comprehensive Survey. Computer science. *IEEE Access: Practical Innovations, Open Solutions.*

Richards, D. (2009). Features and benefits of online counselling: Trinity College online mental health community. *British Journal of Guidance & Counselling, 37*(3), 231–242. doi:10.1080/03069880902956975

Richards, D., & Viganó, N. (2013). Online counseling: A narrative and critical review of the literature. *Journal of Clinical Psychology, 69*(9), 994–1011. doi:10.1002/jclp.21974 PMID:23630010

Rickwood, D., Deane, F. P., Wilson, C. J., & Ciarrochi, J. (2005). Young people's help-seeking for mental health problems. *Advances in Mental Health, 4*(3), 218–251. doi:10.5172/jamh.4.3.218

Rinaldi, C., Franchi, F., Marotta, A., Graziosi, F., & Centofanti, C. (2021). On the Exploitation of 5G Multi-Access Edge Computing for Spatial Audio in Cultural Heritage Applications. *IEEE Access: Practical Innovations, Open Solutions, 9*, 155197–155206. doi:10.1109/ACCESS.2021.3128786

Rindfleisch, A., O'Hern, M., & Sachdev, V. (2017). The digital revolution, 3D printing, and innovation as data. *Journal of Product Innovation Management, 34*(5), 681–690. doi:10.1111/jpim.12402

Riva, G. (2002). Virtual reality for health care: the status of research. CyberPsychology & Behavior, 219-225. doi:10.1089/109493102760147213

Riva, G., Bernardelli, L., Castelnuovo, G., Di Lernia, D., Tuena, C., Clementi, A., Pedroli, E., Malighetti, C., Sforza, F., Wiederhold, B. K., & Serino, S. (2021). A Virtual Reality-Based Self-Help Intervention for Dealing with the Psychological Distress Associated with the COVID-19 Lockdown: An Effectiveness Study with a Two-Week Follow-Up. *International Journal of Environmental Research and Public Health, 18*(15). . doi:10.3390/ijerph18158188

Riva, G. (1997). *Virtual Reality in Neuro-Psycho-Physiology: Cognitive, Clinical and Methodological Issues in Assessment and Treatment* (Vol. 44). IOS Press.

Riva, G. (2003). Applications of virtual environments in medicine. *Methods of Information in Medicine, 42*(5), 524–534. PubMed

Riva, G. (2005). Virtual reality in psychotherapy [Review]. *Cyberpsychology & Behavior, 8*(3), 220–230. doi:10.1089/cpb.2005.8.220 PMID:15971972

Riva, G., Mancuso, V., Cavedoni, S., & Stramba-Badiale, C. (2020). Virtual reality in neurorehabilitation: A review of its effects on multiple cognitive domains. *Expert Review of Medical Devices, 17*(10), 1035–1061. doi:10.1080/17434440.2020.1825939 PMID:32962433

Riva, G., Morganti, F., & Villamira, M. (2004). *Immersive Virtual Telepresence: virtual reality meets eHealth* (Vol. 99). Studies in Health Technology and Informatics. doi:10.3233/978-1-60750-943-1-255

Riva, G., Wiederhold, B. K., & Mantovani, F. (2019). Neuroscience of Virtual Reality: From Virtual Exposure to Embodied Medicine. *Cyberpsychology, Behavior, and Social Networking, 22*(1), 82–96. doi:10.1089/cyber.2017.29099.gri PMID:30183347

Riva, M. T., & Haub, A. L. (2004). Group counseling in the schools. In J. L. DeLucia-Waack, D. A. Gerrity, C. R. Kalodner, & M. T. Riva (Eds.), *Handbook of group counseling and psychotherapy* (pp. 309–321). Sage. doi:10.4135/9781452229683.n22

Rizzo, A. A., Buckwalter, J. G., Bowerly, T., Zaag, C. V., Humphrey, L., Neumann, U., Chua, C., Kyriakakis, C., Rooyen, A. V., & Sisemore, D. (2000). The Virtual Classroom: A Virtual Reality Environment for the Assessment and Rehabilitation of Attention Deficits. *Cyberpsychology, Behavior, and Social Networking*, *3*(3), 483–499. doi:10.1089/10949310050078940

Rizzo, A. A., & Kim, G. J. (2005). A SWOT Analysis of the Field of Virtual Reality Rehabilitation and Therapy. *Presence (Cambridge, Mass.)*, *14*(2), 119–146. doi:10.1162/1054746053967094

Robert, P.H., König, A., Amieva, H., Andrieu, S., Bremond, F., Bullock, R., Ceccaldi, M., Dubois, B., Gauthier, S., Kenigsberg P.A., Nave, S., Orgogozo, J.M., Piano, J., Benoit, M., Touchon, J., Vellas, B., Yesavage, J., & Manera V. (2014). Recommendations for the use of Serious Games in people with Alzheimer's Disease, related disorders and frailty. *Frontiers in Aging Neuroscience*, *6*, 54. . doi:10.3389/fnagi.2014.00054

Roberts, G., Holmes, N., Alexander, N., Boto, E., Leggett, J., Hill, R. M., Shah, V., Rea, M., Vaughan, R., Maguire, E. A., Kessler, K., Beebe, S., Fromhold, T. M., Barnes, G. R., Bowtell, R., & Brookes, M. J. (2019). Towards OPM-MEG in a virtual reality environment. *NeuroImage*, *199*, 408–417. doi:10.1016/j.neuroimage.2019.06.010 PMID:31173906

Rochlen, A. B., Zack, J. S., & Speyer, C. (2004). Online therapy: Review of relevant definitions, debates, and current empirical support. *Journal of Clinical Psychology*, *60*(3), 269–283. doi:10.1002/jclp.10263 PMID:14981791

Rodéhn, C. (2015). Democratization: The performance of academic discourses on democratizing museums. In K. L. Samuels & T. Rico (Eds.), *Heritage Keywords: Rhetoric and Redescription in Cultural Heritage* (pp. 95–110). University Press of Colorado.

Rogers, R. (2018). *Un posto per tutti. Vita, architettura e società giusta* [A Place for Eevryone. Life, architecture and just society]. Johan & Levi.

Rooksby, J., Rost, M., Morrison, A., & Chalmers, M. (2014). Personal tracking as lived informatics. *Proceedings of the SIGCHI Conference on Human Factors in Computing Systems.* 10.1145/2556288.2557039

Roper, K. O. (2017). Facility management maturity and research. *Journal of Facilities Management*, *15*(3), 235–243. doi:10.1108/JFM-04-2016-0011

Rosa, G. (Ed.). (1983). *Semerani+Tamaro. La città e i progetti.* Kappa.

Rosci, C. E., Sacco, D., Laiacona, M., & Capitani, E. (2004). Interpretation of a complex picture and its sensitivity to frontal damage: A reappraisal. *Neurological Sciences*, *25*(6), 322–330. doi:10.100710072-004-0365-6 PMID:15729495

Rose, F. D., Attree, E. A., & Johnson, D. A. (1996). Virtual reality: An assistive technology in neurological rehabilitation. *Current Opinion in Neurology*, *9*(6), 461-467. doi:https://pubmed.ncbi.nlm.nih.gov/9007406/

Rosenthal, B., & Eliane, P. Z. B. (2017). How Virtual Brand Community Traces May Increase Fan Engagement in Brand Pages. *Business Horizons*, *60*(3), 375–384. doi:10.1016/j.bushor.2017.01.009

Rossetti, V., Furfari, F., Leporini, B., Pelegatti, S., & Quarta, A. (2018). Enabling access to cultural heritage for the visually impaired: An interactive 3D model of a cultural site. *Procedia Computer Science*, *130*, 383–391.

Rossi, H. S., Santos, S. M., Prates, R., & Ferreira, R. A. C. (2018). Imaginator: A virtual reality based game for the treatment of sensory processing disorders. *IEEE 6th International Conference on Serious Games and Applications for Health (SeGAH).* 10.1109/SeGAH.2018.8401355

Roussou, M. (2004). Learning by doing and learning through play. *Computers in Entertainment*, *2*(1), 2–10. doi:10.1145/973801.973818

Rubén Alonso, M. B. (2019). *SPHERE: BIM Digital Twin Platform* (Vol. 20). MDPI - Proceeding.

Ruben de Laat, L. v. (2010). Integration of BIM and GIS: The Development of the CityGML GeoBIM Extension. *Advances in 3D Geo-Information Sciences.*

Rubino, D. (2018). *Microsoft HoloLens - Here are the full processor, storage and RAM specs.* https://www.windowscentral.com/microsoft-hololens-processor-storage-and-ram

Rushton, H., & Schnabel, M. A. (2022). Immersive Architectural Legacies: The Construction of Meaning in Virtual Realities. In Visual Heritage: Digital Approaches in Heritage Science. Springer. doi:10.1007/978-3-030-77028-0_13

Rushton, H., Silcock, D., Schnabel, M. A., Moleta, T., & Aydin, S. (2018). Moving images in digital heritage: architectural heritage in virtual reality. *AMPS series, 14,* 29-39.

Russo, M., & De Luca, L. (2021). Semantic-driven analysis and classification in architectural heritage. *Disegnarecon, 14*(26), 1–6.

Saarikko, T., Westergren, U., & Blomquist, T. (2020). Digital transformation: Five recommendations for the digitally conscious firm. *Business Horizons, 63*(6), 825–839. doi:10.1016/j.bushor.2020.07.005

Sahira Banu, K., & Cathrine, L. (2015). General Techniques Involved in Phytochemical Analysis. *International Journal of Advanced Research in Chemical Science,* 2(4), 25–32. www.arcjournals.org

Salama, R., & ElSayed, M. (2018). Basic elements and characteristics of game engine. *Global Journal of Computer Sciences: Theory and Research, 8*(3), 126–131. doi:10.18844/gjcs.v8i3.4023

Salerno, R., & Casonato, C. (2008). Paesaggi culturali. Rappresentazioni, esperienze, prospettive [Cultural landscapes. Representations, experiences, perspectives]. Gangemi.

Saloheimo, T., Kaos, M., Fricker, P., & Hämäläinen, P. (2021). Automatic Recognition of Playful Physical Activity Opportunities of the Urban Environment. *Academic Mindtrek, 2021,* 49–59. doi:10.1145/3464327.3464369

Sanchez, J., & Saenz, M. (2006). 3D sound interactive environments for blind children problem solving skills. *Behaviour & Information Technology, 25*(4), 367–378. doi:10.1080/01449290600636660

Sanchez-Vives, M. V., & Slater, M. (2005). From presence to consciousness through virtual reality. *Nature Reviews. Neuroscience, 6*(4), 332–339. doi:10.1038/nrn1651 PMID:15803164

Sand, A., Rakkolainen, I., Isokoski, P., Kangas, J., Raisamo, R., & Palovuori, K. (2015). Head-mounted display with mid-air tactile feedback. In *Proceedings of the ACM Symposium on Virtual Reality Software and Technology, VRST.* Association for Computing Machinery. 10.1145/2821592.2821593

Santamaría, J., Cordón, O., & Damas, S. (2011). A comparative study of state-of-the-art evolutionary image registration methods for 3D modeling. *Computer Vision and Image Understanding, 115*(9), 1340–1354. doi:10.1016/j.cviu.2011.05.006

Santilli, S., Ginevra, M. C., Di Maggio, I., Soresi, S., & Nota, L. (2021). In the same boat? An online group career counseling with a group of young adults in the time of COVID-19. *International Journal for Educational and Vocational Guidance.* Advance online publication. doi:10.100710775-021-09505-z PMID:34642592

Santoso, T. N. B., & Hastutiningtyas, K. N. (2021). Pengembangan Media Game Edukasi Sebagai Sistem Informasi Alternatif Ice Breaking Pembelajaran Di Masa Pandemi [Development of Educational Game Media as an Alternative Information System for Ice Breaking Learning During a Pandemic]. *Ecodunamika : Jurnal Pendidikan Ekonomi, 4*(1), 1–6.

Santos, V. M. (2009). The relationship and difficulties of students with math: A topic of study. *Revista Zetetiké, 32*(17), 1744.

Sasidharan, S., Chen, Y., Saravanan, D., Sundram, K. M., & Yoga Latha, L. (2011). Extraction, Isolation and Characterization of Bioactive Compounds from Plants' Extracts. *African Journal of Traditional, Complementary, and Alternative Medicines*, *8*(1), 1. doi:10.4314/ajtcam.v8i1.60483 PMID:22238476

Sauer, J., Sonderegger, A., & Schmutz, S. (2020). Usability, user experience and accessibility: Towards an integrative model. *Ergonomics*, *63*(10), 1207–1220. doi:10.1080/00140139.2020.1774080 PMID:32450782

Saurina, N. (2016). Pengembangan Media Pembelajaran Untuk Anak Usia Dini Menggunakan Augmented Reality [Learning Media Development for Early Childhood Using Augmented Reality]. *Jurnal IPTEK*, *20*(1), 95. doi:10.31284/j.iptek.2016.v20i1.27

Sayegh, A. (2020). *Responsive Environments: Defining our Technologically-Mediated Relationship with Space*. Actar D.

Schachter, S., & Singer, J. (1962). Cognitive, social, and physiological determinants of emotional state. *Psychological Review*, *69*(5), 379–399. doi:10.1037/h0046234 PMID:14497895

Schlick, C. (1994). An Inexpensive BRDF Model for Physically Based Rendering. *Computer Graphics Forum*, *13*(3), 233–246. doi:10.1111/1467-8659.1330233

Schmidt, M., Beck, D., Glaser, N., & Schmidt, C. (2017). A Prototype Immersive, Multi-user 3D Virtual Learning Environment for Individuals with Autism to Learn Social and Life Skills: A Virtuoso DBR Update. doi:10.1007/978-3-319-60633-0_15

Schoeb, D. S., Schwarz, J., Hein, S., Schlager, D., Pohlmann, P. F., Frankenschmidt, A., Gratzke, C., & Miernik, A. (2020). Mixed reality for teaching catheter placement to medical students: A randomized single-blinded, prospective trial. *BMC Medical Education*, *20*(1), 510. doi:10.118612909-020-02450-5 PMID:33327963

Schweibenz, W. (1991). The Virtual Museum: New Perspectives for Museums to Present Objects and Information Using the Internet as a Knowledge Base and Communication System. In H. Zimmermann & H. Schramm (Eds.), Knowledge Management und Kommunikationssysteme, Workflow Management, Multimedia, Knowledge Transfer (pp. 185-200). UKV.

Schwitzer, A. M. (2005). Self-development, social support, and student help-seeking. *Journal of College Student Psychotherapy*, *2*(2), 29–52. doi:10.1300/J035v20n02_04

Sdegno, A. (2018). Rappresentare l'opera d'arte con tecnologie digitali: dalla realtà aumentata alle esperienze tattily [Representing the artwork with digital technologies: from augmented reality to tactile experiences]. In Ambienti digitali per l'educazione all'arte e al patrimonio (pp. 256-271). FrancoAngeli.

Sejnowski, T. J. (2018). *The Deep Learning Revolution*. The MIT Press. doi:10.7551/mitpress/11474.001.0001

Self, T., Scudder, R. R., Weheba, G., & Crumrine, D. (2007). A virtual approach to teaching safety skills to children with autism spectrum disorder. *Topics in Language Disorders*, *27*(3), 242–253. doi:10.1097/01.TLD.0000285358.33545.79

Semerani, L. (1983). *Progetti per una città*. FrancoAngeli.

Semerani, L. (1987). *Lezioni di composizione architettonica*. Arsenale.

Semerani, L. (1991). *Passaggio a nord-est: itinerari attorno ai progetti di Luciano Semerani e Gigetta Tamaro*. Electa.

Semerani, L. (2000). *L'altro moderno*. Allemandi.

Semerani, L. (2000). *Luciano Semerani e Gigetta Tamaro: Architetture e progetti*. Skira.

Semerani, L. (2007). *L'esperienza del simbolo: lezioni di teoria e tecnica della progettazione architettonica*. Clean.

Semerani, L. (2013). *Incontri e lezioni: attrazione e contrasto tra le forme*. Clean.

Semerani, L. (2020). *Il ragazzo dell'IUAV.* LetteraVentidue.

Serino, S., Pedroli, E., Tuena, C., De Leo, G., Stramba-Badiale, M., Goulene, K., Mariotti, N. G., & Riva, G. (2017). A Novel Virtual Reality-Based Training Protocol for the Enhancement of the "Mental Frame Syncing" in Individuals with Alzheimer's Disease: A Development-of-Concept Trial. *Frontiers in Aging Neuroscience*, *9*, 240. doi:10.3389/fnagi.2017.00240 PMID:28798682

Serrani, D. (2014). Virtual reality training improves spatial navigation disorientation in dementia patients. *Dementia (London)*, 189–195.

Setyawan, B., Rufii, N., & Fatirul, A. N. (2019). Augmented Reality Dalam Pembelajaran IPA Bagi Siswa SD [Augmented Reality in Natural Science Learning for Elementary School Students]. *Kwangsan: Jurnal Teknologi Pendidikan*, *7*(1), 78–90. doi:10.31800/jtp.kw.v7n1.p78--90

SFPE Handbook of Fire Protection Engineering. (2016). Academic Press.

Shen, J., Xiang, H., Luna, J., Grishchenko, A., Patterson, J., Strouse, R. V., Roland, M., Lundine, J. P., Koterba, C. H., Lever, K., Groner, J. I., Huang, Y., & Lin, E. D. (2020). Virtual Reality–Based Executive Function Rehabilitation System for Children With Traumatic Brain Injury: Design and Usability Study. *JMIR Serious Games*, *8*(3), 8. doi:10.2196/16947 PMID:32447275

Shen, W., Hao, Q., Mak, H., Neelamkavil, J., Xie, H., Dickinson, J., Thomas, R., Pardasani, A., & Xue, H. (2010). Systems integration and collaboration in architecture, engineering, construction, and facilities management: A review. *Advanced Engineering Informatics*, *24*(2), 196–207. https://doi.org/10.1016/j.aei.2009.09.001

Sherman, W. R., & Craig, A. B. (2003). *Understanding Virtual Reality: Interface, Application, and Design.* Morgan Kaufmann Publishers, Inc.

Shine, J. M., Ward, P. B., Naismith, S. L., Pearson, M., & Lewis, S. J. (2011). Utilising functional MRI (fMRI) to explore the freezing phenomenon in Parkinson's disease. *Journal of Clinical Neuroscience*, *18*(6), 807–810. doi:10.1016/j.jocn.2011.02.003 PMID:21398129

Shin, M., & Bryant, D. P. (2015). A synthesis of mathematical and cognitive performances of students with mathematics learning disabilities a synthesis of mathematical and cognitive performances of students with mathematics learning disabilities. *Journal of Learning Disabilities*, *48*(1), 96–112. doi:10.1177/0022219413508324 PMID:24153404

Shi, Y. (2018). Interpreting User Input Intention in Natural Human Computer Interaction. *Proceedings of the 26th Conference on User Modeling, Adaptation and Personalization*, 277–278. 10.1145/3209219.3209267

Siavash H. Khajavi, N. H. (2019). Digital Twin: Vision, Benefits, Boundaries, and Creation for Buildings. *IEEE Access, 7.*

Sicklinger, A. (2020). *Design e Corpo umano. Cenni storici di Ergonomia, Antropometria e Movimento Posturale.* Maggioli Editore.

Siemens Switzerland Ltd. (2022). *Whitepaper. Hospitals harness digitalization to reach new levels of operational performance.* https://new.siemens.com/global/en/markets/healthcare/smart-hospitals/documents-resources/smart-hospitals-whitepaper.html

Sigala, M. (2020). Tourism and COVID-19: Impacts and implications for advancing and resetting industry and research. *Journal of Business Research*, *117*, 312–321. doi:10.1016/j.jbusres.2020.06.015 PMID:32546875

Silcock, D., Rushton, H., Rogers, J., & Schnabel, M. A. (2018). Tangible and intangible digital heritage: creating virtual environments to engage public interpretation. In *Computing for a better tomorrow, 36th annual conference on education and research in computer aided architectural design in Europe, 2*. Lodz University of Technology. 10.52842/conf.ecaade.2018.2.225

Sime, J. D. (1985). Movement toward the Familiar: Person and Place Affiliation in a Fire Entrapment Setting. *Environment and Behavior.*

Sims, D. (1994). Multimedia camp empowers disabled kids. *IEEE Computer Graphics and Applications, 14*(1), 13–14. doi:10.1109/38.250912

Sirilak, S., & Muneesawang, P. (2018). A New Procedure for Advancing Telemedicine Using the HoloLens. *IEEE Access: Practical Innovations, Open Solutions, 6*, 60224–60233. doi:10.1109/ACCESS.2018.2875558

Skiada, R., Soroniati, E., Gardeli, A., & Zissis, D. (2014). EasyLexia: A mobile application for children with learning difficulties. *Procedia Computer Science, 27*, 218–228. doi:10.1016/j.procs.2014.02.025

Skublewska-Paszkowska, M., Milosz, M., Powroznik, P., & Lukasik, E. (2022). 3D technologies for intangible cultural heritage preservation-literature review for selected databases. *Heritage Science, 10*(3), 1–24. doi:10.118640494-021-00633-x PMID:35003750

Slater, M. (2003). A note on presence terminology. *Presence Connect, 3*(3), 1-5.

Slater, M., & Wilbur, S. (1995). Through the looking glass world of presence: A framework for immersive virtual environments. In Five (Vol. 95, pp. 1-20). Academic Press.

Slater, M., & Sanchez-Vives, M. V. (2016). Enhancing Our Lives with Immersive Virtual Reality. *Frontiers in Robotics and AI, 3*, 3–74. doi:10.3389/frobt.2016.00074

Slater, M., & Wilbur, S. (1997). A Framework for Immersive Virtual Environments (FIVE): Speculations on the Role of Presence in Virtual Environments. *Presence (Cambridge, Mass.), 6*(6), 603–616. doi:10.1162/pres.1997.6.6.603

Smith, A., & Peck, B. (2010). The teacher as the 'digital perpetrator': Implementing web 2.0 technology activity as assessment practice for higher education Innovation or Imposition? *Procedia: Social and Behavioral Sciences, 2*(2), 4800–4804. doi:10.1016/j.sbspro.2010.03.773

Solazzo, G., Maruccia, Y., Ndou, V., & Del Vecchio, P. (2022). How to exploit Big Social Data in the Covid-19 pandemic: The case of the Italian tourism industry. *Service Business*. Advance online publication. doi:10.100711628-022-00487-8

Soler, F., Torres, J. C., Leon, A. J., & Luz. (2013). Design of cultural heritage information systems based on information layers. *ACM Journal of Computing and Cultural Heritage, 6*(4), 1-17.

Song, H., Chen, F., Peng, Q., Zhang, J., & Gu, P. (2017). Improvement of userexperience using virtual reality in open-architecture product design. *Proceedings of the Institution of Mechanical Engineers. Part B, Journal of Engineering Manufacture*, 232.

Song, S., Noh, G., Yoo, J., Oakley, I., Cho, J., & Bianchi, A. (2015). Hot & tight: Exploring thermo and squeeze cues recognition on wrist wearables. *Proceedings of the 2015 ACM International Symposium on Wearable Computers - ISWC '15*, 39–42. 10.1145/2802083.2802092

Spence, C., Hobkinson, C., Gallace, A., & Fiszman, B. P. (2013). A touch of gastronomy. *Flavour (London), 2*(1), 14. doi:10.1186/2044-7248-2-14

Spettu, F., Teruggi, S., Canali, F., Achille, C., & Fassi, F. (2021). A hybrid model for the reverse engineering of the Milan Cathedral. Challenges and lesson learnt. *ARQUEOLÓGICA 2.0 - 9th International Congress & 3rd GEORES - GEOmatics and pREServation.*

Spiegel, B. (2020). *VRx: how virtual therapeutics will revolutionize medicine.* Basic Books New York.

Sreelakshmi, M., & Subash, T. D. (2017). Haptic Technology: A comprehensive review on its applications and future prospects. *Materials Today: Proceedings, 2017*(2), 4182–4187. doi:10.1016/j.matpr.2017.02.120

Srivastava, A. (2016). Enriching student learning experience using augmented reality and smart learning objects. *ICMI 2016 - Proceedings of the 18th ACM International Conference on Multimodal Interaction, October 2016*, 572–576. 10.1145/2993148.2997623

Stake, R. E. (2010). *Qualitative research: Studying how things work.* Guilford Press.

Standen, P. J., & Brown, D. J. (2006). Virtual reality and its role in removing the barriers that turn cognitive impairments into intellectual disability. *Virtual Reality (Waltham Cross), 10*(3), 241–252. doi:10.100710055-006-0042-6

Stathopoulou, E. K., & Remondino, F. (2019). Semantic photogrammetry - Boosting image-based 3D reconstruction with semantic labeling. *ISPRS Int. Arch. Photogramm. Remote Sens. Spatial Inf. Sci, XLII-2*(W9), 685–690.

Stephenson, N. (1992). *Snow Crash.* Academic Press.

StilesM. (2019, April 4). Https://medium.com/desn325-emergentdesign/reality-virtuality-continuum-868cb8121680

Stramba-Badiale, C., Mancuso, V., Cavedoni, S., Pedroli, E., Cipresso, P., & Riva, G. (2020). Transcranial Magnetic Stimulation Meets Virtual Reality: The Potential of Integrating Brain Stimulation With a Simulative Technology for Food Addiction. *Frontiers in Neuroscience, 14*, 720. Advance online publication. doi:10.3389/fnins.2020.00720 PMID:32760243

Strickland, D. C., McAllister, D. F., Coles, C., & Osborne, S. (2007). An evolution of virtual reality training designs for children with autism and fetal alcohol spectrum disorders. *PubMed, 27*(3), 226–241. doi:10.1097/01.TLD.0000285357.95426.72 PMID:20072702

Strogatz, S. (2003). *Sincronia. I ritmi della natura, i nostri ritmi.* Rizzoli.

Stupar-Rutenfrans, S., Ketelaars, L. E. H., & van Gisbergen, M. S. (2017). Beat the fear of public speaking: Mobile 360o video virtual reality exposure training in homeenvironment reduces public speaking anxiety. *Cyberpsychology, Behavior, and Social Networking, 20*(10), 624–633. doi:10.1089/cyber.2017.0174 PMID:29039704

Suler, J. R. (2004). The online disinhibition effect. *Cyberpsychology & Behavior, 7*(3), 321–326. doi:10.1089/1094931041291295 PMID:15257832

Suler, J. R. (2005). The online disinhibitation effect. *Journal of Applied Psychoanalytic Studies*, 2(2), 184–188. doi:10.1002/aps.42

Suler, J. R. (2011). The psychology of text relationships. In R. Kraus, G. Stricker, & C. Speyer (Eds.), *Online counseling: A handbook for mental health professionals* (2nd ed., pp. 21–53). Academic Press. doi:10.1016/B978-0-12-378596-1.00002-2

Sutherland, I. E. (1968). A head-mounted three dimensional display. *Proceedings of the December 9-11, 1968, Fall Joint Computer Conference, Part I on - AFIPS '68 (Fall, Part I)*, 757. 10.1145/1476589.1476686

Sutherland, J., Belec, J., Sheikh, A., Chepelev, L., Althobaity, W., & Chow, B. J. W. (2019). Applying Modern Virtual and Augmented Reality Technologies to Medical Images and Models. Journal of Digital Imaging, 1(32), 38–53. doi:10.100710278-018-0122-7

Sutter, C., Drewing, K., & Müsseler, J. (2014). Multisensory integration in action control. *Frontiers in Psychology, 5*, 544. doi:10.3389/fpsyg.2014.00544 PMID:24959154

Suzuki, Y., Suzuki, R., Watanabe, J., Yoshida, A., & Shigeru, S. (2015). Haptic vibrations for hands and bodies. *SIGGRAPH Asia 2015 Haptic Media and Contents Design on - SA '15*, 1–3. doi:10.1145/2818384.2818389

Szabo, V. (2020). Critical and Creative Approaches to Digital Cultural Heritage with Augmented Reality. In L. Hjorth, A. de Souza e Silava & S. K. Lanson (Eds.), The Routledge Companion to Mobile Media Art. Routledge, 1-14.

Tali Hatukaa, H. Z. (2020). From smart cities to smart social urbanism: A framework for shaping the socio-technological ecosystems in cities. *Telematics and Informatics*, 55.

Tancik, M., Mildenhall, B., Wang, T., Schmidt, D., Srinivasan, P. P., Barron, J. T., & Ren, N. (2021). Learned Initializations for Optimizing Coordinate-Based Neural Representations. *Proceedings of the IEEE/CVF Conference on Computer Vision and Pattern Recognition* (CVPR), 2021, 2846-2855.

Tannock, R. (2013). Specific Learning Disabilities in DSM-5: are the changes for better or worse? *International Journal of Research in Learning Disabilities, 1*(2), 2-30. https://eric.ed.gov/?id=EJ1155677

Tanrikulu, İ. (2009). Counselors-in-training students' attitudes towards online counseling. *Procedia: Social and Behavioral Sciences*, *1*(1), 785–788. doi:10.1016/j.sbspro.2009.01.140

Tara, A., Belesky, P., & Ninsalam, Y. (2019). *Towards Managing Visual Impacts on Public Spaces: A Quantitative Approach to Studying Visual Complexity and Enclosure Using Visual Bowl and Fractal Dimension.* Wichmann Verlag. doi:10.14627/537663003

Tarnanas, I., Schlee, W., Tsolaki, M., Müri, R., Mosimann, U., & Nef, T. (2013). Ecological Validity of Virtual Reality Daily Living Activities Screening for Early Dementia: Longitudinal Study. *JMIR Serious Games*, *1*(1), e1. doi:10.2196/games.2778 PMID:25658491

Tcha-Tokey, K., & Loup-Escande, E., Christmann, & O., Richir, S. (2016). A questionnaire to measure the user experience in immersive virtual environments. In *Proceedings of the 2016 Virtual Reality International Conference (VRIC '16)*. Association for Computing Machinery. 10.1145/2927929.2927955

Tcha-Tokey, K., Loup-Escande, E., Christmann, O., Canac, G., Fabien, F., & Richir, S. (2015). Towards a user experience in immersive virtual environment model: A review. In *Proceedings of the 27th Conference on l'Interaction Homme-Machine (IHM 15)*. Association for Computing Machinery.

TechnologiesU. (2022). *Unity*. Retrieved from https://unity.com/

TechRepublic. (2020). *Gartner's top tech predictions for 2021.* CBS Interactive Inc.

Tenenbaum, J. M. (2006). AI Meets Web 2.0: Building the Web of Tomorrow, Today. *AI Magazine*, *27*(4), 47.

Testi, G. (Ed.). (1980). *Progetto Realizzato.* Marsilio Editori.

The Lancet. (2020). COVID-19: protecting health-care workers. *The Lancet, 395*(10228), 922. doi:10.1016/S0140-6736(20)30644-9

The London Charter. (2009). Retrieved on April 27, 2022 from http://www.londoncharter.org/

The PanSurg Collaborative Group. (2020). *The three vital lessons Italian hospitals have learned in fighting covid-19.* Heal Serv J.

Thomas, M. (2011). Technology, Education, and the Discourse of the Digital Native: Between Evangelists and Dissenters. In M. Thomas (Ed.), *Deconstructing Digital Natives: Young People, Technology, and The New Literacies* (pp. 1–14). Routledge. doi:10.4324/9780203818848

Thomas, P. T., Goswami, S. P., & Samasthitha, S. (2015). Developmental coordination disorder (dyspraxia). *Asian Journal of Cognitive Neurology, 3*(1), 41–43. https://www.academia.edu/29076970/developmental_coordination_disorder_dyspraxia

Tian, S., Yang, W., Grange, J. M. L., Wang, P., Huang, W., & Ye, Z. (2019). Smart healthcare: making medical care more intelligent. Global Health Journal, 3(3), 62-65. doi:10.1016/j.glohj.2019.07.001

Tieri, G., Morone, G., Paolucci, S., & Iosa, M. (2018). Virtual reality in cognitive and motor rehabilitation: Facts, fiction and fallacies. *Expert Review of Medical Devices, 15*(2), 107–117. doi:10.1080/17434440.2018.1425613 PMID:29313388

Time Ontology in OWL. (n.d.). Available online: https://www.w3.org/TR/owl-time/

Tinacci, L., Guardone, L., Giusti, A., Pardini, S., Benedetti, C., Di Iacovo, F., & Armani, A. (2022). Distance Education for Supporting "Day One Competences" in Meat Inspection: An E-Learning Platform for the Compulsory Practical Training of Veterinarians. *Education Sciences, 12*(1), 24. Advance online publication. doi:10.3390/educsci12010024

Tokel, S. T., & Topu, F. B. (2016). Üç boyutlu sanal dünyalar [Three dimensional virtual worlds]. In K. Çağıltay & Y. Göktaş (Eds.), *Öğretim Teknolojilerinin Temelleri: Teoriler, Araştırmalar, Eğilimler* [Foundations of Instructional Technologies: Theories, Research, Trends] (2nd ed., pp. 825–844). Pegem Akademi.

Tolentino, L., Birchfield, D., Megowan-Romanowicz, C., Johnson-Glenberg, M. C., Kelliher, A., & Martinez, C. (2009). Teaching and learning in the mixed-reality science classroom. *Journal of Science Education and Technology, 18*(6), 501–1. doi:10.100710956-009-9166-2

Tomasi, S. O., Umana, G. E., Scalia, G., & Winkler, P. A. (2020). In Reply: Rongeurs, Neurosurgeons, and COVID-19: How Do We Protect Health Care Personnel During Neurosurgical Operations in the Midst of Aerosol-Generation From High-Speed Drills? *Neurosurgery, 87*(2), E166–E166. doi:10.1093/neuros/nyaa213 PMID:32385489

Tominaga, S., & Tanaka, N. (2008). Spectral Image Acquisition, Analysis, and Rendering for Art Paintings. *Journal of Electronic Imaging, 17*(4), 13.

Tommasi, C., Achille, C., & Fassi, F. (2016). From point cloud to BIM: A modelling challenge in the Cultural Heritage field. *International Archives of the Photogrammetry, Remote Sensing and Spatial Information Sciences. XLI, B5*, 429–436.

Tong, D., & Canter, D. (1985). The decision to evacuate: A study of the motivations which contribute to evacuation in the event of fire. *Fire Safety Journal, 9*(3), 257–265.

Torrance, K. E., & Sparrow, E. M. (1967). Theory for off-specular reflection from roughened surfaces. *Journal of the Optical Society of America, 57*(9), 1105–1114. doi:10.1364/JOSA.57.001105

Travers, M. F., & Benton, S. (2014). The acceptability of therapist-assisted, Internet-delivered treatment for college students. *Journal of College Student Psychotherapy, 28*(1), 35–46. doi:10.1080/87568225.2014.854676

Trillo, C., Aburamadan, R., Makore, C., Udeaja, C., Moustaka, A., Gyau, K., . . . Mansouri, L. (2021). Towards smart planning conservation of heritage cities: Digital technologies and heritage conservation planning. *Culture and Computing. Interactive Cultural Heritage and Arts. HCII 2021.*

Trillo, C., Aburamadan, R., Mubaideen, S., Salameen, D., & Makore, C. (2020). *Towards a Systematic Approach to Digital Technologies for Heritage Conservation. Insights from Jordan.* Preservation, Digital Technology & Culture. doi:10.1515/pdtc-2020-0023

Trizio, I., Savini, F., Giannangeli, A., Fiore, S., Marra, A., Fabbrocino, G., & Ruggieri, A. (2019). Versatil tools: Digital survey and virtual reality for documentation, analysis and fruition of cultural heritage in seismic areas. *The International Archives of the Photogrammetry, Remote Sensing and Spatial Information Sciences*, *52*(2-3), 377–384. doi:10.5194/isprs-archives-XLII-2-W17-377-2019

Trost, W., Labb, C., & Grandjean, D. (2017). Rhythmic entrainment as a musical affect induction mechanism. *Neuropsychologia*, *96*, 96–110. doi:10.1016/j.neuropsychologia.2017.01.004 PMID:28069444

Tsai, C.-L., Tu, C.-H., Chen, J.-C., Lane, H.-Y., & Ma, W.-F. (2021). Efficiency of an online health-promotion program in individuals with at-risk mental state during the COVID-19 pandemic. *International Journal of Environmental Research and Public Health*, *18*(22), 11875. Advance online publication. doi:10.3390/ijerph182211875 PMID:34831631

Tucci, G., & Bonora, V. (2016). Documenting Syrian Built Heritage to Increase Awareness in the Public Conscience. In M. Silver (Ed.), *Challenges, Strategies and High-Tech Applications for Saving the Cultural Heritage of Syria* (pp. 83–94). Austrian Academy of Sciences Press.

Tucci, G., Bonora, V., Conti, A., & Fiorini, L. (2017). High-quality 3d models and their use in a cultural heritage conservation project. *The International Archives of the Photogrammetry, Remote Sensing and Spatial Information Sciences*, *42*, 687–693.

Tucci, G., Bonora, V., Conti, A., Fiorini, L., & Riemma, M. (2014). Il rilievo digitale del Battistero: dati 3D per nuove riflessioni critiche [Digital survey of the Baptistery: 3D data for new critical reflections]. In F. Gurrieri (Ed.), *Il Battistero di San Giovanni. Conoscenza, diagnostica, conservazione* (pp. 105–117). Mandragora.

Tuena, C., Mancuso, V., Stramba-Badiale, C., Pedroli, E., Stramba-Badiale, M., Riva, G., & Repetto, C. (2021). Egocentric and allocentric spatial memory in mild cognitive impairment with real-world and virtual navigation tasks: A systematic review. *Journal of Alzheimer's Disease*, *79*(1), 95–116. doi:10.3233/JAD-201017 PMID:33216034

Tuena, C., Pedroli, E., Trimarchi, P. D., Gallucci, A., Chiappini, M., Goulene, K., Gaggioli, A., Riva, G., Lattanzio, F., Giunco, F., & Stramba-Badiale, M. (2020). Usability Issues of Clinical and Research Applications of Virtual Reality in Older People: A Systematic Review. *Frontiers in Human Neuroscience*, *14*, 93. doi:10.3389/fnhum.2020.00093 PMID:32322194

Tukiran, Pramudya, A., Wardana, Nurlaila, E., Santi, A. M., & Hidayati, N. (2016). Analisis Awal Fitokimia pada Ekstrak Metanol Kulit Batang Tumbuhan Syzygium (Myrtaceae) [Preliminary Phytochemical Analysis on Methanol Extract of Syzygium (Myrtaceae)]. *Prosiding Seminar Nasional Kimia Dan Workshop*, (September), 2–8.

Tyson, L., & Mischke, J. (2021). *Project Syndicate*. Retrieved from Productivity After the Pandemic: https://www.project-syndicate.org/commentary/productivity-after-the-pandemic-by-laura-tyson-and-jan-mischke-2021-04

Udeaja, C., Trillo, C., Awuah, K., Makore, C., Patel, D., Mansuri, L., & Jha, K. (2020). Urban Heritage Conservation and Rapid Urbanization: Insights from Surat, India. *Sustainability*, *12*(6), 2172. doi:10.3390u12062172

Ugliotti, F. M., Osello, A., Levante, R., & Urbina, E. N. B. (2019). Digital evolution of representation implemented at the Galliera Hospital in Genova [Evoluzione digitale della rappresentazione applicata all'Ospedale Galliera di Genova]. Riflessioni. L'arte del disegno/il disegno dell'arte [Reflections. The art of drawing/the drawing of art] (pp. 1775-1780). Gangemi Editore spa.

Ugliotti, F. M. (2020). Increase the awareness of ALS patients through a Virtual Reality Application. In *EDULEARN20 Proceedings, 12th International Conference on Education and New Learning Technologies* (pp. 4403-4408). IATED Academy. 10.21125/edulearn.2020.1166

Ugliotti, M., De Luca, D., Fonsati, A., Del Giudice, M., & Osello, A. (2021). Students and teachers turn into avatars for online education. In *15th International Technology, Education and Development Conference* (pp. 4556-4565). Valencia: IATED Academy.

Umana, G. E., Pucci, R., Palmisciano, P., Cassoni, A., Ricciardi, L., Tomasi, S. O., Strigari, L., Scalia, G., & Valentini, V. (2022). Cerebrospinal Fluid Leaks After Anterior Skull Base Trauma: A Systematic Review of the Literature. *World Neurosurgery*, *157*, 193–206.e2. doi:10.1016/j.wneu.2021.10.065 PMID:34637942

Umana, G. E., Scalia, G., Yagmurlu, K., Mineo, R., Di Bella, S., Giunta, M., Spitaleri, A., Maugeri, R., Graziano, F., Fricia, M., Nicoletti, G. F., Tomasi, S. O., Raudino, G., Chaurasia, B., Bellocchi, G., Salvati, M., Iacopino, D. G., Cicero, S., Visocchi, M., & Strigari, L. (2021). Multimodal Simulation of a Novel Device for a Safe and Effective External Ventricular Drain Placement. *Frontiers in Neuroscience*, *15*, 690705. Advance online publication. doi:10.3389/fnins.2021.690705 PMID:34194297

UNESCO. (1954). *Convention for the Protection of Cultural Property in the Event of Armed Conflict*. Retrieved on April 27, 2022 from http://portal.unesco.org/en/ev.php-URL_ID=13637&URL_DO=DO_TOPIC&URL_SECTION=201.html

UNESCO. (1976). *Recommendation concerning the Safeguard and Contemporary Role of Historic Areas*. Retrieved on April 27, 2022 from http://portal.unesco.org/en/ev.php-URL_ID=13133&URL_DO=DO_TOPIC&URL_SECTION=201.html

UNESCO. (1994). *The Nara Document on Authenticity*. Retrieved on April 27, 2022 from whc.unesco.org/document/9379

UNESCO. (2003). *Text of the Convention for the Safeguarding of the Intangible Cultural Heritage*. Retrieved April 8, 2022 from https://ich.unesco.org/en/convention

UNESCO. (2003a). *Charter on the Preservation of the Digital Heritage*. Retrieved on April 27, 2022 from http://portal.unesco.org/en/ev.php-URL_ID=17721&URL_DO=DO_TOPIC&URL_SECTION=201.html

UNESCO. (2003b). *Convention for the Safeguarding of Intangible Cultural Heritage*. Retrieved on April 27, 2022 from https://ich.unesco.org/en/convention

UNESCO. (2009). *World Heritage Cultural Landscape*. Retrieved on April 27, 2022 from https://whc.unesco.org/en/culturallandscape/

UNESCO. (2011). *Recommendation on the Historic Urban Landscape*. Retrieved on April 27, 2022 from https://whc.unesco.org/en/activities/638

UNESCO. (2015). *Operational Guidelines for the Implementation of the World Heritage Convention*. Retrieved on April 27, 2022 from https://whc.unesco.org/en/guidelines/

UNESCO. (2020). *Museums around the world in the face of COVID-19*. United Nations Educational, Scientific and Cultural Organization. https://unesdoc.unesco.org/ark:/48223/pf0000373530

UNI EN ISO. (2018). *ISO 9241-11:2018. Ergonomia dell'interazione uomo-sistema - Parte 11: Usabilità: Definizioni e concetti* [Ergonomics of Human-System Interaction – Part 11: Usability: Defintions and Concepts].

UNI. (2017). *UNI 11337-4:2017. Edilizia e opere di ingegneria civile - Gestione digitale dei processi informativi delle costruzioni - Parte 4: Evoluzione e sviluppo informativo di modelli, elaborati e oggetti. Consturction and civil engineering works – Digital Management of Construction Information Processes. Part 4: Evolution and Information of Model.* Designs, and Objects.

Unity Technologies. (2022). https://unity.com/

Unity3D. (n.d.). http://www.unity3d.com

University of Lumar. (2016). *Can Virtual Reality Assist Special Needs Students?* https://degree.lamar.edu/articles/education/can-virtual-reality-assist-special-needs-students

Uohara, M. Y., Weinstein, J. N., & Rhew, D. C. (2020). The Essential Role of Technology in the Public Health Battle Against COVID-19. *Population Health Management, 23*(5), 361–367. doi:10.1089/pop.2020.0187 PMID:32857014

Urech, P. R. W. (2019). *Point-Cloud Modeling: Exploring a Site-Specific Approach for Landscape Design.* Wichmann Verlag. doi:10.14627/537663031

USIBD Level of Accuracy (LOA) Specification Guide v. 3.0-2019. (2019). Academic Press.

Uzun Özer, B., Saçkes, M., & Tuckman, B. W. (2013). Psychometric properties of the Tuckman Procrastination Scale in a Turkish sample. *Psychological Reports, 113*(3), 874–884. doi:10.2466/03.20.PR0.113x28z7 PMID:24693816

Vaia Moustaka, Z. T. (2019). Enhancing social networking in smart cities: Privacy and security bonderlines. *Technological Forecasting and Social Change*, 142.

Vaishnav, R., & Parage, P. (2013). Innovative instructional strategies interactive multimedia instruction and computer aided instruction for teaching biology. *Voice of Research, 2*(2), 1-4. http://www.voiceofresearch.org/doc/sep-2013/sep-2013_1.pdf

Valinejadshoubi, M., Osama, M., & Ashutosh, B. (2022). Integrating BIM into Sensor-based Facilities Management Operations. *Journal of Facilities Management, 20*(3), 385–400. doi:10.1108/JFM-08-2020-0055

Valladares-Rodriguez, S., Fernåndez-iglesias, M., Anido-Rifòn, L., Facal, D., & Pèrez-Rodriguez, R. (2018). Episodix: A serious game to detect cognitive impairment in senior adults. A psychometric study. *PeerJ, 6.* . doi:10.7717/peerj.5478

Valve Corporation. (2022). *The Cooking Game VR.* https://store.steampowered.com/app/857180/The_Cooking_Game_VR/

Van Eerden, J. (2020, January 23). A Davos POV About a 5th Industrial Revolution. *Rea Leaders.* https://real-leaders.com/a-5th-industrial-revolution-what-it-is-and-why-it-matters/

van Ruymbeke, M., Nofal, E., & Billen, R. (2022). 3D Digital Heritage and Historical Storytelling: Outcomes from the Interreg EMR Terra Mosana Project. In *International Conference on Human-Computer Interactio*n (pp. 262-276). Springer.

van Weert, J.C., van Dulmen, A.M., Spreeuwenberg, P.M., Ribbe, M.W., & Bensing, J.M. (2005) Behavioral and mood effects of snoezelen integrated into 24-hour dementia care. *Journal of the American Geriatrics Society, 53*(1), 24-33. . doi:10.1111/j.1532-5415.2005.53006.x

Vancetti, R., & Cereda, R. (2020a). Stazioni metropolitane: la caratterizzazione degli occupanti per le verifiche di esodo ed inclusione con i metodi della fse [Metropolitan stations: characterization of occupants for exodus and inclusion checks with ESF methods]. Antincendio, 58-77.

Vancetti, R., & Cereda, R. (2020b). Progettazione e verifica del sistema di esodo con strumenti alternativi: la realtà virtuale immersiva. In Nuovi orizzonti per l'architettura sostenibile [New horizons for sustainable architecture] (pp. 1522-1530). Academic Press.

Vancetti, R., Cereda, R., & Cosi, F. (2019). Fire Safety Engineering, Universal Design, Realtà Virtuale: nuovi strumenti per una progettazione sempre più smart. Ingegno e costruzione nell'epoca della complessità - Forma urbana e individualità architettonica, 873-882.

Vancetti, R., & Cereda, R. (2019). *La Realtà Virtuale: un nuovo strumento a servizio della progettazione con la Fire Safety Engineering.* Antincendio.

Vandermeulen, B., Hameeuw, H., Watteeuw, L., Van Gool, L., & Proesmans, M. (2018). Bridging Multi-light & Multi-Spectral images to study, preserve and disseminate archival documents. *Archiving Conference, 2018*(1), 64–69. doi:10.2352/issn.2168-3204.2018.1.0.15

Varela, F. J., Thompson, E., & Rosch, E. (1991). *The embodied mind: Cognitive science and human experience*. MIT Press. doi:10.7551/mitpress/6730.001.0001

Vasudevan, S. K., Saravanan, P. G., John, M. K., & Sasidharan, A. (2022). Virtual reality-based real-time solution for children with learning disabilities and slow learners - an innovative attempt. *International Journal of Medical Engineering and Informatics, 14*(2), 165–175. doi:10.1504/IJMEI.2022.121131

Velazquez, R., Bazan, O., & Magana, M. (2009). A shoe-integrated tactile display for directional navigation. *2009 IEEE/RSJ International Conference on Intelligent Robots and Systems*, 1235–1240. 10.1109/IROS.2009.5354802

Ventura, S., Brivio, E., Riva, G., & Baños, R. M. (2019). Immersive Versus Non-immersive Experience: Exploring the Feasibility of Memory Assessment Through 360° Technology. *Frontiers in Psychology, 10*, 2509. Advance online publication. doi:10.3389/fpsyg.2019.02509 PMID:31798492

Vermuyten, H., Beliën, J., De Boeck, L., Reniers, G., & Wauters, T. (2016). A review of optimisation models for pedestrian evacuation and design problems. *Safety Science, 87*, 167–178.

Vichi, M. (2016). *Amyotrophic Lateral Sclerosis in Italy: characteristics and geographical distribution of first hospitalizations in the year 2005*. Academic Press.

VIGIE. (2022). *Final report*. https://op.europa.eu/en/publication-detail/-/publication/dc1c4098-b551-11ec-b6f4-01aa75ed71a1/language-en/format-PDF/source-255964403

Villa, V., Naticchia, B., Bruno, G., Aliev, K., Piantanida, P., & Antonelli, D. (2021). IoT Open-Source Architecture for the Maintenance of Building Facilities, Basel: MDPI AG. *Applied Sciences (Basel, Switzerland), 11*(12), 5374. doi:10.3390/app11125374

Villena-Taranilla, R., Tirado-Olivares, S., C'ozar-Guti'errez, R., & Gonz'alez-Calero, J. A. (2022). Effects of virtual reality on learning outcomes in K-6 education: A meta-analysis. *Educational Research Review, 35*(11), 100434. doi:10.1016/j.edurev.2022.100434

Vito Albino, U. B. (2015). Smart Cities: Definitions, Dimensions, Performance, and Initiatives. *Journal of Urban Technology, 22*.

Vittorio Miori, D. R. (2017). *Improving life quality for the elderly through the Social Internet of Things (SIoT). Global Internet of Things Summit (GIoTS)*. IEEE.

Vogel, D., & Balakrishnan, R. (2004). Interactive public ambient displays: Transitioning from implicit to explicit, public to personal, interaction with multiple users. *Proceedings of the 17th annual ACM symposium on User interface software and technology*, 137–146. 10.1145/1029632.1029656

Vogeley, K., May, M., Ritzl, A., Falkai, P., Zilles, K., & Fink, G. R. (2004). Neural correlates of first-person perspective as one constituent of human self-consciousness. *Journal of Cognitive Neuroscience, 16*(5), 817–827. doi:10.1162/089892904970799 PMID:15200709

Voinescu, A., Fodor, L.-A., Fraser, D. S., Mejias, M., & David, D. (2019). Exploring the Usability of Nesplora Aquarium, a Virtual Reality System for Neuropsychological Assessment of Attention and Executive Functioning. *2019 IEEE Conference on Virtual Reality and 3D User Interfaces (VR)*, 1207–1208. 10.1109/VR.2019.8798191

Walasek, D., & Barszcz, A. (2017). Analysis of the Adoption Rate of Building Information Modeling [BIM] and its Return on Investment [ROI]. *Procedia Engineering, 172*, 1227–1234. https://doi.org/10.1016/j.proeng.2017.02.144

Walter, B., Marschner, S. R., Li, H., & Torrance, K. E. (2007). Microfacet models for refraction through rough surfaces. In *Proceedings of the 18th Eurographics conference on Rendering Techniques*. Eurographics Association.

Wandell, B. A., & Farrell, J. E. (1993). Water into wine: Converting scanner RGB to tristimulus XYZ. Device-Indep. *Color Imaging Imaging Syst. Integr, 1909*, 92–100.

Wang, W., Cheng, J., & Guo, J. L. (2019). Usability of Virtual Reality Application Through the Lens of the User Community: A Case Study. *Extended Abstracts of the 2019 CHI Conference on Human Factors in Computing Systems.*

Wang, J., Zhao, S., Tong, X., Snyder, J., & Guo, B. (2008). Modeling Anisotropic Surface Reflectance with Example-Based Microfacet Synthesis. *ACM Transactions on Graphics, 27*(3), 3. doi:10.1145/1360612.1360640

Wang, X. (2020). Digital Heritage. In H. Guo, M. F. Goodchild, & A. Annoni (Eds.), *Manual of Digital Earth*. Springer. doi:10.1007/978-981-32-9915-3_17

Warburton, S. (2009). Second Life in higher education: Assessing the potential for and the barriers to deploying virtual worlds in learning and teaching. *British Journal of Educational Technology, 40*(3), 414–426. doi:10.1111/j.1467-8535.2009.00952.x

Wattanasoontorn, V., Boada, I., García, R., & Sbert, M. (2013). Serious games for health. *Entertainment Computing, 4*(4), 231–247. doi:10.1016/j.entcom.2013.09.002

Watteeuw, L., Hameeuw, H., Vandermeulen, B., Van der Perre, A., Boschloos, V., Delvaux, L., Proesmans, M., Van Bos, M., & Van Gool, L. (2016). Light, shadows and surface characteristics: The multispectral Portable Light Dome. *Applied Physics. A, Materials Science & Processing, 122*(11), 976. doi:10.100700339-016-0499-4

Weerdmeester, J., van Rooij, M., Harris, O., Smit, N., Engels, R. C. M. E., & Granic, I. (2017). Exploring the Role of Self-efficacy in Biofeedback Video Games. *Extended Abstracts Publication of the Annual Symposium on Computer-Human Interaction in Play*, 453–461. 10.1145/3130859.3131299

Weigl, D. M. (2019). Interweaving and enriching digital music collections for scholarship, performance, and enjoyment. 6th International Conference on Digital Libraries for Musicology. doi.org/10.1145/3358664.3358666.

Welch, J. (2014). *Cultural Heritage, What is it? Why is it important?* Retrieved July 28, 2022 from https://summit.sfu.ca/item/16150

Weng, P. L., & Taber-Doughty, T. (2015). Developing an app evaluation rubric for practitioners in special education. *Journal of Special Education Technology, 30*(1), 43–58. doi:10.1177/016264341503000104

Werbach, K., & Hunter, D. (2015). *The gamification toolkit: dynamics, mechanics, and components for the win*. University of Pennsylvania Press.

Westin, S. H., Arvo, J., & Torrance, K. E. (1992). Predicting reflectance functions from complex surfaces. *Proceedings of the SIGGRAPH 92*, 255–264. 10.1145/133994.134075

Wheeler, J. A. (1989) Information, Physics, Quantum the Search for Links. *The 3rd International Symposium Foundations of Quantum Mechanics*, 310-336.

WHO. (2022). *WHO Coronavirus (COVID-19) Dashboard (2021)*. WHO.

Whyte, J., & Nikolić, D. (2018). *Virtual Reality and the Built Environment*. Routledge. doi:10.1201/9781315618500

Wibawanto, W. (2020). Laboratorium Virtual Konsep Dan Pengembangan Simulasi Fisika LPPM UNNES.

Wiederhold, B. K., & Wiederhold, M. D. (2004). *The future of cybertherapy: improved options with advanced technologies Cybertherapy*. IOS Press.

Wiener, N. (1948). *Cybernetics, or control and communication in the animal and the machine*. Hermann.

Wiewiorra, L., & Godlovitch, I. (2021). The Digital Services Act and the Digital Markets Act – a forward-looking and consumer-centred perspective, Publication for the Committee on the Internal Market and Consumer Protection, Policy Department for Economic, Scientific and Quality of Life Policies. European Parliament.

Wilson, A. J., & Dehaene, S. (2007). *Number sense and developmental dyscalculia, in human behavior, learning and the developing brain: A typical development*. Guilford Press.

Wilson, P. N., Foreman, N., & Stanton, D. (1997). Virtual reality, disability and rehabilitation. *Disability and Rehabilitation*, *19*(6), 213–220. doi:10.3109/09638289709166530 PMID:9195138

Windawati, R., & Koeswanti, H. D. (2021). Pengembangan Game Edukasi Berbasis Android untuk Meningkatkan Hasil Belajar Siswa di Sekolah Dasar [Android-Based Educational Game Development to Improve Student Learning Outcomes in Elementary Schools]. *Journal Basicedu*, *5*(2), 1028–1038.

Windhager, F., Federico, P., Mayr, E., Schreder, G., & Smuc, M. (2016). A Review of Information Visualization Approaches and Interfaces to Digital Cultural Heritage Collections. In *Proceedings of the 9th Forum Media Technology*. St. Pölten, Austria: CEUR-WS. Retrieved March 7, 2022 from https://www.researchgate.net/publication/313442197_A_Review_of_Information_Visualization_Approaches_and_Interfaces_to_Digital_Cultural_Heritage_Collections

Winn, W. (1993). *A Conceptual Basis for Educational Applications of Virtual Reality*. Human Interface Technology Laboratory, Washington Technology Center, University of Washington. http://www.hitl.washington.edu/projects/learning_center/winn/winn-paper.html

Witikon. (n.d.). http://witikon.eu/

Witmer, B. G., Jerome, C. J., & Singer, M. J. (2005). The Factor Structure of the Presence Questionnaire. *Presence (Cambridge, Mass.)*, *14*(3), 298–312. doi:10.1162/105474605323384654

Woodham. R.J. (1980). Photometric Method for Determining Surface Orientation from Multiple Images. *Opt. Eng.*, *19*(1), 191139.

Woodhead, R., Stephenson, P., & Morrey, D. (2018). Digital construction: From point solutions to IoT ecosystem. *Automation in Construction*, *93*, 35–46. https://doi.org/10.1016/j.autcon.2018.05.004

Wu, T.-Y., Chang, Y.-C., Chen, S.-T., & Chiang, I.-T. (2012). A Preliminary Study on Using Augmented Virtuality to Improve Training for Intercollegiate Archers. In *2012 IEEE Fourth International Conference On Digital Game And Intelligent Toy Enhanced Learning* (p. 212-216). Takamatsu, Japan: IEEE Xplore. 10.1109/DIGITEL.2012.58

Wu, Y., Zhang, Y., Shen, J., & Peng, T. (2013). The Virtual Reality Applied in Construction Machinery Industry. In *Virtual, Augmented and Mixed Reality. Systems and Applications* (Vol. 8022, pp. 340–349). Springer Berlin Heidelberg. doi:10.1007/978-3-642-39420-1_36

Wu, C., Shieh, M.-D., Lien, J.-J. J., Yang, J.-F., Chu, W.-T., Huang, T.-H., Hsieh, H.-C., Chiu, H.-T., Tu, K.-C., Chen, Y.-T., Lin, S.-Y., Hu, J.-J., Lin, C.-H., & Jheng, C.-S. (2022). Enhancing Fan Engagement in a 5G Stadium With AI-Based Technologies and Live Streaming. *IEEE Systems Journal*, 1–13. doi:10.1109/JSYST.2022.3169553

Wu, H.-K., Lee, S. W.-Y., Chang, H.-Y., & Liang, J.-C. (2013). Current status, opportunities and challenges of augmented reality in education. *Computers & Education, 62*, 41–49. doi:10.1016/j.compedu.2012.10.024

Xie, Y., Takikawa, T., Saito, S., Litany, O., Yan, S., Khan, N., Tombari, F., Tompkin, J., Sitzmann, V., & Sridhar, S. (2021). Neural Fields in Visual Computing and Beyond. *Computer Graphic Forum*, 41(2), 641-676.

Xie, B., Liu, H., Alghofaili, R., Zhang, Y., Jiang, Y., Lobo, F. D., Li, C., Li, W., Huang, H., Akdere, M., Mousas, C., & Yu, L.-F. (2021). A Review on Virtual Reality Skill Training Applications. *Frontiers in Virtual Reality, 2*, 1–19. doi:10.3389/frvir.2021.645153

X-Tech Blog. (2022). *Let's Start Cooking With Virtual Reality!* http://x-tech.am/lets-start-to-cook-virtual-reality-cooking-lessons/

Xu, Q. Z. S. (2021). Fast Containment of Infectious Diseases With E-Healthcare Mobile Social Internet of Things. IEEE Internet of Things Journal, 8(22).

Yadav, K. (2004). *Development of an IT enabled Instructional Package for Teaching English medium students of Vadodara city* [M.eD Dissertation]. CASE, The MS University of Boroda, Borodara.

Yaman, O., & Karaköse, M. (2016). Development of image processing based methods using augmented reality in higher education. *2016 15th International Conference on Information Technology Based Higher Education and Training (ITHET)*, 1-5. 10.1109/ITHET.2016.7760723

Yang, J., Fricker, P., & Jung, A. (2022). *From Intuition to Reasoning: Analyzing Correlative Attributes of Walkability in Urban Environments with Machine Learning*. Wichmann Verlag. doi:10.14627/537724008

Yanli, X., & Danni, L. (2021). Prospect of Vocational Education under the Background of Digital Age: Analysis of European Union's "Digital Education Action Plan (2021-2027)". In *2021 International Conference on Internet, Education and Information Technology (IEIT)* (pp. 164-167). IEEE.

Yeo, N. L., White, M. P., Alcock, I., Garside, R., Dean, S. G., Smalley, A. J., & Gatersleben, B. (2020). What is the best way of delivering virtual nature for improving mood? An experimental comparison of high definition TV, 360 video, and computer generated virtual reality. *Journal of Environmental Psychology, 72*, 101500. doi:10.1016/j.jenvp.2020.101500 PMID:33390641

Yıldırım, A., & Şimşek, H. (2013). *Sosyal bilimlerde nitel araştırma yöntemleri* [Qualitative research methods in the social sciences] (9th ed.). Seçkin Yayıncılık.

Yildirim, B., Sahin-Topalcengiz, E., Arikan, G., & Timur, S. (2020). Using virtual reality in the classroom: Reflections of STEM teachers on the use of teaching and learning tools. *Journal of Education in Science, Environment and Health, 6*(3), 231–245. doi:10.21891/jeseh.711779

Yoganathan, S., Finch, D. A., Parkin, E., & Pollard, J. (2018). 360o virtual reality video for the acquisition of knot tying skills: A randomised controlled trial. *International Journal of Surgery, 54*, 24–27. doi:10.1016/j.ijsu.2018.04.002 PMID:29649669

Younes, G., Asmar, D., Elhajj, I., & Al-Harithy, H. (2017). Pose tracking for augmented reality applications in outdoor archaeological sites. *Journal of Electronic Imaging, 26*(1), 1–12.

Young, K. S. (2005). An empirical examination of client attitudes towards online counseling. *Cyberpsychology & Behavior, 8*(2), 172–177. doi:10.1089/cpb.2005.8.172 PMID:15938657

Yu, C.-P., Lee, H.-Y., & Luo, X.-Y. (2018). The effect of virtual reality forest and urban environments on physiological and psychological responses. *Urban Forestry and Urban Greening, 35*, 106-114. doi:10.1016/j.ufug.2018.08.013

Yue Pan, L. Z. (2021). A BIM-data mining integrated digital twin framework for advanced project management. *Automation in Construction*, 124.

Zack, J. S. (2011). The technology of online counseling. In R. Kraus, G. Stricker, & C. Speyer (Eds.), *Online counseling: A handbook for mental health professionals* (2nd ed., pp. 67–84). Academic Press. doi:10.1016/B978-0-12-378596-1.00004-6

Zain, N. Z. M., Mahmud, M., & Hassan, A. (2013). Utilization of mobile apps among student with learning disability from Islamic perspective. *5th International Conference Information Community Technology Muslim World*, 1-4. 10.1109/ICT4M.2013.6518889

Zaker, R., & Coloma, E. (2018). Virtual reality-integrated workflow in BIM-enabled projects collaboration and design review: A case study. *Vis. in Eng.*, *6*(1), 4. doi:10.118640327-018-0065-6

Zannoni, M. (2018). *Progetto e interazione. Il design degli ecosistemi interattivi.* Quodlibet. https://www.quodlibet.it/libro/9788822901668

Zannoni, M., & Formia. (2018). *"Geo-media" e Data Digital Humanities.* Academic Press.

Zarei, K. C. (2015). A comprehensive review of amyotrophic lateral sclerosis. *Surgical Neurology International.*

Zarrad, A. (2018). *Game engine solutions. Simulation and gaming.* IntechOpen. doi:10.5772/intechopen.71429

Zeagler, C. (2017). Where to wear it: Functional, technical, and social considerations in on-body location for wearable technology 20 years of designing for wearability. *Proceedings of the 2017 ACM International Symposium on Wearable Computers*, 150–157. 10.1145/3123021.3123042

Zerbi, A., & Mikolajewska, S. (2021). Digital technologies for the virtual reconstruction and projection of lost decorations: the case of the proscenium of the Farnese Theatre in Parma. *DisegnareCon*, *14*(27), 5.1-5.11.

Zhang, Y. R. C. (2020). Developing a visually impaired older people Virtual Reality (VR) simulator to apply VR in the aged living design workflow. *International Conference Information Visualisation (IV).*

Zheng, X., Zhong, T., & Liu, M. (2009). Modeling crowd evacuation of a building based on seven methodological approaches. *Building and Environment*, 437–445.

Zhou, K., Liu, T., & Zhou, L. (2015). Industry 4.0: Towards future industrial opportunities and challenges. *2015 12th International Conference on Fuzzy Systems and Knowledge Discovery (FSKD)*, 2147–2152. doi:10.1109/FSKD.2015.7382284

Zhu, Z. T., Yu, M. H., & Riezebos, P. (2016). A research framework of smart education. *Smart Learning Environments*, *3*(1), 4. Advance online publication. doi:10.118640561-016-0026-2

Ziri, A. E., Bonsma, P., Bonsma, I., Iadanza, E., Maietti, F., Medici, M., Ferrari, F., & Lerones, P. M. (2019). Cultural Heritage sites holistic documentation through Semantic Web technologies. In A. Moropoulou, M. Korres, A. Georgopoulos, C. Spyrakos, & C. Mouzakis (Eds.), *Transdisciplinary Multispectral Modelling and Cooperation for the Preservation of Cultural Heritage* (pp. 347–358). Springer.

Zucchella, C., Sinforiani, E., Tassorelli, C., Cavallini, E., Tost-Pardell, D., Grau, S., Pazzi, S., Puricelli, S., Bernini, S., Bottiroli, S., Vecchi, T., Sandrini, G., & Nappi, G. (2014). Serious games for screening pre-dementia conditions: From virtuality to reality? A pilot project. *Functional Neurology*, *29*(3), 153–158. PMID:25473734

Zyda, M. (2005). From visual simulation to virtual reality to games. *Computer*, *38*(9), 25–32. doi:10.1109/MC.2005.297

Zygouris, S., Giakoumis, D., Votis, K., Doumpoulakis, S., Ntovas, K., Segkouli, S., Karagiannidis, C., Tzovaras, D., & Tsolaki, M. (2015). Can a virtual reality cognitive training application fulfill a dual role? Using the virtual supermarket cognitive training application as a screening tool for mild cognitive impairment. J Alzheimers Dis, 44(4), 1333-47. doi:10.3233/JAD-141260

About the Contributors

Francesca Maria Ugliotti is currently Assistant Professor (RTD-A) of Drawing at Politecnico di Torino, Torino, Italy. She received both M.Sc. and the Ph.D. degrees at Politecnico di Torino in Building Engineering in 2011 and in Urban and Regional Development in 2017, respectively. During the Academic Year 2015/2016, she spent a period in Malaysia as part of the Erasmus Mundus Exchange Program at the University of Kuala Lumpur with the industrial attachment with Telekom Malaysia. She also has a first-level Specializing Master in Real Estate Management at Università degli Studi di Torino in 2012. She is involved in various national funded projects focused on Smart digitalization of the built heritage and has received national awards for research activities. Her research interests concern: i) Building Information Modelling; ii) Smart and Resilient Cities; iii) Facility Management; iv) Multidimensional scenarios for the public and health sector; v) Software solutions for simulating, optimizing and visualizing energy data; vi) VAR for maintenance, communication, awareness-raising.

Anna Osello is full professor at Politecnico di Torino (Italy), and she is in charge of the "Drawing to the Future" research group. Her research interests include virtual and augmented reality as well as building information modeling, and she has studied historical architectures and urban spaces. Osello received a PhD in Drawing and survey of the building heritage at La Sapienza University of Rome (http://www.drawingtothefuture.polito.it/).

* * *

Francesco Alotto is a Biomedical Engineer with experience in the use of ICT techniques to support people health. Passionate about artificial intelligence, autonomous vehicles, data knowledge and healthcare. Born maker with the ability to discover new patterns to improve processes and manage large amount of work without compromising quality.

Fabrizio Ivan Apollonio is Head of Department of Architecture, at Alma Mater Studiorum - University of Bologna. His main research topics lies on Virtual reconstruction, semantic modeling and application in the field of ICT to Cultural Heritage and development of information/cognitive systems aimed to fruition, study and documentation of CH. PhD in Survey of the existing built heritage (University of Ancona), is Full professor in Architectural Representation. He contributed in several research program on architectural representation and surveying and ICT/digital technologies applied to Cultural Heritage. Specialist in urban and architectural surveying with photogrammetric and laser scanning technologies, and in 3D digital modeling mainly focused on Virtual reconstruction, he is fellow of AG Digitale

Rekonstruktion (Germany). Scientific in charge of two competitive MURST-PRIN research projects: in 2000-'02 Struttura urbana e immagine della città'; in 2003-'05 'Atlante dell'iconografia delle città italiane in Età Moderna'. He has been speaker in several national and international conferences, meetings and congress giving reports, presenting papers/posters. Scientific in charge of three DA's Laboratories: "DiMoRe – Lab. of Digital Modeling and representation" (from 2015); "LaMoViDA – Lab. of Modeling and Digital Visualization for Architecture" (from 2010); "LARAC – Lab. of Urban and Architectural Surveying" (from 2005).

Alessandro Basso, RTD B at University of Camerino, SAAD, PhD in Architecture c/o Università degli Studi G.d'Annunzio, Architect, graduated Master II level "Hypergraphics" c/o Quasar institute in Rome, carried out studies and researches on virtual representation, serious game editing for Heritage enhancement, academic collaborations including (PALLADIO GEODATABASE-CISA), 2 PRIN (2007-2009/2011) and an international Italy-China research (Zhongshan project_ scientific coordinator prof.Livio Sacchi). Publications and teaching seminars on the topic of digital survey and 3D visual communication are focused on in-depth knowledge of real time rendering and precomputed animation. Lecturing in Italian and European projects.

Rachele Angela Bernardello is a PhD student at the University of Padova with a thesis in BIM practices for monitoring and management of existing bridges. Rachele graduated in Building Engineering and Architecture and since 2017 she has worked at LIM.lab - Laboratory of Information Modeling in Padua. Her research interests are in BIM and Information exchange referring both to Infrastructures and Buildings, focusing on digital workflows through delivered international standards. Her research also deals with Cultural Heritage Management and Communication based on digital methods.

Maurizio Marco Bocconcino is an Engineer for the Environment and the Territory and PhD in Design and Surveying for the Protection of the Building and Territorial Heritage, he is Associate Professor of Drawing at the Department of Structural, Building and Geotechnical Engineering of the Polytechnic University of Turin (Italy). A member of the R3C | Responsible Risk Resilience Centre of the Politecnico di Torino interdepartmental centre and a member of the Urban and Social Regeneration Group of the Fondazione Sviluppo e Crescita CRT, he deals with information systems and models for the study and representation of the territory and the city and related phenomena.

Peter Bonsma, MSc, is the co-owner of RDF Ltd. and the technical architect of the components developed by RDF Ltd. He has a degree in technical mathematics. After working for many years on mathematical optimization algorithms and compiler development his focus changed to geometry and ICT standardization. Peter developed the basis of several ISO standards/proposals and is part of the Model Support Group of IFC, one of the main standards in the building and construction area. With more than 12 years of experience in geometry standards and development of a commercially used real time CSG algorithm, as well as a long history in Semantic Web, he is now focused as technical architect on complex technical components for other software companies.

Stefano Brusaporci is Full Professor of Graphic Sciences, Surveying, Modeling and Representation of Architecture and Built Heritage at the University of L'Aquila (Italy) – Department of Civil, Construction-Architectural and Environmental Engineering. He is Ph.D. in "Conservation, Planning and Preservation of Settlements and Territorial Contexts of Elevated Environmental and Landscape Value". His research fields are surveying and historical-critical analysis of architecture and historical urban contexts; surveying and documentation of architectural heritage, also with integrated information systems; 3-D modeling and computer-based visualization for architectural and urban interpretation and presentation.

Raffaele Catuogno graduated in Architecture at the University of Naples Federico II. In the same University he obtained a PhD in Technology and Representation of Architecture and the Environment. He was a University Researcher at DiARC, Department of Architecture at the University of Naples Federico II from 2011 to 2016, where he is currently an Associate Professor, covering the courses of Architectural Design, Architectural Surveying and Architectural Representation Techniques. Particularly focused on the topics of analysis and survey of architecture and the city, he turns his research interests to the field of new survey technologies. He has published articles in scientific and A-class journals on digital representation and new technologies for architectural surveying. His main research interests concern the digital representation of architecture, no-contact surveying (in particular laser scanning, lidar and photogrammetric modelling), 3D modelling and integrated surveying, testing new technologies with particular reference to the transition from instrumental surveying to BIM and HBIM. A computer expert in the hardware and software field, he has various skills in programming, graphics, dynamic and interactive representation techniques, photorealistic rendering, photogrammetric modelling and integrated survey.

Emiliano Cereda is a Building Engineer and Research Fellow at the Department of Structural, Geotechnical and Building Engineering, Politecnico di Torino. He received his M.Sc. degree at Politecnico di Torino in 2019, developing a thesis concerning the application of virtual reality tools to Fire Safety Engineering. At drawingTOthefuture lab at Politecnico di Torino, he studies the use of Virtual Reality applied to Fire Safety. His research specialties include evacuation modelling in fire in complex buildings, use of virtual reality in egress simulation, human behaviour in fire, virtual reality for fire safety engineering.

Victoria Andrea Cotella had a double master's degree in Building Engineering-Architecture at the UNISA, Italy and at the UNC, Argentina, developing a thesis on Digital Conservation of Cultural Heritage. Currently, she is carrying out her PhD studies at the University of Naples Federico II, in the field of BIM methodology and automation of graphic representation processes. She has participated in numerous international research projects and conferences in countries such as Portugal, England, USA and Jordan promoting the use of digital technologies, and in particular Heritage Information Modelling, in support of heritage conservation. Also, she has collaborated with the University of Salford (UK) in international research activities and projects aimed at the conservation and dissemination of Jordanian Heritage.

Giorgio Dall'Osso, Designer, PhD. His research focus is the relationship between body and space mediated by technologies and design. Since 2011 he has collaborated in research and teaching at the University of Bologna, the University of San Marino, the IUAV of Venice, the IAAD, and the Mediterranea University of Reggio Calabria. Since 2016 he has been a member of the Advanced Design Unit, the design research group of the University of Bologna. His research interests focus on human body design, interaction design, basic design and behavior design.

Daniela De Luca, born in Turin in 1991, architect and PhD student at Politecnico di Torino. Since 2015, he studies and applies the use of the BIM methodology for different contexts. In particular way, the use of BIM methodology per energy saving, at the European Project: DIMMER (District Infromation Modelling and Management for Energy Reduction). In addition, she has studying new applications of the BIM method in Industry 4.0, for maintenance, training, security and storage using Virtual and Augmented Reality. Since 2016, she has been part of the research group coordinated by Prof. Osello at the drawingTOthefuture and VR@polito laboratory. She is currently involved in the PhD course Urban and Regional development at the Polytechnic of Turin, at the Department of Science, Design and Politics of the Territory. In this area she is developing applications and methodologies related to Virtual Reality in different sectors such as entertainment and culture in order to implement the accessibility of places and content to those with disabilities.

Raffaella De Marco, Ph.D., Engineer and Architect, Research Fellow and Lecturer at DICAr Department of Civil Engineering and Architecture of University of Pavia (Italy), collaborating at DAda-LAB. Young member of Europa Nostra Network. Her research deals with the development of databases and reality-based models on Cultural and Endangered Heritage for conservation, valorisation and management protocols, following international policy requirements. Collaborating in international research missions on UNESCO sites, in particular in Middle East and Palestinian territories, for range-based and photogrammetric documentation at architectural and urban scale. Her research activities focus on the definition of 3D digital databases, structural and urban models and the elaboration of Informative Management systems. Early-Stage Researcher of the European team in the EU project Horizon2020 Marie Skłodowska-Curie Actions (MSCA) Research and Innovation Staff Exchange (RISE) H2020-MSCA-RISE-2018, Project Acronym: PROMETHEUS – Project Number: 821870, and MSCA Fellow (PI) for the Global Post-doctoral Fellowship project "MOEBHIOS" - Project Number: 101064433.

Matteo Del Giudice, born in Turin in 1987, received both M.Sc. in Building Engineering and the Ph.D. degrees in Innovation Technology for the Built Environment at Politecnico di Torino in 2011 and 2016, respectively. He is currently Assistant Professor (RTD-A) at Politecnico di Torino, Department of Structural, Geotechnical and Building Engineering (DISEG). His research activity and publications are aimed at optimizing data management of the existing architectural heritage with BIM and GIS. The themes mentioned are linked to each other in relation to the development of smart cities. Since 2015 he has been part of the research group coordinated by Prof. Osello at the drawingTOthefuture laboratory, where research and training are constantly integrated and developed on the themes of BIM, Virtual and Augmented Reality (http://www.drawingtothefuture.polito.it/).

Sara Ermini is a User Experience and Interaction Designer currently collaborating in research projects with the Laboratory of Virtual Reality and Interaction Design of the University of Siena and teaching VR/AR in higher education institutions.

Purnama Fajri is a Lecturer at Faculty of Pharmacy, Politeknik Kesehatan Kementrian Kesehatan Jakarta II.

Federico Ferrari, Architect, RTDa (2006-2021) at the University of Ferrara. He covered more than 35 teaching positions in Architecture, Industrial Design and University Masters. He is involved in the use of innovative technologies for the enhancement/conservation of Cultural Heritage through 3D/BIM/ XR surveying and modeling, and application of thermography and spectrophotometry. He has authored/ co-authored more than 120 national and international publications and is in the Coordination Group of EU research projects. He is Co-founder of INCEPTION Srl, spin off at the University of Ferrara. He is member of the Scientific Coordination Team (INCEPTION Srl) of the H2020 EU funded project "4CH - Competence Centre for the Conservation of Cultural Heritage".

Riccardo Florio, architect since 1986, carries out his professional, teaching and research activities in Naples. CNR scholarship holder at L'ècole d'Architecture de Versailles, PhD at the University of Palermo, Researcher at the Polytechnic of Bari, he is Associate Professor at the Faculty of Architecture of the University of Naples Federico II, and here, since 2014, he is Full Professor and holder of the Chair of Architectural Design and teaches Urban and Environmental Surveying and Techniques of Representation of Architecture. He has curated and participated in architecture exhibitions in Italy and abroad and has published essays and articles in specialist magazines. He has a long research experience in the field of representation and survey of architecture and the city, which has been followed by numerous important applications in southern Italian cities. From September 2013 to December 2016, he was President of the Urban Planning Commission of the City of Naples.

Arianna Fonsati was born in Turin (TO) on 20.02.1992. She received a graduate Architecture degree at Politecnico di Torino in 2015. She was Research Fellowship at Politecnico di Torino and she is now a PhD Candidate in Urban and Regional Development. Her research activity is related to BIM Standards development and implementation, working both at Drawing TO the Future Lab and Lombardi Ingegneria. She is one of the editors of the book "InfraBIM il BIM per le infrastrutture" (2019).

Fabio Franchi is a research fellow at the Department of Information Engineering, Computer Science and Mathematics of the University of L'Aquila, Italy. He received his M.Sc. degree in Computer Engineering and his Ph.D. in Information and Communications Technology from the University of L'Aquila, in 2012 and 2018, respectively. He performs research on cloud and edge architectures supporting next generation services driven by ultra Reliable and Low Latency Communications, enhanced Mobile Broadband and massive Machine Type Communications. By 2022 he is a Contract Lecturers for "Cloud Architecture and Services" course provided within the Master Degree in Telecommunications Engineering at the University of L'Aquila.

Pia Fricker serves as Professor of Practice and Vice Head of the Department of Architecture at Aalto University, School of Arts, Design and Architecture in Finland. She holds the Professorship of Computational Methodologies in Landscape Architecture and Urbanism and directs the interdisciplinary Urban Studies and Planning Programme at the Department of Architecture. Pia Fricker's research and teaching link urban design and landscape architecture to the field of digital design culture through the lens of emerging technologies. Her research is actively collaborating with other Departments at Aalto University, international academic partners (ETH Zurich, The Singapore University of Technology and Design - SUTD, Hafencity University Hamburg) as well as with professional partners on a national and international level. She is currently leading research projects in data-driven design methods for dynamic urban landscapes and in the field of immersive, collaborative design environments, featuring a new reading and interaction of ecosystem datasets in the realm of Mixed Reality. Pia Fricker holds a PhD in Architecture and a postgraduate degree in Computer Aided Architectural Design from ETH Zurich, a master's and bachelor's degrees in Architecture, specialising in Urban Design and Landscape Architecture from the Technical University of Karlsruhe. Prior to her current position, she was Director of Advanced Studies in Landscape Architecture – Landscape Architecture Design Simulation – at ETH Zurich, Chair of Landscape Architecture, Prof. Christophe Girot from 2009-2015. Pia Fricker is a member of the editorial board of the JoDLA (Journal of Digital Landscape Architecture), the Scientific Program Committee of the DLA conference, the peer review committees of ACADIA (Association for Computer Aided Design in Architecture) and eCAADe (Education and research in Computer Aided Architectural Design in Europe), as well as expert peer reviewer for the International Journal of Architectural Computing, Landscape and Urban Planning Journal, International Journal of Environmental Research and Public Health, and the Journal of Architecture and Urbanism. Pia Fricker has published extensively, and her work has been exhibited at the International Architecture Biennale Venice – EEC, the National Design Centre Singapore, the National Library of Singapore, as well as a part of Helsinki Design Week. In addition to her academic work, she works as a consultant to municipalities and large-scale landscape and urban projects in the area of digitalization, with special focus on adaptive urban development.

Marco Gaiani is a full professor of Architectural Representation and past Director of Department of Architecture and Urban Planning at University of Bologna and INDACO at Politecnico di Milano. A specialist in 3D computer imaging, modeling and visualization for AH and archaeology, and architectural and archaeology DL, he has been working to virtually document, illustrate, and preserve Italy's Heritage for the past 25 years. He was one the first developers/user of laser scanning technology in the CH field. He experimented this technology in detailed surveys, and creating accurate and high resolution 3D models of monuments from the Colosseum to the Caracalla Thermae in Rome, from a number of Roman sepulchers on the via Appia to San Salvatore al Monte church in Florence, from Palladio's buildings and villas, to the houses for rent of beginning of the Twentieth century in Milan, finally to the Pompeii archaeological site. Also, he produced some multimedia visual database widely diffuses, like Andrea Palladio - Le Ville in collaboration with Centro Internazionale di Studi di Architettura Andrea Palladio in Vicenza (CISAAP), with full historical citations and detailed 3D reconstructions of Andrea Palladio's villas. He designed and was responsible of the development of some AH information systems and digital libraries of architecture and archeology including the 3D based IS for the Pompeii archaeological area (in collaboration with Scuola Normale Superiore (SNS) di Pisa) and the PALLADIOlibrary information-communication system of Andrea Palladio, his works and his territory, for CISAAP (2009-2012). He was Scientific Co-ordinator of many Courses for post-degree and post-doctoral students including the two courses "Representation and modeling for CH" of SNS in Volterra (2009 and 2011). Member of the scientific commitee and of the organizing committee of many conferences and workshops like "Museum without walls. New media for news museums", at SIGGRAPH in 1995; "I progetti non realizzati di Palladio" at Centro Internazionale di Studi di Architettura Andrea Palladio, in 1994; he is member of Advisory Board of "Center for Design Visualization of University of California at Berkeley" from april 2001; of Electronic Cultural Atlas Initiative (ECAI.). He was involved in the organization of some International Conferences as General Chair including VSMM 2002, E-Arcom 2007, Simposio Palladiano in 2008 and he is member of scientific committee of some international journals as "Disegnare idee immagini" and "SCIRES-IT".

Alberto Gallace is a cognitive neuroscientist with a special interest in the relationship between the human brain and new technologies. After a PhD in cognitive neuroscience and 4-year of research at University of Oxford in UK, he was appointed as Associate Professor of Psychobiology at University of Milan Bicocca. His research activity is mainly focused on multisensory integration, tactile information processing and on how scientific knowledge about the functioning of our nervous system can allow us to develop and market better and more usable products and technological devices. He is currently director of Mibtec (Mind and Behavior Technological Center – www.mibtec.it), exploring the interaction between human factors and VR/AR technologies. Prof Gallace is also the author of over 80 scientific articles, several chapters and 2 books, one of which (The future of Touch: From Cognitive Neuroscience to Virtual Reality by OUP) is entirely dedicated to tactile interactions with objects and products.

Simone Garagnani is an Architectural Engineer; he holds a PhD in Building and Territorial Engineering from University of Bologna (2010) with a thesis entitled "Digital Models and Design Archives - Integrated systems aimed to architectural documentation". Since 2003 he has been involved in teaching and research activities concerning Drawing, Survey and Digital Representation at the Department of Architecture, University of Bologna. Post doctoral research associate since 2010, Simone was researcher at the Interdepartmental Center for Industrial Research (CIRI) of University of Bologna, developing software and BIM-related methodologies (2012-14). Simone Garagnani was Research Unit Coordinator in the Future in Research project "Kainua. Restituire, percepire, divulgare l'assente. Tecnologie transmediali per la città etrusca di Marzabotto" at the Department of Architecture, a research activity on archaeology and BIM funded by the Italian Ministry of University and Research, in collaboration with CINECA and the Department of History, Cultures and Civilizations of the University of Bologna. Simone Garagnani carries out research concerning computer graphics, digital modelling, architectural 3D laser scanning, BIM and computer archiving technologies applied to Architecture. He developed specific studies on BIM processes related to existing buildings at the University of California Berkeley (Electrical Engineering and Computer Sciences EECS, 2012) as a Visiting Scholar in 2012. In 2015 Simone was Guest Associate Professor at Keio University in Tokyo (Japan). He is currently Assistant Professor at the Department of Architecture, University of Bologna.

Andrea Giordano has a Master's degree in Architecture, PhD in "Survey and Representation of Architecture and Environment" (VI cycle, Napoli). Full Professor (ICAR 17-Drawing) at the University of Padova, where he is Dean of the Department of Civil, Environmental and Architectural Engineering - ICEA and Coordinator of the Master BIM/HBIM. He is responsible of the Laboratory of Drawing and Representation (LDR) and the Laboratory of Information Modeling (LIM) at ICEA. He is also responsible for the survey, representation, and digital rendition of the university's campus. He has published several essayson the theory and history of methods of representation, dealing, most recently, with the use of ICT for research in the field of interoperable and semantic representation. His researches deal with: • Geometric-configurative interpretation of architectural surfaces: the construction of Padua domes; • Representational codes for the verification of landscape design; • New "tools" for the Architectural and Urban Historic Transformations Visualization and Multimedia Representation. • Building Information Modeling (BIM), Scan to BIM & HBIM.

Abdulmenaf Gul received BS degree in Computer Education and Instructional Technology. He received Ph. D. degree from same department at Middle East Technical University in 2016. Currently, he is Assistant Professor at Hakkari University, Faculty of Education, Department of Educational Sciences. His main researcher interests include serious games, gamification, virtual worlds, human-computer interaction and online learning environments.

Nashrul Hakiem is an Associate Professor of Universitas Islam Negeri Syarif Hidayatullah Jakarta, Indonesia. Prior to this, he was a data processing engineer at Indonesian Aerospace and a software engineer in Balicamp (currently known as a TelkomSigma, a subsidiary of PT Telkom Indonesia). His research interests are software engineering, software development, software quality assurance, and cybersecurity.

Kristin Love Huffman is an art and architectural historian of the early modern world. Her scholarship focuses primarily on the material and visual culture of Renaissance Venice, and she is known for her work on architectural spaces and urban systems, printed bird's-eye and orthographic representations of Venice, transcultural exchanges of knowledge, and the digital humanities. She currently serves as Associate Director of Visualizing Venice/Visualizing Cities and as a member of the Digital Advisory Board of the Society of Architectural Historians. For over ten years, she has helped build academic and research programming for the Digital Art History Research Lab at Duke University (formerly the Wired! Lab). Her scholarship has been awarded grants by the National Endowment for the Humanities, the Samuel H. Kress Foundation, the Gladys Krieble Delmas Foundation, and the Center for the Advanced Study of Visual Arts (CASVA). Additional awards include publication subventions from the Renaissance Society of America and the Furthermore Foundation for her forthcoming, edited volume with Duke University Press, A View of Venice: Portrait of a Renaissance City. Recent publications of her digital art history scholarship include: "San Geminiano: 'a Ruby among Pearls,'" JSAH (March 2020) and "Jacopo de' Barbari's View of Venice (1500): 'Image Vehicles' and 'Pathways of Culture' Past and Present," Mediterranea (2019) (http://www.uco.es/ucopress/ojs/index.php/mediterranea/article/view/11530/10681). The latter two publications feature her methods of combining traditional art history (archival work and on-site visual analyses) with digital tools that enhance critical looking and discovery.

Alessandro Innocenti is full professor of economic policy at the University of Siena. He is the coordinator of the Laboratory of Experimental Economics LabSi and of the Laboratory of Virtual Reality of the University of Siena (LabVR UNISI).

Igor Koncar graduated at Medical Faculty University of Belgrade (2002), and since then works at the Clinic for vascular and endovascular surgery of the University Serbian Clinical Center, Belgrade as well as assistant professor at the Belgrade Medical School, University of Belgrade. He published more than 150 Pubmed publications; 1 book, 15 book chapters, more than 100 invited lectures abroad. H index – 16. Section Editor in Journal of Endovascular Therapy and European Journal of Clinical Investigations (EJCI). Reviewer of the European Journal of Vascular and Endovascular Surgery (EJVES) more than 200 reviews and review of 10 ESVS guidelines; Reviewer in Annals of Vascular Surgery (AVS), Journal of Endovascular Therapy (JEVT), Vascular and Endovascular Surgery, Journal of Cardiovascular Surgery (Torino), British Journal of Surgery (BJS). Member of the European Society for Vascular Surgery (Chair of the ESVS Academy since September 2022), European society for Cardiovascular and Endovascular Surgery (Executive committee member), Serbian Society for Cardiovascular Surgery (member of the Executive committee), International Society for Endovascular Specialists (ISES). Organized more than 30 different workshops in the field of Seizing and planning, Writing scientific papers, carotid endarterectomy, EVAR, TEVAR, open aortic repair, anastomosis suturing etc. AWARDS: Serbian Medical Society: For research about "Usage of Recombinant factor VIIa in non hemophiliacs". European Society for Cardiovascular and Endovascular Surgery (ESCVS) - Young vascular surgeon prize for Scientific and Clinical achievements in 2011. School of Medicine, University of Belgrade – Scientific achievements as young researcher SPEAKS English and Portuguese.

Ruth Elenora Kristanty joined Poltekkes Kemenkes Jakarta II and became a lecturer at the Pharmaceutical and Food Analysis Department in 2003. She received the M.Farm degree from the Universitas Indonesia in 2012. Her research interests cover several aspects of phytochemistry, herbal medicines, and learning technology. She has contributed to international conferences and scientific journals.

Pedro Martín Lerones obtained his Degree in Applied Physics in 1997, with further specialization in the design of optical systems for non-contact measurement and 3D imaging. He obtained his PhD in 2007, and worked as Associate Professor in the Technical Hi-School of Engineering - University of Valladolid (Spain) between 2002 and 2011, teaching about 3D digitization and modelling to CAD/CAM. Dr. Lerones is currently Senior Researcher at Fundación CARTIF. Author or co-author of 32 scientific papers and specialized studies (including impact publications), has been participating in 26 national and international RTD projects and commercial contracts mainly oriented to the application of digital technologies to Cultural Heritage and Construction. In addition, he is Evaluator of EU RTD proposals (Code: EX2006C113308); Reviewer for international journals; Core member of the European Focus Area of Cultural Heritage (FACH-ECTP); Member of the Spanish NCP for the JPI Cultural Heritage and Global Change; Collaborator of the initiative: Research for Future Infrastructure Networks in Europe (reFINE); and Founding member of the International Platform: Economic Value of Cultural Heritage (EVoCH).

Ami Licaj is a Designer and PhD in Design, focused on Data Visualization, passionate about processes - and the "designerly" way of dealing with them - applied to digital/social/intangible/future. Since 2015 she has been combining freelance work with research and academic work, previously at the Department of Architecture and Design of the University of Genoa - where she has been for several years a research fellow, teaching tutor, collaborator and contract lecturer - and currently at the Advanced Design Unit of the University of Bologna where she holds a research grant entitled "Design culture and creative practices for sustainable development in the low touch and low travel age".

Federica Maietti M.Arch., PhD, is Associate Professor in the Scientific Sector ICAR/17 at the DIAPReM Centre (Development of Integrated Automatic Procedures for Restoration of Monuments) of the Department of Architecture of the University of Ferrara, where she is responsible for the Heritage Documentation and Diagnostic Survey area. Her interdisciplinary research and skills are focused on the relationship between survey and representation for conservation and restoration, in the field of digital heritage. She is involved in several national and international research activities. She is the author of more than 200 publications in the field of Heritage Documentation, Survey and Representation. Since 2006, she has been a speaker at national and international conferences, in particular on integrated survey methodologies for heritage conservation. She has been the Technical Coordinator of the H2020 project "INCEPTION - Inclusive Cultural Heritage in Europe through 3D semantic modelling", and she is Co-founder of INCEPTION Srl, spin off at the University of Ferrara. She is member of the Scientific

Coordination Team (INCEPTION Srl) of the H2020 EU funded project "4CH - Competence Centre for the Conservation of Cultural Heritage".

Valentina Mancuso is a Ph.D student, Department of Psychology, E-Campus University.

Fabio Manzone, Building Engineer and PhD in Building and Territorial Systems, is a temporary researcher in Building Production at the Department of Structural, Building and Geotechnical Engineering of the Polytechnic of Turin (Italy). Lecturer at the Polytechnic of Turin since 2004 for courses in Project and Construction Management. Since 1998 he has been working as a freelancer, thanks to which he has dealt with ambitious projects and complex construction sites such as the Intesa Sanpaolo Skyscraper in Turin and the Juventus Stadium.

Priyaranjan Maral is an assistant professor at Central University of Rajasthan, India. He obtained his Ph.D. in Cognitive Psychology from University of Allahabad, India. His research interests include Street Children, Resilience, Coping, Addiction, Trauma, Cognition, Emotion, Disasters, Posttraumatic Stress Disorder, Posttraumatic Growth, Women Empowerment, Aging and qualitative research methodology.

Marco Medici, M.Arch. and Ph.D. in Architecture, is Assistant Professor at the University of Ferrara, Department of Architecture. He is part of the DIAPReM-TekneHub research center since 2013, where he has been involved in several research and training projects, as well as technology transfer activities. He developed advanced skills in the digitization of built environment, focusing in particular on BIM modeling applied to Cultural Heritage. On these topics, he took part in international conferences and he is author of national and international scientific papers. In the last years, he's also developing research activities on web-based technologies, virtual environments and algorithm-aided design for architectural modeling. He is Co-founder of INCEPTION Srl, spin off at the University of Ferrara. He is member of the Scientific Coordination Team (INCEPTION Srl) of the H2020 EU funded project "4CH - Competence Centre for the Conservation of Cultural Heritage".

Sandra Mikolajewska graduated in 2017 in Architecture at the University of Parma with a thesis regarding the Church of Santa Maria di Canepanova in Pavia (supervisor: Prof. A. Zerbi, co-supervisor: Prof. B. Adorni). In May 2021, she earned her Pd.D. at the Department of Engineering and Architecture with a dissertation focused on digital technologies for the documentation, conservation and valorization of Cultural Heritage, with an application to the Farnese Theatre in Parma (supervisor: Prof. A. Zerbi). Her research deals with Survey and Representation, with particular attention to the methodologies for the three-dimensional modeling of cultural assets and to the multimedia technologies for the valorization and communication of Cultural Heritage. She is a member of MADLab (Laboratory for the Monitoring, Analysis and Diagnostics of existing buildings) and an adherent member of UID (Unione Italiana del Disegno). She collaborates in many laser scanner and photogrammetric survey campaigns on monumental architectural complexes.

Claudio Mirarchi is researcher at Politecnico di Milano. He has a PhD in Building Engineering achived at Politecnico di Milano and his research is focused on the introduction of digital processes in the construction sector with specific reference to the issues related to the knowledge management area. He is active in the research and practice about the implementation of BIM and digital processes and technologies in the construction sector. He was the project management work package leader of the DigiPLACE project aimed at creating the foundations for the future European platform(s) in the construction sector; he coordinates the development of the Italian standard about common data environment in the context of the UNI 11337 series (working group 4). He has held several courses and speeches at national and international level including the BIMA+ European master, the MSc Building Information Modelling program at the School of Architecture, University of Liverpool, the dissemination activities promoted by the Italian association of construction companies (ANCE) at Italian level, the LC3 conference, the ECPPM conference, the EC3 conference, the BIM world Paris, etc.

Edoardo Montevidoni is a PhD candidate at PoliMI - Department of architecture, built environment and construction engineering (DABC).

Elena Moretti is associate Professor in Applied Biology. She attended University of Reading (UK) and Yale University (USA) before obtaining PhD in "Biology of Germinal Cells". She is member of the Board of Doctoral School in Molecular Medicine (Siena University). She is author of 150 papers in International peer reviewed journals (H index: 32). She is teaching Biology in several degree and master degree courses. She studies human and animal sperm and spermatogenesis, oxidative stress and antioxidants.

Paolo Palmisciano is an Italian medical doctor interested in pursuing a career as neurosurgeon-scientist. Graduated from the University of Catania (Italy) in July 2019, he completed a postdoctoral fellowship in anterior skull base and Artificial Intelligence at the National Hospital of Neurology and Neurosurgery in London (UK) under the mentorship of Dr. Hani J. Marcus. After obtaining his Italian medical license in March 2020, he completed his USMLEs step 1 and step 2 CK. From August 2020, he started a research collaboration with students and residents working at the MD Anderson Cancer Center and the University of Texas Southwestern Medical School, under the mentorship of Dr. Tarek Y. El Ahmadieh and Mr. Ali S. Haider. His research focused on neuro-oncology and skull base oncology. From March 2021, he started a clinical and research fellowship at the Department of Neurosurgery of the Cannizzaro Hospital in Catania (Italy) under the mentorship of Dr. Giuseppe E. Umana. He dedicated his clinical and research work in neuro-oncology and skull base surgery, in spine surgery, and in the application of Mixed Reality in neurosurgery. From March 2022, he is pursuing a postdoctoral research fellowship at the University of Cincinnati under the mentorship of Dr. Mario Zuccarello with a focus on skull base anatomy and surgical approaches.

Sandro Parrinello is an associate professor at DICAr Department of Civil Engineering and Architecture of University of Pavia. PhD in Representation and Survey Sciences with the title of European Research Doctor. Since 2012 he has been visiting professor at Perm National Research Polytechnic University (Russia) and in 2015 he received an honorary degree from the State Academy of Civil Engineering and Architecture in Odessa (Ukraine). Since 2005 he has been a member of UNESCO Red List Forum with the title of expert; in 2011 he was appointed Expert and Voting member as referent for Italy, to the international scientific committee ICOFORT (ICOMOS International Scientific Committee on Fortifications and Military Heritage). In 2016 he was Visiting Professor at Cracow Polytechnic University (Poland) and in 2017 he obtained the National Scientific Qualification as First Level Professor. He is director of Laboratory DAda Lab. and of joint laboratory "Landscape Survey & Design" of University of Pavia. He is responsible for numerous national and international research projects, member of editorial committees of international scientific series and journals and he has organized numerous international conferences on the subject of heritage documentation.

Edoardo Patti is Assistant Professor with tenure track at Politecnico di Torino. He received both M.Sc. and Ph.D. degrees in Computer Engineering at Politecnico di Torino in 2010 and 2014, respectively. His research interests concern: ubiquitous computing and Internet of Things; smart systems, cities and mobility; software architectures with particular emphasis on infrastructure for Ambient Intelligence; software solutions for simulating and optimising energy systems and for energy data visualisation to increase user awareness.

Maurizio Perticarini graduated in 2015 in Architecture with the thesis entitled LACUNA Completing, Rebuilding, Transforming Architecture: Loggia del Capitanio, Andrea Palladio 1571-2015. In October 2018 research fellow at the University of Padua entitled PD-Invisible: PaDova Innovative VISions - views and Imaginings Behind the city Learning. From 2019 PhD student at the Department of Architecture, Industrial Design Architecture, Industrial Design and Cultural Heritage at the Luigi Vanvitelli University.

Stefano Maria Priola is Assistant Professor in Neurosurgery at the Northern Ontario School of Medicine University. He has completed the School of Medicine and Residency in Neurosurgery at the University of Messina, Italy. After that, he moved to Canada where he completed a clinical fellowship in Neuro-oncology and Skull Base Approaches, a clinical fellowship in General Neurosurgery, and a clinical fellowship in Vascular and Endovascular Neurosurgery. His main research interests are Neuro-oncology, Stroke and Traumatic Brain Injury. He is author and co-author of several research papers published on peer-reviewed international journals. He is now staff Neurosurgeon at Health Sciences North, Sudbury (ON), Canada.

Anjana Prusty is currently working as an Assistant Professor in the Humanities and Social Sciences Program at Center for Creative Cognition in the SR University, Warangal, India. Anjana earned her Ph.D. degree from Sikkim University, India with a specialization in Clinical Psychology. She has experience in different NGOs and institutions as a Clinical and School Psychologist. Her main research interest includes disability, affective neuroscience, suicide, depression, substance abuse, antisocial personality disorder, and student's wellbeing.

Claudia Rinaldi received the Laurea degree (cum laude) in electronic engineering from the University of L'Aquila, Italy, in 2005, the master's degree in trumpet from the Conservatory of Music of L'Aquila, in 2006, the Ph.D. degree in electronic engineering from the University of L'Aquila, in 2009, and the bachelor's degree in electronic music from the Conservatory of Music of L'Aquila, in 2013. She is currently an Assistant Professor at the Department of Information Engineering, Computer Science and Mathematics, University of L'Aquila. Her main research interests include digital signal processing algorithms and more in general on the use of technology in artistic fields. Moreover, she is concerned with design, modeling, and optimization of communication algorithms with particular emphasis on the physical layer and software defined radio for the development of transmission systems responding to cognitive radios paradigms.

Gianluca Scalia currently works as Neurosurgeon at ARNAS Garibaldi of Catania.

Roberta Surian was born in 1995, received her Bachelor's degree in Biomedical Engineering in 2018, and then she furthered and completed her specialization in Biomechanics in 2021. She worked as a research fellow to drawingTOthefuturelab at Politecnico di Torino, Italy, where she was part of the research group managed by Prof. Anna Osello at the Department of Structural, Geotechnical and Building Engineering (DISEG). During her research activity, she interfaced with the worlds of Internet of Things (IoT) and Building Information Modeling (BIM) to build technology to support patients with motor neuron degeneration. Currently, she is in charge of Occupant Safety in the Automotive sector.

Saniye Tugba Tokel is an Associate Professor of Computer Education and Instructional Technology at Middle East Technical University. She received B.S. and M.S. degrees in CEIT from METU and Pd.D. degree in Educational Technology from Texas A&M University. Her research interests include virtual worlds, virtual reality, serious games, simulations, presence, human factors, instructional design, online learning, problem solving, scaffolding, metacognition, and cultural issues.

Giuseppe Emmanuele Umana is Consultant Neurosurgeon at Cannizzaro Hospital, Trauma and Gamma-knife center, Catania, Italy. Device developer New technologies beta tester: augmented reality, robotic rehab. The holomedicine association founding member; Advisory board member of ApoQlar Editor and Reviewer for several peer reviewed journals. Visiting Professor at Society for Brain Mapping and Therapeutics (SBMT), Los Angeles, California, USA.

Roberto Vancetti is a PhD in Building Recovery Engineering, is currently Assistant Professor with tenure track (RTI) at the "Department of Structural, Geotechnical and Building Engineering" of Politecnico di Torino. He carries out teaching activities since 2004, first at the Second Faculty of Engineering of the Politecnico di Torino based in Vercelli and subsequently at the Faculty of Building Engineering in Turin. He is currently the holder of the teaching of "Prevention for Fire Safety" and "Fire Safety Engineering" in the Master of Science in Building Engineering. His research interests concern: i) building intervention methodologies for setting standards according to fire safety regulations at different levels of design and execution of the works; ii) Fire Safety Engineering methods applied to building and industrial design with the use of numerical modelling of fire scenarios for fire safety checks; applications of Virtual Reality and Augmented Reality for training on escape routes, evaluations on the effectiveness of emergency signs and further assessments on human behaviour in emergencies; fire safety design through BIM methodology: Interoperability between BIM and Fire Safety Engineering software, Modelling and verification of large-scale escape systems. He is the author of more than 70 articles in national and international scientific journals and monographs in the various disciplinary fields of the research carried out.

Starlight Vattano (1987), PhD, architect, research fellow and adjunct professor at the Università Iuav di Venezia. Since 2017 she has been teaching assistant at the Università di Trento. Her research fields concern the themes of representation in their scientific and communication aspects, digital modelling (in immersive and interactive environments), and visual studies.

Chiara Vernizzi is full professor with the Department of Engineering and Architecture, University of Parma. She received the Laurea degree in Architecture from Politecnico di Milano and the Ph.D. degree in Drawing and Survey of the Built Environment from Università di Ancona, Parma, Italy, in 1991 and 1999, respectively. From 2002 she was assistant and then associate professor (2010) with the Faculty of Engineering, and from 2018 she is full professor with the Department of Engineering and Architecture, University of Parma. She is in member of the "Laboratory of monitoring analysis and diagnostics of the built enviromnent" at the Department of Engineering and Architecture; she is also member of CIDEA (Interdepartmental Centre for Energy and the Environment) anche of CICCREI (Interdepartmental Research Centre for the Conservation, Construction and Regeneration of Buildings and Infrastructures), Università di Parma. The main research interests of C. Vernizzi includes Design and Architectural and Urban Survey, with particular reference to tools and methods of digital representation of the city and the architecture (including BIM and existing BIM); use of IT tools for the graphic restitution of the survey of architectures and historical urban contexts; modern and contemporary graphic representation of the architectural project, from sketch to digital drawing. Vernizzi's work is published in over 120 articles in national and international journals and books, in proceedings of national and international conferences and monographs on the themes of Descriptive Geometry, Survey and Representation of Architecture and the City, Architectural Design, from the historical to the contemporary digital one. She is a Member of the CTS of the Italian Design Union (UID) since 2018, within which he is a member of the Archives Commission. She was from 2018 to present Member of the editorial-scientific board of diségno, biannual magazine of UID - Unione Italiana Disegno. She acts as a technical program committee member for many international conferences. She is the coordinator of the POR-FESR 2014-2020 project eBIM: existing Building Information Modeling per la gestione dell'intervento sul costruito.

Luca Vespasiano is a PhD student at the Department of Uman Science at University of L'Aquila.

Michele Zannoni is Associate Professor at the Department of Architecture of the University of Bologna. His publications include articles in SCIRES-IT journal and Material Design Journal. He is the author of the book Progetto e Interazione (Quodlibet, 2018), which explore the intersection of interaction processes and visual and product design. His scientific research is concerned about digital and physical products and interaction design. From the year 2000, he taught at the IUAV of Venice, the Polytechnic of Milan and the University of the Republic of San Marino where he was the director of the master's degree in Design. Since 1997 he has worked as interaction and visual designer in several design agencies: at Studio Altermedia, at Studio Visuale form 2008 and at last in BDS Design from 2014. In these companies he has worked on national and international projects, winning international awards in the fields of multimedia and interface design and in particular, he has collaborated since 2009 with the Technogym design center, and has designed many of the company's user interfaces.

Andrea Zerbi graduated in 1993 in Architecture at the Faculty of Architecture of the Politecnico di Milano; in 2000 he earned the Ph.D. in Design and Survey of the Building Heritage; in 2009 he was Assistant Professor at the Department of Engineering and Architecture of the University of Parma, where he teaches several courses related to the field of representation; since 2015 he is Associate Professor at the same structure. He mainly deals with survey and representation on urban and architectural scale, areas in which he has worked participating in numerous surveys of important monumental complexes in Parma and other Italian and foreign cities. Over the years he has specialized in new methodologies for surveying and graphic restitution based on the use of increasingly innovative computer tools made available by technological development. He is the author of numerous publications on these topics presented at national and international conferences.

Index

C

D

E

F

G

H

I

L

M

N

O

P

Q

R

S

T

U

V

W